Power Magnetic Devices

IEEE Press
445 Hoes Lane
Piscataway, NJ 08854

IEEE Press Editorial Board
Ekram Hossain, *Editor in Chief*

Jón Atli Benediktsson	Xiaoou Li	Jeffrey Reed
Anjan Bose	Lian Yong	Diomidis Spinellis
David Alan Grier	Andreas Molisch	Sarah Spurgeon
Elya B. Joffe	Saeid Nahavandi	Ahmet Murat Tekalp

Power Magnetic Devices

A Multi-Objective Design Approach

Second Edition

Scott D. Sudhoff

Copyright © 2022 by The Institute of Electrical and Electronics Engineers, Inc. All rights reserved.

Published by John Wiley & Sons, Inc., Hoboken, New Jersey.
Published simultaneously in Canada.

Edition History
John Wiley & Sons, Inc (1e, 2014);

No part of this publication may be reproduced, stored in a retrieval system, or transmitted in any form or by any means, electronic, mechanical, photocopying, recording, scanning, or otherwise, except as permitted under Section 107 or 108 of the 1976 United States Copyright Act, without either the prior written permission of the Publisher, or authorization through payment of the appropriate per-copy fee to the Copyright Clearance Center, Inc., 222 Rosewood Drive, Danvers, MA 01923, (978) 750-8400, fax (978) 750-4470, or on the web at www.copyright.com. Requests to the Publisher for permission should be addressed to the Permissions Department, John Wiley & Sons, Inc., 111 River Street, Hoboken, NJ 07030, (201) 748-6011, fax (201) 748-6008, or online at http://www.wiley.com/go/permission.

Limit of Liability/Disclaimer of Warranty: While the publisher and author have used their best efforts in preparing this book, they make no representations or warranties with respect to the accuracy or completeness of the contents of this book and specifically disclaim any implied warranties of merchantability or fitness for a particular purpose. No warranty may be created or extended by sales representatives or written sales materials. The advice and strategies contained herein may not be suitable for your situation. You should consult with a professional where appropriate. Further, readers should be aware that websites listed in this work may have changed or disappeared between when this work was written and when it is read. Neither the publisher nor author shall be liable for any loss of profit or any other commercial damages, including but not limited to special, incidental, consequential, or other damages.

For general information on our other products and services or for technical support, please contact our Customer Care Department within the United States at (800) 762-2974, outside the United States at (317) 572-3993 or fax (317) 572-4002.

Wiley also publishes its books in a variety of electronic formats. Some content that appears in print may not be available in electronic formats. For more information about Wiley products, visit our web site at www.wiley.com.

Library of Congress Cataloging-in-Publication Data:

Names: Sudhoff, Scott D., author.
Title: Power magnetic devices : a multi-objective design approach / Scott D. Sudhoff.
Description: Second edition. | Hoboken, NJ : Wiley, [2021] | Series: Eee press series on power and energy systems | Includes bibliographical references and index.
Identifiers: LCCN 2021031537 (print) | LCCN 2021031538 (ebook) | ISBN 9781119674603 (hardback) | ISBN 9781119674641 (adobe pdf) | ISBN 9781119674634 (epub)
Subjects: LCSH: Electromagnetic devices. | Power electronics. | Electric machinery.
Classification: LCC TK7872.M25 S83 2022 (print) | LCC TK7872.M25 (ebook) | DDC 621.3–dc23
LC record available at https://lccn.loc.gov/2021031537
LC ebook record available at https://lccn.loc.gov/2021031538

Cover Design: Wiley
Cover Images: (center) Courtesy of Scott D. Sudhoff, (top) © Sam Robinson/Getty Images

Set in 9.5/12.5pt STIXTwoText by Straive, Pondicherry, India

10 9 8 7 6 5 4 3 2 1

To my wife Julie for 38 wonderful years and counting
To my daughter Emily. Hy-potenuse
To my daughter Samantha. Boiler Up!

In memory of my mother, Mary Louise Sudhoff, 1931–2017
To my Father
In memory of my brothers, Doug Sudhoff,
1958–2021, and Brian Sudhoff, 1961–1964
To my brother Steve

Contents

Author Biography *xiii*
Preface *xv*
About the Companion Site *xix*

1 Optimization-Based Design *1*
1.1 Design Approach *1*
1.2 Mathematical Properties of Objective Functions *3*
1.3 Single-Objective Optimization Using Newton's Method *5*
1.4 Genetic Algorithms: Review of Biological Genetics *7*
1.5 The Canonical Genetic Algorithm *10*
1.6 Real-Coded Genetic Algorithms *15*
1.7 Multi-Objective Optimization and the Pareto-Optimal Front *25*
1.8 Multi-Objective Optimization Using Genetic Algorithms *27*
1.9 Formulation of Fitness Functions for Design Problems *31*
1.10 A Design Example *33*
  References *39*
  Problems *40*

2 Magnetics and Magnetic Equivalent Circuits *43*
2.1 Ampere's Law, Magnetomotive Force, and Kirchhoff's MMF Law for Magnetic Circuits *43*
2.2 Magnetic Flux, Gauss's Law, and Kirchhoff's Flux Law for Magnetic Circuits *46*
2.3 Magnetically Conductive Materials and Ohm's Law For Magnetic Circuits *48*
2.4 Construction of the Magnetic Equivalent Circuit *56*
2.5 Translation of Magnetic Circuits to Electric Circuits: Flux Linkage and Inductance *59*
2.6 Representing Fringing Flux in Magnetic Circuits *64*
2.7 Representing Leakage Flux in Magnetic Circuits *68*
2.8 Numerical Solution of Nonlinear Magnetic Circuits *80*
2.9 Permanent Magnet Materials and Their Magnetic Circuit Representation *95*
2.10 Closing Remarks *98*
  References *98*
  Problems *99*

3 Introduction to Inductor Design 103
- 3.1 Common Inductor Architectures 103
- 3.2 DC Coil Resistance 105
- 3.3 DC Inductor Design 108
- 3.4 Case Study 113
- 3.5 Closing Remarks 119
  - References 120
  - Problems 120

4 Force and Torque 123
- 4.1 Energy Storage in Electromechanical Devices 123
- 4.2 Calculation of Field Energy 125
- 4.3 Force from Field Energy 127
- 4.4 Co-Energy 128
- 4.5 Force from Co-Energy 132
- 4.6 Conditions for Conservative Fields 133
- 4.7 Magnetically Linear Systems 134
- 4.8 Torque 135
- 4.9 Calculating Force Using Magnetic Equivalent Circuits 135
  - References 139
  - Problems 139

5 Introduction to Electromagnet Design 141
- 5.1 Common Electromagnet Architectures 141
- 5.2 Magnetic, Electric, and Force Analysis of an Ei-Core Electromagnet 141
- 5.3 EI-Core Electromagnet Design 151
- 5.4 Case Study 155
  - References 162
  - Problems 163

6 Magnetic Core Loss and Material Characterization 165
- 6.1 Eddy Current Losses 165
- 6.2 Hysteresis Loss and the B–H Loop 172
- 6.3 Empirical Modeling of Core Loss 177
- 6.4 Magnetic Material Characterization 183
- 6.5 Measuring Anhysteretic Behavior 188
- 6.6 Characterizing Behavioral Loss Models 197
- 6.7 Time-Domain Loss Modeling: the Preisach Model 201
- 6.8 Time-Domain Loss Modeling: the Extended Jiles–Atherton Model 205
  - References 211
  - Problems 212

7 Transformer Design 215
- 7.1 Common Transformer Architectures 215
- 7.2 T-Equivalent Circuit Model 217
- 7.3 Steady-State Analysis 221
- 7.4 Transformer Performance Considerations 223

7.5	Core-Type Transformer Configuration *231*
7.6	Core-Type Transformer MEC *238*
7.7	Core Loss *244*
7.8	Core-Type Transformer Design *245*
7.9	Case Study *251*
7.10	Closing Remarks *259*
	References *260*
	Problems *260*

8 Distributed Windings and Rotating Electric Machinery *263*

8.1	Describing Distributed Windings *263*
8.2	Winding Functions *271*
8.3	Air-Gap Magneto Motive Force *276*
8.4	Rotating MMF *278*
8.5	Flux Linkage and Inductance *280*
8.6	Slot Effects and Carter's Coefficient *282*
8.7	Leakage Inductance *284*
8.8	Resistance *289*
8.9	Introduction to Reference Frame Theory *290*
8.10	Expressions for Torque *294*
	References *299*
	Problems *299*

9 Introduction to Permanent Magnet AC Machine Design *303*

9.1	Permanent Magnet Synchronous Machines *303*
9.2	Operating Characteristics of PMAC Machines *305*
9.3	Machine Geometry *312*
9.4	Stator Winding *317*
9.5	Material Parameters *320*
9.6	Stator Currents and Control Philosophy *320*
9.7	Radial Field Analysis *321*
9.8	Lumped Parameters *326*
9.9	Ferromagnetic Field Analysis *327*
9.10	Formulation of Design Problem *332*
9.11	Case Study *336*
9.12	Extensions *344*
	References *345*
	Problems *346*

10 Introduction to Thermal Equivalent Circuits *349*

10.1	Heat Energy, Heat Flow, and the Heat Equation *349*
10.2	Thermal Equivalent Circuit of One-Dimensional Heat Flow *352*
10.3	Thermal Equivalent Circuit of a Cuboidal Region *358*
10.4	Thermal Equivalent Circuit of a Cylindrical Region *361*
10.5	Inhomogeneous Regions *367*
10.6	Material Boundaries *373*
10.7	Thermal Equivalent Circuit Networks *376*

10.8	Case Study: Thermal Model of Electromagnet *380*
	References *396*
	Problems *397*

11 Alternating Current Conductor Losses *399*
- 11.1 Skin Effect in Strip Conductors *399*
- 11.2 Skin Effect in Cylindrical Conductors *405*
- 11.3 Proximity Effect in a Single Conductor *409*
- 11.4 Independence of Skin and Proximity Effects *411*
- 11.5 Proximity Effect in a Group of Conductors *413*
- 11.6 Relating Mean-Squared Field and Leakage Permeance *416*
- 11.7 Mean-Squared Field for Select Geometries *417*
- 11.8 Conductor Losses in Rotating Machinery *422*
- 11.9 Conductor Losses in a UI-Core Inductor *426*
- 11.10 Closing Remarks *431*
  - References *431*
  - Problems *432*

12 Parasitic Capacitance *433*
- 12.1 Modeling Approach *433*
- 12.2 Review of Electrostatics *434*
- 12.3 Turn-to-Turn Capacitance *442*
- 12.4 Coil-to-Core Capacitance *446*
- 12.5 Layer-to-Layer Capacitance *449*
- 12.6 Capacitance in Multi-Winding Systems *452*
- 12.7 Measuring Capacitance *455*
  - References *458*
  - Problems *459*

13 Buck Converter Design *461*
- 13.1 Buck Converter Analysis *461*
- 13.2 Semiconductors *469*
- 13.3 Heat Sink *472*
- 13.4 Capacitors *474*
- 13.5 UI-Core Input Inductor *476*
- 13.6 UI-Core Output Inductor *477*
- 13.7 Operating Point Analysis *488*
- 13.8 Design Paradigm *492*
- 13.9 Case Study *495*
- 13.10 Extensions *501*
  - References *501*
  - Problems *501*

14 Three-Phase Inductor Design *503*
- 14.1 System Description *503*
- 14.2 Inductor Geometry *516*
- 14.3 Magnetic Equivalent Circuit *518*
- 14.4 Magnetic Analysis *529*

14.5	Inductor Design Paradigm *533*
14.6	Case Study *537*
	References *541*
	Problems *541*

15 Common-Mode Inductor Design *543*
15.1	Common-Mode Voltage and Current *543*
15.2	System Description *545*
15.3	Common-Mode Equivalent Circuit *546*
15.4	Common-Mode Inductor Specification *552*
15.5	UR-Core Common-Mode Inductor *557*
15.6	UR-Core Common-Mode Inductor Magnetic Analysis *562*
15.7	Common-Mode Inductor Design Paradigm *564*
15.8	Common-Mode Inductor Case Study *566*
	References *571*
	Problems *571*

16 Finite Element Analysis *573*
16.1	Maxwell's and Poisson's Equations *573*
16.2	Finite Element Analysis Formulation *575*
16.3	Finite Element Analysis Implementation *580*
16.4	Closing Remarks *587*
	References *588*
	Problems *588*

Appendix A	**Conductor Data and Wire Gauges** *589*
Appendix B	**Selected Ferrimagnetic Core Data** *593*
Appendix C	**Selected Magnetic Steel Data** *595*
Appendix D	**Selected Permanent Magnet Data** *599*
Appendix E	**Phasor Analysis** *601*
Appendix F	**Trigonometric Identities** *607*

Index *609*

Author Biography

Scott D. Sudhoff received the BS (highest distinction), MS, and PhD degrees in electrical engineering from Purdue University in 1988, 1989, and 1991, respectively. From 1991 to 1993, he served as a consultant for PC Krause and Associates in aerospace power and actuation systems. From 1993 to 1997, he served as a faculty member at the University of Missouri-Rolla, and in 1997, he joined Purdue University where he currently serves as the Michael and Katherine Birck Professor of Electrical and Computer Engineering.

Professor Sudhoff has worked for over 30 years in electric machinery, drives, and power-electronics-based distribution systems. His research sponsors have included the US Navy, Army, Air-Force, DOE, and NASA, as well as many industrial partners. This work has resulted in 6 patents and the graduation of over 30 doctoral students.

He has nearly 100 journal publications and over 100 conference publications including 5 prize papers. He is a co-author of *Analysis of Electric Machinery and Drives Systems*, second and third editions, both published by IEEE Press/Wiley. In 2006, he won the IEEE Power Engineering Society Cyril Veinott Electromechanical Energy Conversion Award for outstanding contributions to the field of electromechanical energy conversion. In 2008, he became an IEEE Fellow. He served as editor-in-chief of *IEEE Transactions on Energy Conversion* from 2007 to 2013 and the editor-in-chief of *IEEE Power and Energy Systems Technology Journal* (which later become *IEEE Open Access Journal of Power and Energy*) from 2013 to 2019.

His interests include electric machinery, power electronics, marine and aerospace power systems, applied control, power magnetics, and evolutionary computing. Much of his current research focuses on genetic algorithms and their application to power electronic converter and electric machine design.

Outside of power magnetics, his interests include running and astronomical imaging where he specializes in planetary nebula and peculiar galaxies.

Preface

This work is intended either as a text book for a senior-level/beginning graduate-level course, or as a resource for the practicing engineer. There are three objectives for the text. The first is to set forth a systematic multi-objective-optimization-based approach for the semi-automated design of power magnet components. A second objective of the text is to discuss physical principals and analysis necessary for the design of power magnetic devices including fields, magnet equivalent circuit analysis, core loss, eddy current losses, thermal analysis, skin and proximity effect, and the like. The third objective of the text is to provide some fundamental background in a variety of devices including inductors, electromagnets, transformers, and rotating electric machinery. It is not the intent to provide a cookbook of design information for all devices. Rather, it is intended a position to leave the reader well poised to start adapting the approach to specific devices of interest to their work—whether it be a transformer or a novel type of rotating machine.

From a pedagogical point of view, the organization of the text is designed to involve the readers in the design process as rapidly as possible. While it might be more efficient to discuss all the relevant physical effects, and then to discuss the design, such an approach is not always satisfying in that it leaves the reader hungry for a meal a long time before dinner is served. For that reason, the first thirteen chapters of the text generally alternate between discussing a physical effect and considering a design problem in which that effect is considered.

Chapter 1 introduces multi-objective optimization using genetic algorithms. This chapter provides a sufficient background to conduct formal multi-objective optimization-based design. Next, Chapter 2 provides a background in the magnetic analysis that is used throughout the book. Formal design is introduced in Chapter 3, wherein a simple inductor design is considered, using the material from Chapters 1 and 2. In Chapter 4, force and torque production are considered – and in Chapter 5 this material is used in simple electromagnet design. Chapter 6 concerns magnetic core loss. In this second edition, this chapter has been greatly expanded to include characterization of magnetic materials, as well as to include more information on modeling of magnetic hysteresis. Chapter 7 uses the magnetic core loss models in the design of a single-phase transformer. The alternating pattern of analysis and design continues, with Chapters 8 and 9 focusing on the rotating machinery.

The text then continues with three primarily analysis chapters to supplement the earlier work, with Chapter 10 focusing on thermal analysis (and revisiting some of the earlier design efforts), Chapter 11 discussing skin and proximity effect losses, and Chapter 12 focusing on parasitic capacitance. The material of these chapters is then tied together in Chapter 13, which considers the design of a dc-to-dc converter. Chapters 12 and 13 are new to the 2nd edition, as are Chapters 14, the design of three-phase inductors, and Chapter 15, which focuses on the design of common-mode inductors. The book concludes with Chapter 16, that is also new, which introduces finite element analysis as a method to validate designs.

The amount of material is adequate for two courses. The author recommends that Chapter 1 through Chapter 6 be covered as a starting point. The remaining chapters could be covered at that discretion of the instructor or the reader.

Chapters 8 and 9 are based on Chapters 2 and 15 of Analysis of Electric Machinery and Drive Systems, 3rd Edition, by Paul Krause, Oleg Wasynczuk, Scott Sudhoff, and Steve Pekarek. This work is also published by IEEE/Wiley.

MATLAB source code to support this book is given in S. D. Sudhoff, "MATLAB codes for Power Magnetic Devices: A Multi-Objective Design Approach, 2nd Edition" [Online]. Available: http://booksupport.wiley.com. This code includes the Genetic Optimization System Engineering Toolbox, a Magnetic Equivalent Circuit Toolbox, a Thermal Equivalent Circuit Toolbox, as well as all the design examples discussed in the book. This should greatly reduce the amount of work needed to either to teach from the book, or to use the principals taught in this text for the reader's own purposes. Partially annotated slides and a solutions manual are also available.

Throughout this work, scalar variables are normally in italic font (for example) while vectors and matrices are bold non-italic (for example). Functions of all dimensionalities are denoted by non-italic non-bold font (for example). Brackets in equations are associated with iteration number in iterative methods.

The author of this book is deeply indebted to many individuals. First, to my parents, who gave me the time, support, and indulgence for me to pursue my interests. While in high-school, I was blessed with many excellent teachers. I am particularly indebted to Sister Thomasita Hayes. As an undergraduate, I was also fortunate to have some outstanding instructors – particularly Stanislaw Zak in control and optimization, Fred Mowle who taught me how to code, and Paul Krause, who taught me electric machinery. I spent the beginning part of my career at the University of Missouri – Rolla where I was fortunate to have a number of excellent mentors including Keith Stanek, Max Anderson, and especially Mariesa Crow and Jim Drewniak.

I would thank many current and former students, post docs, and research scientists who directly or indirectly contributed to this book. These include Benjamin Loop, Chunki Kwon, Jim Cale, Aaron Cramer, Brandon Cassimere, Brant Cassimere, Chuck Sullivan, Ricky Chan, Shengyu Wang, Yonggon Lee, Cahya Harianto, Jacob Krizan, Grant Shane, Omar Laldin, Ahmed Taher, Jamal Alsawhali, Harish Suryanarayana, and Jonathan Crider. Jamal Alsawalhi, Grant Shane, Jonathan Crider, Ahmed Taher, Ruiyang Lin, David Loder, Andrew Kasha, Vinicius Cabral Do Nascimento, Akhil Prasad, and Harshita Singh contributed in performing many of the FEA and/or experimental results in the book. I would also especially thank Dionysios Aliprantis for sparking my interest in genetic algorithms.

A variety of U.S. government agencies have contributed to research efforts which contributed to this book, including the Army, Navy, DOE, Sandia National Labs, NREL, ORNL, and NASA. The Office of Naval Research in particular has provided steady support for my entire career which directly and indirectly supported this work, and without which this work would not have been possible. The support of the Grainger Foundation has also been very important to the program at Purdue.

I would thank my colleagues at Purdue. A key attribute of any institution is the people your work with. With this regard, my colleagues at Purdue, namely Oleg Wasynczuk, Steve Pekarek, and Dionysios Aliprantis, are a great group.

Finally, I had some extra help with the second edition. Special thanks to Dr. Harshita Singh, who thoroughly reviewed Chapters 10 and 12. Dr Paul Ohodnicki (University of Pittsburg) and Drummond Fudge (Continuous Solutions) provided great feedback on Chapters 13, 14, and 15. Dr. Jamal Alsawalhi (Khalifa University) provided detailed reviews on Chapters 6 and 16. Finally, Dr Steve Pekarek (Purdue University) also reviewed Chapter 16. I am greatly indebted to all of these individuals for contributing so much time to the second edition of this book.

About the Companion Site

This book is accompanied by a companion website:

www.wiley.com/go/sudhoff/Powermagneticdevices

The website includes the following materials: Powerpoint slides, Solutions manual, Matlab code

1

Optimization-Based Design

We will begin our study of power magnetic device design with a general consideration of the design process. A case will be made to approach the design process rather formally by converting the design problem into an optimization problem. Next, single-objective optimization is discussed, with particular emphasis on optimization using genetic algorithms (GAs). This is followed by a discussion of multi-objective optimization. Practical aspects of formulating design problems as optimization problems are then considered. The chapter concludes with a design example that focuses on a UI-core inductor.

1.1 Design Approach

It is appropriate to begin this work by considering the design process. Clearly, there are a myriad of different approaches by which components may be designed. For example, a possible manual design process is illustrated in Figure 1.1. In order to consider this process in a more concrete way, suppose that the component we are designing is an electromagnet and that we wish to design an electromagnet so that a certain set of specifications are met.

Using the design process in Figure 1.1, our first step would be to perform a detailed mathematical analysis of the device. Typically, when we analyze a device, our analysis predicts device performance (mass, loss, force) in terms of the device parameters (geometry, materials) rather than directly addressing the design problem by deriving expressions for what the device parameters should be in terms of the device specifications (allowed loss, required force). Therefore, we must manipulate our detailed analysis into a set of design equations that are used to calculate the design parameters as a function of device specifications. However, going from detailed analysis to design equations invariably requires numerous assumptions and approximations, even beyond the ones found in our original "detailed" analysis. As a result, we check our design, either against our original analysis or using some numerical tool such as a finite element analysis. Based on the results from the numerical analysis, we will revise the design and repeat the numerical analysis until specifications are met, at which point we have arrived at a final design. Of course, we often use a more involved design process; for example, another iteration of the design may be made based on physical prototypes.

The manual design process we have been considering involves an engineer in the iteration process. Variations of this process are successfully used ubiquitously throughout the engineering community. However, the process has some significant drawbacks. First, it requires a great deal of engineering time. Second, it requires a great deal of engineering experience. This experience

Power Magnetic Devices: A Multi-Objective Design Approach, Second Edition. Scott D. Sudhoff.
© 2022 The Institute of Electrical and Electronics Engineers, Inc. Published 2022 by John Wiley & Sons, Inc.
Companion website: www.wiley.com/go/sudhoff/Powermagneticdevices

1 Optimization-Based Design

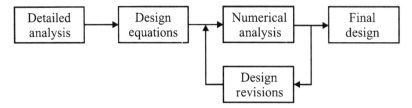

Figure 1.1 A manual design process.

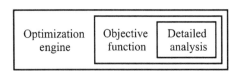

Figure 1.2 Optimization-based design process.

comes into play in the development of the design equations, which often take the form of rules-of-thumb based at least partially on experience. Experience is also a factor in making changes to the design based on the numerical analysis. Finally, while the process has been very successful in yielding working designs, it may not lead to the best design.

An alternate design process is illustrated in Figure 1.2. Therein, an optimization-based design process is shown. In this case, the process is not illustrated in a sequential manner as in Figure 1.1, but rather in an organizational manner. The process again starts with a detailed analysis of the device or component. However, unlike the manual design process, in the optimization-based process, the detailed analysis is not used to formulate design equations. Instead, the detailed analysis is used to calculate design metrics such as mass, cost, and loss. The detailed analysis is also used to check constraints such as achieving some minimum acceptable level of performance. The metrics and constraints are combined into an objective or fitness function. This function is defined so that its optimization results in optimization of the design metrics subject to all design constraints being met.

At the outermost level of this design process, an optimization engine will select the parameters of the design (geometry, materials, etc.) so as to maximize the objective function. In terms of computational algorithm, Figure 1.2 depicts an optimization engine at the outer level. This engine operates on an objective function that is calculated based on the detailed analysis.

There are several advantages of this approach. First, it is unnecessary to formulate design equations. This is beneficial in that it reduces the number of approximations and assumptions made and reduces the amount of design experience needed for a good design. Second, the design is formally optimized with regard to the design metrics, potentially leading to better designs, at least in terms of the design metrics. Third, since the engineer is out of the optimization loop, less engineering time is generally required. There are some disadvantages of the procedure. First, the process can be numerically intense and require significant computing time, sometimes on the order of hours and, in extreme cases, days. Fortunately, computer time is significantly less expensive than engineering time. Second, the quality of the result depends upon the quality of the detailed analysis. In this regard, design experience is still valuable, though not as critical as in the manual design approach.

In order to utilize the optimization-based design process, it is clearly necessary to be able to optimize mathematical functions. For design purposes, we will be optimizing the objective function, which we will also refer to as a fitness function. Optimization is a broad subject, which has been the subject of a strong and sustained interest of a host of researchers over the years. The purpose of this chapter is to introduce the subject to an extent sufficient to enable the reader to utilize an optimization-based design process for power magnetic devices. More thorough study of optimization methods will serve every engineer well; for a good textbook devoted to the subject the reader is referred to Chong and Żak [1].

1.2 Mathematical Properties of Objective Functions

Before discussing optimization algorithms, it is appropriate to discuss some properties of objective functions that are relevant to their optimization, as these properties determine the effectiveness of one optimization approach relative to another.

As we proceed to do this, note that throughout this work, scalar variables are normally in italic font (for example, x) while vector and matrices are bold nonitalic (for example, **x**). Functions of all dimensionalities are denoted by nonitalic nonbold font (for example, x(θ)). Brackets in equations are associated with iteration number in iterative methods.

In considering the properties of the objective function, it is appropriate to begin by defining our parameter vector, which will be denoted as **x**. The domain of **x** is referred to as the search space and will be denoted Ω, which is to say we require $\mathbf{x} \in \Omega$. The elements of parameter vector **x** will include those variables of a design that we are free to select. In general, some elements of **x** will be discrete in nature while others will be continuous. An example of a discrete element might be one that designates a material type from a list of available materials. A geometrical parameter such as the length of a motor would be an example of an element that can be selected from a continuous range. If all members of the parameter vector are discrete, the search space is described as being discrete. If all members of the search space are continuous (in the set of real numbers), the search space is said to be continuous. If the elements of **x** include both discrete and continuous elements, the search space is said to be mixed. It is assumed that the function that we wish to optimize is denoted f(**x**). We will assume that f(**x**) returns a vector of dimension m of real numbers, that is, $f(\mathbf{x}) \in \mathbb{R}^m$, where m is the number of objectives we are considering. For most of this chapter, we will merely consider f(**x**) to be a mathematical function for which we wish to identify the optimizer of; however, in Section 1.9, and in the rest of this book for that matter, we will focus on how to construct f(**x**) so as to serve as an instrument of engineering design.

For this section, let us focus on the case where all elements of **x** are real numbers so that $\mathbf{x} \in \mathbb{R}^n$, where $\mathbb{R}^n$ denotes the set of real numbers of dimension n and where the number of objectives is one (that is, $m = 1$) so that f(**x**) is a scalar function of a vector argument. Finally, let us suppose we wish to minimize f(**x**). A point $\mathbf{x}^*$ is said to be the global minimizer of f over Ω provided that

$$f(\mathbf{x}^*) \leq f(\mathbf{x}) \quad \forall x \in \Omega \setminus \{x^*\} \tag{1.2-1}$$

where $\forall$ is read as "for all" and $\Omega \setminus \{\mathbf{x}^*\}$ denotes the set Ω less the point $\mathbf{x}^*$. If the $\leq$ is replaced by $<$, then $\mathbf{x}^*$ is referred to as the strict global minimizer.

As stated previously, the function f(**x**) can have properties that make it easier or more difficult to find the global minimizer. Some of these properties are depicted in Figure 1.3. An example of a feature that makes it more difficult to find the global minimizer is a discontinuity as shown in Figure 1.3(a). Therein $\Omega = [x_{mn}, x_{mx}]$ and the discontinuity is at $x = x_a$. In this case, the discontinuity results in a point where the function's derivative is undefined. Since many optimization algorithms use the derivative of the function as part of the algorithm, such behavior can be

1 Optimization-Based Design

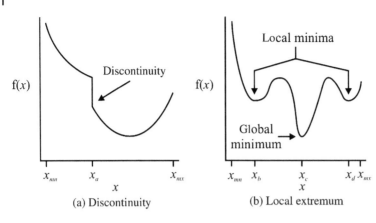

Figure 1.3 Function properties.

problematic. In general, any problem with a discrete or mixed search space will have a discontinuous objective.

Another property that can be problematic is the existence of local minima. These are illustrated in Figure 1.3(b). Therein $x = x_b$, $x = x_c$, and $x = x_d$ are all local minima; however, only the point $x = x_c$ is a global minimizer. Many minimization algorithms can converge to local minimizers and fail to find the global minimizer.

Related to the existence of local extrema is the convexity (or lack thereof) of a function. It is appropriate to begin the discussion of function convexity by considering the definition of a convex set. Let $\Theta \subset \mathbb{R}^n$ denote a set. A set is considered convex if all the points on the line connecting any two points in the set are also in the set. In other words, if

$$\alpha \mathbf{x}_a + (1-\alpha)\mathbf{x}_b \in \Theta \quad \forall \alpha \in [0, 1] \tag{1.2-2}$$

for any two points $\mathbf{x}_a, \mathbf{x}_b \in \Theta$ then Θ is convex. This is illustrated in Figure 1.4 for $\Theta \subset \mathbb{R}^2$.

In order to determine if a function is convex, it is necessary to consider its epigraph. The epigraph of $f(\mathbf{x})$ is simply the set of points greater than or equal to $f(\mathbf{x})$. A function is considered convex if its epigraph is a convex set, as is shown in Figure 1.5. Note that this set will be in $\mathbb{R}^{n+1}$, where n is the dimension of $\mathbf{x}$.

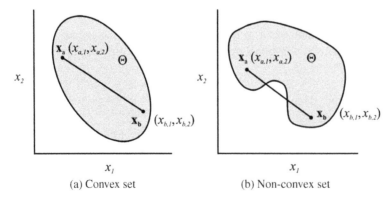

Figure 1.4 Definition of a convex set.

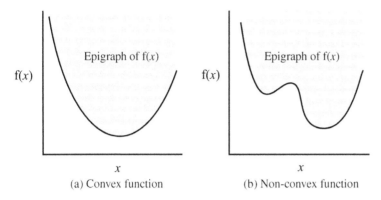

Figure 1.5 Definition of a convex function.

If the function being optimized is convex, the optimization process becomes much easier. This is because it can be shown that any local minimizer of a convex function is also a global minimizer. Therefore the situation shown in Figure 1.3(b) cannot occur. As a result, the minimization of continuous convex functions is straightforward and computationally tractable.

1.3 Single-Objective Optimization Using Newton's Method

Let us consider a method to find the extrema of an objective function f(**x**). Let us focus our attention on the case where $f(\mathbf{x}) \in \mathbb{R}$ and $\mathbf{x} \in \mathbb{R}^n$. Algorithms to solve this problem include gradient methods, Newton's method, conjugate direction methods, quasi-Newton methods, and the Nelder–Mead simplex method, to name a few. Let us focus on Newton's method as being somewhat representative.

In order to set the stage for Newton's method, let us first define some operators. The first derivative or gradient of our objective function is denoted $\nabla f(\mathbf{x})$ and is defined as

$$\nabla f(\mathbf{x}) = \begin{bmatrix} \dfrac{\partial f(\mathbf{x})}{\partial x_1} & \dfrac{\partial f(\mathbf{x})}{\partial x_2} & \cdots & \dfrac{\partial f(\mathbf{x})}{\partial x_n} \end{bmatrix}^T \quad (1.3\text{-}1)$$

The second derivative or Hessian of f(**x**) is defined as

$$F(\mathbf{x}) = \begin{bmatrix} \dfrac{\partial^2 f(\mathbf{x})}{\partial^2 x_1} & \dfrac{\partial^2 f(\mathbf{x})}{\partial x_2 \partial x_1} & \cdots & \dfrac{\partial^2 f(\mathbf{x})}{\partial x_n \partial x_1} \\ \dfrac{\partial^2 f(\mathbf{x})}{\partial x_1 \partial x_2} & \dfrac{\partial^2 f(\mathbf{x})}{\partial^2 x_2} & \cdots & \dfrac{\partial^2 f(\mathbf{x})}{\partial x_n \partial x_2} \\ \vdots & \vdots & \ddots & \vdots \\ \dfrac{\partial^2 f(\mathbf{x})}{\partial x_1 \partial x_n} & \dfrac{\partial^2 f(\mathbf{x})}{\partial x_2 \partial x_n} & \cdots & \dfrac{\partial^2 f(\mathbf{x})}{\partial^2 x_n} \end{bmatrix} \quad (1.3\text{-}2)$$

If $\mathbf{x}^*$ is a local minimizer of f, and if $\mathbf{x}^*$ is in the interior of Ω, it can be shown that

$$\nabla f(\mathbf{x}^*) = \mathbf{0} \quad (1.3\text{-}3)$$

and that

$$F(\mathbf{x}^*) \geq 0 \tag{1.3-4}$$

Note that the statement $F(\mathbf{x}^*) \geq 0$ means that $F(\mathbf{x}^*)$ is positive semi-definite, which is to say that $\mathbf{y}^T F(\mathbf{x}^*)\mathbf{y} \geq 0 \, \forall \, \mathbf{y} \in \mathbb{R}^n$. This second condition verifies that a minimum rather than a maximum has been found. It is important to understand that the conditions (1.3-3) and (1.3-4) are necessary but not sufficient conditions for $\mathbf{x}^*$ to be a mimimizer, unless f is a convex function. If f is convex, (1.3-3) and (1.3-4) are necessary and sufficient conditions for $\mathbf{x}^*$ to be a global minimizer.

At this point, we are posed to set forth Newton's method of finding function minimizers. This method is iterative and is based on a kth estimate of the solution denoted $\mathbf{x}[k]$. Then an update formula is applied to generate a (hopefully) improved estimate $\mathbf{x}[k+1]$ of the minimizer. The update formula is derived by first approximating f as a Taylor series about the current estimated solution $\mathbf{x}[k]$. In particular,

$$f(\mathbf{x}) = f(\mathbf{x}[k]) + \nabla f(\mathbf{x}[k])(\mathbf{x} - \mathbf{x}[k]) + \ldots \\ \frac{1}{2}(\mathbf{x}-\mathbf{x}[k])^T F(\mathbf{x}[k])(\mathbf{x}-\mathbf{x}[k]) + H \tag{1.3-5}$$

where H denotes higher-order terms. Neglecting these higher-order terms and finding the gradient of $f(\mathbf{x})$ based on (1.3-5), we obtain

$$\nabla f(\mathbf{x}) \approx \nabla f(\mathbf{x}[k]) + F(\mathbf{x}[k])(\mathbf{x} - \mathbf{x}[k]) \tag{1.3-6}$$

From the necessary condition (1.3-3), we next take the right-hand side of (1.3-6) and equate it with zero; then we replace $\mathbf{x}$, which will be our improved estimate, with $\mathbf{x}[k+1]$. Manipulating the resulting expression yields

$$\mathbf{x}[k+1] = \mathbf{x}[k] - F(\mathbf{x}[k])^{-1} \nabla f(\mathbf{x}[k]) \tag{1.3-7}$$

which is Newton's method.

Clearly, Newton's method requires $f \subset C^2$, which is to say that f is in the set of twice differentiable functions. Note that the selection of the initial solution, $\mathbf{x}[1]$, can have a significant impact on which (if any) local solution is found. Also note that method is equally likely to yield a local maximizer as a local minimizer.

Example 1.3A Let us apply Newton's method to find the minimizer of the function

$$f(\mathbf{x}) = 2(x_1 - 2)^4 + 3(e^{x_2} - x_1)^2 + 8 \tag{1.3A-1}$$

This example is an arbitrary mathematical function; we will consider how to construct $f(\mathbf{x})$ so as to serve a design purpose in Section 1.9. Inspection of (1.3A-1) reveals that the global minimum is at $x_1 = 2$ and $x_2 = \ln(2)$. However, let us apply Newton's method to find the minimum. Our first step is to obtain the gradient and the Hessian. From (1.3A-1), we have

$$\nabla f(\mathbf{x}) = \begin{bmatrix} 8(x_1-2)^3 - 6(e^{x_2}-x_1) \\ 6(e^{x_2}-x_1)e^{x_2} \end{bmatrix} \tag{1.3A-2}$$

and

$$F(\mathbf{x}) = \begin{bmatrix} 24(x_1-2)^2 + 6 & -6e^{x_2} \\ -6e^{x_2} & 12e^{2x_2} - 6x_1 e^{x_2} \end{bmatrix} \tag{1.3A-3}$$

Table 1.1 Newton's Method Results

k	x[k]	f(x[k])	∇f(x[k])	F(x[k])
1	$\begin{bmatrix} 0 \\ 0 \end{bmatrix}$	43	$\begin{bmatrix} -70 \\ 6 \end{bmatrix}$	$\begin{bmatrix} 102 & -6 \\ -6 & 12 \end{bmatrix}$
2	$\begin{bmatrix} 0.677 \\ -0.162 \end{bmatrix}$	14.2	$\begin{bmatrix} -19.6 \\ 0.888 \end{bmatrix}$	$\begin{bmatrix} 48.0 & -5.10 \\ -5.10 & 5.23 \end{bmatrix}$
3	$\begin{bmatrix} 1.11 \\ 0.0938 \end{bmatrix}$	9.25	$\begin{bmatrix} -5.53 \\ -0.0938 \end{bmatrix}$	$\begin{bmatrix} 24.9 & -6.58 \\ -6.58 & 7.13 \end{bmatrix}$
10	$\begin{bmatrix} 1.95 \\ 0.666 \end{bmatrix}$	8.00	$\begin{bmatrix} -2.23 \cdot 10^{-3} \\ 2.00 \cdot 10^{-3} \end{bmatrix}$	$\begin{bmatrix} 6.07 & -11.7 \\ -11.7 & 22.7 \end{bmatrix}$

Let us arbitrarily take our initial estimate of the solution to be $\mathbf{x}[1] = \begin{bmatrix} 0 & 0 \end{bmatrix}^T$. Table 1.1 lists the numerical results from the repeated application of (1.3-7). As can be seen, during the first three iterations, the value of the function decreases rapidly. However, then the rate of reduction of the function slows. Observe that on the 10th iteration the value of the objective function is the minimum value to three significant digits, though there is still some discrepancy in the estimate of the minimizer. In this problem, the minimum is quite shallow, which reduces the speed of convergence.

Newton's method can be extremely effective on some problems, but prove problematic on others. For example, if f($\mathbf{x}$) is not twice differentiable for some $\mathbf{x}$, difficulties arise since Newton's method requires the function, its gradient, and its Hessian. Many optimization methods require similar information and share similar drawbacks. There are optimization methods that do not require derivative information. One example is the Nelder–Mead simplex method. Even so, this algorithm can still become trapped at local minimizers if the function is not convex.

One feature that makes these methods susceptible to becoming trapped at a local minimum is that they take the approach of starting with a single estimated solution and attempt to refine that estimate. If the single estimate is close to a local extrema, it will tend to converge to that extrema. There is another class of optimization methods that are not based on a single estimate of the solution but on a large number (a population) of estimates. These population-based methods are not as susceptible to convergence to a nonglobal local extrema because there are a multitude of candidate optimizers.

Genetic algorithms (GAs) are a population-based optimization algorithms that have proven very effective in solving design optimization problems. Other population-based optimization methods, such as particle swarm optimization, have also been used successfully. While one can engage in a lengthy debate over which algorithm is superior, such a debate is unlikely to be fruitful. The focus of this text is on posing the design problem as a formal optimization problem; once the problem is so posed, any optimization algorithm can be used. A discussion of GAs is included herein in order to provide the reader with a background in at least one method that can be used for the optimization process.

1.4 Genetic Algorithms: Review of Biological Genetics

In this section, we will set the stage for the use of GAs as optimization engines by reviewing some principles of biological genetics. All living things have a set of instructions on how they are constructed. These instructions are written in the deoxyribonucleic acid (DNA) contained within each

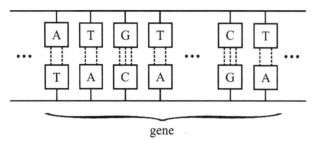

Figure 1.6 Deoxyribonucleic acid (DNA).

cell of that living being. The structure of this molecule was first determined by James Watson and Francis Crick in the 1950s and is depicted in Figure 1.6. Therein, the horizontal strands are made of phosphate and a sugar called deoxyribose. These strands are wound into a helix structure. The short vertical dashed lines in Figure 1.6 indicate weak hydrogen bonds that are instrumental in the duplication of DNA. The letters A and G stand for adenine and guanine, respectively, which are compounds known as purines. The letters T and C designate thymine and cytosine, which are pyrimidines. The combinations AT, TA, GC, and CG form a four-letter alphabet. A sequence of letters from this alphabet forms a gene of a living being. In terms of our discussion on design, we may view the gene as a design parameter of a living organism.

Each DNA molecule in a living organism is known as a chromosome. Living organisms generally have multiple chromosomes. For example, humans have 46 chromosomes per cell. These chromosomes are arranged into 22 pairs (one of each pair contributed by the father and one of each pair contributed by the mother). In addition, there are the two sex chromosomes denoted as X and Y. In humans and many other organisms, the existence of chromosomes in pairs leads to dominant and recessive genes, as discovered by Gregory Mendel, a Roman Catholic monk and botanist, who studied the propagation of traits in pea plants. However, not all living creatures have chromosomes organized in pairs; ants, wasps, and bees are haploid and have only one occurrence of each chromosome, while strawberries are octaploid with eight occurrences of each chromosome. This provides something to contemplate while eating strawberry pie.

In addition to some understanding of genes, chromosomes, and DNA, it is important to consider sexual reproduction and, in particular, the formation of gametes (sperm and egg cells). The formation of gametes is through a process known as meiosis, which is illustrated in Figure 1.7. Let us consider a diploid organism with two pairs of chromosomes, a long set and a short set as shown in Figure 1.7. This is consistent with chromosomes in cells that vary in length. The production of gametes begins with the chromosomes lengthening out within the cell as shown in Figure 1.7 (a). Note there are two copies of this chromosome, one contributed by the father (darkly shaded) and one contributed by the mother (lightly shaded). Next replication of the chromosomes occurs as seen in Figure 1.7(b). The two copies of a chromosome are referred to as chromatids, and they are connected at a point called the centromere. At this point, meiosis starts with the pairing of the chromosomes given by mother and father. In this pairing process, it is possible for the arms of the chromosomes to interchange, thereby leading to a new chromosome that consists of some genes from the father and some genes from the mother. Note that the crossover point is normally between genes; this is because of the large amount of (apparently) nonfunctional DNA in most chromosomes. While one crossover of one chromosome is shown, multiple crossovers in all chromosomes are possible.

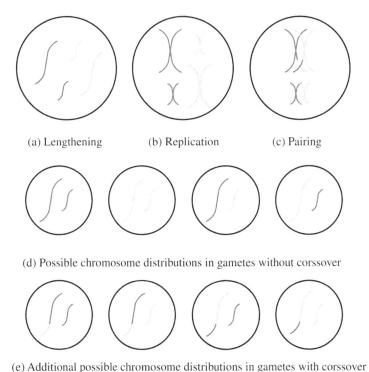

(a) Lengthening (b) Replication (c) Pairing

(d) Possible chromosome distributions in gametes without corssover

(e) Additional possible chromosome distributions in gametes with corssover

Figure 1.7 Meiosis.

After the chromosome pairing and crossover, the cell, which now contains four versions of each chromosome, splits into four cells. The four chromosomes segregate into these four cells. If crossover did not occur, this two-chromosome organism could produce four genotypes (genetically unique) of gametes; therefore, a single mother–father pair could produce sixteen genotypes. However, because of the crossover, many additional genotypes of gametes can be produced. In fact, because of crossover, the number of genotypes that can be produced becomes related to the number of genes, not just the number of chromosomes. Thus, crossover is very important in achieving genetic diversity.

Of course, in the case of humans, chromosome distribution alone with 23 pairs of chromosomes, yields $2^{23} = 8,388,608$ genotypes for the gametes. A set of parents could thus produce $8,388,608^2$ genetically different children, which would seem to be an impressively diverse set, even without crossover. However, in artificial GAs, the number of chromosomes is much smaller, often consisting of a single chromosome.

Beyond increasing the sheer number of genotypes of the gametes, crossover plays another critical role because it allows beneficial genes (traits) on a given chromosome to be decoupled from detrimental genes (traits). Crossover will play a very important role in the operation of GAs.

In addition to gamete diversity due to genetic crossover and chromosomal segregation into gametes, additional diversity is brought about because of mutation. Mutation arises from errors in copying DNA. In mitosis, or cell division, mutation often has little effect, since the mutated cell will often die. However, in meiosis, mutation can have a significant impact since the mutated genetic code will propagate into the genetic code of every cell in the child. Even then, many mutations are not noticeable because they are a part of the genetic code that is unused. When mutation has a noticeable effect, it is generally for the worse. However, occasionally beneficial mutations occur which improve the ability of an individual (and eventually a species) to survive.

A final concept from biology that will serve our needs for an optimization engine is the idea of natural selection and the survival of the fittest, an idea stemming from Charles Darwin's voyages of the H.M.S Beagle, during a period of time roughly contemporary with the work of Mendel and the American Civil War. The idea that the most fit individuals of a population survive to reproduce is directly used in GAs. These algorithms are based on an explicit fitness function, which will be used to determine which individuals "survive" and will be placed into a mating pool.

Clearly, the discussion in this section is at a high level and has been greatly simplified. The interested reader is referred to Crow [2] for a more thorough introduction to the topic.

1.5 The Canonical Genetic Algorithm

A century after the work of Mendel and Darwin, but a mere decade after the work of Watson and Crick, John Holland, a professor at the University of Michigan, proposed using the principles of biological genetics as a computation algorithm for optimization, a concept instantiated by a GA [3]. In this section, we will begin our consideration of GAs with a canonical GA similar to Holland's original vision.

GAs are quite different from traditional optimization algorithms. First of all, GAs operate not on the argument of the function being optimized, but rather on an encoding of the argument. Second, rather than iterating to improve an estimate for an optimizer, GAs iterate to improve a large number of different estimates of the optimizer. This collection of estimates will be referred to as a population. The use of a population of estimated solutions improves the chances of finding a global optimum. Third, GA operations are based only on the values of the objective function—gradients and Hessians are not used, nor even estimated. This property is useful in function with discontinuities or with a discrete or mixed search space. Finally, GA operations are based on probabilistic rather than deterministic computations.

The first concept that must be set forth in a GA is that it, like evolution, operates on a population, not on an individual. We will denote the population within the GA as $\mathbf{P}[k]$, where k is the generation number. The kth generation consists of a number of individuals, that is,

$$\mathbf{P}[k] = \{\boldsymbol{\theta}^1, \boldsymbol{\theta}^2, \cdots \boldsymbol{\theta}^{N_p}\} \tag{1.5-1}$$

where $\boldsymbol{\theta}^i$ is the genetic code for the ith individual in the kth generation of the population and where N_p denotes the number of individuals in the population, which should be an even number. The genetic code for the ith individual may be organized as

$$\boldsymbol{\theta}^i = \begin{bmatrix} \text{chromosome } 1 & \left\{ \begin{matrix} \boldsymbol{\theta}^i_1 \\ \boldsymbol{\theta}^i_2 \end{matrix} \right. \\ \text{chromosome } 2 & \left\{ \begin{matrix} \boldsymbol{\theta}^i_3 \\ \boldsymbol{\theta}^i_4 \\ \boldsymbol{\theta}^i_5 \end{matrix} \right. \\ \vdots & \\ \text{chromosome } N_c & \left\{ \boldsymbol{\theta}^i_{N_g} \right. \end{bmatrix} \tag{1.5-2}$$

where N_c is the number of chromosomes, N_g is the number of genes, and $\boldsymbol{\theta}^i_j$ is the jth gene of the ith individual (and it is understood that we are referring to the kth generation). Each gene is a string sequence. Recall that DNA consists of alphabet AT, TA, CG, and GC. In the case of the canonical GA, the string is most typically a binary sequence. Thus, $\boldsymbol{\theta}^i$ takes the form of a binary number. The significance of the chromosome organization will come into play when we consider reproduction.

The fact that the genes are encoded results in a limitation of the domain of the parameter vector. In other words, the domain of possible values of each element of the parameter vector is inherently limited. In some cases, this property is very convenient, but in other cases this limitation on the domain of the parameter vectors is disadvantageous.

Associated with the genetic code for each population member, we will have a decoding function that translates the genetic code into a parameter vector. In particular,

$$\mathbf{x}^i = \mathrm{d}(\boldsymbol{\theta}^i) \tag{1.5-3}$$

where $\mathbf{x}^i$ is the parameter vector of the ith member of the population and is structured as

$$\mathbf{x}^i = \begin{bmatrix} x^i_1 \\ x^i_2 \\ \vdots \\ x^i_{N_g} \end{bmatrix} \tag{1.5-4}$$

As can be seen, $\mathbf{x}^i$ has one element (denoted with a subscript) for each gene. However, it is not partitioned into chromosomes.

Based on the parameter vector of ith population member, the objective function can be evaluated. In particular,

$$f^i = \mathrm{f}(\mathbf{x}^i) \tag{1.5-5}$$

In the case of a GA, the objective function is referred to as a fitness function. It will be used in a "survival of the fittest" sense to determine which members of the population will mate to form the next generation. In the context of a GA, fitness is viewed in a positive sense, thus it is assumed that we wish to maximize the fitness function. Fortunately, it is a straightforward matter to convert between maximization of a function and the minimization of a function. At this point, we have enough background to discuss the computational aspects of a GA. However, before doing this, it is appropriate to briefly pause in our development and consider an example.

Example 1.5A Suppose the 13th member of the population has the genetic code

$$\boldsymbol{\theta}^{13} = \begin{bmatrix} \underbrace{0 \quad 1 \quad 0}_{\text{gene1}} & \underbrace{1 \quad 1 \quad 0 \quad 1}_{\text{gene2}} & \underbrace{0 \quad 1 \quad 1}_{\text{gene3}} \end{bmatrix}^T \tag{1.5A-1}$$

The decoding algorithm is on a gene-by-gene basis and is of the form

$$x_i = x_{mn,i} + \frac{x_{mx,i} - x_{mn,i}}{2^{l_i} - 1} \sum_{m=1}^{l_i} b_m 2^{m-1} \tag{1.5A-2}$$

where l_i is the number of bits of the ith gene, m is an index ranging from least significant (rightmost) bit to most significant (leftmost) bit for a given gene, b_m is the value of that bit (0 or 1), and $x_{mn,i}$ and $x_{mx,i}$ are the minimum and maximum values of the ith element of the parameter vector. For the problem at hand, we assume $x_{mn,1} = 5$, $x_{mx,1} = 10$, $x_{mn,2} = -2$, $x_{mx,2} = 0$, $x_{mn,3} = 0$, and $x_{mx,3} = 1$. In Section 1.9, we will formally consider the construction of fitness functions, and in Section 1.10, we will consider an engineering example. However, for the moment we will assume a purely mathematical fitness function given by

$$f(\mathbf{x}) = \frac{1}{(x_1 - 1)^2 + (x_2 + 2)^2 + x_3^2 + 1} \tag{1.5A-3}$$

Our goal is to compute the fitness of the 13th member of the population.

The solution to this problem is straightforward. Using (1.5A-2), we obtain $\mathbf{x}_1^{13} = 6.43$, $\mathbf{x}_2^{13} = -0.267$, and $\mathbf{x}_3^{13} = 0.429$. Substitution of these values into (1.5A-3) yields $f^{13} = 0.0297$. Note that while the division of the bits of (1.5A-1) into genes is very important to determine the fitness, the organization of genes into chromosomes is irrelevant for fitness evaluation and so has not been specified in this example.

At this point, we can now consider the primary aspects of a GA. These are illustrated in Figure 1.8. Therein, the first step is initialization that yields an initial population denoted $\mathbf{P}[1]$. Next, the fitness of every member of the population is evaluated. Based on the fitness, a mating pool $\mathbf{M}[k]$ is determined by the selection process. The individuals in this population will mate and genetic operators such as crossover, segregation, and mutation will be used to produce children who will form the next generation $\mathbf{P}[k + 1]$. A stopping criterion is then checked; this can be as simple as checking a generation number. Once the stopping criterion is met, the algorithm concludes by selection of the most fit individual of the final population to be the optimizer. This process is implemented by the argmax() operator, which returns the argument that maximizes its objective (which, in this case, is carried about by inspection of the finite population).

It is now appropriate to consider each of these operations in more detail, beginning with the initialization step. The genetic code of every member of the population is initialized at random. This yields an initial population of designs, denoted $\mathbf{P}[1]$. The next step in the algorithm is to compute

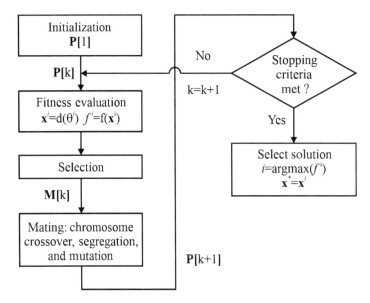

Figure 1.8 Canonical genetic algorithm.

the fitness of every member of the population. This is accomplished by applying (1.5-3) and (1.5-5) to every member of the population. Example 1.5A illustrates this step for a single member of the population.

The next step in the process is selection. In this step, members of the population **P**[k] are placed into the mating pool **M**[k]. Two algorithms to do this are roulette wheel selection and tournament selection. In both methods, the mating pool is initially empty and is filled one member at a time by repeatedly applying the selection mechanism. In roulette wheel selection, members of the population are drawn into the mating pool with a probability proportional to their fitness. In particular, the probability of individual i being drawn into the mating pool on a given draw is given by

$$p^i = \frac{f^i}{\sum_{i=1}^{N_p} f^i} \qquad (1.5\text{-}6)$$

When applying this particular algorithm, it is important that the fitness function be constructed so that $f^i \geq 0$. If this is not the case, it is possible to scale/adjust the fitness so that the condition is satisfied (for example, by adding a constant). Note that once a population member has been copied to the mating pool, it is not removed from the population. Thus, it can be copied to the mating pool multiple times.

In n-way tournament selection, n members of the population are selected at random, and the member of this subset with the highest function is put into the mating pool. Here again, the member placed into the mating pool is not removed from the population. In tournament selection, there is no restriction on the range of the fitness function, which provides a slight simplification.

The next step in process is mating, which is comprised of chromosome crossover, segregation, and mutation. In this step, pairs of parents are used to create pairs of children. Pseudo-code for this step appears in Table 1.2. Therein, underlined text denotes comments. Referring to the

Table 1.2 Pseudo-Code for Mating, Crossover, Segregation, and Mutation

```
for i =1 to N_p/2

          compute element indices
          i₁ = 2i - 1
          i₂ = 2i
          get genetic codes of parents
          θᵖ¹ = i₁th individual in M[k]
          θᵖ² = i₂th individual in M[k]

          apply genetic operators
          apply crossover to {θᵖ¹,θᵖ²} yielding {θᵃ¹,θᵃ²}
          segregate chromosomes of {θᵃ¹,θᵃ²} yielding {θᵇ¹,θᵇ²}
          apply mutation to {θᵇ¹,θᵇ²} yielding {θᶜ¹,θᶜ²}

          place children into next population
          θᶜ¹ becomes the i₁th individual of P[k + 1]
          θᶜ² becomes the i₂th individual of P[k + 1]
end
```

pseudo-code, θ^{p1} and θ^{p2} denote the genetic code from parents taken from the mating pool. A crossover operator is applied to the codes to form intermediate genetic codes θ^{a1} and θ^{a2}. Crossover occurs at random points within a given chromosome. If it occurs, then all elements of the portion of the chromosome strings of the two parents past the crossover point are interchanged. This is similar to biological crossover but not identical: In the case of biological crossover, this process occurs in the formation of the gametes. The net result is the same. Next, the chromosomes of θ^{a1} and θ^{a2} are randomly segregated to form the next stage of intermediate codes θ^{b1} and θ^{b2}, much as the chromosome pairs of a cell are segregated in forming gametes. Genetic codes θ^{b1} and θ^{b2} are then mutated to form θ^{c1} and θ^{c2}, which will be the children. The mutation operator consists of random interchanges of 0 and 1. The final step in the process is that the children θ^{c1} and θ^{c2} are placed into the next generation $\mathbf{P}[k+1]$. It should be noted that in many GAs there is only a provision to have one chromosome, in which case the chromosome segregation process does not come into play and so genetic diversity is brought about solely by crossover and mutation.

Example 1.5B Let us consider the crossover, segregation, and mutation processes on a numerical example. Figure 1.9 illustrates these operators. Therein $g_1 - g_4$ denote genes 1–genes 4 and c_1 and c_2 denote chromosomes 1 and 2, respectively. We start at the left of the diagram with the two parents θ^{p1} and θ^{p2}. The first operator applied is crossover. Assuming the crossover point is between the first and second bits in the first chromosome, and that no crossover occurs in the second chromosome, we arrive at θ^{a1} and θ^{a2}. The portions of the genetic code which have been acted on by the crossover operator are shown in bold. Next, chromosome segregation occurs. In this case, by a random choice the first child inherits the first chromosome from θ^{a2} and the second chromosome from θ^{a1}. The second child inherits the remaining chromosomes. Next, let us assume that mutation acts to change the status of the first bit of gene 3 in θ^{b2}. This yields θ^{c1} and θ^{c2}, where the affected bit is again indicated in boldface.

At this point, the main points of a canonical GA have been described. This canonical form is the type of algorithm first developed by Holland. The canonical GA can also be used to explain the effectiveness of GAs using schema theory. A schema is a pattern of bits; one reason GAs are so effective is that they can be shown to exhibit an implicit parallelism in which all schema that result in higher than average fitness tend to propagate. The interested reader is referred to textbooks on GAs such as those by Goldberg [4,5].

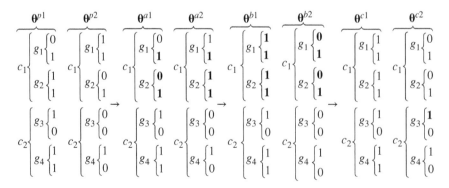

Figure 1.9 Chromosome crossover, segregation, and mutation in Example 1.5B.

At this point, an example would prove useful. Before embarking on such an example, however, it will prove useful to address one weakness of the GA. This weakness is the encoding algorithm. Although the representation of a member of the population as a string similar to that in DNA is intellectually appealing, it is also inconvenient. Further, a binary representation of real numbers suffers from finite resolution and the fact that sometimes large numbers of bits may need to change to accomplish a small change unless, for example, a Gray code scheme is used. As it turns out, it is possible to create a GA without encoding genes as strings and, instead, representing them with real numbers. Real-coded GAs are simpler to write than their canonical counterparts and very effective. They will be our next topic.

1.6 Real-Coded Genetic Algorithms

Real-coded GAs are very similar to canonical GAs except that instead of each gene being represented as a binary string, each gene is represented by a real number. This proves convenient for coding purposes and makes representing each gene to the numerical precision of floating-point numbers for a given machine (computer) straightforward. Beyond the change of the way in which a gene is represented, the algorithm presented in Figure 1.8 is still applicable, though we will need to modify the encoding, crossover, and mutation operators.

Encoding

Let us begin our development by considering the form of the genetic code for the *i*th individual. In the case of the canonical GA, this was given by (1.5-2), where each element of $\boldsymbol{\theta}^i$ was represented by a string. In the real-coded GA, (1.5-2) still applies. However, instead of each element of $\boldsymbol{\theta}^i$ being represented by a string, in a real-coded GA each element is represented by a real number. In fact, a real-coded GA could be written such that $\boldsymbol{\theta}^i = \mathbf{x}^i$. However, it will be convenient to provide a mapping of $\mathbf{x}^i$ to $\boldsymbol{\theta}^i$ so that the gene values of $\boldsymbol{\theta}^i$ fall into the domain [0,1].

The mapping between $\mathbf{x}^i$ and $\boldsymbol{\theta}^i$ is accomplished on a gene-by-gene basis. A simple choice is a linear map. Let x and θ denote a gene (element) of the $\mathbf{x}^i$ and $\boldsymbol{\theta}^i$. For a linear mapping, we have

$$x = \mathrm{d}_j(\theta) \tag{1.6-1}$$

where j denotes the gene number and

$$\mathrm{d}_j(\theta) = x_{mn,j} + (x_{mx,j} - x_{mn,j})\theta \tag{1.6-2}$$

where $x_{mn,j}$ and $x_{mx,j}$ denote the minimum and maximum values of the parameter.

Closely related to linear mapping is integer mapping, wherein x, $x_{mn,j}$, and $x_{mx,j}$ are integers. In this case, (1.6-2) still applies, but we require

$$\theta \in \left\{ \frac{0}{x_{mx,j} - x_{mn,j}}, \frac{1}{x_{mx,j} - x_{mn,j}}, \ldots, \frac{x_{mx,j} - x_{mn,j}}{x_{mx,j} - x_{mn,j}} \right\} \tag{1.6-3}$$

This mapping is useful in choosing, for example, between different types of steel in a design. If the third gene represented the type of steel used, and five types of steels were being considered, then $x_{mn,3} = 1$, $x_{mx,3} = 5$, and $\theta \in \{0.00, 0.25, 0.50, 0.75, 1.00\}$.

In some cases, the domain of a parameter may span many decades in magnitude. If the domain of the parameter is always positive, then a logarithmic mapping is appropriate. In this case,

$$d_j(\theta) = x_{mn,j}\left(\frac{x_{mx,j}}{x_{mn,j}}\right)^\theta \qquad (1.6\text{-}4)$$

Crossover

In the case of the canonical GA, crossover is accomplished by breaking the subject chromosomes of the parents at the same point and interchanging them to form the corresponding chromosomes of the children. This interchange substantially alters the gene where the chromosome is interchanged, and it results in an interchange of genes falling after the interchange point.

There are several algorithms to achieve crossover in real-coded GAs. One of them is single-point crossover, which can be readily generalized to n-point crossover. This method is straightforward and is readily illustrated by the example shown in Figure 1.10. Therein, the genetic code for two parents, $\boldsymbol{\theta}^{p1}$ and $\boldsymbol{\theta}^{p2}$, is shown. Observe that the elements of these vectors are real numbers in the domain [0,1]. In this algorithm, a crossover point is chosen at random and the genes after the crossover points are interchanged (these genes are in bold) to form two new genetic codes, $\boldsymbol{\theta}^{a1}$ and $\boldsymbol{\theta}^{a2}$, which will become children after the application of additional genetic operators such as mutation.

Single-point crossover is very similar to biological crossover. However, note that gene values cannot become altered using this operator. This limits the amount of genetic diversity that can be brought about.

Simple-blend crossover can be used to increase the genetic diversity. Let us consider the jth gene of parents $\boldsymbol{\theta}^{p1}$ and $\boldsymbol{\theta}^{p2}$. If the jth gene is being crossed over, the new gene values are determined by first computing a random number v given by

$$v = \mathrm{U}(\cdot) \qquad (1.6\text{-}5)$$

where $\mathrm{U}(\cdot)$ denotes a random number generator that generates a uniformly distributed random number in the range [0,1]. Next, the gene values of the children are set to

$$\theta_j^{a1} = \frac{1}{2}\left(\theta_j^{p1} + \theta_j^{p2}\right) + \alpha\left(\theta_j^{p2} - \theta_j^{p1}\right)v \qquad (1.6\text{-}6)$$

$$\theta_j^{a2} = \frac{1}{2}\left(\theta_j^{p1} + \theta_j^{p2}\right) - \alpha\left(\theta_j^{p2} - \theta_j^{p1}\right)v \qquad (1.6\text{-}7)$$

where α is an algorithm constant often taken to be 0.5. Observe that in this algorithm the average gene value for the children is the same as for the parents. Canonical GAs also have this property.

Another commonly used crossover algorithm is simulated binary crossover. This algorithm is designed to yield results that would be similar to those obtained by crossover within a gene using

$\boldsymbol{\theta}^{p1}$ = [0.21 0.34 0.89 0.26 **0.01 0.44**]^T

↕ Crossover point

$\boldsymbol{\theta}^{p2}$ = [0.92 0.45 0.02 0.92 **0.84 0.82**]^T

⇓

$\boldsymbol{\theta}^{a1}$ = [0.21 0.34 0.89 0.26 **0.84 0.82**]^T

$\boldsymbol{\theta}^{a2}$ = [0.92 0.45 0.02 0.92 **0.01 0.44**]^T

Figure 1.10 Single-point crossover.

the canonical GA. These properties include the fact that the sum of the gene values is the same for the children as the parents and that the children tend to be similar to the parents (which is also the case in simple-blend crossover). In this algorithm, first a random number in [0,1) is computed as in (1.6-5) from which a factor β is computed as

$$\beta = \begin{cases} (2v)^{\frac{1}{\eta_c+1}} & v \leq \frac{1}{2} \\ \left(\frac{1}{2(1-v)}\right)^{\frac{1}{\eta_c+1}} & v > \frac{1}{2} \end{cases} \qquad (1.6\text{-}8)$$

In (1.6-8), η_c is an algorithm parameter. Once β is found, the genes of the children are computed as

$$\boldsymbol{\theta}_j^{a1} = \frac{1}{2}\left[(1+\beta)\boldsymbol{\theta}_j^{p1} + (1-\beta)\boldsymbol{\theta}_j^{p2}\right] \qquad (1.6\text{-}9)$$

$$\boldsymbol{\theta}_j^{a2} = \frac{1}{2}\left[(1-\beta)\boldsymbol{\theta}_j^{p1} + (1+\beta)\boldsymbol{\theta}_j^{p2}\right] \qquad (1.6\text{-}10)$$

In both simple blend and simulated binary crossover, it is possible to generate gene values outside of [0,1]. For example, suppose the parent gene values are 0.5 and 1, and we use simple-blend crossover with $\alpha = 0.8$. Further, suppose $v = 0.9$. The resulting gene values for the children would be 0.39 and 1.11, the latter of which is outside the allowed set of values. In this case, gene repair becomes necessary. There are several approaches to this. Hard limiting the gene values is one approach. For example, a gene value calculated as 1.2 is simply represented as 1.0. Similarly, a gene value of -2.1 would be set to 0. Alternately, gene values can be ring mapped, which is to say corrected using a modulus 1 operator. In this case, 1.2 would be repaired to 0.2 and -2.1 would be repaired to 0.9. For integer-coded genes, additional repair is necessary in order to make sure that the gene values take on allowed values.

Our discussion of simple blend and simulated binary crossover has focused at the gene level, and so it is now appropriate to consider the application of these processes at the chromosome level. There are several approaches that could be taken. In scalar crossover, each gene in a chromosome that is being operated upon undergoes a crossover operation independently. Each gene in a chromosome uses a different v. In vector crossover, v is the same for all genes in a crossover. This leads to scalar simple-blend crossover, vector simple-blend crossover, scalar simulated binary crossover, and vector simulated binary crossover.

As can be seen, it is perhaps too easy to create new genetic operators. However, before leaving the topic, one final combination will be considered, namely, single-point simple-blend crossover. In this algorithm, a single gene of the chromosome is selected as a crossover point. Unlike single-point crossover, however, the crossover point is a gene, not a position between genes. That gene is operated upon with the simple blend operator. The remaining genes are interchanged as in single-point crossover. The process is illustrated in Figure 1.11. Therein, the third gene is randomly selected as

Figure 1.11 Single-point simple-blend crossover.

$\boldsymbol{\theta}^{p1} = [0.21\ 0.34\ 0.89\ 0.26\ 0.01\ 0.44]^T$
$\boldsymbol{\theta}^{p2} = [0.92\ 0.45\ 0.02\ 0.92\ 0.84\ 0.82]^T$
$\updownarrow$ Crossover point
$\boldsymbol{\theta}^{a1} = [0.21\ 0.34\ 0.10\ 0.92\ 0.84\ 0.82]^T$
$\boldsymbol{\theta}^{a2} = [0.92\ 0.45\ 0.81\ 0.26\ 0.01\ 0.44]^T$

the crossover point, α is taken as 0.5, and υ was 0.81 (determined at random). Single-point simple-blend crossover is somewhat similar to the inheritance of quantitative traits wherein multiple genes are involved in determining the extent of an attribute. One could also devise single-point simulated binary crossover or multipoint simple-blend crossovers in a straightforward fashion.

Mutation

In the process of the duplication of chromosomes to form gametes, occasionally errors in DNA duplication occur, which are referred to as mutations. Normally, particularly in mature species, these mutations have either little effect or a harmful effect. However, occasionally mutations yield beneficial traits. This is particularly true in the early evolution of a species. Mutation in GAs helps to explore the parameter space; in doing so, many mutations are harmful, but occasionally mutations occur which are very beneficial.

Referring to the pseudo-code in Table 1.2, genes are selected at random for mutation. Many chromosomes will have no mutations, and others may have several. There are several approaches that can be applied to mutate a gene. One approach, referred to as total mutation, is to simply reinitialize the mutated gene at random in the interval [0,1]. Figure 1.12 illustrates the total mutation operator applied to the third gene of a chromosome (which could be either *a1* or *a2* with respect to Table 1.2).

A second method is partial absolute mutation. In this method, if the *j*th gene is selected for mutation, the mutated value is given by

$$\theta_j^b = \theta_j^a + \sigma N(\cdot) \tag{1.6-11}$$

where $N(\cdot)$ denotes a zero-mean, unity variance Gaussian random number generator and σ is a desired standard deviation. A related method is partial relative mutation wherein the modified gene value may be expressed as

$$\theta_j^b = \theta_j^a(1 + \sigma N(\cdot)) \tag{1.6-12}$$

In partial absolute mutation, the size of the mutation is independent of the value of the gene; in relative mutation, the size of the perturbation tends to increase with the magnitude of the coded gene value. In both of these methods, gene repair is necessary because either method could result in a gene value outside of [0,1].

There are also vector-based mutation operators. In absolute vector mutation, we have

$$\boldsymbol{\theta}^b = \boldsymbol{\theta}^a + \sigma N(\cdot)\mathbf{V}(\cdot) \tag{1.6-13}$$

where $\mathbf{V}(\cdot)$ denotes a unit vector with the same dimensionality as the chromosome in a random direction. For relative vector mutation, we have

$$\boldsymbol{\theta}_j^b = \boldsymbol{\theta}_j^a(1 + \sigma N(\cdot)\mathbf{V}_j(\cdot)) \quad \forall j \in [1, 2, ..., N_g] \tag{1.6-14}$$

where N_g is the number of genes in the chromosome in question. Both of these operators require gene repair.

$\boldsymbol{\theta}^a$ = [0.21 0.34 **0.89** 0.26 0.84 0.82]^T
⇓ Mutation
$\boldsymbol{\theta}^b$ = [0.92 0.45 **0.25** 0.25 0.84 0.82]^T

Figure 1.12 Total mutation.

A final mutation operator we will consider is integer mutation. Recall that in integer coding, we are representing integers as real numbers and mapping them to a discrete set of values within [0,1]. In the case of integer-coded genes, the mutation operators just described are not appropriate. Thus, it is convenient to use a total mutation of these genes with the result discretized to the allowed values.

The mutation operators just described all have uniform and nonuniform versions. In uniform versions, the parameters of the algorithms (the mutation rates, and standard deviations of mutation amount) are constant with respect to generation number. In nonuniform mutation, these rates vary. Normally, high mutation rates and large standard deviations are used at the beginning of evolution while smaller rates and standard deviations are used toward the end of the study when the population is mature from an evolutionary point of view.

Example 1.6A Let us illustrate the use of an elementary GA based on a real-coded version of Figure 1.8. In particular, let us attempt to find the value of **x** that maximizes the function

$$f(\mathbf{x}) = \frac{1}{(x_1 x_2 - 6)^2 + 4(x_2 - 3)^2 + 1} \qquad (1.6A\text{-}1)$$

For the purposes of solving this problem, we will use linear coding with $x_{mn,1} = x_{mn,2} = 0$ and $x_{mx,1} = x_{mx,2} = 5$. In order to form the mating pool, we will use two-way tournament selection. We will also assume simple-blend crossover with $\alpha = 0.5$ and a probability of crossover of 0.5. We will use total mutation with the probability of a gene mutation of 0.1. In this example, we will consider three generations with a population size of 8. The code for this example can be obtained from Sudhoff [6].

Table 1.3 lists the data for the evolution. The first population's parameters were determined by random initialization. Next, the parameter vector for each member of the population was decoded yielding $\mathbf{x}^i$ for every member of the population. This is then evaluated by calculating $f^i = f(\mathbf{x}^i)$, where the fitness is given by (1.6A-1). At this point, the maximum, median, and mean fitness values of the population are 0.2213, 0.04123, and 0.06940, respectively. The next step is the creation of the mating pool using tournament selection. Applying simple-blend crossover, mutation, and associated gene repair operators yields the second population. Decoding to obtain the parameter vector for each element and then evaluating the fitness, it can be seen that maximum and mean fitness values of the population have increased to 0.4886 and 0.09160 while the median fitness is decreased to 0.03661. Repeating the process a third time, the maximum, median, and mean fitness increase to 0.7719, 0.08486, and 0.1774, respectively. After the third generation, taking the individual with the highest fitness yields $x^* = [1.942 \quad 3.233]^T$. This is starting to approach the correct solution, which is $x = [2 \quad 3]^T$. Note, however, that a different run will produce different results since a population size of 8 over three generations is inadequate to find a consistent solution.

Before concluding this example, it is interesting to repeat it with a more substantial population size and over a larger number of generations. Figure 1.13 illustrates an optimization run over eight generations with a population size of 25 individuals. In Figure 1.13(a), the best fitness in the population, median fitness of the population, and mean fitness of the population are shown versus generation. All three of these quantities can be seen to increase with generation number, though not monotonically. Figure 1.13(b) depicts a plot of the individuals in terms of their parameter vectors. Therein, an o, x, +, ∗, star, diamond, triangle, and pentagram depict members of the first through eighth generations, respectively. In addition, the darkness of the shading of each point is

Table 1.3 Example of Real-Coded Genetic Algorithm Evolution

				Generation 1				
P[1]	0.3210	0.8222	0.5718	0.6991	0.4416	0.4657	0.6754	0.9085
	0.8296	0.5707	0.2860	0.7963	0.4462	0.2790	0.9037	0.7472
$\mathbf{x}^i$	1.6051	4.1109	2.8591	3.4957	2.2079	2.3283	3.3769	4.5426
	4.1478	2.8534	1.4301	3.9813	2.2311	1.3952	4.5183	3.7360
f^i	0.1492	0.0295	0.0689	0.0148	0.2213	0.0530	0.0104	0.0081
M[1]	0.5718	0.4657	0.8222	0.3210	0.4657	0.8222	0.4416	0.3210
	0.2860	0.2790	0.5707	0.8296	0.2790	0.5707	0.4462	0.8296
				Generation 2				
P[2]	0.5718	0.1355	0.8975	0.6534	0.4657	0.5279	0.3627	0.8467
	0.2860	0.3321	0.7424	0.6579	0.2790	0.0321	0.6971	0.5786
$\mathbf{x}^i$	2.8591	0.6774	4.4874	3.2668	2.3283	2.6394	1.8133	4.2336
	1.4301	1.6606	3.7118	3.2895	1.3952	0.1604	3.4857	2.8932
f^i	0.0689	0.0313	0.0086	0.0419	0.0530	0.0155	0.4886	0.0249
M[2]	0.1355	0.5718	0.3627	0.5718	0.8467	0.8467	0.5718	0.6534
	0.3321	0.2860	0.6971	0.2860	0.5786	0.5786	0.2860	0.6579
				Generation 3				
P[3]	0.1355	0.5718	0.5462	0.3883	0.8467	0.8467	0.6235	0.6017
	0.3321	0.2860	0.3365	0.6467	0.5786	0.5786	0.5216	0.4223
$\mathbf{x}^i$	0.6774	2.8591	2.7308	1.9417	4.2336	4.2336	3.1174	3.0085
	1.6606	1.4301	1.6824	3.2334	2.8932	2.8932	2.6082	2.1114
f^i	0.0313	0.0689	0.1008	0.7719	0.0249	0.0249	0.1625	0.2335

proportional to generation number. Thus, it can be seen that as the evolution progresses, the population moves toward the objective function maximizer. At the end of the run, the estimate of the solution is $x^* = [2.133 \quad 2.755]^T$, which yields $f(x^*) = 0.797$. With the given limited number of generations and small population size, the results vary from run to run. This variance could be

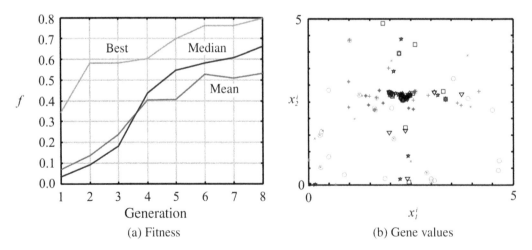

Figure 1.13 Fitness and gene values for Example 1.6A.

eliminated with a larger population and more generations, but it would make the resulting figures (particularly Figure 1.13(b)) hard to trace.

While Example 1.6A helped to illustrate the operation of a real-coded GA, practical GAs typically have several additional features to increase their efficacy. We will now briefly consider some of these additional operators.

Scaling

Scaling the fitness values of a population can result in improved algorithm performance. Scaling algorithms are only used in the context of roulette wheel selection. Consider a situation early in an evolution. Suppose individual A is significantly more fit than the remainder of the population. In this case, many copies of individual A will become part of the mating pool. In fact, without scaling, copies of individual A can rapidly dominate the population, leading to premature convergence and a failure to fully explore the search space. In this case, the scaling algorithm could be used to reduce the fitness of individual A so that it does not become as common as quickly, permitting the evolution to explore other avenues.

Conversely, late in the evolution, suppose that individual B is slightly better than the rest of the population. Because the fitness of most individuals is quite good, the chances of individual B being put into the mating pool are not much more than that of an average individual. In this case, it is appropriate to scale the fitness of population member B so as to increase its likelihood of being put into the mating pool. Another purpose of scaling is that for roulette wheel selection, all fitness values need to be positive.

Many scaling algorithms take the form of (1.6-15). Therein, a and b are coefficients, which vary by algorithm, and f' denotes the scaled fitness. Expressions for a and b are listed by method in Table 1.4. In this table, f_{min}, f_{max}, f_{avg}, and f_{med} denote the minimum, maximum, average, and median fitness of the population.

$$f' = \max(0, af + b) \tag{1.6-15}$$

Another scaling method approach is quadratic scaling. In this algorithm, the scaled fitness is calculated as

$$f' = af^2 + bf + c \tag{1.6-16}$$

Table 1.4 Linear Scaling Methods

Method	a	b	Comments
Offset scaling	1	$-f_{min}$	Ensures positive fitness
Standard linear scaling	$\dfrac{(k-1)f_{avg}}{f_{max} - f_{avg}}$	$f_{avg}(1-a)$	Most fit individual k times more likely to be in mating pool than average
Modified linear scaling	$\dfrac{(k-1)f_{med}}{f_{max} - f_{med}}$	$f_{med}(1-a)$	Most fit individual k times more likely to be in mating pool than median
Mapped linear scaling	$\dfrac{(k-1)f_{avg}}{f_{max} - f_{min}}$	$-af_{min} + 1$	Minimum fitness mapped to 1; maximum fitness to k
Sigma truncation	1	$-(f_{avg} - kf_{std})$	Average fitness maps to kf_{std}

where a, b, and c are given by

$$\begin{bmatrix} a \\ b \\ c \end{bmatrix} = \begin{bmatrix} f_{max}^2 & f_{max} & 1 \\ f_{avg}^2 & f_{avg} & 1 \\ f_{min}^2 & f_{min} & 1 \end{bmatrix}^{-1} \begin{bmatrix} k_{max} \\ 1 \\ k_{min} \end{bmatrix} \qquad (1.6\text{-}17)$$

In (1.6-17), k_{max} and k_{min} are algorithm constants selected such that $k_{max} > 1$ and $0 < k_{min} < 1$. In this approach, an individual with a fitness equal to the population average is scaled to 1, the most fit individual has a scaled fitness of k_{max}, and the least fit individual has a scaled fitness of k_{min}. Thus, the most fit individual is k_{max} more likely as an average fit individual to go into the mating pool, and the least fit individual is k_{min} times more likely as the average fit individual to be put in the mating pool (which is less likely than average since $k_{min} < 1$).

Diversity Control

The point of a population-based optimization is to search for the minimizer or maximizer using a large number of candidate solutions. Having large numbers of identical members of the population defeats the purpose of having a population in the first place. Diversity control algorithms examine the closeness of the solutions and compute a penalty for identical or nearly identical individuals; the penalized fitness is then used in the selection process.

Elitism

In biological systems, there is no guarantee that the most fit individual in a population will survive long enough to procreate because of environmental influences. For example, the most fit grizzly bear in Alaska could be killed by a hunter. While examples of this occur every day in the natural world, as engineers we may not want to lose the best induction motor design in our population. This has led to the elitism operator. This operator compares the most fit individual in a new population to the most fit individual in the previous population. If the best fitness in the new population is lower than the previous population, the most fit member of the previous population replaces a member of the new population. In this way, the best fitness in the population cannot go down. While this operator is distinctly nonbiological, it normally results in improved algorithm performance.

Migration

In biological populations, it sometimes occurs that a population becomes geographically dispersed, and then members of these geographically isolated regions continue to evolve on separate evolutionary tracks. Sometimes, because of storms or random events, a few members of one region migrate into another region. These individuals then have children with members of the region which they have migrated into, often yielding very fit offspring.

It is possible to create GAs where such migration occurs. In such a GA, each individual is associated with a region, and for the most part, only individuals within a given region interact. Each region acts as a separate population. Occasionally, however, a few individuals are transferred between regions.

At first consideration, one might conclude that incorporating such behavior would have little effect on the algorithm. However, in Sudhoff [7] it is shown that at least in the case of one motor design example, the migration operator had a profound effect on the performance of the algorithm.

That paper included a comparison between GA optimization and particle swarm optimization, and it found that choice of algorithm was less important than the use of migration in the GA or a somewhat analogous feature known as neighborhoods in particle swarm optimization.

Death

In the canonical GA, the entire population is replaced by children. Those individuals who are selected in the mating pool but do not undergo any crossover, segregation, or mutation "survive" to the next generation unchanged. However, in some GAs, rather than forming a mating pool of the same size as the population, a mating pool that is a fraction of the size of the population is formed, thereby generating a population of children a fraction of the size of the population. The children replace members of the existing population, so that a new population is formed which has the children, plus some members of the previous population. In order to make room for the children (to hold the population size constant), it becomes necessary to decide which members of the current population will "die" to make room for the introduction of children. The selection of those individuals to be replaced can be made on criteria other than those used for selection—for example, based on diversity (or more particularly lack thereof) or some other attribute. In some sense, these approaches are reminiscent of biological populations, where the boundaries between generations are gradual.

Local Search

After the initial generations of a GA, it is often the case that the most fit individuals are near a solution. In this case, some GAs will initiate local searches around the most fit individual. One way to accomplish this is to create a set of slight mutations of the most fit individual. Let us refer to the most fit individual as individual A, and refer to the population of mutations of this individual as $\mathbf{P}_A$. If the most fit individual in $\mathbf{P}_A$ is more fit than individual A, then the most fit individual of $\mathbf{P}_A$ replaces individual A in the population.

Deterministic Search

Many classical optimization methods are very effective if they initialized to be close to the solution. This is because many functions are "locally" convex. At the same time, while GAs are often very effective at getting close to a global optimum, they may not always converge rapidly from a good approximation of a solution to the exact solution. This suggests a combination of the two approaches, wherein the best individual of the final population is used to initialize a classical optimization method. To this end, the Nelder–Mead simplex method [1] is particularly attractive because it does not require gradients or Hessians. In performing such an optimization, it is normally the case that in problems with a mixed search space, those genes that represent discrete choices are held fixed.

Enhanced Real-Coded Genetic Algorithm

Before concluding this section, it is interesting to place all of the aforementioned operators into a single block diagram so that the relationships between the operators can be made clear. This is illustrated in Figure 1.14. Therein, initialization and the first fitness evaluation are as in the canonical GA diagramed in Figure 1.8. Next, the scaling algorithm is (optionally) employed. Note that the input and output of the scaling operator is the population $\mathbf{P}[k]$; the same symbol for the input and output is used since the scaling operator does not operate on the population; it merely adjusts

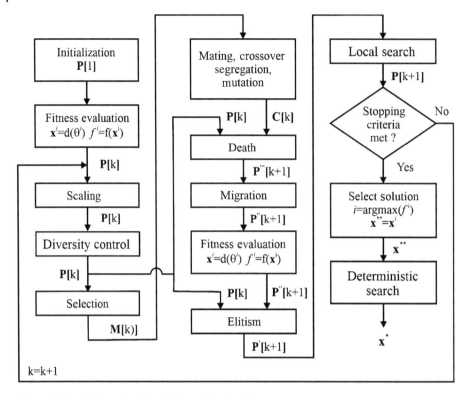

Figure 1.14 Enhanced real-coded genetic algorithm.

fitness values. The same is true of the diversity control operator, if utilized. Next, the selection operator creates a mating pool $\mathbf{M}[k]$. Based on the mating pool, the mating, crossover, segregation, and mutation operators yield a population of children $\mathbf{C}[k]$.

The death algorithm selects a member of the current population to be replaced by the children, yielding the beginnings of the next generation, $\mathbf{P}'''[k+1]$. The primes are used because this population will be modified before becoming the next generation. First the migration operator will be applied, which may occasionally move individuals from one region to another. This yields $\mathbf{P}''[k+1]$. Next, the gene values are decoded and the fitness evaluated. The result is again denoted $\mathbf{P}''[k+1]$ since the population itself does not change. The elitism operator is then applied, which compares the most fit individual of the previous population $\mathbf{P}[k]$ to the most fit individual in $\mathbf{P}''[k+1]$ and replaces that individual if appropriate. This yields population $\mathbf{P}'[k+1]$. The local search operator is employed to generate $\mathbf{P}[k+1]$.

Once the next generation is formed, the stopping criterion is checked. Often, the stopping criterion is a check to see if a sufficient number of generations have passed. If the stopping criterion has not been met, the process repeats, starting with scaling. If the stopping criterion has been met, then an estimate of the solution x^{**} is selected as the most fit individual of the population. This estimate can then be passed to a deterministic optimization algorithm for further refinement, yielding x^*.

The algorithm depicted in Figure 1.14 is probably more involved than is typical, because many optional operators have been used. For example, diversity control does not have to be used. Scaling is not necessary when using tournament or when the fitness values are always positive. A separate death algorithm is not necessary if the entire population is replaced by children. Random and deterministic search routines are often not used. Of the optional algorithms, elitism is fairly important. In addition, although not as commonly employed as elitism, the migration operator has been shown to have a significant impact on performance. Clearly, there is a wide variety of GA operators, and many books have been written on this subject. The reader is referred to references [4,5,8,9] for just a few of these. The good news is that almost all variations are effective optimization engines— they vary primarily in how quickly they converge and their probability of finding global solutions.

1.7 Multi-Objective Optimization and the Pareto-Optimal Front

Much of our focus in this chapter has focused on the single-objective optimization of an objective or fitness function f(x). However, in the case of design problems, it is normally the case that there are multiple objectives of interest. In this section, we begin our consideration of the multi-objective optimization problem.

Let us begin our discussion by the consideration of some designs of an electric motor. Suppose we are designing a motor and wish to minimize material cost and also to minimize loss. Further, suppose that we had eight designs available, each of which met all specifications for the machine, but each of which had a different cost and a different loss, as illustrated in Figure 1.15. Therein, each number listed is an index of a design. Thus, the x above point 1 has coordinates given by f($\mathbf{x}^1$).

Now let us consider the question of which design is the best. Let us first compare design 5 to design 7 (again using our shorthand notation of using a 5 to designate $\mathbf{x}^5$). Design 7 has lower loss and lower cost than design 5, and so it is clearly better than design 5. Using similar reasoning, we can see that design 8 is better than designs {4,5,7}. Now let us compare design 8 to design 6. Design 8 has lower loss, but design 6 has lower cost. At this point, it is difficult to say which is better. This leads to the concept of dominance.

Consider two parameter vectors $\mathbf{x}^a$ and $\mathbf{x}^b$. The parameter vector $\mathbf{x}^a$ is said to dominate $\mathbf{x}^b$ if f($\mathbf{x}^a$) is no worse than f($\mathbf{x}^b$) in all objectives and f($\mathbf{x}^a$) is strictly better than f($\mathbf{x}^b$) in at least one objective. The statement "$\mathbf{x}^a$ dominates $\mathbf{x}^b$" is equivalent to the statement that "$\mathbf{x}^b$ is dominated by $\mathbf{x}^a$." Returning to the example of the previous paragraph, we can say that design 8 dominates designs {4,5,7}.

Consider a set of parameter vectors denoted $\mathbf{X}$. The nondominated set $\mathbf{X}_{nd} \subset \mathbf{X}$ is a set of parameter vectors that are not dominated by any other member parameter vector in $\mathbf{X}$. In our example, $\mathbf{X} = \{1, 2, ..., 8\}$ and $\mathbf{X}_{nd} = \{1, 6, 8\}$. The nondominated set can be viewed as the set of best designs.

If we consider the set of parameters to be the entire domain for the parameter vector (i.e., Ω) then the set of nondominated designs is referred to as the Pareto-optimal set. The boundary defined by the objectives over this set is known as the Pareto-optimal front. Pareto was an Italian engineer,

Figure 1.15 Motor performance objective space.

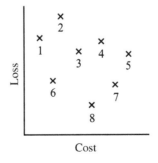

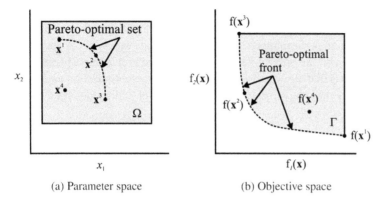

Figure 1.16 Pareto-optimal set and front.

economist, and philosopher who lived in the period of 1848–1923 (overlapping Mendel and Darwin) and who, besides his work in multi-objective optimization, formulated the Pareto principle that 80% of effects stem from 20% of causes (and that 80% of wealth is owned by 20% of the population).

The concepts of a Pareto-optimal set and Pareto-optimal front are illustrated in Figure 1.16 for a system with a two-dimensional parameter space (i.e., a two-dimensional parameter vector) and a two-dimensional objective space. As an example, if we were designing an inductor, the variables of the parameter space may consist of the number of turns and diameter of the wire, variables in the objective space might be inductor volume and inductor power loss. Returning to Figure 1.16(a) we see the parameter space defined over a set Ω (shaded). Members of this set include $\mathbf{x}^1, \mathbf{x}^2, \mathbf{x}^3$, and $\mathbf{x}^4$. Every point in Ω is mapped to a point in objective space Γ (also shaded) shown in Figure 1.16(b). For example, $\mathbf{x}^4$ in parameter space maps to $f(\mathbf{x}^4)$ in objective space. Points along the dashed line in Ω in Figure 1.16(a) are nondominated and form the Pareto-optimal set. They map to the dashed line in Figure 1.16(b) in objective space where they form the Pareto-optimal front.

The goal of multi-objective optimization is to calculate the Pareto-optimal set and the associated Pareto-optimal front. One may argue that in the end, only one design is chosen for a given application. This is often the case. However, to make such a decision requires knowledge of what the tradeoff between competing objectives is—that is, the Pareto-optimal front.

There are many approaches to calculating the Pareto-optimal set and the Pareto-optimal front. These approaches include the weighted sum method, the ε-constraint method, and the weighted metric method, to name a few [10]. In each of these methods, the multi-objective problem is converted to a single-objective optimization, and that optimization is conducted to yield a single point on the Pareto-optimal front. Repetition of the procedure while varying the appropriate parameters yields the Pareto-optimal front.

In order to illustrate this procedure, let us consider the ε-constraint method for a system with two objectives, $f_1(\mathbf{x})$ and $f_2(\mathbf{x})$, which we wish to minimize. In order to determine the Pareto-optimal set and front, the problem

$$\begin{aligned} \text{minimize} \quad & f_2(\mathbf{x}) \\ \text{subject to} \quad & f_1(\mathbf{x}) \leq \varepsilon \end{aligned} \qquad (1.7\text{-}1)$$

is repeatedly solved for different values of ε. This minimization can be carried out numerically using any single-objective minimization method. Solving (1.7-1) with $\varepsilon = \varepsilon_{1a}$ yields $\mathbf{x}^{1a}$ with objective

Figure 1.17 Calculation of Pareto-optimal front with ε-constraint method.

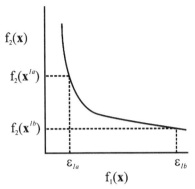

function values $(\varepsilon_{1a}, f_2(\mathbf{x}^{1a}))$ which is a point on the Pareto-optimal front, as shown in Figure 1.17. Solving (1.7-2) with $\varepsilon = \varepsilon_{1b}$ yields $\mathbf{x}^{1b}$ with objective function values $(\varepsilon_{1b}, f_2(\mathbf{x}^{1b}))$. Repeating the procedure over a range of ε values, the Pareto-optimal set and front can be determined.

In more concrete terms, suppose we wished to minimize volume and loss for some electromagnetic device. If our first objective was volume and our second objective was loss, we would repeatedly minimize loss with different constraints on the volume. In this case, each ε value would correspond to a different volume. For each numerically different volume constraint, we would get the corresponding loss and thereby—one single-objective optimization at a time—build up a Pareto-optimal front between volume and loss.

As it turns out, GAs are well-suited to compute Pareto-optimal sets. This is because they operate on a population. In multi-objective optimizations, the population of designs can be made to conform to the Pareto-optimal set. This allows a GA to determine the Pareto-optimal set and Pareto-optimal front in a single analysis without requiring the solution of a separate optimization for every point on the front as is needed in the weighted sum method, ε-constraint method, and weighted metric methods.

1.8 Multi-Objective Optimization Using Genetic Algorithms

GAs are well-suited for multi-objective optimization, and there are a large number of approaches that can be taken. The goal in all of these methods is to evolve the population so that it becomes a Pareto-optimal set.

Multi-objective GAs fall into two classes, nonelitist and elitist. The elitist strategies are particularly effective because they explicitly identify and preserve, when possible, the nondominated individuals. Elitist strategies include the elitist nondominated sorting GA (NSGA-II), distance-based Pareto GA, and the strength Pareto GA [11]. In order to keep the present discussion limited that we might soon start discussing device design, we will somewhat arbitrarily focus our attention on the elitist NSGA-II.

Before setting forth this algorithm, we will first consider the problem of finding the nondominated individuals in a population. Since finding the Pareto-optimal front is the goal of any multi-objective optimization, being able to identify those candidate solutions which are nondominated will be an important task. In a generic sense, the nondominated solutions will be given a favored status in many multi-objective methods. To identify the nondominated solutions, we will consider Kung's method. Consider a population $\mathbf{P}$. We wish to find the nondominated subset of $\mathbf{P}$,

Table 1.5 Pseudo-Code for Kung's Method

```
function R=front(S)
   if |S|=1
      R=S
   else
      i=floor(|S|/2)
      T=front(S^{1:i})
      B=front(S^{i+1:|S|})
      N=solutions of B not dominated
        by any solution of T
      R=T ∪ N
   end
end
```

which we will denote **E**. The first step in Kung's method is to sort the members of **P** from best to worst in terms of the first objective. In the event that two or more members have identical values of the first objective, this subset is ordered by the second objective (and by subsequent objectives if necessary). This sorted population will be denoted $\mathbf{P}_s$. The nondominated set **E** is then calculated as

$$\mathbf{E} = \text{front}(\mathbf{P}_s) \tag{1.8-1}$$

where front() is a recursively defined function. In order to describe this function, let $|\cdot|$ denote the number of elements in a set, and let $\mathbf{P}^{a:b}$ denote the ath to bth individuals in a population or set. With this notation, the pseudo-code for Kung's algorithm is listed in Table 1.5. Therein, **S** and **R** denote the input and output of the routine. The terms **T**, **B**, and **N** refer to intermediate variables that represent sets. The variable i is a scalar integer index.

In order to illustrate this algorithm, let us apply Kung's algorithm to the set of designs illustrated in Figure 1.15. For this problem, $\mathbf{P} = \{1, 2, 3, 4, 5, 6, 7, 8\}$. Considering cost to be our first objective, $\mathbf{P}_s = \{1, 6, 2, 3, 8, 4, 7, 5\}$. Figure 1.18 illustrates the calls to the front routine. Therein, the order of the calls and the input and output arguments to each call are indicated. The basis of this method is that whenever the front is called, its return argument **R** will never have any dominated members. This is a result of the following: (a) The sorting of the first objective ensures that no member of a return argument **T** can be dominated by a member of **B**, (b) every member of **B** is checked against every member of **T** before returning to the next higher level, and (c) a set of one is inherently a nondominated set. As a result, **T**, **B**, and **R** are all nondominated sets.

In executing this example, calls 4, 5, 7, 8, 11, 12, 14, and 15 return nondominated sets because their input and output argument set size is 1. In calls 3, 6, 10, and 13, no member of an input set **T** can be dominated by the corresponding set **B** because of the ordering with respect to the first objective. The return arguments of calls 3, 6, 10, and 13 cannot have dominated members because the input sets to these calls (**T** and **B**) were nondominated, no member of **T** can be dominated by a member of **B**, and every member of **B** is checked against every member of **T**. Thus, proceeding from the bottom of Figure 1.18 to the top as subroutine calls are made, we see that the process involves a gradual combining and sifting of smaller nondominated sets to yield a larger nondominated set.

One question that arises in this example is why we need function calls beyond call 1 as it generates the nondominated set. It must be remembered that the calls are recursive in nature and so the algorithm of call 1 cannot be completed without calls 2–15. Another question that is often asked is if

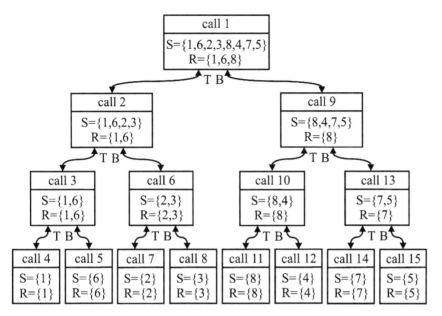

Figure 1.18 Application of Kung's method.

we really need to make calls to front when the number of elements in the input set is one. With some additional if-then constructs, calls to front with an input set size of one could in fact be avoided.

In addition to Kung's method, two more processes will be needed in order to set forth the elitist NSGA-II. The first of these is nondominated sorting (NDS). In order to understand NDS, let us consider again Figure 1.15 with population $\mathbf{P} = \{1, 2, 3, 4, 5, 6, 7, 8\}$. As we have already discussed, the nondominated set is $\{1,6,8\}$, which was found using Kung's algorithm. Let us consider this set front 1, denoted $\mathbf{F}_1$. Now suppose we remove the members of $\mathbf{F}_1$ from $\mathbf{P}$, and find the nondominated set of the remaining population by again applying Kung's algorithm. This will yield the second front, given by $\mathbf{F}_2 = \{2, 3, 7\}$. Now let us once more consider the population $\mathbf{P}$ but now less the members in $\mathbf{F}_1$ and $\mathbf{F}_2$. Finding the nondominated set of the remaining population by again applying Kung's algorithm yields the third front, $\mathbf{F}_3 = \{4, 5\}$. In this way, it is possible to associate with every member of the population a rank with regard to which front they are associated. This ranking will be used to compare different members of the population in order to determine which members will become a part of the mating pool.

Although a population member on the first front can be said to be superior to a member on the second front, the question arises as to how to compare solutions on the same front. This issue will be resolved using the concept of crowding distance. The crowding distance associated with a solution $\mathbf{x}^i$ is defined as

$$c^i = \sum_{o=1}^{O} \frac{f_o^{N_L(o,i,F_n)} - f_o^{N_S(o,i,F_n)}}{\max_{j \in P}\left(f_o^j\right) - \min_{j \in P}\left(f_o^j\right)} \quad i \in \mathbf{F}_n \quad (1.8\text{-}2)$$

where $\mathbf{F}_n$ is the set of indices in front n, we assume i is in front $\mathbf{F}_n$, o is an index of objective number, O is the number of objectives, $N_L(o, i, \mathbf{F}_n)$ returns the index of the individual with the next largest fitness (greater than individual i) in the oth objective in front $\mathbf{F}_n$, and $N_S(o, i, \mathbf{F}_n)$ returns the index of the individual with the next smallest fitness (less than individual i) in the oth objective in front F_n. In the case where there is no next larger or smaller individual in some objective—that is, it is a maximum or minimum in some objective—the crowding distance is taken as a very large number (pseudo-infinity).

Crowding tournament selection uses the concepts of front rank and crowding distance in order to decide which individuals to put into the mating pool. In this method, individuals $\mathbf{x}^{c1}$ and $\mathbf{x}^{c2}$ are randomly drawn from the current population. If one of these solutions has a better front rank than the other, it is copied into a mating pool. If the two individuals have the same front rank, then the one with the better (larger) crowding distance goes into the mating pool, as it has greater diversity in terms of objectives.

Example 1.8A The concept of crowding distance is illustrated in Figure 1.19. Suppose we wish to calculate the crowding distance of population member 7. Using (1.8-2), we have

$$c^7 = \frac{6-2}{8-1} + \frac{7-3}{11-3} = 1.07 \qquad (1.8A\text{-}1)$$

At this point, the remaining elements of the NSGA-II can be set forth. The process is illustrated in Figure 1.20. Therein, we start with a population $\mathbf{P}[k]$. Using crowded tournament selection (CTS), a mating pool $\mathbf{M}$ is created which is in turn used to create a population of children $\mathbf{C}$. Next, the current population $\mathbf{P}[k]$ is combined with the children $\mathbf{C}$ to form an enlarged population $\mathbf{R}$. Non NDS is used to divide this population into groups based on front, yielding sets $\mathbf{F}_1$, $\mathbf{F}_2$, and so on. Next, these sets are used to create the next generation $\mathbf{P}[k+1]$. This is done by filling $\mathbf{P}[k+1]$ with sets of fronts from first to last. At some point, shown as $\mathbf{F}_3$ in Figure 1.20, the full front will have more members

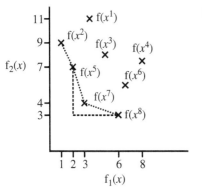

Figure 1.19 Crowding distance.

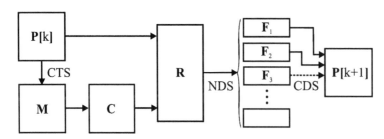

Figure 1.20 Elitist nondominated sorting genetic algorithm (NSGA-II).

than remain empty in $\mathbf{P}[k+1]$. At this point, crowding distance selection (CDS) is used to determine which members of the final included front are included in $\mathbf{P}[k+1]$. In particular, the members of $\mathbf{F}_3$ with the best crowding distance are used to complete the next population.

Using this algorithm, as the population evolves, it will come closer and closer to approaching the Pareto-optimal set, which will lead to a family of designs. We will use multi-objective optimization extensively in this book for the design of power magnetic devices.

1.9 Formulation of Fitness Functions for Design Problems

The previous sections of this chapter have focused on the design process, and some general-purpose single- and multi-objective optimization techniques. In this section, we consider methodologies to construct fitness functions.

In constructing the fitness function, we will have a variety of metrics, such as mass and loss, and also a number of constraints related to the appropriate operation and construction of the device of interest. It will often be the case that we will have to perform multiple analyses in order to evaluate metrics, and that some of these analyses may be computationally expensive. The fact that a variety of analyses of varying computational intensity will be required will be a significant consideration in the construction of the fitness function.

Let us begin our discussion of the construction of the fitness function with consideration of the constraints. Let us assume that we have C constraints, and use c_i to denote the status of the ith constraint. If the constraint is satisfied, we will set $c_i = 1$. If the constraint is not satisfied, we will have $0 \leq c_i < 1$. It is convenient to define c_i so that it approaches 1 as the constraint becomes closer to becoming satisfied.

In order to test constraints, it is convenient to define the less-than-or-equal-to and greater-than-or-equal-to functions as

$$\text{lte}(x, x_{mx}) = \begin{cases} 1 & x \leq x_{mx} \\ \dfrac{1}{1 + x - x_{mx}} & x > x_{mx} \end{cases} \tag{1.9-1}$$

$$\text{gte}(x, x_{mn}) = \begin{cases} 1 & x \geq x_{mn} \\ \dfrac{1}{1 + x_{mn} - x} & x < x_{mn} \end{cases} \tag{1.9-2}$$

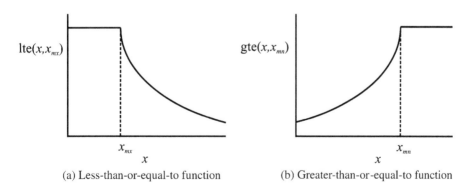

(a) Less-than-or-equal-to function   (b) Greater-than-or-equal-to function

Figure 1.21 Constraint functions.

These functions are illustrated in Figure 1.21.

As an example, we may require the height of a transformer to be less than a maximum value h_{mx}. If this is the first constraint and if we use $h(\mathbf{x})$ to represent the height in terms of our design parameters, our first constraint could be calculated as

$$c_1 = \text{lte}(h(\mathbf{x}), h_{mx}) \tag{1.9-3}$$

Next, let us consider our design metrics. The design metrics will also be a function of the parameter vector $\mathbf{x}$. In this text, common metrics will be mass and loss. Let the number of metrics be denoted M, and the value of the ith metric be denoted m_i. It will henceforth be assumed that all metric values are greater than zero.

As we will see, metrics and objectives are closely related but not synonymous. Metrics will be based on attributes of interest. Objectives will be based on the metrics, but also be influenced by constraints.

Let us now discuss the definition of the objective or fitness function. In keeping with the usual practice of GAs, we will assume that our fitness function (which is synonymous with the objective function) is of a form to be maximized. One approach to creating a fitness function begins with first forming a combined constraint. This can be done by averaging the constraints as

$$\bar{c} = \frac{1}{C} \sum_{i=1}^{C} c_i \tag{1.9-4}$$

Next, the elements of the fitness function are defined as either

$$f_i = \begin{cases} \varepsilon(\bar{c} - 1) & \bar{c} < 1 \\ m_i & \bar{c} = 1 \end{cases} \quad \text{(maximize metric)} \tag{1.9-5}$$

or

$$f_i = \begin{cases} \varepsilon(\bar{c} - 1) & \bar{c} < 1 \\ \dfrac{1}{m_i} & \bar{c} = 1 \end{cases} \quad \text{(minimize metric)} \tag{1.9-6}$$

where ε is a small positive number. The value has no impact on results but is convenient for plotting the fitness versus generation. It is commonly chosen to be on the order of 10^{-10}.

With this definition, designs that do not meet all constraints have negative fitness, while designs that do meet all constraints have positive fitness. If the constraints are not met, the fitness increases (becomes less negative) as the constraints become closer to being met. Note that the metric in the denominator in (1.9-6) does not pose a risk of a singularity because for most cases the metrics cannot be zero if all constraints are met. For example, a workable inductor design will have nonzero mass. Also, observe that all elements of the fitness function are negative (and equal) in the case where one or more constraints are not met; all elements of the fitness function are positive (and generally speaking not equal) if all of the constraints are met.

While the use of (1.9-5) and (1.9-6) is convenient for many design problems, there are cases when certain calculations are nonsensical unless constraints are met. There are also cases where certain calculations are computationally expensive, and there is a desire to avoid them if possible. For example, if predicting loss were computationally expensive, and that it is already known that not all constraints are passed, then there is little motivation to proceed with the loss calculation. Table 1.6 contains pseudo-code to treat this case. Therein, C_I and C_s denote the number of constraints imposed and satisfied, respectively, thus far in the current evaluation of the fitness function.

Table 1.6 Approach for Treating Constraints

```
calculate constraint c_i
```
$C_I = C_I + 1$
$C_S = C_S + c_i$

if ($C_S < C_I$)

$$\mathbf{f} = \varepsilon\left(\frac{C_S - C}{C}\right)\begin{bmatrix} 1 & 1 & \cdots & 1 \end{bmatrix}^T$$

```
    return
end
```

Both of these quantities are initialized to zero. After the constraint in question is calculated, the number of constraints imposed and number of constraints satisfied are updated. If the number of constraints satisfied falls below the number of constraints imposed, the fitness is calculated based on the number of constraints satisfied, and no further action (calculation of further constraints or metrics) is taken. This can lead to a significant computational savings for some problems.

The code block shown in Table 1.6 is repeated sequentially for every constraint. After all constraints are tested, the metrics may be calculated and the fitness assigned as in (1.9-5) and (1.9-6) for the case when all constraints are met. There are also variations of this procedure. For example, it may be convenient to calculate and test several constraints at a time.

1.10 A Design Example

In this section, we will conclude this chapter with a design example. In particular, we will endeavor to design a UI-core inductor using an optimization-based design process. At this point, we have not studied any magnetics or magnetic devices. This will be the subject of the remainder of this book. Since we are not yet in a position to derive or understand the relationships needed, an elementary analysis will be provided and the reader is asked to simply accept these relationships at face value for the time being. It should be observed that the analysis used to derive the needed relationships is very simplistic, but this does not matter because our purpose here is only to look at the design process. We will conduct a much more detailed analysis and design in subsequent chapters.

A UI-core inductor is depicted in Figure 1.22. Therein, the dark region is the magnetic core. The magnetic core conducts magnetic flux and is made of a U-shaped piece (the U-core) and an I-shaped piece (the I-core). Conductors (wires) pass through the middle of the U, a region called the slot, and also around the outside of the U-core to form a winding. The slot has a width denoted w_s and a height denoted d_s. The width of the core (both U- and I-cores) is denoted the core width w_c, and the length of the core pieces is l_c. The two cores are separated by an air gap g. The cross-sectional drawing (a) is the view that one would obtain by looking to the right from the left side of the side view (b) if the side view were cut in half (in the direction into the page). The light region indicates a winding comprised of N turns of wire.

Our goal in this design will be to design an inductor that has an inductance of at least L_{mn}, a flux density below B_{mx}, and a current density below J_{mx} at rated dc current i_{rt}. It is desirable to minimize

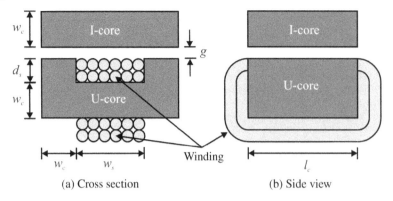

Figure 1.22 UI-core inductor.

the mass M of the inductor and to minimize the power loss at rated current, denoted P_{rt}. We will also constrain our designs to have a mass below M_{mx} and a power loss less than P_{mx}.

The free parameters in our design are the number of turns N, the slot depth d_s, the slot width w_s, the core thickness w_c, the core depth l_c, and the air gap g. Thus, our parameter vector may be expressed as

$$\mathbf{x} = \begin{bmatrix} N^* & d_s & w_s & w_c & l_c & g \end{bmatrix}^T \quad (1.10\text{-}1)$$

In (1.10-1), N^* is the desired number of turns rather than the actual number of turns N because, as a design parameter, we will let the number of turns be represented as a real number rather than an integer. This is because for large N this variable acts in a more continuous rather than discrete fashion. The actual number of turns is calculated from the desired number as

$$N = \text{round}(N^*) \quad (1.10\text{-}2)$$

In order to perform the optimization, we will need to analyze the device. It is assumed that windings occupy the entire slot, that the core is infinitely permeable, and that the fringing and leakage flux components are negligible. Again, for the reader unfamiliar with these terms, the analysis can be taken as a set of arbitrary mathematical equations; we will spend the rest of the book defining and developing more accurate expressions for these quantities.

With these assumptions, the mass of the design may be expressed as

$$M = 2(2w_c + w_s + d_s)l_c w_c \rho_{mc} + (2l_c + 2w_c + \pi d_s)d_s w_s k_{pf} \rho_{wc} \quad (1.10\text{-}3)$$

In (1.10-3), ρ_{mc} and ρ_{wc} denote the mass density of the magnetic core and wire conductor, respectively, and k_{pf} is the fraction of the U-core window occupied by conductor. Ideally, it would be 1, but 0.7 is a very high number in practice.

The next step is the computation of loss. The power dissipation of the winding at rated current may be expressed as

$$P_{rt} = \frac{(2l_c + 2w_c + \pi d_s)N^2 i_{rt}^2}{d_s w_s k_{pf} \sigma_{wc}} \quad (1.10\text{-}4)$$

In (1.10-4), σ_{wc} denotes the conductivity of the wire conductor.

There are constraints both on the inductance, flux density at rated current, and current density at rated current. These quantities may be expressed as

$$L = \frac{\mu_0 l_c w_c N^2}{2g} \tag{1.10-5}$$

$$B_{rt} = \frac{\mu_0 N i_{rt}}{2g} \tag{1.10-6}$$

$$J_{rt} = \frac{Ni}{w_s d_s k_{pf}} \tag{1.10-7}$$

In (10.1-5) and (1.10-6), μ_0 is the magnetic permeability of free space, a constant equal to $4\pi 10^{-7}$ H/m.

In order to formulate a fitness function, expressions (1.10-1)–(1.10-7) can be sequentially evaluated. Then constraint functions can be evaluated as

$$c_1 = \text{gte}(L, L_{mn}) \tag{1.10-8}$$
$$c_2 = \text{lte}(B_{rt}, B_{mx}) \tag{1.10-9}$$
$$c_3 = \text{lte}(J_{rt}, J_{mx}) \tag{1.10-10}$$
$$c_4 = \text{lte}(P_{rt}, P_{mx}) \tag{1.10-11}$$
$$c_5 = \text{lte}(M, M_{mx}) \tag{1.10-12}$$

Keeping with (1.9-4), we find the aggregate constraint

$$\bar{c} = \frac{1}{5}(c_1 + c_2 + c_3 + c_4 + c_5) \tag{1.10-13}$$

We will consider both single- and multi-objective optimization. For the single-objective case, we will minimize mass and our fitness is given by

$$f = \begin{cases} \varepsilon(\bar{c}-1) & \bar{c} < 1 \\ \dfrac{1}{M} & \bar{c} = 1 \end{cases} \tag{1.10-14}$$

For the multi-objective case, the fitness function will be taken as

$$f = \begin{cases} \varepsilon(\bar{c}-1)[1\ \ 1]^T & \bar{c} < 1 \\ \left[\dfrac{1}{M}\ \ \dfrac{1}{P_{rt}}\right]^T & \bar{c} = 1 \end{cases} \tag{1.10-15}$$

In (1.10-14) and (1.10-15), we will take $\varepsilon = 10^{-10}$.

For our design, let us consider a ferrite material for the core with $B_{mx} = 0.617$ T and $\rho_{mc} = 4680$ kg/m^3, and consider copper for the wire with $\rho_{wc} = 8890$ kg/m^3 and $J_{mx} = 7.5$ A/mm^2. We will take

Table 1.7 Domain of Design Parameters

Parameter	N	d_s (m)	w_s (m)	w_c (m)	l_c (m)	g (m)
Min. value	1	10^{-3}	10^{-3}	10^{-3}	10^{-3}	10^{-5}
Max. value	10^3	10^{-1}	10^{-1}	10^{-1}	10^{-1}	10^{-2}
Encoding	log	log	log	log	log	log
Chromosome	1	1	1	1	1	1

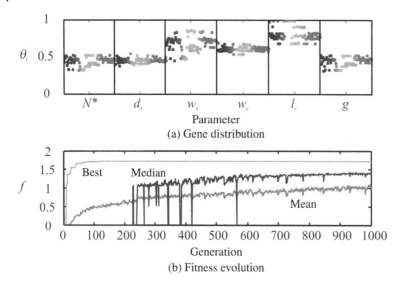

Figure 1.23 Single-objective optimization study.

rated current to be 10 A and take the minimum inductance L_{mn} to be 1 mH. Finally, let us take the maximum allowed mass as $M_{mx} = 1$kg, and the maximum allowed loss to be $P_{mx} = 1$W.

The next step in the design process is to determine the parameter space Ω. This is tabulated in Table 1.7. Some level of engineering estimation is required to select a reasonable range. However, situations where a range is incorrectly set are usually easy to detect by looking at the population distribution. We will return to this point.

We have now set forth a fitness function and a domain for the parameter vector, and so we can proceed to conduct an optimization. We will begin with a single-objective case. To conduct this study, a MATLAB-based genetic optimization toolbox known as GOSET was used. This open-source code and the code for this particular example are available at no cost in Sudhoff [6].

Figure 1.23 illustrates the progression of the study, which was conducted with a population size of 1000 over 1000 generations. Therein, Figure 1.23(a) shows the gene distribution at the end of the optimization. Recall that θ^i is the normalized value of the ith gene. Each design is shown in encoded parameter space as a series of dots, each with its own shade (for example, a certain dark shade may correspond to design 37 of the population). Because of the large numbers of designs, it is not possible to pick one design among all the designs. However, a sense of the distribution of the gene (parameter) values in the population can readily be obtained. The horizontal coordinate of each design within its parameter window is proportional to its ranked fitness—with lower ranks toward the left side of the window for a given parameter and higher ranks toward the right side of a given window. Considerable information can be discerned from the distribution plot. For example, there seems to be more sensitivity to d_s (which is tightly clustered) than to w_s (which is less tightly clustered). A distribution of a gene (parameter) value at the bottom or top of the range indicates that it may be appropriate to adjust the domain of that parameter.

Figure 1.23(b) depicts the fitness versus generation. The best fitness in the population, the median fitness of the population, and the mean fitness of the population are shown. Note that for a few generations, the best fitness is zero (actually slightly < 0), but then the best fitness increases rapidly until generation 150 or so, after which the fitness climes more slowly. The median and mean fitness rise more slowly than the fitness of the best individual. Observe that there are large rapid changes in

Turns $N = 25$ ($N^* = 25.3$)
Slot depth d_s (mm) = 8.39
Slot width w_s (mm) = 25.5
Core width w_c (mm) = 15.0
Core length l_c (mm) = 43.3
Air gap g (mm) = 0.255

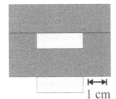

Figure 1.24 UI-core design.

the median fitness because this is a fitness of the median individual which changes from generation to generation. The mean fitness of the population is more stable. As can be seen, the mean fitness of the population occasionally goes down; this does not happen in the case of the fitness of the most fit individual in the population because of the elitism operator.

The most fit individual in the final population is illustrated in Figure 1.24, which lists the design parameters as well as a cross-sectional diagram. Note that $N^* = 25.3$ maps to $N = 25$ from (1.10-2). The design's mass is 0.578 kg, and the power loss at rated current is $0.9\overline{9}$ W. The inductance is right at 1 mH and the flux density is at 0.617 T. The current density of the design is 1.67 A/mm^2. It would appear that our design is against all constraint limits except those on mass and current density.

At this point, the question arises regarding how we know that our design is optimal. Unfortunately, we do not. There is not an optimization algorithm known that can guarantee convergence to the global optimum for a generic problem without known mathematical properties. However, in the GOSET code used for this example, a traditional optimization method (Nelder–Mead simplex) is used to optimize the design starting from the endpoint of the GA run, and this helps to ensure a local optimum. Still, there is no guarantee that a global optimum is obtained. Therefore, the prudent designer will re-run the optimization several times in order to gain confidence in the results. The runs can then be inspected to see if all runs converged to the same fitness. If significant variation in fitness has occurred, the use of more generations and/or a larger population size is indicated.

For our single-objective optimization problem, the optimization was re-run a multitude of times in order to investigate the variability of the design obtained from one run to the next. We will view variation of parameters and metrics in terms of normalized standard deviations. For example, the normalized standard deviation of the number of turns is the standard deviation of the number of turns divided by the median value of the number of turns for each design, interpreted as a percentage. Conducting the optimization process 100 times yielded the following normalized standard deviations: N with a 11%, d_s with a 4.4%, w_s with a 17%, w_c with a 6.4%, l_c with a 14%, and g with a 11% standard deviation. These may seem relatively large. However, it is interesting that normalized standard deviation in mass is only 1.0%. This indicates that there is a family of designs with equally good performance. It is interesting to observe that while appreciable design variation was found, every solution determined was viable (and not that different in terms of performance metrics).

It may seem objectionable to the reader that the results are not repeatable, which arises from the use of a random set of initial designs, and stochastic operators in the GA. However, even Newton's method will generate random variation in the solution of an optimization problem if the initial condition is selected at random. In Newton's method, providing a consistent initial condition will of course produce a consistent final answer; however, being consistent can merely mean being

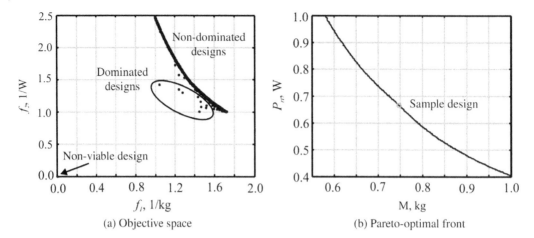

Figure 1.25 Multi-objective optimization results.

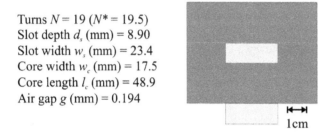

Turns $N = 19$ ($N^* = 19.5$)
Slot depth d_s (mm) = 8.90
Slot width w_s (mm) = 23.4
Core width w_c (mm) = 17.5
Core length l_c (mm) = 48.9
Air gap g (mm) = 0.194

Figure 1.26 Sample design from Pareto-optimal front.

consistently incorrect—which can happen if the algorithm becomes consistently trapped at a same local minimizer while missing the global minimizer.

Let us now turn our attention to a multi-objective optimization of the UI-core inductor. The only change in our approach is the replacement of (1.10-14) by (1.10-15). Using 2000 generations with a population size of 1000 yields the results shown in Figure 1.25. Figure 1.25(a) illustrates the objective space at the end of the optimization run. Each point in Figure 1.25(a) represents the objectives of a complete design. Nonviable designs have fitness values close to the origin (and are slightly negative in each axis). Viable, but dominated, designs are also apparent. The remaining designs are nondominated. In Figure 1.25(b) only the nondominated designs are shown, and the fitness elements are reciprocated so that mass and loss can be plotted directly. Again, each point represents the mass and loss of an individual design. Note the tradeoff between mass and loss. This will be a recurrent theme in this text. A sample design on the front is also indicated; this design is illustrated in more detail in Figure 1.26. The sample design has a mass of 0.75 kg, and a loss at rated current of 0.67 W. The inductance is just over 1 mH, and flux density at rated current is 0.617 T. The current density at rated current is 1.3 A/mm^2.

Figure 1.27 illustrates the gene distribution of the final population of designs. In Figure 1.27(a), the genes are sorted by objective 1. This means that the genes of designs with higher mass are toward the left of the parameter window, and genes of designs with lower mass are toward the right.

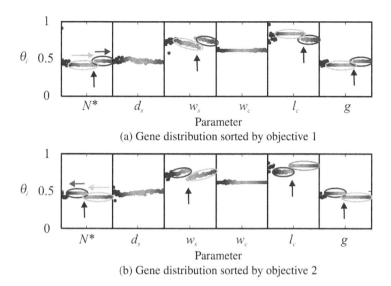

Figure 1.27 Sample design from Pareto-optimal front.

In Figure 1.27(b), the genes are sorted by objective 2, so that designs with the most loss are toward the left, and designs with the least loss are toward the right.

Unlike the case of single-objective optimization, the clustering of all values of a gene to approximately the same value is not expected in multi-objective optimization because the parameters will vary along the front. In order to illustrate this, consider the slot depth d_s. Observe that in Figure 1.27 (a) it has a slightly downward slope while in Figure 1.27(b) it has a slightly upward slope. This is because as we move from a low-mass high-loss design to a high-mass low-loss design the slot depth decreases. The core depth w_c can be seen to be approximately constant.

The remaining parameters undergo more interesting behaviors. Consider N^*, for example. Observe that in Figure 1.27(a), the nondominated solutions fall into two groups, which are indicated with a darker shaded and lighter shaded ellipses for lower and higher mass, respectively. The direction of decreasing mass is indicated with an arrow. Observe that the designs undergo a bifurcation indicated by a black vertical arrow. This can also be seen in Figure 1.27(b), wherein the sets are again circled. Note that the direction of decreasing mass is now to the left. The designs that are not in the two groups are dominated solutions. The bifurcation of the design space is also readily apparent in the slot width w_s, core length l_c, and air gap g. Such bifurcations in the design space can be the result of the change of a discrete variable, or the result of the design space moving into or out of a constraint. In this example, if we replace (10.1-1) with $N = N^*$, the bifurcation disappears. Of course, in doing this, our problem becomes strictly mathematical in nature since N must be an integer in practice.

References

1 E. K. Chong and S. H. Zak, *An Introduction to Optimization*, third edition. Hoboken: John Wiley & Sons, Inc., 2008.

2 J. F. Crow, *Genetics Notes*. New York: Macmillan Publishing Company, 1983.
3 J. H. Holland, *Adaptation in Natural and Artificial Systems*, Boston: MIT Press, 1992.
4 D. E. Goldberg, *Genetic Algorithms in Search, Optimization, & Machine Learning*. Boston: Addison Wesley Longman, Inc., 1989.
5 D. E. Goldberg, *The Design of Innovation*. Boston: Kluwer Academic Publishers, 2002.
6 S. D. Sudhoff, MATLAB codes for Power Magnetic Devices: A Multi-Objective Design Approach, second edition. Available: http://booksupport.wiley.com.
7 B. N. Cassimere and S. D. Sudhoff, Population based design of a permanent magnet synchronous machine, *IEEE Transactions on Energy Conversion*, vol. **24**, no. 2, pp. 347–357, 2009.
8 Z. Michalewicz, *Genetic Algorithms + Data Structures = Evolution Programs*, third, revised and extended edition. Berlin: Springer-Verlag, 1999.
9 A. Osyczka, *Evolutionary Algorithms for Single and Multicriteria Design Optimization*. Heidelberg: Physica-Verlag, 2002.
10 G. P. Liu, J. B. Yang and J. F. Whidborne, *Multiobjective Optimization and Control*. Exeter: Research Studies Press Ltd., 2003.
11 K. Deb, *Multi-Objective Optimization using Evolutionary Algorithms*. Chichester: John Wiley & Sons, Ltd., 2001.

Problems

1 It is desired to minimize the function

$$g(x_1, x_2) = (x_1 - 1)^2 + 3(x_2 - 4)^2 - 200$$

What is a possible fitness function (the answer is not unique) if using a canonical GA?

2 The fitness values of the members of a population are 23, 96, 42, 12, 8, 7, and 47. What is the expected number of times the individual with a fitness of 42 will appear in the mating pool? Use roulette wheel selection.

3 Consider two parents, with $x_1 = 0.2$ and $x_2 = 0.5$. Consider simulated binary crossover with $\eta_c = 1$. Form 10^5 children, and plot a histogram of the children arranged in 20 equally spaced bins on the interval 0 to 1. Implement gene repair using ring mapping.

4 Repeat Problem 3 with $x_2 = 0.25$.

5 Suppose a design problem was coded with five genes on a single chromosome, and single-point crossover was used. During the crossover between two parents, how many different children (in terms of genotype) could be produced?

6 A logarithmically mapped gene has a range from 10^{-3} to 10^6. If the value for this gene for a particular individual is 37.6, what is its' normalized value? What would its' normalized value be if it were linearly mapped.

7 During an evolution, the minimum, maximum, and average fitness of a population is 1.2, 270, and 40, respectively. If the most fit individual is to be three times more likely than the average fit individual to be selected, and the least fit individual is 1/3 as likely as the average fit individual to be selected, what is the scaled fitness of an individual whose raw fitness is 58.

8 Below are the objective function values for an inductor design. It is desired to minimize both objectives.

Individual	1	2	3	4	5	6
Mass (kg)	5	3	1	4	2	4
Loss (kW)	2	3	6	5	4	1

Use Kung's algorithm to identify the nondominated individuals.

9 Consider Problem 8. Compute the crowding distance of all nondominated individuals.

2

Magnetics and Magnetic Equivalent Circuits

The purpose of this chapter is to set forth the basic background needed to analyze electromagnetic and electromechanical systems. This will include a review of magnetics, particularly magnetostatics. We will see how a current i (in amperes, A) relates to an magnetomotive force (MMF) source F (in amperes, A or ampere-turns, A-t) that results in field intensity H (in amperes/meter, A/m) and in turn a flux density B (in Tesla, T). Associated with a flux density we will have flux Φ (in webers, Wb) and flux linkage λ (in webers, Wb, or weber-turns, Wb-t or equivalently volt seconds, Vs). The time rate of flux linkage λ will close the loop and create a voltage that affects the current. Throughout this work, SI units will be used. The reader is referred to [1] for a discussion of other unit systems. As a reminder, scalar variables are normally in italic font (for example, x) while vector and matrices are bold nonitalic (for example, **x**). The one exception to this in this chapter is that the flux, which is a scalar, is denoted by a nonitalic font. Functions of all dimensionalities are denoted by a nonitalic nonbold font (for example, x(θ)). Brackets in equations are associated with iteration number in iterative methods.

After the review of magnetics, the notion of magnetic equivalent circuits (MECs) will be explored as an approximate but often fairly accurate technique for the analysis of magnetic systems. This technique was introduced in references [2–5] to analyze electric machinery. Using this technique, we will see how to translate a geometrical description of a magnetic device consisting of coils of wire, a magnetic conductor such as iron, and permanent magnet material into an electrical circuit description. Much of the chapter will be devoted to showing how this technique can be made to be reasonably accurate because it will be heavily used for design later in this book.

2.1 Ampere's Law, Magnetomotive Force, and Kirchhoff's MMF Law for Magnetic Circuits

We begin our analysis of magnetic systems by developing the concept of kirchhoff's voltage law for magnetic systems—in particular the idea that the sum of the MMF drops around any closed loop is equal to the sum of the MMF sources. This is directly analogous to Kirchhoff's voltage law in an electric circuit, namely that the sum of the voltages (sources or drops) around any closed loop is zero. To formalize the idea for magnetic circuits, we begin with Ampere's law, which states that the line integral of the field intensity over a closed path is equal to the current enclosed by that path. This may be stated mathematically as

$$\oint_{l_\Gamma} \mathbf{H} \cdot d\mathbf{l} = i_{\text{enc},\Gamma} \tag{2.1-1}$$

Power Magnetic Devices: A Multi-Objective Design Approach, Second Edition. Scott D. Sudhoff.
© 2022 The Institute of Electrical and Electronics Engineers, Inc. Published 2022 by John Wiley & Sons, Inc.
Companion website: www.wiley.com/go/sudhoff/Powermagneticdevices

44 | *2 Magnetics and Magnetic Equivalent Circuits*

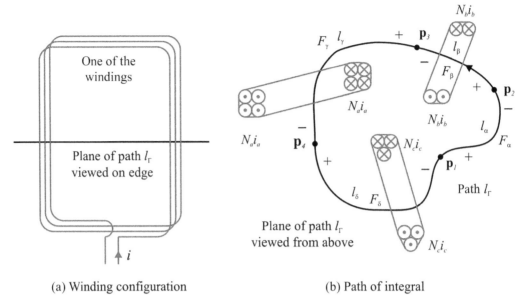

Figure 2.1 Ampere's law.

In (2.1-1), **H** is the field intensity (a vector field), **dl** is an incremental segment in the path (again a vector), l_Γ denotes a path, and $i_{enc,\Gamma}$ is the total current enclosed by that path, where the sign is determined in accordance with the right-hand rule.

In order to understand (2.1-1), let us consider Figure 2.1, which depicts a path in a plane intersected by several loops of wire known as windings. In Figure 2.1(a), one of three looped windings consisting of N turns of conductor (wire) is shown. In this portion of the figure, we view the plane of interest on edge. Figure 2.1(b) depicts the plane with all three windings from above. Therein, the ⊗ symbol is used to denote a conductor whose direction is into the page, and the ⊗ symbol is used to denote a conductor whose direction is out of the page. As can be seen, the path encloses one side of each winding.

Let us begin our discussion of (2.1-1) with consideration of the enclosed current $i_{enc,\Gamma}$. Consider Figure 2.1(b). Starting with the "a" phase, there are N_a conductors carrying a current i_a into the page within the path (and the same number coming out of the page outside of the path). The path also encloses N_b conductors carrying a current i_b out of the page, and N_c conductors carrying a current i_c into the page.

The positive direction for current flow is determined by the direction of the path and can be found using the right-hand rule. In particular, curling the fingers of one's right hand, the thumb of that hand will point in the direction considered to be positive. Thus, for our example, the "b" current is defined in the positive direction while the "a" and "c" currents are in the negative direction. The total current enclosed by the path in this example is thus given by

$$i_{enc,\Gamma} = -N_a i_a + N_b i_b - N_c i_c \qquad (2.1\text{-}2)$$

It is convenient to define MMF source as a product of conductors and current. We will symbolize MMF with a F. For this system, the MMF sources for the three windings are expressed as

$$F_a = -N_a i_a \qquad (2.1\text{-}3)$$

$$F_b = N_b i_b \tag{2.1-4}$$

$$F_c = -N_c i_c \tag{2.1-5}$$

Thus, for this example and in general,

$$i_{\text{enc},\Gamma} = \sum_{s \in S_\Gamma} F_s \tag{2.1-6}$$

In (2.1-6), S_Γ denotes the set of names (subscripts) for the MMF sources associated with path Γ. For this example, $S_\Gamma = \{\text{'}a\text{'},\text{'}b\text{'},\text{'}c\text{'}\}$.

Now let us turn our attention to the left-hand side of (2.1-1), which involves the integral of vector-valued field intensity around a closed path Γ. Before proceeding, it is worthwhile to remember that (2.1-1) applies to any closed path that one might imagine; further, the starting point (which is also the ending point) of the integration is irrelevant. To proceed, it is convenient to define the concept of an MMF drop. An MMF drop is defined as the line integral of vector-valued field intensity on an open path. In particular, MMF drop y is defined as

$$F_y = \int_{l_y} \mathbf{H} \cdot d\mathbf{l} \tag{2.1-7}$$

In (2.1-7), l_y denotes the path associated with F_y. As an example, in Figure 2.1(b), the path l_α for MMF drop F_α has a starting point $\mathbf{p}_1$ and ending point $\mathbf{p}_2$. The path l_δ for MMF drop F_δ has a starting point $\mathbf{p}_4$ and an ending point $\mathbf{p}_1$.

The integration around a closed path can be broken into the sum of integrations around open paths. For the present example, we have

$$\oint_{l_\Gamma} \mathbf{H} \cdot d\mathbf{l} = \int_{l_\alpha} \mathbf{H} \cdot d\mathbf{l} + \int_{l_\beta} \mathbf{H} \cdot d\mathbf{l} + \int_{l_\gamma} \mathbf{H} \cdot d\mathbf{l} + \int_{l_\delta} \mathbf{H} \cdot d\mathbf{l} \tag{2.1-8}$$

Each term in the right-hand side of (2.1-1) can be viewed as an MMF drop. Thus,

$$\oint_{l_\Gamma} \mathbf{H} \cdot d\mathbf{l} = F_\alpha + F_\beta + F_\gamma + F_\delta \tag{2.1-9}$$

or, in general

$$\oint_{l_\Gamma} \mathbf{H} \cdot d\mathbf{l} = \sum_{d \in D_\Gamma} F_d \tag{2.1-10}$$

where D_Γ denotes the set of names (subscripts) for the MMF drops associated with path Γ. For our example, $D_\Gamma = \{\text{'}\alpha\text{'},\text{'}\beta\text{'},\text{'}\gamma\text{'},\text{'}\delta\text{'}\}$.

Combining (2.1-1), (2.1-6), and (2.1-10), we arrive at Kirchhoff's MMF law:

$$\sum_{s \in S_\Gamma} F_s = \sum_{d \in D_\Gamma} F_d \tag{2.1-11}$$

which may be stated as follows:

The sum of the MMF drops around a closed loop is equal to the sum of the MMF sources for that loop.

This statement will prove crucial to our development of the MEC.

Example 2.1A Suppose there exists an H field given by

$$\mathbf{H} = x\mathbf{a}_x + (x^2 + y)\mathbf{a}_y \qquad (2.1\text{A-1})$$

where $\mathbf{a}_x$ and $\mathbf{a}_y$ are unit vectors in the x- and y-directions, respectively. What is the MMF drop starting at the point (1,1) and going directly to the point (5,2)?

To solve this problem, note that moving from (1,1) to (5,2) along a direct path the coordinates may be expressed as

$$x = 1 + 4\alpha \qquad (2.1\text{A-2})$$

$$y = 1 + \alpha \qquad (2.1\text{A-3})$$

As α goes from 0 to 1 our point on the (x, y) plane moves from (1,1) to (5,2). Our incremental path $d\mathbf{l}$ may be expressed as

$$d\mathbf{l} = \mathbf{a}_x dx + \mathbf{a}_y dy \qquad (2.1\text{A-4})$$

From (2.1A-2) and (2.1-3), we have that

$$dx = 4 d\alpha \qquad (2.1\text{A-5})$$

$$dy = d\alpha \qquad (2.1\text{A-6})$$

Combining (2.1A-4)–(2.1A-6), we obtain

$$d\mathbf{l} = 4\mathbf{a}_x d\alpha + \mathbf{a}_y d\alpha \qquad (2.1\text{A-7})$$

Substitution of (2.1A-1) and (2.1A-7) into (2.1-7) yields

$$F = \int_{\alpha=0}^{1} 4x + (x^2 + y) d\alpha \qquad (2.1\text{A-8})$$

Substitution of (2.1A-2) and (2.1A-3) into (2.1A-8) and evaluating yields an MMF drop of 23.83 A.

2.2 Magnetic Flux, Gauss's Law, and Kirchhoff's Flux Law for Magnetic Circuits

MMF in magnetic circuits is analogous to electromotive force in electric circuits. We will next turn our attention to magnetic flux, which is analogous to electric current. Magnetic flux is always associated with a surface through which it flows. In particular, magnetic flux is defined as the integral of the normal component of the flux density over a surface. Stated mathematically, the flux through open surface $\mathbf{S}_m$ may be expressed as

$$\Phi_m = \int_{S_m} \mathbf{B} \cdot d\mathbf{S} \qquad (2.2\text{-}1)$$

In (2.2-1), $\mathbf{S}_m$ denotes a surface through which the flux flows. Often, but not always, the surface is associated with a coil of wire (a winding) that forms the periphery of the open surface.

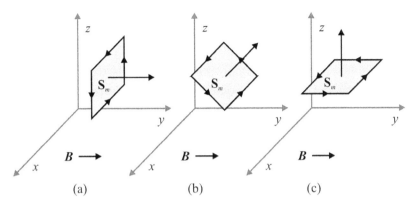

Figure 2.2 Calculation of flux.

Example 2.2A In order to understand (2.1-1), let us consider Figure 2.2. Therein, x, y, and z-axis of a Cartesian coordinate system are shown and $\mathbf{S}_m$ denotes a surface. Suppose there exists a uniform flux density field of $\mathbf{B} = 1.5\mathbf{a}_y$ T, where $\mathbf{a}_y$ is a unit vector in the direction of the y-axis. We wish to calculate the flux through open surface $\mathbf{S}_m$, whose periphery is a coil of wire wound in the indicated direction.

The direction of a surface is normal to its tangent plane at any point. When the surface is associated with a winding, as it is in this case, the direction of the surface is further determined by the direction of the winding using the right-hand rule. In particular, wrapping the fingers of one's right hand in the direction of the winding (the direction considered positive for current flow), one's thumb points in the direction of the surface. This requirement indicates that the surface $\mathbf{S}_m$ to be directed to the right rather than to the left in Figure 2.2(a). Let us suppose that the area of surface $\mathbf{S}_m$ is 2 m². In the case of Figure 2.2a, the direction of the surface and the flux density are parallel, and since the surface is a plane and the flux density is constant, we can readily solve (2.2-1) to yield a flux of 3 Tm² or 3 webers (Wb). In Figure 2.2(b), the surface is still a plane and the flux density is constant; however because the surface is tilted $\pi/4$ radians with respect to the flux density, the dot product operator causes a reduction of flux to $3\cos(\pi/4)$ Wb. In Figure 2.2(c), the surface and flux density are everywhere perpendicular; in this case the flux is reduced to zero.

Our definition of magnetic flux considers the integration of flux density over an open surface. Gauss's law addresses the integration of flux density over a closed surface (i.e., the surface bounding some volume). In particular, Gauss's law states that the surface integral of flux density over any closed volume is zero. In particular, we have

$$\oint_{S_\Gamma} \mathbf{B} \cdot d\mathbf{S} = 0 \qquad (2.2\text{-}2)$$

where $\mathbf{S}_\Gamma$ denotes a chosen closed surface. Any closed surface can be broken into a number of open surfaces. For example, a can (a closed volume) has a surface that can be broken into a top, a bottom, and the cylindrical side. Note that the direction of these surfaces must be consistently defined to be in or out. Let us break up $\mathbf{S}_\Gamma$ into a set of open surfaces denoted $\{\mathbf{S}_\alpha, \mathbf{S}_\beta, \mathbf{S}_\gamma...\}$ which together make up the closed surface $\mathbf{S}_\Gamma$. Further, let $\mathbf{O}_\Gamma$ be a set of the names (subscripts) for the components of the set of surfaces. For this case, we obtain

$$\mathbf{O}_\Gamma = \{'\alpha','\beta','\gamma'...\} \qquad (2.2\text{-}3)$$

From (2.2-2) and (2.2-3), we have

$$\sum_{o \in O_\Gamma} \int_{S_o} \mathbf{B} \cdot d\mathbf{S} = 0 \qquad (2.2\text{-}4)$$

where o denotes a member of the set O_Γ. Comparing (2.2-4) to (2.2-1), we obtain

$$\sum_{o \in O_\Gamma} \Phi_o = 0 \qquad (2.2\text{-}5)$$

In (2.2-5), Φ_o is the flux flowing through surface $\mathbf{S}_o$.

The expression (2.2-5) states that the sum of the fluxes flowing into any closed volume must be zero. Equation (2.2-5) is the basis for Kirchhoff's current law for magnetic circuits. However, before we can proceed, we must introduce the idea of a node in a magnetic circuit. In an electric circuit, we think of a node as an equipotential region; for example, it could be a bus that we consider to be at a uniform electric potential relative to ground. A magnetic node will have a similar property: it will be a (usually small) volume over which any MMF drop will be neglected. Since a node will be volume (even if small), (2.2-5) must apply. Thus, we have a version of Kirchhoff's current law for magnetic circuits:

The sum of the fluxes into (or out of) any node must be zero.

We will refer to this law as Kirchhoff's flux law.

2.3 Magnetically Conductive Materials and Ohm's Law For Magnetic Circuits

We have just derived Kirchhoff's MMF law for magnetic circuits and Kirchhoff's flux law for magnetic circuits. In this section, we will derive the last result we will need for our study of magnetic circuits, which is Ohm's Law.

Magnetic Materials

The interaction of materials with magnetic fields stems from two principal effects, the first of which arises because of the magnetic moment of the electron as it moves in its shell (inaccurately but usefully viewed as the moment due to the electron moving in an orbit around the nucleus), and the second effect is due to the magnetic moment that arises because of electron spin. The interaction of these two effects produces six different types of materials: diamagnetic, paramagnetic, ferromagnetic, antiferromagnetic, ferrimagnetic, and superparamagnetic.

The diamagnetic effect is present in all materials, and it has a net effect of acting against an applied magnetic field. Consider a magnetic moment caused by the electron moving within its shell. An applied field in the direction of the magnetic moment will create a force on the electron. However, the Coulomb force on the electron is unaffected. Since the effective radius of the electron's orbit (as viewed in crude terms) is fixed by the shell, the electron must slow, thereby reducing the moment of the electron moving in its shell and thereby causing a change in the moment that opposes the field. Considering the case where the applied field is opposed to the magnetic moment yields a similar conclusion. Diamagnetic materials are those materials in which the diamagnetic

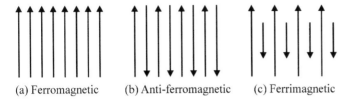

(a) Ferromagnetic   (b) Anti-ferromagnetic   (c) Ferrimagnetic

Figure 2.3 Atomic magnetic moment arrangements.

effect dominates other mechanisms so that the material's field tends to act against the applied field. In these materials, the net magnetic moment of the material is either zero or very small. Diamagnetic materials include hydrogen, helium, copper, gold, and silicon.

Let us next consider paramagnetic materials. In these materials, there is a small net magnetic moment. The moments due to electron spin do not quite cancel the moments of electrons within their shells. In this case, the moments will align with an applied field, which tends to aid the applied field. Materials in which the tendency to aid the field overcomes the diamagnetic effect are called paramagnetic. Paramagnetic materials include oxygen and tungsten.

In ferromagnetic, antiferromagnetic, ferrimagnetic, and superparamagnetic materials, the individual atoms possess a strong net moment. The arrangement of the atomic moments is shown in Figure 2.3. In ferromagnetic materials, the atoms possess a large net moment, and interatomic forces are such that they cause these moments to line up over regions called domains. In the absence of an external field, the moments of the domains cancel out; however, in the presence of an external field the domain walls will move in such a way that domains aligned with the applied field will grow; and unaligned domains will shrink, resulting in a strong interaction with external fields. These materials are very important in engineering applications. Materials that possess this property at room temperature include iron, nickel, and cobalt.

Antiferromagnetic materials possess the property that interatomic forces cause the magnetic moments of adjacent atoms to be aligned in an antiparallel fashion. The net magnetic moment of a group of atoms is therefore zero. These materials are not greatly affected by the presence of an external field.

Ferrimagnetic materials are similar to antiferromagnetic materials in that the magnetic moments of adjacent atoms are aligned in an antiparallel fashion; however, in this case, the moments between directions are unequal. These materials exhibit a significant response to an external field. While their response is, in general, less than that of ferromagnetic materials, one class of ferrimagnetic materials (ferrites) has a conductivity that is much lower than ferromagnetic materials, making it well suited for high-frequency applications. Examples of ferrites include iron oxide magnetite (Fe_3O_4), nickel–zinc ferrite ($Ni_{1/2},Zn_{12}Fe_2O_4$), and nickel ferrite ($NiFe_2O_4$).

A final class of materials that is useful in magnetic recording medium is superparamagnetic materials that consist of ferromagnetic materials in a nonferromagnetic matrix, which blocks domain wall movement.

For a more extensive discussion of the different types of materials, the reader is referred to Hayt [6]. However, our interest will primarily be in ferromagnetic and ferrimagnetic materials; other materials we will simply consider have the same response to a magnetic material as a vacuum.

In considering ferromagnetic and ferrimagnetic materials, let us assume that B and H are parallel so that we may treat these quantities as scalars. The relationship between flux density B and field intensity H is depicted in Figure 2.4. As can be seen, the value of flux density B cannot be uniquely established from the value of field intensity H for values in the vicinity of the origin. This is because

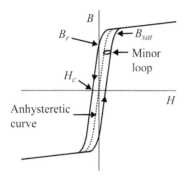

Figure 2.4 B–H characteristic of ferromagnetic and ferrimagnetic materials.

of an effect known as magnetic hysteresis which results from domain wall motion. In particular, as the domain walls move, they become "stuck" at pinning sites that are defects in the atomic arrangement. The movement of the domain walls past the pinning sites requires an amount of nonrecoverable energy; this loss results in hysteresis. The direction of the traversal of the loop is indicated with arrows in Figure 2.4. While large excitations of the core results in the complete traversal of the loop, it is also possible to enter the central region of the loop. Such a traversal is known as a minor loop. Parameters of interest that characterize the loop include the coercive force H_c and residual flux density B_r. Hysteresis is difficult to model; detailed trajectories of B versus H are normally calculated using either a variant of the Jiles–Atherton model [7–10] or the Preisach model [11]. However, even these techniques have limited accuracy. Estimates of hysteresis losses are often accomplished using the Steinmetz relationship [12]; however, this is only a behavioral representation. We will discuss all of these models in some detail in Chapter 6.

Because of the complexity of predicting or even describing the behavior of the B versus H characteristic, for the present we will confine our attentions to a very simple model—simply ignoring the hysteresis effect. While such an approach may appear rash, it is generally the case that most of the energy storage in electromagnetic systems occurs in air, not in the magnetic material. As a result, for the majority of designs, hysteresis losses are modest.

Neglecting hysteresis yields the anhysteretic B versus H characteristic, which is also shown in Figure 2.4. Note that even this characteristic is not trivial; after the flux density reaches a value of B_{sat} the slope of the anhysteretic curve decreases markedly as the material is said to saturate. Normally, a high degree of saturation is to be avoided because it can result in large currents and high losses.

In addition to saturating at high flux density, the anhysteretic curve is subject to an additional nonlinearity at very low flux levels (not shown in Figure 2.4). In particular, the slope of the B versus H characteristic also drops at very low flux levels. This can be an important effect when employing parameter identification using small-signal perturbations.

The anhysteretic relationship between flux density B and field intensity H can be captured in many different forms. In the absence of magnetic materials, the relationship between B and H may be expressed as

$$B = \mu_0 H \tag{2.3-1}$$

where μ_0 is the magnetic permeability of free space which is equal to $4\pi \cdot 10^{-7}$ Henries per meter (H/m). In magnetic materials, the magnetic moments of the material alter (2.3-1) and introduce another term. In particular, the relationship between B and H becomes

$$B = \mu_0(H + M) \tag{2.3-2}$$

or

$$B = \mu_0 H + M \tag{2.3-3}$$

where M is referred to as the magnetization and has units of A/m in (2.3-2) and T in (2.3-3). Unfortunately, both (2.3-2) and (2.3-3) are in common use. The magnetization represents the flux density due to the material.

In the linear part of the anhysteretic relationship, the magnetization is related to the field intensity by

$$M = \chi H \tag{2.3-4}$$

if M is in A/m or

$$M = \mu_0 \chi H \tag{2.3-5}$$

if M is in T. In (2.3-4) and (2.3-5), χ is known as susceptibility and is dimensionless. Substitution of (2.3-4) into (2.3-2) or (2.3-5) into (2.3-3) yields

$$B = \mu_0 (1 + \chi) H \tag{2.3-6}$$

which may be written as

$$B = \mu_0 \mu_r H \tag{2.3-7}$$

where μ_r is the relative magnetic permeability (i.e., relative to free space) and may be expressed as

$$\mu_r = 1 + \chi \tag{2.3-8}$$

For ferromagnetic and ferrimagnetic materials, values of susceptibility and relative permeability commonly range from the tens to tens of thousands.

Returning to (2.3-7), we note it may be written more simply as

$$B = \mu H \tag{2.3-9}$$

where μ is the total magnetic permeability, which may be expressed as

$$\mu = \mu_0 \mu_r \tag{2.3-10}$$

In order to represent the nonlinearity in the magnetic material, we may represent the permeability as being a function of either the field intensity or the flux density so that

$$B = \mu_H(H) H \tag{2.3-11}$$

or

$$B = \mu_B(B) H \tag{2.3-12}$$

In (2.3-11) and (2.3-12), the H or B subscript of the permeability function is used as a reminder that the function in question is written for a specific argument.

Let us briefly consider the anhysteretic properties of an actual material. Figure 2.5 depicts the anhysteretic $B - H$ characteristic for a sample of M47 silicon steel. This was measured using the procedure set forth in Shane and Sudhoff [13]. Observe that the relationship between B and H is linear for field intensities less than about 100 A/m. For field intensities greater than about 100 A/m, saturation begins to occur as all domains become aligned. The corresponding permeability functions, normalized to the permeability of free space, are shown in Figure 2.6. For low field values, the relative permeability is nearly 10^4. However, the permeability function can be seen to

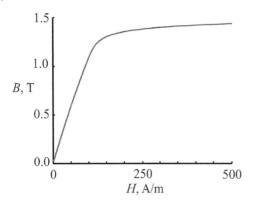

Figure 2.5 B - H Characteristic of a M47 silicon steel.

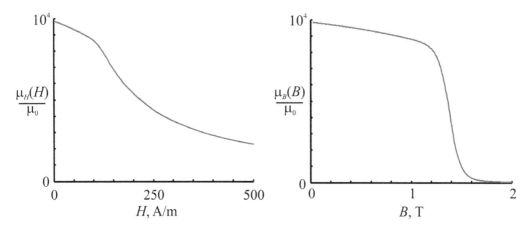

Figure 2.6 Permeability functions for a M47 silicon steel.

fall off with H or B. The fall-off of permeability with B is particularly steep as the material saturates magnetically. It should be noted that the magnetic characteristics of a given steel can vary significantly with processing.

Now let us consider the construction of the permeability functions. One approach is to simply view these as interpolated functions of tabulated data. However, analytical expressions are also useful. To this end, let us consider the relationship between B and H given by (2.3-3). With this form, one approach to representing the magnetization as a function of field intensity is

$$M(H) = \text{sgn}(H) \sum_{k=1}^{K} \frac{m_k |H/h_k|}{1 + |H/h_k|^{n_k}} \qquad (2.3\text{-}13)$$

wherein m_k, h_k, and n_k are coefficients, and K represents the number of terms. It is required that $n_1 = 1$, and $n_k > \forall k \in [2...K]$, and that $h_k > \forall k \in [1...K]$. The sgn() function is defined by

$$\text{sgn}(x) = \begin{cases} 1, & x > 0 \\ 0, & x = 0 \\ -1, & x < 0 \end{cases} \qquad (2.3\text{-}14)$$

2.3 Magnetically Conductive Materials and Ohm's Law For Magnetic Circuits

Let us consider the first term in the series for which $n_1 = 1$. From (2.3-13), we can see that the magnetization will rise nearly linearly with H for small H, but as H/h_1 becomes significant relative to 1, the rate of increase in the first term will decay. As $H \to \infty$, the first term goes to m_1 which is the saturated value of magnetization.

The remaining terms in (2.3-13) are somewhat similar in behavior: they increase linearly with H for a time; but instead of going to a constant, they eventually decay to zero because of the requirement that $n_k > 1$ for $k > 1$. These terms are primarily used to shape the initial part of the magnetization characteristic.

From (2.3-3), (2.3-11), and (2.3-13), one obtains

$$\mu_H(H) = \mu_0 + \sum_{k=1}^{K} \frac{m_k}{h_k} \frac{1}{1 + |H/h_k|^{n_k}} \qquad (2.3\text{-}15)$$

For the purposes of solving nonlinear MECs, the derivative of the permeability with respect to the field intensity is occasionally needed. From (2.3-15), we obtain

$$\frac{d\mu_H(H)}{dH} = -\operatorname{sgn}(H) \sum_{k=1}^{K} \frac{m_k n_k}{h_k^2} \frac{|H/h_k|^{n_k - 1}}{(1 + |H/h_k|^{n_k})^2} \qquad (2.3\text{-}16)$$

Next, let us characterize permeability as a function of B. Unfortunately, the form of this expression is difficult to capture analytically. However, one approach is suggested in Shane and Sudhoff [13]. Manipulating (2.3-3), we may write

$$\mu_B(B) = \mu_0 \frac{r(B)}{r(B) - 1} \qquad (2.3\text{-}17)$$

where $r(B)$ is the ratio of flux density to magnetization (in T)

$$r(B) \triangleq \frac{B}{M(B)} \qquad (2.3\text{-}18)$$

The function $r(B)$ is constant for small values of flux density and then begins to rise in an almost linear fashion as B increases. Figure 2.7 illustrates this function for the sample of M47 steel considered earlier.

The function $r(B)$ can be written as

$$r(B) = \frac{\mu_r}{\mu_r - 1} + g(B) \qquad (2.3\text{-}19)$$

Figure 2.7 r(B) for a M47 silicon steel sample.

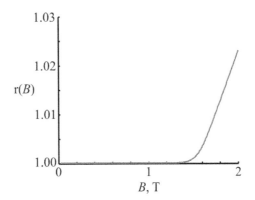

where μ_r is the relative permeability in the magnetically linear region, and g(B) is a function that is essentially zero in the magnetically linear range but eventually begins to rise nearly linearly once saturation occurs. The reader can verify that upon setting g(B) = 0 in (2.3-19) then combining (2.3-17), (2.3-19), and (2.3-3), one obtains (2.3-7), which is our expression relating B and H for magnetically linear conditions.

From (2.3-19) observe that g(B) only differs from r(B) by a constant. From Figure 2.5, we infer that dg(B)/dB is a "soft" step function: it starts at zero, rises, and then becomes a constant. This is suggestive of a sigmoid function. Thus

$$\frac{dg(B)}{dB} = \text{sgn}(B) \sum_{k=1}^{K} \frac{\alpha_k}{1 + e^{-\beta_k(|B| - \gamma_k)}} \tag{2.3-20}$$

where the summation is used in order to gain generality, if needed. In (2.3-20), if $K = 1$, then α_1 sets the slope at which g(B) rises after the transition point, γ_1 sets the transition point, and β_1 sets the "stiffness" of the step function. For $K > 1$, each term in the summation has this form, and α_k, γ_k, and β_k set the slope, transition point, and stiffness of the kth term, respectively.

From (2.3-20), it follows that

$$g(B) = \sum_{k=1}^{K} \alpha_k |B| + \delta_k \ln\left(\varepsilon_k + \zeta_k e^{-\beta_k |B|}\right) \tag{2.3-21}$$

where

$$\delta_k = \frac{\alpha_k}{\beta_k} \tag{2.3-22}$$

$$\varepsilon_k = \frac{e^{-\beta_k \gamma_k}}{1 + e^{-\beta_k \gamma_k}} \tag{2.3-23}$$

$$\zeta_k = \frac{1}{1 + e^{-\beta_k \gamma_k}}. \tag{2.3-24}$$

For simplicity, and numerical prudence, it is advantageous to rewrite (2.3-20) as

$$\frac{dg(B)}{dB} = \text{sgn}(B) \sum_{k=1}^{K} \frac{\eta_k}{\theta_k + e^{-\beta_k |B|}} \tag{2.3-25}$$

where

$$\eta_k = \alpha_k e^{-\beta_k \gamma_k} \tag{2.3-26}$$

$$\theta_k = e^{-\beta_k \gamma_k}. \tag{2.3-27}$$

As we solve nonlinear MECs, we will also need to calculate the derivative of the permeability with respect to flux density. From (2.3-17) and (2.3-19), we have

$$\frac{d\mu_B(B)}{dB} = -\frac{\mu_0}{(r(B) - 1)^2} \frac{dg(B)}{dB} \tag{2.3-28}$$

where dg(B)/dB is found from (2.3-20).

The steps to calculate $\mu_B(B)$ are as follows. First, g(B) is calculated using (2.3-21); then r(B) is found from (2.3-19), and $\mu_B(B)$ is found from (2.3-17). Evaluating dg(B)/dB from (2.3-26) and making use of r(B), $d\mu_B(B)/dB$ is readily found from (2.3-28).

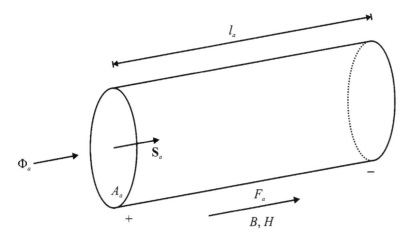

Figure 2.8 Fields in a material sample "a."

Ohm's Law

Thus far in this chapter, we have considered (a) the relationship between MMF sources and MMF drops in the form of Kirchhoff's MMF law and (b) the relationship between fluxes going into a common volume or node that is governed by Kirchhoff's flux law. We will now use the material relationships that we have just set forth to determine the relationship between MMF drop and flux flow, which is captured by Ohm's law for MECs.

To this end, consider Figure 2.8, which shows a piece of magnetic material "a." We are interested in relating the flux going into the material, denoted Φ_a, through surface $\mathbf{S}_a$ to the MMF drop across the length of the material F_a. The length of the material is l_a; the area of the uniform cross section of the material is A_a. Since the flux is assumed to go into one end and come out of the other, with no flux leaving along the length, this situation is often described as a flux tube.

Assuming that the field intensity is uniform, then from (2.1-7) the MMF drop across the material may be expressed as

$$F_a = H_a l_a \qquad (2.3\text{-}29)$$

where H_a is the field intensity $\mathbf{H}$ in the indicated direction (parallel to the length). Next, assuming that the flux density $\mathbf{B}$ is uniform and parallel with the surface S_a (and can thus be represented with a scalar B_a whose direction is the same as the surface), from 2.1-1) the flux into the surface may be expressed as

$$\Phi_a = B_a A_a \qquad (2.3\text{-}30)$$

Manipulating (2.3-9), (2.3-29), and (2.3-30), we arrive at

$$F_a = R_a \Phi_a \qquad (2.3\text{-}31)$$

where R_a is referred to as a reluctance of sample "a" (or the flux tube "a") which may be expressed as

$$R_a = \frac{l_a}{A_a \mu_a} \qquad (2.3\text{-}32)$$

In (2.3-32), μ_a is the permeability of the material "a." Reluctance is analogous to resistance and has units of 1/H. Equation (2.3-31) represents Ohm's law for a magnetic circuit element.

As an alternative to (2.3-31) and (2.3-32), we may also write the equivalent relationship as

$$\Phi_a = P_a F_a \qquad (2.3\text{-}33)$$

where P_a is referred to as the permeance of sample "a," which may be expressed as

$$P_a = \frac{A_a \mu_a}{l_a} \qquad (2.3\text{-}34)$$

Permeance has units of H and is analogous to conductance.

For nonlinear magnetic systems, the permeability is expressed as a function of either flux density or field intensity. Formally indicating this dependence, Ohm's law may be expressed as

$$F_a = R_a(\xi)\Phi_a \qquad (2.3\text{-}35)$$

where

$$R_a(\xi) = \begin{cases} \dfrac{l_a}{A_a \mu_H(F_a/l_a)} & \text{Form 1}: \xi = F_a \\ \dfrac{l_a}{A_a \mu_B(\Phi_a/A_a)} & \text{Form 2}: \xi = \Phi_a \end{cases} \qquad (2.3\text{-}36)$$

Alternately, in terms of permeance, (2.3-33) may be expressed as

$$\Phi_a = P_a(\xi) F_a \qquad (2.3\text{-}37)$$

where

$$P_a(\xi) = \begin{cases} \dfrac{A_a \mu_H(F_a/l_a)}{l_a} & \text{Form 1}: \xi = F_a \\ \dfrac{A_a \mu_B(\Phi_a/A_a)}{l_a} & \text{Form 2}: \xi = \Phi_a \end{cases} \qquad (2.3\text{-}38)$$

In our derivation of Ohm's law for magnetic circuits, we assumed a straight piece of material with a uniform cross section and uniform length. Other geometries will also yield the relationships (2.3-31) and (2.3-33); however, the expression for the reluctances and permeances values will change. We will consider many examples of these situations in Sections 2.6 and 2.7.

Before concluding this section, note that in both magnetic and electric circuits, Kirchhoff's laws are in fact physical laws: they stem directly from Maxwell's equations. However, in electric circuits, a third commonly quoted physical law—in particular Ohm's law—is not a physical law at all but is instead a property of some types of conducting materials. The same can be said of Ohm's law for magnetic circuits; it is not a physical law, merely a property that some materials possess.

2.4 Construction of the Magnetic Equivalent Circuit

In this section, the procedure whereby we translate a geometrical description of a device into an MEC is considered by way of a relatively simple example—a UI-core inductor. Inductors are commonly used in power electronics circuits to reduce current ripple. They are also used in passive filters. In this beginning analysis of an inductor, we will take a very basic approach—one not sufficiently accurate to perform a design. However, it will be useful in illustrating the procedure for obtaining the MEC. It will also be useful in the next section wherein the translation of magnetic circuits to electric circuits is considered. Of course, the ultimate goal of this analysis is to provide an

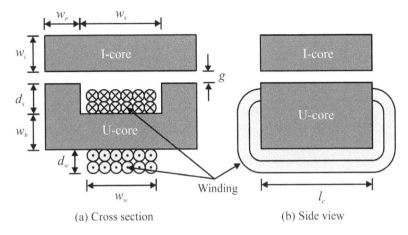

Figure 2.9 UI-core inductor.

analysis that is of sufficient accuracy to facilitate design. To this end, we will improve our MEC in Sections 2.6 and 2.7.

A UI-core inductor is illustrated in Figure 2.9. The geometry is similar to that discussed in Section 1.10; however, some additional aspects have been introduced. For example, the winding bundle is assumed to have a cross section of the winding depth d_w by winding width w_w instead of being assumed to occupy the entire slot that has dimensions of the slot depth d_s by the slot width w_s. This feature will accommodate spacers and incomplete slot fill. In addition, the width of the I-core, the width of the ends of the U-core, and the width of the base of the U-core, denoted w_i, w_e, and w_b, respectively, are not assumed to be equal as they were in Section 1.10. The air gap $\Phi_{s,i}$ is defined in Section 1.10 and shown in Figure 2.9.

The first step in the construction of the MEC is to select node locations. In doing this, it is important to realize that there is no unique choice for nodes. The role of the nodes is to break the circuit into regions that may be treated as lumped reluctances or permeances, either linear or nonlinear. These nodes should have the property that there is negligible (preferably zero) MMF drop across them, because they will be treated as equipotential points in the equivalent circuit. In general, there is a tradeoff between accuracy and the number of nodes. The main factor in selecting nodes is that they usually mark the endpoints of regions that can be treated as lumped magnetic circuit elements.

One possible selection of nodes is illustrated in Figure 2.10. Therein, Figure 2.10(a) shows the location of the nodes on the inductor cross section, where they are shown as thin volumes with various shapes. Figure 2.10(b) depicts the corresponding MEC. Nodes are numbers 0–6 and labeled N_0 through N_6. Node 2 is missing; we will add this node in a later section when we increase the accuracy of the MEC. The "shape" of the node is a function of the behavior of the flux in the region of the node. For example, node 3 is a thin rectangle that is perpendicular to the direction of flux. Nodes 4 and 5 are shaped like an upside down "T." Flux is assumed to enter node 4 from the bottom of the "T" traveling upward, and then leave from the vertical part of the node traveling horizontally. In this case, the MMF drop from the bottom of the I-core to the center of the I-core, being a short distance, is neglected. Node 0 and node 1 are both shaped like "+" signs. The flux path makes a right angle turn at these nodes. We will often neglect the subtleties of the flux paths at the corners; since the lengths of these regions are normally short, they have little effect on the overall circuit. Node 0 will be taken as a ground node with a defined potential of 0 A.

2 Magnetics and Magnetic Equivalent Circuits

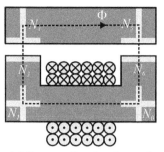
(a) Cross-section showing nodes

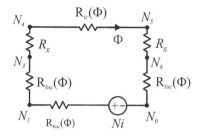
(b) Magnetic equivalent circuit

Figure 2.10 Construction of the MEC.

The next step in developing the MEC is to connect the nodes with circuit elements representing the reluctance between nodes and to represent the MMF sources. In Figure 2.10(b), the reluctances include:

$R_{ic}(\Phi)$	The reluctance of the I-core
$R_{buc}(\Phi)$	The reluctance along the base of the U-core
$R_{luc}(\Phi)$	The reluctance of a leg of the U-core
R_g	The reluctance of the air gap

The reluctances $R_{ic}(\Phi)$, $R_{buc}(\Phi)$, and $R_{luc}(\Phi)$ are functions of flux, because they correspond to a magnetically nonlinear material. As an alternative, they could have been made a function of MMF drop. Since the air gap reluctance R_g corresponds to a linear magnetic material (air), it is a constant.

In order to determine these reluctances, the region between nodes is represented as a rectangular section of material wherein the length of the rectangular section is given by the mean distance between nodes. This approach is known as the mean path approximation. This allows the reluctances to be expressed using (2.3-36). In particular, we have

$$R_{ic}(\Phi) = \frac{w_s + w_e}{w_i l_c \mu_B\left(\frac{\Phi}{w_i l_c}\right)} \tag{2.4-1}$$

$$R_{buc}(\Phi) = \frac{w_s + w_e}{w_b l_c \mu_B\left(\frac{\Phi}{w_b l_c}\right)} \tag{2.4-2}$$

$$R_{luc}(\Phi) = \frac{2d_s + w_b}{2w_e l_c \mu_B\left(\frac{\Phi}{w_e l_c}\right)} \tag{2.4-3}$$

$$R_g = \frac{g}{w_e l_c \mu_0} \tag{2.4-4}$$

Note that all parentheses in (2.4-1)–(2.4-4) denote function arguments, not multiplication.

The flux path shown in Figure 2.10 is in the counterclockwise direction. Therefore, within this path, positive current is defined to be into the page, which is consistent with the direction of the winding. Thus, the MMF source in Figure 2.10 may be expressed as

$$F_s = Ni \tag{2.4-5}$$

Note the placement of the MMF in Figure 2.10(b) is not unique since it merely has to be in the series path, though with the proper polarity.

Figure 2.11 Simplified magnetic equivalent circuit for a UI-core inductor.

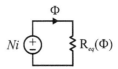

From Kirchhoff's MMF law, the sum of the MMF drops around a closed loop is equal to the sum of the MMF sources for that loop. Thus, in Figure 2.10, traversing the nodes in the order 0, 1, 3, 5, 6, 0 the sum of the MMF drops is equal to the source MMF and so

$$Ni = \left(R_{ic}(\Phi) + R_{buc}(\Phi) + 2R_{luc}(\Phi) + 2R_g\right)\Phi \tag{2.4-6}$$

All network reduction techniques the reader is accustomed to for electric resistive circuits also apply to magnetic circuits. These network reductions include series combinations, parallel combinations, voltage division, and current division. This is because these techniques are all based on Kirchhoff's laws, which are basically identical for electric and magnetic circuits. The only caution that needs to be taken is when the reluctances are a nonlinear function of flux through the branch; in this case, we need to develop appropriate expressions for the flux in each branch. In this UI-core example, we have a common flux through all reluctances. Hence, we can combine the series resistances to form the simple network shown in Figure 2.11. Therein,

$$R_{eq}(\Phi) = R_{ic}(\Phi) + R_{buc}(\Phi) + 2R_{luc}(\Phi) + 2R_g \tag{2.4-7}$$

From Figure 2.11, and Ohm's law we have the implicit equation for flux

$$\Phi = \frac{Ni}{R_{eq}(\Phi)} \tag{2.4-8}$$

Example 2.4A Let us consider a UI-core inductor with the following parameters: $w_i = 25.1$ mm, $w_b = w_e = 25.3$ mm, $w_s = 51.2$ mm, $d_s = 31.7$ mm. $l_c = 101.2$ mm, $g = 1.00$ mm, $w_w = 38.1$ mm, and $d_w = 31.7$ mm. The winding is composed of 35 turns. In this case, the wire used was a sheet conductor with a cross section of 38.1 mm by 0.450 mm. Suppose the magnetic core material has a constant relative permeability of $\mu_r = 7700$ (see (2.3-7)), and that a current of 25.0 A is flowing. Compute the flux through the magnetic circuit and the flux density in the I-core.

We begin by computing the reluctances of the equivalent circuit. From (2.4-1)–(2.4-4), we obtain $R_{ic} = 3.11 \cdot 10^3 \text{H}^{-1}$, $R_{buc} = 3.09 \cdot 10^3 \text{H}^{-1}$, $R_{luc} = 1.79 \cdot 10^3 \text{H}^{-1}$, and $R_g = 3.11 \cdot 10^5 \text{H}^{-1}$. Note that the air gap reluctance dominates the other terms. From (2.4-7), we obtain $R_{eq} = 6.31 \cdot 10^5 \text{H}^{-1}$. From Figure 2.11, $\Phi = 1.39$ mWb. Dividing the flux by the cross-sectional area of the I-core, the flux density in the I-core is 0.546 T.

At this point, the reader may be curious as to the accuracy of our computations. We will address this point in the next section.

2.5 Translation of Magnetic Circuits to Electric Circuits: Flux Linkage and Inductance

The MEC is a tool that we can use to relate flux Φ to a MMF source Ni. We can also use the MEC to determine field strengths within the circuit. Often, however, it is our desire to develop an electrical circuit model that will represent an electromagnetic or electromechanical device. In this section, we will consider the relationship between the MEC and the corresponding electric equivalent circuit.

Flux Linkage

The basic concept of flux as the integral of flux density over a finite surface was set forth in (2.2-1). The concept of flux linkage is closely related to the concept of flux. In particular, let us assume that a winding is wound around the periphery of a surface "x," and that the flux linking this surface is Φ_x. Then, the flux linking this winding is denoted λ_x and is related to the flux by

$$\lambda_x = N_x \Phi_x \qquad (2.5\text{-}1)$$

where N_x is the number of turns. It should be noted that for (2.5-1) to hold, the defined direction for positive current must be counterclockwise around the surface as viewed from a vantage wherein the defined direction of positive flux is toward the observer. Alternately, stated in terms of the "right-hand rule," if the fingers of one's right-hand wrap around the periphery of a surface in the defined direction for positive current, then one's right thumb (extended away from one's hand) will be pointing in the defined direction for positive flux. If this convention is not followed, then a minus sign is introduced into (2.5-1).

Inductance

Inductance is a metric associated with the flux linkage versus current characteristic of a magnetic device. Let us consider a system with a single winding "x." Neglecting hysteresis, the relationship between flux linkage through winding "x" and the current into winding "x" is often as depicted in Figure 2.12. We will use this figure to introduce the concept of inductance. In particular, the absolute inductance at an operating point i_{x1}, λ_{x1} is defined as

$$L_{x,\text{abs}} = \left.\frac{\lambda_x}{i_x}\right|_{i_{x1},\lambda_{x1}} \qquad (2.5\text{-}2)$$

Also of interest is the incremental inductance that is defined as the slope of the flux linkage versus current characteristic at a given operating point. In particular,

$$L_{x,\text{inc}} = \left.\frac{\partial \lambda_x}{\partial i_x}\right|_{i_{x1},\lambda_{x1}} \qquad (2.5\text{-}3)$$

Several observations are in order. First, the incremental inductance is generally lower than the absolute inductance. Second, in a magnetically linear system, the absolute and incremental

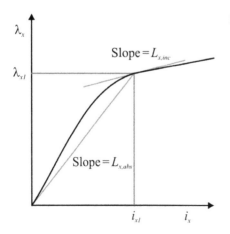

Figure 2.12 Flux linkage versus current.

inductances are the same. Finally, it should be noted that the "abs" or "inc" subscripts are normally not used; these will have to be understood from the context of the discussion.

In order to understand how inductance is related to our MEC, consider the simple MEC shown in Figure 2.10. Therein flux in winding "x" would be proportional to the MMF as

$$\Phi_x = \frac{N_x i_x}{R_x} \tag{2.5-4}$$

From (2.5-1), we have

$$\lambda_x = \frac{N_x^2 i_x}{R_x} \tag{2.5-5}$$

Thus

$$\frac{\lambda_x}{i_x} = L_x \tag{2.5-6}$$

where

$$L_x = \frac{N_x^2}{R_x} \tag{2.5-7}$$

Since R_x is constant, this system is magnetically linear and there is no reason to distinguish between absolute and incremental inductance.

In a magnetic system with a single winding, the inductance is a self-inductance. In other words, the inductance relates how much flux links a winding due to the amount of current in that winding. In systems with multiple windings, self- inductance is still of interest. However, there is also the concept of mutual inductance that relates how much flux will appear in a given winding due to a current in a different winding.

Let us consider a system with several windings. In linear magnetic systems, or when discussing incremental inductance, the mutual inductance $L_{x,y}$ between winding "x" and winding "y" describes how much the flux linking winding "x" changes due to a change in the current in winding "y." In particular,

$$L_{x,y} = \frac{\partial \lambda_x}{\partial i_y} \tag{2.5-8}$$

Ordinarily, the concept of absolute self- and mutual inductances is not terribly useful in multi-winding nonlinear magnetic systems. Instead, it is more common to define flux linkages in terms of a leakage component and magnetizing component and then relate the magnetizing flux linkage to a magnetizing current in such a way that the analysis can be essentially treated as a single-winding problem. We will see an example of this when we consider a transformer in Chapter 7.

Faraday's Law

Consider a winding "x" on an electromechanical device. Denoting the voltage across the winding as v_x, Faraday's law states that the voltage across the winding is equal to the time rate of change of the flux linkage. If we allow for an Ohmic drop $r_x i_x$ where r_x is the resistance of the winding, we have that

$$v_x = r_x i_x + \frac{d\lambda_x}{dt} \qquad (2.5\text{-}9)$$

Equation (2.5-9) is extremely important; it is the basic voltage equation for nearly every device considered throughout this work.

Before concluding this section, let us briefly return to the characteristic shown in Figure 2.12. From (2.5-6), we have that

$$\lambda_x = L_x i_x \qquad (2.5\text{-}10)$$

Substitution of (2.5-10) into (2.5-9) yields

$$v_x = r_x i_x + L_x \frac{di_x}{dt} \qquad (2.5\text{-}11)$$

Note that this simple form is only possible if L_x is constant; otherwise, there will be additional terms.

Before concluding this section, we will now consider a two-part example. First, in Example 2.5A, we will predict the $\lambda - i$ characteristic of a UI-core inductor. Next, in Example 2.5B, we will discuss the measurement of this characteristic.

Example 2.5A In this example, let us predict the flux linkage versus current characteristics of an inductor. To this end, consider an inductor with the parameters set forth in Example 2.4A. The permeability function of the material is given by (2.3-17), (2.3-18), and (2.3-21), and where $\mu_r = 7701$ H/m and the remaining parameters are given by Table 2.1.

From (2.4-8) and (2.5-1), we write the expression

$$\lambda = \frac{N^2 i}{\mathrm{R}_{eq}\left(\frac{\lambda}{N}\right)} \qquad (2.5\text{A-}1)$$

where $\mathrm{R}_{eq}()$ is a function representing the equivalent reluctance, that is given by (2.4-7). We wish to use 2.5A-1 to generate λ versus i. This will be done by generating and then plotting a set of points λ_k and i_k where $k \in [1, 2, ...K]$ and K is the number of point to plot. We could approach this by substituting in $i = i_k$ and solving (2.5A-1) to generate the corresponding λ_k. This approach is suggested by the fact that we normally consider i_k as the independent variable. However, it is much easier to start with flux linkage value λ_k and to generate current value i_k. This is because we may rearrange (2.5A-1) as

$$i = \frac{\lambda \mathrm{R}_{eq}\left(\frac{\lambda}{N}\right)}{N^2} \qquad (2.5\text{A-}2)$$

Table 2.1 Permeability Function Parameters

$\Phi_b(k+1)$	1	2	3	4
α_k (1/T)	1.6254	0.052254	0.0065337	0.0053922
β_k (1/T)	105.4353	21.34779	4.84803	911.9827
γ_k (T)	0.49258	0.51834	0.85782	0.41737

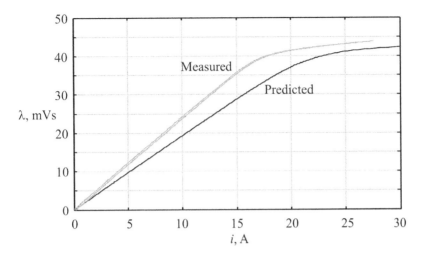

Figure 2.13 Measured and predicted $\lambda - i$ characteristics.

Using (2.5A-2) and starting with flux linkages values λ_k to generate current points i_k circumvents the need to iteratively solve a nonlinear equation. Taking this approach yields the predicted $\lambda - i$ characteristic shown in Figure 2.13. Also shown in this figure is a measured $\lambda - i$ characteristic obtained using the procedure outlined in the next example. As can be seen, there is a considerable difference. Improving our MEC to reduce this discrepancy will be the focus of the next section.

Example 2.5B Let us consider the measurement of flux linkage versus current. It is often the case that we wish to measure the $\lambda - i$ characteristics of an inductor or other device. One approach to this measurement is to apply a zero-mean periodic ac voltage to the inductor and allow the system to reach steady state. An ac source (in this case a power electronic converter) is connected to the inductor via either an isolation transformer or a series capacitor, either of which ensure there is no dc component in the applied voltage waveform. Then, the voltage and current waveforms are recorded. Let v_m and i_m denote the measured voltage and current. The measured quantities are often susceptible to dc offsets in the measurement devices. The dc offset may be removed by using

$$v(t) = v_m(t) - \frac{1}{T} \int_{t=0}^{T} v_m(t) dt \qquad (2.5\text{B-1})$$

$$i(t) = i_m(t) - \frac{1}{T} \int_{t=0}^{T} i_m(t) dt \qquad (2.5\text{B-2})$$

In (2.5B-1) and (2.5B-2), T is the period of the waveform and $v(t)$ and $i(t)$ denote processed waveforms.

The voltage, current, and flux linkage are related by

$$v = ri + \frac{d\lambda}{dt} \qquad (2.5\text{B-3})$$

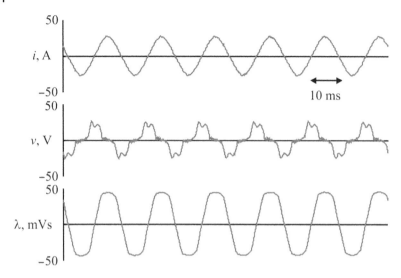

Figure 2.14 Measurement of $\lambda - i$ characteristics.

where r is the resistance of the winding (14.3 mΩ for the inductor used). Assuming that the flux linkage waveform will not have a dc offset, we may solve (2.5B-3) for flux linkage as

$$\lambda(t) = \int_{\tau=0}^{t} v(\tau) - ri(\tau)\,d\tau - \frac{1}{T} \int_{t=0}^{T} v(t) - ri(t)\,dt \qquad (2.5\text{B-4})$$

In the first integral in (2.5B-4), τ is a dummy variable for time, since the end point of our integration is time, t. The second integral calculates the time zero value of flux linkage so that there is no dc offset in the flux linkage waveform. Note that this is appropriate for this inductor; it is not appropriate for systems involving, for example, permanent magnets. The resulting current, voltage, and flux linkage waveforms are shown in Figure 2.14. The ripple in the voltage waveform has to do with the interaction of capacitances and inductances in the power converter output filter. Plotting $\lambda(t)$ versus $i(t)$ yields the characteristic shown in Figure 2.13. Observe the narrow hysteresis loop.

2.6 Representing Fringing Flux in Magnetic Circuits

In Example 2.5A, we found that the accuracy of our MEC is wanting. In this section, we will begin to improve the accuracy of our MEC.

One reason that our MEC was inaccurate had to do with fringing flux. Fringing flux is associated with air gaps in a magnetic circuit. In our UI-core example, this occurs where the U-core legs come together with the I-core as shown in Figure 2.15. In analyzing the UI-core inductor reluctance, we only considered the flux path labeled direct flux. However, flux will also take the path on the inner corner and outer corner of the gap, which is labeled fringing flux in Figure 2.15. In addition, fringing flux will also occur on a path shaped like the path on the outside corners, but physically on the front and back faces of the UI-core. The permeances associated with the fringing flux paths can markedly reduce the total air gap reluctance, causing an increase in air gap permeance.

Figure 2.15 Fringing Flux.

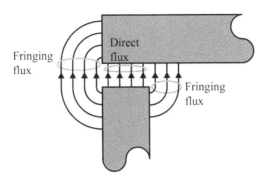

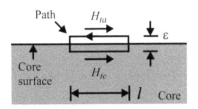

(a) Tangential field component

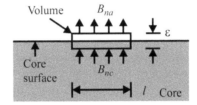

(b) Normal field component

Figure 2.16 Field components at air-core interface.

Fringing flux is readily incorporated into an MEC. To do so, the first observation made is that the flux will enter and leave the core material in a direction normal to the surface of the core. This can be seen by considering Figure 2.16. First let us consider Figure 2.16(a). Therein, let us apply Ampere's law to the indicated path. Traversing the path indicated, which does not enclose any current, and letting the height of the path ε approach zero, we have

$$H_{tc}l - H_{ta}l = 0 \tag{2.6-1}$$

where H_{tc} and H_{ta} are the tangential components of the field intensity in the core and air, respectively. From (2.6-1), we see that these two field intensities are equal. It follows that

$$\frac{B_{ta}}{\mu_0} = \frac{B_{tc}}{\mu_r \mu_0} \tag{2.6-2}$$

where B_{ta} and B_{tc} are the tangential components of the flux density in air and the core, respectively, and μ_r is the relative permeability of the core. From (2.6-2), we obtain

$$B_{ta} = \frac{B_{tc}}{\mu_r} \tag{2.6-3}$$

Assuming that the relative permeability of the core is large, the tangential component of the flux density in air above the core will be a small fraction of the tangential component of the flux density just below the core's surface.

Now consider Figure 2.16(b) wherein the normal component of the flux density is depicted. Therein, the rectangle is a cross section of a rectangular volume. The sides of this volume have a height ε that will approach 0. Since no flux will be associated from these sides (since they have no height) it follows from Gauss's law that

$$B_{nc}ld - B_{na}ld = 0 \tag{2.6-4}$$

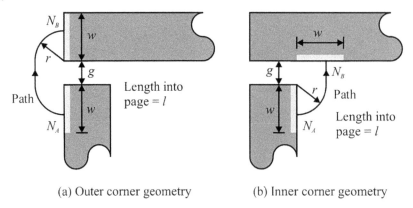

(a) Outer corner geometry (b) Inner corner geometry

Figure 2.17 Calculation of fringing flux.

where B_{nc} and B_{na} are the normal components of the flux density in the core and air, respectively, and where d is the depth of the rectangular volume into the page. From (2.6-4), we obtain

$$B_{na} = B_{nc} \qquad (2.6\text{-}5)$$

Thus, the normal component of the flux density in air above the core is the same as the normal component just under the surface. From (2.6-3), since the tangential component of the flux density in air is highly attenuated from that of the core, it follows that the flux density at the surface of a magnetic material will predominately be in normal direction, unless the material is saturated to such an extent that the permeability falls substantially. We conclude that flux will enter and leave magnetic material in a direction normal to the surface. With this observation, we can now calculate the effects of the fringing flux.

Although the calculation of fringing flux is highly geometry dependent, the two geometries shown in Figure 2.17 are both very common (and very relevant to our UI-core example). Let us first consider Figure 2.17(a). Therein, the lightly shaded region indicates assumed locations for nodes of our MEC. Our goal is to be able to calculate the flux between the two nodes. Assuming that the flux leaves and enters core material in the normal direction, we assume that the field follows the indicated path, which consists of two quarter-circles connected by a line segment of length g. The MMF drop between the two paths may be expressed as

$$F = \int_{N_A}^{N_B} \mathbf{H} \cdot dL \qquad (2.6\text{-}6)$$

Assuming the field intensity is uniform along the length of the path and in the direction of the path, from (2.6-6) we find that the field intensity in the direction of the path is given by

$$H = \frac{F}{g + \pi r} \qquad (2.6\text{-}7)$$

It is important to observe that while H is uniform along the path, it varies with r. Since the fringing flux is flowing through air, from (2.6-7) the flux density may be expressed as

$$B = \frac{\mu_0 F}{g + \pi r} \qquad (2.6\text{-}8)$$

Now let us calculate the flux leaving node A. From our definition of flux, the fringing flux leaving node A may be expressed as

$$\Phi = \int_{S_A} \mathbf{B} \cdot d\mathbf{S} \qquad (2.6\text{-}9)$$

where S_A denotes the surface of node A. Observe that the direction of the flux density given by (2.6-8) is normal to the node surface. Making use of this fact and taking

$$d\mathbf{S} = l\,dr \qquad (2.6\text{-}10)$$

where l is the depth of the geometry into the page, we have

$$\Phi = \int_0^w \frac{\mu_0 F}{g + \pi r} l\,dr \qquad (2.6\text{-}11)$$

Carrying out the indicated integration, the fringing flux associated with the indicated path may be expressed as

$$\Phi = P_{fa} F \qquad (2.6\text{-}12)$$

where P_{fa} is the permeance associated with the geometry of Figure 2.17(a) and is expressed as

$$P_{fa} = \frac{\mu_0 l}{\pi} \ln\left(1 + \pi \frac{w}{g}\right) \qquad (2.6\text{-}13)$$

Repeating this procedure for the geometry of Figure 2.17(b), one obtains a similar result; in particular

$$\Phi = P_{fb} F \qquad (2.6\text{-}14)$$

where P_{fb} is the permeance associated with fringing geometry in Figure 2.17(b) and is given by

$$P_{fb} = \frac{2\mu_0 l}{\pi} \ln\left(1 + \frac{\pi}{2} \frac{w}{g}\right) \qquad (2.6\text{-}15)$$

Let us now apply the expressions for fringing permeance to our UI-core example. Previously, our expression for the reluctance of the air gap was given by (2.4-4). This corresponds to the direct flux path with a permeance we will now denote as

$$P_{gd} = \frac{\mu_0 l_c w_e}{g} \qquad (2.6\text{-}16)$$

Applying the expression for fringing permeance of the outside corner to the UI-core inductor, note that $l = l_c$ and that $w = \min(w_i, d_s + w_b)$. Thus the outside corner fringing permeance is given by

$$P_{foc} = \frac{\mu_0 l_c}{\pi} \ln\left(1 + \frac{\pi}{g} \min(w_i, d_s + w_b)\right) \qquad (2.6\text{-}17)$$

To apply (2.6-15) to the inside corner, we will again take $l = l_c$. However, the choice for the appropriate value of w requires consideration. It would seem that w should not exceed half the slot width or it will overlap the node from the other side. Further, it should not exceed the slot depth. Thus, one approach is to take

$$w = \min(d_s, w_s/2) \qquad (2.6\text{-}18)$$

whereupon the fringing permeance of the inside corner becomes

$$P_{fic} = \frac{2\mu_0 l_c}{\pi} \ln\left(1 + \frac{\pi}{2} \frac{\min(d_s, w_s/2)}{g}\right) \quad (2.6\text{-}19)$$

It should be observed that in (2.6-18), this choice of w means that some of the fringing flux will intersect the winding and thus not be fully coupled by it. This effect is neglected. One reason why this approximation is usually acceptable is that the flux density is highest on the innermost region of the corner where the portion of the winding not coupled is small.

Fringing flux will also occur on the front and back face of the UI-core. The path for this flux is similar to Figure 2.17(a) except that the path for the front face extends out of the page rather than out to the left. The expression (2.6-13) is applicable with $w = \min(w_i, d_s + w_b)$ and the length $l = w_e$, whereupon the fringing face permeance becomes

$$P_{ff} = \frac{\mu_0 w_e}{\pi} \ln\left(1 + \pi \frac{\min(w_i, d_s + w_b)}{g}\right) \quad (2.6\text{-}20)$$

Parallel permeances, like conductances in electrical circuits, add. Thus, the total air gap permeance of the UI-core inductor may be expressed as

$$P_g = P_{gd} + P_{foc} + P_{fic} + 2P_{ff} \quad (2.6\text{-}21)$$

where the factor of 2 arises because there is fringing on both the front and back face of the UI-core arrangement. The reluctance may be calculated as

$$R_g = \frac{1}{P_g} \quad (2.6\text{-}22)$$

Example 2.6A In this example, let us consider the effects of including fringing flux on the air gap permeance of the UI-core example of Example 2.4A and 2.5A. Using (2.6-16), (2.6-17), (2.6-19), and (2.6-20), we obtain P_{gd}, P_{foc}, P_{fic}, and P_{ff} of 3.22 μH, 0.177 μH, 0.301 μH, and 44.3 nH, respectively. These terms comprise 85, 4.7, 8.0, and 2.3% to the total air gap permeance, so that the total fringing permeance is roughly 15% of the total air gap permeance—a modestly significant contribution. The resulting $\lambda - i$ characteristic is shown in Figure 2.18. Comparing this to Figure 2.13, it is clear that our MEC has become considerably more accurate.

In Example 2.6A, we saw that fringing flux plays an important role in the performance of a magnetic device and should be represented in the MEC. However, it is not the only important secondary effect. Leakage permeance also plays an important role. We will consider this in the next section and create an even more accurate magnetic model.

2.7 Representing Leakage Flux in Magnetic Circuits

Leakage flux refers to flux that is flowing outside of the intended path in a magnetic circuit. In some sense, this would include fringing flux, but we have treated that aspect of the flux separately because it is associated with the air gap. However, flux can leak from the intended path in other places, and this "leak" significantly affects the circuit. In fact, in Example 2.6A, the dominant remaining error in the analysis is the failure to incorporate the effects of leakage flux. In this section, we will look at several situations that give rise to leakage flux and derive permeances useful in calculating these flux terms. We will then apply the resulting expression to the UI- core inductor.

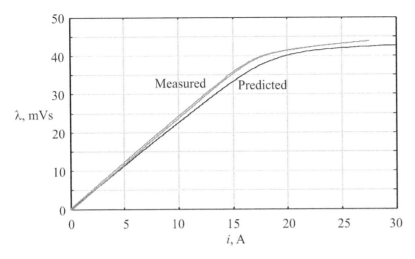

Figure 2.18 Measured and predicted $\lambda - i$ characteristics including fringing flux.

Figure 2.19 Leakage flux.

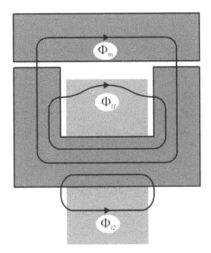

Figure 2.19 illustrates the leakage flux path in the context of a UI-core inductor. Therein the main component of the flux is denoted Φ_m, where the m stands for magnetizing flux, the main component of the inductor flux. In the previous section, when we considered the fringing flux components, these flux components could be viewed as portions of the magnetizing flux since they follow the intended path except for around the air gap. The quantities Φ_{l1} and Φ_{l2} in Figure 2.19 denote the leakage flux paths. These components of flux never cross the air gap. Another feature of the leakage flux is that in many situations, leakage flux passes through the winding so that not all the flux links all of the turns. This complicates the calculation of the leakage flux and the permeances used to represent the leakage flux.

The reader may question why flux would tend to travel through the indicated leakage paths given the low permeability of air. The short answer is because it can. In electrical circuits, the conductivity of copper and the conductivity of air are different by 23 orders of magnitude (10^{23}). As a result, the leakage of electric current into air can usually be neglected unless the electric field strength exceeds the breakdown voltage of air. In magnetic circuits, however, the difference between the permeability of the magnetic materials and the permeability of air is only three or four orders of magnitude, and the air paths often have a larger cross section. As a result, it is often the case that if a high accuracy is to be obtained, then leakage flux must be considered.

The reader may also question why the path of Φ_{l1} deflects as it passes through the slot. This is because the slot has both horizontal and vertical field components. We will consider the flux path in more detail later in this section.

Our development will proceed as follows. First, we will consider the energy stored in a magnetically linear inductor and the energy stored in a linear magnetic material. While this may seem off topic, we will make some observations that will prove very helpful in calculating leakage permeance. Next, we will consider the leakage permeance associated with several geometries. Finally, we will apply our results to our running case study of a UI-core inductor.

Energy Storage in Magnetically Linear Systems

We will now make some observations on energy storage, which will prove useful in calculating leakage permeances. Let us first consider the energy stored in a magnetically linear inductor. Such an inductor is defined by having a flux linkage λ that is proportional to current i, that is,

$$\lambda = Li \tag{2.7-1}$$

where L is not a function of the current. The electrical power flowing into the winding is given by

$$p = iv \tag{2.7-2}$$

where v is the voltage across the winding. Treating the resistive loss of the inductor to be external to the inductor, so that the voltage to which we are referring is the voltage after any ohmic drop has been removed, the voltage is equal to the time derivative of flux linkage, so we obtain

$$p = i\frac{d\lambda}{dt} \tag{2.7-3}$$

Substitution of (2.7-1) into (2.7-3) yields

$$p = Li\frac{di}{dt} \tag{2.7-4}$$

Let us suppose that at time $t = 0$ the inductor current, and stored energy are zero. Let us further suppose that at time $t = t_1$, the inductor current is $i = i_1$. The energy that has flowed into the inductor (and hence stored in the inductor) may be expressed as

$$E = \int_{t=0}^{t_1} p \, dt \tag{2.7-5}$$

Substitution of (2.7-4) into (2.7-5) yields

$$E = \int_{i=0}^{i=i_1} Li \, di \tag{2.7-6}$$

which reduces to

$$E = \frac{1}{2}Li_1^2 \tag{2.7-7}$$

Figure 2.20 Toroidal geometry.

Since this expression is valid for any value of i_1, we may more simply write the familiar expression as

$$E = \frac{1}{2}Li^2 \qquad (2.7\text{-}8)$$

which may also be expressed as

$$E = \frac{1}{2}PN^2i^2 \qquad (2.7\text{-}9)$$

where P is the effective permeance of the inductor.

Equations (2.7-8) and (2.7-9) give an expression for the energy stored in a magnetically linear inductor from a circuit perspective. Let us now derive a similar result from a field perspective. To this end, let's start by applying Ampere's law to the geometry shown in Figure 2.20. This figure shows a top view of a toroid of rectangular cross section that is uniformly wound. Assume that the inner and outer radius are very large and that the outer radius is only slightly larger than the inner radius. Applying Ampere's law, we have that

$$Hl = Ni \qquad (2.7\text{-}10)$$

where l is the circumference at the mean radius, and H is the field intensity in the clockwise direction. Again, with our assumption that the inner and outer radius differ only slightly the flux through the toroid may be expressed as

$$\Phi = AB \qquad (2.7\text{-}11)$$

where A is the cross section of the toroid given by $(r_o - r_i)d$ where r_o, r_i, and d are, respectively, the outer toroid radius, inner toroid radius, and the depth of the toroid into the page. The flux linking the winding is N times the flux, so we have

$$\lambda = NAB \qquad (2.7\text{-}12)$$

2 Magnetics and Magnetic Equivalent Circuits

Manipulating (2.7-3), (2.7-10), and (2.7-12), the power into the core's magnetic field may be expressed as

$$p = \underbrace{\frac{Hl}{N}}_{i} \underbrace{NA\frac{dB}{dt}}_{\frac{d\lambda}{dt}} \tag{2.7-13}$$

which reduces to

$$\frac{p}{V} = H\frac{dB}{dt} \tag{2.7-14}$$

where V is the volume of the toroid. Integrating (2.7-14) with respect to time, the energy density of the toroid (i.e., energy per unit volume) may be expressed as

$$\frac{E}{V} = \int_0^{t_1} H\frac{dB}{dt} dt \tag{2.7-15}$$

In (2.7-15), t_1 is some arbitrary point in time. From (2.7-15), it follows that

$$\frac{E}{V} = \int_0^{B_1} H dB \tag{2.7-16}$$

where B_1 is the circumferential flux density at time t_1. Assuming a magnetically linear material, substitution of (2.3-9) into (2.7-16) yields

$$\frac{E}{V} = \mu \int_0^{H_1} H dH \tag{2.7-17}$$

which reduces to

$$\frac{E}{V} = \frac{1}{2}\mu H^2 \tag{2.7-18}$$

where we have dropped the subscript 1 from H since H could represent the field intensity at any point in time.

Expression (2.7-18) was derived for a toroidal geometry with a linear magnetic material. The right-hand side of (2.7-18) is the energy density. However, from a physical point of view, if (2.7-18) yields the energy density within the toroid, the question arises as to whether the energy density would change if the geometry were different. Because an infinitesimally small amount of material does not know the geometry in which it is located, it follows that the expression for energy density, namely $\mu H^2/2$, is generally applicable to linear magnetic materials, with H being the magnitude of the field at some point. Thus, in a general situation with magnetically linear material but with spatially varying fields, we have

$$E = \frac{1}{2}\mu \int_V H^2 dV \tag{2.7-19}$$

We now have two expressions for energy stored. In (2.7-9), we have energy related to permeance. In (2.7-19), energy stored is related to fields and geometry. By relating these two expressions, we will

Figure 2.21 Horizontal and vertical slot leakage flux.

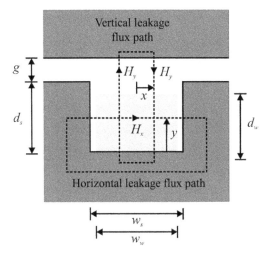

be able to compute permeance in situations where flux penetrates into a winding. This will be our next topic.

Slot Leakage Flux Permeance

A major contributor to leakage flux is slot leakage flux. This corresponds to Φ_{l1} in Figure 2.19. One component of slot leakage flux is associated with a horizontal field component labeled H_x in Figure 2.21. This figure is certainly relevant to the UI-core inductor, but is equally relevant to many other situations. For example, if the I-core was removed, it approximates the slot of a motor winding (we will consider this in Chapter 8). In order to calculate this component, consider the indicated path. Applying Ampere's law to the horizontal flux path, we have

$$H_x w_s = \begin{cases} Ni \dfrac{y}{d_w}, 0 \leq y \leq d_w \\ Ni, d_w \leq y \leq d_s \end{cases} \quad (2.7\text{-}20)$$

In (2.7-20), there are several points of interest. First, we have neglected the MMF drop across all portions of the path in the core material. This is because the permeability of the core is much higher than that of air, so for most situations the MMF drop around the path is dominated by $H_x w_s$ term. Another point of interest is on the right-hand side of (2.7-20). Observe that the y/d_w factor is present in order to describe the fraction of the winding enclosed by the path.

Solving (2.7-20) for the field intensity and substituting the result into (2.7-19) with $\mu = \mu_0$ and with $dV = w_s l_c dy$ yields

$$E = \frac{1}{2}\mu_0 \int_{y=0}^{d_w} \left(\frac{Niy}{w_s d_w}\right)^2 w_s l dy + \frac{1}{2}\mu_0 \int_{y=d_w}^{d_s} \left(\frac{Ni}{w_s}\right)^2 w_s l dy \quad (2.7\text{-}21)$$

where l is the length of the core into the page. Evaluating (2.7-21) we have the energy stored by the horizontal component of the field in the slot is given by

$$E = \frac{1}{6}\mu_0 \frac{d_w l}{w_s} N^2 i^2 + \frac{1}{2}\mu_0 \frac{l}{w_s}(d_s - d_w)N^2 i^2 \quad (2.7\text{-}22)$$

From (2.7-9), the energy stored by the field may be expressed in terms of permeance as

$$E = \frac{1}{2} P_{hsl} N^2 i^2 \tag{2.7-23}$$

where P_{hsl} represents the permeance associated with the horizontal slot leakage flux path. Equating (2.7-22) with (2.7-23), we have

$$P_{hsl} = \frac{\mu_0 l (3d_s - 2d_w)}{3 w_s} \tag{2.7-24}$$

Note that this is a generic expression for the permeance associated with the horizontal field in a slot; we will consider the application of this to our UI core later in this section.

The expression for horizontal leakage permeance includes the region of the slot above the winding. In some cases, it may be advantageous to represent this term separately. Normally, however, d_s and d_w are very close so that the resulting permeance term is very small and the extra complication involved in representing it separately is not justified.

Another point of interest relevant to the horizontal leakage permeance is that it describes the amount of flux linking the winding. This will not be exactly the same as the total amount of flux that crosses the slot in the horizontal direction. The permeance related to the total flux across the slot in the horizontal direction can be obtained from the definition of flux and (2.7-20), which yields

$$P'_{hsl} = \frac{\mu_0 l}{w_s} \left(d_s - \frac{1}{2} d_w \right) \tag{2.7-25}$$

where the prime is used to differentiate this estimate from that of (2.7-24). The computation of (2.7-24) is left as an exercise (see Problem 16). Comparing (2.7-25) to (2.7-24) one can see that $P'_{hsl} > P_{hsl}$, which makes sense since the permeance given by (2.7-25) represents total horizontal leakage flux while the component (2.7-24) is only that portion of the horizontal leakage flux that would couple the winding. This leaves the question as to which value to use (2.7-24) or (2.7-25).

The simplest approach is to simply use (2.7-24) in the equivalent circuit. This will normally result in the appropriate leakage flux contribution as seen by the primary winding. Since the flux contribution due to the horizontal leakage permeance is rather small, the error arising from such an approach is generally acceptable. Another approach is to use (2.7-25) as the permeance in the MEC, but to compensate for the difference by placing a permeance of $P_{hsl} - P'_{hsl}$ in parallel with the source, as a correction. However, this only results in a slight improvement in accuracy.

Before concluding our conversation of the horizontal flux leakage, it will be useful to find the height up the slot of which one-half the horizontal leakage flux is above this point and one-half the horizontal slot leakage flux is below this point. From (2.7-20) and the definition of flux, the half-way point from the bottom of the slot may be readily expressed as

$$d_h = \begin{cases} \sqrt{d_w(d_s - d_w/2)}, & d_w \geq 2d_s/3 \\ d_s/2 + d_w/4, & d_w \leq 2d_s/3 \end{cases} \tag{2.7-26}$$

The height d_h will prove useful in the formulation of the MEC.

Next let us consider the leakage permeance associated with the vertical component of the field. Applying Ampere's law to vertical leakage flux path, we have

$$2H_y(d_s + g) = \begin{cases} \dfrac{2x}{w_w} Ni, & 0 \leq x \leq \dfrac{w_w}{2} \\ Ni, & \dfrac{w_w}{2} \leq x \leq \dfrac{w_s}{2} \end{cases} \tag{2.7-27}$$

Applying the same procedure to find the energy as for the horizontal leakage flux, one obtains an expression for a permeance associated with the vertical slot leakage flux. In particular,

$$P_{vsl} = \frac{\mu_0 l(3w_s - 2w_w)}{12(d_s + g)} \quad (2.7\text{-}28)$$

The derivation of this result is left as an exercise (see Problem 17). Note that in applying the procedure, some care is involved to accommodate the fact that x only varies over half a slot while we wish to compute the energy in the entire slot.

Before considering another common geometry that involves leakage flux, it is interesting to briefly reconsider Figure 2.19. Note the deflection drawn into the path of Φ_{l1}. This deflection is due to the combination of the horizontal and vertical field components shown in Figure 2.21.

Another point of interest is that when we computed the two leakage permeances, we essentially applied superposition, treating the field separately. This is valid in this case because the field components considered are orthogonal. Had they not been orthogonal, the two calculations could not have been separated because the energy calculation involves nonlinear operations.

At this point, let us consider another situation that results in leakage flux. In particular, we will consider exterior flux permeance, which is associated with the part of a winding or coil that is not within the slot. We will study this source of leakage flux for two situations: one in which the bundle of conductors is adjacent to a magnetically conductive material and one in which a bundle of conductors is in isolation.

Exterior Adjacent Conductor Leakage Flux Permeance

Let us now consider the leakage flux associated with a bundle of conductors adjacent to a magnetically conductive material as shown in Figure 2.22. Therein, the darkly shaded region is magnetic core and the lightly shaded region represents N conductors directed into the page. Using the field path shown, Ampere's law yields

$$Hl_p = \begin{cases} Ni\dfrac{r}{d_w}, 0 \leq r \leq r_1 \\ Ni, r_1 \leq r \leq r_2 \end{cases} \quad (2.7\text{-}29)$$

In (2.7-29), l_p is the length of the path which may be expressed as

$$l_p = w_w + \pi r \quad (2.7\text{-}30)$$

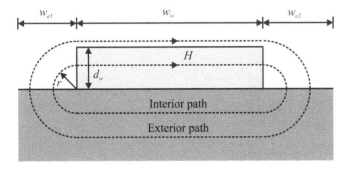

Figure 2.22 Exterior adjacent leakage flux (approach 1).

and r_1 and r_2 are given by

$$r_1 = \min(w_{e1}, w_{e2}, d_w) \tag{2.7-31}$$

$$r_2 = \min(w_{e1}, w_{e2}) \tag{2.7-32}$$

Applying (2.7-19), with a differential volume of $dV = l_p l dr$ and equating the result to (2.7-9), yields

$$P_{eal} = \frac{\mu_0 l_c}{\pi d_w^2} \left(\frac{r_1^2}{2} - \frac{w_w r_1}{\pi} + \frac{w_w^2}{\pi^2} \ln\left(\frac{w_w + \pi r_1}{w_w}\right) + d_w^2 \ln\left(\frac{w_w + \pi r_2}{w_w + \pi r_1}\right) \right) \tag{2.7-33}$$

In deriving this result, the right-hand side of (2.7-30) must replace l_p within the integral since it is a function of the variable of integration.

It should be observed that the expression (2.7-33) is highly approximate. However, since its contribution to the overall behavior is modest, the overall result will not be highly sensitive to the error.

In situations in which the end widths w_{e1} or w_{e2} are small relative to d_w, other approaches to calculating the leakage permeance may be more appropriate. One approach suggested in Cale and Sudhoff [14] is illustrated in Figure 2.23. In this approach, the leakage permeance is calculated in two parts. The first part is associated with flux flowing around the winding. This is calculated by using a series combination of three simple reluctances each calculated using (2.3-32). In particular, the reluctance associated with the series combination of reluctance is given by

$$R_{eale} = \underbrace{\frac{d_w + d_e/2}{\mu_0 w_{e1} l_c}}_{R_{e1}} + \underbrace{\frac{w_w + w_{e1}/2 + w_{e2}/2}{\mu_0 d_e l_c}}_{R_w} + \underbrace{\frac{d_w + d_e/2}{\mu_0 w_{e2} l_c}}_{R_{e2}} \tag{2.7-34}$$

Unfortunately, d_e is unknown. Flux, like current, favors the easiest path. Therefore, let us choose d_e so as to minimize reluctance. Setting the partial derivative of R_{eali} with respect to d_e to zero (to find the value of d_e which minimizes the reluctance) and substituting the resulting value of d_e back into (2.7-34) yields

$$R_{eale} = \frac{d_w(w_{e1} + w_{e2}) + \sqrt{(2w_w + w_{e1} + w_{e2})(w_{e1} + w_{e2})w_{e1} w_{e2}}}{\mu_0 l w_{e1} w_{e2}} \tag{2.7-35}$$

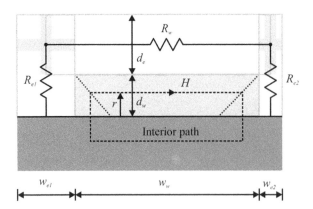

Figure 2.23 Exterior adjacent leakage flux (approach 2).

To calculate the portion of the permeance associated with the flux flowing within the winding bundle, we will assume the rectangular path indicated in Figure 2.23. Neglecting the MMF drop in the core material, we have

$$Hl_p = Ni\frac{a_p}{d_w w_w} \tag{2.7-36}$$

where l_p and a_p are the length of the path and area associated with the path, respectively, and are given by

$$l_p = |w_w - 2d_w| + 4r \tag{2.7-37}$$

and

$$a_p = r(|w_w - 2d_w| + 2r) \tag{2.7-38}$$

We will use (2.7-19) to obtain the energy stored in the region internal to the conductor. To this end, the differential volume may be expressed as

$$dV = l_p l\, dx \tag{2.7-39}$$

Using (2.7-36)–(2.7-39) in conjunction with (2.7-19), the energy stored in the conductor region may be expressed as

$$E = \frac{1}{2}\mu_0 \frac{N_p^2 i_p^2 l}{w_w^2 d_w^2} \int_0^{r_{mx}} \frac{r^2(|w_w - 2d_w| + 2r)^2}{|w_w - 2d_w| + 4r} dr \tag{2.7-40}$$

Evaluating (2.7-40) with $r_{mx} = \min(d_w, w_w/2)$ yields

$$P_{eali} = \frac{\mu_0 l}{256 w_w^2 d_w^2}\left[\begin{array}{l} 16k_2^4 + 16\sqrt{2}k_1 k_2^3 + 4k_1^2 k_2^2 \\ -2\sqrt{2}k_1^3 k_2 + k_1^4 \ln\left(1 + \frac{2\sqrt{2}k_2}{k_1}\right) \end{array}\right] \tag{2.7-41}$$

where

$$k_1 = |w_w - 2d_w| \tag{2.7-42}$$

and

$$k_2 = \frac{1}{\sqrt{2}} \min(2d_w, w_w) \tag{2.7-43}$$

From a coding point of view, it can be shown that the natural log term goes to zero as k_1 goes to zero.

In the explanation thus far, it has been implicitly assumed that $w_w > 2d_w$. However, with the given definitions of k_1 and k_2 the result is valid even if this inequality does not hold.

After P_{eali} and R_{eale} have both been calculated, the net permeance may be calculated as

$$P_{eal} = \frac{1}{R_{eale}} + P_{eali} \tag{2.7-44}$$

In order to apply this result to our UI-core inductor, we would take

$$w_{e1} = w_{e2} = w_e + \frac{1}{2}(w_s - w_w) \tag{2.7-45}$$

and

$$l = l_c + 2w_b \qquad (2.7\text{-}46)$$

In (2.7-46), the $2w_b$ term occurs because this leakage path will occur on the front and back faces in addition to on the bottom of the U-core.

Exterior Isolated Conductor Leakage Flux Permeance

If a winding bundle is far from a magnetic material, the field pattern will be different. A crude approximation to this is depicted in Figure 2.24. Note that this situation does not occur in our UI-core example; it will occur when we consider transformer design and rotating electric machinery. In this figure, the darkly shaded region represents the core material assumed to be distant from the winding bundle. The lightly shaded region is a cross section of the winding which is directed into the page. Let us first consider the permeance arising from the flux flow within the winding. Assuming that the field lines are parallel to the indicated interior path, we may write

$$Hl_p = \frac{a_p}{w_w d_w} Ni \qquad (2.7\text{-}47)$$

where l_p and a_p are the length of the path and area enclosed by the path, respectively, which may be expressed as

$$l_p = 2|w_w - d_w| + 8x \qquad (2.7\text{-}48)$$

and

$$a_p = 2x|w_w - d_w| + 4x^2 \qquad (2.7\text{-}49)$$

We will use (2.7-19) to obtain the energy stored in the region internal to the conductor. To this end, the differential volume may be expressed as

$$dV = l_p l dx \qquad (2.7\text{-}50)$$

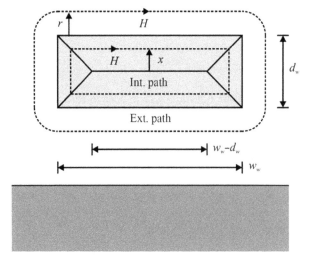

Figure 2.24 Exterior isolated leakage flux.

where l is the length into the page. Using (2.7-47)–(2.7-50) in conjunction with (2.7-19), we obtain

$$E = \frac{1}{2}\mu_0 \frac{N_p^2 i_p^2 l}{w_w^2 d_w^2} \int_0^{x_{mx}} \frac{(2x|w_w - d_w| + 4x^2)^2}{2|w_w - d_w| + 8x} dx \tag{2.7-51}$$

For the situation shown in Figure 2.24, wherein $w_w > d_w$, we would have $x_{mx} = d_w/2$. However, if the aspect ratio were reversed, then the orientation of our path changes and $x_{mx} = w_w/2$. Thus, in general $x_{mx} = \min(w_w, d_w)/2$. Evaluating (2.7-51) and equating the result with (2.7-9), we obtain

$$P_{eili} = \frac{\mu_0 l}{128 w_w^2 d_w^2} \left(4k_2^4 + 8k_1 k_2^3 + 2k_1^2 k_2^2 - 2k_1^3 k_2 + k_1^4 \ln\left(1 + \frac{2k_2}{k_1}\right) \right) \tag{2.7-52}$$

where

$$k_1 = |w_w - d_w| \tag{2.7-53}$$

and

$$k_2 = \min(w_w, d_w) \tag{2.7-54}$$

As in (2.7-41), from a coding point of view, it can be shown that the natural log term goes to zero as k_1 goes to zero.

Consideration of the exterior path yields a permeance associated with the exterior path given by

$$P_{eile} = \frac{\mu_0 l}{2\pi} \ln\left(1 + \frac{\pi r_{mx}}{d_w + w_w}\right) \tag{2.7-55}$$

The derivation of (2.7-55) is left as an exercise (see Problem 20). In (2.7-55), r_{mx} is the maximum allowed radius for the flux path. For example, r_{mx} could represent the nearest distance to the magnetic material in Figure 2.24.

Adding the two permeance components together, we have

$$P_{eil} = P_{eili} + P_{eile} \tag{2.7-56}$$

Incorporation of Leakage Flux Permeances into UI-Core Inductor

At this point, let us return to our UI-core inductor example and consider how we may incorporate leakage flux permeances/reluctances into the equivalent circuit. One approach is shown in Figure 2.25. Therein, many changes have been made from our initial equivalent circuit shown in Figure 2.10.

The first change made is that the shape of nodes 3, 4, 5, and 6 has been altered. This is not a matter of representing leakage flux, but rather to indicate the extended surface of the nodes made to accommodate fringing, which is represented by using (2.6-16)–(2.6-21) to calculate the air gap reluctance, rather than (2.4-4).

Next, nodes 2 and 7 have been added to the equivalent circuit, dividing the legs into lower and upper sections. Between these nodes the reluctance $R_{hsl} = 1/P_{hsl}$ has been placed to represent the horizontal slot leakage. The expression for the horizontal slot leakage permeance is given by (2.7-24). The path of this leakage flux flows from leg to leg in the horizontal direction. Splitting the legs into an upper and lower portion facilitates incorporating this flux component. This leaves the question of determining the height where the legs are split. We will take this point to be at the height

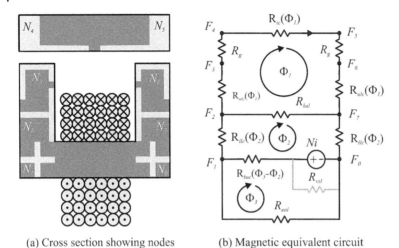

(a) Cross section showing nodes (b) Magnetic equivalent circuit

Figure 2.25 Magnetic equivalent circuit with leakage flux permeances.

above the bottom of the slot which splits the horizontal leakage flux in two equal parts, as given by (2.7-26), whereupon the reluctance $R_{luc}(\Phi)$ of (2.4-3) becomes split into two reluctances

$$R_{llc}(\Phi) = \frac{2d_h + w_b}{2w_el_c\mu_B\left(\frac{\Phi}{w_el_c}\right)} \tag{2.7-57}$$

$$R_{ulc}(\Phi) = \frac{d_s - d_h}{w_el_c\mu_B\left(\frac{\Phi}{w_el_c}\right)} \tag{2.7-58}$$

The remaining core branch reluctances are unchanged, though because of the added branches, we have multiple flux components. In Figure 2.25, Φ_1, Φ_2, and Φ_3 all denote mesh fluxes that we will use to analyze the circuit. The core reluctances all become a function of the appropriate branch flux or fluxes. For example, R_{ic} is a function of Φ_1 because that is the flux that flows through it; R_{buc} is a function of $\Phi_3 - \Phi_2$ because it is this difference in mesh fluxes that flows through the branch.

Turning our attention to the lower portion of the circuit, the leakage reluctance R_{eal} is placed between nodes 0 and 1 because this flux component makes use of the lower (base) of the U-core. The vertical slot leakage R_{vsl} is placed directly across the source MMF because it makes use of relatively little core material. This reluctance is shown as being lightly shaded in Figure 2.25(b) in order to indicate that it will not be included in the network solution of the nonlinear circuit. Since it is linear, and directly across the MMF source, the flux flowing through this branch can be explicitly calculated separately from the remainder of the MEC.

At this point, the leakage inductances have been added to the MEC, which raises the questions of what improvement in accuracy we have achieved by including these paths. We will answer this question. However, before doing so, we will have to consider the problem of solving the nonlinear network equations corresponding to Figure 2.25, which is significantly more complicated than our previous attempt at an MEC shown in Figure 2.10(b). This is the topic of the next section.

2.8 Numerical Solution of Nonlinear Magnetic Circuits

The analysis of magnetic circuits such as the one shown in Figure 2.25(b) is directly analogous to the solution of electric circuits. Any technique applicable to the solution of resistive electric circuits can be used for magnetic circuits. Although there are many analysis approaches that can be used, when faced with a complicated magnetic circuit, mesh analysis and nodal analysis are most conveniently applied. In order to set the stage for these approaches, this section begins with a definition of a standard branch that will be used to describe the connection between two nodes. Then linear magnetic circuit analysis using nodal and mesh analysis will be discussed. The nonlinear case is then considered. It will be shown that mesh analysis can have a significant advantage over nodal analysis for nonlinear magnetic circuits. An algorithm for the solution of nonlinear magnetic circuits based on mesh analysis is then set forth. The section concludes by applying this algorithm to the solution of the UI-core example considered throughout this chapter.

Standard Branch

In formulating the solution to the network (circuit) equations, it is useful to define a standard branch, which is shown in Figure 2.26. Therein, a standard branch for nodal analysis and a standard branch for mesh analysis are shown. The two branches are almost identical, except that for nodal analysis we formulate the branch using a permeance (which can be set to zero) and in the mesh case we formulate the branch using a reluctance (which can be set to zero). In both cases, $\Phi_{b,i}$ denotes the flux going into the branch, $\Phi_{s,i}$ denotes a flux source associated with the branch, $F_{s,i}$ denotes an MMF source associated with the branch, and $F_{b,i}$ denotes the MMF drop across the branch. For the mesh branch, the reluctance is denoted $R_{b,i}$ and the flux through this reluctance is denoted $\Phi_{R,i}$; for the nodal branch the permeance is denoted $P_{b,i}$ and the flux through this permeance is denoted $\Phi_{P,i}$. Of course, the two branches are entirely equivalent with $R_{b,i} = 1/P_{b,i}$ for the case when neither the reluctance nor the permeance is zero. Often, the MMF source will represent a winding; we will often use the flux source to represent a permanent magnet material, as will be discussed in Section 2.9.

From Figure 2.26, we see that we may write the expressions

$$\Phi_{b,i} = P_{b,i}(F_{b,i} - F_{s,i}) + \Phi_{s,i} \tag{2.8-1}$$

for a nodal branch or

$$F_{b,i} = R_{b,i}(\Phi_{b,i} - \Phi_{s,i}) + F_{s,i} \tag{2.8-2}$$

for the mesh branch. Our next step will be to see how the idea of a standard branch is useful in solving magnetic circuits.

Figure 2.26 A standard branch.

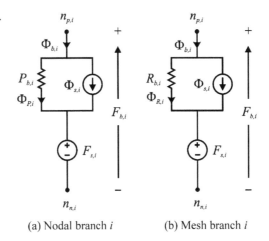

(a) Nodal branch i (b) Mesh branch i

Nodal Analysis

Nodal analysis of MECs is based on Kirchhoff's flux law—that is, that the sum of the flux into every node must be equal to zero. In this approach, every node of a magnetic circuit is labeled 0 through N_n, where N_n is the number of node potentials to be determined, and the potential of node 0 is defined to be zero (or ground in an electrical circuit). The magnetic circuit may then be described by a list of branches, labeled 1 through N_b, where N_b is the number of branches. Using this information, it is possible to form a set of equations of the form

$$\mathbf{P}_N \mathbf{F}_N = \mathbf{\Phi}_N \tag{2.8-3}$$

where $\mathbf{P}_N \in \mathbb{R}^{N_n \times N_n}$ is a network permeance matrix, $\mathbf{\Phi}_N \in \mathbb{R}^{N_n}$ is the network flux vector, and $\mathbf{F}_N \in \mathbb{R}^{N_n}$ is a vector of network node potentials. Solving (2.8-3) yields the node potentials of the MEC.

In order to systematically formulate the nodal equations, our branch description for the ith branch will include all circuit elements shown in Figure 2.26(a) and also include $n_{p,\,i}$ and $n_{n,\,i}$, which represent the positive and negative node number, respectively. Using this nomenclature, the MMF drop across the branch may be expressed as

$$F_{b,i} = F_{n_{p,i}} - F_{n_{n,i}} \tag{2.8-4}$$

so we may write (2.8-1) as

$$\Phi_{b,i} = P_{b,i}\left(F_{n_{p,i}} - F_{n_{n,i}} - F_{s,i}\right) + \Phi_{s,i} \tag{2.8-5}$$

Now let us apply Kirchhoff's flux law to the jth node (>0) of our MEC. Requiring that the sum of the flux leaving the jth node is zero, we have

$$\sum_{i \in L_{p,j}} \Phi_{b,i} - \sum_{i \in L_{n,j}} \Phi_{b,i} = 0 \tag{2.8-6}$$

In (2.8-6), $L_{p,\,j}$ is the list (set) of all branches that have node j as the positive node, and $L_{n,\,j}$ is a list of all branches that have node j as the negative node.

Equation (2.8-6) will be the basis for the jth row of the matrix/vector equation (2.8-3). This is not a requirement of the mathematics or physics, but making the jth row of (2.8-3) to correspond to the jth nodal equation will prove convenient.

Let us suppose that some branch k is in the positive node list, that is, $k \in L_{p,\,j}$. Using (2.8-1), we may write (2.8-6) as

$$P_{b,k}\left(F_{n_{p,k}} - F_{n_{n,k}} - F_{s,k}\right) + \Phi_{s,k} + \sum_{i \in L_{p,j}/k} \Phi_{b,i} - \sum_{i \in L_{n,j}} \Phi_{b,i} = 0 \tag{2.8-7}$$

where $L_{p,\,j}/k$ denotes the set $L_{p,\,j}$ less the element k. Rearranging (2.8-7), we obtain

$$P_{b,k} F_{n_{p,k}} - P_{b,k} F_{n_{n,k}} + \sum_{i \in L_{p,j}/k} \Phi_{b,i} - \sum_{i \in L_{n,j}} \Phi_{b,i} = P_{b,k} F_{s,k} - \Phi_{s,k} \tag{2.8-8}$$

The implications of (2.8-8) are that branch k, which has its positive terminal at node j, will affect the jth row and $n_{p,\,k}$ column of $\mathbf{P}_N$ with a contribution of $P_{b,\,k}$, the jth row and $n_{n,\,k}$ column of $\mathbf{P}_N$

Table 2.2 Nodal Analysis Formulation Algorithm (NAFA)

$P_N = 0$
$\Phi_N = 0$
for $i = 0$ to $i = N_b$
  if $n_{p,\,i} > 0$
    $P_{N:n_{p,i},n_{p,i}} = P_{N:n_{p,i},n_{p,i}} + P_{b,i}$
    $\Phi_{N,n_{p,i}} = \Phi_{N,n_{p,i}} + P_{b,i}F_{s,i} - \Phi_{s,i}$
  end
  if $n_{n,\,i} > 0$
    $P_{N:n_{n,i},n_{n,i}} = P_{N:n_{n,i},n_{n,i}} + P_{b,i}$
    $\Phi_{N,n_{n,i}} = \Phi_{N,n_{n,i}} - P_{b,i}F_{s,i} + \Phi_{s,i}$
  end
  if $n_{p,\,i} > 0$ and $n_{n,\,i} > 0$
    $P_{N:n_{p,i},n_{n,i}} = P_{N:n_{p,i},n_{n,i}} - P_{b,i}$
    $P_{N:n_{n,i},n_{p,i}} = P_{N:n_{n,i},n_{p,i}} - P_{b,i}$
  end
end

---

with a contribution of $-P_{b,k}$, and the jth element of $\mathbf{\Phi}_N$ with a contribution of $P_{b,k}F_{s,k} - \Phi_{s,k}$. Observe that $j = n_{p,k}$ since $n_{p,k} \in L_{p,\,j}$. Note that if $n_{n,\,k}$ is zero, then there is no contribution to the zero column, which is not represented since the potential of node zero is taken as zero.

Now let us suppose that some branch l is in the negative node list, that is, $l \in L_{n,\,j}$. Using (2.8-1), we may write (2.8-6) as

$$-P_{b,l}\left(F_{n_{p,l}} - F_{n_{n,l}} - F_{s,l}\right) - \Phi_{s,l} + \sum_{i \in L_{p,j}} \Phi_{b,i} - \sum_{i \in L_{n,j}/l} \Phi_{b,i} = 0 \qquad (2.8\text{-}9)$$

Note the change in signs relative to (2.8-7) because the chosen node is from the negative list. Rearranging (2.8-9), we have

$$-P_{b,l}F_{n_{p,l}} + P_{b,l}F_{n_{n,l}} + \sum_{i \in L_{p,j}} \Phi_{b,i} - \sum_{i \in L_{n,j}/l} \Phi_{b,i} = -P_{b,l}F_{s,l} + \Phi_{s,l} \qquad (2.8\text{-}10)$$

The implications of (2.8-10) are that branch l, which is attached with its negative node to node j, will affect the jth row and $n_{p,\,l}$ column of $\mathbf{P}_N$ (with a contribution of $-P_{b,\,l}$), the jth row and $n_{n,\,l}$ column of $\mathbf{P}_N$ with a contribution of $P_{b,\,l}$ and the jth element of $\mathbf{\Phi}_N$ with a contribution of $-P_{b,\,l}F_{s,\,l} + \Phi_{s,\,l}$. Note that $j = n_{n,\,l}$ and if $n_{n,\,l}$ is zero, then there is no contribution to the zero column, which is not represented since the potential of node zero is taken as zero.

Together (2.8-8) and (2.8-10) suggest the Nodal Analysis Formulation Algorithm (NAFA) described in pseudo-code form in Table 2.2. Therein, the procedure to establish the $\mathbf{P}_N$ and $\mathbf{\Phi}_N$ is set forth.

In order to illustrate this procedure, let us now consider an example.

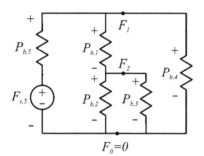

Figure 2.27 Sample MEC.

Table 2.3 Branch List for Nodal Analysis

$d_w w_w$	$n_{p,i}$	$n_{n,i}$	$P_{b,i}$	$F_{s,i}$	$\Phi_{s,i}$
1	1	2	2	0	0
2	2	0	4	0	0
3	2	0	5	0	0
4	1	0	6	0	0
5	1	0	4	100	0

Example 2.8A In this example, let us formulate the network equations for the MEC shown in Figure 2.27 and solve them to determine the flux coming out of the MMF source. Therein, we have two nonzero node potentials and five branches. The first four branches are all permeances. We will take $P_{b,1} = 2$ H, $P_{b,2} = 4$ H, $P_{b,3} = 5$ H, and $P_{b,4} = 6$ H. The fifth branch is an MMF source $F_{s,5} = 100$ with a series permeance $P_{s,5} = 4$. Our circuit is described by the branch list presented in Table 2.3.

Observe that in coming up with the branch list, the assignment of what is a positive node versus a negative node is only important in branch 5, which has a signed source.

Let us now apply the algorithm in Table 2.3. We start with $\mathbf{P}_N = 0$ and $\mathbf{\Phi}_N = 0$. Processing the first branch, we have

$$\mathbf{P}_N = \begin{bmatrix} 2 & -2 \\ -2 & 2 \end{bmatrix}, \mathbf{\Phi}_N = \begin{bmatrix} 0 \\ 0 \end{bmatrix} \quad (2.8A\text{-}1)$$

Processing the second branch, we obtain

$$\mathbf{P}_N = \begin{bmatrix} 2 & -2 \\ -2 & 6 \end{bmatrix} \quad \mathbf{\Phi}_N = \begin{bmatrix} 0 \\ 0 \end{bmatrix} \quad (2.8A\text{-}2)$$

After processing the third and fourth branch, we get

$$\mathbf{P}_N = \begin{bmatrix} 8 & -2 \\ -2 & 11 \end{bmatrix} \quad \mathbf{\Phi}_N = \begin{bmatrix} 0 \\ 0 \end{bmatrix} \quad (2.8A\text{-}3)$$

Finally, after processing the fifth branch,

$$\mathbf{P}_N = \begin{bmatrix} 12 & -2 \\ -2 & 11 \end{bmatrix}, \mathbf{\Phi}_N = \begin{bmatrix} 400 \\ 0 \end{bmatrix} \quad (2.8A\text{-}4)$$

Solving (2.8-3) with $\mathbf{P}_N$ and $\mathbf{\Phi}_N$ given by (2.8A-3) yields $F_1 = 34.4$ A and $F_2 = 6.25$ A. Using this value of F_1 in (2.8-5) for branch 5, we obtain the flux leaving the branch of 262.5 Wb.

Mesh Analysis

Nodal analysis is based on the application of Kirchhoff's flux law at every nonzero node of the MEC. Mesh analysis is based on the application of Kirchhoff's MMF law around every independent loop. Using mesh analysis, we will form a system of equations of the form

$$\mathbf{R}_N \mathbf{\Phi}_N = \mathbf{F}_N \tag{2.8-11}$$

where the network reluctance matrix $\mathbf{R}_N \in \mathbb{R}^{N_m \times N_m}$, the network MMF $\mathbf{F}_N \in \mathbb{R}^{N_m}$, and the vector of network mesh fluxes is $\mathbf{\Phi}_N \in \mathbb{R}^{N_m}$ where N_m is the number of mesh equations. Although we have used some of the same symbols as in our nodal analysis, the definitions and dimensionality are different.

In order to systematically formulate the mesh equations, two lists will be associated with each branch. The first list will be a set of the indices of the network mesh fluxes which enter the positive node of the branch. The set will be denoted $N_{p,i}$. The second list will be set of indices of the network mesh fluxes which enter the negative node of the branch. This set will be denoted $N_{n,i}$. The flux going into the ith branch may be expressed as

$$\Phi_{b,i} = \sum_{m \in N_{p,i}} \Phi_{N,m} - \sum_{m \in N_{n,i}} \Phi_{N,m} \tag{2.8-12}$$

In order to formulate the mesh equations, let us sum the MMF across each standard branch for the jth loop of an MEC. We have

$$\sum_{i \in L_{p,j}} F_{b,i} - \sum_{i \in L_{n,j}} F_{b,i} = 0 \tag{2.8-13}$$

In (2.8-13), $L_{p,j}$ is the list (set) of all branches for which the jth mesh flux, denoted $\mathbf{\Phi}_{N,j}$, enters the positive node. Similarly, $L_{n,j}$ is the list (set) of all branches for which $\mathbf{\Phi}_{N,j}$ enters the negative node.

Although we could order the equations differently, let us use (2.8-13) to develop the jth row of the matrix/vector equation (2.8-11). As in the case of nodal analysis, we will consider one branch at a time. Suppose the kth branch is in the set $L_{p,j}$. From (2.8-2) and (2.8-12), we have

$$F_{b,k} = R_{b,k} \left(\sum_{m \in N_{p,k}} \Phi_{N,m} - \sum_{m \in N_{n,k}} \Phi_{N,m} - \Phi_{s,k} \right) + F_{s,k} \tag{2.8-14}$$

Substitution of (2.8-14) into (2.8-13)

$$R_{b,k} \left(\sum_{m \in N_{p,k}} \Phi_{N,m} - \sum_{m \in N_{n,k}} \Phi_{N,m} - \Phi_{s,k} \right) + F_{s,k} + \sum_{i \in L_{p,j}/k} F_{b,i} - \sum_{i \in L_{n,j}} F_{b,i} = 0 \tag{2.8-15}$$

which can be rearranged as

$$R_{b,k} \left(\sum_{m \in N_{p,k}} \Phi_{N,m} - \sum_{m \in N_{n,k}} \Phi_{N,m} \right) + \sum_{i \in L_{p,j}/k} F_{b,i} - \sum_{i \in L_{n,j}} F_{b,i} = R_{b,k} \Phi_{s,k} - F_{s,k} \tag{2.8-16}$$

Table 2.4 Mesh Analysis Formulation Algorithm (MAFA)

$\mathbf{R}_N = 0$

$\mathbf{F}_N = 0$

```
for  i = 0  to  i = N_b
  if  |N_{p, i}| > 0
    R_{N:N_{p,i},N_{p,i}} = R_{N:N_{p,i},N_{p,i}} + R_{b,i}
    F_{N,N_{p,i}} = F_{N,N_{p,i}} + R_{b,i}Φ_{s,i} - F_{s,i}
  end
  if  |N_{n, i}| > 0
    R_{N:N_{n,i},N_{n,i}} = R_{N:N_{n,i},N_{n,i}} + R_{b,i}
    F_{N,N_{n,i}} = F_{N,N_{n,i}} - R_{b,i}Φ_{s,i} + F_{s,i}
  end
  if  |N_{p, i}| > 0  and  |N_{n, i}| > 0
    R_{N:N_{p,i},N_{n,i}} = R_{N:N_{p,i},N_{n,i}} - R_{b,i}
    R_{N:N_{n,i},N_{p,i}} = R_{N:N_{n,i},N_{p,i}} - R_{b,i}
  end
end
```

In (2.8-16), the fact that branch $R_{luc} = 1.79 \cdot 10^3$ H^{-1} is in the positive branch list for mesh equation j means that j is a member of $N_{p, k}$. Thus (2.8-16) will hold for all $j \in N_{p, k}$. This means that the net effect of branch k on the mesh equations is

i) A contribution of $R_{b,k}$ to $R_{N:N_{p,k},N_{p,k}}$. Here the notation $\mathbf{A}_{N:b, c}$ would indicate, if $b = \{2, 3\}$ and $c = \{1, 3\}$, referral to the (2,1), (2,3), (3,1), and (3,4) elements of $\mathbf{A}_N$. The colon separates the subscript from the indices.

ii) A contribution of $-R_{b,k}$ to the matrix elements $R_{N:N_{p,k},N_{n,k}}$.

iii) A contribution of $R_{b, k}Φ_{s, k} - F_{s, k}$ to the vector elements $\mathbf{F}_{N,N_{p,k}}$. Here the notation $\mathbf{A}_{N : b}$ would indicate, if $\lambda_x = N_x Φ_x$, referral to the second and third elements of $\mathbf{A}_N$.

If we repeat this process but take branch k to be in the negative branch list, we conclude that the affect will be

i) A contribution of $R_{b,k}$ to the matrix elements $R_{N:N_{n,k},N_{n,k}}$.

ii) A contribution of $-R_{b,k}$ to the matrix elements $R_{N:N_{n,k},N_{p,k}}$.

iii) A contribution of $-R_{b,k}Φ_{s, k} + F_{s, k}$ to the vector elements $\mathbf{F}_{N,N_{n,k}}$.

Based on these observations, one obtains the Mesh Analysis Formulation Algorithm (MAFA) shown in Table 2.4. Therein, "$r_x i_x$," when applied to a set, yields the number of elements in that set. When utilizing this algorithm, it should be remembered that the operations are set based. For example, if processing the seventh branch $R_{b, 7} = 2$ and $N_{p, 7} = \{3, 4\}$, then the (3,3), (3,4), (4,3), and (4,4) elements of $\mathbf{R}_N$ would be incremented by 2.

Before proceeding further, let us reconsider Example 2.8A, only this time using mesh analysis.

Figure 2.28 Sample MEC

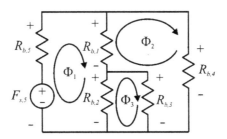

Table 2.5 Branch List for Mesh Analysis

$i_k k \in [1, 2, \cdots K]$	$N_{p,i}$	$N_{n,i}$	$R_{b,i}$	$F_{s,i}$	$\Phi_{s,i}$
1	{1}	{2}	0.500	0	0
2	{1}	{3}	0.250	0	0
3	{3}	{2}	0.200	0	0
4	{2}	{}	0.167	0	0
5	{}	{1}	0.250	100	0

Example 2.8B Let us reconsider the example of Example 2.8A using mesh analysis. Figure 2.28 illustrates the same circuit as Example 2.8A, but indicates the meshes used for analysis. The positive and negative node for each branch is also shown (these were selected arbitrarily). The first step in the analysis is to establish the branch list (Table 2.5). This is similar to the nodal case, with the exception of instead of listing the positive and negative nodes, positive and negative mesh lists are provided instead.

Let us now apply the MAFA algorithm. We start with $\mathbf{R}_N = 0$ and $\mathbf{F}_N = 0$. Processing the first branch, we have that

$$\mathbf{R}_N = \begin{bmatrix} 0.5 & -0.5 & 0 \\ -0.5 & 0.5 & 0 \\ 0 & 0 & 0 \end{bmatrix}, \quad \mathbf{F}_N = \begin{bmatrix} 0 \\ 0 \\ 0 \end{bmatrix} \tag{2.8B-1}$$

Next, after processing the second branch, we obtain

$$\mathbf{R}_N = \begin{bmatrix} 0.75 & -0.5 & -0.25 \\ -0.5 & 0.5 & 0 \\ -0.25 & 0 & 0.25 \end{bmatrix}, \quad \mathbf{F}_N = \begin{bmatrix} 0 \\ 0 \\ 0 \end{bmatrix} \tag{2.8B-2}$$

Processing branch 3, we obtain

$$\mathbf{R}_N = \begin{bmatrix} 0.75 & -0.5 & -0.25 \\ -0.5 & 0.7 & -0.2 \\ -0.25 & -0.2 & 0.45 \end{bmatrix}, \quad \mathbf{F}_N = \begin{bmatrix} 0 \\ 0 \\ 0 \end{bmatrix} \tag{2.8B-3}$$

and branch 4, we get

$$\mathbf{R}_N = \begin{bmatrix} 0.75 & -0.5 & -0.25 \\ -0.5 & 0.867 & -0.2 \\ -0.25 & -0.2 & 0.45 \end{bmatrix}, \quad \mathbf{F}_N = \begin{bmatrix} 0 \\ 0 \\ 0 \end{bmatrix} \quad (2.8\text{B-4})$$

Finally, processing branch 5,

$$\mathbf{R}_N = \begin{bmatrix} 1.00 & -0.5 & -0.25 \\ -0.5 & 0.867 & -0.2 \\ -0.25 & -0.2 & 0.45 \end{bmatrix}, \quad \mathbf{F}_N = \begin{bmatrix} 100 \\ 0 \\ 0 \end{bmatrix} \quad (2.8\text{B-5})$$

Solving (2.8-11) with $\mathbf{R}_N$ and $\mathbf{F}_N$ given by (2.8B-5) yields $\mathbf{\Phi}_N = [262 \ \ 206 \ \ 237]^T$. Thus the flux leaving branch 1 is 262 Wb, which is in agreement with Example 2.8A.

The nodal and mesh algorithms have both been presented in the context of a standard branch with linear elements. Our next main analysis objective is to see how to modify these so that they can accommodate nonlinear magnetic circuits. Before doing this, however, it will be appropriate to briefly compare nodal and mesh analysis to see which might be the more fruitful approach.

Comparison of Nodal and Mesh Analysis

For linear magnetic systems, it is difficult to clearly justify nodal analysis versus mesh analysis. The answer would, no doubt, depend on the number of meshes versus the number of nodes, the range of the elements considered, and the purpose for which the analysis is being conducted. For nonlinear analysis, however, it has been suggested that mesh analysis has a distinct advantage [15]. In this section, we will briefly consider why this might be true by way of a simple example.

Example 2.8C In this example, let us consider the simple nonlinear MEC shown in Figure 2.29. Therein, both R_1 and R_2 will be nonlinear elements. Our goal in this example will be to explore the relative merits of mesh versus nodal analysis in solving nonlinear MECs.

For this example, we will assume that the magnetic material characteristics of both reluctances are given by

$$B = \mu_0 H + M_{sat} \frac{H}{\delta H + h} \quad (2.8\text{C-1})$$

where

$$\delta = \begin{cases} 1, & B, H \geq 0 \\ -1, & B, H < 0 \end{cases} \quad (2.8\text{C-2})$$

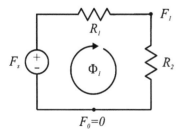

Figure 2.29 Simple nonlinear MEC.

We will take $M_{sat} = 1$ T, and $h = 159.2$ A/m. This form is convenient because it behaves roughly similar to measured characteristics and can be used to conveniently express permeability as a function of either B or H. If we wish to calculate permeability as a function of H, then substitution of H into (2.8C-2) and (2.8C-1) allows B to be calculated whereupon permeability may be expressed as

$$\mu = \frac{B}{H} \qquad (2.8C\text{-}3)$$

In the case that we wish to calculate permeability as a function of flux density B, it can be shown that the sequence of calculations

$$a = \mu_0 \delta \qquad (2.8C\text{-}4)$$

$$b = \mu_0 h + M_{sat} - \delta B \qquad (2.8C\text{-}5)$$

$$c = -Bh \qquad (2.8C\text{-}6)$$

$$H = \frac{-b + \sqrt{b^2 - 4ac}}{2a} \qquad (2.8C\text{-}7)$$

yields H whereupon (2.8C-3) can again be used to calculate permeability.

Let us begin our analysis using nodal analysis. In this case, since we are solving for the nodal MMFs, which are closely associated with the field intensity, we will calculate the reluctances using permeability as a function of field intensity. The reluctance terms may thus be expressed as

$$R_1 = \frac{l_1}{A_1 \mu_H((F_1 - F_s)/l_1)} \qquad (2.8C\text{-}8)$$

$$R_2 = \frac{l_2}{A_2 \mu_H(F_1/l_2)} \qquad (2.8C\text{-}9)$$

For this example, we will take $l_1 = l_2 = 1$ cm, $A_1 = 1.2$ cm^2, and $A_2 = 0.8$ cm^2.

Writing the nodal equation for node 1, we have the nonlinear equation

$$\frac{F_1 - F_s}{R_1\left(\frac{F_1 - F_s}{l_1}\right)} + \frac{F_1}{R_2\left(\frac{F_1}{l_2}\right)} = 0 \qquad (2.8C\text{-}10)$$

In order to solve (2.8C-10), we introduce the residual function

$$g(F_1) = \frac{F_1 - F_s}{R_1\left(\frac{F_1 - F_s}{l_1}\right)} + \frac{F_1}{R_2\left(\frac{F_1}{l_2}\right)} \qquad (2.8C\text{-}11)$$

The value of F_1 that results in $g(F_1) = 0$ is the solution to (2.8C-10).

Figure 2.30 illustrates the residual function (solid line) as a function of F_1. Solving for the point where $g(F_1) = 0$ can prove problematic. Suppose we iteratively solve the nonlinear equation (2.8C-11) using a Newton–Raphson method. In this approach, the nonlinear equation of interest is linearized, and an update is found based on the zero crossing of the linearized representation of the nonlinear equation of interest. Consider the case where we start at Point 1, which consists of the pair $(F_1, g(F_1))$. Linearizing (2.8C-11) yields an anticipated solution at Point 2'. However, because the equation is nonlinear, the updated solution estimate actually corresponds to Point 2. Linearizing about Point 2 yields a solution that would be at Point 3', except again for the nonlinearity, which causes us to instead be at Point 3. As we continue to iterate, we find that we are not in fact approaching the solution, but rather circling around it.

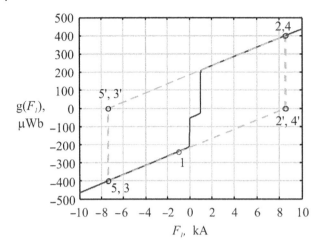

Figure 2.30 Residual for nodal analysis.

Now let us consider a mesh-based approach. In this case, the reluctances are expressed with the permeability as a function of flux density, since it is closely tied to flux. We have

$$R_1 = \frac{l_1}{A_1 \mu_B(\Phi_1/A_1)} \tag{2.8C-12}$$

$$R_2 = \frac{l_2}{A_2 \mu_B(\Phi_1/A_2)} \tag{2.8C-13}$$

Writing the single mesh equation, we have

$$-F_s + (R_1(\Phi_1) + R_2(\Phi_2))\Phi_1 = 0 \tag{2.8C-14}$$

To solve this problem, we define the residual function

$$f(\Phi_1) = -F_s + (R_1(\Phi_1) + R_2(\Phi_1))\Phi_1 \tag{2.8C-15}$$

We will solve (2.8C-14) by finding the value of Φ_1, which yields $f(\Phi_1) = 0$. This residual function is plotted in Figure 2.31. Therein, the nearly horizontal segment of the residual from -0.1 mT $< \Phi_1 < 0.1$ mT has a slightly negative though increasing residual value, which is difficult to see because of the scale.

Let us consider the problem of solving (2.8C-15) using a Newton–Raphson approach as we did in the case of Figure 2.30. Let us start at Point 1. Linearizing at this point, we anticipate the solution to be at Point 2′ but, because of the nonlinearity, end up at Point 2. After another iteration, we end up at Point 3. The slope at Point 3 is low, causing the anticipated solution at Point 4′ to be far from the solution and to be quite different from the value at Point 4. However, once we arrive at Point 4, we find that the residual is fairly linear all the way through the value of the true solution. As a result, anticipated solution Point 5′ is quite close to Point 5. Past this point the method converges rapidly. The reader is referred to Derbas et al. [15] for additional discussion.

Nonlinear Analysis of Magnetic Equivalent Circuits

In this section, our attention will be focused on solving a set of network equations. As it turns out, this will not be very different from solving the network equations in a linear system.

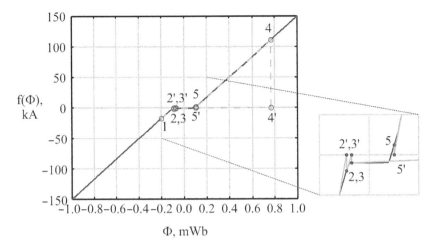

Figure 2.31 Residual using mesh analysis.

Suppose that we have a vector-valued function f(**x**) where f() and x are in $\mathbb{R}^n$, and we wish to solve

$$f(\mathbf{x}) = 0 \tag{2.8-17}$$

One method to solve this problem is the Newton–Raphson method. In this method, we start with the present estimate of a solution $\mathbf{x}_{pe}$ and generate a new (and hopefully improved) estimate $\mathbf{x}_{ne}$.

To generate an improved estimate, we express f(**x**) as a Taylor series. In particular, f(**x**) may be approximated in the region near the present estimate $\mathbf{x}_{pe}$ as

$$f(\mathbf{x}) = f(\mathbf{x}_{pe}) + \left.\frac{\partial f(\mathbf{x})}{\partial \mathbf{x}}\right|_{\mathbf{x}=\mathbf{x}_{pe}} (\mathbf{x} - \mathbf{x}_{pe}) + \mathbf{H} \tag{2.8-18}$$

where **H** represents higher-order terms. Neglecting these terms (which should go to zero in the region of $\mathbf{x}_{pe}$), setting **x** to $\mathbf{x}_{ne}$, and setting f(**x**) = 0 to be zero, (2.8-18) may be manipulated to yield

$$\mathbf{x}_{ne} = \mathbf{x}_{pe} - \left[\left.\frac{\partial f(\mathbf{x})}{\partial \mathbf{x}}\right|_{\mathbf{x}=\mathbf{x}_{pe}}\right]^{-1} f(\mathbf{x}_{pe}) \tag{2.8-19}$$

Iteratively applying (2.8-19) followed by setting $\mathbf{x}_{pe} = \mathbf{x}_{ne}$ forms the basis of the Newton–Raphson method. The iteration continues until some metric is satisfied, for example,

$$\|\mathbf{x}_{ne} - \mathbf{x}_{pe}\|_2 < \varepsilon \tag{2.8-20}$$

or

$$\|f(\mathbf{x}_{ne})\|_2 < \varepsilon \tag{2.8-21}$$

where $\|\ \|_2$ denotes the 2-norm of a vector and ε is the required tolerance.

While (2.8-19) is a useful formulation for many problems, we observe that the basis for the Newton–Raphson method is linearization of the nonlinear set of equations. In applying the Newton–Raphson algorithm to the solution of an MEC, one approach is to explicitly formulating a set of nonlinear equation (2.8-18) and performing an update via (2.8-19). However, organizing and formulating the set of nonlinear equations is inconvenient. A more convenient but entirely equivalent approach is to linearize all the standard branch equations about the estimated solution and then solve the MEC for an improved solution using the NAFA or MAFA algorithm. This will be our approach herein. Based on our comparison of the convergence properties of nodal and mesh analysis presented in the previous subsection, we will concentrate our efforts on using mesh analysis.

Referring to Figure 2.26(b) and repeating equation (2.8-2), though making the reluctance an explicit function of the flux, we may express the branch equation as

$$F_{b,i} = R_{b,i}(\Phi_{R,i})(\Phi_{b,i} - \Phi_{s,i}) + F_{s,i} \tag{2.8-22}$$

where

$$\Phi_{R,i} = \Phi_{b,i} - \Phi_{s,i} \tag{2.8-23}$$

We will take $F_{s,i}$ and $\Phi_{s,i}$ as constants, so that the nonlinearity is associated with $R_{b,i}(\Phi_{R,i})$. Substitution of (2.3-37) into (2.8-22) yields

$$F_{b,i} = \frac{l_{b,i}}{A_{b,i}\mu_{B,i}(\Phi_{R,i}/A_{b,i})}(\Phi_{b,i} - \Phi_{s,i}) + F_{s,i} \tag{2.8-24}$$

where $l_{b,i}$, $A_{b,i}$, and $\mu_{B,i}$ are the length, magnetic cross-sectional area, and permeability functions associated with branch i.

Linearizing (2.8-24) will yield the form

$$F_{b,i} \approx F_{b0,i} + \left.\frac{\partial F_{b,i}}{\partial \Phi_{b,i}}\right|_{\Phi_{b0,i}}(\Phi_{b,i} - \Phi_{b0,i}) \tag{2.8-25}$$

where $F_{b0,i}$ is the branch MMF drop at a branch flux of $\Phi_{b0,i}$. The subscript 0 subscript will indicate values about that the linearization is occurring.

From (2.8-24) and (2.8-25), we obtain

$$F_{b0,i} = R_{b0,i}(\Phi_{b0,i} - \Phi_{s,i}) + F_{s,i} \tag{2.8-26}$$

where $R_{b0,i}$ is the reluctance at branch flux $\Phi_{b0,i}$ given by

$$R_{b0,i} = \frac{l_{b,i}}{A_{b,i}\mu_{B,i}(B_{0,i})} \tag{2.8-27}$$

In (2.8-27), $B_{0,i}$ is the flux density in the resistive branch at a branch flux of $\Phi_{b0,i}$ which may be expressed as

$$B_{0,i} = \frac{\Phi_{b0,i} - \Phi_{s,i}}{A_{b,i}} \tag{2.8-28}$$

From (2.8-24), it can be shown that the partial derivative of the branch MMF drop with respect to the branch flux may be expressed as

$$\frac{\partial F_{b,i}}{\partial \Phi_{b,i}} = R_{b0,i}(1 - S_{0,i}) \tag{2.8-29}$$

where $S_{0,i}$ is an intermediate variable defined by

$$S_{0,i} = \frac{1}{l_{b,i}} R_{b0,i} \left.\frac{\partial \mu_B}{\partial B}\right|_{B0,i}(\Phi_{b0,i} - \Phi_s) \tag{2.8-30}$$

Combining (2.8-25), (2.8-26), and (2.8-29), we obtain

Table 2.6 Nonlinear MEC Solution Algorithm

```
Φ_b[1] = 0

k = 1
e = 0
while ( k = 1 )or(( k < k_mx ) and ( e > 0 ))
  linearize branch models
  execute MAFA algorithm
  solve (2.8-11) to yield  Φ_b(k + 1)
  e = ‖Φ_b[k + 1] - Φ_b[k]‖_2 - K_r‖Φ_b[k + 1] + Φ_b[k]‖_2 - K_a
  k=k+1;
end
```

$$F_{b,i} \approx F_{seff,i} + R_{beff,i}\Phi_{b,i} \tag{2.8-31}$$

where $F_{seff,\,i}$ and $R_{beff,\,i}$ are effective source MMF and effective reluctance of the branch based on the linearized branch model. The effective flux source is zero. These terms are given by

$$F_{seff,i} = F_{s,i} + R_{b0,i}(S_{0,i}\Phi_{b0,i} - \Phi_{s,i}) \tag{2.8-32}$$

and

$$R_{beff,i} = R_{b0,i}(1 - S_{o,i}) \tag{2.8-33}$$

Equation (2.8-31) represents an effective branch equation based on a linearization of the nonlinear branch. The sequence of calculations to determine the effective branch parameters is (2.8-28) for $B_{0,\,i}$; (2.8-27) for $R_{b0,\,i}$; (2.8-30) for $S_{0,\,i}$; (2.8-32) for $F_{seff,\,i}$; and finally (2.8-33) for $R_{beff,\,i}$. At this point, we have a linearized model of the branch.

After linearizing all branch equations, we can use the MAFA algorithm to solve for an update for the mesh fluxes. Next, we relinearize all the branches and repeat to obtain an even better estimate of the branch fluxes. This algorithm is summarized in the pseudo-code in Table 2.6.

There are several points in the pseudo-code of Table 2.6 that warrant further explanation and comments. First, there is no need to keep track of the branch flux vector at every iteration. To this end, if desired $\Phi_b[k + 1]$ and $\Phi_b[k]$ could be replaced by $\Phi_{b,ne}$ and $\Phi_{b,ce}$, where the subscripts ne and ce indicate new estimate and current estimate, respectively. If this is done, then the statement $\Phi_b[1]$ should be replaced by $\Phi_{b,ce} = 0$ and the statement $\Phi_{b,ce} = \Phi_{b,ne}$ should be added immediately before the end statement of the while loop. The variable k is an iteration counter; if k becomes equal to the maximum allowed iteration count k_{mx}, the loop is exited. Typically, convergence occurs in less than a half dozen iterations.

The error metric e is a measure of whether convergence has been obtained. Let us consider the definition of this metric that is given by

$$e = \|\Phi_b[k+1] - \Phi_b[k]\|_2 - K_r\|\Phi_b[k+1] + \Phi_b[k]\|_2 - K_a \tag{2.8-34}$$

The error metric is dominated by the two norm of the difference between the current and previous branch flux estimated, which is represented by the first term. From the pseudo-code in Table 2.6, we see that that loop execution will terminate when the error metric $P_{hsl} = \frac{\mu_0 l_c}{w_s}\left(d_s - \frac{2}{3}d_w\right)$ becomes negative. However, the first term in (2.8-34) could never become negative and, because of numerical errors, can never even become zero. This is the reason for remaining terms. The remaining term with the K_r factor can be seen to be proportional to the norm of the average of the new and previous branch flux estimate. The K_r factor will be a small positive number. The net effect of this term be that e will become negative if the difference between the branch flux estimates is very small relative to the average of the branch flux estimates. The K_r factor is known as the relative error tolerance. This term achieves the desired effect on its own unless the branch flux estimates are zero. In this case, the absolute error tolerance term (the K_a) makes it possible to drive e negative. Typical values of K_r, K_a, and k_{mx} are 10^{-12}, 10^{-14}, and 50, respectively.

Application to UI-Core Inductor

At this point, we are ready to bring our UI-core example to a conclusion. We have established a nonlinear MEC to represent the UI-core (Figure 2.25(b)) and have algorithms to solve the nonlinear MEC in Tables 2.4 and 2.6. We will now put this circuit and these algorithms to use.

Example 2.8D In this example, let us now investigate the impact of including leakage inductance in the UI-core inductor explored in Examples 2.4A, 2.5A, and 2.6A. Using the MEC of Figure 2.25 (b), and solving this nonlinear circuit using the nonlinear MEC solution algorithm of Tables 2.4 and 2.6, one obtains Figure 2.32. Therein, the measured and predicted λ versus i characteristics are shown. Clearly, the inclusion of leakage flux significantly improved the accuracy of the MEC over what we observed in Example 2.6A, which was in turn a significant improvement over Example 2.5A. The predicted characteristics are in reasonable agreement with the observed behavior.

With the improvements to our MEC, our methodology seems reasonably accurate, but the reader may have some concern that the method is rather involved to use. Fortunately, computer codes are available to formulate and solve MECs. These codes are similar to codes used to solve electrical circuits and involve describing the system in terms of nodes and branch descriptions. One open-source no-cost code is available in Sudhoff [16].

Between Chapter 1 and the material in this chapter, enough background has been set forth for us to start to design some magnetic components such as inductors and transformers. However, before we do so, we will consider the properties and modeling of hard magnetic materials or permanent

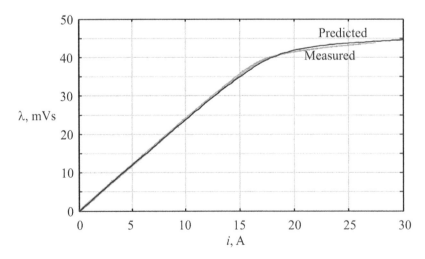

Figure 2.32 Measured and predicted $\lambda - i$ characteristics including fringing and leakage flux.

magnets. Permanent magnets are widely used in rotating electric machinery, but have other uses as well—including inductor design [17].

2.9 Permanent Magnet Materials and Their Magnetic Circuit Representation

Permanent magnet materials are useful in many situations. They are often used in electric machinery, for example. Permanent magnet materials have a $B-H$ characteristic similar to ferrimagnetic and ferromagnetic materials, except that the hysteresis loop is very wide—so wide that permanent magnet materials retain significant magnetization even when no external field is applied. For this reason, permanent magnets are referred to "hard" magnetic materials whereas ferrimagnetic and ferromagnetic materials are referred to as "soft" magnetic materials

Permanent magnet parameters are illustrated in terms of the magnet $B-H$ and $M-H$ characteristic in Figure 2.33. Therein, the $B-H$ relationship is often referred to as the material's normal characteristic, while the $M-H$ relationship is referred to as the intrinsic characteristic. In Figure 2.33 B_r is the residual flux density of the material. This is the flux density or magnetization when the field intensity is zero. The field intensity where the flux density goes to zero is known as the coercive force H_c. The intrinsic coercive force is the point where the magnetization goes to zero and is designated H_{ci}. The susceptibility of the material is denoted χ_m.

Permanent magnet material is generally operated in the 2nd quadrant if it is positively magnetized or 4th quadrant if it is negatively magnetized. It is important to make sure that

$$H \geq H_{\lim} \quad \text{positively magnetized} \tag{2.9-1}$$
$$H \leq -H_{\lim} \quad \text{negatively magnetized} \tag{2.9-2}$$

in order to avoid demagnetization, where $H_{\lim}$ is a minimum allowed field intensity to avoid demagnetization, which is a negative number whose magnitude is less than that of H_{ci}, and which is a function of magnet material and often of operating temperature. It should also be noted that while the shape of the $M-H$ characteristic is fairly consistent between materials, the shape of the $B-H$ curve is not; indeed $B-H$ may take on the slanted shape shown in Figure 2.33, or appear relatively square.

The parameters B_r, H_{ci}, H_c, and χ_m are a function of not only the material, but how it is processed; this include the degree to which the material is magnetized. In many types of permanent magnet

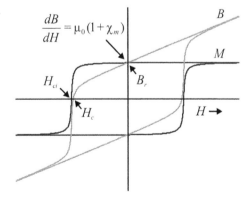

Figure 2.33 B-H and M-H Characteristics of PM Material.

material, the parameters are a strong function of temperature. Above the Curie temperature, denoted T_C, a permanent magnet will lose its magnetization; thus this temperature is critical, depending upon the magnet type and application.

There are a great variety of permanent magnet materials. Metallic metals such as iron, nickel, and cobalt are an example (since they exhibit hysteresis), but these metals are not generally used as the magnitude of the coercive force is quite low. Ceramic or ferrite magnetic materials are made of powdered iron oxide and barium/strontium carbonate ceramic. Ceramic magnets are relatively inexpensive and are used in electric machinery. Alnico (aluminum, nickel, cobalt) is another common class of magnet, though it is somewhat easily demagnetized (its coercive force is relatively small). Rare-earth combinations such as samarium–cobalt (SmCo) and neodynmium–iron–boron (NdFeB) magnets are very important in electric machinery. These magnets have both a high residual flux density and a large coercive force.

Within the classes of magnets, depending on the material, they can be processed many ways. For example, they can be cast wherein the constituent material is melted and poured into a mold. This often results in excellent magnetic properties at the sacrifice of mechanical properties. Magnet materials are also often sintered, wherein a powdered form of the material is compressed under heat (but below the melting point). Diffusion causes the material to bond. Sintering sometimes reduces the magnetic performance slightly, but often yields good mechanical properties. Bonded magnets combine powdered material with a resin or polymer to form a material that can be injection molded.

A list of some properties of common classes of magnets are listed in Table 2.7 [18]. Therein, Alnico, SmCo, Ferrite, bonded NdFeB, and sintered NdFeB are compared. In terms of Curie temperature, Alnico and SmCo magnets can have a Curie temperature in excess of 700°C; NdFeB materials have a more modest Curie temperature of 300–400°C.

One may be tempted to think that the strongest magnets will yield the "best" electromechanical device. However, we will see that the situation is more subtle than this because flux can become concentrated in devices creating a choke point for flux. In addition, it is often the case that a high magnitude of coercive force is often not needed. These issues are explored in Krizan and Sudhoff [18].

Let us now consider how to model a permanent magnet material in the context of an MEC. We will consider a sample of material shaped as in Figure 2.8, wherein we were considering Ohm's law, but in this case, we will consider permanent magnet material. We will assume that the material is operating in the second or fourth quadrants of the $B-H$ characteristic, in a region where $|H| < |H_{ci}|$. In this region, the relationship between M and H is approximated as being affine, so we may model the magnetization in a material sample as

Table 2.7 Selected Permanent Magnet Properties

Class	B_r (T)	H_{ci} (kA/m)	χ_m	ρ (g/cm³)
Alnico	0.55 to 1.4	−38 to −150	0.60 to 6.8	6.9 to 7.3
SmCo	0.87 to 1.2	−1400 to −2800	0.023 to 0.096	8.3 to 8.4
Ferrite	0.23 to 0.41	−200 to −320	0.045 to 0.58	4.8 to 4.9
Bonded NdFeB	0.47 to 0.66	−470 to −1300	0.066 to 0.40	4.9 to 5.7
Sintered NdFeB	1.2 to 1.4	−960	0.092 to 0.36	7.5

2.9 Permanent Magnet Materials and Their Magnetic Circuit Representation

$$M = \mu_0 \chi_m H \pm B_r \tag{2.9-3}$$

In (2.9-3), the plus sign applies to operation in the 2nd quadrant and the minus sign applies to operation in the 4th quadrant. Substitution of (2.9-3) into (2.3-3) yields

$$B = \mu_0(1 + \chi_m)H \pm B_r \tag{2.9-4}$$

Going back to our work of Section 2.3, (2.3-30) and (2.3-31) still apply. Manipulating (2.3-30), (2.3-31), and (2.9-4), we may write

$$F_a = R_a(\Phi_a - \Phi_{r,a}) \tag{2.9-5}$$

where the subscript a refers to material sample a and

$$R_a = \frac{l_a}{A_a \mu_0 (1 + \chi_{m,a})} \tag{2.9-6}$$

and

$$\Phi_{r,a} = A_a B_{r,a} \tag{2.9-7}$$

It can also be shown that

$$\Phi_a = P_a F_a + \Phi_{r,a} \tag{2.9-8}$$

where

$$P_a = \frac{A_a \mu_0 (1 + \chi_{m,a})}{l_a} \tag{2.9-9}$$

This suggests the equivalent circuit element shown in Figure 2.34. Note that this equivalent circuit model only holds in the region wherein (2.9-3) holds; however, we will normally have design constraints such that this is the case.

Example 2.9A Let us briefly consider the factors leading to demagnetization. Consider the situation shown in Figure 2.35. Therein, a piece of permanent magnet material (PM) of length l_m is in series with an air gap of length l_g and a core, which we will take to be infinitely permeable. The magnet is magnetized in the indicated directions. Denoting the field intensity and flux density in the upper upward direction in the magnet as B_m and H_m, from (2.9-4) we have that

$$B_m = \mu_0(1 + \chi_m)H_m + B_r \tag{2.9A-1}$$

In the air gap, denoted the field intensity and flux density as H_g and B_g in the downward direction, we have

$$B_g = \mu_0 H_g \tag{2.9A-2}$$

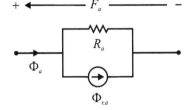

Figure 2.34 Permanent magnet MEC element.

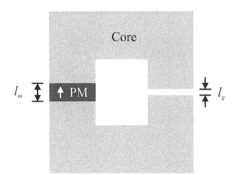

Figure 2.35 Permanent magnet configuration.

Applying Ampere's law, and assuming the core to be infinitely permeable, we deduce that

$$H_m l_m + H_g l_g = 0 \tag{2.9A-3}$$

Assuming the air gap and magnet to have identical cross-sectional areas, we know that from Gauss's Law $B_g = B_m$. From this, coupled with (2.9A-1)–(2.9A-3), we may solve for the field intensity in the magnet material. In particular,

$$H_m = -\frac{B_r}{\mu_0 \left(1 + \chi_m + \frac{l_m}{l_g}\right)} \tag{2.9A-4}$$

Observe the field intensity in the permanent magnet is negative. Also note that as the magnet becomes shorter relative to the air gap, the field intensity becomes more negative and more inclined to demagnetization.

2.10 Closing Remarks

In our work thus far, we have focused on the use of MECs to analyze magnetic systems. In the next chapter, the use of this technique in optimization-based inductor design will be introduced, thus tying our work in Chapters 1 and 2 together.

References

1 J. K. Watson, *Applications of Magnetism*, 2nd edition, Published by author, 1985.
2 G. R. Slemon, Equivalent circuits for transformers and machines including non-linear effects, *Proceedings IEE*, vol. **100**, pp. 129–143, July 1953.
3 J. Fiennes, New approach to general theory of electric machines using magnetic equivalent circuits, *Proceedings IEEE*, vol. **120**, pp. 94–104, 1973.
4 V. Ostovic, A method for evaluation of transient and steady state performance in saturated squirrel cage induction machines, *IEEE Transactions on Energy Conversion*, vol. **1**, no. 3, pp. 190–197, 1986.
5 V. Ostovic, Magnetic equivalent circuit presentation of electric machines, *Electric Machines and Power Systems*, vol. **22**, no. 2, pp. 407–432, 1987
6 J. W. H. Hayt, *Engineering Electromagnetics*. New York: McGraw-Hill, 1981.

7 D. C. Jiles and D. L. Atherton, "Ferromagnetic hysteresis," *IEEE Transactions on Magnetics*, vol. **19**, no. 5, pp. 2183–2185, 1983.
8 K. H. Carpenter and S. Warren, A wide bandwidth dynamic hysteresis model for magnetization in soft ferrites, *IEEE Transactions on Magnetics*, vol. **28**, no. 5, pp. 2037–1441, 1992.
9 D. Jiles, *Introduction to Magnetism and Magnetic Materials*, 2nd edition, London: Chapman & Hall, 1998.
10 H. Singh and S. Sudhoff, Reconsideration of energy balance in Jiles–Atherton model for accurate prediction of B–H trajectories in ferrites, *IEEE Transactions on Magnetics*, vol. **56**, no. 7, pp. 1–8, 2020.
11 F. Preisach, Über die magnetische Nachwirkung, *Zeitschrift für Physik*, vol. **94**, pp. 277–302, 1935.
12 B. S. Guru and H. R. Hiziroğlu, *Electric Machinery and Transformers*, 2nd edition, Harcourt Brace & Company, 1995.
13 G. M. Shane and S. D. Sudhoff, Refinements in anyhysteretic characterization and permeability modeling, *IEEE Transactions on Magnetics*, vol. **46**, no. 11, pp. 3834–3843, 2010.
14 J. L. Cale and S. D. Sudhoff, Accurately modeling EI core inductors using a high fidelity magnetic equivalent circuit approach, *IEEE Transactions on Magnetics*, vol. **42**, no. 1, pp. 40–46, 2006.
15 H. W. Derbas, J. M. Williams, A. C. Koenig, and S. D. Pekarek, A comparison of nodal- and mesh-based magnetic equivalent circuit models, *IEEE Transactions on Energy Conversion*, vol. **24**, no. 2, pp. 388–396, 2009.
16 S. D. Sudhoff, MATLAB Codes for Power Magnetic Devices: A Multi-Objective Design Approach, 2nd edition. Available: http://booksupport.wiley.com.
17 G. M. Shane and S. D. Sudhoff, *Design and optimization of permanent magnet inductors*, in 2012 Applied Power Electronics Conference, Orlando, FL, 2012.
18 J. Krizan and S. D. Sudhoff, *Theoretical performance boundaries for permanent magnet machines as a function of magnet type*, in 2012 Power Engineering Society General Meeting, San Diego, CA, 2012.

Problems

1 Consider Example 2.1A. What is the MMF drop traveling in a straight line from (5,2) to (5,1).

2 Consider Example 2.1A. What is the MMF drop traveling in a straight line from (5,1) to (1,1).

3 Consider Example 2.1A. Consider the path traveling in a straight lines from (5,2) to (5,1) to (1,1) and back to (5.2). How much current going into the page does the path enclose?

4 Consider a Cartesian coordinate system on a plane (the plane lies on the x- and y-axis). A current of 100 A is flowing up out of the plane at the origin. Find the MMF drop between the point $x = 1$, $y = 0$ and the point $x = 3$, $y = 2$.

5 Consider a Cartesian coordinate system (x, y, z). Suppose a uniform H-field of 1 A/m exists in the direction of the x-axis. Calculate the MMF drop from the point $(1,1,1)$ to the point $(2,5,-1)$.

6 Consider a Cartesian coordinate system (x, y, z). A uniform B-field of 1 T exists along the direction of the x-axis. Now consider the rectangular plane bounded by the points $(1,0,0)$, $(0,1,0)$, $(0,1,1)$, $(1,0,1)$. How much flux is traveling through the rectangular plane?

7 Consider a cone-shaped device with a large end with a radius of 0.1 m, and a small end with a radius of 0.03 m. The two ends are 2 m apart. A flux of 3 mWb enters the large end. A flux of 2 mWb enters the solid through the side of the cone. The B-field leaving the small end of the cone exists in a direction normal to the small end of the cone and is uniform. What is the flux density leaving the small end?

8 The $B - H$ characteristic of a certain magnetic material may be expressed as

$$B(H) = \mu_0 H + \frac{M_{sat} H}{H_b + |H|}$$

where $\mu_0 = 4\pi 10^{-7}$ A/m, $H_b = 200$ A/m, and $M_{sat} = 1.0$ T. Plot B versus H for $0 \leq H \leq 10000$, μ_B/μ_0 versus B for $0 \leq B \leq$ (value of B corresponding to $H = 10000$ A/m), and μ^H/μ_0 versus H for $0 \leq H \leq 10000$.

9 Consider a toroid with a rectangular cross section of 1 cm by 1 cm. The mean radius is 8 cm. Into this toroid, there is cut a 1 mm air gap. The toroid has 100 turns of wire. The magnetic material has a permeability of

$$\mu_B(B) = \mu_0 \frac{\sqrt[n]{B^n + a^n}}{\sqrt[n]{B^n + a^n} - 1}$$

where $n = 5$ and $a = 1.001$. This form looks strange mathematically but has the correct shape. Plot the flux linkage versus current for flux linkages in the range of 0–5 mVs. Ignore leakage inductance.

10 The flux linkage versus current characteristic of a device is given by

$$\lambda = \frac{10(1 - e^{-5i}) + 5i}{1000}$$

Find the absolute and incremental inductance at a nominal current of 250 mA.

11 Consider the toroid of Problem 9, with the exception that the magnetic material is replaced by a different material wherein the permeability is constant and equal to 2000 times the permeability of free space. Two windings are placed on the device such that positive current in one winding will tend to produce negative flux in the other winding. Suppose winding 1 has 50 turns; winding 2 has 20 turns. What is (a) the self-inductance of winding 1, (b) the self-inductance of winding 2, and (c) the mutual inductance between the windings. Finally (d), if the current in the winding 1 is 5 A, and the current in winding 2 is 3 A, what is the flux linkage in winding 1? Ignore leakage inductance.

12 Consider a UI-core with the following parameters: $w_b = w_e = w_i = 1$ cm; $w_s = w_w = 5$ cm; $d_s = d_w = 2$ cm; $l_c = 5$ cm; $g = 1$ mm; $N = 100$. Suppose the material used is such that for a flux density less than 1.3 T (the saturation point), the magnetic material is linear and has a permeability 1500 times that of free space (i.e., a relative permeability of 1500). Neglecting fringing and leakage, what current will result in a flux density of 1.3 T? In other words, if we call this the point of magnetic saturation, what current will saturate the core.

13 Consider Problem 12. What is the field intensity in the core and in the air gap?

14 Consider Problem 12. What is the inductance?

15 Consider Problem 12. For the stated conditions, how much energy is stored in the air gap? How much energy is stored in the core?

16 Using (2.7-20) and the definition of flux, derive (2.7-25).

17 Using (2.7-27) derive (2.7-28).

18 Using (2.7-34) derive (2.7-35).

19 From (2.7-36)-(2.7-39) derive (2.7-40)-(2.7-43).

20 From Figure 2.24 derive (2.7-55).

21 Consider the MEC shown in Figure 2.25(b). Suppose that all reluctances have a value of 1 H^{-1}, and that $Ni=100$. Order the branches in ascending order wherein branch number increases as one moves from top to bottom and from left to right. Write the branch list for the circuit. Then, using the NAFA, compute $\mathbf{P}_N$ and $\mathbf{\Phi}_N$.

22 Consider the MEC shown in Figure 2.25(b). Suppose that all reluctances have a value of 1 H^{-1}, and that Ni=100. Using the MAFA, compute $\mathbf{R}_N$ and $\mathbf{F}_N$.

23 Consider Example 2.9A. Suppose the cross section of the magnet is different from the cross section of the air gap. Denoting the magnet and air gap cross sections as A_m and A_g, derive an expression for the field intensity in the magnet similar to (2.9A-4).

3

Introduction to Inductor Design

In Chapter 1, we focused our attention on the design process and formulating design problems as single or multiobjective optimization problems. In Chapter 2, our attention shifted dramatically as we reviewed magnetics, magnetic equivalent circuits, and means to solve magnetic equivalent circuits. In this chapter, we will use the results from Chapters 1 and 2 in order to consider the problem of inductor design.

The placement of this chapter in this book is in some sense early: We have not yet covered many important topics relevant to inductor design including capacitance, AC losses, or thermal analysis, to name a few. While we will consider these topics later in this work, it is hoped that by engaging in the design process early, the reader will find the text more enjoyable and, at the same time, be better poised to accommodate other factors into the design framework. In Chapter 13, we will revisit the design of the UI-core inductor in a much more comprehensive fashion.

This chapter will proceed as follows. First, common inductor architectures will be briefly reviewed. Next, the calculation of coil resistance will be considered. The formulation of an inductor design problem as an optimization problem will then be set forth. Finally, a case study in inductor design will be presented.

3.1 Common Inductor Architectures

There are many different ways to construct an inductor. A few of these are shown in Figure 3.1. Therein, we have already seen the UI-core inductor shown in Figure 3.1(a). However, if we close the windings on the front and the back of the inductor rather than on the bottom, we arrive at the alternate UI-core architecture shown in Figure 3.1(b). An EI-core inductor is shown in Figure 3.1(c). In this arrangement, the conductors close around the front and back of the core to form complete turns. An MEC for the EI-core inductor is set forth in Cale and Sudhoff [1] and in Chapter 5. The center of the "E" is approximately twice as wide as the end of the "E" so that the flux density is approximately uniform.

A solenoid inductor is shown in Figure 3.1(d). In this figure, the view of the core is a longitudinal cross-section of the cylindrical core. The axial cross-section is normally circular, though, is occasionally rectangular. It is often the case that solenoid inductors are made without any core at all, in which case such inductors are referred to as air-core inductors. One disadvantage of a solenoid inductor is that the fields produced by the device go far from the device since they are not constrained to within the core.

Power Magnetic Devices: A Multi-Objective Design Approach, Second Edition. Scott D. Sudhoff.
© 2022 The Institute of Electrical and Electronics Engineers, Inc. Published 2022 by John Wiley & Sons, Inc.
Companion website: www.wiley.com/go/sudhoff/Powermagneticdevices

104 | *3 Introduction to Inductor Design*

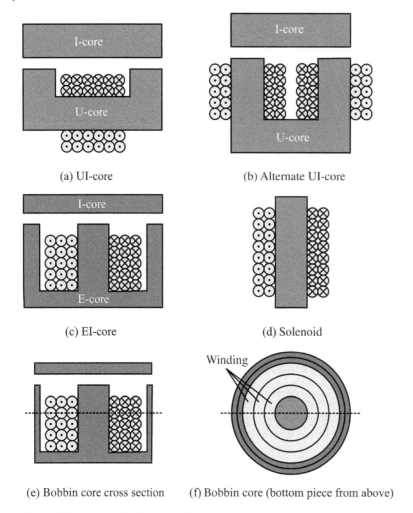

Figure 3.1 Common inductor architectures.

In Figure 3.1(e) and Figure 3.1(f) a bobbin-core inductor architecture is shown. In these figures, the view in Figure 3.1(e) is the cross-section at the point of the dashed line in Figure 3.1(f) as viewed from the side, and the view in Figure 3.1(f) is a top view of the cross-section indicated by the dashed line in Figure 3.1(e) as viewed from above. One advantage of the arrangement is in accommodating the bending radius of the conductor and that it can be very easy to wind (the winding can be wound on a spool and then placed in the core). A disadvantage of the arrangement is thermal transfer; for this reason, the outer ring of the core is often left incomplete to facilitate heat transfer.

Note that for the purposes of construction, windings are generally not wrapped directly onto the core, but rather a liner often called a slot liner, is placed on the core first, in order to provide abrasion resistance and in some cases to increase the electrical insulation between the core and winding. As this is normally thin, we will not explicitly consider it in this chapter. It can become important, however, particularly in motor design, where its thickness can significantly decrease the number of conductors that can be placed in the slots. Another practical detail is potting material, which is a resin-like material impregnated into the coil in order to provide for additional mechanical durability and increased thermal transfer and, occasionally, to improve electrical insulation properties. However, this has little effect on the magnetic analysis.

The best way to select the architecture for an inductor is to compare the Pareto-optimal fronts of the objectives relevant to a given application across a set of inductor types. Thus, our goal in this chapter will be to show how to determine the Pareto-optimal front for a given inductor architecture. In particular, we will establish the Pareto-optimal front between mass and loss in the case of the UI-core inductor arrangement shown in Figure 3.1(a).

3.2 DC Coil Resistance

In our study of inductor design, the DC winding resistance will be very important. In this section, a general approach to finding the DC resistance of a generic coil of wire will be set forth. This approach will then be applied to some specific configurations.

A typical coil is shown in Figure 3.2, and consists of a conductor wound around an open area N times. In Figure 3.2(a), a top view of the coil is shown. A cross-section of the coil is shown in Figure 3.2(b). This is precisely the geometry of coil that we would have in, for example, a UI-core inductor. Relating the coil in Figure 3.2 to the UI-core in Figure 3.1, the bottom of the "U" would pass through the center of the coil in Figure 3.2, the thickness of the coil t_c would be equal to the depth of the winding d_w, and the depth of the coil d_c would be equal to the width of the winding w_w. The reason that our generic coil in Figure 3.1 is dimensioned differently from our UI-core example is so it can be applied to other inductor geometries.

To calculate the resistance of a coil, we begin with the resistance of an electrical conductor, which may be expressed as

$$R = \frac{l}{a\sigma} \tag{3.2-1}$$

where l is the length of the electrical conductor, a is the cross-section of the conductor, and σ is the conductivity of the conductor. The difficulty in using (3.2-1) is in knowing the length of the conductor which is wound around the coil. Note that different turns around the center of the coil will have different lengths depending upon if an individual turn is toward the inside of the coil or toward the outside of the coil.

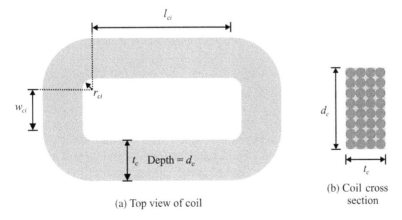

(a) Top view of coil

(b) Coil cross section

Figure 3.2 Example of a coil.

In order to calculate the length of the conductor, denote the volume of the coil as V_{cl}. Note that not all of the space occupied by a coil will be filled with conductive material. This is because the conductive material will not pack perfectly and because the conductor insulation and slot liner will occupy volume. Let us define the packing factor k_{pf} as the ratio of the total volume of conductor, denoted V_{cd}, to the total volume of the coil, V_{cl}, or

$$k_{pf} = \frac{V_{cd}}{V_{cl}} \tag{3.2-2}$$

In general, it is desirable to have as high a packing factor as possible. High packing factors lead to compact windings with better thermal conductivity than windings with low packing factor. The packing factor is primarily a function of the manufacturing process. It can be shown that the packing factor of round conductors cannot exceed 0.907 (see Problem 1). However, the packing factor is often much less: Values of around 0.45 are not uncommon.

The volume of the conductor may be related to the length of the conductor by

$$V_{cd} = la \tag{3.2-3}$$

Manipulating (3.2-1), (3.2-2), and (3.3.3), the coil resistance is given by

$$R = \frac{V_{cl} k_{pf}}{a^2 \sigma} \tag{3.2-4}$$

Let us denote the cross-sectional area of the coil as A_{cl} and denote the total cross-section of all the turns of a conductor in a coil as A_{cd}. In (3.2-2), the packing factor relates the conductor and coil volume. The packing factor also relates the conductor and coil areas to a good approximation, thus

$$k_{pf} = \frac{A_{cd}}{A_{cl}} \tag{3.2-5}$$

The total cross-section of conductor may be related to the cross-section of a single conductor as

$$A_{cd} = Na \tag{3.2-6}$$

where N is the number of turns in a coil. Incorporating (3.2-5) and (3.2-6) into (3.2-4) yields

$$R = \frac{V_{cl} N^2}{k_{pf} A_{cl}^2 \sigma} \tag{3.2-7}$$

For design purposes, (3.2-7) allows the winding resistance to be calculated based on the volume allotted for the coil region, the cross-section allotted for the coil, and the expected packing factor. Included in (3.2-7) is an implicit calculation of the conductor cross-sectional area.

Example 3.2A In order to clarify some of the definitions that have been introduced, consider a cross-section of a coil shown in Figure 3.3. Let us compute A_{cd}, A_{cl}, and k_{pf} in terms of the conductor radius r. By inspection, $A_{cd} = 6\pi r^2$, $A_{cl} = (4r)(6r) = 24r^2$, and thus, from (3.2-5), $k_{pf} = 0.785$.

In many cases, it is useful to treat the conductor cross-sections as if it is continuously variable. However, it is also often desirable to use standard conductor sizes (gauges). In this case, the conductor cross-section is based on the coil area A_{cl} and the assumed packing factor k_{pf}^*. From (3.2-5) and (3.2-6) the desired conductor cross-sectional area is first found as

$$a^* = \frac{k_{pf}^* A_{cl}}{N} \tag{3.2-8}$$

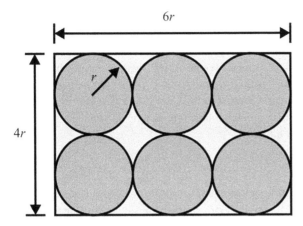

Figure 3.3 Cross-section of coil.

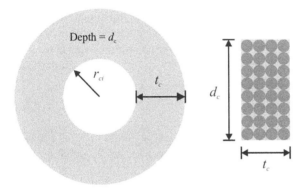

(a) Top view of coil (b) Cross-section of coil

Figure 3.4 Bobbin wound coil.

and the actual conductor area is then based on available wire sizes (selecting the gauge with larger area less than a^*), whereupon the resistance can be calculated using (3.2-4). Standard wire sizes are listed in Appendix A.

Let us now consider the coil arrangement shown in Figure 3.2. Using geometrical arguments, we have

$$A_{cl} = t_c d_c \tag{3.2-9}$$

$$V_{cl} = d_c \left(\pi (t_c + r_{ci})^2 - \pi r_{ci}^2 + (2l_{ci} + 2w_{ci})t_c \right) \tag{3.2-10}$$

A bobbin wound coil is depicted in Figure 3.4. For this coil, (3.2-9) still applies and the expression for the coil volume becomes

$$V_{cl} = \pi d_c \left((t_c + r_{ci})^2 - r_{ci}^2 \right) \tag{3.2-11}$$

3.3 DC Inductor Design

In this section, we will consider the problem of designing an inductor for an application in which we desire to have a required incremental inductance L_{inc} at a given DC bias current i_{dc0}. Normally, this sort of inductor would be commonly used for filtering purposes in power electronics applications, such as (a) a DC link inductor in an electric drive system [2], or (b) an inductor in a DC/DC converter [3]. We will be interested in minimizing mass and in minimizing loss, and so we will be faced with a multiobjective optimization problem.

Problem Formulation

As discussed in Section 3.1, there are many different inductor architectures. However, in this section, we will focus on just one of these: the UI-core inductor that we considered in Chapter 2 and that is shown in Figure 3.5.

Our design approach will be to pose the design problem as a formal multiobjective optimization problem. This will involve four steps: the identification of the search space, compiling a list of design constraints, formulation of the design metrics, and finally the synthesis of a fitness function.

Let us begin with the formulation of the search space. The design space must include, in some form, all the geometrical parameters shown in Figure 3.5, the magnetic material used, and the type of conductor used. Also, the number of turns of conductor N and the cross-sectional area of the wire must be specified.

One choice for the parameter vector to characterize the design is

$$\boldsymbol{\theta} = \begin{bmatrix} T_{cr} & T_{cd} & g & l_c & w_e & r_{ie} & r_{be} & a^* & N^* & N_w^* & N_d^* & c_w & c_d \end{bmatrix} \tag{3.3-1}$$

In (3.3-1), T_{cr} and T_{cd} are integers that will be used to designate the type of core material and type of conductor material from a list of allowed materials. The parameters g, l_c, and w_e are all geometrical in nature and are defined by Figure 3.5.

The next parameter r_{ie} is the ratio of w_i to w_e so that

$$w_i = r_{ie} w_e \tag{3.3-2}$$

In essence, r_{ie} is being used as a design parameter rather than w_i. The reason for this is to speed convergence of the optimization algorithm. Nominally, we would expect w_i to be very similar to w_e, but perhaps not exactly equal. By making r_{ie} a parameter rather than w_i, and by using a narrow

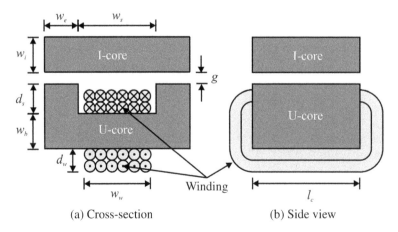

Figure 3.5 UI-Core inductor.

range for r_{ie}, say $0.9 \leq r_{ie} \leq 1.1$, we can cause w_i to be appropriately sized relative to w_e, but still not make any assumption on the exact relationship between the two. Using a similar argument, we will take

$$w_b = r_{be}w_e \tag{3.3-3}$$

We would expect w_b to be very similar to w_i and so we may specify $0.9 \leq r_{be} \leq 1.1$ when setting up the parameter range.

The next element in the parameter vector is a^*, which will be used to denote the desired conductor cross-sectional area. This is different from the actual cross-sectional area, because we will assume that the conductor size must conform to standard sizes. Thus, the actual conductor size is calculated in accordance with

$$a = \text{round}_{WG}(a^*) \tag{3.3-4}$$

where $\text{round}_{WG}()$ is a function that rounds the argument not to the nearest integer, but to the closest available wire gauge. Note that there are several wire gauge standards. Information on the American Wire Gauge is set forth in Appendix A.

From the conductor area, note that the conductor radius (w/o insulation) may be expressed as

$$r_c = \sqrt{\frac{a}{\pi}} \tag{3.3-5}$$

The next parameter specified is N^*, which is the desired number of turns. From this, the actual number of turns is calculated as

$$N = \text{round}(N^*) \tag{3.3-6}$$

where the round() function rounds to the next nearest integer. The reason for the use of N^* as a real number and applying (3.3-6) rather than making N an integer parameter directly is that, at a fundamental level, the number of turns is more of a quantitative trait rather than a qualitative trait.

The next parameters relate to the coil dimensions and are denoted N_w^* and N_d^*. These parameters are the desired number width and depth of the coil in terms of the number of conductors. Both of these numbers will be specified as real numbers and then rounded as

$$N_w = \text{round}(N_w^*) \tag{3.3-7}$$

$$N_d = \text{round}(N_d^*) \tag{3.3-8}$$

Referring to Figure 3.5, $N_w = 6$ and $N_d = 2$. The reason for specifying N_w^* and N_d^* rather than N_w and N_d follows similar reasoning to specifying N^* rather than N as discussed above.

Next, the winding width and depth are computed as

$$w_w = 2r_c k_b N_w \tag{3.3-9}$$

$$d_w = 2r_c k_b N_d \tag{3.3-10}$$

In (3.3-9) and (3.3-10), r_c is the conductor radius and k_b is a build factor that increases the effective radius of the conductor to account for insulation and to give some margin between conductors. A typical value is in the range of 1.05–1.4 that corresponds to packing factors from 71% to 40%. The calculation of w_w and d_w by (3.3-7)–(3.3-10), when coupled with the constraint that $N_d N_w > N$, will ensure that the winding can be physically constructed.

The final parameters of $\boldsymbol{\theta}$ are clearances c_w and c_d that relate the slot depth and width to the winding depth and width. In particular,

$$w_s = w_w + 2c_w \tag{3.3-11}$$

$$d_s = d_w + c_d \tag{3.3-12}$$

Ideally, one would think that the optimal value of the clearances is zero. We will see if this is so in the context of a case study in the next subsection.

At this point, it can be seen that the parameter vector $\boldsymbol{\theta}$ is in fact sufficient to allow all the parameters of the design to be calculated. Let us now turn our attention to the calculation of design metrics from the parameter vector. For this problem formulation, we will be interested in the mass of the inductor and the loss.

Let us first consider the calculation of mass. From basic geometry, the mass of the core is readily expressed as

$$M_{cr} = \rho_{cr}((w_b + w_i)(w_s + 2w_e) + 2d_s w_e)l_c \tag{3.3-13}$$

where ρ_{cr} is the mass density of the core material.

In order to compute the mass of the coil, we note that from (3.2-10), taking $t_c = d_w$, $l_{ci} = l_c$, $w_{ci} = w_b$, and $r_{ci} = 0$ the volume of the coil may be expressed as

$$V_{cl} = w_w(\pi d_w^2 + (2l_c + 2w_b)d_w) \tag{3.3-14}$$

The next step is to compute the packing factor of the coil. This is given by

$$k_{pf} = \frac{Na}{d_w w_w} \tag{3.3-15}$$

Using V_{cl} and k_{pf} and (3.3-2), the conductor volume may be calculated as

$$V_{cd} = k_{pf} V_{cl} \tag{3.3-16}$$

where upon the conductor mass may be expressed as

$$M_{cd} = \rho_{cd} V_{cd} \tag{3.3-17}$$

The total mass of the inductor is then expressed as

$$M_L = M_{cr} + M_{cd} \tag{3.3-18}$$

This expression neglects the mass of the wire insulation and any structural elements. We will refer to the mass calculated in this way as the electromagnetic mass since it does not include structural elements.

The next metric of interest is loss. From (3.2-7) and noting that

$$A_{cl} = d_w w_w \tag{3.3-19}$$

the resistance is readily expressed as

$$R = \frac{V_{cl} N^2}{k_{pf} d_w^2 w_w^2 \sigma_{cd}} \tag{3.3-20}$$

The loss of the inductor at the nominal operating current may then be expressed as

$$P_L = R i_{dc0}^2 \tag{3.3-21}$$

Note that this is the DC loss—that is the loss associated with the DC component of the current. As we will see in Chapters 6 and 11 there are many other loss mechanisms. However, for the moment we will consider (3.3-21) to adequately capture the losses.

Now that we have expressions for the metrics of interest, we will consider the constraints on the design. For our first constraint, we will require

$$c_1 = \text{gte}(N_d N_w, N) \tag{3.3-22}$$

where gte(,) is the greater-than-or-equal-to function defined in (1.9-1).

We will also require that the packing factor be limited to $k_{pf,\,mx}$. Thus, our second constraint is

$$c_2 = \text{lte}(k_{pf}, k_{pf,mx}) \tag{3.3-23}$$

where lte (,) is the less-than-or-equal-to function defined by (1.9-2). Typical values $k_{pf,\,mx}$ range from 0.35 to 0.65. This constraint is highly related to the build factor. It can be readily shown that

$$k_{pf} < \frac{\pi}{4k_b^2} \tag{3.3-24}$$

In the design of electromagnetic and electromagnet devices, thermal constraints are often very important in low-mass designs. We will consider thermal equivalent circuits in Chapter 10. In the meantime, one approach to ensuring achievable behavior is by imposing a current density limit. The current density at rated current will be given by

$$J = \frac{i_{dc0}}{a} \tag{3.3-25}$$

We will impose the constraint

$$c_3 = \text{lte}(J, J_{mx}) \tag{3.3-26}$$

where J_{mx} is the maximum allowed current density. For a passively air-cooled component, a current density limit of $J_{mx} = 7.5 \cdot 10^6$ A/m^2 is sometimes achievable in copper. Again, it is recognized that a better formulation is to compute the temperature profile of the device and limit the temperature, but the approach (3.3-25)–(3.3-26) will be adequate for the present purposes.

It is often desirable that inductors do not have highly exaggerated aspect ratios. In our case, for example, an extremely long stick-like inductor may prove objectionable. Thus, we may wish to limit the aspect ratio. To this end, we may express the overall height, width, and length of our inductor as

$$h_L = d_w + w_b + d_s + g + w_i \tag{3.3-27}$$

$$w_L = 2w_e + w_s \tag{3.3-28}$$

$$l_L = 2d_w + l_c \tag{3.3-29}$$

We will define the aspect ratio as

$$\alpha_L = \frac{\max(h_L, w_L, l_L)}{\min(h_L, w_L, l_L)} \tag{3.3-30}$$

and enforce the constraint

$$c_4 = \text{lte}(\alpha_L, \alpha_{mx}) \tag{3.3-31}$$

Another limit that we may wish to place on the device is the allowed mass and allowed loss. While these quantities are both metrics, it is often the case that there is a point beyond which the mass is clearly not tolerable, likewise with loss. Thus, we will impose the constraints that

$$c_5 = \text{lte}(M_L, M_{Lmxa}) \tag{3.3-32}$$

$$c_6 = \text{lte}(P_L, P_{Lmxa}) \tag{3.3-33}$$

where M_{Lmxa} and P_{Lmxa} are the maximum allowed mass and loss, respectively. Imposing these limits narrows the search effort to the region of interest.

The next constraint we will place on the design is on the incremental inductance. The incremental inductance is most straightforwardly found from the numerical linearization of the $\lambda - i$ characteristic. In particular, the incremental inductance may be calculated as

$$L_{inc} = \frac{\lambda|_{i_{dc0} + \Delta i} - \lambda|_{i_{dc0} - \Delta i}}{2\Delta i} \qquad (3.3\text{-}34)$$

where $\lambda|_{i_{dc0} + \Delta i}$ is the flux linkage when the current is $i_{dc0} + \Delta i$, where Δi is a perturbation in current from its nominal value. The flux linkage is calculated using the MEC that was developed in Chapter 2. We will require that the MEC converges, which will be constraint 7, or c_7, which will be 1 if convergence is obtained at both flux-linkage points, 0.5 if the analysis only converges at one point, or zero if it doesn't converge at either point. We will require that the incremental inductance be greater than some minimum required inductance L_{rqi}, which is the value of the inductance we are trying to achieve. The inductance constraint is

$$c_8 = \text{gte}(L_{inc}, L_{rqi}) \qquad (3.3\text{-}35)$$

The final constraint we will place on the design is designed to make sure that the MEC on which the design is based is reasonably accurate. This constraint is formed by first defining a magnetizing flux ratio as

$$r_\Phi = \left|\frac{N\Phi_1|_{i = i_{dc0} + \Delta i}}{\lambda|_{i = i_{dc0} + \Delta i}}\right| \qquad (3.3\text{-}36)$$

In (3.3-36), Φ_1 refers to mesh flux 1 of our MEC (see Figure 2.25). It is the flux through the I-core. The quantity λ/N is the flux through the winding. For "normal" designs, this ratio should be fairly close to 1; it is a measure of the flux through the intended path relative to the total flux. Low values of flux ratio indicate that the leakage flux dominates, which should not be the case. In our design space, if the magnetizing flux ratio is low but our model is correct, then in a sense it does not matter. However, if the leakage flux dominates the design, the MEC loses accuracy. In order to maintain accuracy (and our concept of operation of the inductor), we will impose the constraint

$$c_9 = \text{gte}(r_\Phi, r_{\Phi rq}) \qquad (3.3\text{-}37)$$

where $r_{\Phi rq}$ is the minimum required magnetizing flux ratio.

The next step is to construct a fitness function. We have two metrics of interest, along with $C = 9$ constraints. The first six of these constraints are computationally straightforward; the last three involve the solution of an MEC and are more computationally involved. This suggests the calculation of the fitness function in accordance with the pseudo-code shown in Table 3.1. Therein, ε is a small positive number as discussed in Section 1.9. The basic philosophy represented in this code is to avoid the most involved computations if possible. This would be particularly important if a more involved magnetic analysis such as FEA was used.

It should be noted that the formulation of the design problem as posed has deficiencies. For example, we have not included losses associated with current ripple, nor have we included temperature effects, nor mechanical issues such as the bending radius of the wire. In spite of this, the approach is a good first start and will enable us to start our efforts in design now, rather than waiting for several chapters.

Table 3.1 Calculation of Fitness Function

---

calculate M_L, P_L

calculate constraints c_1 through c_6

$C_s = c_1 + c_2 + c_3 + c_4 + c_5 + c_6$

$C_I = 6$

if (C_s C_I)

$$\mathbf{f} = \varepsilon\left(\frac{C_s - C}{C}\right)\begin{bmatrix}1\\1\end{bmatrix}$$

return

end

construct MEC and calculate c_7, c_8, and c_9

$C_s = C_I + c_7 + c_8 + c_9$

$C_I = C_I + 3$

if ($C_s < C_I$)

$$\mathbf{f} = \varepsilon\left(\frac{C_s - C}{C}\right)\begin{bmatrix}1\\1\end{bmatrix}$$

return

end

$$\mathbf{f} = \begin{bmatrix}\frac{1}{M_L}\\ \frac{1}{P_L}\end{bmatrix}$$

return

3.4 Case Study

Now let us consider a case study in the design of a UI-core inductor. The application is for a DC inductor for a power electronics application. Our choice for conductor materials will be copper or aluminum. Material properties of these are listed in Appendix A. We will also restrict our attention to the use of American wire gauges, also listed in Appendix A. Although we have not studied AC power losses in the core material, we will restrict our design to ferrite materials that have relatively low high-frequency losses. The specific materials considered have properties listed in Appendix B. Specific requirements are listed in Table 3.2. Our goal is to design an inductor to satisfy all of these requirements. We will also be interested in minimizing the mass, and minimizing the loss. In order to determine the performance tradeoff between these competing objectives, we will perform a multiobjective optimization.

As discussed in Chapter 1, it is necessary to establish the domain of the design parameters. This is listed in Table 3.3. Therein, the minimum and maximum value for each design parameter is listed, as is the encoding scheme. In the case of the materials, the integer range corresponds to the materials in Appendix A as previously mentioned. Note that of the conductor material types, only copper and aluminum are considered. Establishing the range of the remaining parameters requires an educated guess. However, the logarithmic encoding used for the genes facilitates the use of a wide range. A poorly chosen range will be evident in plot of the gene distribution.

Table 3.2 Inductor Specifications

Symbol	Description	Value
i_{dc0}	Nominal DC operating current	10 A
L_{rqi}	Required incremental inductance at nominal current	5 mH
J_{mxa}	Maximum allowed current density at nominal current	7.6 MA/m^2
α_{mxa}	Maximum allowed aspect ratio	3
M_{Lmxa}	Maximum allowed mass	5 kg
P_{Lmxa}	Maximum allowed loss at nominal current	100 W
k_{pf-mxa}	Maximum allowed packing factor	0.7
k_b	Winding build factor	1.05
Δi	Perturbation in current for inductance calculation	0.1 A
$r_{\Phi rq}$	Minimum required magnetizing flux ratio	0.9

Table 3.3 Design Space

Par.	Description	Min.	Max.	Enc.	Gene
T_{cr}	Core material	1	5	Int	1
T_{cd}	Conductor material	1	2	Int	2
g	Air gap (m)	10^{-4}	10^{-2}	Log	3
l_c	Core length (m)	10^{-2}	1	Log	4
w_e	U-core end width (m)	2×10^{-3}	10^{-1}	Log	5
r_{ie}	w_i to w_e ratio	0.5	1.5	lin	6
r_{be}	w_b to w_e ratio	0.5	1.5	lin	7
a_c	Cross-sectional conductor area (m^2)	10^{-9}	10^{-4}	log	8
N^*	Desired number of turns	1	10^3	log	9
N_w^*	Desired slot width in conductors	1	10^3	log	10
N_d^*	Desired slot depth in conductors	1	10^3	log	11
c_w	Slot clearance in width (m)	10^{-6}	10^{-2}	log	12
c_d	Slot clearance in depth (m)	10^{-6}	10^{-2}	log	13

The result of a genetic-algorithm based optimization run is shown in Figure 3.6. The code to conduct this run is available in reference [4]. This run involved a population size of 3000 individuals over an evolution of 3000 generations. Repeated runs produced the same Pareto-optimal front. In Figure 3.6, each point marks the loss and mass of a specific nondominated design. The tradeoff between mass and loss is one that reverberates in the design of electromagnetic and electromechanical devices: We will always have the ability to trade between loss and mass (or volume or cost). Also indicated in Figure 3.6 is a specific design, labeled Design 50 (it is in 50th place in a list of designs ordered by mass). We look at that particular design more closely after examining some of the more general trends.

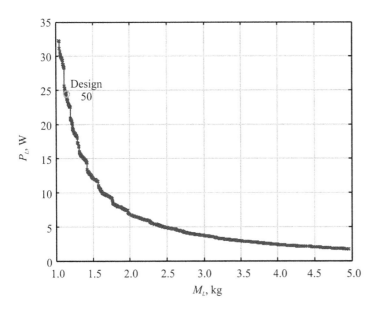

Figure 3.6 UI-Core inductor Pareto-optimal front.

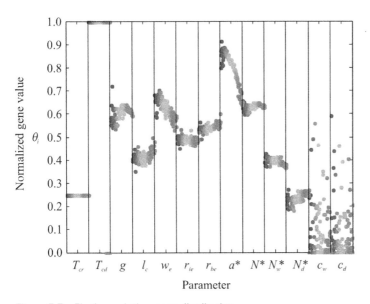

Figure 3.7 Final population gene distribution.

The gene distribution plot for the final population is shown in Figure 3.7. Recall from Chapter 1, that each design is shown as a circle with a unique shade. While it is not possible to make out individual designs because of limited shade resolution, the trends of normalized parameter values are clear. In Figure 3.7, θ_i denotes the normalized gene value that falls in [0,1]. Here $i = 1$ corresponds to T_{cr}, $i = 2$ to T_{cd}, and so on. Each vertical window shows all the gene values for every member of the population. In this figure, designs are plotted in order of the first element of the fitness function, so the genes (normalized parameters) of designs with the highest mass (or unsatisfied constraints) are to the left and the genes of designs with the lowest mass are to the right.

In regard to the gene distribution plot, note that the type of core material is tightly clustered at 0.25, which corresponds to Ferrite MN60LL. Relative to the other materials, this material doesn't have the highest initial permeability, the highest saturation flux density, or the lowest density. However, reviewing the materials $B - H$ characteristics in Appendix B, it can be seen that over a certain range of H this material yields the largest flux density, and so the choice makes some sense. The conductor type, T_{cd} is 2 (normalized value of 1), which corresponds to aluminum. Note that while copper is a better conductor on a per volume basis, aluminum is a better conductor on a per mass basis. Of the remaining parameters, while patterns are evident, it appears that none of the parameters are against the limits of the gene range, and so the ranges are appropriate, except for the clearances c_w and c_d. In these cases, the distribution plot indicates that the best designs are mostly clustered around small values. Because any clearance would just take up extraspace (and indirectly add mass), this makes sense. Note that some degree of scattering is indicated in the clearances, but the range was such that even a large value was small in an absolute sense. Given this tendency, a refinement to the define formulation would be to set the clearances to correspond to the thickness of a slot liner.

It is interesting to see how the designs change as we move along the Pareto-optimal front from low-mass high-loss designs to high-mass low-loss designs. To explore this, let us first consider Figure 3.8. In this figure, each point represents the total mass of a design on the Pareto-optimal front in the horizontal axis and the current density at the nominal DC current for that design in the vertical axis. The designs are sorted in terms of increasing mass, and a line is drawn connecting all the points. This resulting characteristic suggests that the dominant design change as one move along the Pareto-optimal front is the current density. Low-mass high-loss designs have a high current density while high-mass low-loss designs have a relatively low current density. It is the direct or

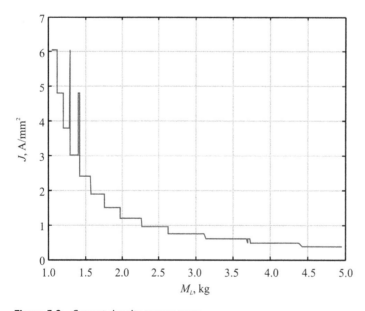

Figure 3.8 Current density versus mass.

indirect manipulation of the current density which is the dominant differentiator in the designs along the front. We will find current density to be a dominant factor in controlling the placement along the Pareto-optimal front throughout this book.

A feature of interest in Figure 3.8 is that the current density changes in discrete steps. This behavior is a direct result of picking the conductor from a finite set of wire gauges. As a result, only a finite set of current density values can result. Also, the maximum allowed current density of 7.5 A/mm^2 is not reached; this is because the next lower diameter wire size would actually be above this limit. The highest current density below the limit is slightly over 6 A/mm^2.

The reader will notice that the current density decreases almost monotonically with increasing mass. The reason for the deviation from a monotonic relationship may be related to the nature of the design problem. In particular, there are sufficient degrees of freedom that nearly identical performance can be achieved by more than one design. This can cause the predicted Pareto-optimal front near the end of an optimization run to approach the true Pareto-optimal front in an asymptotic way—it can be very close but it can never exactly reach the true front. In this case the result is a fluctuation in the current density with mass. Another optimization run would produce qualitatively similar but not identical fluctuation. Observe that this fluctuation is not associated with a visible fluctuation in the Pareto-optimal front because there are other design changes that largely negate the effect in terms of performance metrics. It is this attribute that reduces the speed of convergence to begin with.

Figure 3.9 illustrates the number of turns N and the arrangement of conductors in terms of the slot shape. Here again, the traces are formed by plotting the parameter of interest for a given design versus the mass of that design and then connecting the points to form a contiguous trace (before plotting, the designs are ordered in terms of increasing mass). Recall from (3.3-9) and (3.3-10) that N_w and N_d are the width and depth of the slot shape in terms of the number of conductors.

Comparing Figures 3.9 to 3.8, observe that for every decrease in current density (increase in wire size), there is a step increase in the number of turns. To understand this, consider two designs, one on either side of the step. Being adjacent on the Pareto-optimal front, the two designs have similar

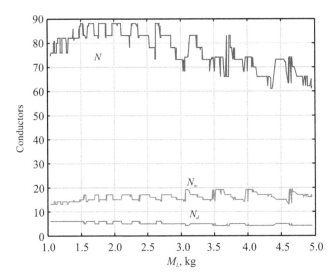

Figure 3.9 Conductor count versus mass.

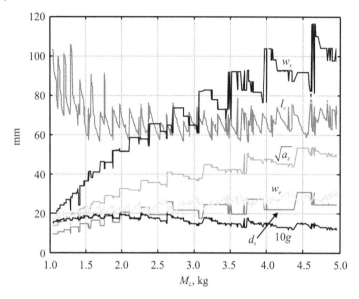

Figure 3.10 Dimensions versus mass.

masses and similar losses. After a step change to the next larger wire size, significant design changes are needed to accommodate to keep the loss roughly the same. We will see an increase in the number of turns, along with an increase in the slot dimensions. Observe that for a given wire size (region of constant current density) the number of turns goes down with increasing mass. The behaviors are irregular; much of this is due to the stiff nonlinearities associated with discretization of wire size, the number of turns, and the arrangement of conductors.

Figure 3.10 depicts many of the dimensions of the inductor designs as a function of the mass of those designs. Again, each point corresponds to a specific design. The designs have been sorted in terms of increasing mass, and a line is drawn between the points to form a continuous trace. The variables shown are, in order on the right-hand side of the figure, the slot width w_s, the core length l_c, the square root of the slot area $\sqrt{a_s} = \sqrt{w_s d_s}$, the width of the ends of the core w_e, the slot depth d_s, and 10 times the air gap 10g. By comparing Figure 3.10 with Figure 3.8, we can observe that the fluctuation of these variables is correlated with current density and conductor size.

The next step in the design process is to choose a design from the Pareto-optimal front. There are several ways this could be done. For example, data from the Pareto-optimal front could be used in conjunction with design data for other components to perform a higher-level optimization. However, for now let us simply pick one design from the front based on what would seem to be a desirable combination of mass and loss. In this case, let us choose Design 50, so-named because it is the 50th design in a list of designs ranked in order of increasing mass. Design data for this design are listed in Table 3.4, and a cross-section of the design appears in Figure 3.11.

The data in Table 3.4 are divided into three columns: one for the core, one for the winding, and one for design metrics that include mass and loss, but also include some other information of interest. Reviewing the data in Table 3.4, we see that the mass in this design is dominated by the core. We also see that N is only slightly less than $N_w N_d$, implying that our "grid" of conductors in the winding is almost full—which we would expect. From the metric column, we see that

Table 3.4 Design 50 Data

UI-Core Data	Winding Data	Metrics
Material = MN60LL	Material = Aluminum	
$w_e = 1.72$ cm	AWG = 14	$M_L = 1.14$ kg
$w_i = 1.67$ cm	$a_c = 2.08$ mm^2	$P_L = 24.5$ W
$w_b = 1.82$ cm	$N = 76$	$J = 4.80$ A/mm^2
$w_s = 2.22$ cm	$N_w = 13$	$L_{inc} = 5.00$ mH
$d_s = 1.03$ cm	$N_d = 6$	$R_L = 245$ mΩ
$l_c = 9.24$ cm	$d_w = 1.03$ cm	$h_L = 5.69$ cm
$M_{cr} = 1.03$ kg	$w_w = 2.22$ cm	$w_L = 5.66$ cm
$g = 1.58$ mm	$M_{cd} = 0.108$ kg	$l_L = 11.3$ cm

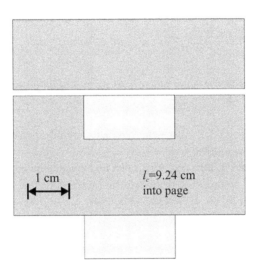

Figure 3.11 Design 50.

the incremental inductance is the minimum allowed value. Looking at the overall height, width, and length, it would appear that the aspect ratio constraints did not come into play, at least in this design.

3.5 Closing Remarks

The goal of this chapter was to provide a first example of electromagnetic component design using a multiobjective approach. While only DC loss has been thus far considered, in a DC application this will normally be the dominant loss term. We will consider AC losses, capacitance, and thermal modeling in subsequent chapters and in Chapter 13 revisits the design of the UI-core inductor considering all of these factors. While there is considerable literature in inductor design, there are a few works that may be of particular interest. A discussion of some high-frequency effects is set forth in Wallmeier et al. [5]. Examples of inductor design using some alternate optimization approaches are set forth in Pahner et al. [6] and in Njiende and Frohleke [7]. An interesting work using an evolutionary algorithm to control the details of the inductor shape is set forth in Watanabe et al. [8]. A multiobjective design process is very similar to the one in this chapter, but for an EI-core inductor is set forth in Cale and Sudhoff [9]. Although not discussed in this chapter, another interesting class of the inductor is permanent magnet inductor (PMI), which uses a permanent magnet to partially offset the flux due to the winding, allowing the core size to be reduced. A multiobjective design approach, again similar to the one presented in this chapter, but applied to a PMI, is set forth in Shane and Sudhoff [10]. Therein, a PMI has shown to have only about 60% of the mass of an EI-core inductor for the same loss.

References

1 J. L. Cale and S. D. Sudhoff, Accurately modleing EI core inductors using a high fidelity magnetic equivalent circuit approach, *IEEE Transactions on Magnetics*, vol. **42**, no. 1, pp. 40–46, 2006.
2 P. C. Krause, O. Wasynczuk, and S. D. Sudhoff, *Analysis of Electric Machinery and Drive Systems*, 2nd edition. New York: IEEE Press/Wiley, 2002.
3 N. Mohan, T. M. Undeland, and W. P. Robbins, *Power Electronics*, 2nd edition. New York: John Wiley & Sons, Inc, 1995.
4 S. D. Sudhoff, *MATLAB codes for Power Magnetic Devices: A Multi-Objective Design Approach* [Online], 2nd edition. Hoboken, NJ: John Wiley & Sons Inc. Available at http://booksupport.wiley.com.
5 P. Wallmeier, N. Frohleke, and H. Groststollen, Automated optimization of high frequency inductors, in *IECON*, Aachen, Germany, 1998.
6 U. Pahner, K. Hameyer and R. Belmans, A parallel implementation of a parametric optimization environment—Numerical optimization of an inductor for traction drive systems, *IEEE Transactions on Energy Conversion*, vol. **14**, no. 4, pp. 1329–1334, 1999.
7 H. D. Njiende and N. Frohleke, Optimization of inductors in power convers feeding high power piezoelectric motors, *Journal of Electrical & Electronics Engineering*, vol. **22**, no. 1, pp. 25–31, 2003.
8 K. Watanabe, F. Campelo, Y. Iijima, K. Kawano, T. Matsuo, T. Mifune, and H. Igarashi, Optimization of inductors using evolutionary algorithms and its experimental validation, *IEEE Transactions on Magnetics*, vol. **46**, no. 8, pp. 3393–3396, 2010.
9 J. Cale and S. D. Sudhoff, Ferrimagnetic inductor design using population-based design algorithms, *IEEE Transactions on Magnetics*, vol. **45**, no. 2, pp. 726–734, 2009.
10 G. Shane and S. D. Sudhoff, Design and optimization of permanent magnet inductors, in *Applied Power Electronics Conference and Exposition*, Orlando, 2012.

Problems

1 Show that the packing factor of round conductors cannot exceed 0.907.

2 Consider the UI-core inductor arrangement shown in Figure 3.1(b). Develop an expression for the winding resistance assuming the two coils are (*a*) in series, (*b*) in parallel. You may neglect provision for bending radius. Would you anticipate any practical advantage of one configuration versus the other?

3 The code for the design study is set forth in Sudhoff [4]. Plot the Pareto-Optimal fronts using different numbers of generations and population size. In particular, run with the following combinations (generation, population): (100, 100), (200, 200), (400, 400), (800, 800), (1600, 1600). Make a recommendation for the number of generations and population size.

4 The code for the design study is set forth in Sudhoff [4]. Modify the code assuming that the conductor area is as desired; that is the conductor size is continuously variable rather than be restricted to discrete sizes. How does the Pareto-optimal front change?

5 The code for the design study is set forth in Sudhoff [4]. Modify the code to minimize volume and loss, where the volume is the volume of the circumscribing box.

4

Force and Torque

In Chapter 1, we considered the design process and formulated design problems as optimization problems. We then reviewed magnetics and magnetic equivalent circuit analysis in Chapter 2. In Chapter 3, we used the results from Chapters 1 and 2 in order to design an electromagnetic device—an inductor. Our next step will be to proceed from strictly electromagnetic devices to ones that are also electromechanical. In electromechanical devices, we are typically interested in force or torque production. Thus, the topic of this chapter is the prediction of electromagnetic force and torque.

4.1 Energy Storage in Electromechanical Devices

In this chapter, an energy-based approach to the calculation of force and torque is set forth. We will find that if we can derive an expression for the amount of energy stored, the calculation of the force or torque will be straightforward. We will consider two variations of the approach. In the first variation, a macroscopic view of the device is taken, and it is assumed that the relationship between current, flux linkage, and position—that is, the device flux linkage equation—is known. From this information, a method to derive an expression for force will be set forth. The same method can readily be used to calculate the torque. The second variation of the approach will view the device microscopically (in a figurative sense) and will not require the system flux linkage equations to be formulated. This variation may be used to calculate the force or torque from a magnetic equivalent circuit.

Let us now consider the first approach where the relationship between the flux linkage, current, and mechanical position is known. It will be assumed that this relationship can be expressed in one of two forms. In the first form, the current $\mathbf{i}$ is expressed as a function of flux linkage λ and position x, that is,

$$\mathbf{i} = \mathbf{i}(\lambda, x) \qquad (4.1\text{-}1)$$

The second form of the flux linkage equation is used considerably more often than the first. In this case, the flux linkage is expressed in terms of current and position as

$$\lambda = \lambda(\mathbf{i}, x) \qquad (4.1\text{-}2)$$

Let us now consider the energy stored by an electromechanical device. Generally, any such device will involve a magnetic field. The energy stored in this magnetic field will be referred to as the field energy and denoted W_f. The energy that is stored in the magnetic field has two possible sources: the

Power Magnetic Devices: A Multi-Objective Design Approach, Second Edition. Scott D. Sudhoff.
© 2022 The Institute of Electrical and Electronics Engineers, Inc. Published 2022 by John Wiley & Sons, Inc.
Companion website: www.wiley.com/go/sudhoff/Powermagneticdevices

electrical system or the mechanical system. The energy entering the stored field from the electrical system is denoted W_e; energy entering the stored field from the mechanical system is denoted W_m. Assuming that the coupling field is lossless, we have that

$$W_f = W_e + W_m \tag{4.1-3}$$

The assumption of a lossless field does not mean that it is assumed that the device is lossless. Many sources of losses are external to the coupling field (resistive drops, eddy current losses, friction, etc.). However, there is one important source of loss that the assumption of a lossless core does preclude: magnetic hysteresis. Nevertheless, the analysis set forth will prove extremely useful and sufficiently accurate in the majority of cases.

As it turns out, the field energy and a related quantity referred to as the co-energy will play a critical role in determining force. In order to determine the field energy, let us consider the jth winding of the electromechanical device. The current in this winding is denoted i_j and the voltage associated with the time rate of flux across the winding is e_j, where

$$e_j = \frac{d\lambda_j}{dt} \tag{4.1-4}$$

The resistive loss across the coil is not considered; this is not because resistive losses are neglected, but rather that they will simply be considered to be external to the energy conversion process.

The energy entering the coupling field through the jth winding from time t_0 to time t_f may be expressed as

$$W_{ej} = \int_{t_0}^{t_f} e_j i_j \, dt \tag{4.1-5}$$

Substitution of (4.1-4) into (4.1-5) yields

$$W_{ej} = \int_{t_0}^{t_f} i_j \frac{d\lambda_j}{dt} \, dt \tag{4.1-6}$$

which reduces to

$$W_{ej} = \int_{\lambda_{j,0}}^{\lambda_{j,f}} i_j \, d\lambda_j \tag{4.1-7}$$

where $\lambda_{j,0}$ is λ_j at t_0 and $\lambda_{j,f}$ is λ_j at t_f.

Summing the electrical input energy overall J windings, we have that

$$W_e = \sum_{j=1}^{J} \int_{\lambda_{j,0}}^{\lambda_{j,f}} i_j \, d\lambda_j \tag{4.1-8}$$

Mechanical energy may be expressed as the integral of force over a distance; thus we have

$$W_m = -\int_{x_0}^{x_f} f_e \, dx \tag{4.1-9}$$

where $x = x_0$ at t_0 and $x = x_f$ at $t = t_f$. Therein, it is assumed that the electromagnetic force f_e and displacement x are defined to be in the same direction. The negative sign in (4.1-9) arises from the fact that positive force over a positive distance will cause positive work being done on the mechanical system, which is a negative contribution to the coupling field.

Substitution of (4.1-8) and (4.1-9) into (4.1-3) yields the primary results of this section; which is that the energy in the coupling field may be expressed as

$$W_f = \sum_{j=1}^{J} \int_{\lambda_{j,0}}^{\lambda_{j,f}} i_j d\lambda_j - \int_{x_0}^{x_f} f_e dx \tag{4.1-10}$$

4.2 Calculation of Field Energy

In the previous section, an expression for the energy in the coupling field was set forth (4.1-10). In this section, we focus on the evaluation of this expression. To this end, we will concentrate on evaluating (4.1-10) when the system flux linkage equations are of the form wherein current is a function of flux linkage and position as in (4.1-1). If this is not the case, it is much easier to calculate co-energy, an alternative but equally useful concept that will be introduced in Section 4.4.

The field energy represents energy stored in a lossless and therefore conservative field. As a result, the energy stored in the field is a function of the state—and not how that state was reached. In other words, the field energy is a function of present conditions, not past history.

Our chief difficulty in evaluating (4.1-10) is that it involves force, which is unknown. It is, after all, what we are trying to calculate. To circumvent this difficulty, we will perform a mathematical experiment. In this experiment, we will specify the trajectory of our state variables (in this case the flux linkage and position) in such a way as to make (4.1-10) as easy to evaluate as possible. Since the field is conservative, the actual trajectory is irrelevant; thus we pick the easiest way to evaluate the trajectory.

To this end, one common trajectory is to pick t_0 to correspond to a point in time wherein the field energy is zero, and the electrical system is unexcited. At this point, since there is no magnetic field, there can be no force due to the magnetic field. Hence, $f_e = 0$. Under these conditions, we position the mechanical system, varying x from its initial value of x_0 at $t = t_0$ to a value of x_f at $t = t_f$. This part of the trajectory does not contribute to W_f since f_e is zero. Thus the expression for field energy reduces from (4.1-10) to

$$W_f(\lambda, x) = \sum_{j=1}^{J} \int_{\lambda_{j,0}}^{\lambda_{j,f}} i_j(\lambda, x) d\lambda_j \tag{4.2-1}$$

In (4.2-1), the explicit dependence of the current on flux linkage and position is shown in accordance with (4.1-1), as is the resulting dependence of field energy on flux linkage and position. These functional dependencies will prove important.

The expression (4.2-1) cannot be blindly evaluated. It is a trajectory integration requiring significant care, and so we will consider its evaluation in a series of examples. In each example, we perform a numerical experiment to bring the system from its initial state to an arbitrary final state.

Example 4.2A In this example, consider a single input electrical system for which

$$i = (5 + 2x)\lambda^2 \tag{4.2A-1}$$

From (4.2-1) we have

$$W_f = \int_0^{\lambda_f} i(\lambda, x) \, d\lambda \tag{4.2A-2}$$

Substitution of (4.2A-1) into (4.2A-2) with x fixed at x_f yields

$$W_f = \frac{1}{3}(5 + 2x_f)\lambda_f^3 \tag{4.2A-3}$$

In (4.2A-3), we have written the field energy as a variable rather than as a function since the right-hand side makes the functional dependence clear. Since (4.2A-2) is valid for any final time, it is convenient to let t_f be the present time t whereupon λ_f is the present value of flux linkage λ, and x_f is the present value of position x. Thus

$$W_f = \frac{1}{3}(5 + 2x)\lambda^3 \tag{4.2A-4}$$

Example 4.2B Let us now consider a multi-input system, wherein we have

$$i_1 = 5x\lambda_1 + (10 + 2x)e^{2\lambda_1 + 2\lambda_2} \tag{4.2B-1}$$

$$i_2 = 7\lambda_2 + (10 + 2x)e^{2\lambda_1 + 2\lambda_2} \tag{4.2B-2}$$

Again, we will consider the initial flux linkages to be zero, and the final flux linkages to be $\lambda_1 = \lambda_{1,f}$ and $\lambda_2 = \lambda_{2,f}$. From (4.2-1), we obtain

$$W_f = \int_0^{\lambda_{1,f}} i_1(\lambda, x)d\lambda_1 + \int_0^{\lambda_{2,f}} i_2(\lambda, x)d\lambda_2 \tag{4.2B-3}$$

To evaluate (4.2B-3), we will position the mechanical system at $x = x_f$. For this example, it is convenient to observe that the field energy given by (4.2B-3) may be expressed as

$$W_f = W_{f,step1} + W_{f,step2} \tag{4.2B-4}$$

where

$$W_{f,step1} = \int_0^{\lambda_{1,f}} i_1(\lambda, x)|_{\lambda_2 = 0} d\lambda_1 + \int_0^0 i_2(\lambda, x)|_{\lambda_2 = 0} d\lambda_2 \tag{4.2B-5}$$

and where

$$W_{f,step2} = \int_{\lambda_{1,f}}^{\lambda_{1,f}} i_1(\lambda, x)|_{\lambda_1 = \lambda_{1,f}} d\lambda_1 + \int_0^{\lambda_{2,f}} i_2(\lambda, x)|_{\lambda_1 = \lambda_{1,f}} d\lambda_2 \tag{4.2B-6}$$

The second term in (4.2-B5) and the first term in (4.2-B6) are clearly zero. This formulation corresponds to a mathematical experiment wherein the first step λ_1 is brought from 0 to $\lambda_{1,f}$ with λ_2

fixed at zero. The second step, λ_2 goes from 0 to $\lambda_{2,f}$ with λ_1 fixed at $\lambda_{1,f}$. Evaluating (4.2B-5) and (4.2B-6) yields

$$W_{f,\text{step 1}} = \frac{5}{2}x_f\lambda_{1f}^2 + \frac{1}{2}(10 + 2x_f)\left(e^{2\lambda_1} - 1\right) \tag{4.2B-7}$$

and

$$W_{f,\text{step 2}} = \frac{7}{2}\lambda_{2f}^2 + \frac{1}{2}(10 + 2x_f)\left(e^{2\lambda_{1f} + 2\lambda_{2f}} - e^{2\lambda_{1f}}\right) \tag{4.2B-8}$$

Summing (4.2B-7) and (4.2B-8) in accordance with (4.2B-4) we have

$$W_f = \frac{5}{2}x_f\lambda_{1f}^2 + \frac{1}{2}(10 + 2x_f)\left(e^{2\lambda_{1f} + 2\lambda_{2f}} - 1\right) + \frac{7}{2}\lambda_{2f}^2 \tag{4.2B-9}$$

Since (4.2B-9) holds for any final time t_f, we will choose $t_f = t$, whereupon $\lambda_1 = \lambda_{1f}$, $\lambda_2 = \lambda_{2f}$, and $x = x_f$. Thus

$$W_f = \frac{5}{2}x\lambda_1^2 + \frac{1}{2}(10 + 2x)\left(e^{2\lambda_1 + 2\lambda_2} - 1\right) + \frac{7}{2}\lambda_2^2 \tag{4.2B-10}$$

Clearly, the process of bringing the flux linkages up one at a time can be extended to systems with any number of inputs.

It should be observed that in both of these examples, the field energy was found as a function of position x and flux linkages. A method to find the field energy in terms of the currents can be found using the considerably more involved procedure set forth in Krause et al. [1]. However, as previously mentioned, this process is never really necessary because if the flux linkage equations are in the second form (4.1-2) then it is possible to find an alternative quantity known as co-energy in a straightforward fashion. The co-energy will be discussed in Section 4.4 and is closely related to the field energy.

4.3 Force from Field Energy

Our ultimate objective is the calculation of force, not the calculation of field energy. However, once an expression for field energy has been derived an expression for force is readily obtained. To see this, let us first take the total derivative of (4.1-10). By the fundamental theorem of calculus, we have

$$dW_f = \sum_{j=1}^{J} i_j d\lambda_j - f_e dx \tag{4.3-1}$$

Now let us suppose that using the method of Section 4.2 we have an expression for the field energy in terms of the flux linkages, $W_f(\lambda, x)$. Taking the total derivative of the field energy yields

$$dW_f = \sum_{j=1}^{J} \frac{\partial W_f(\lambda, x)}{\partial \lambda_j} d\lambda_j + \frac{\partial W_f(\lambda, x)}{\partial x} dx \tag{4.3-2}$$

Equating (4.3-1) and (4.3-2), we obtain

$$\sum_{j=1}^{J} i_j d\lambda_j - f_e dx = \sum_{j=1}^{J} \frac{\partial W_f(\lambda, x)}{\partial \lambda_j} d\lambda_j + \frac{\partial W_f(\lambda, x)}{\partial x} dx \tag{4.3-3}$$

Equation (4.3-3) holds for all infinitesimally small values of $d\lambda_j$ and dx. In other words, for every combination of small $d\lambda_j$ and dx values, (4.3-3) holds. Since zero is infinitesimally small, let us set all $d\lambda_j$ equal to zero in (4.3-3). This yields the simple and powerful result,

$$f_e = -\frac{\partial W_f(\lambda, x)}{\partial x} \tag{4.3-4}$$

This result should be committed to memory. It is important to observe that for (4.3-4), the expression for field energy must be expressed exclusively in terms of flux linkage and position.

Example 4.3A In order to demonstrate the utility of (4.3-4), let us reconsider the magnetic system of Example 4.2A, and attempt to calculate the force. Applying (4.3-4) to (4.2A-4) we have

$$f_e = -\frac{2}{3}\lambda^3 \tag{4.3A-1}$$

Example 4.3B Let us reconsider the magnetic system set forth in Example 4.2B. In particular, applying (4.3-4) to (4.2B-10), we obtain

$$f_e = -\left(e^{2\lambda_1 + 2\lambda_2} - 1\right) - \frac{5}{2}\lambda_1^2 \tag{4.3B-1}$$

Before concluding this section, the importance of the result should again be contemplated. In particular, using the methods of field energy, once an expression for the flux linkages is found, it is straightforward to first find the field energy and then an expression for force. In other words, the flux linkage equations are sufficient information to derive the expression for force. No other additional information (such as geometry, etc.) is needed.

4.4 Co-Energy

In Sections 4.2 and 4.3, it was assumed that the flux linkage equations were in the form of (4.1-1); in particular, it was assumed that the currents could be expressed as an explicit function of flux linkage and position. In this section, we consider the more common case where flux linkage is expressed as a function of current and position. This corresponds to the formulation set forth in (4.1-2).

Our basic approach to finding an expression for torque, in this case, will be through the use of a concept known as co-energy. To introduce this concept, let us reconsider the evaluation of the field energy given by (4.2-1). In particular, consider the jth term in the summation that we will define as

$$W_{f,j} = \int_{\lambda_{j,0}}^{\lambda_{j,f}} i_j d\lambda_j \tag{4.4-1}$$

Comparing (4.2-1) and (4.4-1), we have

$$W_f = \sum_{j=1}^{J} W_{f,j} \tag{4.4-2}$$

Figure 4.1 Field energy and co-energy.

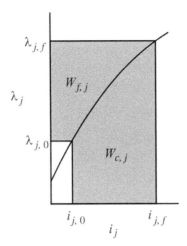

The evaluation of (4.4-1) may be viewed geometrically in Figure 4.1. Clearly, (4.4-1) corresponds to the indicated area. The corresponding component of co-energy $W_{c,j}(\mathbf{i}, x)$ is the complementary area also indicated in Figure 4.1 and expressed mathematically as

$$W_{c,j} = \int_{i_{j,0}}^{i_{j,f}} \lambda_j di_j \tag{4.4-3}$$

where $i_{j,0}$ and $i_{j,f}$ correspond to $\lambda_{j,0}$ and $\lambda_{j,f}$, respectively. Observe that while we represented the field energy as a function of flux linkage and position, we will calculate the co-energy as a function of current and position.

The total co-energy may then be expressed as

$$W_c = \sum_{j=1}^{J} W_{c,j} \tag{4.4-4}$$

Combining (4.4-3) and (4.4-4), we obtain

$$W_c(\mathbf{i}, x) = \sum_{j=1}^{J} \int_{i_{j,0}}^{i_{j,f}} \lambda_j(\mathbf{i}, x) di_j \tag{4.4-5}$$

which is analogous to (4.2-1). In (4.4-5) the functional dependence of co-energy and flux linkage on current and position has been emphasized. Observe that, as in the evaluation of field energy, the integral in (4.4-5) is integration over a trajectory and must be evaluated accordingly.

It is apparent from Figure 4.1 that

$$W_{c,j} + W_{f,j} = \lambda_{j,f} i_{j,f} - \lambda_{j,0} i_{j,0} \tag{4.4-6}$$

Throughout Sections 4.2 and 4.3, the use of the f subscript on the currents and flux linkages is to allow us to use either current or flux linkage as a variable of integration since we should not use the variable of integration to also be a limit of integration. Outside of the integration operator, we can consider $\lambda_{j,f}$ and $i_{j,f}$ to be the present values of interest λ_j and i_j. For this reason, as we have done earlier, we drop the f subscript in (4.4-6) which may then be written as

$$W_{c,j} + W_{f,j} = \lambda_j i_j - \lambda_{j,0} i_{j,0} \tag{4.4-7}$$

From (4.4-2), (4.4-5), and (4.4-7), it is readily shown that

$$W_c + W_f = \sum_{j=1}^{J} \left(\lambda_j i_j - \lambda_{j,0} i_{j,0} \right) \tag{4.4-8}$$

Equation (4.4-8) has some important physical ramifications. In particular, recall the field energy is a conservative field in that it is only a function of the state, not of how that state was achieved. Since the field energy is only a function of the state, and the term on the right-hand side of the equal sign in (4.4-8) is only a function of the state, it follows from (4.4-8) that the co-energy is only a function of state. In other words, co-energy is a conservative field. Thus (4.4-5) may be evaluated along any trajectory and the same result will be achieved.

Example 4.4A Let us find the co-energy associated with the electromechanical system described by the flux linkage equations

$$\lambda_1 = 2i_1 + \frac{1}{2+x}(i_1 + i_2)^{0.5} \tag{4.4A-1}$$

$$\lambda_2 = 5i_2^{0.4} + \frac{1}{2+x}(i_1 + i_2)^{0.5} \tag{4.4A-2}$$

In evaluating the co-energy, we will take an approach similar to that in Example 4.2B. In particular, we will utilize a two-step process, wherein in the first step we bring up the first current (in this case from zero) while holding the second current at zero. Then, we hold the first current constant while we bring up the second current. In particular, starting with (4.4-5) we have

$$W_c(\mathbf{i}, x) = \int_0^{i_{1,f}} \lambda_1(\mathbf{i}, x) di_1 + \int_0^{i_{2,f}} \lambda_2(\mathbf{i}, x) di_2 \tag{4.4A-3}$$

with x held fixed at x_f. Equation (4.4A-3) can be broken up as

$$W_c = W_{c,\text{step}1} + W_{c,\text{step}2} \tag{4.4A-4}$$

where

$$W_{c,\text{step}1} = \int_0^{i_{1,f}} \lambda_1(\mathbf{i}, x)\big|_{i_2=0} di_1 + \int_0^{0} \lambda_2(\mathbf{i}, x)\big|_{i_2=0} di_2 \tag{4.4A-5}$$

and

$$W_{c,\text{step}2} = \int_{i_{1,f}}^{i_{1,f}} \lambda_1(\mathbf{i}, x)\big|_{i_1=i_{1,f}} di_1 + \int_0^{i_{2,f}} \lambda_2(\mathbf{i}, x)\big|_{i_1=i_{1,f}} di_2 \tag{4.4A6}$$

As we did in the case of field energy, the functional dependence of co-energy on current and position is not explicitly indicated after (4.4A-3).

Clearly, the second term in (4.4A-5) and the first term in (4.4A-6) are zero. Substitution of (4.4A-1) into (4.4A-5) with $i_2 = 0$ yields

$$W_{c,step\,1} = \int_0^{i_{1,f}} 2i_1 + \frac{1}{2+x_f} i_1^{0.5} \, di_1 \qquad (4.4\text{A}7)$$

which evaluates to

$$W_{c,step\,1} = i_{1,f}^2 + \frac{2}{3} \frac{1}{2+x_f} i_{1,f}^{1.5} \qquad (4.4\text{A}8)$$

For the second step, we bring up the second current while we hold the first current equal to its final value from step 1. Substituting (4.4A-2) into (4.4A-6) yields

$$W_{c,step\,2} = \int_0^{i_{2,f}} 5i_2^{0.4} + \frac{1}{2+x_f}(i_{1,f} + i_2)^{0.5} \, di_2 \qquad (4.4\text{A}9)$$

that evaluates to

$$W_{c,step\,2} = \frac{5}{1.4} i_{2,f}^{1.4} + \frac{2}{3}\frac{1}{2+x_f}(i_{1,f} + i_{2,f})^{1.5} - \frac{2}{3}\frac{1}{2+x_f}(i_{1,f})^{1.5} \qquad (4.4\text{A}10)$$

Adding (4.4A-8) to (4.4A-10) yields

$$W_c = i_{1,f}^2 + \frac{2}{3}\frac{1}{2+x_f}(i_{1,f} + i_{2,f})^{1.5} + \frac{5}{1.4}i_{2,f}^{1.4} \qquad (4.4\text{A}11)$$

As a final note, we observe that t_f, $i_{1,f}$, $i_{2,f}$, and x_f could represent any value of time, current, and position; so we drop the subscript f leading to our final expression for this example:

$$W_c = i_1^2 + \frac{2}{3}\frac{1}{2+x}(i_1 + i_2)^{1.5} + \frac{5}{1.4}i_2^{1.4} \qquad (4.4\text{A}12)$$

Example 4.4B In this example, we consider the same magnetic system as in Example 4.4A, but we will solve the problem in a different way. In particular, in evaluating (4.4-5) we will assume that the currents follow the trajectory

$$i_1 = i_{1,f}\alpha \qquad (4.4\text{B-}1)$$
$$i_2 = i_{2,f}\alpha \qquad (4.4\text{B-}2)$$

where α varies from 0 to 1. Observe that from (4.4-B1) and (4.4-B2) we have that

$$di_1 = i_{1,f} d\alpha \qquad (4.4\text{B-}3)$$
$$di_2 = i_{2,f} d\alpha \qquad (4.4\text{B-}4)$$

Incorporating (4.4B-1)–(4.4B-4) into (4.4-8), we have

$$W_c = \int_0^1 \left(2i_{1,f}\alpha + \frac{1}{2+x}(i_{1,f} + i_{2,f})^{0.5}\alpha^{0.5}\right) i_{1,f} d\alpha + \int_0^1 \left(5i_{2,f}^{0.4}\alpha^{0.4} + \frac{1}{2+x}(i_{1,f} + i_{2,f})^{0.5}\alpha^{0.5}\right) i_{2,f} d\alpha$$

$$(4.4\text{B-}5)$$

4.5 Force from Co-Energy

At this point, while the reader may feel comfortable calculating co-energy, the reader may be questioning its use. In answer to this, we will show that co-energy provides a useful vehicle in the calculation of force.

We begin our development by rearranging (4.4-8) such that

$$W_c = \sum_{j=1}^{J} (\lambda_j i_j - \lambda_{j,0} i_{j,0}) - W_f \tag{4.5-1}$$

From which

$$dW_c = \sum_{j=1}^{J} (\lambda_j di_j + i_j d\lambda_j) - dW_f \tag{4.5-2}$$

Substitution of (4.5-1) into (4.5-2)

$$dW_c = \sum_{j=1}^{J} (\lambda_j di_j + i_j d\lambda_j) - \sum_{j=1}^{J} i_j d\lambda_j + f_e dx \tag{4.5-3}$$

which reduces to

$$dW_c = \sum_{j=1}^{J} \lambda_j di_j + f_e dx \tag{4.5-4}$$

In our next step, let us suppose that we have the co-energy as a function of current and position, that is, $W_c(\mathbf{i}, x)$. Taking the total derivative yields

$$dW_c = \sum_{j=1}^{J} \frac{\partial W_c(\mathbf{i}, x)}{\partial i_j} di_j + \frac{\partial W_c(\mathbf{i}, x)}{\partial x} dx \tag{4.5-5}$$

Equating (4.5-4) and (4.5-5), we obtain

$$\sum_{j=1}^{J} \lambda_j di_j + f_e dx = \sum_{j=1}^{J} \frac{\partial W_c(\mathbf{i}, x)}{\partial i_j} di_j + \frac{\partial W_c(\mathbf{i}, x)}{\partial x} dx \tag{4.5-6}$$

Equation (4.5-6) must hold for all infinitesimally small values of di_j and dx. Since zero qualifies as an infinitesimally small value, it is convenient to set $di_j = 0$ for all j. This yields our desired result, namely that

$$f_e = \frac{\partial W_c(\mathbf{i}, x)}{\partial x} \tag{4.5-7}$$

Equation (4.5-7) is a very important result—it is the workhorse result in terms of the calculation of electromagnetic force. In evaluating (4.5-7), it is critically important that the co-energy is strictly a function of current and position; if it is expressed as a function of flux linkage, (4.5-7) will yield erroneous results.

Example 4.5A-1. Let us reconsider Example 4.4A. In particular, we will find the electromagnetic force for the magnetic system specified therein. Applying (4.5-7) to (4.4A-12) yields

$$f_e = -\frac{2}{3}\frac{1}{(2+x)^2}(i_1+i_2)^{1.5} \qquad (4.5\text{A-}1)$$

4.6 Conditions for Conservative Fields

An important assumption in deriving our results in this chapter was that the field energy and co-energy are conservative fields. We have argued that from a physical viewpoint, this amounts to neglecting magnetic hysteresis. However, it also places a mathematical restriction on how we describe the system flux linkage equations. In particular, it can be shown that for a multi-input electrical system of the form (4.1-1) we must have that

$$\frac{\partial i_j(\boldsymbol{\lambda},x)}{\partial \lambda_k} = \frac{\partial i_k(\boldsymbol{\lambda},x)}{\partial \lambda_j} \qquad (4.6\text{-}1)$$

and that for a system of the form (4.1-2) we must have

$$\frac{\partial \lambda_j(\mathbf{i},x)}{\partial i_k} = \frac{\partial \lambda_k(\mathbf{i},x)}{\partial i_j} \qquad (4.6\text{-}2)$$

for all $j, k \in [1 \cdots J]$. A discussion of this result as well as the implications of violating these constraints is set forth in references [2–4]. Thus, when using (4.3-4) and (4.5-7) the reader is advised to make sure that the flux linkage equations obey (4.6-1) or (4.6-2), respectively.

Example 4.6A-1. Let us consider the system represented by the flux linkage equations (4.6A-1) and (4.6A-2):

$$\lambda_1 = 2xi_1^2 + 4i_1 i_2 \qquad (4.6\text{A-}1)$$

$$\lambda_2 = \frac{7i_2}{3+x} + 5i_1 \qquad (4.6\text{A-}2)$$

Could this system represent a conservative field?
From (4.6A-1) and (4.6A-2), we have

$$\frac{\partial \lambda_1}{\partial i_2} = 4i_1 \qquad (4.6\text{A-}3)$$

$$\frac{\partial \lambda_2}{\partial i_1} = 5 \qquad (4.6\text{A-}4)$$

Clearly,

$$\frac{\partial \lambda_1}{\partial i_2} \neq \frac{\partial \lambda_2}{\partial i_1} \qquad (4.6\text{A-}5)$$

and so we conclude that (4.6A-1) and (4.6A-2) do not represent a system with a conservative field.

4.7 Magnetically Linear Systems

As a useful special case, let us consider the co-energy and field energy produced by a magnetically linear system with J inputs in which

$$\boldsymbol{\lambda} = \mathbf{L}\mathbf{i} \tag{4.7-1}$$

and where $\mathbf{L}$ is not a function of $\boldsymbol{\lambda}$ or $\mathbf{i}$. With these conditions, a magnetic system that can be described by (4.7-1) is referred to as being magnetically linear. Note that such a system is not conservative unless $\mathbf{L}$ is also symmetric.

We will first consider the calculation of co-energy. Our initial condition for calculation of the co-energy is that $\mathbf{i} = 0$ and that our final condition is $\mathbf{i} = \mathbf{i}_f$. It is convenient to utilize a trajectory as in Example 4.4B wherein

$$\mathbf{i} = \alpha \mathbf{i}_f \tag{4.7-2}$$

where α will vary from 0 to 1. Clearly,

$$d\mathbf{i}_j = \mathbf{i}_{j,f} d\alpha \tag{4.7-3}$$

Utilizing (4.4-5) in conjunction with (4.7-1)–(4.7-3) we have that

$$W_c = \sum_{j=1}^{J} \int_0^1 \left. (\mathbf{L}\mathbf{i}_f \alpha) \right|_{j\text{'th row}} i_{j,f} d\alpha \tag{4.7-4}$$

which evaluates to

$$W_c = \sum_{j=1}^{J} \frac{1}{2} \left. (\mathbf{L}\mathbf{i}_f) \right|_{j\text{'th row}} i_{j,f} \tag{4.7-5}$$

Rearranging (4.7-5), we obtain

$$W_c = \sum_{j=1}^{J} \frac{1}{2} \left. (\mathbf{L}\mathbf{i}_f)^T \right|_{j\text{'th column}} i_{j,f} \tag{4.7-6}$$

which may be written more simply as

$$W_c = \frac{1}{2} \mathbf{i}_f^T \mathbf{L} \mathbf{i}_f \tag{4.7-7}$$

Dropping the subscript f, the co-energy for a magnetically linear conservative system may be expressed as

$$W_c = \frac{1}{2} \mathbf{i}^T \mathbf{L} \mathbf{i} \tag{4.7-8}$$

Following an analogous procedure for the field energy, it can be shown that

$$W_f = \frac{1}{2} \boldsymbol{\lambda}^T \mathbf{L}^{-1} \boldsymbol{\lambda} \tag{4.7-9}$$

As a point of interest, using the flux linkage equations (4.7-1), it is readily shown that for this magnetically linear conservative system the field energy and co-energy are equal, that is,

$$W_c = W_f \tag{4.7-10}$$

This result does not hold for magnetically nonlinear or nonconservative systems. Thus, if magnetic saturation occurs, then (4.7-10) will not hold.

Occasionally, there is confusion involving the implications of a system being magnetically linear versus the implications of a system being conservative. If a system is not conservative, then the field- and co-energy approaches to computing force via (4.3-4) and (4.5-7) are not valid. If the system is conservative, then (4.3-4) and (4.5-7) are valid whether the system is magnetically linear or not. If a system is conservative and magnetically linear, (4.7-8) and (4.7-9) hold, allowing field and co-energy to be calculated more simply than in the nonlinear case. If a system is magnetically linear, but not conservative, then (4.3-4), (4.5-7), (4.7-8), and (4.7-9) do not hold.

4.8 Torque

Thus far in our development, all of our results have focused on translational systems with a mechanical position x and an electromagnetic force f_e defined to be positive in the same direction as x. We will also be very interested in rotational systems. In this case, we will denote the angular position as θ_{rm}, and we will be interested in calculating torque, denoted T_e, which will be defined such that it tends to cause angular acceleration in the positive θ_{rm} direction. With regard to the subscript rm, the r designates rotor since we will be typically applying this result to an electric motor or generator, and the m in the subscript denotes mechanical position (as opposed to electrical position, but we needn't worry about that now).

The expression for mechanical work in a mechanically translational system is the integral of force over a distance as given by (4.1-9). For rotational systems, work is the integral of torque over angular displacement. Thus our expression for mechanical work becomes

$$W_m = - \int_{\theta_{rm,0}}^{\theta_{rm,f}} T_e \, d\theta_{rm} \tag{4.8-1}$$

Equation (4.8-1) is of the same form as (4.1-9) so all of our results for translation systems also hold for rotational systems with the exception that electromagnetic torque T_e replaces force f_e and rotational position θ_{rm} replaces translational position x.

4.9 Calculating Force Using Magnetic Equivalent Circuits

In the previous sections of this chapter, we considered the field and co-energy approach to calculate force or torque, wherein the $\lambda - i$ characteristics of the device are used to formulate field energy or co-energy which is then used to calculate force or torque. In a sense, the approach represented a macroscopic device view. In this section, we consider a microscopic device view in which the flux linkage equations need not be explicitly formulated. This approach facilitates the calculation of force or torque when using magnetic equivalent circuit analysis.

We will begin our development by defining $\mathbf{\Phi}_B$ to be a vector of branch fluxes. We have

$$\mathbf{\Phi_B} = [\Phi_{b,1} \ \Phi_{b,2} \ \cdots \ \Phi_{b,N_b}]^T \tag{4.9-1}$$

We can express our flux linkages in terms of the branch fluxes as

$$\lambda = \mathbf{N}_\lambda \mathbf{\Phi}_B \tag{4.9-2}$$

where $\mathbf{N}_\lambda$ is a constant matrix with elements corresponding to the number of turns, with a number of rows equal to the number of flux linkages and with a number of columns corresponding to the number of branches. Substitution of (4.9-2) into (4.3-1) yields

$$dW_f = \sum_{j=1}^{J} i_j \mathbf{N}_\lambda \big|_{j\text{th row}} d\mathbf{\Phi}_B - f_e dx \tag{4.9-3}$$

We will now leave this result and consider another approach to calculating the differential field energy. By comparing the two, we will arrive at an expression for force.

In order to calculate a second expression for differential field energy, let us consider how much field energy is stored by a single branch. Consider a system in which we apply a voltage v to a single-turn coil, which in turn causes a current i to flow. This current then acts as an MMF source for the MEC shown in Figure 4.2. The power going into the electrical circuit associated with branch b may be expressed as

$$P_b = iv \tag{4.9-4}$$

and thus, assuming that the mechanical system is already positioned, the field energy stored by the branch between the time of the beginning of our experiment t_0 and the end of our experiment t_f may be written as

$$W_{f,b} = \int_{t_0}^{t_f} iv \, dt \tag{4.9-5}$$

Since the coil has a single turn, the current is equal to the branch MMF and the voltage is equal to the time rate of change of flux into the branch. Thus we may write

$$W_{f,b} = \int_{t_0}^{t_f} F_b \frac{d\Phi_b}{dt} dt \tag{4.9-6}$$

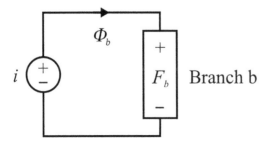

Figure 4.2 Excitation of a single branch.

where F_b and Φ_b are the branch MMF drop and branch flux into the bth branch. From (4.9-6) we have

$$W_{f,b} = \int_{\Phi_{b,0}}^{\Phi_{b,f}} F_b d\Phi_b \qquad (4.9\text{-}7)$$

where $\Phi_{b,0}$ is the branch flux at $t = t_0$ and $\Phi_{b,f}$ is the branch flux at $t = t_f$. It follows that the total field energy stored in all elements of the MEC may be expressed as

$$W_f = \sum_{b \in \mathbf{B}} \int_{\Phi_{b,0}}^{\Phi_{b,f}} F_b d\Phi_b \qquad (4.9\text{-}8)$$

where $\mathbf{B}$ is a set of the indices of all branches in the MEC.

It is convenient to break our field energy into two terms as

$$W_f = W_{fI} + W_{fD} \qquad (4.9\text{-}9)$$

where W_{fI} is the field energy due to those branches independent of any mechanical degree of freedom and W_{fD} is the field energy associated with those branches which are dependent on the mechanical degree of freedom x. This dependence can be taken to mean that some parameter of the branch is a function of the mechanical degree of freedom. We may write W_{fI} as

$$W_{fI} = \sum_{b \in \mathbf{B}_I} \int_{\Phi_{b,0}}^{\Phi_{b,f}} F_b d\Phi_b \qquad (4.9\text{-}10)$$

where $\mathbf{B}_I$ is the set of branch indices of branches which are independent of the mechanical degree of freedom x (i.e., their shape is constant). Likewise, W_{fD} may be expressed as

$$W_{fD} = \sum_{b \in \mathbf{B}_D} \int_{\Phi_{b,0}}^{\Phi_{b,f}} F_b d\Phi_b \qquad (4.9\text{-}11)$$

where $\mathbf{B}_D$ is the set of branch indices of those branches whose shape depends on the mechanical degree of freedom x.

We will assume that those branches whose shape is a function of position are magnetically linear and have permeance which, while a function of mechanical position, is not a function of flux. Typically, these branches represent flux paths through air. For such branches, we may write

$$F_b = \frac{\Phi_b}{P_b(x)} \qquad (4.9\text{-}12)$$

From (4.9-11) and (4.9-12) we obtain

$$W_{fD} = \frac{1}{2} \sum_{b \in \mathbf{B}_D} \frac{\Phi_b^2}{P_b(x)} \qquad (4.9\text{-}13)$$

The stage is now set to calculate our second expression for differential field energy. From (4.9-9) we have

$$dW_f = dW_{fI} + dW_{fD} \qquad (4.9\text{-}14)$$

Since W_{fI} does not involve our mechanical degree of freedom x, dW_{fI} will not involve dx and so will be of the form

$$dW_{fI} = \mathbf{A}_W d\mathbf{\Phi}_B \tag{4.9-15}$$

With some work, we could establish $\mathbf{A}_W$, but we will not because it will not prove necessary. From (4.9-13) we have

$$dW_{fD} = \frac{1}{2} \sum_{b \in B_D} \frac{2\Phi_b}{P_b(x)} d\Phi_b - \frac{\Phi_b^2}{P_b^2(x)} \frac{\partial P_b}{\partial x} dx \tag{4.9-16}$$

Combining (4.9-14)–(4.9-16), we have our second expression for differential field energy:

$$dW_f = \mathbf{A}_W d\mathbf{\Phi}_B + \frac{1}{2} \sum_{b \in B_D} \left[\frac{2\Phi_b}{P_b(x)} d\Phi_b - \frac{\Phi_b^2}{P_b^2(x)} \frac{\partial P_b(x)}{\partial x} dx \right] \tag{4.9-17}$$

We are now close to the desired result. Equating our first expression for differential field energy (4.9-3) to our second (4.9-17), we have

$$\sum_{j=1}^{J} i_j \mathbf{N}_\lambda |_{j\text{th row}} d\mathbf{\Phi}_B - f_e dx = \mathbf{A}_W d\mathbf{\Phi}_B + \frac{1}{2} \sum_{b \in B_D} \left[\frac{2\Phi_b}{P_b(x)} d\Phi_b - \frac{\Phi_b^2}{P_b^2(x)} \frac{\partial P_b(x)}{\partial x} dx \right] \tag{4.9-18}$$

Some care is required to interpret (4.9-18). From (4.9-1), recall that

$$d\mathbf{\Phi}_B = [d\Phi_{b,1} \quad d\Phi_{b,2} \quad \cdots \quad d\Phi_{b,N_b}]^T \tag{4.9-19}$$

Thus (4.9-18) can be thought of as being a linear equation in terms of the $d\Phi_b$ and dx. It must hold true for any allowable combination of $d\Phi_b$ and dx. The phrase "allowable combination" refers to the fact that the $d\Phi_b$ are not all independent variables, though they are independent of dx. Dependencies will arise from Kirchoff's flux law stating that the sum of the fluxes into a node must be zero. Still, setting all $d\Phi_b$ equal to zero is an allowable combination, and from this, we conclude that

$$f_e = \frac{1}{2} \sum_{b \in B_D} \frac{\Phi_b^2}{P_b^2(x)} \frac{\partial P_b(x)}{\partial x} \tag{4.9-20}$$

or, alternately, that

$$f_e = \frac{1}{2} \sum_{b \in B_D} F_b^2 \frac{\partial P_b(x)}{\partial x} \tag{4.9-21}$$

which is the desired result.

The expression (4.9-21) will be very valuable because it will allow us to use a MEC to calculate the force (or torque) without formally identifying a device's $\lambda - i$ characteristic, using this to find field energy or co-energy, and then differentiating (with the appropriate sign) to find force or torque.

In Chapter 5, we will use this result extensively as we use a MEC to calculate the force in an electromagnet.

References

1 P. C. Krause, O. Wasynczuk, and S. D. Sudhoff, *Analysis of Electric Machinery and Drive Systems*, 2nd edition. NJ: IEEE Press/Wiley-Interscience, 2003.
2 J. A. Melkebeek and J. L. Willems, Reciprocity relations for the mutual inductances between orthogonal axis windings in saturated salient-pole machines, *IEEE Trnasactions on Industry Applications*, vol. **26**, no. 1, pp. 104–110, 1990.
3 P. W. Sauer, Constraints on saturated modeling in AC machines, *IEEE Transactions on Energy Conversion*, vol. **7**, no. 1, pp. 161–167, 1992.
4 D. C. Aliprantis, *Advances in Electric Machine Modeling and Evolutionary Parameter Identification*, Ph. D. thesis. West Lafayette: Purdue University, 2003.
5 J. L. Cale and S. D. Sudhoff, Accurately modeling EI core inductors using a high fidelity magnetic equivalent circuit approach, *IEEE Transactions on Magnetics*, vol. **42**, no. 1, pp. 40–46, 2006.

Problems

1 An electromechanical device has the following flux linkage equations:

$$i_1 = \frac{0.3\lambda_1}{2+x} + \frac{0.7\,\text{sgn}\,(\lambda_1+\lambda_2)\left(1-e^{-|\lambda_1+\lambda_2|}\right)}{4+x^2}$$

$$i_2 = 0.5\lambda_2 + \frac{0.7\,\text{sgn}\,(\lambda_1+\lambda_2)\left(1-e^{-|\lambda_1+\lambda_2|}\right)}{4+x^2}$$

Note $\text{sgn}(u)$ is 1 if $u > 0$ and -1 if $u < 0$ and 0 if $u = 0$. However, you need only consider the positive current/flux linkage case. Compute the field energy as a function of λ_1 and λ_2.

2 Suppose that the field energy is given by

$$W_f = \frac{\lambda_1 + 0.3(\lambda_1+\lambda_2)^2 + 5\lambda_2^3}{7+x^2+x}$$

Derive an expression for the electromagnetic force.

3 Consider the following system:

$$\lambda_1 = 5i_1 + \frac{2\left(1-e^{-(i_1+2i_2+i_3)}\right)}{7+x}$$

$$\lambda_2 = 2i_2 + \frac{4\left(1-e^{-(i_1+2i_2+i_3)}\right)}{7+x}$$

$$\lambda_3 = i_3 + \frac{2\left(1-e^{-(i_1+2i_2+i_3)}\right)}{7+x}$$

Compute the co-energy, assuming that all currents are positive.

4 The flux linkage of a certain rotational electromechanical device may be expressed as

$$\lambda = (5+2\sin 8\theta_{rm})i$$

where θ_{rm} is the rotor position and i is the current. What is the electromagnetic torque?

4 Force and Torque

5 Is the following system magnetically linear? Could it represent a conservative field?

$$\lambda_1 = 5i_1 + \frac{7i_2}{5+x^2}$$

$$\lambda_2 = 2i_2 + \frac{3i_1}{5+x^2}$$

6 The flux linkage equations of a magnetically linear system may be expressed as

$$\begin{bmatrix} \lambda_1 \\ \lambda_2 \\ \lambda_3 \end{bmatrix} = \begin{bmatrix} 4 & \frac{2}{1+x} & 1 \\ \frac{2}{1+x} & 8 & 3 \\ 1 & 3 & 9 \end{bmatrix} \begin{bmatrix} i_1 \\ i_2 \\ i_3 \end{bmatrix}$$

Compute the force.

7 Find an expression for force from field energy analogous to (4.3-4) if there are two mechanical degrees of freedom.

8 Find an expression for force from co-energy analogous to (4.5-7) if there are two mechanical degrees of freedom.

9 Derive (4.7-9).

10 From (4.7-1), (4.7-8), and (4.7-9) derive (4.7-10).

11 Consider the expression for the reluctance on the direct flux path across the air gap of a UI-core inductor given by (2.4-4). Use this to derive an expression for force due to this branch in terms of the MMF drop across the air gap.

12 Consider the expression for the reluctance on the direct flux path across the air gap of a UI-core inductor given by (2.4-4). Use this to derive an expression for force due to this branch in terms of the flux density in the air gap.

13 Consider the UI-core from Chapter 2. Using the parameters of Example 2.4A and the magnetic equivalent circuit discussed in Section 2.6, plot the electromagnetic force on the I-core as a function of current as the current is varied from 0 to 30 A. In addition, plot the force due to the direct air gap path versus the force due to some of the fringing paths.

14 Consider the UI-core from Chapter 2. Using the parameters of Example 2.4A and the magnetic equivalent circuit of Figure 2.25, plot the electromagnetic force on the I-core as a function of current as the current is varied from 0 to 30 A.

5

Introduction to Electromagnet Design

In Chapter 4, an approach to calculate force and torque for electromechanical devices was set forth. We will use this approach extensively in our consideration of electromechanical devices. In this chapter, we will consider the design of perhaps the simplest electromechanical device—an electromagnet.

This chapter will proceed as follows. First, common electromagnet configurations will be briefly reviewed. Next, we will set the stage for the design of an EI-core electromagnet. The EI-core arrangement is selected because it is readily built and the magnetic model can be used to study the EI-core inductor in addition to the electromagnet. The first step is to develop a MEC for this architecture. This is followed by the formulation of the EI-core electromagnet design problem as an optimization problem. Finally, a case study in electromagnet design is set forth.

5.1 Common Electromagnet Architectures

There are a wide variety of electromagnet arrangements. For example, any inductor arrangements we considered in Chapter 3 (see Figure 3.1) could be used as an electromagnet, as shown in Figure 5.1. Therein, in Figure 5.1(a) and (b) we see the two U-core arrangements. Therein, note that a bottom piece is an object we are trying to attract, which may or may not be an I-core. This is similar to the case in Figure 5.1(c) where an E-core arrangement is shown. The arrangement is shown in Figure 5.1(d), the solenoid, is particularly appropriate when a relatively long travel distance is desired. This is essentially identical to the inductor arrangement shown in Fig 3.1(d), except that the winding is not wound directly on the steel member so that it can move relative to the winding. The cross-section of the movable member is often circular, but could be of any cross-section. The bobbin arrangement is shown in Figure 3.1(e) and (f) is also very commonly used, particularly in holding applications. The U-core and E-core arrangements are also effective in holding applications.

5.2 Magnetic, Electric, and Force Analysis of an Ei-Core Electromagnet

In this section, we will set the stage for the design of an EI-core electromagnet. This topology has been selected for several reasons. First, it is relatively easy to construct using laminated steel. Second, the analysis is a straightforward extension of our analysis of the UI-core inductor. Third, the

Power Magnetic Devices: A Multi-Objective Design Approach, Second Edition. Scott D. Sudhoff.
© 2022 The Institute of Electrical and Electronics Engineers, Inc. Published 2022 by John Wiley & Sons, Inc.
Companion website: www.wiley.com/go/sudhoff/Powermagneticdevices

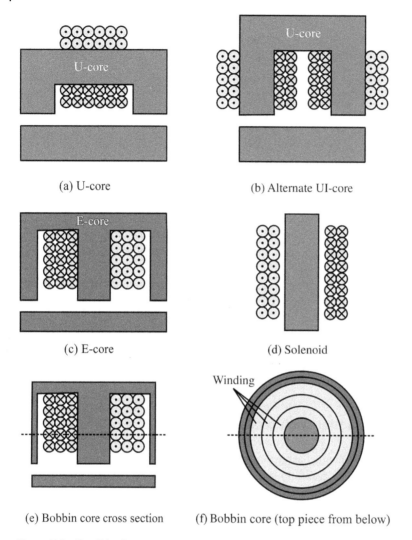

Figure 5.1 Possible electromagnet arrangements.

magnetic analysis that we will perform will be immediately useful to those with an interest in designing an EI-core inductor.

A diagram of the EI-core electromagnet is shown in Figure 5.2. Therein, most of the dimensions are denoted as in the case of the UI-core inductor, with the exception that we now have a center post of width w_c. The device is assumed to have N turns of the conductor. Figure 5.2(a) depicts a cross-section. As can be seen, conductors are going into the left slot and coming out of the right slot. The turns close in front of and in the back of the E-core, as can be seen in the side view in Figure 5.2(b).

Our analysis will be broken into three parts. First, we will perform an electric analysis that will be used to predict the current. Second, we will require a magnetic analysis in order to determine the system fluxes and MMF drops. Finally, we will require a force analysis so that we can predict the force between the E-core and the I- core.

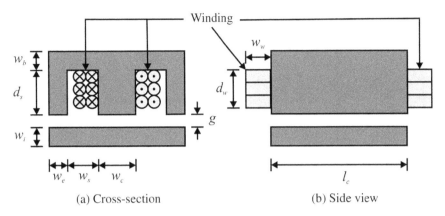

Figure 5.2 EI-core electromagnet.

Electrical Analysis

We will begin our analysis with consideration of the coil resistance. We will follow the approach set forth in Section 3.2. We begin by observing that the cross-sectional area and volume of the core may be expressed as

$$A_{cl} = w_w d_w \tag{5.2-1}$$

and

$$V_{cl} = d_w \left(\pi w_w^2 + (2l_c + 2w_c) w_w \right) \tag{5.2-2}$$

respectively. From (3.2-7) we have

$$R_{cl} = \frac{V_{cl} N^2}{k_{pf} A_{cl}^2 \sigma} \tag{5.2-3}$$

where σ is the conductivity of the conductor and k_{pf} is the packing factor as described in Section 3.2.

The steady-state DC current may be expressed as

$$i = \frac{v}{R_{cl}} \tag{5.2-4}$$

where v is the DC voltage across the coil.

Magnetic Analysis

Let us next perform a magnetic analysis of the device by formulating a magnetic equivalent circuit. A possible structure of a magnetic equivalent circuit is shown in Figure 5.3. Therein, reluctances with parentheses are a function of the branch flux going through them; we will formally identify these arguments subsequently.

We will begin our development by specifying the reluctance of the core pieces. From our work in Chapter 2, we may write

$$R_b(\Phi) = \frac{2w_s + w_c + w_e}{2w_b l_c \mu_B(\Phi/(w_b l_c))} \tag{5.2-5}$$

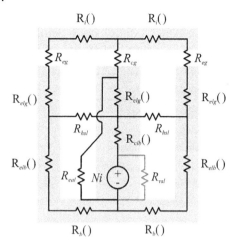

Figure 5.3 EI-core magnetic equivalent circuit.

$$R_i(\Phi) = \frac{2w_s + w_c + w_e}{2w_i l_c \mu_B(\Phi/(w_i l_c))} \qquad (5.2\text{-}6)$$

$$R_{elb}(\Phi) = \frac{2d_h + w_b}{2w_e l_c \mu_B(\Phi/(w_e l_c))} \qquad (5.2\text{-}7)$$

$$R_{elg}(\Phi) = \frac{d_s - d_h}{w_e l_c \mu_B(\Phi/(w_e l_c))} \qquad (5.2\text{-}8)$$

$$R_{clb}(\Phi) = \frac{2d_h + w_b}{2w_c l_c \mu_B(\Phi/(w_c l_c))} \qquad (5.2\text{-}9)$$

$$R_{clg}(\Phi) = \frac{d_s - d_h}{w_c l_c \mu_B(\Phi/(w_c l_c))} \qquad (5.2\text{-}10)$$

where in each case Φ denotes the flux through the appropriate branch. In (5.2-8) and (5.2-9), d_h is the halfway point where half the horizontal leakage flux is below this depth and half the leakage flux are above this point. The derivation of this distance culminates in (2.7-26), which is repeated here for the reader's convenience:

$$d_h = \begin{cases} \sqrt{d_w(d_s - d_w/2)} & d_w \geq 2d_s/3 \\ d_s/2 + d_w/4 & d_w \leq 2d_s/3 \end{cases} \qquad (5.2\text{-}11)$$

Calculation of fringing permeances is discussed in Section 2.6. Using that approach, the end air-gap permeance may be expressed as

$$P_{eg} = P_{egd} + P_{efoc} + P_{efic} + 2P_{eff} \qquad (5.2\text{-}12)$$

In (5.2-12), P_{egd} is the end air-gap permeance through the direct path given by

$$P_{egd} = \frac{\mu_0 w_e l_c}{g} \qquad (5.2\text{-}13)$$

and P_{efoc} is the end-leg fringing permeance associated with an outer corner. This permeance is calculated using

$$P_{efoc} = \frac{\mu_0 l_c}{\pi} \ln\left(1 + \frac{\pi}{g} \min(w_i, d_s + w_b)\right) \tag{5.2-14}$$

Next in (5.2-12), the quantity P_{efic} is the end-leg fringing permeance associated with an inner corner. This term can be shown to be given by

$$P_{efic} = \frac{2\mu_0 l_c}{\pi} \ln\left(1 + \frac{\pi}{4} \frac{\min(2d_s, w_s)}{g}\right) \tag{5.2-15}$$

Finally in (5.2-12), P_{eff} is the end-leg fringing face permeance computed from

$$P_{eff} = \frac{\mu_0 w_e}{\pi} \ln\left(1 + \pi \frac{\min(w_i, d_s + w_b)}{g}\right) \tag{5.2-16}$$

A discussion and derivation of these terms were given in Section 2.6 in the context of the UI-core inductor.

The permeance of the center gap may be expressed as

$$P_{cg} = P_{cgd} + 2P_{efic} + 2P_{cff} \tag{5.2-17}$$

where P_{cgd} is the permeance of the center leg directly across the gap given by

$$P_{cgd} = \frac{\mu_0 w_c l_c}{g} \tag{5.2-18}$$

P_{efic} is the fringing permeance associated with an inner corner and is the same as that for the end leg as given by (5.2-15) and P_{cff} is the center leg fringing formula associated with the face of the center leg which may be expressed as

$$P_{cff} = \frac{\mu_0 w_c}{\pi} \ln\left(1 + \pi \frac{\min(w_i, d_s + w_b)}{g}\right) \tag{5.2-19}$$

The horizontal and vertical slot leakage permeances may be calculated as in Section 2.7. In particular,

$$P_{hsl} = \frac{\mu_0 l_c (3d_s - 2d_w)}{3w_s} \tag{5.2-20}$$

$$P_{vsl} = \frac{\mu_0 l_c (3w_s - 2w_w)}{6(d_s + g)} \tag{5.2-21}$$

Note that (5.2-21) has twice the value given by (2.7-28). This is because the same permeance is used to represent two slots. The doubling does not occur in (5.2-20) since this permeance appears twice in the MEC. In addition, in (5.2-21) it is assumed that the fact that the winding is not centered in the slot has negligible impact on the permeance.

A slight improvement to (5.2-20) and (5.2-21) is suggested in Cale and Sudhoff [1], involving the addition of a term to each of these expressions for leakage flux over the front and back faces. Consider the horizontal leakage inductance. In deriving (5.2-20), a path similar to the one shown in Figure 5.4(a) was used, where the flux travels straight across the air gap. However, some flux will follow a path that looks the same as Figure 5.4(a) in a front view, but which flairs from the front face as it travels across the width of the slot in the air as shown in Figure 5.4(b).

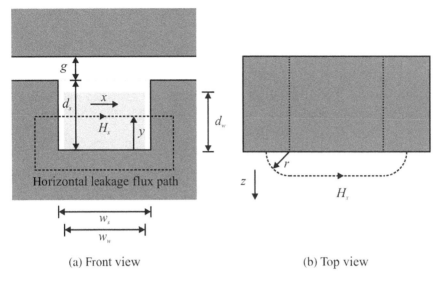

Figure 5.4 Slot leakage path.

Using our work in Chapter 2, we can find that the field intensity in the direction of the path (which is labeled H_x since it is mostly in the x-direction but includes a z-direction component in part of the path) is given by

$$H_x = \begin{cases} \dfrac{Ni}{\pi r + w_s} \dfrac{y}{d_w} & y \leq d_w \\ \dfrac{Ni}{\pi r + w_s} & d_w \leq y \leq d_s \end{cases} \quad (5.2\text{-}22)$$

From (2.7-19), the energy associated with this path may be expressed as

$$E = \frac{1}{2}\mu_0 \int_0^{r_{mx}} \int_0^{d_s} H_x^2 (\pi r + w_s) \, dy \, dr \quad (5.2\text{-}23)$$

where r_{mx} is the maximum radius used; for this case, taking

$$r_{mx} = \min(w_e, w_c/2) \quad (5.2\text{-}24)$$

is appropriate. Substitution of (5.2-22) and (5.2-24) into (5.2-23), evaluating, and comparing the result to (2.7-9), we conclude that the horizontal leakage permeance for one face is

$$P_{hslf} = \frac{\mu_0 (3d_s - 2d_w)}{3\pi} \ln\left(1 + \frac{\pi \min(w_e, w_c/2)}{w_s}\right) \quad (5.2\text{-}25)$$

Adding twice this value (as there are two faces) to (5.2-20), we obtain a revised expression for the horizontal leakage inductance which includes face horizontal leakage. In particular,

$$P_{hsl} = \frac{\mu_0 (3d_s - 2d_w)}{3}\left(\frac{l_c}{w_s} + \frac{2}{\pi} \ln\left(1 + \frac{\pi \min(2w_e, w_c)}{2w_s}\right)\right) \quad (5.2\text{-}26)$$

Incorporating the vertical face slot leakage into (5.2-21) one obtains

$$P_{vsl} = \frac{\mu_0(3w_s - 2w_w)}{6}\left(\frac{l_c}{d_s + g} + \frac{2}{\pi}\ln\left(1 + \frac{\pi \min(w_i, w_b)}{d_s + g}\right)\right) \quad (5.2\text{-}27)$$

The derivation of (5.2-27) is left as an exercise.

The last leakage permeance term is associated with the leakage flux through the end turns over the faces of the E-core. Following the derivation in Section 2.7, this term may be expressed

$$P_{eal} = \frac{2}{R_{eale}} + 2P_{eali} \quad (5.2\text{-}28)$$

where the factor of 2 comes into play because there is a front and back face. In (5.2-28), R_{eale} may be expressed

$$R_{eale} = \frac{w_w(w_b + w_{e2}) + \sqrt{(2d_w + w_b + w_{e2})(w_b + w_{e2})w_b w_{e2}}}{\mu_0 w_c w_b w_{e2}} \quad (5.2\text{-}29)$$

where

$$w_{e2} = d_s - d_w \quad (5.2\text{-}30)$$

The permeance P_{eali} is given by

$$P_{eali} = \frac{\mu_0 w_c}{256 w_w^2 d_w^2}\left[\begin{array}{l} 16k_2^4 + 16\sqrt{2}k_1 k_2^3 + 4k_1^2 k_2^2 - \cdots \\ 2\sqrt{2}k_1^3 k_2 + k_1^4 \ln\left(1 + \frac{2\sqrt{2}k_2}{k_1}\right) \end{array}\right] \quad (5.2\text{-}31)$$

where

$$k_1 = |d_w - 2w_w| \quad (5.2\text{-}32)$$

and

$$k_2 = \begin{cases} \sqrt{2}w_w & d_w > 2w_w \\ d_w/\sqrt{2} & d_w \leq 2w_w \end{cases} \quad (5.2\text{-}33)$$

In comparing (5.2-29)–(5.2-33) to (2.7-34)–(2.740) note that the l in Section 2.7 is w_c for the EI-core geometry and that the role of w_w and d_w are interchanged.

At this point, all elements of our MEC have been defined. Before concluding this section, note that using circuit reduction techniques, the MEC shown in Figure 5.3 can be reduced to the MEC of Figure 5.5. Therein, when calling the reluctance functions, note that the flux argument is divided by two since the mesh fluxes will represent the value going into both end legs. Also note that the reluctance term R_{vsl} is not formally included in the MEC since it is across an MMF source. It is shown as being lightly shaded since the flux into this branch is calculated separately. That is why it is not associated with a mesh flux.

Force Analysis

The next step in our analysis will be to determine an expression for force. From (4.9-20) to (4.9-21) we may write

$$f_e = -\frac{1}{2}\left(\frac{1}{2}\frac{1}{P_{eg}^2}\frac{\partial P_{eg}}{\partial g} + \frac{1}{P_{cg}^2}\frac{\partial P_{cg}}{\partial g}\right)\Phi_1^2 - \frac{1}{2}N^2 i^2 \frac{\partial P_{vsl}}{\partial g} \quad (5.2\text{-}34)$$

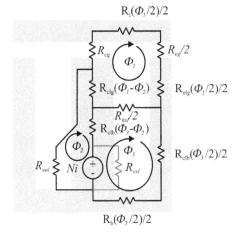

Figure 5.5 Reduced magnetic equivalent circuit.

There are several points to note in (5.2-34). First, a minus sign has been introduced so that f_e is an attractive force and will be positive. Otherwise, f_e, which is nominally defined to be in the direction of increasing g, would be negative. Secondly, there is a factor of 2 associated with the P_{eg} term; this is because there are two end gaps. Finally, the first two terms have been written in terms of flux, while the third term has been expressed in terms of MMF. This was done purely for convenience when using the MEC.

From (5.2-12)–(5.2-16) we have

$$\frac{\partial P_{eg}}{\partial g} = \frac{\partial P_{egd}}{\partial g} + \frac{\partial P_{efoc}}{\partial g} + \frac{\partial P_{efic}}{\partial g} + 2\frac{\partial P_{eff}}{\partial g} \tag{5.2-35}$$

where

$$\frac{\partial P_{egd}}{\partial g} = -\frac{\mu_0 w_e l_c}{g^2} \tag{5.2-36}$$

$$\frac{\partial P_{efoc}}{\partial g} = -\frac{\mu_0 l_c}{g} \frac{\min(w_i, d_s + w_b)}{g + \pi \min(w_i, d_s + w_b)} \tag{5.2-37}$$

$$\frac{\partial P_{efic}}{\partial g} = -\frac{2\mu_0 l_c}{g} \frac{\min(2d_s, w_s)}{4g + \pi \min(2d_s, w_s)} \tag{5.2-38}$$

$$\frac{\partial P_{eff}}{\partial g} = -\frac{\mu_0 w_e}{g} \frac{\min(w_i, d_s + w_b)}{g + \pi \min(w_i, d_s + w_b)} \tag{5.2-39}$$

Similarly, from (5.2-17)–(5.2-19) we have

$$\frac{\partial P_{cg}}{\partial g} = \frac{\partial P_{cgd}}{\partial g} + 2\frac{\partial P_{efic}}{\partial g} + 2\frac{\partial P_{cff}}{\partial g} \tag{5.2-40}$$

where

$$\frac{\partial P_{cdg}}{\partial g} = -\frac{\mu_0 w_c l_c}{g^2} \qquad (5.2\text{-}41)$$

$\partial P_{efic}/\partial g$ is given by (5.2-38), and

$$\frac{\partial P_{cff}}{\partial g} = -\frac{\mu_0 w_c}{g} \frac{\min(w_i, d_s + w_b)}{g + \pi \min(w_i, d_s + w_b)} \qquad (5.2\text{-}42)$$

From (5.2-21) we have

$$\frac{\partial P_{vsl}}{\partial g} = -\frac{\mu_0 (3w_s - 2w_w)}{6(d_s + g)} \left(\frac{l_c}{d_s + g} + \frac{2\min(w_i, w_b)}{d_s + g + \pi \min(w_i, w_b)} \right) \qquad (5.2\text{-}43)$$

Note that some of the partial derivatives have a singularity as g goes to zero. This will not result in an infinite force, however, since the flux will be finite and because the reluctances approach zero as g goes to zero. The net result is that the force will take on a finite value as g tends to zero. In order to illustrate this easily, let us temporarily neglect fringing flux in the central leg. From (5.2-18) we see that

$$\frac{1}{P_{cgd}^2} \frac{\partial P_{cgd}}{\partial g} = -\frac{1}{\mu_0 w_c l_c} \qquad (5.2\text{-}44)$$

Note that the flux density in the central leg may be expressed

$$B_c = \frac{\Phi_1}{w_c l_c} \qquad (5.2\text{-}45)$$

From the P_{cgd} term in (5.2-18), (5.2-44), and (5.2-45) we have that, neglecting fringing, the force in the direction of g due to the central leg may be expressed as

$$f_e|_{\text{central leg}} = -\frac{w_c l_c B_c^2}{2\mu_0} \qquad (5.2\text{-}46)$$

We see that the force generated is proportional to the area that the flux goes through (i.e., the magnetic cross-section) and the square of the flux density through the magnetic cross-section.

Example 5.2A As an example, let us predict the performance characteristics of an EI-core arrangement. The parameters of the core are as follows: $d_s = 48.8$ mm, $w_s = 37$ mm, $w_b = 27.3$ mm, $w_e = 28.7$ mm, $w_c = 57.4$ mm, $w_i = 30.5$ mm, $l_c = 119.8$ mm. The air gap $g = 2.96$ mm. The core is made of 3C90 ferrite whose parameters are listed in Appendix B. The winding was the following parameters: $w_w = 15.8$ mm, $d_w = 44.4$ mm, $N = 40$, $a_c = 11$ µm². The conductor is rectangular in shape with rounded edges.

Let us begin our assessment with a calculation of resistance. Using (5.2-1)–(5.2-3), we obtain $R_{cl} = 24.7$ mΩ. The measured value is 30.5 mΩ. The rather significant discrepancy arises from the fact that in the hardware device the shape of the winding deviates from the assumed shape: The large-diameter conductor results in significant bowing in the slot and on the ends. Indeed, the inside of the coil was approximately 9 mm from the core at either of the ends and approximately 9.5 mm from the center post of the E-core at the middle of the center post of the E-core.

The flux linkage versus current characteristic, though not strictly needed for operation as an electromagnet, is of interest because it indicates the accuracy of the MEC. From Figure 5.5, the flux linking the primary winding may be expressed as

$$\lambda = N(\Phi_3 - \Phi_2 + NiP_{vsl}) \qquad (5.2\text{A-}1)$$

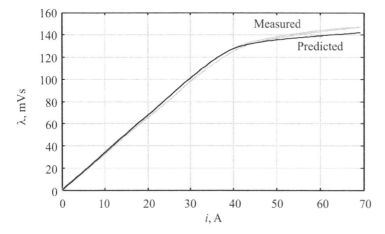

Figure 5.6 Flux linkage versus current.

The predicted and measured $\lambda - i$ characteristic is shown in Figure 5.6. The predicted and measured results are reasonably consistent, though do not match as well as in the case of the UI-core inductor we studied in Chapter 2. One reason for this may be the distortion in the shape of the coil caused by the use of a rather large conductor size.

Next, let us consider force production. Using (5.2-34), we obtain the force versus current characteristic shown in Figure 5.7. As can be seen, force rises vary rapidly with current at first, but once magnetic saturation occurs, the force rises much more slowly. Figure 5.8 depicts the ratio of force, f_e, to power loss, P_l, which is calculated based on the current and the measured coil resistance. As can be seen, initially the ratio of force to power loss is constant; however, as the magnetic material saturates the ratio of force to power loss decreases markedly. This illustrates one penalty of heavy magnetic saturation in the design of magnetic devices.

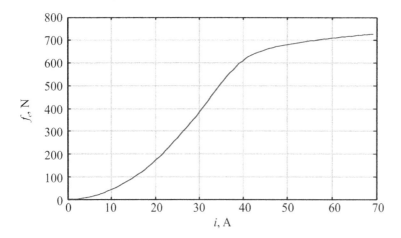

Figure 5.7 Force versus current.

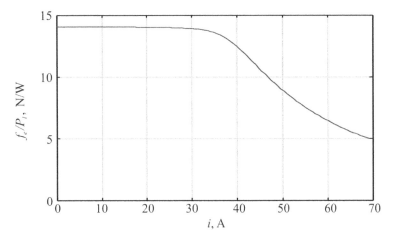

Figure 5.8 Force over power loss versus current.

5.3 El-Core Electromagnet Design

In this section, we will use the results from Section 5.2 in order to design an electromagnetic latch. In particular, our system will be as shown in Figure 5.9. Therein, when the switch is closed, a voltage is applied to the coil of an electromagnet. This will provide an attractive force on the I-core attached to the door. This force will be sufficient to prevent the door from being opened. The diode present is needed in order to provide a continuous path for current to flow if the switch is opened, in order to prevent an electrical arc across the switch. A chief drawback of this arrangement is that power must be continuously applied to lock the door. However, in a situation where we wish for the door to be unlocked in a failure condition (with a priority of safety over security), the arrangement can be beneficial.

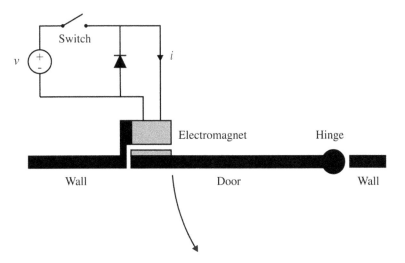

Figure 5.9 Electromagnetic door latch.

Our goal in our design will be to generate a prescribed amount of force at a given air gap. The specified air gap will be small, but nonzero, in order to accommodate imperfections of the contact surface. The current into the electromagnet will not be specified, but the voltage source will. It will be somewhat arbitrarily required that the EI-core have a square cross-section. In this application, it will be desired to minimize the electromagnet volume and power consumption.

Problem Formulation

Our design approach, as in the case of the inductor design, will be to pose the design problem as a formal multiobjective optimization problem. This will involve four steps: the identification of the search space, compiling a list of design constraints, formulation of the design metrics, and finally the synthesis of a fitness function.

Let us begin with the formulation of the search space. The design space must include, in some form, all the geometrical parameters shown in Figure 5.2, the magnetic material used, and the type of conductor used. Also, the number of turns of conductor N and the cross-sectional area of the wire must be specified.

One choice for parameter vector to characterize the design is

$$\boldsymbol{\theta} = \begin{bmatrix} T_{cr} & T_{cd} & w_c & r_{ec} & r_{ic} & r_{bc} & a_c^* & N^* & N_w^* & N_d^* \end{bmatrix} \tag{5.3-1}$$

In (5.3-1), T_{cr} and T_{cd} are integers that will be used to designate the type of core material and type of conductor material from a list of allowed materials. The parameter w_c is the width of the core center leg. The parameters r_{ec}, r_{ic}, and r_{bc} are ratios defined such that

$$w_e = \frac{1}{2} r_{ec} w_c \tag{5.3-2}$$

$$w_i = \frac{1}{2} r_{ic} w_c \tag{5.3-3}$$

$$w_b = \frac{1}{2} r_{bc} w_c \tag{5.3-4}$$

The use of ratios as design parameters rather than the actual parameters was discussed in our inductor design work in Section 3.3. The factor of 1/2 is introduced because the flux through the center leg will be twice that through the end legs.

The next element in the parameter vector is a_c^*, which will be used to denote the desired conductor cross-sectional area. This is different from the actual cross-sectional area, because we will assume that the conductor size must conform to standard sizes. Thus, as in our inductor design work, the actual conductor size is calculated in accordance with

$$a_c = \text{round}_{WG}(a_c^*) \tag{5.3-5}$$

where $\text{round}_{WG}()$ is a function that rounds the argument not to the nearest integer, but the closest available wire gauge. Note that there are several wire gauge standards. Information on the American Wire Gauge is set forth in Appendix A.

From the conductor area, note that the conductor radius (w/o insulation) may be expressed as

$$r_c = \sqrt{\frac{a_c}{\pi}} \tag{5.3-6}$$

The next parameter specified is N^*, which is the desired number of turns. From this, the actual number of turns is calculated as

$$N = \text{round}(N^*) \tag{5.3-7}$$

where the round() function rounds to the nearest integer. The reason for the use of N^* as a real number and applying (5.3-7) rather than making N an integer parameter directly is that N may be thought of as a discrete representation of a continuous quantity.

The next parameters relate to the coil dimensions and are denoted N_w^* and N_d^*. These parameters are the desired width and depth of the coil in terms of the number of conductors. Both of these numbers will be specified as real numbers and then rounded as

$$N_w = \text{round}(N_w^*) \tag{5.3-8}$$
$$N_d = \text{round}(N_d^*) \tag{5.3-9}$$

The reason for specifying N_w^* and N_d^* rather than N_w and N_d follows a similar reasoning to specifying N^* rather than N as discussed above.

Next, the winding width and depth are computed as

$$w_w = 2r_c k_b N_w \tag{5.3-10}$$
$$d_w = 2r_c k_b N_d \tag{5.3-11}$$

In (5.3-9) and (5.3-10), k_b is a build factor that increases the effective radius of the conductor to account for insulation and to give some margin between conductors. Typical values range from 1.05 to 1.4 which corresponds to packing factors from 71% to 40%. The calculation of w_w and d_w by the method (5.3-5)–(5.3-11), when added to the constraint that $N_d N_w > N$, will ensure that the winding can be physically constructed.

In the case of our work in inductor design, the parameter vector included clearances in the winding width and depth denoted c_w and c_d. Our intuition was that there was little motivation to make these more than a minimal value, and this was borne out by our studies. Thus in this example, we will not use them as a design variable, but we will include them as design specifications. They could be specified based on the intended slot liner. In terms of these clearances,

$$w_s = w_w + c_w \tag{5.3-12}$$
$$d_s = d_w + c_d \tag{5.3-13}$$

The final geometrical variable we need is the length of the core, l_c. As mentioned, we will assume that a square profile is desired. Therefore,

$$l_c = 2w_s + 2w_e + w_c \tag{5.3-14}$$

At this point, it can be seen that the parameter vector $\boldsymbol{\theta}$ is sufficient to allow all the parameters of the design to be calculated. Let us now turn our attention to the calculation of design metrics from the parameter vector. For this problem formulation, we will be interested in electromagnet volume and loss.

Let us first consider the calculation of volume. The overall height, width, and length of our electromagnet may be expressed as

$$h_E = w_i + g + d_s + w_b \tag{5.3-15}$$
$$w_E = 2w_s + 2w_e + w_c \tag{5.3-16}$$

$$l_E = l_c + 2w_w \tag{5.3-17}$$

The volume of the circumscribing box may be expressed as

$$V_E = h_E w_E l_E \tag{5.3-18}$$

In order to calculate the power loss, we note that an expression for the coil resistance is given by (5.2-3). The power loss of the electromagnet may be expressed

$$P_E = \frac{v^2}{R_{cl}} \tag{5.3-19}$$

Now that we have expressions for the metrics of interest, we will now consider the constraints on the design. For our first constraint, we will require

$$c_1 = \text{gte}(N_d N_w, N) \tag{5.3-20}$$

where gte(,) is the greater-than-or-equal-to function defined in (1.9-1).

We will also require that the packing factor be limited to $k_{pf,mx}$. The packing factor is given by

$$k_{pf} = \frac{N a_c}{w_w d_w} \tag{5.3-21}$$

Thus, our second constraint is

$$c_2 = \text{lte}(k_{pf}, k_{pf,mx}) \tag{5.3-22}$$

where lte(,) is the less-than-or-equal-to function defined by (1.9-2). One might question the need to constrain the packing factor given the use of the build factor. Strictly speaking, it is not. However, one reason to retain a constraint on the packing factor is that a given coil may not layout as neatly as assumed in developing the build factor.

In the design of electromagnetic and electromagnet devices, thermal constraints are often very important in low-mass designs. We will consider thermal equivalent circuits in Chapter 10. In the meantime, one approach to ensuring achievable behavior is by imposing a current density limit. The current density at rated current will be given by

$$J = \frac{i}{a_c} \tag{5.3-23}$$

where the current i is calculated using (5.2-4). We will impose the constraint

$$c_3 = \text{lte}(J, J_{\text{mx}}(T_{cd})) \tag{5.3-24}$$

where $J_{\text{mx}}(T_{cd})$ is the maximum allowed current density, which, for this study, will be a function of the type of conductor. For copper, the maximum current density will be taken to be 7.6 A/mm², while for aluminum 6.52 A/mm² will be used. In Chapter 10, we will rework this example and incorporate a formal thermal analysis.

For construction purposes, it is often (but not always) desirable that components do not have highly exaggerated aspect ratios. To this end, we will define the aspect ratio as

$$\alpha_L = \frac{\max(h_E, w_E, l_E)}{\min(h_E, w_E, l_E)} \tag{5.3-25}$$

and enforce the constraint

$$c_4 = \text{lte}(\alpha_L, \alpha_{mx}) \tag{5.3-26}$$

Another limit that we may wish to place on the device is the allowed volume and allowed loss. While these quantities are both metrics, it is often the case that there is some volume beyond which the design is clearly not tolerable; likewise with loss. Thus we will impose the constraints that

$$c_5 = \text{lte}(V_E, V_{Emxa}) \tag{5.3-27}$$

$$c_6 = \text{lte}(P_E, P_{Emxa}) \tag{5.3-28}$$

where V_{Emxa} and P_{Emxa} are the maximum allowed volume and loss, respectively. Imposing these limits narrows the search effort to the region of interest.

The calculation of force will require the use of the MEC. This will lead to two constraints. The first of these is a convergence of the analysis. In particular, c_7 will be set to 1 if the analysis converges and 0 otherwise. The next constraint we will place on the design is the force. The calculation of the force was discussed in Section 5.2 using (5.2-34)–(5.2-43). We impose the constraint

$$c_8 = \text{gte}(f_e, f_{Emnr}) \tag{5.3-29}$$

where f_{Emnr} is the minimum required electromagnet force. This force will be at the current calculated by (5.2-4) and at a specified air gap.

It is appropriate to briefly pause in our development and to consider how to address the issue of inaccuracies in our analysis of the design process. Error in the force calculation or the resistance calculation will affect the force. One way to address this is to simply require more force than is required by the problem specification. Of course, this requires some experience so that the accuracy of the analysis may be assessed. In the case of this problem, we will assume that the specified force already includes such a margin.

In the case of the inductor design in Chapter 3, we placed a constraint on the percentage of flux flowing through the magnetizing path. In that case, there was a design motivation for leakage flux in that it contributed to the inductance, yet we were hesitant to permit it as it contributed to inaccuracy (and to proximity effect loss, though we haven't discussed that yet). In the electromagnet example, it is essentially only the flux crossing the air gap that contributes to force, so there is no motivation to favor designs with high leakage permeances. As a result, we will not need this constraint.

The next step is to construct a fitness function. We have two metrics of interest and $C = 8$ constraints. The first six of these constraints are computationally straightforward; the last involves the solution of a MEC and is somewhat more involved. This suggests the calculation of the fitness function in accordance with the pseudo-code shown in Table 5.1. The idea is to avoid the most involved computation if possible.

It should be noted that the formulation of the design problem as posed has deficiencies. For example, we have not included temperature effects outside of current density, nor mechanical issues such as the bending radius of the wire. Still, the problem as posed is sufficient to be of interest.

5.4 Case Study

Let us now consider a numerical case study. The design specifications for this study are listed in Table 5.2. The required force is approximately 560 lb$_f$, which should be adequate to keep most people from opening the door.

As discussed in Chapter 1, it is necessary to establish the domain of the design parameters. This is listed in Table 5.3. Therein, the minimum (Min.) and maximum (Max.) allowed values for each

Table 5.1 Calculation of Fitness Function

```
calculate V_E, P_E
calculate constraints c_1 through c_6
C_s = c_1 + c_2 + c_3 + c_4 + c_5 + c_6
C_I = 6
if (C_s < C_I)
```

$$\mathbf{f} = \varepsilon\left(\frac{C_s - C}{C}\right)\begin{bmatrix}1\\1\end{bmatrix}$$

```
    return
end
construct MEC and calculate c_7 and c_8
C_s = C_s + c_7 + c_8
C_I = C_I + 2
if (C_s < C_I)
```

$$\mathbf{f} = \varepsilon\left(\frac{C_s - C}{C}\right)\begin{bmatrix}1\\1\end{bmatrix}$$

```
    return
end
```

$$\mathbf{f} = \begin{bmatrix}\dfrac{1}{V_E}\\ \dfrac{1}{P_E}\end{bmatrix}$$

```
return
```

Table 5.2 Electromagnet Specifications

Symbol	Description	Value
v	Applied voltage	12 V
g	Air gap	1 mm
a_{mxa}	Maximum allowed aspect ratio	3
V_{Emxa}	Maximum allowed volume	1 L
P_{Emxa}	Maximum allowed loss	50 W
k_{pf-mxa}	Maximum allowed packing factor	0.7
k_b	Winding build factor	1.05
f_{Emnr}	Minimum required electromagnetic force	2500 N
c_w	Winding clearance in width	2 mm
c_d	Winding clearance in depth	2 mm

Table 5.3 Design Space

Par.	Description	Min.	Max.	Enc.	Gene
T_{cr}	Core material	1	5	int	1
T_{cd}	Conductor material	1	2	int	2
w_c	Center leg width	2×10^{-3}	10^{-1}	log	3
r_{ec}	Twice w_e to w_c ratio	0.5	1.5	lin	4
r_{ic}	Twice w_i to w_c ratio	0.25	1.5	lin	5
r_{bc}	Twice w_b to w_c ratio	0.25	1.5	lin	6
a_c^*	Cross-sectional conductor area (m²)	10^{-9}	10^{-4}	log	7
N^*	Desired number of turns	1	10^3	log	8
N_w^*	Desired slot width in conductors	1	10^3	log	9
N_d^*	Desired slot depth in conductors	1	10^3	log	10

design parameter are listed, as is the encoding (Enc.) scheme. In the case of the materials, the integer range corresponds to the materials in the appendices. Note that of the conductor material types, only copper and aluminum are considered. For core materials, we will consider the silicon steel materials listed in Appendix C. Establishing the range of the remaining parameters requires an educated guess. However, logarithmic encoding facilitates the use of a wide range. A poorly chosen range will be evident in the gene distribution.

The result of a genetic algorithm based optimization run is shown in Figure 5.10. The code to conduct this run is available in Sudhoff [2]. This run involved a population size of 2000 individuals over evolution of 2000 generations. Repeated runs produced the same Pareto-optimal front. In Figure 5.10, each x marks the loss and mass of a specific nondominated design. As in the case

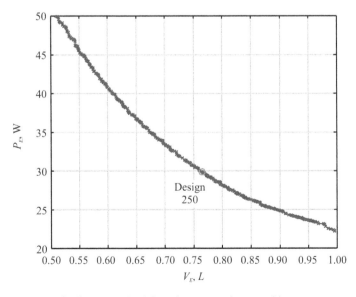

Figure 5.10 Pareto-optimal front between volume and loss.

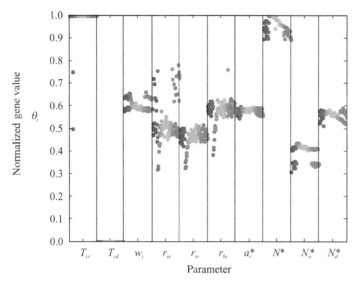

Figure 5.11 Gene distribution plot for the electromagnet.

of the inductor design in Chapter 3, we see a fundamental tradeoff between size and loss. Also indicated in Figure 5.10 is a specific design, labeled Design 250, which is the 250th design in order of increasing volume. We will look at that particular design more closely after examining some of the more general trends.

A gene distribution plot is shown in Figure 5.11. Therein, the ordering from gene values from left to right is in order of decreasing volume. Each set of points of a unique shade (or a unique distance from the left side of each window) represents an individual design. As can be seen, the same core material with a normalized gene value of 1 is used in all designs. This value maps to $T_{cr} = 5$ or Hyperco 50, which has the highest saturation flux density of all the materials considered. It can also be seen that the normalized gene value for the conductor type is zero, which maps to $T_{cd} = 1$; which corresponds to copper, and which has the best conductivity of the materials considered on a per-volume basis. The ratio r_{ce} has normalized values of approximately 0.5 which maps to 1, indicating that the width of the ends will be approximately one-half the width of the center, as we would expect. Observe that r_{ic}, is somewhat smaller than r_{bc}, indicating that the width of the base of the E-core is wider than the width of the I-core. The desired conductor area a_c^* can be seen to be fairly consistent. The number of turns varies across the population as the volume changes.

In Figure 5.12, each point represents the current density and volume of a given design. Points for all designs on the Pareto-optimal front are shown. The designs are sorted in terms of increasing volume, and a line connecting the points is drawn in order to form a continuous trace. As one would expect, the current density is highest for low-volume designs and lowest for large-volume designs, but the relationship is not monotonic. The rapid changes in current density are associated with the discretization of the conductor size.

Figure 5.13 is constructed in a fashion similar to Figure 5.12 but illustrates conductor counts versus volume. Note that the square root of the number of turns has been shown rather than the number of turns in order to better scale the results for plotting. As can be seen, the number of turns generally increases with increasing volume until the maximum allowed number of turns is reached. Irregularities in the characteristics are driven by toggling between the two wires sizes used in the designs.

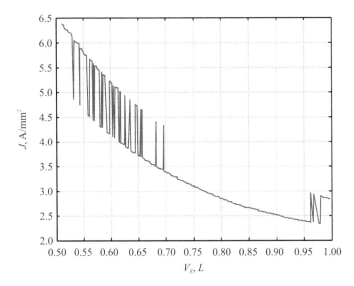

Figure 5.12 Current density versus volume for electromagnet.

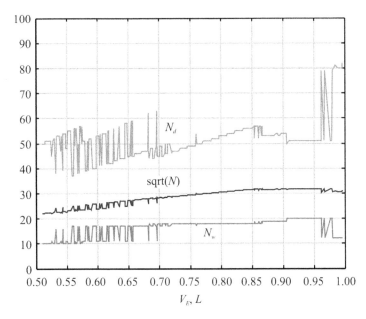

Figure 5.13 Conductor count versus volume for electromagnet.

The core dimensions of the designs are shown as a function of the design volumes in Figure 5.14 (using the same procedure as for Figure 5.12). Therein, we see that $w_c \approx 2w_e \approx 2w_b$ for volumes between 0.75 L and 0.85 L. We also see that w_i tends to be less than w_b. This may be because this portion of the core does not carry any leakage flux. Additional geometrical parameters are shown in Figure 5.15. Of particular note is the fluttering between wire gauges at low volumes (seen by the variation in r_c) and at high volumes.

5 Introduction to Electromagnet Design

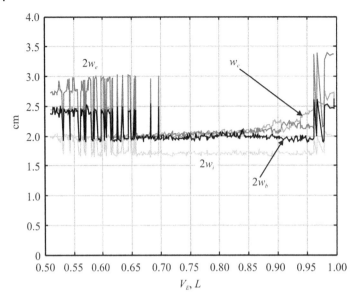

Figure 5.14 Core widths versus volume for electromagnet.

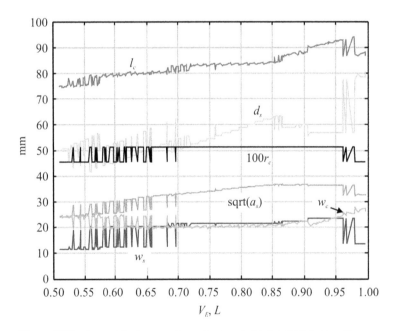

Figure 5.15 Geometrical parameters versus volume for electromagnet.

Figure 5.16 Geometrical parameters versus volume for electromagnet.

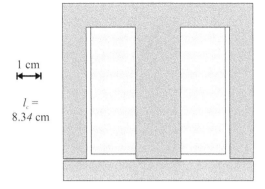

Table 5.4 Design 250 Data

El-Core Data	Winding Data	Metrics
Material = Hiperco50	Material = Copper	V_E = 764 mL
w_c = 2.02 cm	AWG = 18	P_E = 29.9 W
w_e = 1.03 cm	a_c = 0.823 mm²	i_E = 2.49 A
w_i = 8.36 mm	N = 882	J_E = 3.03 A/mm²
w_b = 9.93 mm	N_w = 18	R_E = 4.82 Ω
w_s = 2.13 cm	N_d = 50	h_E = 7.50 cm
d_s = 5.57 cm	d_w = 5.37 cm	w_E = 8.34 cm
l_c = 8.34 cm	w_w = 1.93 cm	l_E = 12.2 cm

Figure 5.16 depicts a scaled drawing of Design 250 from the Pareto-optimal front, with parameters listed in Table 5.4. Again, this is the 250th design in order of increasing volume from an optimization run.

Before concluding this section, it is interesting to investigate the question of the impact of wire size discretization. Figure 5.17 illustrates a Pareto-optimal front obtained using the formulation described in this chapter with one exception: The round$_{WG}$() operator in (5.3-5) is not used, rather $a_c = a_c^*$. As can be seen, the obtainable performance improves slightly. This is because the design space is larger and (possibly) because it is easier for the optimization algorithm to converge. In Figure 5.18, the current density of each design as a function of the volume of each design is shown. In this case, the large rapid variation in current density is not seen, and current density decreases gradually with increasing volume.

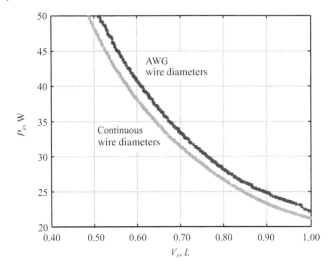

Figure 5.17 Effects of wire size descritization on Pareto-optimal front.

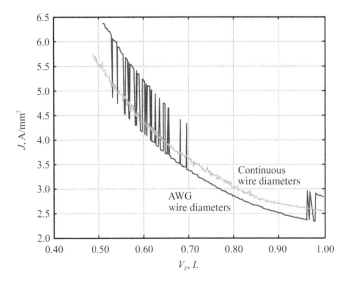

Figure 5.18 Effects of wire size discretization on current density.

References

1 J. L. Cale and S. D. Sudhoff, Accurately modleing EI core inductors using a high fidelity magnetic equivalent circuit approach, *IEEE Transactions on Magnetics*, vol. **42**, no. 1, pp. 40–46, 2006.
2 S. D. Sudhoff, *MATLAB codes for Power Magnetic Devices: A Multi-Objective Design Approach* [Online], 2nd edition. Hoboken, NJ : John Wiley & Sons Inc. Available: http://booksupport.wiley.com.

Problems

1. Using (5.2-22)–(5.2-24) obtain (5.2-25).

2. Derive (5.2-26).

3. Consider the force associated with the center post and the fringing permeances. Show that if Φ is the total flux going from the center post across the air gap, the force associated with the center post region remains finite as the air gap goes to zero.

4. Consider Design 250 discussed in Section 5.4. What is the flux density in various parts of the core?

5. Consider Design 250 discussed in Section 5.4. Plot the electromagnet force versus air gap, as the air gap varies from nearly 0 to 5 mm.

6. Consider the design code given in Sudhoff [2]. Modify the code so that the electromagnet does not have to be square. Plot the Pareto-optimal front between volume and loss, superimposing the results on top of those for the case study in Section 5.4. Create figures analogous to Figures 5.11–5.15 showing the final parameter distribution and the variation of the current density and parameters along the front. What is the lowest volume device that can be obtained? Report the parameters of the lowest volume device.

7. Consider the design code given in Sudhoff [2]. Modify the code so that: (i) a square wire of continuously variable size is used assuming that rounding of the corners will reduce the conductor area to 95% of that perfectly square wire, (ii) there is no limit on packing factor, (iii) the build constant k_b is kept the same, and (iv) the maximum power dissipation is increased to 200 W. Plot the Pareto-optimal front between volume and loss, superimposing the results on top of those for the case study in Section 5.4. Create figures analogous to Figures 5.10–5.16 showing the final parameter distribution and the variation of the current density and parameters along the front. What is the lowest volume device that can be obtained? Report the parameters of the lowest volume device.

6

Magnetic Core Loss and Material Characterization

The focus of this chapter is on (i) calculating magnetic material losses and (ii) characterizing magnetic materials, especially with regard to loss. Magnetic material losses arise from several causes. Eddy currents induced by the time rate of change of flux in the material are one source of loss. A second source of loss is magnetic hysteresis, which is associated with the nonuniquely valued relationship between flux density and field intensity. The loss mechanism here involves several components including overcoming energy associated with pinning sites, localized eddy currents associated with domain wall movement, and the dynamics associated with the magnetic material itself. While the loss mechanisms are intertwined, it is convenient to think of them separately. In this chapter, first eddy current loss and then hysteresis loss will be considered, and behavioral models to predict these losses will be set forth. Next, experimental procedures to characterize the material anhysteretic $B-H$ characteristic and losses will also be discussed. Before concluding, time-domain models for modeling hysteresis will be considered.

6.1 Eddy Current Losses

Eddy current losses can occur in any conductive medium, magnetic or not. Eddy current losses refer to the resistive losses associated with induced currents. Let us begin our consideration of eddy current losses by studying the situation in Figure 6.1. Therein, a conductive material is shown (lightly shaded). As can be seen, the material is rectangular with a cross section of depth d by width w. The length of the material into the page is l. The diagonal line segments are at an angle of $\pm 45°$ relative to horizontal, and so the length of the horizontal line segment in the center of the sample is $|w - d|$ ($w > d$ in the figure). Let us suppose that the flux density in the material is uniform into the page.

The darkly shaded region in Figure 6.1 depicts a differential current path. It is convenient to define

$$k_1 = \min(w, d) \tag{6.1-1}$$
$$k_2 = |w - d| \tag{6.1-2}$$

Observe that the area enclosed by the path at position x may be expressed as

$$a_p = 2k_2 x + 4x^2 \tag{6.1-3}$$

and that the length of the path is given by

$$l_p = 2k_2 + 8x \tag{6.1-4}$$

Power Magnetic Devices: A Multi-Objective Design Approach, Second Edition. Scott D. Sudhoff.
© 2022 The Institute of Electrical and Electronics Engineers, Inc. Published 2022 by John Wiley & Sons, Inc.
Companion website: www.wiley.com/go/sudhoff/Powermagneticdevices

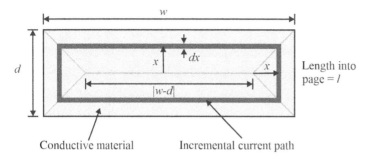

Figure 6.1 Calculation of eddy current losses.

Given that the path is short-circuited, the voltage equation describing the circuit formed by the path is

$$0 = r_p i_p + \frac{d\lambda_p}{dt} \tag{6.1-5}$$

where r_p is the Ohmic resistance of the path, i_p is the current in the differential path, and λ_p is the flux linking the differential path. The resistance of the path may be expressed as

$$r_p = \frac{l_p}{\sigma l dx} \tag{6.1-6}$$

where σ is the materials' conductivity and l is the length of the conductor into the page.

We will assume the existence of a uniform time-varying flux density B directed along the axis perpendicular to the page, and with the direction into the page considered positive. Since there is only a single turn, the flux linkage is given by

$$\lambda_p = a_p B \tag{6.1-7}$$

The instantaneous power dissipated by the current in the differential path is given by

$$dP_p = r_p i_p^2 \tag{6.1-8}$$

Combining (6.1-5)–(6.1-8), the instantaneous power dissipated by the current in the differential path may be expressed as

$$dP_p = \frac{\sigma l a_p^2}{l_p} \left(\frac{dB}{dt}\right)^2 dx \tag{6.1-9}$$

The time-average power associated with the path is thus given by

$$d\overline{P}_p = \frac{\sigma l a_p^2}{l_p} \frac{1}{T} \int_0^T \left(\frac{dB}{dt}\right)^2 dt\, dx \tag{6.1-10}$$

where T is the period of the flux density waveform and the overbar indicates temporal average. Considering all paths from $x = 0$ to $x = k_1/2$, the total average power dissipated may be expressed as

$$\overline{P} = \sigma l \int_0^{k_1/2} \frac{a_p^2}{l_p} \frac{1}{T} \int_0^T \left(\frac{dB}{dt}\right)^2 dt\ dx \tag{6.1-11}$$

Substitution of (6.1-3) and (6.1-4) into (6.1-11) and integrating, one obtains

$$\overline{P} = \frac{\sigma l}{128}\left(4k_1^4 + 8k_2k_1^3 + 2k_2^2k_1^2 - 2k_2^3k_1 + k_2^4 \ln\left(1 + \frac{2k_1}{k_2}\right)\right)\frac{1}{T}\int_0^T \left(\frac{dB}{dt}\right)^2 dt \qquad (6.1\text{-}12)$$

Recall that k_1 and k_2 are given by (6.1-1) and (6.1-2).

From (6.1-12), we observe that the eddy current loss will be a function of the geometry of the device. At a fundamental level, it can be seen to be driven by the time rate of change of the flux density.

The expression (6.1-12) is approximate. First, we assumed a path that is not very realistic at the ends. In addition, associated with the incremental path is inductive energy storage; this will reduce the eddy current in the path, especially as frequency increases. The presence of the eddy current will also alter the flux distribution within the sample. We will examine the impact of these effects on the accuracy of (6.1-12) in the following example.

Example 6.1A In this example, we will use (6.1-12) to predict the power loss in an aluminum alloy and compare the losses to those measured. First, let us define excitation-normalized power loss density as

$$\hat{p} = \frac{\overline{P}}{V\dfrac{1}{T}\displaystyle\int_0^T \left(\dfrac{dB}{dt}\right)^2 dt} \qquad (6.1\text{A-}1)$$

where V is the volume of a sample. The quantity $\hat{p}$ represents the time-averaged power loss per unit volume and per excitation. Substitution of (6.1-12) into (6.1A-1) yields

$$\hat{p} = \frac{\sigma\left(4k_1^4 + 8k_2k_1^3 + 2k_2^2k_1^2 - 2k_2^3k_1 + k_2^4\ln\left(1 + \dfrac{2k_1}{k_2}\right)\right)}{128wd} \qquad (6.1\text{A-}2)$$

which is independent of frequency and the shape of the flux density waveform.

Now, let us apply (6.1A-2) to the sample shown in Figure 6.2. Therein, the material sample is a toroid. The configuration has a magnetizing (excitation) winding consisting of $N_m = 24$ conductors in the center of the toroid, which have a distributed return path outside of the toroid, approximating a cylindrical current sheet. This arrangement yields a very ideal field arrangement within the sample. A sense test winding of $N_s = 300$ turns is used to sense the time rate of flux within the sample. No current is present in the sense winding. Details on the arrangement are set forth in [1], which uses the same test fixture.

The inner and outer radii of the toroid will be denoted r_{in} and r_{out}, respectively, and the depth of the toroid into the page will be designated d. The cross section of our sample is w by d, where

$$w = r_{out} - r_{in} \qquad (6.1\text{A-}3)$$

The dimensions will be chosen such that w and d are small relative to the mean toroidal radius. Two samples will be considered, the first with an approximately square cross section and the second with a rectangular cross section.

Parameters of the two aluminum alloy samples are listed in Table 6.1. Note that the conductivity of the two alloys was based on an average of published estimates and is rather different from that of pure aluminum. Also note that the permeability, obtained using low-frequency flux measurements, is significantly higher than that of pure aluminum, though it is still low. Both alloys have iron

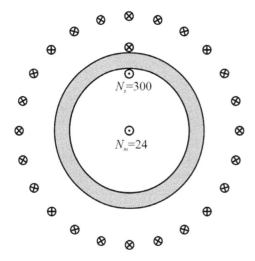

Figure 6.2 Top view of test configuration.

Table 6.1 Aluminum Alloy Test Samples

Material	σ (MS/m)	μ_r	r_{in} (mm)	r_{out} (mm)	w (mm)	d (mm)
6061T6	25.3	1.23	140	150	10	12.7
6013	23.3	1.20	140	150	10	27.2

content. We will refer to the first sample listed as the square sample because of its almost square cross section; the second sample will be referred to as the rectangular sample.

Figure 6.3 illustrates the normalized power loss for the two samples. This was obtained using (6.1A-2), which yields the "predicted" traces in the figure.

The traces labeled "measured loss, predicted B" are obtained experimentally in conjunction with (6.1A-1). The procedure to measure the power loss is as follows. First, observe that the portion of the power into the magnetizing winding, which is not being resistively dissipated, is given by

$$P_m = i_m(v_m - r_m i_m) \tag{6.1A-4}$$

where v_m is the primary voltage and r_m is the resistance of the magnetizing winding. Thus,

$$P_m = i_m \frac{d\lambda_m}{dt} \tag{6.1A-5}$$

The primary flux linkage can be divided into two terms as

$$\lambda_m = \lambda_{me} + N_m \Phi_s \tag{6.1A-6}$$

where λ_{me} is the flux linkage associated with flux paths which are external to the sample, and Φ_s is the flux in the sample. Substituting (6.1A-6) into (6.1A-5),

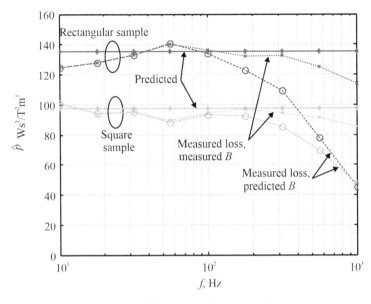

Figure 6.3 Excitation-normalized power loss versus frequency.

$$P_m = \underbrace{i_m \frac{d\lambda_{me}}{dt}}_{P_e} + \underbrace{N_m i_m \frac{d\Phi_s}{dt}}_{P_s} \quad (6.1A\text{-}7)$$

Considering (6.1A-7), the first term P_e is power going into the spatial region external to the sample, and the second term P_s is the power going into the sample. Note that the open-circuited secondary voltage may be expressed as

$$v_s = N_s \frac{d\Phi_s}{dt} \quad (6.1A\text{-}8)$$

From (6.1A-7) and (6.1A-8), we conclude that the instantaneous power into the sample is given by

$$P_s = \frac{N_m}{N_s} i_m v_s \quad (6.1A\text{-}9)$$

and so the average power loss in the sample may be measured as

$$\overline{P}_s = \frac{N_m}{N_s} \frac{1}{T} \int_0^T v_s i_m \, dt \quad (6.1A\text{-}10)$$

In order to evaluate the normalized power loss, we will also need the flux density waveform. We will have two approaches to calculating this. In the first method, we will calculate the flux density waveform. From the test geometry and applying Ampere's law, the flux density in the sample ignoring eddy currents and using a mean path length of $\pi(r_{in} + r_{out})$ is given by

$$B = \frac{\mu_0 \mu_r N_m i_m}{\pi(r_{in} + r_{out})} \quad (6.1A\text{-}11)$$

Observe that in order to evaluate the normalized power loss, it is not the flux density, but rather the time derivative of flux density, which is needed. Assuming that the current is sinusoidal, the time derivative of the flux density can be determined from the time derivative of the current, which is found based on the representation of its fundamental component in order to avoid numerical noise. Thus,

$$\frac{dB}{dt} = \frac{\mu_0 \mu_r N_m}{\pi(r_{in} + r_{out})} \frac{di_m}{dt} \quad (6.1\text{A-}12)$$

This approach to obtaining the time derivative of the flux density will be referred to as the calculated method.

A second (and more direct and accurate) means of obtaining the flux density waveform is noting that the secondary voltage may be expressed as

$$v_s = wdN_s \frac{dB}{dt} \quad (6.1\text{A-}13)$$

Thus,

$$\frac{dB}{dt} = \frac{v_s}{wdN_s} \quad (6.1\text{A-}14)$$

Hence, the secondary voltage can be used to obtain the spatial mean of the time derivative of the flux density. Again, this will be based on the fundamental component of the waveform in question. Evaluating the time derivative of the flux density in this way will be referred to as using the measured flux density.

Referring back to Figure 6.3, observe that the predicted normalized power loss density is reasonably consistent with the value obtained based on the measured loss and measured flux density (i.e., using (6.1A-1), (6.1A-10), and (6.1A-14)) for frequencies less than 300 Hz. As the frequency increases beyond this point, the flux density varies significantly within the sample as a result of the distributed nature of the eddy current. Observe that the predicted normalized loss is only consistent with the value obtained from the measured power loss and the calculated flux density (i.e., using (6.1A-1), (6.1A-10), and (6.1A-12)) until 100 Hz. The reason for the greater deviation is that the eddy current reduces the spatial mean value of the time derivative of the flux density from the calculated value. This error, which amounts to an error in the flux density waveform, will also affect the calculation of other losses in the device, such as hysteresis loss.

Let us now return to our attention to (6.1-12). Let us suppose the flux density is sinusoidal so that

$$B = B_{pk} \cos(\omega_e t) \quad (6.1\text{-}13)$$

where B_{pk} and ω_e designate the peak amplitude and radian frequency of the flux density waveform. In this case, we can readily show that

$$\frac{1}{T}\int_0^T \left(\frac{dB}{dt}\right)^2 dt = \frac{1}{2}\omega_e^2 B_{pk}^2 \quad (6.1\text{-}14)$$

6.1 Eddy Current Losses

Thus, for sinusoidal excitation, we would expect the eddy current power loss within a given material to increase with the peak flux density squared and with frequency squared.

As a generalization of this result, if we express the flux density waveform as a Fourier series of the form

$$B = \sum_{k=1}^{K} B_{a,k} \cos(k\omega_e t) + B_{b,k} \sin(k\omega_e t) \tag{6.1-15}$$

it is straightforward to show that

$$\frac{1}{T}\int_0^T \left(\frac{dB}{dt}\right)^2 dt = \frac{1}{2}\omega_e^2 \sum_{k=1}^{K} k^2 \left(B_{a,k}^2 + B_{b,k}^2\right) \tag{6.1-16}$$

where K is the number of harmonic components used to represent the flux density waveform.

Let us now consider the problem of reducing eddy current losses. From (6.1-12), one option to accomplish this is by decreasing the conductivity of the material. It is for this reason that ferrimagnetic materials have an advantage over ferromagnetic materials – their conductivity is much lower. Within the class of ferromagnetic materials, silicon is added to the steel to decrease conductivity. Another effective method to reduce loss is to build devices out of laminations. The laminations are insulated (to varying degrees, depending on the method) so as to prevent current flow between them.

In order to see how this might be effective, let us again consider the situation shown in Figure 6.1 and let dimension d denote the depth of a lamination, which will be assumed to be very thin compared to the other dimensions. From (6.1-1) and (6.1-2), we would expect $k_2 \gg k_1$, which will allow us to simplify (6.1-12).

To this end, recall the Taylor series expansion of $\ln(1 + y)$. In particular, for small values of y,

$$\ln(1+y) = y - \frac{1}{2}y^2 + \frac{1}{3}y^3 - \frac{1}{4}y^4 + h_5(y) \tag{6.1-17}$$

where $h_5(y)$ denotes a function composed of terms or order 5 and higher. This suggests that

$$\ln\left(1 + \frac{2k_1}{k_2}\right) = \frac{2k_1}{k_2} - \frac{1}{2}\left(\frac{2k_1}{k_2}\right)^2 + \frac{1}{3}\left(\frac{2k_1}{k_2}\right)^3 - \frac{1}{4}\left(\frac{2k_1}{k_2}\right)^4 + h_5\left(\frac{2k_1}{k_2}\right) \tag{6.1-18}$$

Using (6.1-18) in (6.1-12)

$$\overline{P} = \frac{\sigma l}{128}\left(\frac{32}{3}k_2 k_1^3 + k_2^4 h_5\left(\frac{2k_1}{k_2}\right)\right)\frac{1}{T}\int_0^T \left(\frac{dB}{dt}\right)^2 dt \tag{6.1-19}$$

In (6.1-19), observe that $h_5(2k_1/k_2)$ is made up of terms of $2k_1/k_2$ to the fifth power and higher. It follows that for small ratios of k_1/k_2, the $k_2^4 h_5(2k_1/k_2)$ term is negligible compared to the $k_2 k_1^3$ term and so we have

$$\overline{P} = \frac{\sigma l}{12} k_2 k_1^3 \frac{1}{T}\int_0^T \left(\frac{dB}{dt}\right)^2 dt \tag{6.1-20}$$

Replacing k_1 by d, and approximating k_2 with w, we have

$$\overline{P} = \frac{\sigma w d l}{12} d^2 \frac{1}{T}\int_0^T \left(\frac{dB}{dt}\right)^2 dt \tag{6.1-21}$$

so that the time average of the volumetric power loss density defined as

$$\bar{p} = \frac{\bar{P}}{wdl} \tag{6.1-22}$$

becomes

$$\bar{p} = \frac{\sigma d^2}{12} \frac{1}{T} \int_0^T \left(\frac{dB}{dt}\right)^2 dt \tag{6.1-23}$$

Beyond this point, the expression (6.1-23) is interesting from several respects. First, observe that if the dimensions are large compared to the thickness of the lamination (now denoted d) and the time varying flux density is perpendicular to the plane of the laminations, then the power loss is not a function of the sample shape. Also, observe that the power loss density is a function of the lamination thickness. Clearly, (6.1-23) indicates the thinner the lamination, the lower the eddy current power loss will be. The thickness of a lamination is governed by structural and manufacturing limits, with values between 0.05 and 0.5 mm being common. Using a large number of thin laminations will clearly yield lower losses than a smaller number of thick laminations but will normally be more expensive. In addition, the use of thinner laminations results in a lower stacking factor. The stacking factor relates the volume of steel in the laminated structure relative to the volume of the structure. As the number of laminations increases, the amount of insulating material and gaps (due to surface roughness) increase, and so the material becomes less effective at conducting magnetic flux.

As a final note to this section, the use of the overbar to designate the time-average value of power loss densities is generally not used. It has been done in this section to make it easier to differentiate between instantaneous and average power. However, normally this difference is understood by context, and in most cases, "power loss density" should be taken to mean "time-average power loss density."

6.2 Hysteresis Loss and the *B–H* Loop

The focus of this section is magnetic hysteresis loss. The source of this loss is domain wall movement. Recall that in ferromagnetic materials, the magnetic moments of adjacent atoms are aligned, and in ferrimagnetic materials, groups of atoms can be thought of as having a net magnetic moment and that these aggregate groups are aligned. However, the alignment of moments does not extend through the entire material, but instead only through regions referred to as domains. Figure 6.4 illustrates a set of domains for a rectangular section of material (note that in practice, the domains are on small scale, not a large scale as shown in the figure). In Figure 6.4(a), the situation is shown when there is no externally applied field. The domains are numbered 1–10. If an external field is applied, the domain walls shift, and domains aligned with the direction of the applied field become larger, while those not aligned with the applied field become smaller. This is illustrated in Figure 6.4 (b), wherein domains 5, 7, and 8 have become larger in response to the applied field. The fact that these aligned domains become larger when an external field is applied is the reason why there is a relatively large change in flux in response to a given change in field intensity. Increasing the field intensity further will cause further shift in the domain wall until the aligned domains occupy the entire sample.

The domain walls separating the magnetic domains do not move freely. Rather they are held in place by irregularities in the material structure referred to as pinning sites. Energy must

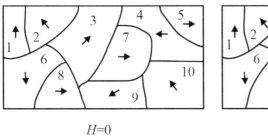
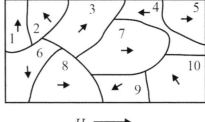

(a) Domains without applied field (b) Domains with applied field

Figure 6.4 Magnetic domains.

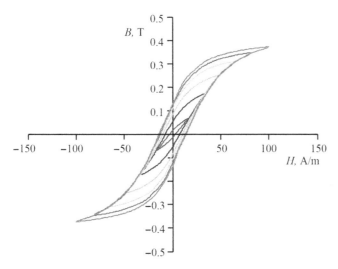

Figure 6.5 B-H characteristics of a ferrite material MN80C.

be expended to move the walls. The movement of a wall will induce localized eddy currents, which are also associated with loss. As a result, if one removed the external field shown in Figure 6.4(b), the domain wall locations would move somewhat, but not all the way back to their positions in Figure 6.4(a). As a result, the $B - H$ characteristic of a magnetic material is not single-valued.

Figure 6.5 illustrates the $B - H$ characteristics of a ferrite material MN80C. As can be seen, a number of $B - H$ trajectories are shown. Each of these trajectories is shaded differently, and each has a different range over which the field intensity is varied. While the largest of the $B - H$ trajectories (or loops) is usually considered to be "the" $B - H$ characteristic of the material, the actual $B - H$ trajectory is a function of the excitation and forms a loop well inside the main loop, as we see in the figure. Unfortunately, every time we traverse a $B - H$ loop, we lose energy.

In order to establish the total energy loss, consider the toroidal material sample shown in Figure 6.6. Therein, it is assumed that the toroid is uniformly wound around its circumference. We will also assume that the difference between the inner and outer radii is small so that the field is uniform within the material. Let l_T denote the length of the circumference of the toroid measured

Figure 6.6 Large uniformly wound toroid.

at the mean radius, and let a_T denote the cross-sectional area of the toroid. The volume of the toroid is given by

$$V_T = l_T a_T \tag{6.2-1}$$

The number of turns of wire around the toroid will be denoted N. We will assume that our wire has negligible resistance. Applying a voltage v to the winding, the instantaneous power into the winding may be expressed as

$$P = vi \tag{6.2-2}$$

The voltage may be expressed as

$$v = \frac{d\lambda}{dt} \tag{6.2-3}$$

Combining (6.2-2) and (6.2-3)

$$P = i\frac{d\lambda}{dt} \tag{6.2-4}$$

From (6.2-4), the energy that has flowed into the winding from time 0 to T is given by

$$E = \int_0^T i\frac{d\lambda}{dt} dt \tag{6.2-5}$$

or

$$E = \int_{\lambda_0}^{\lambda_T} i\, d\lambda \tag{6.2-6}$$

where λ_0 and λ_T denote the flux linkage at times 0 and T, respectively. The energy E is not energy-stored; rather, it is energy-transferred to the material.

From Ampere's law, we have that the current and field intensity in the toroid are related by

$$i = \frac{l_T H}{N} \tag{6.2-7}$$

The flux linking our winding may be expressed as

$$\lambda = Na_T B \tag{6.2-8}$$

Substitution of (6.2-7) and (6.2-8) into (6.2-6) and using (6.2-1), we have

$$E = V_T \int_{B_0}^{B_T} H dB \tag{6.2-9}$$

where B_0 and B_T denote the flux density at times 0 and T, respectively. The energy transferred to the material per unit volume is defined as

$$e = \frac{E}{V_T} \tag{6.2-10}$$

Thus, the energy density transfer within the magnetic material is given by

$$e = \int_{B_0}^{B_T} H dB \tag{6.2-11}$$

Note that while we derived this result for a specific situation, the material in question does not "realize" that it is in the specified toroid; hence, (6.2-11) is actually a generic result for the change in energy density.

Now, let us conduct a second experiment. Consider the $B - H$ characteristic shown in Figure 6.7. Suppose that at time zero, we are at point (H_0, B_0), and then as time goes on we move to point (H_1, B_1) and eventually to (H_2, B_2) as we follow the rightmost arc of the $B - H$ characteristic. Then, the field intensity decreases and we move to (H_3, B_3) and eventually back to the starting point (H_0, B_0) along the leftmost arc of the $B - H$ characteristic. Our goal is to track the energy density transfer during this trajectory. Applying (6.2-11), but breaking the integral into a number of sub-trajectories, we have

$$e = \underbrace{\int_{B_0}^{B_1} H dB}_{\text{Term 1}} + \underbrace{\int_{B_1}^{B_2} H dB}_{\text{Term 2}} + \underbrace{\int_{B_2}^{B_3} H dB}_{\text{Term 3}} + \underbrace{\int_{B_3}^{B_0} H dB}_{\text{Term 4}} \tag{6.2-12}$$

Term 1 corresponds to the vertically hashed area in Figure 6.7. Since H is negative and dB is positive, the area under the curve represented by the integral is a negative contribution to the energy transfer density. The second term, with a positive H and positive dB, is a positive contribution and equal to the area in the lightly shaded region (with or without horizontal hash marks). The third term corresponds to the horizontally hashed region. This term is negative because H is positive and dB is negative. The final contribution to the integral, term 4, corresponds to the darkly shaded region (with or without vertical hash marks). Since H is negative and dB is also negative, this is a positive area and positive contribution to the overall integral. The two positive areas contributed by terms 2 and 4, less the two negative areas from terms 1 and 3, yield the shaded (either lightly or darkly) nonhashed (either vertically or horizontally) area. In other words, the energy density transfer associated with one traversal of the $B - H$ characteristics is the area enclosed by the $B - H$ trajectory. It follows that the average power loss density in the material due to hysteresis may be expressed as

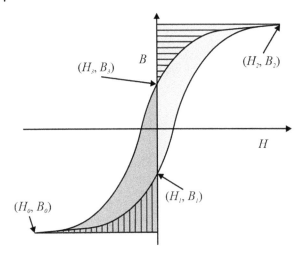

Figure 6.7 Hysteresis characteristic.

$$\bar{p}_h = f e_{BH} \tag{6.2-13}$$

where f is frequency, that is, the number times we traverse the characteristic per second, and e_{BH} is the area enclosed by the $B - H$ characteristic. From Figure 6.5, it is easy to see that this area will be a function of the amplitude of the field intensity. It will also change with frequency and the shape of the field intensity and flux density waveforms.

The $B - H$ trajectories we have looked at so far have been symmetrical, but in general, this need not be the case. Let us consider the situation of an inductor in a buck converter. The interested reader may want to briefly skip ahead and read the first two pages of Section 13.1 for an overview of such a converter. The relevant factor in this application is that the fields will have a large bias with a ripple component. This situation is illustrated in Figure 6.8. Therein, one extremum of the trajectory is the point (H_0, B_0). Then, the field decreases until the point (H_1, B_1) is reached. At this point, the field intensity is increased until the point (H_0, B_0) is again reached, completing a cycle. This situation is referred to as traversing a minor loop.

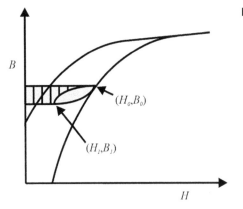

Figure 6.8 Minor loop behavior.

Let us consider the energy loss in this situation. The energy density transferred to the device during this trajectory may be expressed as

$$e = \underbrace{\int_{B_0}^{B_1} H dB}_{\text{Term 1}} + \underbrace{\int_{B_1}^{B_0} H dB}_{\text{Term 2}} \qquad (6.2\text{-}14)$$

Evaluating term 1 yields the vertically hashed area in Figure 6.8. Since H is positive and dB is negative, this counts as negative area or energy density. Evaluating term 2 yields the shaded area, which includes the vertically hashed area. Since H is positive and dB is positive, this is positive area or energy density. The sum of these areas is the shaded, nonhashed area. Thus, the area enclosed by the trajectory represents the per cycle loss.

6.3 Empirical Modeling of Core Loss

In this section, we consider a variety of empirical or behavioral approaches to modeling core loss. These are not based on a detailed physical model but rather on representing observed behavior in a mathematical form using curve fitting.

The Steinmetz Equation

In the previous section, we found that the power loss density is proportional to the area enclosed by the $B - H$ characteristic (6.2-13). In the Steinmetz approach, which is meant for sinusoidal excitation, the area of the loop is expressed as

$$e_{BH} = \frac{k_h}{f_b}\left(\frac{f}{f_b}\right)^{\alpha-1}\left(\frac{B_{pk}}{B_b}\right)^{\beta} \qquad (6.3\text{-}1)$$

where k_h, α, and β are the parameters associated with the model obtained from curve-fitting, f_b and B_b are the base frequency and base flux density, which we will take to be 1 Hz and 1 T, respectively, and B_{pk} is the peak flux density. The use of f_b and B_b allows k_h to have units independent of α and β. From (6.3-1), the loop area goes up with either frequency or peak flux density; the use of $\alpha - 1$ in (6.3-1) rather than α is so that when (6.3-1) is substituted into (6.2-13), we obtain the simple form

$$\overline{p}_h = k_h \left(\frac{f}{f_b}\right)^{\alpha}\left(\frac{B_{pk}}{B_b}\right)^{\beta} \qquad (6.3\text{-}2)$$

While (6.3-2) has proven useful, it has a number of significant drawbacks. First, note that as B_{pk} increases, the power loss density is predicted to grow without bound. Physically, we note that because of saturation, at some point, the area of the $B - H$ characteristic will cease to expand with increasing B. This is not the case with eddy current loss. Second, it must be realized that this is a behavioral model used to fit to measured data – and with only three model parameters, the quality of the fit is limited. Third, for (6.3-2) to yield useful predictions, the flux-density waveform must be ac and essentially sinusoidal. In order to overcome these difficulties, alternate methods have been proposed. These include the modified Steinmetz equation (MSE) and the generalized Steinmetz equation, which will both be discussed henceforth.

The Modified Steinmetz Equation

One of the goals of the MSE [2] is to enable accurate predictions for nonsinusoidal waveforms. From a physical point of view, in this approach, it is assumed that hysteresis loss is a form of eddy current loss. In particular, when a domain wall moves, there is an associated localized eddy current induced by a rapid change of magnetization in the region of the material where the wall has moved. Eddy current losses are driven by the time rate of change of flux density, so the average rate of change of flux density is of interest. This average rate of change may be expressed as

$$\dot{B} = \frac{1}{\Delta B} \oint \frac{dB}{dt} dB \qquad (6.3\text{-}3)$$

In (6.3-3), the closed path integral refers to integration over a periodic cycle, and

$$\Delta B = B_{mx} - B_{mn} \qquad (6.3\text{-}4)$$

where B_{mx} and B_{mn} are the maximum and minimum flux densities, respectively, within the cycle. Observe that the average is an average with respect to the flux density trajectory, not an average with respect to time. Alternatively, (6.3-3) may be expressed as

$$\dot{B} = \frac{1}{\Delta B} \int_0^T \left(\frac{dB}{dt}\right)^2 dt \qquad (6.3\text{-}5)$$

where T is the period of the flux density waveform.

In [2], it is proposed that $\dot{B}$ be used to define an equivalent frequency. In particular, the equivalent frequency is defined as

$$f_{eq} = \frac{2}{\Delta B^2 \pi^2} \int_0^T \left(\frac{dB}{dt}\right)^2 dt \qquad (6.3\text{-}6)$$

and that the energy density loss per cycle be defined as

$$e_{BH} = \frac{k_h}{f_b} \left(\frac{f_{eq}}{f_b}\right)^{\alpha-1} \left(\frac{\Delta B}{2B_b}\right)^{\beta} \qquad (6.3\text{-}7)$$

where, as in the case of the Steinmetz model, we will take f_b and B_b to be 1 Hz and 1 T, respectively. The motivation behind this is to simultaneously accomplish the following: (i) the energy lost per cycle is now related to $\dot{B}$ and (ii) for a sinusoidal waveform, (6.3-6) yields $f_{eq} = f$, the fundamental frequency, and $\Delta B = 2B_{pk}$ so, (6.3-7) reduces to the Steinmetz case. Substitution of (6.3-7) into (6.2-13) yields a power loss density of

$$\bar{p}_h = k_h \left(\frac{f_{eq}}{f_b}\right)^{\alpha-1} \left(\frac{\Delta B}{2B_b}\right)^{\beta} \frac{f}{f_b} \qquad (6.3\text{-}8)$$

To assess the power loss density, (6.3-4), (6.3-6), and (6.3-8) are sequentially evaluated. Again, for sinusoidal flux density, this reduces to the Steinmetz equation (6.3-2), but for nonsinusoidal flux density waveforms, the power loss density could be higher or lower. One would expect the accuracy of the method to be reduced for nonmonotonic waveforms (i.e., waveforms with minor loops). Although the method is most often applied to ac waveforms, a modification of the method for dc-biased waveforms is also set forth in [2].

6.3 Empirical Modeling of Core Loss

Example 6.3A For the case of a sinusoidal flux density waveform, the MSE reduces to the Steinmetz equation with $B_{pk} = B_{mx} = -B_{mn}$. Let us compare the power loss density for a sinusoidal waveform to that of a triangular waveform in which the flux density starts at $-B_{pk}$, increases linearly in time to B_{pk}, and then falls linearly in time to $-B_{pk}$. From (6.3-1), we first have

$$\Delta B = 2B_{pk} \tag{6.3A-1}$$

Next, from the description of the waveform, we have

$$\frac{dB}{dt} = \pm 4B_{pk}f \tag{6.3A-2}$$

Substitution of (6.3A-1) and (6.3A-2) into (6.3-6), we obtain an equivalent frequency of

$$f_{eq} = \frac{8}{\pi^2}f \tag{6.3A-3}$$

which is slightly lower than the fundamental frequency. Substitution of (6.3A-1) and (6.3A-3) into (6.3-8), we obtain

$$\bar{p}_h = \left(\frac{8}{\pi^2}\right)^{\alpha-1} k_h \left(\frac{f}{f_b}\right)^{\alpha} \left(\frac{B_{pk}}{B_b}\right)^{\beta} \tag{6.3A-4}$$

Comparing (6.3A-3) to (6.3-2), which can be shown to match the MSE model for the sinusoidal case, we can see that the MSE model predicts that this triangular waveform would be associated with a lower loss density than a sinusoidal waveform with the same peak value and fundamental frequency.

The Series Modified Steinmetz Equation

The MSE is an empirical model and will, like all empirical models, or all models in general, have an error associated with it. One method to reduce this error is to use a series based on the Steinmetz form. In particular, in the series modified Steinmetz equation (or sMSE model), the loss density is expressed as

$$\bar{p}_h = \frac{f}{f_b} \sum_{k=1}^{K} k_{h,k} \left(\frac{f_{eq}}{f_b}\right)^{\alpha_k - 1} \left(\frac{\Delta B}{2B_b}\right)^{\beta_k} \tag{6.3-9}$$

where K is the number of terms. This approach is useful when a single term cannot be well fit to measured loss data.

The Generalized Steinmetz Equation

Another method for predicting core losses is the generalized Steinmetz equation. This method is based on the premise that the instantaneous power loss density is a function of the time rate of change in flux density and the value of flux density. In other words,

$$p = f\left(\frac{dB}{dt}, B\right) \tag{6.3-10}$$

where p is the instantaneous power loss density. The loss density function chosen is

$$p = k \left| \frac{1}{f_b B_b} \frac{dB}{dt} \right|^\alpha \left| \frac{B}{B_b} \right|^{\beta-\alpha} \tag{6.3-11}$$

This form can be made compatible with the original Steinmetz equation (6.3-2) for sinusoidal excitation. Like the MSE, loss is a function of the time rate of change of flux density. Using (6.3-11), the average power loss density may be expressed as

$$\overline{p}_h = \frac{1}{T} k \int_0^T \left| \frac{1}{f_b B_b} \frac{dB}{dt} \right|^\alpha \left| \frac{B}{B_b} \right|^{\beta-\alpha} dt \tag{6.3-12}$$

It can be shown that by selecting

$$k = \frac{k_h}{(2\pi)^{\alpha-1} \int_0^{2\pi} |\cos\theta|^\alpha |\sin\theta|^{\beta-\alpha} d\theta} \tag{6.3-13}$$

the loss model reduces to the Steinmetz model for sinusoidal excitation.

It has been reported that the accuracy of the GSE model deteriorates for nonmonotonic waveforms. A method to correct this, the improved GSE model, or iGSE, is reported in [3]. An improved–improved GSE model, denoted the i²GSE, appears in [4], and an expanded GSE model is proposed in [5]. Besides the MSE model, the GSE model, and its variants, the literature is rich with core loss models. For example, the natural Steinmetz extension is set forth in [6]. A method to calculate loss particularly designed for the dc-biased case is reported in [7].

Example 6.3B Let us compare the loss predictions of the MSE and GSE core loss models for MN60LL ferrite. This material has $\alpha = 1.034$, $\beta = 2.312$, and $k_h = 40.08 \text{W/m}^3$. We will consider excitation of the form

$$B = B_1 \cos(2\pi f t) + B_3 \cos(6\pi f t) \tag{6.3B-1}$$

We will consider three cases. For case 1, $B_1 = 0.50$ and $B_3 = -0.05$T. For case 2, $B_1 = 0.45$ and $B_3 = 0$ T. For case 3, $B_1 = 0.409$ and $B_3 = 0.0409$T. In each case, the peak flux density is 0.45 T, and the frequency f will be taken to be 10 kHz. The waveforms for the three cases are illustrated in Figure 6.9. As can be seen, case 1 is a rather triangular-looking waveform, case 2 is a sinusoid,

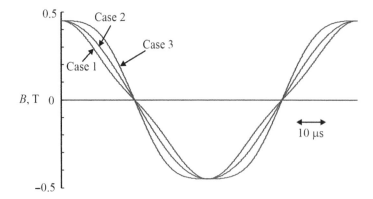

Figure 6.9 Flux density waveforms.

and case 3 is a flat-topped waveform. Computing the losses using the MSE by sequentially evaluating (6.3-4), (6.3-6), and (6.3-8) yields 87.4, 86.5, and 86.2 kW/m³ for the three cases, respectively. Evaluating (6.3-13) yields $k = 21.8$ W/m³. Applying (6.3-12) to the three cases, we obtain power loss densities of 86.9, 86.5, and 86.8 kW/m³, respectively, which are similar to the predictions of the MSE model. Thus, for modestly nonsinusoidal waveforms, this brief study suggests that the differences between the model predictions are limited.

Combined Loss Modeling

In ferrimagnetic materials, the hysteresis loss models discussed in this section may be used by themselves. In the case of ferromagnetic materials, the losses due to eddy currents will be substantial. For this case, we can combine the hysteresis loss and core loss models.

To this end, note that for the case of a thin lamination, from (6.1-23), we may express the power loss density in a thin lamination due to eddy currents as

$$\bar{p}_e = k_e \frac{1}{T} \int_0^T \left(\frac{dB}{dt}\right)^2 dt \qquad (6.3\text{-}14)$$

where, from (6.1-23), k_e is a function of the conductivity and lamination thickness. Indeed, comparing (6.3-14) with (6.1-23), we might conclude

$$k_e = \frac{\sigma d^2}{12} \qquad (6.3\text{-}15)$$

where d is the lamination thickness. However, for reasons we will soon discuss (6.3-15) is only a rough approximation.

Combining the eddy current loss density (6.3-14) with, for example, the sMSE hysteresis loss model, we have that the total loss density in a ferromagnetic material may be expressed as

$$\bar{p}_t = \frac{f}{f_b} \sum_{k=1}^{K} k_{h,k} \left(\frac{f_{eq}}{f_b}\right)^{\alpha_k - 1} \left(\frac{\Delta B}{2B_b}\right)^{\beta_k} + k_e \frac{1}{T} \int_0^T \left(\frac{dB}{dt}\right)^2 dt \qquad (6.3\text{-}16)$$

There are several points to note in (6.3-16). First, as has been mentioned, hysteresis and eddy current loss mechanisms are intricately coupled. Because of this, it is better to view (6.3-16) as a behavioral loss model. By measuring power loss density for a variety of flux density waveforms, the coefficients k_h, α, β, and k_e can be simultaneously determined using curve fitting techniques. Thus, it is unlikely the value of k_e given by (6.3-15) will yield good results.

A second point of interest in (6.3-16) is that this expression for power loss density is not for a particular material but rather for a particular material with a particular lamination thickness. This is because the eddy current loss will be a function of the thickness of the lamination. In other words, samples of the same material but with different lamination thicknesses will have different values of k_e.

Finally, as was noted in Section 6.1, the use of the overbar notation in (6.3-16) is generally not used. Rather, it is generally understood that p_t is the time-average power loss density. The overbar will be omitted later in this text.

Example 6.3C In this example, let us consider the contribution of core loss to eddy current loss for a sample of laminated M19 steel. The MSE parameters of the material are $k_h = 50.7 \, \text{W/m}^3$, $\alpha = 1.34$, $\beta = 1.82$, and $k_e = 27.5 \cdot 10^{-3} \, \text{Am/V}$. Let us plot the power loss density due to hysteresis (the term with k_h) and the power loss density due to bulk eddy currents (the term with k_e) versus frequency. We will assume a sinusoidal flux density given by

$$B = B_{pk} \cos(2\pi f t) \tag{6.3C-1}$$

where $B_{pk} = 1.1$ T. Since the flux density is sinusoidal, the MSE model reduces to the Steinmetz model and so the power loss density due to hysteresis is given by

$$\bar{p}_h = k_h \left(\frac{f}{f_b}\right)^\alpha \left(\frac{B_{pk}}{B_b}\right)^\beta \tag{6.3C-2}$$

From (6.3C-1) and (6.3-14), the loss density due to eddy currents is given by

$$\bar{p}_e = 2\pi^2 k_e B_{pk}^2 f^2 \tag{6.3C-3}$$

The total power loss density is given by

$$\bar{p}_t = \bar{p}_h + \bar{p}_e \tag{6.3C-4}$$

The resulting power loss density components are illustrated in Figure 6.10. Therein, it can be seen that at low frequencies, the hysteresis losses are the dominant loss mechanism. However, at high frequencies, bulk eddy current losses dominate. The high loss past a few hundred Hertz explains why silicon steel materials are not used at elevated frequencies.

Temperature Dependence

Before concluding this section, it should be mentioned that losses are temperature-dependent and, in fact, tend to decrease with increasing temperature, at least in some materials. A straightforward modification to the Steinmetz, MSE, and GSE models is set forth in [8]. In this approach, a saturation factor of the form

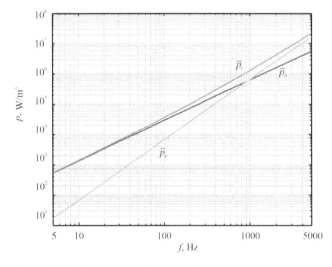

Figure 6.10 Loss components versus frequency.

$$c_s = \frac{1}{1 + k_s \left(\frac{B_{mx}}{B_b}\right)^\beta} \quad (6.3\text{-}17)$$

and a temperature factor of the form

$$c_T = \frac{1}{1 + e^{\gamma(T - T_0)}} \quad (6.3\text{-}18)$$

are introduced into the expressions for power loss density, where k_s, γ, and T_o are additional material parameters. With these factors, the Steinmetz model (6.3-2) becomes

$$\overline{p}_h = k_h \left(\frac{f}{f_b}\right)^\alpha \left(\frac{B_{pk}}{B_b}\right)^\beta c_s c_T \quad (6.3\text{-}19)$$

or

$$\overline{p}_h = k_h \frac{(B_{pk}/B_b)^\beta}{1 + k_h (B_{pk}/B_b)^\beta} \left(\frac{f}{f_b}\right)^\alpha \frac{1}{1 + e^{\gamma(T - T_0)}} \quad (6.3\text{-}20)$$

The approach may be similarly applied to other behavioral loss models. In [8], these factors are shown to drastically improve the accuracy of the models. Interestingly, the two factors work much better in tandem than separately.

6.4 Magnetic Material Characterization

In Sections 6.1–6.3, we considered various ways of estimating losses in magnetic materials. In all cases, these models were presented in terms of material parameters. In the remaining sections of this chapter, we will consider how such material parameters are measured. The development will be as follows. First, in this section, different test fixtures for magnetic material characterization will be set forth. Next, in Section 6.5, the process of measuring the anhysteretic material properties is considered. Then, we will address the problem of experimentally obtaining loss coefficients for empirical core loss models, such as the MSE model. We will begin our examination of test fixtures by considering the Epstein frame.

Epstein Frame

Figure 6.11 depicts an Epstein frame, which can be used to test both directional and nondirectional materials. This frame is constructed of a stack of laminations, which have length l, width b, and total depth into the page d. The laminations are staggered so that the seams where laminations come to together in one layer are covered by the next layer. Often, these dimensions are selected in accordance with standards such as IEC_60404 [9]. For example, therein, it is specified that for the 25-cm Epstein frame, 280 mm $\leq l \leq$ 320 mm, $b = 30$ mm, and that the total mass of the laminations comes to at least 0.24 kg (which effectively provides a minimum bound on d).

The Epstein frame utilizes two sets of windings: the excitation winding and the sense winding. The excitation winding is made up of the four outside coils, which are connected in series. Exciting this winding results in flux. By measuring the excitation current, multiplying it by the total number of turns in the four coils, and dividing by the mean path length of the core yield an approximate value for field intensity. The sense winding is made up of the four inner coils, which are also

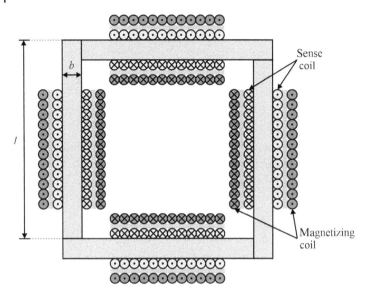

Figure 6.11 Epstein frame.

connected in series. The voltage across the coils is measured; this voltage can be integrated to determine the flux linkage; the flux linkage is divided by the product of turns and magnetic cross section to determine the flux density waveform in the sample.

Not shown in Figure 6.11, an Epstein frame is often equipped with a mutual inductor, which is physically placed in the center of the frame. The value of mutual inductance and polarities of connection are such that when magnetizing winding is excited without a sample, essentially no voltage is measured on the sense winding. This allows M vs H to be directly measured rather than B vs H. Second, it corrects for the impact of any space between the secondary winding and the sample (to facilitate a bobbin, e.g.). In a sense, this is a leftover from the past; a similar correction could be made mathematically without the need to fabricate a mutual inductor.

While the Epstein frame is a classic method to characterize laminated steel, there are some issues with the test. In particular, the calculation of the field intensity is based on the mean path length approximation (essentially the same approximation we made when creating MECs). While this does work well, some error is involved. Likewise, some error is involved in assuming a uniform distribution of flux density within the material. It should be noted that it is probably best to stamp the samples from a die and anneal them in the same way as their intended use, as compared to, for example, laser cutting, which can locally alter material properties.

Single- and Double-Sheet Testers

One alternative to Epstein frames is single- and double-sheet testers. These are designed for the characterization of a single lamination (or a pair of laminations for the double-sheet tester). These testers can again be used with either oriented or nonoriented materials. One advantage of these testers is that they require much less material than is required for the Epstein test and, accordingly, much less MMF for excitation.

Single- and double-sheet testers are shown in Figure 6.12. Let us focus our attention first on the single-sheet tester. In this arrangement, there is a magnetic core associated with the

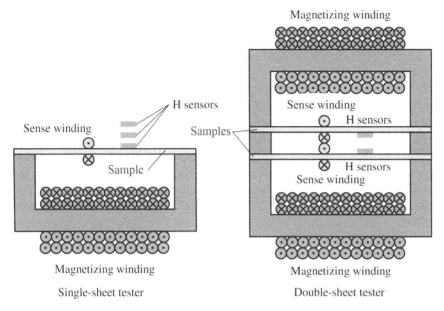

Figure 6.12 Single- and double-sheet testers.

magnetization winding, which is not necessarily the same material as being tested. The sample is placed over the core, essentially forming a UI-core path, where the I-core is the material sample being characterized. A sense winding is used to determine the flux density in the sample, in a manner similar to that of the Epstein tester. However, the measurement of the field intensity is more involved.

Because of the nonuniform path, it is not easy to analytically estimate the field intensity in the sample. Such an estimate is possible but is subject to significant error, and so a more direct approach is desired. To this end, it is noted that using Ampere's law, it can be shown that the tangential field intensity in the sample is the same as that in the free space immediately outside of the sample. So, any device that can be used to measure field in free space could be used to measure the field intensity just outside of the sample, so that the field intensity in the sample is known.

The caveat to this is that all physical sensors occupy space, and so, it is not possible to measure H exactly at the material interface; it can only be measured some distance from it. Unfortunately, this field is not uniform and so this introduces an error. This problem can be overcome by using multiple field sensors so that the spatial gradient of H can be determined, whereupon H in the sample can be deduced.

In the double-sheet tester, an alternate approach is used. In this setup, two back-to-back single-sheet testers are mirrored, as shown in Figure 6.12. The use of two testers in this configuration creates a symmetry that reduces the field intensity gradient around the sample, thereby establishing a more uniform field, which can be more easily measured.

As mentioned previously, when characterizing some laminations, the use of laser cutting can cause significant changes in the magnetic performance. Single- and double-sheet testers, if used with sufficiently large and squarish samples, are less susceptible to this because most of the material is away from the edges where the material degradation takes place.

Toroidal Tester

Another common testing arrangement is a toroidal tester shown in Figure 6.13. This arrangement can be used to test toroidal ferrite cores, toroidal tape-wound cores, and laminated toroidal cores constructed from isotropic magnetic steels.

The instantiation of the toroidal tester is shown in Figure 6.13; the sense winding is wound directly on the core, just as in the case of the Epstein, single-, and double-sheet testers. In this case, the primary or magnetizing winding is kept relatively far from the core to avoid local variations in the field, though this is not always done. One advantageous property of the toroidal tester is that because of the axial symmetry, given the magnetizing winding MMF, it is easier to accurately compute the field intensity as a function of radius within the core. Of course, the fact that the field intensity is a function of radius will introduce complications if there is a large difference between the inner and outer radii. Fortunately, this can be accounted for, as we will see in the next section.

Considerations When Characterizing Laminated and Tape-Wound Samples

Before proceeding, some comments are on the order of the preparation of laminated and tape-wound samples. The first of these concerns stacking factor. If we take a stack of laminations, not all the volume within the stack is the material being characterized. The laminations generally have an insulating layer, and there is effectively dead space between the laminations because of surface roughness. Clearly, the fraction of dead space/insulation is a function of the thickness

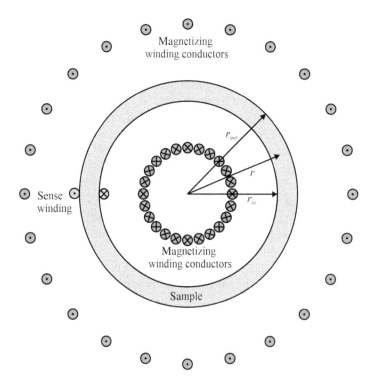

Figure 6.13 Toroidal tester.

or gauge of the steel. Herein, we will take the approach of aggregating the stack and treating it as a homogenized material. This goes for not only the magnetic properties but also the physical properties such as mass density. The advantage of this approach is that it is simple and reflects how the laminations will be used. However, it also means that if we have the same material with two different finishes, we would need to characterize the material for each finish. Needless to say, this also goes for changing the lamination thickness – but in this case, we would have had to recharacterize anyway because the eddy current losses will change.

However, it should be noted that there is another school of thought in which the fact that there is inert space between laminations is characterized by what is known as a stacking factor, in an attempt to characterize the magnetic material at a more fundamental level. The stacking factor is the fraction of a stack occupied by active material. This approach will result in a larger magnitude of flux density for a given field intensity than the former approach. Clearly, the stacking factor also needs to be accounted for when using the material. In any case, it is important to realize that depending upon the approach, viewing the material as homogenized versus using a stacking factor will yield a different material characteristic; however, when both approaches are applied correctly and consistently, both yield identical predictions of device performance.

Another aspect of a sample preparation has to do with the shaping of samples. One method of forming samples for characterization is to stamp them using a die, though a die can be expensive. After the laminations are cut, annealing may be required in order to restore the magnetic properties caused by the deformation. Alternately, samples may be laser-cut. However, this also causes local degradation in the material properties, which is hard to repair.

To illustrate this, the B–H characteristics of three toroidal samples of an M15 steel are shown in Figure 6.14. The samples were laser-cut from the same stock. The 3-mm sample had an inner radius of 97 mm, an outer radius of 100 mm, and a height of 7.3 mm; the 10-mm sample had an inner radius of 49 mm, an outer radius of 59 mm, and a height of 5.9 mm; and the 80-mm sample had an inner radius of 20 mm, an outer radius of 100 mm, and a height of 7.1 mm. The mean flux density versus mean field intensity is plotted. As can be seen, the 3-mm sample and, to a lesser extent, the 10-mm sample have a significant "distortion" near the origin, which is attributed to the laser-cutting process. As the sample becomes wider, and more of the material is further away from a laser cut, the effect becomes less pronounced. The interested reader is referred to [10–12].

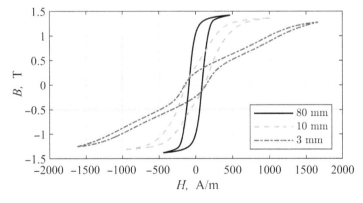

Figure 6.14 Impact of laser cutting.

6.5 Measuring Anhysteretic Behavior

When characterizing magnetic materials, the first step is to characterize the anhysteretic B–H characteristic. Here, we will focus our attention on how this is done using the toroidal tester, though we will discuss characterization using other methods as well.

Excitation and Data Collection

In order to characterize the anhysteretic behavior using any of the test fixtures we discussed, we will begin by injecting a current into the magnetizing winding. The current waveform need not be sinusoidal but should be (i) relatively low frequency so that eddy currents are negligible, (ii) periodic, and (iii) odd half-wave symmetric. This latter point precludes dc offsets. DC offsets can be avoided through the use of a series capacitor or through the use of a parallel isolation transformer.

When using either the Epstein frame or a toroidal tester, we will measure the magnetizing current $i_m(t)$. When referring to $i_m(t)$, we will be referring to a preprocessed waveform such that dc measurement offsets have been removed and that, optionally, the waveform has been low-pass-filtered (though with a sufficiently high cutoff frequency that only noise is removed). Recall that our source should be such that dc offsets cannot exist – however, our measured data may include a dc offset because of current probe offsets, and these offsets must be removed. In discrete time, we will denote $i_{m,i} = i_m(t_i)$, where t_i is the ith sampling instant. We will use the magnetizing current to compute the field intensity when using the Epstein frame or the toroidal tester. In the case of the single- or double-sheet tester, we will assume we have measured H "directly," already taking into effect the gradient calculation in the case of the single-sheet tester. In this case, we have $H_i = H(t_i)$. Again, we will assume that the measured data has probe-induced dc bias and noise removed.

For all the text fixtures considered, we will measure the voltage across the sense winding. The measured secondary voltage at $t = t_i$ is denoted $v_{s,i} = v_s(t_i)$. Here again, it is assumed that the measured secondary voltage is preprocessed such that any dc offset due to measurement has been removed, as has any high-frequency noise. From the voltage waveform, the sense winding waveform is integrated to determine the sense winding flux linkage as

$$\lambda_s(t) = \int_{t_0}^{t} v_s(\tau)d\tau + \lambda_s(t_0) \tag{6.5-1}$$

where $\lambda_s(t_0)$ is determined such that $\lambda_s(t)$ has no dc component.

Anhysteretic Data

In the case of both the Epstein frame and toroidal tester, following the procedure described above, we can construct a waveform for the magnetizing current versus sense winding flux linkage. However, this data includes the impact of hysteresis, whereas, for the moment, we wish to obtain anhysteretic data. To this end, we will "horizontally" average the lower and upper arcs of the hysteretic data, where we have plotted magnetizing current (or field intensity) on the horizontal axis and sense winding flux linkage on the vertical axis.

The process for obtaining anhysteretic data is illustrated in Figure 6.15. The first step is to divide our data set into two sets. The first set will correspond to data along the lower arc of the hysteresis curve, wherein the sense voltage is positive and the sense flux linkage is increasing. In particular,

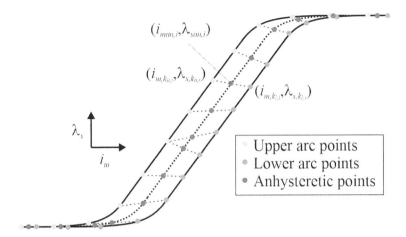

Figure 6.15 Determining the anhysteretic characteristic.

we define a set of indices corresponding to the lower arc of the hysteresis loop, wherein the flux linkage is increasing as

$$\mathbf{K}_l = i : v_{s,i} > 0 \tag{6.5-2}$$

The second set of indices corresponds to the upper arc of the hysteresis curve, wherein the sense voltage is negative and the sense flux linkage is decreasing. We define

$$\mathbf{K}_u = i : v_{s,i} < 0 \tag{6.5-3}$$

Next, for each measured flux-linkage value, we determine indices

$$k_{l,i} = \underset{j \in \mathbf{K}_l}{\operatorname{argmin}}\left(\left|\lambda_{s,i} - \lambda_{s,j}\right|\right) \tag{6.5-4}$$

$$k_{u,i} = \underset{j \in \mathbf{K}_u}{\operatorname{argmin}}\left(\left|\lambda_{s,i} - \lambda_{s,j}\right|\right) \tag{6.5-5}$$

The indices $k_{l,i}$ and $k_{u,i}$ represent those indices of the sensor flux linkage, which are closest $\lambda_{s,i}$ on the lower and upper arcs, respectively. Of course, one of these will be an exact match, which is to say that either $k_{u,i} = i$ or $k_{l,i} = i$ for any value of i. Then, we create an anhysteretic data point by averaging the flux linkage and magnetizing currents for the two indices. In particular,

$$\lambda_{san,i} = \frac{1}{2}\left(\lambda_{s,k_{l,i}} + \lambda_{s,k_{u,i}}\right) \tag{6.5-6}$$

$$i_{man,i} = \frac{1}{2}\left(i_{m,k_{l,i}} + i_{m,k_{u,i}}\right) \tag{6.5-7}$$

At this point, in the case of the Epstein frame, the anhysteretic field intensity is given by

$$H_{an,i} = \frac{N_m i_{man,i}}{l_{mp}} \tag{6.5-8}$$

and the magnetization is given by

$$M_{an,i} = \frac{\lambda_{san,i}}{N_s a_{mcs}} \tag{6.5-9}$$

where l_{mp}, a_{mcs}, N_m, and N_s are the effective magnetic path length, magnetic cross section, total number of magnetizing turns, and total sense winding turns, respectively. In (6.5-9), magnetization is found rather than flux density assuming that a compensating mutual inductor has been employed (either physically or virtually), in accordance with [9].

In the case of a single- or double-sheet tester, the field intensity is processed analogously to the magnetizing current in the case of the Epstein or toroidal testers; thus,

$$H_{an,i} = \frac{1}{2}\left(H_{k_{l,i}} + H_{k_{u,i}}\right) \tag{6.5-10}$$

Similar to (6.5-9),

$$B_{an,i} = \frac{\lambda_{san,i}}{N_s a_{mcs}} \tag{6.5-11}$$

At this point, we have data points that represent the anhysteretic magnetization curve for either the Epstein frame or single- or double-sheet testers. This data can be used verbatim with an interpolation routine, or fit to a curve, a process we will discuss later in this section.

In the case of the toroid tester, if the difference between the inner and outer radii is small, we can apply (6.5-8) and (6.5-11) to determine B and H, with

$$a_{mcs} = (r_{out} - r_{in})d \tag{6.5-12}$$

and

$$l_{mp} = \pi(r_{in} + r_{out}) \tag{6.5-13}$$

where d is the depth of the toroid into the page in Figure 6.13. Again, this leaves us with numerical data representing the anhysteretic magnetization characteristic. However, while the results may sometimes be acceptable, it is possible to improve the results by considering the impact of radial field variation. This is considered in the next subsection.

Impact of Radial Field Variation

In the case of the toroidal tester, using Ampere's law, the field intensity within the sample may be expressed as

$$H = \frac{N_m i_m}{2\pi r} \tag{6.5-14}$$

where the direction of H is circumferential around the toroid. Because of the radial field variation, knowledge of the flux through the sense winding is not sufficient in itself to calculate the flux density. A means of circumventing this problem was set forth in [1, 13].

Taking the approach of [13], we will assume that the anhysteretic magnetization may be expressed as

$$M_{an}(H) = \delta_H \sum_{k=1}^{K} \frac{m_k |H/h_k|}{1 + |H/h_k|^{n_k}} \tag{6.5-15}$$

where $\delta_H = 1$ if $H > 0$, $\delta_H = -1$ if $H < 0$, and $\delta_H = 0$ if $H = 0$. Further, we will take $n_1 = 1$, $n_k > 1$ $\forall k \in [2 \cdots K]$, and $h_k > 0$ $\forall k \in [1 \cdots K]$. From (6.5-14), (6.5-15), the definitions of flux (2.2-1) and flux linkage (2.5-1), we obtain

$$\lambda_{scan}(\boldsymbol{\theta}_M, i_m) = \frac{N_m N_s d}{2\pi} \left[\mu_0 \ln\left(\frac{r_{out}}{r_{in}}\right) + \sum_{k=1}^{K} \frac{m_k}{h_k n_k} \ln\left(\frac{|N_m i_m|^{n_k} + (2\pi r_{out} h_k)^{n_k}}{|N_m i_m|^{n_k} + (2\pi r_{in} h_k)^{n_k}}\right) \right] i_m \quad (6.5\text{-}16)$$

where the subscript "scan" stands for secondary computed anhysteretic and where $\boldsymbol{\theta}_M$ is a vector of parameters given by

$$\boldsymbol{\theta}_M = [m_1 \cdots m_K \ h_1 \cdots h_K \ n_2 \cdots n_K] \quad (6.5\text{-}17)$$

Using (6.5-16), we will find $\boldsymbol{\theta}_M$ such that the fitted anhysteretic flux linkage versus magnetizing current characteristic matches the measured characteristic as closely as possible. In other words, we will create a parameter identification problem and apply optimization techniques to minimize the error, thus identifying $\boldsymbol{\theta}_M$ and effectively identifying the M versus H characteristic.

In order to accomplish this, we will define the error at the jth data point to be

$$e_j = \frac{|\lambda_{san,j} - \lambda_{scan}(\boldsymbol{\theta}_m, i_{m,j})|}{\|\lambda_{san}\|_\infty} \quad (6.5\text{-}18)$$

where $\lambda_{san,j}$ is the jth measured anhysteretic flux linkage determined from (6.5-6), $\lambda_{scan}(\boldsymbol{\theta}_m, i_{m,j})$ is the jth computed anhysteretic flux linkage found using (6.5-16) with model parameters $\boldsymbol{\theta}_m$ and with $i_{m,j}$ being the jth measured anhysteretic magnetizing current found from (6.5-7), and finally with $\|\lambda_{san}\|_\infty$ being the maximum absolute value of all the $\lambda_{san,j}$ points. Associated with the error at point j is a weighting function suggested in [13] as

$$w_j = \left(\frac{|\lambda_{san,j}|}{\overline{|\lambda_{san}|}}\right)^3 \quad (6.5\text{-}19)$$

where $\overline{|\lambda_{san}|}$ is the mean of the absolute value of the elements of λ_{san}. This weighting factor is used to improve accuracy in the knee of the B–H curve. Taking the elements of e_j and w_j to form column vectors, the total error metric is taken as

$$E_M(\boldsymbol{\theta}_M) = \mathbf{e}^T \mathbf{w} \quad (6.5\text{-}20)$$

where E_M is a function of $\boldsymbol{\theta}_M$ since the error at each point is a function of $\boldsymbol{\theta}_M$, though this explicit dependence is not shown on the right-hand side of (6.5-20).

Before proceeding, it will be useful to consider some constraints. First, again following [13], it is desirable that

$$\frac{dM}{dH} > 0 \quad \forall H \in [0, \infty) \quad (6.5\text{-}21)$$

and that

$$\frac{d^2 M}{dH^2} < 0 \quad \forall H \in [0, \infty) \quad (6.5\text{-}22)$$

which requires that magnetization increases with increasing field intensity, but that the rate of that increase decreases as field intensity increases.

To implement these constraints, it is readily shown that

$$\frac{dM}{dH} = \sum_{k=1}^{K} \frac{m_k}{h_k} \frac{1 + (H/h_k)^{n_k}(1-n_k)}{(1+(H/h_k)^{n_k})^2} \quad H > 0 \quad (6.5\text{-}23)$$

and that

$$\frac{d^2M}{dH^2} = \sum_{k=1}^{K} \frac{m_k n_k}{h_k^2} \left(\frac{H}{h_k}\right)^{n_k-1} \frac{(n_k-1)(H/h_k)^{n_k} - (1+n_k)}{(1+(H/h_k)^{n_k})^3} \quad H > 0 \qquad (6.5\text{-}24)$$

As suggested in [13], in order to test the constraints (6.5-21) and (6.5-22), it is convenient to create a vector of test points $\mathbf{H}_t$ that are logarithmically distributed between a minimum value H_{mn} and a maximum value H_{mx}, perhaps from 1 to 10^7 A/m. Then, we create test functions

$$t_{M1}(\boldsymbol{\theta}_M) = \min\left(\left.\frac{dM}{dH}\right|_{H \in \mathbf{H}_t}\right) \qquad (6.5\text{-}25)$$

$$t_{M2}(\boldsymbol{\theta}_M) = \max\left(\left.\frac{d^2M}{dH^2}\right|_{H \in \mathbf{H}_t}\right) \qquad (6.5\text{-}26)$$

and corresponding constraint functions

$$c_{M1}(\boldsymbol{\theta}_M) = \begin{cases} 1 & t_{M1}(\boldsymbol{\theta}_M) > 0 \\ t_{M1}(\boldsymbol{\theta}_M) & t_{M1}(\boldsymbol{\theta}_M) \leq 0 \end{cases} \qquad (6.5\text{-}27)$$

$$c_{M2}(\boldsymbol{\theta}_M) = \begin{cases} 1 & t_{M2}(\boldsymbol{\theta}_M) < 0 \\ -t_{M2}(\boldsymbol{\theta}_M) & t_{M2}(\boldsymbol{\theta}_M) \geq 0 \end{cases} \qquad (6.5\text{-}28)$$

and

$$c_M(\boldsymbol{\theta}_M) = \frac{1}{2}(c_{M1}(\boldsymbol{\theta}_M) + c_{M2}(\boldsymbol{\theta}_M)) \qquad (6.5\text{-}29)$$

At this point, a fitness function can be defined. In particular, one possibility is

$$f_M(\boldsymbol{\theta}_M) = \begin{cases} \dfrac{1}{E_M(\boldsymbol{\theta}_M) + \varepsilon} & c_M(\boldsymbol{\theta}_M) = 1 \\ \varepsilon(c_M(\boldsymbol{\theta}_M) - 1) & c_M(\boldsymbol{\theta}_M) < 1 \end{cases} \qquad (6.5\text{-}30)$$

where ε is any small positive number, such as 10^{-3}. The desired parameter vector is obtained from

$$\boldsymbol{\theta}_M = \mathrm{argmax}(f_M(\boldsymbol{\theta}_M)) \qquad (6.5\text{-}31)$$

Note that once the parameters are obtained, it is very easy to find the permeability as a function of H. In particular, from (2.3-3) and (6.5-15), we have

$$\mu_h(H) = \mu_0 + \sum_{k=1}^{K} \frac{m_k}{h_k} \frac{1}{1+|H/h_k|^{n_k}} \qquad (6.5\text{-}32)$$

Example 6.5A Let us now consider characterizing a toroidal sample of MN80C ferrite. In particular, the sample has an inner radius of 44.53 mm, an outer radius of 69.98 mm, and a height of 25.57 mm. The number of magnetizing and sense turns are 24 and 4, respectively, and we will excite the sample magnetizing winding with a 100-Hz voltage waveform, which result in the magnetizing current and sense winding voltage shown in the first and second traces of Figure 6.16. In order to filter the waveforms, they were expressed as a Fourier series using the first 20 odd harmonics, and this series was used to represent the waveform. This series was based on five cycles of measured data (not all shown). The restriction to represent the data with only odd-harmonics enforces that desired

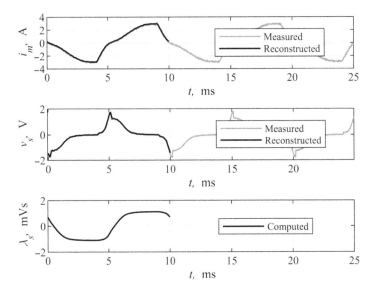

Figure 6.16 Characterization waveforms in example 6.5A.

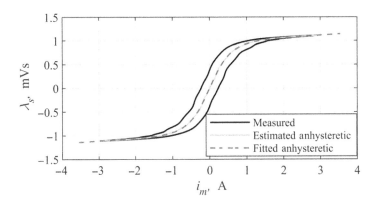

Figure 6.17 Sense winding flux linkage versus magnetizing current in example 6.5A.

symmetry. The original and reconstructed data are shown in Figure 6.16. Then, (6.5-1) was used to determine the sense flux linkage.

The resulting sense flux linkage versus magnetizing current plot is shown in Figure 6.17. Then, carrying out the procedure of (6.5-2)–(6.5-7), the anhysteretic flux linkage versus magnetizing characteristic was determined (which is also shown in Figure 6.16). Finally, (6.5-31) is solved using a single-objective genetic algorithm as an optimization engine. A two-term series was found to fit the data well (6.5-15), and so $\theta_M = [m_1 \; m_2 \; h_1 \; h_2 \; n_2]$, where the parameter range for the search was taken to be $0 \leq m_1 \leq 2\,\text{T}$, $-1 \leq m_2 \leq 2\text{T}$, $1 \leq h_k \leq 251\,\text{A/m}$, and $1.1 \leq n_2 \leq 5$. The magnetization terms were chosen to be linearly encoded genes; the remaining variables were logarithmically encoded. Running the optimization yielded $m_1 = 575.9\,\text{mT}$, $m_2 = 387.7\,\text{mT}$, $h_1 = 164.3\,\text{A/m}$, $h_2 = 58.70\,\text{A/m}$, and $n_2 = 1.952$. Recall that it is required that $n_1 = 1$. The resulting $\lambda_{s,an} - i_m$ characteristic is shown in Figure 6.17 and the B–H characteristic is shown in Figure 6.18. Therein, a box is shown, which

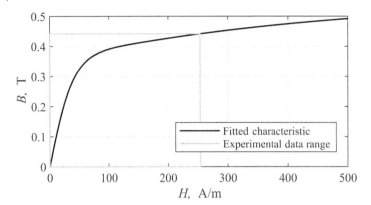

Figure 6.18 B–H characteristic in example 6.5A.

breaks apart the region of interpolation versus extrapolation. Clearly, there is always a danger in extrapolating, and so it is best to rigorously excite the sample. However, this can be overdone, in that representing extremely hard saturation can reduce the fidelity at lower flux levels for a given number of terms in (6.5-15).

Characterization of $\mu_B(B)$

It is often required that the permeability function be expressed as a function of flux density instead of field intensity, as we have done previously. To this end, we will assume the form described in Section 2.3 and (2.3-17)–(2.3-28). For this form, from (2.3-17), (2.3-19), and (2.3-20), one can see that the parameters to be identified may be arranged into a vector as

$$\boldsymbol{\theta}_\mu = [\mu_r \; \alpha_1 \cdots \alpha_K \; \beta_1 \cdots \beta_K \; \gamma_1 \cdots \gamma_K] \tag{6.5-33}$$

The remaining parameters of the model, namely, the values of δ_k, ε_k, and γ_k, may be expressed in terms of the parameters comprising (6.5-33) through (2.3-22)–(2.3-24).

We will also assume the knowledge of the anhysteretic B–H curve in terms of data points B_j and H_j, where $j \in [1 \cdots J]$ and J is the number of points. In the case of data acquired from an Epstein frame, or a single-sheet tester, these points could be acquired rather directly. In the case of the toroid tester, these points could be obtained from (2.3-11) and (6.5-32), and interpreting all field quantities as being anhysteretic.

Our approach to characterizing $\mu_B(B)$ will again be one of curve fitting through optimization, as set forth in [13]. Let us define the jth element of an error vector as

$$e_{\mu,j}(\boldsymbol{\theta}_\mu) = \frac{\|H_j - B_j/\mu_B(B_j)\|}{|H_j|} \tag{6.5-34}$$

In (6.5-34), the functional dependence of $\mu_B()$ on the parameter vector $\boldsymbol{\theta}_\mu$ is left implicit. Also, we will avoid any sample point wherein $H_j = 0$ in order to avoid a singularity. The overall error metric is expressed as

$$E_\mu(\boldsymbol{\theta}_\mu) = \frac{\|\mathbf{e}_\mu(\boldsymbol{\theta}_\mu)\|_2}{J} \tag{6.5-35}$$

Before minimizing $E_\mu(\boldsymbol{\theta}_\mu)$, however, we will introduce some constraints. First, it will be required that

$$\mu_B(B) > 0 \quad \forall B \tag{6.5-36}$$

Secondly, it is required that $\mu_B(B)$ be such that

$$\frac{dB}{dH} > \mu_0 \quad \forall B \in [0, \infty) \tag{6.5-37}$$

which physically implies that the incremental magnetization with respect to field intensity must be positive. Finally, we require

$$\frac{d^2B}{dH^2} < 0 \quad \forall B \in [0, \infty) \tag{6.5-38}$$

which is to say that the slope of the B–H characteristic decreases as H increases.

In order to test these constraints, we create a vector of test points $\mathbf{B}_t$, which are distributed between a minimum value B_{mn} to a maximum value B_{mx}. For example, we may take $B_{mn} = 0$ and $B_{mx} = 5T$ for a silicon steel. Next, we create test functions

$$t_{\mu 1}(\boldsymbol{\theta}_\mu) = \min\left(\mu_B(B)|_{B \in \mathbf{B}_t}\right) \tag{6.5-39}$$

$$t_{\mu 2}(\boldsymbol{\theta}_\mu) = \min\left(\left[\frac{\mu_B^2(B)}{\mu_B(B) - \frac{d\mu_B(B)}{dB}B} - \mu_0\right]\bigg|_{B \in \mathbf{B}_t}\right) \tag{6.5-40}$$

$$t_{\mu 3}(\boldsymbol{\theta}_\mu) = \max\left(\left[2\mu_B(B)\frac{d\mu_B(B)}{dB} + B\left(-2\left(\frac{d\mu_B(B)}{dB}\right)^2 + \mu_B(B)\frac{d^2\mu_B(B)}{dB^2}\right)\right]\bigg|_{B \in \mathbf{B}_t}\right) \tag{6.5-41}$$

These functions are used to test (6.5-36), (6.5-37), and (6.5-38), respectively. In (6.5-40), the term in brackets is dB/dH less μ_0 (see problem 9). In (6.5-41), the term in brackets is not d^2B/dH^2 but has the same sign (see problem 10). Note that $d\mu_B(B)/dB$ is given by (2.3-28). In (6.5-41), from (2.3-28) and (2.3-25)

$$\frac{d^2\mu_B(B)}{dB^2} = \frac{\mu_0}{(r(B)-1)^2}\left[\frac{2}{r(B)-1}\left(\frac{dg(B)}{dB}\right)^2 - \frac{d^2g(B)}{dB^2}\right] \tag{6.5-42}$$

where $r(B)$ is given by (2.3-19), $dg(B)/dB$ is given by (2.3-20), and

$$\frac{d^2g(B)}{dB^2} = \sum_{k=1}^{K} \frac{\eta_k \beta_k e^{-\beta_k|B|}}{(\theta_k + e^{-\beta_k|B|})^2} \tag{6.5-43}$$

We now introduce constraint functions that utilize the three test functions. In particular, we define the three constraints

$$c_{\mu 1}(\boldsymbol{\theta}_\mu) = \begin{cases} 1 & t_{\mu 1}(\boldsymbol{\theta}_\mu) > 0 \\ t_{\mu 1}(\boldsymbol{\theta}_\mu) & t_{\mu 1}(\boldsymbol{\theta}_\mu) \leq 0 \end{cases} \tag{6.5-44}$$

$$c_{\mu 2}(\boldsymbol{\theta}_\mu) = \begin{cases} 1 & t_{u2}(\boldsymbol{\theta}_\mu) > 0 \\ t_{u2}(\boldsymbol{\theta}_\mu) & t_{u2}(\boldsymbol{\theta}_\mu) \leq 0 \end{cases} \qquad (6.5\text{-}45)$$

$$c_{\mu 3}(\boldsymbol{\theta}_\mu) = \begin{cases} 1 & t_{u3}(\boldsymbol{\theta}_\mu) < 0 \\ -t_{u3}(\boldsymbol{\theta}_\mu) & t_{u3}(\boldsymbol{\theta}_\mu) \geq 0 \end{cases} \qquad (6.5\text{-}46)$$

and combine them with

$$c_\mu(\boldsymbol{\theta}_\mu) = \frac{1}{3}\left(c_{\mu 1}(\boldsymbol{\theta}_\mu) + c_{\mu 2}(\boldsymbol{\theta}_\mu) + c_{\mu 3}(\boldsymbol{\theta}_\mu)\right) \qquad (6.5\text{-}47)$$

Next, we define our fitness function as

$$f_u(\boldsymbol{\theta}_\mu) = \begin{cases} \dfrac{1}{E_\mu(\boldsymbol{\theta}_\mu) + \varepsilon} & c_\mu(\boldsymbol{\theta}_\mu) = 1 \\ \varepsilon\left(c_\mu(\boldsymbol{\theta}_\mu) - 1\right) & c_\mu(\boldsymbol{\theta}_\mu) < 1 \end{cases} \qquad (6.5\text{-}48)$$

whereupon

$$\boldsymbol{\theta}_\mu = \mathrm{argmax}\left(f_u(\boldsymbol{\theta}_\mu)\right) \qquad (6.5\text{-}49)$$

Example 6.5B In this example, we will extend Example 6.5A by finding $\mu_B(B)$. In particular, we will carry out (6.5-49). Here again, we will use a single-objective genetic algorithm. The first step is to sample the anhysteretic B–H characteristic. Logarithmically distributing 500-point H values between 1 and 1000 A/m and generating corresponding B values using (2.3-11) and (6.5-32) yield the B–H characteristic shown in Figure 6.19 and labeled as "original." Our next step will be to fit to this data. Using a bit of trial-and-error, we will arrive at using two terms so that $\boldsymbol{\theta}_\mu = [\mu_r \ \alpha_1 \ \alpha_2 \ \beta_1 \ \beta_2 \ \gamma_1 \ \gamma_2]$. The gene range for the parameters is taken to be $10 \leq \mu_r \leq 10^6$, $10^{-6} \leq \alpha_k \leq 10^3 1/T$, $10^{-1} \leq \beta_k \leq 10^3 1/T$, and $10^{-2} \leq \gamma_k \leq 10^1$ T. These ranges are quite large and so all genes were logarithmically encoded. The suitability of the ranges was deemed acceptable through inspection of the gene distribution of the final plots. The genetic algorithm used a population size of 1000 over 1000 generations. Carrying out the optimization yields $\mu_r = 8024$, $\alpha_1 \leq 27.31 \cdot 10^{-3} 1/T$, $\alpha_2 \leq 163.3 \cdot 10^{-6} 1/T$, $\beta_1 = 27.631/T$, $\beta_2 = 26.731/T$, $\gamma_1 = 454.1$ mT, and $\gamma_2 = 86.35$ mT. The resulting B–H characteristic is also shown in Figure 6.19 and labeled as "Fit." As can be seen, the representation of the original data set seems reasonable.

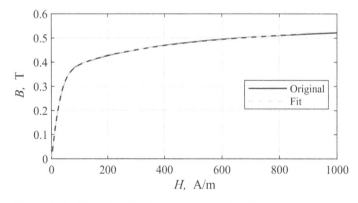

Figure 6.19 Example 6.5B characterization of $\mu_B(B)$.

6.6 Characterizing Behavioral Loss Models

Measuring Power Loss Density

We know from our work in Section 2.7 and specifically (2.7-14) or by taking the time derivative of (6.2-11) that the instantaneous power density into magnetic core may be expressed as

$$p = H \frac{dB}{dt} \tag{6.6-1}$$

Thus, the average power loss density with periodic excitation may be expressed as

$$\bar{p} = \frac{1}{t_f - t_0} \int_{t_0}^{t_f} H \frac{dB}{dt} dt \tag{6.6-2}$$

where t_0 is a time such that the system is in the periodic steady state and t_f is a time integer number of cycles after t_0.

In test stands wherein B is assumed to be spatially uniform across the magnetic cross section

$$v_s = N_s a_{mcs} \frac{dB}{dt} \tag{6.6-3}$$

so, (6.6-2) becomes

$$\bar{p} = \frac{1}{N_s a_{mcs}} \frac{1}{t_f - t_0} \int_{t_0}^{t_f} H v_s dt \tag{6.6-4}$$

which is the expression for power loss density that would be used in the case of a single- or double-sheet tester, where both H and v_s are measured waveforms (with instrumentation dc offsets removed).

In the case of an Epstein frame, we know that

$$H = \frac{N_m i_m}{l_{mp}} \tag{6.6-5}$$

so, (6.6-4) becomes

$$\bar{p} = \frac{N_m}{N_s a_{mcs} l_{mp}} \frac{1}{t_f - t_0} \int_{t_0}^{t_f} i_m v_s dt \tag{6.6-6}$$

again removing dc offsets from all measured waveforms since our source will be designed such that these cannot exist.

In the case of the toroidal tester, the power loss density varies radially. From (6.1A-10), we have that the spatial mean of the temporal-average power loss density is given by

$$\langle \bar{p} \rangle = \frac{1}{\pi (r_o^2 - r_i^2) d} \frac{N_m}{N_s} \frac{1}{t_f - t_0} \int_{t_0}^{t_f} i_m v_s dt \tag{6.6-7}$$

We could also have obtained this result by taking the spatial average of (6.6-2) along with (6.5-15), the definition of flux linkage, and Faraday's law.

Loss Characterization Assuming Uniform Fields

Now, let us consider the problem of determining the parameters of a loss model, which we will accomplish by curve fitting. The first step in this approach will be to collect loss data under a wide variety of conditions, wherein we vary fundamental frequency, amplitude, and waveshape. We will assume that we have N power loss measurements and denote the power loss density of the nth measurement as $\bar{p}_n$. For each waveform obtained, we will remove any instrumentation-induced dc offsets in the waveforms and then use (6.6-4), (6.6-6), or (6.6-7) to determine the loss associated with a given flux density waveform.

Let $\mathbf{B}_n$ denote the time history of the nth flux density waveform (so formally, it includes temporal corresponding flux density data), and $p_x(\mathbf{B}, \boldsymbol{\theta}_x)$ be a function that calculates power loss density of model x in terms of a flux density waveform $\mathbf{B}$ and model parameters $\boldsymbol{\theta}_x$. For example, we may have $x = $ "sMSE" model, whereupon

$$\boldsymbol{\theta}_{sMSE} = \begin{bmatrix} k_{h,1} \cdots k_{h,K} & \alpha_{h,1} \cdots \alpha_{h,K} & \beta_{h,1} \cdots \beta_{h,K} & k_e \end{bmatrix}^T \tag{6.6-8}$$

Next, we will define the error associated with the nth data point as

$$e_n(\boldsymbol{\theta}_x) = w_n \left| \frac{\bar{p}_n - p_x(\mathbf{B}_n, \boldsymbol{\theta}_x)}{\bar{p}_n} \right| \tag{6.6-9}$$

where w_n is a weight associated with the nth point. Clearly, this is not the only choice. However, it has been found to be effective.

An algorithm to determine the weights is as follows. First, the waveforms are sorted into N_b bins based on the peak absolute value of the flux density waveforms. Let o_j be the number of operating points that are classified into the jth bin. Now, suppose that the nth operating point falls into bin $b(n)$. Then, the weight of nth operating point is taken to be

$$w_n = \frac{1}{o_{b(n)} \sum_{j=1}^{N_b} \frac{1}{o_j}} \tag{6.6-10}$$

Essentially, the weighting provided by (6.6-10) makes it seem to the error function that the distribution of operating point samples is uniform with regard to flux density level. It is readily shown that the sum of the weights over all operating points is 1.

Once the error of all operating points is found, the fitness functions are defined as

$$f(\boldsymbol{\theta}_x) = \frac{1}{\sqrt{\frac{1}{N} \sum_{n=1}^{N} e_n^2(\boldsymbol{\theta}_x)}} \tag{6.6-11}$$

whereupon the optimal parameter vector is given by

$$\boldsymbol{\theta}_x^* = \operatorname{argmin}(f(\boldsymbol{\theta}_x)) \tag{6.6-12}$$

Example 6.6A Let us again consider the MN80c toroid whose parameters are listed in Example 6.5A. In this example, we will take loss measurements and use the procedure of this section, culminating with (6.6-12), to determine the MSE parameters for the material. The first aspect is to collect data. In particular, 50 waveforms were collected with fundamental frequencies ranging from 100 to 17.36 kHz and with peak flux densities ranging from 66 to 390 mT. Over these 50 waveforms, the power loss density varied from 0.207 to 182 W/l. A genetic algorithm was used to perform the

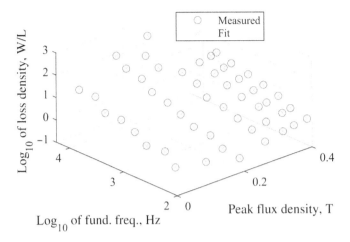

Figure 6.20 Example 6.8A measured and fitted losses.

optimization (6.6-12), using a population size of 1000 and 1000 generations. Attempting a single term, the parameter ranges selected were $-10^4 \leq k_{h,1} \leq 10^4$, $0.5 \leq \alpha_{h,1} \leq 5$, $0.1 \leq \beta_{h,1} \leq 5$, and $10^{-10} \leq k_e \leq 10^2$. The reader may question the negative part of the domain of $k_{h,1}$; this is sometimes useful when using more than one term. The first three parameters were linearly encoded and the last logarithmically. Again, it is appropriate to check the gene distribution at the end of a run to make sure that there is no clustering against a limit. Carrying out the optimization yields $k_{h,1} = 0.1663 \text{J/l}$, $\alpha_{h,1} = 1.005$, $\beta_{h,1} = 2.129$, and $k_e = 1.050 \cdot 10^{-13}$ Js/l. This last value is negligible. The maximum loss error in representing any point of the data is 7.6%; the mean error in predicting loss over all points is 2.5%. This loss can be reduced somewhat by increasing the number of terms in the sMSE model. Figure 6.20 depicts the measured and fitted losses.

Loss Characterization with Spatially Varying Fields

When using the toroid tester, the loss density is clearly a function of radius. To account for this, we will break the toroid into S sub-toroids by radius. These regions are arranged such that they do not overlap, but the union of the S sub-toroid regions incapsulates the entire toroid. The subscript s will be used to denote region s. We will denote the inner, mean, and outer radii of the sth region as $r_{in,s}$, $r_{mn,s}$, and $r_{out,s}$, respectively. One way to assign these radii is

$$r_{in,s} = r_{in} \kappa^{s-1} \qquad (6.6\text{-}13)$$

$$r_{out,s} = r_{in} \kappa^s \qquad (6.6\text{-}14)$$

where

$$\kappa = \sqrt[S]{\frac{r_{out}}{r_{in}}} \qquad (6.6\text{-}15)$$

where r_{in} and r_{out} are the inner and outer radii, respectively, as shown in Figure 6.14. Alternate divisions are possible; however, the proposed selection keeps the ratio of the outer path length to the inner magnetic path length the same for each segment. The mean radius, magnetic path length, magnetic cross section, and the magnetic volume of the sth segment are given by

$$r_{mn,s} = \frac{1}{2}(r_{in,s} + r_{out,s}) \tag{6.6-16}$$

$$l_{mp,s} = 2\pi r_{mn,s} \tag{6.6-17}$$

$$a_{mc,s} = d(r_{out,s} - r_{in,s}) \tag{6.6-18}$$

$$V_{m,s} = \pi d(r_{out,s}^2 - r_{in,s}^2) \tag{6.6-19}$$

where d is the depth of the toroid into the page.

For a given power loss measurement, we will need to determine how much power is lost in each segment for a given loss model. To determine this, we must determine the flux density in each segment. On approach to do this is as follows. Let us assume that we have already characterized the anhysteretic $\mu_H()$. The spatial mean of the field intensity waveform in the sth sub-toroid is given by

$$\langle \mathbf{H}_s \rangle = \frac{N_m \mathbf{i}_m}{l_{mp,s}} \tag{6.6-20}$$

whereupon an estimate for the spatial mean of the flux density is given by

$$\langle \mathbf{B}_{est,s} \rangle = \mu_H(\langle \mathbf{H}_s \rangle) \langle \mathbf{H}_s \rangle \tag{6.6-21}$$

In evaluating (6.6-21), recall that $\mu_H()$ is a function and so the parentheses do not imply multiplication. Here, the flux density is an estimate because of any error in the permeability function, not to mention the impact of hysteresis. Thus, the flux in the sth sub-toroid may be estimated as

$$\Phi_{est,s} = \langle \mathbf{B}_{est,s} \rangle a_{mc,s} \tag{6.6-22}$$

It follows that the fraction of the flux in the sth segment is given by

$$\mathbf{w}_s = \frac{\Phi_{est,s}}{\sum_{s=1}^{S} \Phi_{est,s}} \tag{6.6-23}$$

It should be kept in mind that $\mathbf{w}_s$ is a time-varying quantity. In the absence of hysteresis and saturation, it would be time-independent.

Next, we will assume that $\mathbf{w}_s$, which is based on an estimate of the flux density, is a reasonable estimate of the fraction of the measured flux in the sth sub-toroid. Thus, we form a "corrected" estimate of the flux density as

$$\langle \mathbf{B}_{cest,s} \rangle = \frac{\mathbf{w}_s \lambda_s}{a_{mc,s} N_s} \tag{6.6-24}$$

Note that the multiplications in (6.6-24) are pointwise (element-by-element). Also, the subscript "s" on λ and N denotes "sense" winding, whereas the other occurrences are indices.

The reason that $\langle \mathbf{B}_{cest,s} \rangle$ is used rather than $\langle \mathbf{B}_{est,s} \rangle$ is several-fold. First, consider a magnetically linear case. In this case, one can show if one used the incorrect permeability, $\langle \mathbf{B}_{cest,s} \rangle$ would be calculated exactly, whereas $\langle \mathbf{B}_{est,s} \rangle$ would have an error. Secondly, when magnetic saturation and hysteresis are present, $\langle \mathbf{B}_{cest,s} \rangle$ will increase and decrease whenever λ_s increases or decreases, as one would expect. The same cannot be said of $\langle \mathbf{B}_{est,s} \rangle$. Thus, $\langle \mathbf{B}_{cest,s} \rangle$ is a much more robust estimate of the flux density in the sub-toroids than $\langle \mathbf{B}_{est,s} \rangle$.

At this point, the power loss in the entire toroid may be calculated as

$$\overline{P}_{x,n}(\mathbf{B}_{cest,s,n}, \boldsymbol{\theta}_x) = \sum_{s=1}^{S} \overline{p}_x(\langle \mathbf{B}_{cest,s,n} \rangle, \boldsymbol{\theta}_x) V_{m,s} \tag{6.6-25}$$

where $\overline{p}_x(\mathbf{B}, \boldsymbol{\theta}_x)$ is our power loss density for loss model "x," $\mathbf{B}$ is the time history of the flux density waveform, and $\boldsymbol{\theta}_x$ is the vector of parameters associated with the model. We can then formulate the error associated with a loss measurement $\mathbf{n}$ as

$$e_n(\boldsymbol{\theta}_x) := \left| \frac{\overline{P}_n - \overline{P}_{x,n}(\mathbf{B}_{cest,s,n}, \boldsymbol{\theta}_x)}{\overline{P}_n} \right| \tag{6.6-26}$$

which is analogous to (6.6-9) except it uses power loss instead of power loss density and forgoes the use of a weighting function, which could be included if desired. In this expression, $\mathbf{B}_{cest,s,n}$ is the flux density waveform for sub-toroid s and power loss measurement n.

Finally, we can calculate a fitness function using (6.6-11) and find a model parameter vector using (6.6-12).

Example 6.6B In this example, we reconsider Example 6.6A, except that we apply the wide toroid method. Breaking up the toroid into 12 sub-toroids, we arrive at $k_{h,1} = 0.1664$ J/l, $\alpha_{h,1} = 1.005$, $\beta_{h,1} = 2.123$ $k_e = 1.000 \cdot 10^{-13}$ Js/l. This last value is negligible. The maximum loss error in representing any point of the data is 7.7%; the mean error in predicting loss over all points is 2.5%. To three significant digits, these are the same results as obtained in Example 6.6A. Although the subject toroid is not particularly wide (the ratio of the outer radius to inner radius is 1.57), one would have expected a bigger difference. One reason for this may be that the error that is present is dominated by other factors, such as temperature variation, when collecting the samples. As one would expect, on a very wide toroid, the difference is more pronounced.

6.7 Time-Domain Loss Modeling: the Preisach Model

While the approaches of Section 6.3 are very useful and are the most often used approaches for loss modeling, there are other approaches to the modeling of hysteresis behavior that are more accurate, especially in the case of complicated excitation waveforms. These include the Preisach model and the Jiles–Atherton model. These models are quite different from those we have been discussing in that they are time-domain models that predict the specific B–H trajectory. They offer significantly more information than the empirical models discussed in Section 6.3. They are inherently able to accommodate different input waveforms, including losses resulting from minor loops.

We will begin with the Preisach model. In this approach, a magnetic material is represented by a number of elements referred to as hysterons. Each hysteron has two magnetization states, $+M_h$ or $-M_h$, as illustrated in Figure 6.21. Each hysteron is characterized by the value of two parameters, U and V. The magnetization state of a hysteron is $+M_h$ if H becomes greater than U. Likewise, the magnetization state becomes $-M_h$ if H becomes less than V. Again, each hysteron may have unique values for U and V.

The material magnetization is then taken to be equal to

$$M = \int_{-\infty}^{\infty} \int_{-\infty}^{U} Q(U,V) P(U,V) dV dU \tag{6.7-1}$$

In (6.7-1), $Q(U,V)$ describes the current status of hysterons with parameters U and V as 1 if positively magnetized and -1 if negatively magnetized. The function $P(U,V)$ describes the hysteron

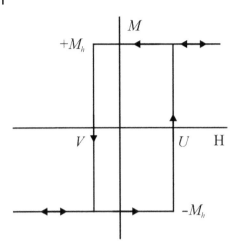

Figure 6.21 Hysteron behavior.

density. Note that M_h does not factor into the magnetization calculation. This is because the density function offers enough degrees of freedom to eliminate the need for another parameter.

With our definition of a hysteron, P(U, V) is zero for $V > U$. Assuming all the hysterons are in a positively magnetized state, the saturated magnetization may be expressed as

$$M_s = \int_{-\infty}^{\infty} \int_{-\infty}^{U} P(U,V) dV dU \qquad (6.7\text{-}2)$$

It will be convenient to define the normalized magnetization as

$$m = \frac{M}{M_s} \qquad (6.7\text{-}3)$$

Similarly, the normalized hysteron density function may be expressed as

$$p(U,V) = \frac{1}{M_s} P(U,V) \qquad (6.7\text{-}4)$$

In order to understand the Preisach model, consider the situation in Figure 6.22. Therein, the system is shown in four different states. In each case, the magnetization is shown as a function of U and V. Light shading indicates negatively magnetized hysterons and dark shading indicates positively magnetized hysterons.

Consider the system in state 1. Therein, we assume that $H = -\infty$ and system has all hysterons in the negative state. Thus, the normalized magnetization for state 1 is $m_{s1} = -1$.

Now, let us suppose that H increases from $-\infty$ to a value H_1. All hysterons for which $U < H$ will change state. The normalized magnetization for state 2 may be expressed as

$$m_{s2} = m_{s1} + 2 \int_{-\infty}^{H_1} \int_{-\infty}^{U} p(U,V) dV dU \qquad (6.7\text{-}5)$$

The factor of 2 in (6.7-5) arises since each hysteron changes from a negative state to a positive state.

Now, suppose the field intensity decreases to a value $H_2 < H_1$. All hysterons with $V > H$ will change state. Thus, the magnetization for state 3 may be expressed as

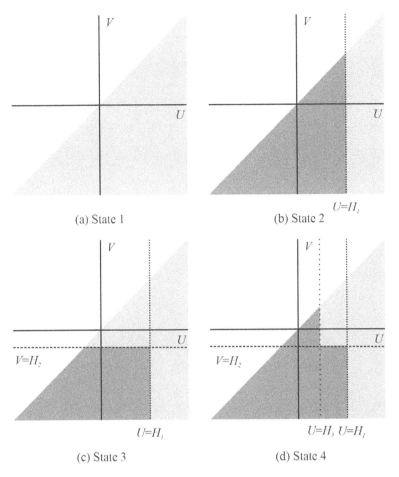

Figure 6.22 Example magnetization states.

$$m_{s3} = m_{s2} - 2 \int_{H_2}^{H_1} \int_{V}^{H_1} p(U,V) dU dV \tag{6.7-6}$$

To end our experiment, let us suppose that we next increase H to a value H_3 such that $H_2 < H_3 < H_1$. As H increases, all hysterons for which $U < H$ will change state. Thus, the magnetization increases to

$$m_{s4} = m_{s3} + 2 \int_{H_2}^{H_3} \int_{H_2}^{U} p(U,V) dV dU \tag{6.7-7}$$

As can be seen, recording of transition points is a key part of the method. To proceed, let us now re-examine the transition from state 3 to state 4. An enlarged view of this transition is shown in Figure 6.23(a). Therein, the white triangle is the region that is changing magnetization, and the very dark region indicates incremental area. After state 3, as the field increases, the magnetization is expressed as

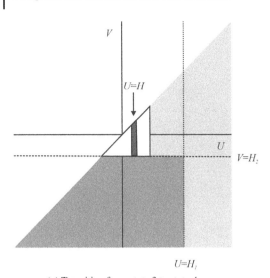

Figure 6.23 Incremental magnetization.

(a) Transition from state 3 to state 4

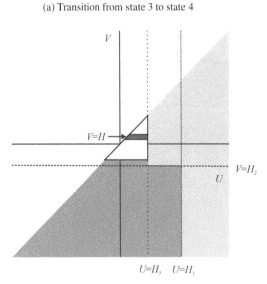

(b) Decrease in H after state 4

$$m = m_{s3} + 2 \int_{H_2}^{H} \int_{H_2}^{U} p(U,V) dV dU \qquad (6.7\text{-}8)$$

If we were to set $H = H_3$, (6.7-8) becomes (6.7-7). Taking the derivative of (6.7-8) with respect to H, we have that

$$\frac{dm}{dH} = 2 \int_{H_2}^{H} p(H,V) dV \qquad (6.7\text{-}9)$$

We may generalize (6.7-9) to apply whenever H is increasing provided we interpret H_2 as the largest field intensity minimum less than the current value of H.

Figure 6.23(b) shows the situation if the field intensity were to decrease after reaching state 4. For this situation, the magnetization would be given by

$$m = m_{s4} - 2 \int_H^{H_3} \int_V^{H_3} p(U,V)\,dU\,dV \qquad (6.7\text{-}10)$$

Again differentiating, we have

$$\frac{dm}{dH} = 2 \int_H^{H_3} p(U,H)\,dU \qquad (6.7\text{-}11)$$

As with (6.7-9), we can generalize (6.7-11) by interpreting H_3 as the minimum of the previous maxima less than the current value of H. In our example, if H were to fall less than H_3, then it would be replaced by H_1 in our integral.

In order to use this method, we may determine the B–H trajectory by solving the differential equation

$$\frac{dm}{dt} = \frac{dm}{dH}\frac{dH}{dt} \qquad (6.7\text{-}12)$$

where dm/dH is found using (6.7-9) or (6.7-11) depending upon the sign of dH/dt. However, in so doing, a list of the relevant maxima and minima of the field intensity needs to be addressed, so that the proper limits of integration can be evaluated in the cited equations. From the normalized magnetization, the actual magnetization and flux density may be calculated as

$$M = M_s m \qquad (6.7\text{-}13)$$
$$B = \mu_0 H + M \qquad (6.7\text{-}14)$$

From the B–H trajectory, the power loss density may be ascertained using (6.6-2).

Before concluding this section, the information presented here with regard to the Preisach model is intended to give the reader a flavor of the method, which involves numerous subtleties. The interested reader is referred to a book devoted to the subject [14].

6.8 Time-Domain Loss Modeling: the Extended Jiles–Atherton Model

An alternate approach to the time-domain modeling of magnetic behavior is the Jiles–Atherton model. This approach is very attractive in that it represents hysteresis with a first-order ordinary differential equation. The classical Jiles–Atherton model and numerous variations have been developed and are described in [15–21]. However, one drawback of the method is that some parts of the model derivation are questionable [22], and the model can predict nonphysical behavior.

As an alternative, herein we will study a variation known as the extended Jiles–Atherton model [23]. Before starting this discussion, it should be mentioned that in the derivation of the classical Jiles–Atherton model, magnetization M is expressed in units of A/m, so that

$$B = \mu_0(H + M) \qquad (6.8\text{-}1)$$

Herein, and in the extended Jiles–Atherton model, M will take on units of T so that

$$B = \mu_0 H + M \tag{6.8-2}$$

The basic premise of the classical Jiles–Atherton model is that

$$\begin{pmatrix} \text{Energy Into} \\ \text{Lossy Sample} \end{pmatrix} = \begin{pmatrix} \text{Energy Stored in} \\ \text{Lossless Sample} \end{pmatrix} + \text{Loss} \tag{6.8-3}$$

which, on the surface, seems reasonable.

Suppose we have a magnetic material and increase the flux density from an initial value of B_i to a final value of B_f. From (6.2-11), the energy density transferred into a lossy sample may be expressed as

$$e_{in} = \int_{B_i}^{B_f} H\, dB \tag{6.8-4}$$

From (6.2-11), the energy density that would have been stored in a lossless sample is given by

$$e_{s,lossless} = \int_{B_i}^{B_f} H_{an}\, dB \tag{6.8-5}$$

where H_{an} is the anhysteretic field intensity – that is, the field intensity that would be present at a given value of B in the absence of magnetic hysteresis. In particular, from (2.3-11),

$$B = \mu_B(B) H_{an} \tag{6.8-6}$$

or, alternately, from (2.3-12)

$$B = \mu_H(H_{an}) H_{an} \tag{6.8-7}$$

Also, from (6.8-2),

$$B = \mu_0 H_{an} + M_{an} \tag{6.8-8}$$

In a mathematically minor but physically very significant difference between the classical Jiles–Atherton model and the extended Jiles–Atherton model, the extended model assumes the energy stored in a lossy sample to be different from that stored in a lossless sample. Thus, our energy balance equation becomes

$$\begin{pmatrix} \text{Energy Into} \\ \text{Lossy Sample} \end{pmatrix} = \begin{pmatrix} \text{Energy Stored in} \\ \text{Lossy Sample} \end{pmatrix} + \text{Loss} \tag{6.8-9}$$

Basically, the premise of the extended Jiles–Atherton model is that the presence of magnetic hysteresis fundamental changes the amount of energy that can be stored in a material.

In [23], the energy stored in a lossy sample is expressed as

$$e_{s,lossy} = \int_{B_i}^{B_f} s H_{an}\, dB \tag{6.8-10}$$

where s is referred to as the energy storage factor, which is itself a function of the level of magnetization. Symbolically, (6.8-9) may be expressed as

$$e_{in} = e_{s,lossy} + e_{loss} \tag{6.8-11}$$

To proceed, further, we must derive an expression for the energy lost, which is the same for the classical and extended Jiles–Atherton models. In the Jiles–Atherton models, it is assumed that the magnetization has two components – an irreversible component that is set by pinning sites, which hold the domain walls in place, and a reversible component, which is associated with the bending of the domain walls. The irreversible and reversible components of the magnetization are denoted M_{irr} and M_{rev}, respectively, and the total magnetization is the sum of these components. Thus,

$$M = M_{irr} + M_{rev} \tag{6.8-12}$$

Now, let us consider irreversible magnetization. Let us consider a sample of magnetic material of volume V. For the sake of simplicity, let us further assume that there are only two directions for the magnetization of the domains – aligned and counter-aligned. Let us suppose that the total volume of the aligned domains is V_a and the total volume of the counter-aligned domains is V_c. The spatial average of the irreversible component of magnetization may be expressed as

$$\langle M_{irr} \rangle = \frac{1}{V}(V_a M_s - V_c M_s) \tag{6.8-13}$$

Actually, in general, all of the field quantities we have been working with can be viewed as spatial averages, but here, we are highlighting that fact. Note that because we are only considering two domain orientations,

$$V = V_a + V_c \tag{6.8-14}$$

Now, suppose that we move some domain walls and thus change the volume of aligned domains by ΔV_a. We would then also change the volume of the counter-aligned domains by $-\Delta V_a$. The change in irreversible magnetization is given by

$$\Delta \langle M_{irr} \rangle = 2M_s \frac{\Delta V_a}{V} \tag{6.8-15}$$

In the Jiles–Atherton model, it is assumed that the domain walls are held in place by pinning sites. If the volumetric density of the pinning sites is ρ_p, and the energy required to move the domain wall off a pinning site is e_p, then the energy density loss associated with our experiment is

$$e_{loss} = \frac{\Delta V_a}{V} \rho_p e_p \delta \tag{6.8-16}$$

In (6.8-16), δ is positive if ΔV_a is positive, negative if ΔV_a is negative, and zero otherwise. This ensures that the loss is always positive. Combining (6.8-15) and (6.8-16) yields

$$\frac{e_{loss}}{\Delta \langle M_{irr} \rangle} = \frac{\rho_p e_p}{2M_s} \delta \tag{6.8-17}$$

Interpreting (6.8-17) in differential form, we conclude that the loss may be expressed as

$$e_{loss} = \int_{M_{irr,i}}^{M_{irr,f}} k \delta \, dM_{irr} \tag{6.8-18}$$

where

$$k = \frac{p_p e_p}{2M_s} \tag{6.8-19}$$

In (6.8-17), δ is positive if the field is increasing, negative if it is decreasing, and zero otherwise. We will refer to k as the pinning site factor. Like the energy storage factor, it will also prove to be a function of magnetization.

At this point, from (6.8-4), (6.8-10), (6.8-11), and (6.8-18),

$$\int_{B_i}^{B_f} H\,dB = \int_{B_i}^{B_f} sH_{an}\,dB + \int_{M_{irr,i}}^{M_{irr,f}} k\delta\,dM_{irr} \tag{6.8-20}$$

Differentiating (6.8-20) with respect to flux density yields

$$H = sH_{an} + k\delta\frac{dM_{irr}}{dB} \tag{6.8-21}$$

Further manipulation yields

$$\frac{dM_{irr}}{dt} = \frac{H - sH_{an}}{k}\left|\frac{dB}{dt}\right| \tag{6.8-22}$$

As can be seen, the irreversible magnetization is governed by a first-order differential equation. To proceed, to use this relationship, we need to address several points. First, we need to address the pinning site factor. In [23], the following empirical form is proposed:

$$k = a_k e^{-M_{an}^2/b_k} \tag{6.8-23}$$

where a_k and b_k are material constants. Note that in these terms, the subscript k is not an index but rather indicates that these constants are associated with the calculation of the variable k.

Likewise, the energy storage factor is also represented empirically as

$$s = c_s + a_s e^{-M_{an}^2/b_s} \tag{6.8-24}$$

where a_s, b_s, and c_s are material constants. Here again, the subscript s is to indicate an association with the variable s rather than an index. It is shown in [23] that the model works reasonably well if k and s are taken to be constants. However, the accuracy is significantly improved if the forms (6.8-23) and (6.8-24) are used. While not yet evident, we will find that by making k and s a function of anhysteretic magnetization, rather than some other quantity (such as the irreversible magnetization), considerable computational advantage is obtained.

While (6.8-22) governs irreversible magnetization, we must also address reversible magnetization, so-named because it can be changed without loss, since, while domain walls are bent, the active pinning sites are not moved. In the Jiles–Atherton models, it is assumed that the reversible component of magnetization is expressed as

$$M_{rev} = c(M_{an} - M) \tag{6.8-25}$$

Thus, from (6.8-25), if the magnetization is less than the anhysteretic magnetization, the domain walls "bend" to increase the reversible magnetization. Manipulating (6.8-12) and (6.8-25), we obtain

$$M_{rev} = \frac{c}{1+c}(M_{an} - M_{irr}) \tag{6.8-26}$$

At this point, we can simplify our model. Through algebraic manipulation of (6.8-8), (6.8-22), and (6.8-26), it can be shown that

$$\frac{dM_{irr}}{dt} = aM_{irr} + b \tag{6.8-27}$$

where

$$a = -\frac{1}{(1+c)k\mu_0}\left|\frac{dB}{dt}\right| \tag{6.8-28}$$

and

$$b = \frac{\left(s - \frac{c}{1+c}\right)M_{an} + (1-s)B}{k\mu_0}\left|\frac{dB}{dt}\right| \tag{6.8-29}$$

The relationship (6.8-27) is essentially a state-space form for a first-order system with unity input. However, the coefficients a and b are not constants and instead given by somewhat involved expressions. However, from a numerical solution point of view, this is irrelevant.

To see this, suppose we have vectors to represent B and its time derivative dB/dt at points of time t stored in a time vector. From (6.8-6) and (6.8-8), we could then determine the time history of M_{an} and then, using (6.8-23) and (6.8-24), of k and s. Finally, we could establish the time history of a and b using (6.8-28) and (6.8-29), all a priori to solving the ordinary differential equation (6.8-27). What we are left with is to solve a linear, albeit time-varying, first-order ordinary differential equation.

While (6.8-27) could be solved with any simulation engine, for the purpose of rapid evaluation, it is convenient to combine the differential equation with a solution algorithm. To this, let t_i denote the ith time instant of discretized data. We will denote the ith time step as

$$h_i = t_{i+1} - t_i \tag{6.8-30}$$

Note that while in some time-domain simulations, the time step is fixed. However, utilization of a variable time step is often highly advantageous, especially in switched systems (in our case, the switching will arise from the switching of power semiconductors in power electronic applications).

If one applies the backward trapezoidal method to (6.8-27), one obtains the discrete-time equation

$$M_{irr,i+1} = a_{D,i}M_{irr,i} + b_{D,i} \tag{6.8-31}$$

where

$$a_{D,i} = \frac{2 + h_i a_i}{2 - h_i a_{i+1}} \tag{6.8-32}$$

and

$$b_{D,i} = h_i \frac{b_i + b_{i+1}}{2 - h_i a_{i+1}} \tag{6.8-33}$$

In (6.8-32) and (6.8-33), a_i and b_i are values of a and b given by (6.8-28) and (6.8-29), respectively, at $t = t_i$. The initial condition $M_{irr,1}$ is generally unknown. However, if one approximates $M_{irr,1}$ by B_1, convergence is generally quite rapid, allowing a steady-state waveform to be achieved.

Once the waveform for M_{irr} is known, corresponding values of M_{rev}, M, and H can be found from (6.8-26), (6.8-12), and (6.8-2), respectively. Then, the power loss density may be found by numerical integration of (6.6-2), where t_0 is an initial time such that the system is in quasi steady state, and t_f is a final time such that an integer number of cycles have passed since t_0. In many power electronic applications, there is no true periodic steady state and so these definitions must be applied somewhat loosely. Note that it has been assumed that dB/dt is known as an input waveform to our analysis and so numerical differentiation is not carried out as part of (6.8-33) – otherwise, it would be better to reformulate the integral.

Because we will be using this model in Chapter 15 for the design of common-mode inductors, it is convenient to encapsulate it as

$$[p, \mathbf{H}] = f_{EJA}(\mathbf{t}, \mathbf{B}, p\mathbf{B}, \mathbf{D}_{cr}) \qquad (6.8\text{-}34)$$

where $\mathbf{t}$, $\mathbf{B}$, and $p\mathbf{B}$ are the input vectors of time, flux density, and the time derivative of flux density, respectively, $\mathbf{D}_{cr}$ is the structure of core parameter (including those of the EJA model), $\mathbf{H}$ is an output vector with the time history of the field intensity, and p is the power loss density.

It should be noted that while the EJA model has been found to perform well when predicting single-looped waveforms, it has been found to sometimes perform poorly in situations where both major and minor loops are simultaneously present.

Example 6.8A In this example, we will use (6.8-34) to predict core loss and B–H trajectories for a grade of ferrite (MN8CX) using the parameters set forth in [23]. In particular, $a_k = 11.424$ A/m, $b_k = 0.156$ T^2, $c = 12498$, $a_s = 0.187$, $b_s = 0.058$ T^2, and $c_s = 0.946$. The anhysteretic parameters are listed in Appendix B. Let us consider two excitations. For the first case,

$$B = 0.4\cos(2\pi ft) \qquad (6.8\text{A-}1)$$

and for the second case

$$B = 0.1\cos(2\pi ft) \qquad (6.8\text{A-}2)$$

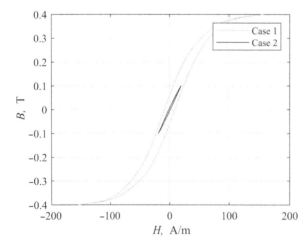

Figure 6.24 Trajectories predicted by the extended Jiles–Atherton model.

where we will take $f = 10$ kHz. Expressions for the time derivative of the flux density are readily established using (6.8A-1) and (6.8A-2). Using the extended Jiles–Atherton model, the resulting trajectories are shown in Figure 6.24. Observe that in case 2, the "effective" permeability of the material drops in an interior loop situation. The power loss densities for cases 1 and 2 are 117 and 3.33 kW/m^3, respectively, a difference of a factor of 35.

References

1 J. L. Cale and S. D. Sudhoff, An Improved Magnetic Characterization Method for Highly Permeable Materials, *IEEE Transactions on Magnetics*, vol. **42**, no. 8, pp. 1974–1981, 2006.
2 J. Reinert, A. Brockmeyer and R. De Doncker, Calculation of Losses in Ferro- and Ferrimagnetic Materials Based on the Modified Steinmetz Equation, *IEEE Transactions on Industry Applications*, vol. **37**, no. 4, pp. 1055–1061, 2001.
3 K. Venkatachalam, C. R. Sullivan, C. R. Abdallah and H. Tacca, "Accurate prediction of ferrite core loss with nonsinusoidal waveforms using only Steinmetz parameters," in 2002 *IEEE Workshop on Computers in Power Electronics,* 2002. *Proceedings*, 2002, pp. 36–41, doi: 10.1109/CIPE.2002.1196712.
4 J. Muhlethaler, J. Biela, J. Kolar and A. Ecklebe, Improved Core-Loss Calculation for Magnetic Components Employed in Power Electronic Systems.
5 K. Chen, "Iron-Loss Simulation of Laminated Steels Based on Expanded Generalized Steinmetz Equation," 2009 Asia-Pacific Power and Energy *Engineering Conference*, 2009, pp. 1–3, doi: 10.1109/APPEEC.2009.4918455.
6 A. Van den Bossche, V. Valchev and G. Georglev, "Measurement and Loss Model of Ferrites with Non-sinusoidal Waveforms,"" in **2004** Power Electronics Specialists Conference, Aachen, 2004.
7 J. Cale and S. Sudhoff, A Field-Extrema Hysteresis Loss Model for High-Frequency Ferrimagnetic Materials, *IEEE Transactions on Magnetics*, vol. **44**, no. 7, pp. 1728–1736, 2008.
8 J. Alsawalhi and S. Sudhoff, Saturable Thermally-Representative Steinmetz-Based Loss Models, *IEEE Transactions on Magnetics*, vol. **49**, no. 11, pp. 5438–5445, 2013.
9 IEC, "IEC60604-2 Magnetic materials - Part 2: Methods of measurement of the magnetic properties of electric steel strip and sheet by means of an Epstein Frame," International Electrotechnical Commission, Geneva, Switzerland, 2008.
10 E. G. Araujo, J. Schneider, K. Verbeken, G. Pasquarella and Y. Houbaert, Dimensional Effects on Magnetic Properties of Fe-Si Steels Due to Laser and Mechanical Cutting, *IEEE Transactions on Magnetics*, vol. **46**, no. 2, pp. 213–216, 2010.
11 R. Siebert, J. Schneider and E. Beyer, Laser Cutting and Mechanical Cutting of Electrical Steels and its Effect on the Magnetic Properties, *IEEE Transactions on Magnetics*, vol. **50**, no. 4, pp. 1–4, 2014.
12 K. Bourchas, Manufacturing Effects on Iron Losses in Electric Machines, KTH (Degree Project), Stockholm, Sweden, 2015.
13 G. M. Shane and S. D. Sudhoff, Refinements in Anhysteretic Characterization and Permeability Modeling, *IEEE Transactions on Magnetics*, vol. **46**, no. 11, pp. 3834–3843, 2010.
14 E. D. Torre, Magnetic Hysteresis, Piscataway, NJ: IEEE Press, 1999.
15 D. C. Jiles, Ferromagnetic Hysteresis, *IEEE Transactions on Magnetics*, vol. *IEEE Transactions on Magnetics*, vol. **19**, no. 5, pp. 2183–2185, 1983.
16 D. C. Jiles and D. L. Atherton, Theory of Ferromagnetic Hysteresis, *Journal of Magnetism and Magnetic Materials*, vol. **61** no. (1–2), pp. 48–60, 1986.

17 D. C. Jiles and J. B. Thoelke, Theory of Ferromagnetic Hysteresis: Determination of Model Parameters From Experimental Minor Loops, *IEEE Transactions on Magnetics*, vol. **25**, no. 5, pp. 3928–3930, 1989, doi: 10.1109/20.42480.

18 D. C. Jiles, A Self Consistent Generalized Model for the Calculation of Minor Loop Excursions in the Theory of Hysteresis, *IEEE Transactions on Magnetics*, vol. **28**, no. 5, pp. 2602–2604, 1992.

19 D. C. Jiles, J. B. Thoelke and M. K. Devine, Numerical Determination of Hysteresis Parameters for the Modeling of Magnetic Properties Using the Theory of Ferromagnetic Hysteresis, *IEEE Transactions on Magnetics*, vol. **28**, no. 1, pp. 27–35, 1992

20 K. H. Carpenter and S. Warren, A Wide Bandwidth, Dynamic Hysteresis Model for Magnetization in Soft Ferrites, *IEEE Transactions on Magnetics*, vol. **28**, no. 5, pp. 2037–2040, 1992.

21 D. C. Jiles, Frequency Dependence Of Hysteresis Curves in "Non-Conducting" Magnetic Materials, *IEEE Transactions on Magnetics*, vol. **29**, no. 6, pp. 3490–3492, 1993.

22 S. E. Zirka, Y. I. Moroz, R. G. Harrison and K. Chwastek, On Physical Aspects of the Jiles-Atherton Hysteresis Models, *Journal of Applied Physics*, vol. **112**, p. 043916, 2012. https://doi.org/10.1063/1.4747915.

23 H. Singh and S. Sudhoff, Reconsideration of Energy Balance in Jiles–Atherton Model for Accurate Prediction of B-H Trajectories in Ferrites, *IEEE Transactions on Magnetics*, vol. **56**, no. 7, pp. 1–8, 2020.

24 S. D. Sudhoff, "*MATLAB codes for Power Magnetic Devices: A Multi-Objective Design Approach*, 2nd Edition," [Online]. Available: http://booksupport.wiley.com

Problems

1 Derive an expression for eddy current loss in a conductive medium similar to (6.1-12) but based on the path shown in Figure 6.25.

2 Consider the rectangular toroidal sample of Example 6.1A. Compute the power loss in the sample if the flux density waveform is given by

$$B(t) = (\cos(1000t))^3$$

3 Suppose the field intensity and flux density waveforms in a material are given by

$$H = H_{pk} \cos(\omega t)$$
$$B = B_{pk} \cos(\omega t - \phi)$$

Derive an expression for the power loss density in terms of B_{pk}, H_{pk} and ϕ.

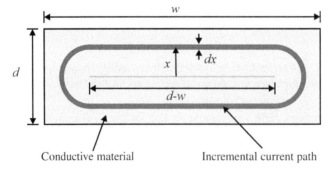

Figure 6.25 Alternate path for eddy current derivation.

4 Consider Example 6.3B. Compare the loss predictions of the MSE model and GSE model if the flux density waveform is given by

$$B = 0.45(\cos(2\pi ft))^3$$

5 Using the material parameters in Appendix C, plot the loss density of M19, M36, M43, and M47 steel samples versus frequency for the waveform

$$B = 1.1 \cos(2\pi ft) - 0.1 \cos(6\pi ft)$$

Plot the loss over a frequency range from 5 to 5000 Hz using a log–log plot. Use the MSE model.

6 Using the data in Appendix C, compare the value of k_e predicted by (6.3-15) to the value listed for M19.

7 Using (6.5-14) and (6.5-15), derive (6.5-16).

8 Given (6.5-15), verify (6.5-23) and (6.5-24).

9 In (6.5-40), verify whether the term in brackets is dB/dH less μ_0.

10 In (6.5-41), verify whether the term in brackets has the same sign as d^2B/dH^2.

11 Starting with (2.3-28), verify (6.5-42).

12 Starting with (2.3-25), verify (6.5-43).

13 Let us reconsider Example 6.6A. In particular, recall that for that example, the sample has an inner radius of 44.53 mm, an outer radius of 69.98 mm, and a height of 25.57 mm. Let us consider pretend the existence of a material with the same loss data, but for a different toroid, one with the same mean radius, and same volume, but one for which the ratio of outer to inner radius is 5. Using code from [24], determine the combined loss parameters as in Example 6.6A.

14 Using code from [24], repeat Example 6.6B, using the fictitious toroidal dimensions for problem 13.

15 Using (6.8-8), (6.8-22), and (6.8-26), we obtain (6.8-27), (6.8-28), and (6.8-29).

16 From (6.8-27), and assuming backward trapezoidal integration, we obtain (6.8-31), (6.8-32), and (6.8-33).

17 Modify the code for Example 6.8A (available at [24]) to plot the B–H trajectory and determine the power loss density for the excitation given by

$$B = 0.2 - 0.2 \cos(2\pi ft)$$

when the frequency f is 20 kHz.

7

Transformer Design

Transformers are extremely important. They are used to change ac voltage levels, as well as to provide galvanic isolation between circuits. Single- and three-phase transformers are extensively employed in the world's power distribution system. In this chapter, we will consider the design of single-phase power transformers. This will be the first device we have considered with multiple windings, and also the first device for which we include core loss in the design process.

Our development will begin with a description of common single-phase transformer architectures. Next, we will review the classic transformer T-equivalent circuit and will consider its use in steady-state phasor analysis. With this brief review complete, we will next discuss transformer performance considerations such as the calculation of transformer parameters, regulation, magnetizing current, operating point analysis, and inrush current, all in general terms. Next, we will focus our attention on one specific class of transformer, develop an Magnetic Equivalent Circuit (MEC), and ultimately a design approach. A case study in transformer design is then presented. The chapter will conclude with comments on additional factors relevant to the design process.

7.1 Common Transformer Architectures

In this chapter, we will focus our attention on single-phase transformers. Single-phase transformers are often classified as being either core type or shell type. These two types are shown in Figure 7.1. In the core-type transformer, the transformer windings surround the core, while in the shell-type transformer, the core surrounds the windings.

Let us first consider the core-type transformer as depicted in Figure 7.1(a). Therein, the darkest shaded region could be thought of as a UI-core with no gap. The moderately shaded region is a low-voltage or LV winding, which consists of two coils that could be connected in series or parallel. The lightly shaded region is a high-voltage or HV winding. The placement of the HV winding at the outside reduces insulation requirements. Often, the HV winding is referred to as the primary winding, and the LV winding is referred to as the secondary winding. It might occur to the reader that it would be simpler to simply place one winding on one leg and place the other winding on the other leg. Indeed, we often think of a transformer in these terms. However, the arrangement shown in Figure 7.1, while more complicated, has an advantage in that what would otherwise be leakage flux couples both windings. As a result, this arrangement has lower leakage flux. Leakage flux will be shown to be problematic in terms of transformer operation.

A shell-type transformer is shown in Figure 7.1(b), and is also a commonly used arrangement, particularly at high powers. It can be thought of as an EI-core with no gap. In this arrangement, the

---

Power Magnetic Devices: A Multi-Objective Design Approach, Second Edition. Scott D. Sudhoff.
© 2022 The Institute of Electrical and Electronics Engineers, Inc. Published 2022 by John Wiley & Sons, Inc.
Companion website: www.wiley.com/go/sudhoff/Powermagneticdevices

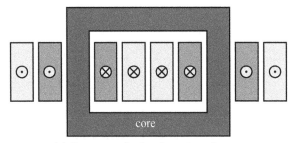

(a) Core-type single-phase transformer

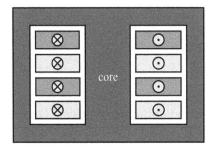

(b) Shell-type single-phase transformer

Figure 7.1 Transformer types.

LV and HV windings are again indicated by moderately and lightly shaded regions. Again, multiple coils are used, and the coils are alternated between LV and HV windings in order to reduce leakage flux. This arrangement will require more insulation space than a core-type transformer. However, the space between the windings can be useful in cooling the windings.

While it has just been stated that the two arrangements can be thought of as being UI- and EI-core arrangements without gaps, there is a subtlety involved in this. Consider Figure 7.2, which shows a cross section of one of the core legs shown in Figure 7.1(a), but seen from above. As can be seen, the coils in this case are circular to facilitate an easy bending radius with large conductors. In this case, it can be seen that the core legs have an approximately circular cross section. This is accomplished

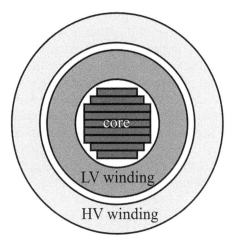

Figure 7.2 Cross section of one leg of core type transformer.

in a laminated structure by having the end laminations shaped differently than the central laminations. This allows what might otherwise be wasted space to conduct flux and will also help reduce leakage flux. This idea could also be applied to a magnetic cross section that was rectangular by profiling the ends as semicircles. An extended discussion of the construction of single-phase transformers is set forth in Pansini [1].

7.2 T-Equivalent Circuit Model

We will begin our study of transformer design with a review of the single-phase transformer circuit model, in particular the T-equivalent circuit model. To this end, consider the elementary transformer shown in Figure 7.3. Note that while this depiction of a transformer will be convenient for derivation of an equivalent circuit model, transformer construction is generally as described in Section 7.1. In Figure 7.3, a closed core is shown (no air gap) and there are two windings: a primary winding consisting of N_p turns around the core and a secondary winding with N_s turns around the core. Unlike the inductor studied in Chapter 3, a transformer is not an energy storage device.

In order to derive a T-equivalent circuit for the transformer, let us begin with a generic magnetic equivalent circuit for the device. This is shown in Figure 7.4. There are many quantities of interest therein. The currents into the primary and secondary winding are denoted as i_p and i_s, respectively. The flux through the primary and secondary windings are designated Φ_p and Φ_s. The magnetizing

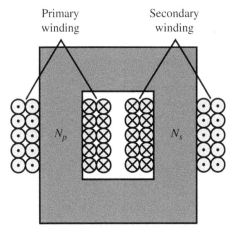

Figure 7.3 Elementary transformer.

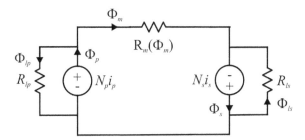

Figure 7.4 Transformer magnetic equivalent circuit.

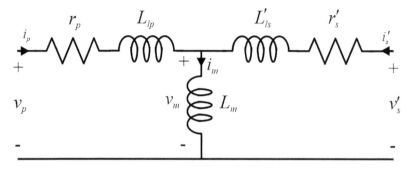

Figure 7.5 T-equivalent circuit.

flux, Φ_m, is that component of the flux common to the primary and secondary winding. Leakage flux components associated with the primary and secondary winding are labeled Φ_{lp} and Φ_{ls}. The reluctance of the magnetic core is $R_m(\Phi_m)$. This includes the complete circuit around the core. Leakage reluctances R_{lp} and R_{ls} are associated with flux paths that do not make use of the full winding. The MEC shown in Figure 7.4 is rather elementary and will be used as a basis to derive the transformer T-equivalent circuit; we will consider a much more detailed MEC when we embark upon transformer design.

The transformer T-equivalent circuit model is shown in Figure 7.5. Therein, the primed quantities are scaled by a factor related to the transformer turns ratio. We will discuss this in detail in our circuit derivation. Many transformer requirements are stated in terms of the T-equivalent circuit. For that reason, understanding the T-equivalent circuit is important.

To begin our development, from Figure 7.4 the flux into the primary and secondary windings may be expressed as

$$\Phi_p = \Phi_{lp} + \Phi_m \tag{7.2-1}$$

$$\Phi_s = \Phi_{ls} + \Phi_m \tag{7.2-2}$$

From Figure 7.4, the leakage and magnetizing fluxes may be expressed

$$\Phi_{lp} = \frac{N_p i_p}{R_{lp}} \tag{7.2-3}$$

$$\Phi_{ls} = \frac{N_s i_s}{R_{ls}} \tag{7.2-4}$$

$$\Phi_m = \frac{N_p i_p + N_s i_s}{R_m} \tag{7.2-5}$$

The voltage equations for the primary and secondary windings are given by

$$v_p = r_p i_p + \frac{d\lambda_p}{dt} \tag{7.2-6}$$

$$v_s = r_s i_s + \frac{d\lambda_s}{dt} \tag{7.2-7}$$

where the primary and secondary flux linkages may be expressed as

$$\lambda_p = N_p \Phi_p \tag{7.2-8}$$

$$\lambda_s = N_s \Phi_s \tag{7.2-9}$$

Let us consider the primary side of the circuit. Substitution of (7.2-3) into (7.2-1) and the result into (7.2-8) yields

$$\lambda_p = L_{lp} i_p + \lambda_m \qquad (7.2\text{-}10)$$

where L_{lp} is the primary side leakage inductance given by

$$L_{lp} = \frac{N_p^2}{R_{lp}} \qquad (7.2\text{-}11)$$

and λ_m is the magnetizing flux linkage seen by the primary winding, which may be expressed as

$$\lambda_m = N_p \Phi_m \qquad (7.2\text{-}12)$$

Substituting the flux linkage equation (7.2-10) into the voltage equation (7.2-6) we arrive at

$$v_p = r_p i_p + L_{lp}\frac{di_p}{dt} + v_m \qquad (7.2\text{-}13)$$

where

$$v_m = \frac{d\lambda_m}{dt} \qquad (7.2\text{-}14)$$

Observe that (7.2-13) is the expression we obtain by writing the loop equation of the left-hand side of Figure 7.5.

Now let us consider the magnetizing flux linkage at greater length. From (7.2-5) and (7.2-12), the magnetizing flux linkage may be written

$$\lambda_m = \frac{N_p^2}{R_m}\left(i_p + \frac{N_s}{N_p} i_s\right) \qquad (7.2\text{-}15)$$

It is convenient to identify the term in parenthesis in (7.2-15) as the magnetizing current:

$$i_m = i_p + \frac{N_s}{N_p} i_s \qquad (7.2\text{-}16)$$

However, from Figure 7.5, we have

$$i_m = i_p + i_s' \qquad (7.2\text{-}17)$$

These two definitions can be made consistent by defining the referred secondary current as

$$i_s' = \frac{N_s}{N_p} i_s \qquad (7.2\text{-}18)$$

The definition of a referred (scaled) secondary current and voltage is critical to being able to develop a T-equivalent circuit. Without this, it would not be possible to develop such a circuit model.

The magnetizing flux linkage can now be expressed as

$$\lambda_m = L_m (i_p + i_s') \qquad (7.2\text{-}19)$$

where

$$L_m = \frac{N_p^2}{R_m} \qquad (7.2\text{-}20)$$

Note that R_m and hence L_m can be a function of the magnetizing flux (or flux linkage), but that dependency is not explicitly shown.

Thus far, we have developed the relationships we need for the primary side and the magnetizing branch of our T-equivalent circuit. Next, let us consider the secondary side voltage equations. We may write

$$v_s = r_s i_s + \frac{d\lambda_s}{dt} \tag{7.2-21}$$

Substitution of (7.2-4) into (7.2-2) and the result into (7.2-9) gives

$$\lambda_s = L_{ls} i_s + N_s \Phi_m \tag{7.2-22}$$

where

$$L_{ls} = \frac{N_s^2}{R_{ls}} \tag{7.2-23}$$

Next, we combine (7.2-22) and (7.2-21) to yield

$$v_s = r_s i_s + L_{ls} \frac{di_s}{dt} + N_s \frac{d\Phi_m}{dt} \tag{7.2-24}$$

Earlier in this development, we replaced the secondary current i_s by a referred secondary current i'_s. Thus, we must express (7.2-24) in terms of the referred current if we are to avoid representing the same current two different ways. This yields

$$v_s = r_s \frac{N_p}{N_s} i'_s + L_{ls} \frac{N_p}{N_s} \frac{di'_s}{dt} + N_s \frac{d\Phi_m}{dt} \tag{7.2-25}$$

which does not correspond to Figure 7.5. However, defining the referred secondary voltage as

$$v'_s = \frac{N_p}{N_s} v_s \tag{7.2-26}$$

(7.2-25) becomes

$$v'_s = r'_s i'_s + L'_{ls} \frac{di'_s}{dt} + v_m \tag{7.2-27}$$

which exactly corresponds to Figure 7.5. In (7.2-27),

$$r'_s = \left(\frac{N_p}{N_s}\right)^2 r_s \tag{7.2-28}$$

$$L'_{ls} = \left(\frac{N_p}{N_s}\right)^2 L_{ls} \tag{7.2-29}$$

Comparing (7.2-26) and (7.2-18), we can see that voltage and current are referred by reciprocal ratios. As a result, the power expressed in terms of referred variables is identical to power in terms of our original variables.

Referring to our equivalent circuit, for an ideal transformer we would have r_p, r'_s, L_{lp}, and L'_{ls} all be zero, and have L_m be infinite. In such a case, we would have $v_s = N_s v_p / N_p$ and $i_p = -N_s i_s / N_p$. Many of our transformer metrics will be aimed at achieving these conditions in practice to the extent possible.

7.3 Steady-State Analysis

It is assumed that the reader is adept at steady-state phasor analysis of electric circuits. If this is not the case, the reader is referred to Appendix E. For the purposes of setting forth notation, consider a variable of the form

$$f_x(t) = \sqrt{2} F_x \cos\left(\omega_e t + \phi_{fx}\right) \tag{7.3-1}$$

In (7.3-1), f may denote a voltage, current, or flux linkage and x will be a subscript associated with the specific voltage, current, or flux linkage of interest. The quantity F_x will be used to denote the root-mean-square (rms) amplitude of the fundamental component of the quantity of interest, and ϕ_{fx} will be used to denote the phase angle. The phasor representation of the signal will be denoted

$$\tilde{F}_x = F_x e^{j\phi_{fx}} \tag{7.3-2}$$

and is also often written in the form $F_x \angle \phi_{fx}$.

From (7.3-2), observe

$$F_x = |\tilde{F}_x| \tag{7.3-3}$$

and

$$\phi_{fx} = \text{angle}\left(\tilde{F}_x\right) \tag{7.3-4}$$

Note that the rms value of a time signal is defined as

$$F_{x-rms} = \sqrt{\frac{1}{T}\int_0^T f_x^2(t) dt} \tag{7.3-5}$$

Substitution of (7.3-1) into (7.3-5) yields $F_{x-rms} = F_x$. However, this is not the case for a nonsinusoidal signal.

In order to demonstrate this notation as well as the steady-state equivalent circuit of Section 7.2, let us now consider an example.

Example 7.3A Let us consider the use of the equivalent circuit model. Suppose we have a transformer with the following parameters: $r_p = r'_s = 50$ mΩ, $L_{lp} = L'_{ls} = 1$ mH, $L_m = 100$ mH, $N_p = 50$, and $N_s = 10$. The primary voltage is given by

$$v_p(t) = 200\sqrt{2} \cos(\omega_e t) \tag{7.3A-1}$$

where $\omega_e = 2\pi 50$ rad/s. A load resistance of $r_l = 1$ Ω is placed on the secondary. We wish to find the secondary voltage. To solve this problem, we will use steady-state circuit analysis.

Referring to Figure 7.6, it is convenient to first find the impedance of the primary branch (the primary resistance and leakage inductance) as

$$Z_p = r_p + j\omega_e L_{lp} = 0.318 \angle 81.0° \tag{7.3A-2}$$

Next, the impedance of the secondary branch (the referred secondary resistance and leakage inductance) may be found from

7 Transformer Design

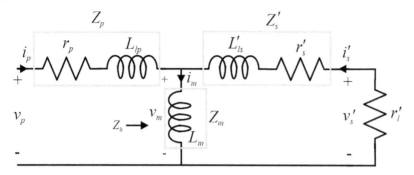

Figure 7.6 Circuit for Example 7.3A.

$$Z'_s = r'_s + j\omega_e L'_{ls} = 0.318\angle 81.0° \tag{7.3A-3}$$

The impedance of the magnetizing branch is given by

$$Z_m = j\omega_e L_m = 31.4\angle 90° \tag{7.3A-4}$$

In order to compute the load impedance, it is critical to remember that our equivalent circuit is in terms of referred variables. The load may be represented by

$$v_s = -r_l i_s \tag{7.3A-5}$$

where the minus sign is due to the fact that the secondary current is defined to be positive into the transformer. Using (7.2-18) and (7.2-26), it is easy to show that in terms of referred variables (7.3A-5) becomes

$$v'_s = -r'_l i'_s \tag{7.3A-6}$$

where

$$r'_l = \left(\frac{N_p}{N_s}\right)^2 r_l \tag{7.3A-7}$$

Evaluating (7.3A-7)

$$r'_l = 25 \; \Omega \tag{7.3A-8}$$

Now that the branch impedances have been calculated, observe that the impedance Z_b looking back into the circuit at the indicated point may be expressed

$$Z_b = \frac{Z_m(Z'_s + r'_l)}{Z_m + Z'_s + r'_l} = 19.5\angle 39° \tag{7.3A-9}$$

The magnetizing voltage may now be expressed using voltage division as

$$\tilde{V}_m = \frac{Z_b}{Z_b + Z_p}\tilde{V}_p = 198\angle -0.618° \tag{7.3A-10}$$

Applying voltage division, we obtain

$$\tilde{V}'_s = \frac{r'_l}{r'_l + Z'_s}\tilde{V}_m = 197\angle -1.34° \tag{7.3A-11}$$

Converting back to original (nonreferred) form and expressing in the time domain, (7.3A-11) becomes

$$v_s(t) = 39.4\sqrt{2}\cos(\omega_e t - 0.0233) \qquad (7.3A\text{-}12)$$

where 0.0233 radians corresponds to 1.34 degrees.

7.4 Transformer Performance Considerations

In the previous section, the T-equivalent circuit of a single-phase transformer was set forth. The stage is now set to consider some performance characteristics in terms of the T-equivalent circuit, and at the same time to set forth some of the analysis required to design the transformer. In order to do this, we will assume that we have access to a magnetic analysis to calculate the primary and secondary flux linkages as a function of the primary and secondary current, as well as transformer parameters. We will set forth such a magnetic analysis for a core type transformer in Section 7.6.

Calculation of Lumped Parameters

In designing a transformer, we will find it useful to determine the electrical parameters, which is to say the parameters of the T-equivalent circuit. The first parameters to calculate will be the number of turns. It is assumed here that the primary and secondary winding are both made up of a number of identical coils and that these coils will be connected in series or parallel. Let us denote the number of turns in one coil of winding x as N_{xcl}, where x is p for primary or s for secondary. The number of turns will be given by

$$N_x = N_{xcl} N_{xcs} \qquad (7.4\text{-}1)$$

where N_{xcs} is the number of coils in series. Observe that parallel windings do not factor into the calculation of number of turns—in a sense a parallel winding just adds to the conductor area.

Let us now consider the calculation of the inductances associated with the T-equivalent circuit. First, we will consider the calculation of the magnetizing inductance. Because of magnetic saturation, this inductance is a function of the magnetizing current. However, at low magnetizing currents, it is linear. In this region, let us denote the magnetizing inductance as L_{m0}, which, from the T-equivalent circuit may be calculated as

$$L_{m0} = \frac{\lambda_s|_{i_p = i_{p,t},\, i_s = 0} N_p}{N_s i_{p,t}} \qquad (7.4\text{-}2)$$

In (7.4-2), $i_{p,t}$ denotes a test current that should be small relative to the expected magnetizing current (~10%).

The leakage inductances do not vary with current nearly as much as the magnetizing inductance since they are associated with magnetic paths that are dominated by reluctances through nonferrous paths. For normal current levels, the leakage inductances can be calculated from architecture of the T-equivalent circuit and our magnetic analysis as

$$L_{lp} = \frac{\lambda_p|_{i_p = i_{p,t},\, i_s = 0}}{i_{p,t}} - L_{m0} \qquad (7.4\text{-}3)$$

$$L'_{ls} = \left(\frac{N_p}{N_s}\right)^2 \frac{\lambda_s|_{i_p=0, i_s=i_{s,t}}}{i_{s,t}} - L_{m0} \tag{7.4-4}$$

In (7.4-3) and (7.4-4), $i_{p,t}$ and $i_{s,t}$ are again test currents that should map to a small magnetizing current (considering the turns ratio in the case of the secondary current).

In addition to the inductances, it will be desirable to calculate the winding resistances. It will be assumed that each transformer winding consists of a number of identical coils. Using (3.2-7), the resistance of a winding may be expressed as

$$r_x = \frac{V_{xcl} N_{xcl}^2 N_{xcs}}{k_{xpf} A_{xcl}^2 \sigma_{xc} N_{xcp}} \tag{7.4-5}$$

In (7.4-5), x is a generic subscript that is p for the primary winding and s for the secondary winding. The symbols V_{xcl}, A_{xcl}, N_{xcl}, and k_{xpf} denote the volume, area, turns, and packing factor of one of the primary or secondary coils, as described in Section 3.2. The terms N_{xcs} and N_{xcp} denote the number of coils in series, and number of coils in parallel, respectively. The variable σ_{xc} denotes the conductivity of the winding conductor. In the case of the secondary winding, we have

$$r'_s = \left(\frac{N_p}{N_s}\right)^2 r_s \tag{7.4-6}$$

Regulation

The output voltage of an ideal transformer is independent of load current. Because of the winding resistance, and more importantly the leakage reactances, this does not occur. The regulation of a transformer is defined as the difference between the no-load voltage magnitude and the full-load voltage magnitude relative to the no-load voltage magnitude. In particular, the full-load regulation is generally defined as

$$\chi = \left| \frac{|\tilde{V}_{s,fl}| - |\tilde{V}_{s,nl}|}{|\tilde{V}_{s,nl}|} \right| \tag{7.4-7}$$

In evaluating (7.4-7), it should be realized that the load (secondary) current must be specified in terms of magnitude and phase.

An alternate definition of regulation is

$$\chi = \left| \frac{\tilde{V}_{s,fl} - \tilde{V}_{s,nl}}{\tilde{V}_{s,nl}} \right| \tag{7.4-8}$$

One advantage of (7.4-8) over (7.4-7) is that it is primarily a function of the magnitude of the load current. To see this, from the T-equivalent circuit we have that the no-load and full-load voltages are given by

$$\tilde{V}_{s,nl} = \frac{j\omega_e L_m}{r_p + j\omega_e (L_{lp} + L_m)} \frac{N_s}{N_p} \tilde{V}_p \tag{7.4-9}$$

and

$$\tilde{V}_{s,fl} = \frac{j\omega_e L_m}{r_p + j\omega_e (L_{lp} + L_m)} \frac{N_s}{N_p} \tilde{V}_p - \left(r'_s + j\omega_e L'_{ls} + \frac{j\omega_e L_m (r_p + j\omega_e L_{lp})}{r_p + j\omega_e (L_{lp} + L_m)} \right) \left(\frac{N_s}{N_p} \right)^2 \tilde{I}_{s,fl} \tag{7.4-10}$$

Substitution of (7.4-9) and (7.4-10) into (7.4-8) and manipulating, we obtain

$$\chi = \left| (r'_s + j\omega_e L'_{ls}) \left(1 + \frac{L_{lp}}{L_m} - \frac{jr_p}{\omega_e L_m}\right) + r_p + j\omega_e L_{lp} \right| \frac{I'_{s,fl}}{V_p} \qquad (7.4\text{-}11)$$

Since $L_{lp} \leq L_m$ and $r_p \ll \omega_e L_m$ for typical parameters (7.4-11) is often approximated as

$$\chi = \left| r_p + r'_s + j\omega_e (L_{lp} + L'_{ls}) \right| \frac{I'_{s,fl}}{V_p} \qquad (7.4\text{-}12)$$

As can be seen, (7.4-11) and (7.4-12) are only functions of the magnitude of the secondary current, not the phase.

As it turns out, our design effort on the transformer will be considerably more involved than our previous efforts. For this reason, it is important to organize and partition some of the different types of analysis we will conduct. To this end, in order to summarize the series of calculations associated with the electrical parameters and regulation, let us denote the vector of electrical parameters as

$$\mathbf{E} = \begin{bmatrix} N_p & N_s & r_p & L_{lp} & L_{m0} & L'_{ls} & r'_s & \chi \end{bmatrix} \qquad (7.4\text{-}13)$$

The sequence of calculations, (7.4-1)–(7.4-6) and (7.4-11), can be represented in functional form as

$$\mathbf{E} = \mathrm{F}_E(\mathbf{T}, \mathbf{D}) \qquad (7.4\text{-}14)$$

where $\mathbf{T}$ is a vector containing the transformer description (dimensions and materials) as required for the analysis and $\mathbf{D}$ is a vector of design information, which includes the numerical values of the test currents as well as other information. These vectors will be formally defined in Section 7.5 and Section 7.8, respectively.

Magnetizing Characteristics

Observe that our MEC analysis is assumed to be of the form wherein the currents are input variables and the flux linkages are output variables. In order to analyze transformer operating points, which will be our next topic, it will prove useful to be able to create a function that describes the magnetizing flux linkage as an input variable and the magnetizing current as an output variable.

To this end, suppose we create a vector of test currents wherein $i_{m,j}$ is the jth test magnetizing current. Setting the primary current to zero and the referred secondary current equal to the magnetizing current, the corresponding magnetizing flux linkage may be expressed as

$$\lambda_{m,j} = \lambda_p \big|_{i_p = 0, i_{s,j} = N_p i_{m,j}/N_s} \qquad (7.4\text{-}15)$$

Repeating this calculation for different test current values one builds up a range of magnetizing flux linkage versus magnetizing current values. From these values, we construct a function

$$i_m = \mathrm{F}_{im}(\lambda_m, \mathbf{T}) \qquad (7.4\text{-}16)$$

using linear or quadratic interpolation. The maximum value of primary current used in constructing this relationship should be a multiple of the expected value so that extrapolation is unnecessary.

Operating Point Analysis

Let us now consider the analysis of an operating point. In order to analyze an operating point, we will use the equivalent circuit developed in Section 7.2, with a modification of a resistance placed in

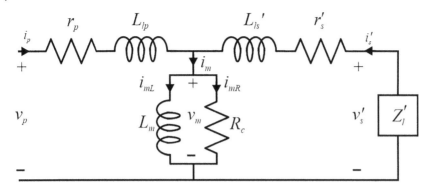

Figure 7.7 Modified T-equivalent circuit.

parallel with the magnetizing branch in order to represent core loss, as shown in Figure 7.7. Here the focus will be on analysis during normal conditions with a secondary current between no load and full load. Under these conditions, we would expect the leakage inductances to be linear but the magnetizing inductance and core loss resistance elements of the equivalent circuit to change modestly as a function of load.

To begin our analysis, we will assume a fixed primary voltage $\tilde{V}_p$ and a fixed (referred) secondary load impedance Z'_l. An algorithm to compute the steady-state operating point is as follows. This algorithm is based on a combination of (a) a T-equivalent electrical circuit discussed in Section 7.2 and (b) a magnetic equivalent circuit. It is an approximate method, but relatively accurate. One reason that it is approximate is that the leakage inductances share some of the same magnetic path as the magnetizing inductance. A second reason the method is approximate is because of core loss. The iterative algorithm will be presented in steps. The variable k will denote the iteration number.

Step 1: Initialization In this step, the electrical circuit is initialized. The iteration index k is set to one, and the core resistance used to represent core loss and magnetizing inductance are initialized as

$$R_c^1 = \infty \tag{7.4-17}$$

$$L_m^1 = L_{m0} \tag{7.4-18}$$

In (7.4-17) and (7.4-18), the superscript denotes the iteration number k, which is one in this initialization step. We will not use the bracket notation for iteration as in the rest of text, because it would prove awkward when used in conjunction with parenthesis that we will also need.

Step 2: Circuit Solution The next step will be to solve the T-equivalent circuit. To this end, it is convenient to define the magnetizing branch impedance as

$$Z_m^k = \frac{j\omega_e L_m^k R_c^k}{R_c^k + j\omega_e L_m^k} \tag{7.4-19}$$

the primary branch impedance as

$$Z_p = r_p + j\omega_e L_{lp} \tag{7.4-20}$$

the secondary branch impedance as

$$Z'_s = r'_s + j\omega_e L'_{ls} \tag{7.4-21}$$

and the loaded secondary branch impedance as

$$Z'_{sl} = Z'_s + Z'_l \tag{7.4-22}$$

Next, we define the magnetizing-secondary-load impedance as

$$Z^k_{msl} = \frac{Z^k_m Z'_{sl}}{Z^k_m + Z'_{sl}} \tag{7.4-23}$$

The magnetizing voltage may then be expressed as

$$\widetilde{V}^k_m = \frac{Z^k_{msl}}{Z^k_{msl} + Z_p} \widetilde{V}_p \tag{7.4-24}$$

and the secondary voltage as

$$\widetilde{V}'^k_s = \frac{Z'_l}{Z'_l + Z'_s} \widetilde{V}^k_m \tag{7.4-25}$$

Step 3: Nonlinear Magnetizing Current Our next step will be to calculate the detailed waveform of the magnetizing current. There is a bit of a paradox in this calculation. We will assume that the magnetizing voltage is sinusoidal, but not assume that the magnetizing current is sinusoidal. Suppose that the magnetizing voltage was in fact a perfect sinusoid. Because of the nonlinear magnetics, it follows that the magnetizing current would still not be sinusoidal. The nonsinusoidal current component, in conjunction with the winding resistances and leakage inductances, causes the magnetizing voltage to be nonsinusoidal. However, because the magnetizing current is small, and because the winding resistances and leakages are small, the amount of distortion in this voltage is very small, and so we will treat the magnetizing voltage as if it is sinusoidal.

Converting the phasor representation of the magnetizing voltage to the time domain and noting that

$$v_m = \frac{d\lambda_m}{dt} \tag{7.4-26}$$

for steady-state conditions one obtains

$$\lambda^k_m(t) = \frac{\sqrt{2} V^k_m}{\omega_e} \sin(\omega_e t + \phi^k_{vm}) \tag{7.4-27}$$

Using the magnetizing current function defined by (7.4-16), we can define the resulting magnetizing current waveform by

$$\hat{i}^k_{mL}(t) = F_{im}(\lambda^k_m(t)) \tag{7.4-28}$$

Computationally, it is necessary to compute the waveform over nominally one-half a cycle. Note that the magnetizing current waveform given by (7.4-28) does not exactly coincide with its phasor description since it includes harmonics. This is the reason for the "^" notation.

Step 4: Lumped Circuit Parameter Update The next step of the analysis is focused on obtaining a revised magnetizing inductance and core resistance.

From the magnetizing current waveform and the secondary current, the primary current waveform is obtained from

$$\hat{i}_p^k(t) = \hat{i}_{mL}^k(t) - i_s'^k(t) \tag{7.4-29}$$

The secondary current waveform for this iteration, $i_s^k(t)$, exactly corresponds to its phasor representation. As in the case of the magnetizing waveform, a "^" notation is used because the time-domain waveform will not exactly correspond to its phasor representation. This is because of two reasons: (a) the nonlinearity in the magnetics and (b) (7.4-29) does not include a component of current to represent core loss. This is intentional because we will use the primary current waveform in the MEC to calculate field waveforms, and the MEC does not include this term. Again, all time-domain waveforms need to be represented over one-half of a cycle.

Next, the primary and secondary currents waveforms will be passed to a function that will be used to calculate the flux density waveforms, which in turn will be used to calculate core loss. This will be functionally represented as

$$P_{cl}^k = F_{cl}\left(\hat{i}_p^k(t), i_s^k(t), \mathbf{T}\right) \tag{7.4-30}$$

where $\mathbf{T}$ is a variable (a vector or structure) that includes all geometrical and material data for the transformer. It is configuration dependent, but will be set forth for a specific case in Section 7.5. The function itself is of course dependent on the details of the device and its magnet representation. It will be discussed for a particular case in Sections 7.6 and 7.7.

Once the core loss is calculated, an updated expression for the core loss resistance is computed as

$$R_c^{k+1} = \frac{(V_m^k)^2}{P_{cl}^k} \tag{7.4-31}$$

In the next step, the fundamental component of the magnetizing current is computed. Using Fourier series techniques and the property of odd half-wave symmetry, the Fourier series coefficient are defined as

$$a_{m1}^k = \frac{4}{T}\int_0^{T/2} i_{mL}^k(t) \cos(\omega_e t) dt \tag{7.4-32}$$

$$b_{m1}^k = \frac{4}{T}\int_0^{T/2} i_{mL}^k(t) \sin(\omega_e t) dt \tag{7.4-33}$$

from which rms value of the fundamental component of the magnetizing component is computed as

$$\hat{I}_{mL}^k = \frac{1}{\sqrt{2}}\sqrt{(a_{m1}^2 + b_{m1}^2)} \tag{7.4-34}$$

Again the "^" notation is used, because the rms magnitude given by (7.4-34) does not correspond exactly to time-domain waveform (7.4-28), which may have harmonic content.

Next, the magnetizing inductance is computed as

$$L_m^{k+1} = \frac{V_m^k}{\omega_e \hat{I}_{mL}^k} \tag{7.4-35}$$

Note that this inductance is not an incremental inductance nor an absolute inductance; rather it is an inductance relating a fundamental component of flux linkage to a fundamental component of current.

Step 5: Convergence Evaluation In this step, it is determined whether or not the algorithm has converged. One method to do this is to define the error metric

$$e = \max\left(\left|\frac{R_c^{k+1} - R_c^k}{R_c^{k+1}}\right|, \left|\frac{L_c^{k+1} - L_c^k}{L_c^{k+1}}\right|\right) \tag{7.4-36}$$

If the error metric is greater than an allowed value and the number of iterations does not exceed a limit, the iteration count is incremented by one and the algorithm returns to Step 2. Otherwise, the analysis is complete. Generally, the algorithm converges very rapidly.

Step 6: Final Calculations After convergence, the power dissipation can be calculated. Dropping the iteration superscript, the resistive and total power loss may be calculated as

$$P_{rl} = r_p I_p^2 + r_s' I_s'^2 \tag{7.4-37}$$

and

$$P_l = P_{rl} + P_{cl} \tag{7.4-38}$$

respectively. Observe that (7.4-37) approximates the magnetizing current by its fundamental component.

Another variable of interest is the total magnetizing current. This may be expressed as

$$i_m(t) = \frac{v_m(t)}{R_c} + i_{mL}(t) \tag{7.4-39}$$

The current density in each winding is given by

$$J_x = \frac{|I_x|}{a_{xc} N_{xpr} N_{xcp}} \tag{7.4-40}$$

For the purposes of supporting our subsequent design efforts, it will be convenient to summarize the operating point calculations in functional form. Representing our operating point inputs as

$$\mathbf{O}_I = \begin{bmatrix} V_p & Z_l & \omega_e \end{bmatrix} \tag{7.4-41}$$

and our desired operating point outputs as

$$\mathbf{O}_D = \begin{bmatrix} I_p & V_s & P_{rl} & P_{cl} & P_l & R_c & L_m & \theta_e(t) & B_{bl}(t) & B_{el}(t) & i_m(t) & J_p & J_s \end{bmatrix} \tag{7.4-42}$$

we represent the procedures of this section in functional form as

$$\mathbf{O}_D = \mathbf{F_O}(\mathbf{E}, \mathbf{T}, \mathbf{O}_I) \tag{7.4-43}$$

In (7.4-42), the time-domain waveforms represent vectors of elements containing a time history of the indicated variable over one-half a cycle.

The reader may have serious reservations about the usefulness of the T-equivalent circuit, given the somewhat involved procedure to use it presented herein. However, most of the time, a significantly simpler approach can be used—which is simply to consider L_m and R_c to be constant, whereupon the T-equivalent circuit can be solved as any circuit. For a given primary voltage at a given frequency, these parameters do not vary tremendously under load, since they are both determined by the magnetizing voltage, which is fairly close to the primary voltage. For design purposes, however, L_m and R_c are not known, and so a more involved procedure is needed.

No Load, Full Load, and Overload Analysis

The algorithm set forth in the previous section is generic to any operating point. However, there are several operating points that are of particular interest. These include no-load, full-load, and overload conditions.

No-load conditions are of interest because a transformer will draw current (magnetizing current) and produce loss even if the secondary winding is open circuited. The loss is a concern because it can occur over long periods of time. Even without the loss, the magnetizing current of the transformer draws reactive power that will result in power system losses and contribute to voltage sag. For these reasons, the no-load primary current is often constrained. We will denote this current as $I_{p,nl}$ and calculate it using the procedure of the previous subsection. We will typically also wish to calculate the no-load secondary voltage, denoted as $V_{s,nl}$, to see if it falls within the desired range.

Full-load conditions are also of interest, particularly for checking the regulation of the transformer. In addition, it is often the case that the transformer be required to operate (at least for a period of time) under overloaded conditions. All of these conditions may be analyzed using the work of the previous section.

Inrush Current

Thus far, most of the transformer characteristics we have considered have been steady-state quantities. Some behaviors of interest are transient in nature, however. One of the most important the transient characteristics is transformer inrush current.

Consider an unloaded transformer disconnected from a source. Suppose that at time $t = t_c$ the unloaded transformer is connected to a primary source. Neglecting the primary resistance in (7.2-6), we have

$$v_p = \frac{d\lambda_p}{dt} \qquad (7.4\text{-}44)$$

Thus the primary flux linkage may be expressed

$$\lambda_{p,nl}(t) = \int_{t_c}^{t} v_p(t)dt \qquad (7.4\text{-}45)$$

where the initial flux linkage at time $t = t_c$ is zero.

The primary voltage may be expressed in the time domain as

$$v_p = \sqrt{2}V_p \cos\left(\omega_e t + \phi_{vp}\right) \qquad (7.4\text{-}46)$$

Substitution of (7.4-46) into (7.4-45) yields

$$\lambda_p(t) = \frac{\sqrt{2}V_p}{\omega_e}\left(\sin\left(\omega_e t + \phi_{vp}\right) - \sin\left(\omega_e t_c + \phi_{vp}\right)\right) \qquad (7.4\text{-}47)$$

The first term in (7.4-47) is the steady-state response. The second term represents an initial dc offset. In the worst-case scenario, t_c is such that

$$\omega_e t_c + \phi_{vp} = -\frac{\pi}{2} \qquad (7.4\text{-}48)$$

which yields a worst-case peak magnetizing flux linkage of

$$\lambda_{p,wci} = \frac{2\sqrt{2}v_p}{\omega_e} \qquad (7.4\text{-}49)$$

which is twice the normal value.

The fact that the primary flux linkage will, in the transient case, reach a value twice its normal peak is problematic. It follows that the magnetizing flux linkage will have a peak value nearly twice its steady-state value. If, for normal conditions, the flux density in the transformer is slightly below the value where the magnetic core saturates, during an inrush transient the core will become highly saturated, which will cause the magnetizing inductance to collapse, whereupon a large primary current (the inrush current) will result even though the transformer is unloaded. This primary current could be many times the normal magnetizing current. If the transformer is loaded, the primary current will be even larger. It follows that inrush current should be addressed in the design process. One approach to doing this will be set forth in Section 7.8.

7.5 Core-Type Transformer Configuration

In this section, we will set the stage for an initial transformer design by setting forth a configuration for our design. Figure 7.8 illustrates a cross section of the core-type configuration we will consider. The darkest shaded region is the core, the next lighter level of shading indicates the secondary winding, and the lightest level of shading indicates the primary winding. As can be seen, the primary and secondary windings are each comprised of two coils. It will be assumed that the base and top legs of the core are of rectangular cross section. However, the end legs may have a rounded cross section, as shown in Figure 7.9.

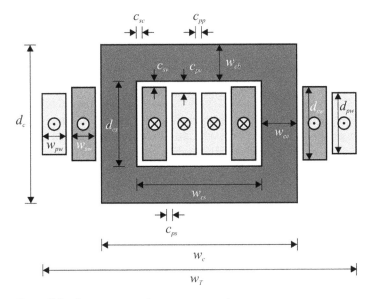

Figure 7.8 Core-type transformer cross section.

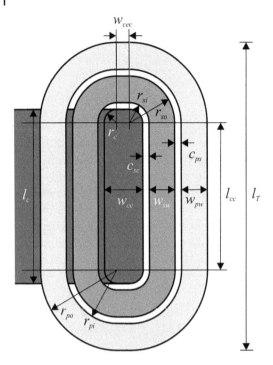

Figure 7.9 End leg cross section with coils.

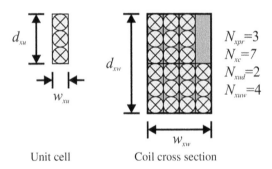

Figure 7.10 Coil construction.

Figure 7.10 depicts a cross section of one of the coils. Therein, the letter x in the subscript is replaced by either p for primary winding or s for secondary winding. In Figure 7.10, we see that the coil cross section may be thought of as being constructed from a unit cell arrangement of N_{xpr} parallel conductors in a coil, which are arranged next to each other on a single layer. These parallel conductors only produce one turn. The unit cell is then stacked N_{xud} units deep and N_{xuw} units wide. Note that the fill does not have to be complete. The number of turns in a coil is N_{xcl}. The number of coils in series and the number of coils in parallel will be denoted as N_{xcs} and N_{xcp}, respectively. Since we only have two coils per winding, either $N_{xcs} = 1$ and $N_{xcp} = 2$ or $N_{xcs} = 2$ and $N_{xcp} = 1$. The former case is generally preferred as it will allow smaller conductor sizes, which facilitates easier bending and in addition is preferable in terms of ac losses, though we won't be in a position to study this until Chapter 11.

Because of the extensive number of variables involved, a listing is provided in Table 7.1, which focuses on core dimensions and clearances, and in Table 7.2 which concerns winding parameters.

Table 7.1 Core and Clearance Dimensions

Symbol	Description
m_c	Material of core
l_{cc}	Length of core center
r_c	Rounding radius of core end leg
w_{cec}	Width of core end leg center
w_{cb}	Width of core base/top leg
ρ_c	Mass density of core
μ_r	Parameter of $\mu_B(B)$
$\boldsymbol{\alpha}_\mu$	Parameter vector of $\mu_B(B)$
$\boldsymbol{\beta}_\mu$	Parameter vector of $\mu_B(B)$
$\boldsymbol{\gamma}_\mu$	Parameter vector of $\mu_B(B)$
k_h	MSE loss parameter
k_e	MSE loss parameter
α	MSE loss parameter
β	MSE loss parameter
d_{cs}	Depth of core slot
w_{cs}	Width of core slot
d_c	Depth of core
w_c	Width of core
l_c	Length of core
w_{ce}	Width of core end leg
A_{cbl}	Area of core base leg
l_{cbl}	Magnetic path length of core base leg
A_{cel}	Area of core end leg
l_{cel}	Magnetic path length of core end leg
V_c	Volume of core
M_c	Mass of core
c_{sc}	Horizontal clearance of secondary winding to core
c_{ps}	Clearance between primary and secondary windings
c_{pp}	Horizontal clearance between primary winding coils
c_{pv}^*	Required vertical clearance between primary and core
c_{sv}^*	Required vertical clearance between secondary and core
c_{pv}	Vertical clearance between primary and core
c_{sv}	Vertical clearance between secondary winding and core

Table 7.2 Winding Parameters

Symbol	Description
m_p	Material for primary conductor
a^*_{pt}	Desired total area of all primary coil parallel conductors
k_{pb}	Primary winding build factor
N_{ppr}	Number of primary conductors in parallel
N_{pud}	Number of primary coil unit cells in depth direction (vertical)
N_{puw}	Number of primary unit cells in width direction (horizontal)
N_{pcl}	Number of primary turns per coil
N_{pcs}	Number of primary coils in series
N_{pcp}	Number of primary coils in parallel
ρ_{pc}	Mass density of primary conductor material
σ_{pc}	Conductivity of primary conductor material
$J_{p,mxa}$	Primary winding maximum allowed current density
a_{pc}	Area of a primary coil conductor
r_{pc}	Radius of a primary coil conductor
t_{pins}	Thickness of primary conductor insulation
d_{pu}	Depth of primary winding unit cell
w_{pu}	Width of primary winding unit cell
d_{pw}	Depth of a primary winding coil
w_{pw}	Width of a primary winding coil
r_{pi}	Inner radius of primary winding
r_{po}	Outer radius of primary winding
l_{pc}	Straight length of primary winding
w_{pc}	Straight width of primary winding
k_{ppf}	Primary winding packing factor
A_{pcl}	Primary winding coil cross section
V_{pcl}	Primary winding coil volume
M_p	Mass of primary winding
k_{pbd}	Primary winding bending constant
m_s	Material for secondary conductor
a^*_{st}	Desired total area of all secondary coil parallel conductors
k_{sb}	Secondary winding build factor
N_{spr}	Number of secondary condors in parallel
N_{sud}	Number secondary coil unit cells in depth direction (vertical)
N_{suw}	Number of secondary until cells in width direction (horizontal)
N_{scl}	Number of secondary turns per coil
N_{scs}	Number of secondary coils in series
N_{scp}	Number of secondary coils in parallel
ρ_{sc}	Mass density of secondary conductor material

Table 7.2 (Continued)

Symbol	Description
σ_{sc}	Conductivity of secondary conductor material
$J_{s,\,mxa}$	Secondary winding maximum allowed current density
a_{sc}	Area of a secondary coil conductor
r_{sc}	Radius of a secondary coil conductor
t_{sins}	Thickness of secondary conductor insulation
d_{su}	Depth of secondary unit cell
w_{su}	Width of secondary unit cell
d_{sw}	Depth of a secondary winding coil
w_{sw}	Width of a secondary winding coil
r_{si}	Inner radius of secondary winding
r_{so}	Outer radius of secondary winding
l_{sc}	Straight length of secondary winding
w_{sc}	Straight width of secondary winding
k_{spf}	Secondary winding packing factor
A_{scl}	Secondary coil cross section
V_{scl}	Secondary coil volume
M_s	Mass of secondary winding
k_{sbd}	Secondary winding bending constant

Although there are a significant number of variables involved in the description of this transformer configuration, many of these are not independent. In order to help us organize the description of the transformer, let us define

$$\mathbf{C}_I = \begin{bmatrix} m_c & l_{cc} & r_c & w_{cec} & w_{cb} \end{bmatrix} \tag{7.5-1}$$

$$\mathbf{G}_I = \begin{bmatrix} c_{sc} & c_{ps} & c_{pp} & c_{pv}^* & c_{sv}^* \end{bmatrix} \tag{7.5-2}$$

$$\mathbf{P}_I = \begin{bmatrix} m_p & a_{pt}^* & k_{pb} & t_{pins} & N_{ppr} & N_{pud} & N_{puw} & N_{pcl} & N_{pcs} & N_{pcp} \end{bmatrix} \tag{7.5-3}$$

$$\mathbf{S}_I = \begin{bmatrix} m_s & a_{st}^* & k_{sb} & t_{sins} & N_{spr} & N_{sud} & N_{suw} & N_{scl} & N_{scs} & N_{scp} \end{bmatrix} \tag{7.5-4}$$

In (7.5-1)–(7.5-3), the $\mathbf{C}_I, \mathbf{G}_I, \mathbf{P}_I$, and $\mathbf{S}_I$ denote core, gap (or clearance), primary winding, and secondary winding variables, respectively, and the subscript "I" denotes independent variables. We will also define a corresponding set of dependent variables:

$$\mathbf{C}_D = \begin{bmatrix} \mathbf{P}_c & d_{cs} & w_{cs} & d_c & w_c & l_c & w_{ce} & A_{cbl} & l_{cbl} & A_{cel} & l_{cel} & V_c & M_c \end{bmatrix} \tag{7.5-5}$$

$$\mathbf{G}_D = \begin{bmatrix} c_{pv} & c_{sv} \end{bmatrix} \tag{7.5-6}$$

$$\mathbf{P}_D = \begin{bmatrix} \mathbf{P}_p & a_{pc} & r_{pc} & d_{pu} & w_{pu} & d_{pw} & w_{pw} & r_{pi} & r_{po} & l_{pc} & w_{pc} & k_{ppf} & A_{pcl} & V_{pcl} & M_p & k_{pbd} \end{bmatrix} \tag{7.5-7}$$

$$\mathbf{S}_D = \begin{bmatrix} \mathbf{P}_s & a_{sc} & r_{sc} & d_{su} & w_{su} & d_{sw} & w_{sw} & r_{si} & r_{so} & l_{sc} & w_{sc} & k_{spf} & A_{scl} & V_{scl} & M_s & k_{sbd} \end{bmatrix} \tag{7.5-8}$$

In (7.5-5), (7.5-6), and (7.5-8), $\mathbf{P}_c, \mathbf{P}_p, \mathbf{P}_s$ are vector or data structures that contain information on the core, primary winding, and secondary winding materials, respectively. For our purposes, we have

$$\mathbf{P}_c = \begin{bmatrix} \rho_c & \mu_r & \alpha_\mu & \beta_\mu & \gamma_\mu & k_h & k_e & \alpha & \beta \end{bmatrix} \tag{7.5-9}$$

$$\mathbf{P}_p = \begin{bmatrix} \rho_{pc} & \sigma_{pc} & J_{p,mxa} \end{bmatrix} \tag{7.5-10}$$

$$\mathbf{P}_s = \begin{bmatrix} \rho_{sc} & \sigma_{sc} & J_{s,mxa} \end{bmatrix} \tag{7.5-11}$$

The elements of these structures are described in Tables 7.1 and 7.2. In the case of the core material, the elements μ_r, α_μ, β_μ, and γ_μ are all associated with the anhysteretic permeability function as described in Section 2.3. The subscript μ has been added to differentiate these quantities from the Modified Steinmetz Equation (MSE) loss model parameters (k_h, k_e, α, and β) that use some of the same symbols. We will characterize the anhysteretic magnetization curve based on $\mu_B()$ of Section 2.3 and use the MSE loss model to calculate core loss.

From the independent variables, we will readily be able to calculate all the dependent fields using geometrical relationships, or, in the case of the material information, using table lookup. Once the dependent variables are calculated, it will be convenient to merge the dependent and independent variables as

$$\mathbf{C} = \begin{bmatrix} \mathbf{C}_I & \mathbf{C}_D \end{bmatrix} \tag{7.5-12}$$

$$\mathbf{G} = \begin{bmatrix} \mathbf{G}_I & \mathbf{G}_D \end{bmatrix} \tag{7.5-13}$$

$$\mathbf{P} = \begin{bmatrix} \mathbf{P}_I & \mathbf{P}_D \end{bmatrix} \tag{7.5-14}$$

$$\mathbf{S} = \begin{bmatrix} \mathbf{S}_I & \mathbf{S}_D \end{bmatrix} \tag{7.5-15}$$

Note that in (7.5-1)–(7.5-15), we have organized our variables into vectors. From a programming point of view, however, these quantities are best viewed as data structures. For example, $\mathbf{C}$ would be a structure with fields corresponding to all elements in (7.5-1) and (7.5-5); those elements in (7.5-1) could be viewed as independent or input fields, while those corresponding to (7.5-5) are viewed as dependent or output fields.

In order to calculate our geometrically dependent variables from our independent variables, it is easiest to start with the windings. In particular, the following sequence of calculations yields the dependent fields. The sequence is first executed for the secondary winding (with x replaced by s) and then executed for the primary winding (with x replaced by p). For the most part, the calculations will not be explained as they follow directly from the geometry. However, there are a few points of note. In (7.5-16), the round operator picks the next closest conductor area from a listing of standard wire gauges. In arranging the conductors, the thickness of the insulation is used in (7.5-18) and (7.5-19) to create some pressure to choose the minimum number of parallel conductors. In (7.5-26), the radius calculation differs depending upon whether the sequence is applied to the primary or secondary winding. It should also be noted that the sequence of calculations uses the independent variables of the core and clearances, not those variables associated with the windings. Finally, in (7.5-30), the bending constant of each winding is the ratio of the bending radius of the conductors to the conductor diameter. This value will be limited to practically achievable values.

$$a_{xc} = \text{round}_{SWG}\left(\frac{a^*_{xt}}{N_{xpr}}\right) \tag{7.5-16}$$

$$r_{xc} = \sqrt{\frac{a_{xc}}{\pi}} \tag{7.5-17}$$

$$d_{xu} = 2(r_{xc} + t_{xins})N_{xpr}k_{xb} \tag{7.5-18}$$

$$w_{xu} = 2(r_{xc} + t_{xins})k_{xb} \tag{7.5-19}$$

$$d_{xw} = d_{xu}N_{xud} \tag{7.5-20}$$

$$w_{xw} = w_{xu}N_{xuw} \tag{7.5-21}$$

$$l_{xc} = l_{cc} \tag{7.5-22}$$

$$w_{xc} = w_{cec} \tag{7.5-23}$$

$$A_{xcl} = w_{xw}d_{xw} \tag{7.5-24}$$

$$k_{xpf} = \frac{a_{xc}N_{xpr}N_{xcl}}{A_{xcl}} \tag{7.5-25}$$

$$r_{xi} = \begin{cases} r_c + c_{sc} & x = \text{'s'} \\ r_{so} + c_{ps} & x = \text{'p'} \end{cases} \tag{7.5-26}$$

$$r_{xo} = r_{xi} + w_{xw} \tag{7.5-27}$$

$$V_{xcl} = d_{xw}\left(\pi\left(r_{xo}^2 - r_{xi}^2\right) + 2w_{xw}(l_{xc} + w_{xc})\right) \tag{7.5-28}$$

$$M_x = 2V_{xcl}k_{xpf}\rho_{xc} \tag{7.5-29}$$

$$k_{xbd} = \frac{r_{xi}}{2r_{xc}} \tag{7.5-30}$$

After the winding calculations are completed, core and clearance calculations can be carried out. Again using geometry, one obtains

$$d_{cs} = \max\left(d_{sw} + 2c_{sv}^*, d_{pw} + 2c_{pv}^*\right) \tag{7.5-31}$$

$$c_{pv} = \frac{1}{2}(d_{cs} - d_{pw}) \tag{7.5-32}$$

$$c_{sv} = \frac{1}{2}(d_{cs} - d_{sw}) \tag{7.5-33}$$

$$w_{cs} = 2w_{sw} + 2w_{pw} + 2c_{sc} + 2c_{ps} + c_{pp} \tag{7.5-34}$$

$$w_{ce} = w_{cec} + 2r_c \tag{7.5-35}$$

$$l_c = l_{cc} + 2r_c \tag{7.5-36}$$

$$d_c = d_{cs} + 2w_{cb} \tag{7.5-37}$$

$$w_c = w_{cs} + 2w_{ce} \tag{7.5-38}$$

$$A_{cbl} = w_{cb}l_c \tag{7.5-39}$$

$$l_{cbl} = w_{cs} + w_{ce} \tag{7.5-40}$$

$$A_{cel} = \pi r_c^2 + l_{cc}w_{ce} + 2w_{cec}r_c \tag{7.5-41}$$

$$l_{cel} = d_{cs} + w_{cb} \tag{7.5-42}$$

$$V_c = 2\left(A_{cbl}w_c + A_{cel}d_{cs} - \pi r_c^2 w_{cb}\right) \tag{7.5-43}$$

$$M_c = V_c\rho_c \tag{7.5-44}$$

Finally, there are some calculations relevant to the overall transformer, including total length, total width, and total electromagnetic mass. The total depth is equal to the core depth.

$$l_T = l_{cc} + 2(r_c + c_{sc} + w_{sw} + c_{ps} + w_{pw}) \tag{7.5-45}$$

$$w_T = w_c + 2(c_{sc} + w_{sw} + c_{ps} + w_{pw}) \tag{7.5-46}$$

$$M_T = M_c + M_p + M_s \tag{7.5-47}$$

The sequence of calculations represented by (7.5-16)–(7.5-30), utilized first for the secondary winding and then for the primary winding, followed by (7.5-31)–(7.5-47) can be thought of in function form as

$$\mathbf{T} = F_{CGPS}(\mathbf{C}_I, \mathbf{G}_I, \mathbf{P}_I, \mathbf{S}_I) \tag{7.5-48}$$

where

$$\mathbf{T} = [\mathbf{C} \ \mathbf{G} \ \mathbf{P} \ \mathbf{S} \ l_T \ w_T \ M_T] \tag{7.5-49}$$

7.6 Core-Type Transformer MEC

In order to design a transformer, we will need a magnetic analysis. Here again, we will use a magnetic equivalent circuit. A possible MEC of the core-type transformer is depicted in Figure 7.11. Therein, the MMF associated with the primary and secondary coils is $N_{pc}i_{pc}$ and $N_{sc}i_{sc}$, respectively, where i_{pc} and i_{sc} are the currents in a primary or secondary coil, which are related to the winding currents by

$$i_{xc} = \frac{i_x}{N_{xcp}} \tag{7.6-1}$$

where x may be p or s. The reluctances $R_{bl}()$ and $R_{el}()$ are associated with the base/top leg and end leg of the core, respectively, and are a function of the flux passing through them. The primary winding coil leakage reluctance is denoted as R_{pl}, which we will assume to be constant. The secondary coil leakage is broken into two terms, R_{sli} and R_{sle}, which correspond to terms that are independent and dependent on using the core as part of the magnetic path. As we proceed in our discussion, we will freely go between permeances and reluctances where P_x and R_x are reciprocals of each other and x is the appropriate subscript.

Before proceeding to discuss each of these terms in detail, it is convenient to set forth a reduced magnetic equivalent circuit. Using circuit reduction techniques yields the circuit shown in

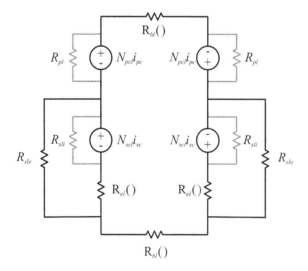

Figure 7.11 Core type transformer magnetic equivalent circuit.

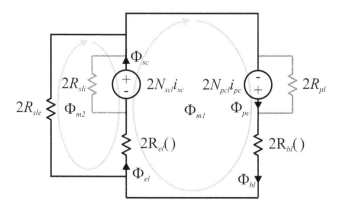

Figure 7.12 Reduced magnetic equivalent circuit.

Figure 7.12. Therein, Φ_{m1} and Φ_{m2} denote mesh fluxes, and Φ_{pc} and Φ_{sc} denote the flux associated with one primary and one secondary coil, respectively. Fluxes Φ_{el} and Φ_{bl} are the flux through the end leg and base/top leg, respectively, and will be useful for calculating core loss. Observe that the R_{sli} and R_{pl} reluctances are shown as being lightly shaded since they will not be formally a part of our MEC as they are placed directly across MMF sources.

The flux linking the primary and secondary windings may be expressed

$$\lambda_p = N_{pcl}N_{pcs}\left(\Phi_{m1} + \frac{N_{pcl}i_{pc}}{R_{pl}}\right) \tag{7.6-2}$$

$$\lambda_s = N_{scl}N_{scs}\left(\Phi_{m1} - \Phi_{m2} + \frac{N_{scl}i_{sc}}{R_{slci}}\right) \tag{7.6-3}$$

Computation of the reluctances of the magnetizing paths is straightforward. In particular,

$$R_{el}(\Phi_{m1} - \Phi_{m2}) = \frac{l_{el}}{A_{el}\mu_B((\Phi_{m1} - \Phi_{m2})/A_{el})} \tag{7.6-4}$$

$$R_{bl}(\Phi_{m2}) = \frac{l_{bl}}{A_{bl}\mu_B(\Phi_{m2}/A_{bl})} \tag{7.6-5}$$

Next, let us consider the leakage paths. A general indication of the paths is depicted in Figure 7.13. Therein, the leakage flux paths are divided into two types: (a) those in the slot and thus in the interior of the core and (b) and those on the outside of the slot and exterior to the core.

Clearly, the nature of the leakage path changes substantially depending upon whether the section of the coil in question is inside or outside of the core window. It will prove convenient to formally define which part of the coil is interior to the core window and which part is exterior to the core window. To this end, consider Figure 7.14. Therein, the centerline of each coil is indicated, the portion of the center coil considered to be interior to the core window is shown as a solid line, and that portion considered to be exterior to the core window is shown as a dashed line.

We will take the division of the coil into interior and exterior portions to occur where the centerline of the coil intersects the edge of the core. The angle (defined in Figure 7.14) where this occurs is at

$$\alpha_x = \operatorname{asin}\left(\frac{2r_c}{r_{xi} + r_{xo}}\right) \tag{7.6-6}$$

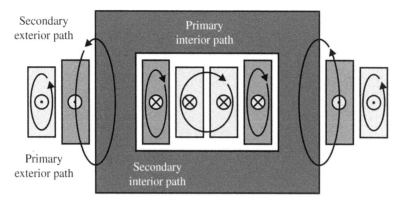

Figure 7.13 Core type transformer leakage paths.

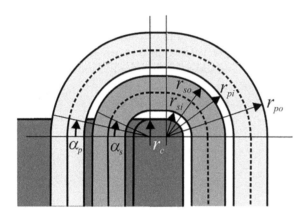

Figure 7.14 Division of winding into interior and exterior portions.

The length of the interior and exterior portions of each coil may then be found as

$$l_{xi} = l_{xc} + \alpha_x(r_{xi} + r_{xo}) \tag{7.6-7}$$

$$l_{xe} = l_{xc} + 2w_{cec} + (\pi - \alpha_x)(r_{xi} + r_{xo}) \tag{7.6-8}$$

Now let us consider the leakage paths of a primary coil. The leakage permeance of the primary coil will be denoted as P_{pl} and will be broken into two parts: that resulting from the leakage flux path in the interior portion of the core window, P_{pli}, and that resulting from the leakage flux path in the exterior portion of the core window, denoted as P_{ple}. These terms are related by

$$P_{pl} = P_{pli} + P_{ple} \tag{7.6-9}$$

Let us first consider the interior core window term. The situation in this case is rather different than in the case of a UI- or EI-core with a gap. To see this, consider Figure 7.15, which depicts a generic conductor bundle inside a core, and let us assume the core is infinitely permeable. If we assume that we have a uniform vertical field, based on Path 1 we conclude that the vertical component of the field at the indicated position, denoted as H_y, is given by

$$H_y = \frac{Ni}{d_s} \tag{7.6-10}$$

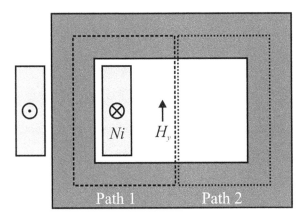

Figure 7.15 Consideration of a vertical leakage path.

However, if we perform the same calculation based on Path 2, we conclude

$$H_y = 0 \qquad (7.6\text{-}11)$$

Given the discrepancy in our results, we conclude that the assumption of a uniform vertical or horizontal field pattern is a poor one in this situation. Further, the assumption that the MMF drop in the magnetic material is negligible in this situation is a bit problematic in this case. If there were truly no MMF drop across the path, then the inductance presented by the windings would be infinite. As a result, the current in (7.6-10) would be zero. While this would make (7.6-10) and (7.6-11) at least consistent, the assumption is not very conducive in helping us compute the leakage inductance.

As an alternative, we will assume that the leakage flux traveling within the core window predominately stays in the core window, and so the leakage paths that remain do not utilize the core. In the case of the primary winding, we will assume the leakage flux paths P1 and P2 indicated in Figure 7.16. Therein, it can be seen that when computing the permeance, we will consider the path that includes both coils, which we will treat as a single winding bundle with the same outside dimension. The permeance associated with P1 is based on our work in Section 2.7 and in particular

Figure 7.16 Leakage flux paths.

(2.7-52)–(2.7-54). The permeance associated with P2 is based on a series connection of reluctances that surround the combined winding bundle. The vertical reluctance terms will be assumed to occupy one-half the space between the primary and secondary windings.

The result of such an analysis yields

$$P_{pli} = \begin{cases} \dfrac{\mu_o l_{pi} \left(4k_{pi2}^4 + 8k_{pi1} k_{pi2}^3 + 2k_{pi1}^2 k_{pi2}^2 - 2k_{pi1}^3 k_{pi2}^1 + k_{pi1}^4 \ln\left(1 + \dfrac{2k_{pi2}}{k_{pi1}}\right) \right)}{64(2w_{pw} + c_{pp})^2 d_{pw}^2} \\ + \dfrac{\mu_o l_{pi}}{\left(\dfrac{d_{cs} + d_{pw}}{c_{ps}} + \dfrac{(2w_{pw} + c_{ps} + c_{pp})}{c_{pv}} \right)} \end{cases} \quad (7.6\text{-}12)$$

where

$$k_{pi1} = |2w_{pw} + c_{pp} - d_{pw}| \quad (7.6\text{-}13)$$

$$k_{pi2} = \min(2w_{pw} + c_{pp}, d_{pw}) \quad (7.6\text{-}14)$$

In (7.6-12), the upper term corresponds to path P1 and the lower term corresponds to path P2. Note that the derivation of (7.6-12) includes a multiplication by a factor of two so that it becomes a single-coil equivalent.

The derivation of the primary winding exterior leakage terms is not as burdened with subtleties. Using paths P3 and P4, and our work in Section 2.7 discussed in the section on exterior isolated conductor leakage flux and culminating in (2.7-52)–(2.7-55), we obtain

$$P_{ple} = \begin{cases} \dfrac{\mu_o l_{pe} \left(4k_{pe2}^4 + 8k_{pe1} k_{pe2}^3 + 2k_{pe1}^2 k_{pe2}^2 - 2k_{pe1}^3 k_{pe2}^1 + k_{pe1}^4 \ln\left(1 + \dfrac{2k_{pe2}}{k_{pe1}}\right) \right)}{128 w_{pw}^2 d_{pw}^2} \\ + \dfrac{\mu_o l_{pe}}{2\pi} \ln\left(1 + \dfrac{\pi(w_{sw} + 2c_{ps})}{2d_{pw} + 2w_{pw}}\right) \end{cases} \quad (7.6\text{-}15)$$

where

$$k_{pe1} = |w_{pw} - d_{pw}| \quad (7.6\text{-}16)$$

$$k_{pe2} = \min(w_{pw}, d_{pw}) \quad (7.6\text{-}17)$$

In (7.6-15), the upper term corresponds to P3 and the lower term corresponds to P4. In the term corresponding to P4, the maximum distance for flux to flow from the coil—that is r_{mx} in (2.7-55)—is taken as the distance from the primary coil to the center of the secondary coil. Beyond this point, the flux density due to this path will be considered to couple the secondary coil.

Let us now consider the secondary winding. For the interior leakage inductance, we will consider paths P5 and P6. The permeance associated with path P5 is calculated using the development of Section 2.7 and in particular (2.7-52)–(2.7-54) with appropriate changes in variable names. The permeance associated with path P6 is determined by using a series combination of reluctances around the winding; the reluctance corresponding to the region between the primary and secondary winding is based on a width of one-half the distance between the coils, the other half being allocated to path P2. This yields

Figure 7.17 Exterior secondary leakage flux paths.

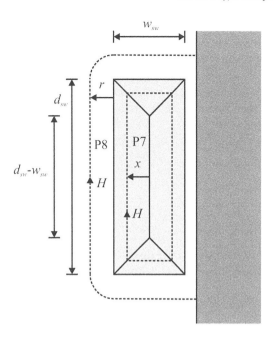

$$P_{sli} = \begin{cases} \dfrac{\mu_0 l_{si} \left(4k_{si2}^4 + 8k_{si1}k_{si2}^3 + 2k_{si1}^2 k_{si2}^2 - 2k_{si1}^3 k_{si2}^1 + k_{si1}^4 \ln\left(1 + \dfrac{2k_{si2}}{k_{si1}^4}\right)\right)}{128 w_{sw}^2 d_{sw}^2} \\ + \dfrac{\mu_0 l_{si}}{\left((d_{cs} + d_{sw})\left(\dfrac{1}{2c_{sc}} + \dfrac{1}{c_{ps}}\right) + \dfrac{(2w_{sw} + c_{sc} + c_{ps})}{c_{sv}}\right)} \end{cases} \quad (7.6\text{-}18)$$

where

$$k_{si1} = |w_{sw} - d_{sw}| \tag{7.6-19}$$

$$k_{si2} = \min(w_{sw}, d_{sw}) \tag{7.6-20}$$

The exterior portion of the secondary leakage inductance is based on paths P7 and P8 in Figure 7.16, which are highlighted in Figure 7.17. Using the techniques of Section 2.7, one obtains

$$P_{sle} = \begin{cases} \dfrac{\mu_0 l_{se}\left(16k_{se2}^4 + 16\sqrt{2}k_{se1}k_{se2}^3 + 4k_{se1}^2 k_{se2}^2 - 2\sqrt{2}k_{se1}^3 k_{se2} + k_{se1}^4 \ln\left(1 + \dfrac{2\sqrt{2}k_{se2}}{k_{se1}}\right)\right)}{256 w_{sw}^2 d_{sw}^2} \\ + \dfrac{\mu_0 l_{se}}{\pi} \ln\left(1 + \dfrac{\pi \min(2c_{ps} + w_{pw}, d_c - d_{sw})}{4w_{sw} + 4c_{sc} + 2d_{sw}}\right) \end{cases}$$

$$(7.6\text{-}21)$$

$$k_{se1} = |d_{sw} - 2w_{sw}| \tag{7.6-22}$$

$$k_{se2} = \dfrac{1}{\sqrt{2}} \min(2w_{sw}, d_{sw}) \tag{7.6-23}$$

where first term corresponds to P7 and the second term to P8. In the second term, r varies from 0 to the point corresponding to the center of the primary coil.

In general, the inputs to our MEC will be the primary and secondary currents, i_p and i_s, and the transformer parameters **T**. The outputs are the primary and secondary flux linkages, λ_p and λ_s. In addition, it will be useful to return the flux density in the base and end legs, which we will use to calculate core loss; this will be our next topic. The base leg and end leg flux may readily be expressed as

$$B_{bl} = \frac{\Phi_{bl}}{A_{bl}} \tag{7.6-24}$$

and

$$B_{el} = \frac{\Phi_{el}}{A_{el}} \tag{7.6-25}$$

where the base leg and end leg flux are available as branch fluxes in the MEC, as seen in Figure 7.12. With these input and output definitions, our MEC may be functionally represented as

$$[\lambda_p, \lambda_s, B_{bl}, B_{el}] = F_{MEC}(i_p, i_s, \mathbf{T}) \tag{7.6-26}$$

In (7.6-26), the flux linkages, flux densities, and currents could be scalars or correspondingly dimensioned vectors of points representing waveforms.

7.7 Core Loss

Core loss is a significant contributor to overall transformer loss and dominates no-load losses. The calculation of the transformer core loss will utilize our transformer MEC, as well as our work in Chapter 6. The approach will be to use the MEC of the transformer to compute the flux density waveforms in the base/top leg and the end legs, and then from these waveforms we will use the MSE model presented in Chapter 6 to compute the core loss.

Specifically, our input to the core loss calculation will be primary and secondary current waveforms, $i_p(t)$ and $i_s(t)$. In determining a transformer operating point for a given load, these variables are determined in Steps 3 and 4 of the algorithm used to analyze a steady-state operating point described in Section 7.4. Computationally, these waveforms are stored as vectors with points corresponding to discrete points in time. Using our MEC of Section 7.6, we may calculate the flux density in the base leg and end leg. In particular,

$$[\lambda_p(t),\ \lambda_s(t),\ B_{bl}(t),\ B_{el}(t)] = F_{MEC}(i_p(t), i_s(t), \mathbf{T}) \tag{7.7-1}$$

Here again, computationally $B_{bl}(t)$ and $B_{el}(t)$ are represented by a vector of points. Based on these waveforms, the power loss density in the base leg and end leg may be computed by first determining the maximum flux density as

$$B_{x,mx} = \max(|B_x(t)|) \tag{7.7-2}$$

then the equivalent frequency as

$$f_{x,eq} = \frac{1}{B_{x,mx}^2 \pi^2} \int_0^{T/2} \left(\frac{dB_x}{dt}\right)^2 dt \tag{7.7-3}$$

and finally the power loss density as

$$p_x = k_h \left(\frac{f_{x,eq}}{f_b}\right)^{\alpha-1} \left(\frac{B_{x,mx}}{B_b}\right)^{\beta} \left(\frac{f}{f_b}\right) + k_e \frac{2}{T} \int_0^{T/2} \left(\frac{dB_x}{dt}\right)^2 dt \qquad (7.7\text{-}4)$$

In (7.7-1)–(7.7-4), x is either bl for the base leg or el for the end leg. These equations are the combined loss model (using the MSE form) presented in Section 6.4. The parameters k_h, α, β, and k_e are parameters of the core material as described in Section 6.3. The equations have been modified under that the assumption that the flux density waveforms are odd half-wave symmetric, so that it is only necessary to compute them for one-half of an electrical cycle. The differentiation of the flux density waveform is accomplished numerically.

After the power loss density in the base leg and end leg is computed, the core loss is readily expressed as

$$P_{cl} = 2p_{bl}A_{bl}l_{bl} + 2p_{el}A_{el}l_{el} \qquad (7.7\text{-}5)$$

Functionally, we may represent the sequence of calculations (7.7-1)–(7.7-5) as

$$P_{cl} = F_{CL}(i_p(t), i_s(t), \mathbf{T}) \qquad (7.7\text{-}6)$$

which is the form required by the calculation of the transformer operating point in Section 7.4.

7.8 Core-Type Transformer Design

At this point, we are now poised to formulate transformer design as an optimization problem. In this section, one possible formulation will be proposed. In Section 7.9, we will carry out a case study.

Design Space

The first step in formulating the design problem as an optimization problem will be to identify the design space. Here, we will take the design space to be

$$\boldsymbol{\theta} = \begin{bmatrix} m_c & l_{cc} & r_c & w_{cec} & w_{cb} & m_p & a_{pt}^* & N_{ppr} & N_{pcl}^* & R_{pdw}^* & m_s & a_{st}^* & N_{spr} & R_{Nps}^* & R_{sdw}^* \end{bmatrix} \qquad (7.8\text{-}1)$$

Observe that (7.8-1) includes many, but not all, of the independent variables associated with the core $\mathbf{C}_I$, gaps (clearances) $\mathbf{G}_I$, primary winding $\mathbf{P}_I$ and secondary winding $\mathbf{S}_I$ as defined by (7.5-1)–(7.5-4). Variables with an asterisk are so marked because the parameter listed is a desired value; the actual value will be determined to be close to this value subject to discretization of allowable values. Previously undefined variables in (7.8-1) include R_{pdw}^*, the desired primary coil depth to width ratio, R_{sdw}^*, the desired secondary coil depth to width ratio, and R_{Nps}^*, the desired primary to secondary turns ratio.

The remaining independent variables will be incorporated into $\mathbf{D}_{fp}$ as

$$\mathbf{D}_{fp} = \begin{bmatrix} c_{sc} & c_{ps} & c_{pp} & c_{pv}^* & c_{sv}^* & N_{pcs} & N_{pcp} & k_{pb} & t_{pins} & N_{scs} & N_{scp} & k_{sb} & t_{sins} \end{bmatrix} \qquad (7.8\text{-}2)$$

The variable $\mathbf{D}$ will be a vector of design information, which computationally is most conveniently viewed as a structure. The subscript fp denotes fixed parameters. Hence, we are assuming that most of the clearances will be fixed in advance based on the clearance necessitated by the insulation breakdown voltage. As can be seen, the configuration of each winding in terms of the number of coils in series and the number of coils in parallel, as well as the assumed build factor, is also specified in advance.

Using $\boldsymbol{\theta}$ and $\mathbf{D}_{fp}$ all the independent variables associated with $\mathbf{C}_I$, $\mathbf{G}_I$, $\mathbf{P}_I$, and $\mathbf{S}_I$ are known except for the primary winding variables N_{pcl}, N_{pud}, and N_{puw} and secondary winding variables N_{scl}, N_{sud}, and N_{suw}. The primary variables are readily found from the sequence of calculations

$$N_{pcl} = \text{round}\left(N_{pcl}^*\right) \tag{7.8-3}$$

$$N_{pud} = \text{ceil}\left(\sqrt{\frac{N_{pcl}R_{pdw}^*}{N_{ppr}}}\right) \tag{7.8-4}$$

$$N_{puw} = \text{ceil}\left(\frac{N_{pcl}}{N_{pud}}\right) \tag{7.8-5}$$

The expression (7.8-3) rounds the number of turns to an integer; the expressions (7.8-5) and (7.8-6) set the number of unit cells in such a way as to approximately achieve the correct aspect ratio.

For the secondary winding, the remaining independent variables are calculated as

$$N_{scl} = \max\left(1, \text{round}\left(\frac{N_{pcl}N_{pcs}}{R_{Nps}^* N_{scs}}\right)\right) \tag{7.8-6}$$

$$N_{sud} = \text{ceil}\left(\sqrt{\frac{N_{scl}R_{sdw}^*}{N_{spr}}}\right) \tag{7.8-7}$$

$$N_{suw} = \text{ceil}\left(\frac{N_{scl}}{N_{sud}}\right) \tag{7.8-8}$$

The expression (7.8-6) is designed to obtain the desired turns ratio as closely as possible given the discretization of the turn counts. Equations (7.8-7) and (7.8-8) determine the number of unit cells so as to approximately obtain the desired coil aspect ratio.

Metrics and Constraints

Let us now consider a set of constraints on the transformer. First, we will have constraints on the total length, width, and depth of the transformer. This yields the set of constraints

$$c_1 = \text{lte}(l_T, l_{Tmxa}) \tag{7.8-9}$$

$$c_2 = \text{lte}(w_T, w_{Tmxa}) \tag{7.8-10}$$

$$c_3 = \text{lte}(d_c, d_{Tmxa}) \tag{7.8-11}$$

where l_{Tmxa}, w_{Tmxa}, and d_{Tmxa} are the maximum allowed length, width, and depth, respectively. Related to the overall dimensions, we will constrain the aspect ratio of the transformer. We will define the aspect ratio as

$$\alpha_T = \frac{\max(l_T, w_T, d_c)}{\min(l_T, w_T, d_c)} \tag{7.8-12}$$

The constraint on the aspect ratio will be given by

$$c_4 = \text{lte}(\alpha_T, \alpha_{Tmxa}) \tag{7.8-13}$$

where α_{Tmxa} is the maximum allowed aspect ratio.

In order to implement the curvature at the ends of the core, the end laminations will be different from the laminations in the center region. We will define the lamination factor as

$$k_l = \frac{2r_c}{w_{ce}} \qquad (7.8\text{-}14)$$

This factor is the fraction of end laminations needed to implement the curvature. This will be subject to the constraint

$$c_5 = \text{lte}(k_l, k_{lmxa}) \qquad (7.8\text{-}15)$$

where k_{lmxa} is the maximum allowed lamination fraction.

In order to focus the range of the objective space, we will constrain the electromagnetic mass, leading to the constraint

$$c_6 = \text{lte}(M_T, M_{Tmxa}) \qquad (7.8\text{-}16)$$

We will require that the depth of the primary winding, which is stacked on top of the secondary winding, be less than that of the secondary winding. Thus

$$c_7 = \text{lte}(d_{pw}, d_{sw}) \qquad (7.8\text{-}17)$$

Large conductors do not bend easily. Therefore, it is appropriate to limit the bending radius. This leads to the constraints

$$c_8 = \text{gte}(k_{pbd}, k_{bdmnr}) \qquad (7.8\text{-}18)$$

$$c_9 = \text{gte}(k_{sbd}, k_{bdmnr}) \qquad (7.8\text{-}19)$$

where k_{bdmnr} is the minimum required bending radius constant. Typical values for the bend radius constant are in the range of 6–20.

The remaining constraints are related to electrical performance. The first of these is a constraint on the transformer regulation. The transformer regulation is calculated based on (7.4-11). This will lead to two constraints. The first of these, c_{10}, is related to the convergence of the MEC because it is used to calculate the electrical parameters. If this constraint is not met, the fitness is calculated based on the constraints passed thus far. If this constraint is met, then a constraint on regulation is imposed:

$$c_{11} = \text{lte}(\chi, \chi_{mxa}) \qquad (7.8\text{-}20)$$

In order to limit the inrush current, we will utilize the transformer MEC with a primary current equal to the maximum allowed inrush current and zero secondary current. The execution of the MEC will lead to a constraint variable c_{12}, which is 1 if the MEC converges and 0 if not. If this constraint is passed, we will require that the primary flux linkage satisfy

$$c_{13} = \text{gte}(\lambda_{pi}, \lambda_{pwci}) \qquad (7.8\text{-}21)$$

where λ_{pi} is the primary flux linkage at a primary current equal to the maximum allowed peak inrush current, denoted as i_{pimxa}, and λ_{pwci} is the worst-case inrush primary flux linkage which may be found using (7.4-49).

Constraints will be imposed based on specific operating points. The first requirement in doing this is to perform the MEC analysis required to formulate the $F_{im}()$ function given by (7.4-16). The constraint c_{14} will be calculated based on convergence of the MEC in this regard. If convergence fails, then the fitness will be calculate based on the number of constraints satisfied. Otherwise, execution will proceed to analyze operating points.

To this end, we will let $\mathbf{V}_p$, $\mathbf{Z}_l$, and $\boldsymbol{\omega}_e$ be vectors of primary voltage, secondary impedance, and radian frequency values. Each primary voltage and secondary impedance value correspond to an operating point of interest. Thus $V_{p,k}$, $Z_{l,k}$, and $\omega_{e,k}$ are the primary voltage, load impedance, and operating frequency for the kth operating point. For each operating point, we will solve the equivalent circuit as discussed in Section 7.4 under the section on operating point analysis. We will let $\mathbf{I}_p$, $\mathbf{V}_s$, $\mathbf{I}_s$, $\mathbf{P}_{prl}$, $\mathbf{P}_{srl}$, $\mathbf{P}_{cl}$, $\mathbf{P}_l$, $\mathbf{J}_p$, $\mathbf{J}_s$, $\mathbf{R}_c$, and $\mathbf{L}_m$ be vectors whose elements describe the primary current, secondary voltage, primary winding resistive loss, secondary winding resistive loss, core loss, total loss, primary winding current density, secondary winding current density, core resistance, and magnetizing inductance at each operating point.

We will organize $\mathbf{Z}_l$ such that the first element is essentially infinite so that it represents the no-load case. The constraint c_{15} will be based on the convergence of the operating point analysis. For the no-load case, we impose a constraint on the no-load primary current

$$c_{16} = \text{lte}(|I_{p,1}|, I_{p,nlmxa}) \tag{7.8-22}$$

as well as constraints on the range of the secondary voltage

$$c_{17} = \text{gte}(|V_{s,1}|, V_{snl,mnr}) \tag{7.8-23}$$

$$c_{18} = \text{lte}(|V_{s,1}|, V_{snl,mxa}) \tag{7.8-24}$$

On the remaining operating points, we will have a constraint c_{16+3k} on the convergence of the operating point analysis, where k is the operating point number. We will also constrain the current density in the primary and secondary windings at each operating point using

$$c_{17+3k} = \text{lte}(J_{p,k}, J_{p,mxa}) \tag{7.8-25}$$

$$c_{18+3k} = \text{lte}(J_{s,k}, J_{s,mxa}) \tag{7.8-26}$$

After the loss at every operating point is computed, we can compute the weighted loss as

$$P_l = \mathbf{w}^T \mathbf{P}_l \tag{7.8-27}$$

Each element of $\mathbf{w}$ corresponds to a loss weight for the corresponding operating point. The sum of the elements of $\mathbf{w}$ should be one. Observe that this measure of power loss will also be one of our metrics of interest. Denoting the number of operating points as K, a constraint is also placed on the total loss:

$$c_{19+3K} = \text{lte}(P_l, P_{lmxa}) \tag{7.8-28}$$

This concludes our list of constraints.

The design vector $\mathbf{D}_{fp}$ given by (7.8-2) includes fixed parameters of the designs. It is also convenient to organize the design specifications. To this end, we define

$$\begin{aligned}\mathbf{D}_{ds} = [&f \ \omega_e \ V_{p0} \ V_{snlmnr} \ V_{snlmxa} \ I_{pnlmxa} \ I_{sfl} \ \chi_{mxa} \ i_{pimxa} \ \lambda_{pwci} \ \ldots \\ &\mathbf{V}_p \ \mathbf{Z}_l \ \boldsymbol{\omega}_e \ \mathbf{w} \ P_{lmxa} \ M_{Tmxa} \ l_{Tmxa} \ w_{Tmxa} \ d_{Tmxa} \ k_{lmxa} \ k_{bdmnr} \ k_{lmxa} \ \alpha_{mxa} \ \ldots \\ &i_{p,t} \ i_{s,t} \ N_{impt} \ N_{tpt} O_{emxa} \ O_{lmxa}] \end{aligned}$$
$$\tag{7.8-29}$$

In (7.8-29), f is the fundamental frequency and ω_e is the corresponding radian frequency. Most of the remaining variables have been defined. Of those that have not, the currents $i_{p,t}$ and $i_{s,t}$ are the test currents used to determine the transformer parameters as described in Section 7.4. The quantities N_{impt} and N_{tpt} are the number of data points used in describing the magnetizing flux linkage characteristic and the number of time points used in representing the time domain waveforms when calculating transformer operating points, respectively. The variables O_{emxa} and O_{imxa} represent the maximum allowed error and maximum allowed iterations of the operating point analysis described in Section 7.4.

Before proceeding, it will be convenient to define our aggregate design information vector **D** as

$$\mathbf{D} = \begin{bmatrix} \mathbf{D}_{fp} & \mathbf{D}_{ds} \end{bmatrix} \tag{7.8-30}$$

From the information in our parameter vector $\boldsymbol{\theta}$ and our design information vector **D** our goal is to formulate a fitness function $\mathbf{f}(\boldsymbol{\theta}, \mathbf{D})$ for use in transformer design.

If all constraints are met, the fitness will be calculated as

$$\mathbf{f}(\boldsymbol{\theta}, \mathbf{D}) = \begin{bmatrix} \dfrac{1}{M_T} & \dfrac{1}{P_l} \end{bmatrix}^T \tag{7.8-31}$$

If the constraints are not met, then the fitness is calculated based on the percentage of constraints satisfied. This will be described in the next section.

Calculation of Fitness

Now that we have discussed the design constraints and metrics in general terms, it is appropriate to discuss the calculation of fitness algorithmically. One approach is set forth in pseudo-code form in Table 7.3.

Therein, Step 1 is the calculation of the transformer geometry. This begins with the initialization of the number of constraints based on the number of operating points K. Next the fields of $\mathbf{C}_I$, $\mathbf{G}_I$, $\mathbf{P}_I$, and $\mathbf{S}_I$ which can be determined based on the parameter vector $\boldsymbol{\theta}$ and design information **D** are determined. The remaining fields can be assigned using the cited expressions. Next, the dependent geometrical parameters are performed. None of these calculations are computationally intense. After the computations are made, a test is made based on the first nine constraints, which are all geometrical in nature. If not all constraints are passed, the fitness is calculated. Otherwise, execution proceeds to Step 2. Pseudo-code for the testing of constraints is listed in Table 7.4.

In Step 2, the transformer parameters and regulation are determined. This begins with setting up the MEC by calculating the permeances and processing the branch lists. Since the MEC will be called several times, it is expedient to carry out and store the results of that part of the analysis which can be used several times. Next, the first use of the MEC will be to compute the electrical parameters of the transformer using the sequence of calculations represented by (7.4-14). Next, the convergence of the MEC calls used to determine the electrical parameters is tested, and, if successful, the regulation is tested.

In Step 3, a function that returns the magnetizing current as a function of magnetizing flux linkage is created. This is described in Section 7.4 under magnetizing characteristics. It is needed for the operating point solution. The convergence of the MEC analysis used for this task is again tested.

Step 4 is focused on addressing inrush current. Here again, this involves first a test on the MEC analysis, and then a test based on the results of the analysis. Failure of either of the tests (constraints) results in a negative fitness as described in the pseudo-code of Table 7.4.

The no-load operating point is analyzed in Step 5. This will include of convergence test (which requires convergence of the MEC as well as the operating point algorithm), as well as constraints on the no-load primary current and no-load secondary voltage.

Table 7.3 Pseudo-code for Calculation of the Fitness Function

1. Calculate transformer geometry and constraints
   initialize number of constraints to $N_C = 19 + 3K$
   assign fields of $\mathbf{C}_I$, $\mathbf{G}_I$, $\mathbf{P}_I$ and $\mathbf{S}_I$ based on $\boldsymbol{\theta}$ and $\mathbf{D}$
   find N_{pcl}, N_{pud}, N_{puw} using (7.8-3) - (7.8-5)
   find N_{scl}, N_{sud}, N_{sud} using (7.8-6) - (7.8-8)
   calculate $\mathbf{C}_D$, $\mathbf{G}_D$, $\mathbf{P}_D$ and $\mathbf{S}_D$ using (7.5-49)
   create $\mathbf{T}$ using (7.5-50)
   evaluate c_1 - c_9 using (7.6-11) - (7.6-22)
   test constraints (Table 7.4)

2. Determine transformer parameters and regulation
   set up MEC
   determine transformer parameters (7.4-14)
   evaluate c_{10} (convergence of MEC for parameters)
   test constraints (Table 7.4)
   evaluate c_{11} (regulation) using (7.8-20)
   test constraints (Table 7.4)

3. Consider inrush flux linkage
   compute inrush flux linkage using MEC
   determine c_{12} (convergence of MEC for inrush)
   test constraints (Table 7.4)
   evaluate c_{13} (inrush) using (7.8-21)
   test constraints (Table 7.4)

4. Create magnetizing current function (7.4-16)
   determine c_{14} based on convergence of MEC
   test constraints (Table 7.4)

5. Analyze no load operating point
   determine c_{15} based on convergence of MEC
   test constraints (Table 7.4)
   evaluate c_{16} - c_{18} using (7.8-22) - (7.8-24)
   test constraints (Table 7.4)

6. Analyze remaining operating points
   k=2
   $P_l = w_1 \mathbf{P}_{l,\,1}$
   while (k≤K) and ($C_S = C_I$)
       analyze operating point k
       determine $c_{16\,+\,3k}$ based on convergence
       calculate $c_{17\,+\,3k}$-$c_{18\,+\,3k}$ using (7.8-25) - (7.8-26)
       update constraints C_S and C_I
       $P_l = P_l + \mathbf{P}_{l,\,k} w_k$
       k=k+1
   end
   test constraints (Table 7.4)
   compute $c_{19\,+\,3K}$ using (7.8-28)
   test constraints (Table 7.4)

7. Compute fitness
   compute fitness using (7.8-31)
   return

Table 7.4 Pseudo-code for Check of Constraints Satisfied Against Imposed

```
update C_s
update C_I
if (C_s < C_I)
```
$$\mathbf{f} = \varepsilon\left(\frac{C_S - C}{N_C}\right)\begin{bmatrix}1\\1\end{bmatrix}$$
```
    return
end
```

The remaining operating points are considered in Step 6. This begins by setting the operating point index k to 2, and initializing the aggregate loss P_l to $\mathbf{P}_{l,\,k}$ where $\mathbf{P}_l$ will be an array of losses at each operating point and $\mathbf{w}$ is a vector with the loss weight of each operating point. A while loop is used to cycle through remaining operating points. Constraints tested are based on convergence of the MEC/operating point algorithm and current density in the two windings. Note that the while loop terminates after all operating points are considered or if the number of constraints satisfied falls below the number of constraints imposed. Once the while loop terminates, the constraints are tested. The constraint on aggregate loss is then tested.

In Step 7, the fitness is computed and the routine terminates. As can be seen, the procedure is systematic. In the next section, we will use this code to design a single-phase core-type transformer.

7.9 Case Study

In this section, let us now consider a case study in the design of a single-phase core-type transformer. We will somewhat arbitrarily choose the transformer ratings. In this case, we design for a rated or base primary voltage of $V_{pb} = 480$ V and a base secondary voltage of $V_{sb} = 240$ V. We will select a base rated load of $S_b = 5$ kVA. The transformer will be designed for 60 Hz.

The rated base primary and secondary currents are given by

$$I_{pb} = \frac{S_b}{V_{pb}} \tag{7.9-1}$$

and

$$I_{sb} = \frac{S_b}{V_{sb}} \tag{7.9-2}$$

which yields 10.4 and 20.8 A, respectively. The rated load impedance is given by

$$Z_{lb} = \frac{V_{sb}}{I_{sb}} \tag{7.9-3}$$

which yields 11.5 Ω.

The assumed specifications are set forth in Table 7.5. Many of the quantities listed are in terms of rated or base values. From the operating point $\mathbf{V}_p$, $\mathbf{Z}_l$, and $\boldsymbol{\omega}_e$ vectors observe that half-load and full-load operating points are both considered, in addition to the no-load point which is automatically treated. Both the half-load and full-load impedances are taken to be resistive. The dimensionality of the power loss weight vector $\mathbf{w}$ is one element larger than the operating

Table 7.5 Transformer Specifications

Symbol	Description	Value
f	Fundamental frequency	60 Hz
ω_e	Fundamental frequency	377 rad/s
V_{p0}	Nominal primary voltage	V_{pb}
V_{snlmnr}	Minimum required no-load secondary voltage	$0.98 V_{sb}$
V_{snlmxa}	Maximum allowed no-load secondary voltage	$1.02 V_{sb}$
I_{pnlmxa}	Maximum allowed no-load secondary current	$0.1 I_{pb}$
I_{sfl}	Full-load secondary current	I_{sb}
χ_{mxa}	Maximum allowed regulation	0.05
i_{pimxa}	Maximum instantaneous primary inrush current	$2\sqrt{2} I_{pb}$
λ_{pwci}	Worst case instantaneous primary flux linkage	$2\sqrt{2} V_{pb}/\omega_e$
$\mathbf{V}_b$	Operating point voltages	$V_{pb}[1\ \ 1]$ V
$\mathbf{Z}_l$	Operating point impedances	$Z_{lb}[2\ \ 1]$ Ω
$\boldsymbol{\omega}_e$	Operating point frequencies	$\omega_e[1\ \ 1]$ rad/s
$\mathbf{w}$	Operating point loss weights	[0.1 0.4 0.5]
P_{lmxa}	Maximum allowed aggregate loss	150 W
M_{Tmxa}	Maximum allowed total electromagnetic mass	40 kg
l_{Tmxa}	Maximum allowed total length	1 m
w_{Tmxa}	Maximum allowed total width	1 m
d_{Tmxa}	Maximum allowed total depth	1 m
k_{lmxa}	Maximum allowed lamination fraction	1
k_{bdmnr}	Minimum required bend radius ratio	8
α_{mxa}	Maximum allowed aspect ratio	3
i_{pt}	Primary winding test current	1 nA
i_{st}	Secondary winding test current	1 nA
N_{impt}	Number of magnetizing current points	25
N_{tpt}	Number of time points per half-cycle	25
O_{emxa}	Maximum allowed operating point calculation error	10^{-3}
O_{imxa}	Maximum allowed operating point iterations	10

point vectors since the first element corresponds to the no-load condition (which could be viewed as another operating point). The last six entries of Table 7.5 relate to algorithm parameters rather than true specifications. Table 7.6 itemizes fixed design parameters constituting the elements of $\mathbf{D}_{fp}$. This includes primarily clearances and information on the winding configuration. Finally, the domain of $\boldsymbol{\theta}$ defined by (7.8-2), which comprises the design space, is summarized in Table 7.7.

The multi-objective optimization yields the Pareto-optimal front shown in Figure 7.18, which shows the tradeoff between aggregate loss and electromagnetic mass for the transformer. Design 100 is highlighted as we will consider this design in some detail later on in the development.

7.9 Case Study

Table 7.6 Transformer Fixed Parameters

Symbol	Description	Value
c_{sc}	Secondary to core horizontal clearance	2.5 mm
c_{ps}	Primary to secondary clearance	2.5 mm
c_{pp}	Primary to primary clearance	2.5 mm
c_{pv}^*	Desired primary vertical clearance	2.5 mm
c_{sv}^*	Desired secondary vertical clearance	2.5 mm
N_{pcs}	Number of primary coils in series	1
N_{pcp}	Number of primary coils in parallel	2
k_{pb}	Primary winding build factor	1.2
t_{pins}	Primary conductor insulation thickness	30 µm
N_{scs}	Number of secondary coils in series	1
N_{scp}	Number of secondary coils in parallel	2
k_{sb}	Secondary winding build factor	1.2
t_{sins}	Secondary conductor insulation thickness	30 µm

Table 7.7 Design Space

Par.	Description	Min.	Max.	Enc.	Gene
m_c	Core material	1	4	int	1
l_{cc}	Length of core center (m)	10^{-4}	1	log	2
r_c	Rounding radius of core end leg (m)	10^{-6}	10^1	log	3
w_{cec}	Width of core end leg center (m)	10^{-4}	10^1	log	4
w_{cb}	Width of core base top leg (m)	10^{-3}	1	log	5
m_p	Primary material	1	2	int	6
a_{pt}^*	Desired total area of all primary coil parallel conductors	10^{-6}	10^{-3}	log	7
N_{ppr}	Number of primary conductors in parallel	1	5	int	8
N_{pcl}^*	Desired number of primary turns per coil	10	10^3	log	9
R_{pdw}^*	Desired primary coil depth to width ratio	0.2	20	log	10
m_s	Secondary material	1	2	int	11
a_{st}^*	Desired total area for all secondary coil parallel conductors (m²)	10^{-6}	10^{-3}	log	12
N_{spr}	Number or secondary conductors in parallel	1	5	int	13
R_{Nps}^*	Desired primary to secondary turns ratio	1.9	2.1	lin	14
R_{sdw}^*	Desired secondary depth to width ratio	0.2	20	log	15

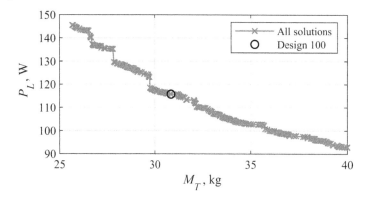

Figure 7.18 Transformer design Pareto-optimal front.

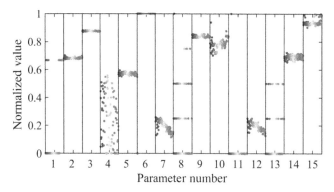

Figure 7.19 Parameter distribution.

The parameter distribution of the final population is depicted in Figure 7.19. The parameter of interest corresponds to the element of **θ** in (7.8-1). For example, parameter 1 is m_c and parameter 2 is l_{cc}. For each parameter, parameter values of each design are ordered within each window in order of increasing fitness in the first objective. Thus, points to the right in each window correspond to the lowest mass designs. Looking at the normalized parameter values, we can make several observations. First, from parameter 1, it is evident that the high-mass low-loss designs use $m_c = 1$ (normalized value = 0) which corresponds to M19, while many of the low-mass high-loss designs used $m_c = 4$ (normalize value = 0.67), which corresponds to M43. This is because for the particular materials considered, M19 has lower losses than M43; but M43 had a higher permeability.

The length of core center l_{cc} (parameter 2) and core radius r_c (parameter 3) can be seen to be consistent across all the designs. The width of the core end leg center region, w_{cec} (parameter 4) is widely dispersed amount lower values; however, in all cases the value is small relative to other dimensions. The width of the core base, w_{cb} (parameter 5), appears constant across all designs.

The primary and secondary winding materials, m_p (parameter 6) and m_s (parameter 11), are almost always chosen to be aluminum and copper, respectively. Copper is a better conductor than aluminum on a per volume basis; aluminum is better on a per mass basis. Since the secondary winding is inside the primary, there is considerable pressure to reduce the volume of the secondary winding; for the primary winding there is more pressure on the mass.

We will next consider a_{pt}^* (parameter 7), which is the desired total area of all paralleled conductors of the primary winding. It is clearly highly correlated with mass. Once again, small conductors and higher current density lead to lower mass but less efficient designs. The corresponding variable for the secondary, a_{st}^* (parameter 12) follows a similar trend.

Let us next consider the number of parallel conductors in the primary and secondary coils, denoted as N_{ppr} (parameter 8) and N_{spr} (parameter 13), respectively. There is a little optimization pressure on these variables. At some level, insulation thickness favors fewer conductors in parallel, and bending radius favors a higher number of conductors in parallel. If ac losses were considered, this would provide additional pressure for higher number of conductors in parallel because it would reduce skin and proximity effects, as we will see in Chapter 11. As it is, both windings generally utilize between 1 and 3 conductors in parallel.

Somewhat unexpectedly, the desired primary coil aspect ratio R_{pdw}^* (parameter 10) seems to be correlated with mass. This does not seem to be the case with the desired secondary coil aspect ratio R_{sdw}^* (parameter 15), except at very low masses. Proceeding to consider the number of turns per primary coil, it can be seen that N_{pcl}^* (parameter 9) is relatively constant across all designs, as is the primary to secondary turns ratio R_{Nps}^* (parameter 14).

Let us now consider a specific design from the front. We will consider Design 100 from the front, wherein the designs are labeled in terms of increasing mass. This design has a total electromagnetic mass of 30.9 kg and an aggregate loss of 116 W. The parameters for this design are listed in Table 7.8. Cross sections of the design are given in Figure 7.20.

The resulting electrical parameters of the design as calculated by the design code are listed in Table 7.9. Given the assumptions made in developing the magnetic equivalent circuit, the accuracy of the leakage inductance is of interest. To this end, results were obtained using 3D finite element analysis. Finite element analysis is a common method for numerical computation of fields. This topic will be introduced, at least in the 2D case, in Chapter 16.

In order to determine the nominal parameters using FEA, the core was represented as being magnetically linear so as to calculate the unsaturated inductance. As a first test, a primary current of $i_p = 1$ A and a secondary current of $i_s = -N_p/N_s$ A was injected into the transformer, and the energy stored by the magnetic fields was found. From Figure 7.6, under these conditions, for which there is no magnetizing current, the energy stored is $(L_{lp} + L'_{ls})/2$; equating the energies yielded that the sum of the primary and referred secondary leakage inductances is 3.7 mH, whereas our MEC yields 4.4 mH, an error of 18%. The fact that the MEC overestimated the sum of the leakages means the regulation will be somewhat better than predicted. Injecting a primary current of $i_p = 1$A with $i_s = 0$ A and again using energy arguments yields $L_{lp} + L_{m0} = 74.3$ H. From our MEC, we obtained 64.2 H, an error of 14%. Here again, our MEC is pessimistic. Finally, injecting $i_p = 0$ A and $i_s = -N_p/N_s$ A into our FEA model, we obtain $L'_{ls} + L_{m0} = 74.3$ H, whereas the MEC predicted 64.2 H, again for an error of 14%.

The reader may wonder why sums of parameters were examined rather than finding the parameters individually. The reason is that in the case of the FEA, the large difference between the leakage inductances which are on the order of mH, and magnetizing inductances that are on the order of tens of H, prohibits making meaningful mathematical operations on these quantities. In other words, a very small numerical error in energy would lead to a very large numerical error in leakage inductance. However, the sums, as posed, do not have this issue.

Table 7.8 Design 100 Parameters

Core		Primary Winding		Secondary Winding	
Symbol	Value	Symbol	Value	Symbol	Value
Steel	**M19**	**Material**	**Al**	**Material**	**Cu**
d_c	26.09 cm	AWG	12	AWG	12
w_c	17.55 cm	r_{pc}	1.026 mm	r_{sc}	1.026 mm
l_c	10.09 cm	N_{pcs}	1	N_{scs}	1
d_{cs}	14.95 cm	N_{pcp}	2	N_{scp}	2
w_{cs}	7.334 cm	N_{ppr}	1	N_{spr}	1
l_{cc}	5.243 cm	N_{puw}	8	N_{suw}	4
w_{cec}	2.560 mm	N_{pud}	57	N_{sud}	57
r_c	2.425 cm	N_{pcl}	455	N_{scl}	223
w_{cb}	5.568 cm	w_{pw}	2.028 cm	w_{sw}	1.014 cm
A_{ccb}	56.21 cm^2	d_{pw}	14.49 cm	d_{sw}	14.45 cm
A_{cel}	46.50 cm^2	r_{pi}	3.939 cm	r_{si}	2.675 cm
V_c	3.157 L	r_{po}	5.967 cm	r_{so}	3.689 cm
M_c	23.37 kg	l_{pc}	5.243 cm	l_{sc}	5.243 cm
		w_{pc}	2.560 mm	w_{sc}	2.560 mm
		k_{ppf}	51.37	k_{spf}	50.36
		k_{pb}	19.19	k_{sb}	13.04
		M_p	3.430 kg	M_s	4.065 kg

While the results are acceptable, there are some ways to improve the calculations. One source of error for the leakage inductance calculation is that the primary side leakage estimate is susceptible to error because it is far from the core which makes the leakage path hard to estimate. A computationally efficient semi-numerical approach using Biot–Savart Law and the method of images is set forth in [2]. Additional discussion on prediction of MEC leakage inductances, and different ways to calculate these inductances both analytically and using FEA is set forth in [3]. While the prior two references focused on the leakage inductance, it is also possible to improve the estimate of the magnetizing inductance. This could be done by representing the core with parallel magnetic paths from the inside to the outside of the core. This is more important in the case of a nongapped structure such as a transformer than in a gapped structure such as an inductor because in the case of a gapped structure the gap typically dominates the core reluctance, and so less accuracy is needed in representing the balance of the core.

The primary flux linkage versus primary current (with no secondary current) is depicted in Figure 7.21 as calculated using the MEC. Also shown is the location of assumed worst-case primary flux linkage λ_{pwci} and allowed peak inrush current i_{pimxa}, and the worst case inrush current based on the MEC i_{pimx}, which is 29.42 A.

One factor that may reduce the primary current somewhat during the inrush event is the primary resistance. This could be included in our analysis by incorporating a time-domain simulation as part of our fitness function. However, we will not take this approach here.

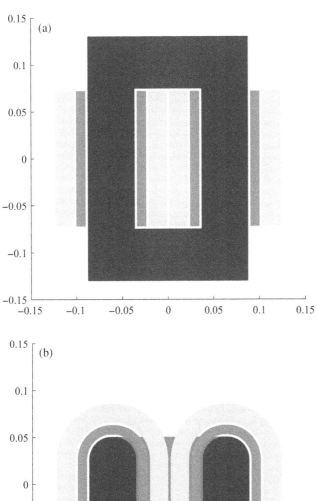

Figure 7.20 Design 100 cross sections.

Table 7.9 Electrical Parameters

$N_p/N_s = 455/223$	$r_p = 0.7675\ \Omega$	$r'_s = 0.7298\ \Omega$
$L_{m0} = 64.2$ H	$L_{lp} = 3.435$ mH	$L'_{ls} = 0.9194$ mH

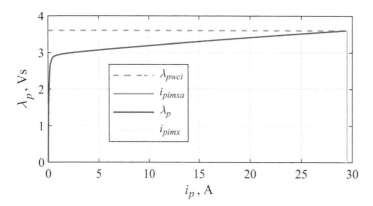

Figure 7.21 Primary flux linkage versus current.

Table 7.10 Operating Point Performance

Variable	No load	Half load	Full load
V_p (V)	480 ∠ 0°	480 ∠ 0°	480 ∠ 0°
f (Hz)	60	60	60
Z_l (Ω)	∞	23.0 ∠ 0°	11.5 ∠ 0°
I_p (A)	0.07732 ∠ −33.98°	4.990 ∠ −1.457°	9.762 ∠ −2.150°
V_s (V)	235.2 ∠ −0.005953°	231.6 ∠ −0.9712°	228.0 ∠ −1.907°
I_s (A)	0	10.05 ∠ 179.0°	19.79 ∠ 178.1°
P_{pr} (W)	0.004588	19.11	73.14
P_{sr} (W)	0	17.71	68.64
P_c (W)	30.77	30.31	29.84
P_l (W)	30.77	67.12	171.6
R_c (Ω)	7485	7478	7473
L_m (H)	29.47	29.71	29.95
J_p (A/mm²)	0.01169	0.7542	1.475
J_s (A/mm²)	0	1.519	2.991

Table 7.10 lists the operating characteristics of the design for no-load, half-load, and full-load conditions. As can be seen, core loss varies little with loading. The resistive losses are significantly impacted by the transformer load. It is interesting that even under full-load conditions, the transformer current density is rather modest. This may be due to the fact that the regulation of the transformer is limited, which in turn causes the resistance to be bounded, which in turn limits the current density.

Core flux density waveforms for no-load conditions are given in Figure 7.22. Therein, the flux density in the base leg and end leg, defined as

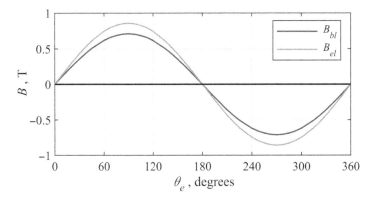

Figure 7.22 No-load flux density waveforms.

$$B_{bl} = \frac{\Phi_{bl}}{A_{bl}} \tag{7.9-4}$$

and

$$B_{el} = \frac{\Phi_{el}}{A_{el}} \tag{7.9-5}$$

respectively, are shown. It is interesting that the flux density in the end leg is higher than the flux density in the base leg. Note that increasing the magnetic cross section of the end leg increases the path length of the primary and secondary coil, so in a sense this cross section is more expensive than the cross section on the base/top legs of the core. This may lead to designs in which the flux density on the end legs has a higher peak value than on the top and bottom legs.

7.10 Closing Remarks

In this introductory chapter on transformer design, we have established a procedure for designing a transformer. Of course, there is room left for significant enhancement. In the case of almost any transformer, an addition of a thermal analysis would be extremely beneficial. We will discuss thermal analysis of electromechanical devices in Chapter 10. Another aspect of transformer behavior we did not address is the short circuit performance. In the case of very high-power transformers, fault considerations are important because the forces on the conductors can become very significant when a large transformer's secondary becomes shorted. Another aspect of transformer performance that is sometimes important is capacitance. The capacitance between the layers of a winding can affect how the transformer responds to high-frequency transients. The capacitance between windings is also important, especially with regard to common-mode current path through the device. Related to the capacitance are the details of the insulation systems. This is clearly an important topic in medium voltage and HV transformers. In the case of high-frequency transformers, high-frequency conductor losses such as skin effect and proximity effect can become important. We will discuss these loss mechanisms in Chapter 11. For additional information specifically related to optimization-based design of transformers, the reader is referred to references [4–6].

References

1 A. J. Pansini, *Electrical Transformers and Power Equipment*. Englewood Cliffs, NJ: Prentice Hall, 1988.
2 V. S. Duppalli and S. Sudhoff, Computationally efficient leakage inductance calculation for a high-frequency core-type transformer, *in IEEE Electric Ship Technologies Symposium (ESTS), Arlington, VA*, 2017.
3 A. Taher, S. Sudhoff, and S. Pekarek, Calculation of a tape-wound transformer leakage inductance using the MEC Model, *IEEE Transactions on Energy Conversion*, vol. **30**, no. 2, pp. 541–549, 2015.
4 W. G. Hurley, W. H. Wolfle, and G. B. Breslin, Optimized transformer design: Inclusive of high-frequency effects, *IEEE Transactions on Power Electronics*, vol. **13**, no. 4, pp. 651–659, 1998.
5 R. A. Jabr, Application of geometric programming to transformer design, *IEEE Transactions on Magnetics*, vol. **41**, no. 11, pp. 4261–4269, 2005.
6 E. I. Amoiralis, P. S. Georgilakis, M. A. Tsili, and A. G. Kladas, Global transformer optimization method using evolutionary design and numerical field computation, *IEEE Transactions on Magnetics*, vol. **45**, no. 3, pp. 1720–1723, 2009.
7 S. D. Sudhoff, "*MATLAB Codes for Power Magnetic Devices: A Multi-Objective Design Approach*, 2nd Edition," [Online]. Available: http://booksupport.wiley.com.

Problems

1 Consider the transformer parameters and excitation given in Example 7.3 A. Suppose a load consists of an inductor that presents a 1 Ω reactive load. Compute the actual (nonreferred) secondary voltage.

2 Consider the transformer parameters and excitation given in Example 7.3 A. Suppose a load consists of a capacitor that presents a 1 Ω reactive load. Compute the actual (nonreferred) secondary voltage.

3 Consider the transformer parameters and excitation given in Example 7.3 A. Suppose the secondary is short-circuited. What will be the magnitude of the short-circuited primary current?

4 Consider the transformer parameters and excitation given in Example 7.3 A. Suppose the secondary is open circuited. What will be the magnitude of the no-load primary current?

5 Consider the transformer corresponding to Design 29 whose parameters are given in Table 7.8. Suppose the primary voltage is 480 V, and the full load referred secondary current is 10.4 A. Compare the regulation predicted by (7.4-11) to the regulation predicted by (7.4-12), to the regulation predicted by (7.4-12) with the resistances neglected.

6 Derive the portion of P_{pli} associated with path P2 in Figure 7.16.

7 Derive the portion of P_{ple} associated with path P4 in Figure 7.16.

8 Derive the portion of P_{sli} associated with path P6 in Figure 7.16.

9 Derive the portion of P_{sle} associated with path P8 in Figure 7.16.

10 Derive (7.8-4).

11 Modify the code used in the case study (available in Sudhoff [7]) to use only 500 generations and a population size of 500. How much does the Pareto-optimal front change?

12 Modify the code used in the case study (available in Sudhoff [7]) to eliminate the requirement on inrush current. Compare the Pareto-Optimal front obtained with that presented in the case study. What is the design impact of this requirement?

13 Rerun the code used in the case study (available in Sudhoff [7]), but allow the regulation to increase from 5 to 10%. Compare the Pareto-optimal front obtained with that presented in the case study. What is the design impact allowing greater regulation?

8

Distributed Windings and Rotating Electric Machinery

Thus far, we have studied several devices which used coils of a conductor. In each of these cases, the winding is a simple lumped coil or pair of coils. In this chapter, we study a more complicated winding arrangement known as a distributed winding, which is often used in rotating electric machinery. In these machines, the goal is to establish a continuously rotating set of north and south poles on the stator (the stationary part of the machine), which interacts with an equal number of north and south poles on the rotor (the rotating part of the machine), to produce uniform torque. Several concepts are needed to study this type of device. These concepts include distributed windings, winding functions, rotating magneto motive force (MMF) waves, and inductances and resistances of distributed windings. We will also study a mathematical analysis technique known as a reference frame theory. The amalgamation of these ideas and concepts will set the stage for us to design a permanent magnet AC machine in the next chapter.

8.1 Describing Distributed Windings

A photograph of the stator (stationary part) of a 3.7 kW, 1800 rpm induction motor is shown in Figure 8.1, where the stator core can be seen inside the stator housing. The core includes the stator slots between the stator teeth. The slots are filled with slot conductors that, along with the end-turns, form complete coils. The windings of the machine are termed distributed because they are not wound as simple coils, but are rather wound in a spatially distributed fashion.

Figure 8.2 depicts a generic electrical machine. The stationary stator and rotating rotor are labeled, but details such as the stator slots, windings, and rotor construction are omitted. The stator reference axis may be considered to be mechanically attached to the stator, and the rotor reference axis of the rotor. Angles defined in Figure 8.2 include position measured relative to the stator, denoted ϕ_{sm}, position measured relative to the rotor, denoted ϕ_{rm}, and the position of the rotor relative to the stator, denoted θ_{rm}. The mechanical rotor speed is the time derivative of θ_{rm} and is denoted ω_{rm}.

The position of a given feature can be described using either ϕ_{sm} or ϕ_{rm}; however if we are describing the same feature using both of these quantities, these two measures of angular position are related by

$$\theta_{rm} + \phi_{rm} = \phi_{sm} \tag{8.1-1}$$

Much of our analysis may be conducted either in terms of ϕ_{sm} or ϕ_{rm}. As such, we will use ϕ_m as a generic symbol to stand for either quantity, as appropriate.

Power Magnetic Devices: A Multi-Objective Design Approach, Second Edition. Scott D. Sudhoff.
© 2022 The Institute of Electrical and Electronics Engineers, Inc. Published 2022 by John Wiley & Sons, Inc.
Companion website: www.wiley.com/go/sudhoff/Powermagneticdevices

Figure 8.1 Distributed winding stator.

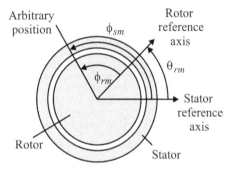

Figure 8.2 Definition of position measurements.

The goal of a distributed winding is to create a set of uniformly rotating poles on the stator which interact with an equal number of poles on the rotor. The number of poles on the stator will be designated P, and must be an even number (it is the number of poles, not the number of pole pairs). The number of poles largely determines the relationship between the rotor speed and the ac electrical frequency. Figure 8.3 illustrates the operation of 2-, 4-, and 6-pole machines. Therein, N_s, S_s, N_r, and S_r denote north stator, south stator, north rotor, and south rotor poles, respectively. A north pole is where positive flux leaves a magnetic material and a south pole is where flux enters a magnetic material. Electromagnetic torque production results from the interaction between the stator and rotor poles.

When analyzing machines with more than two poles, it is convenient to define equivalent "electrical" angles of the positions and speed. In particular, define

$$\phi_s = P\phi_{sm}/2 \qquad (8.1\text{-}2)$$
$$\phi_r = P\phi_{rm}/2 \qquad (8.1\text{-}3)$$

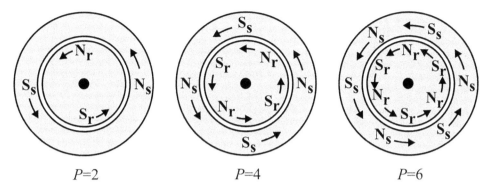

Figure 8.3 P-pole machines.

$$\theta_r = P\theta_{rm}/2 \tag{8.1-4}$$

$$\omega_r = P\omega_{rm}/2 \tag{8.1-5}$$

In terms of electrical position (8.1-1) becomes

$$\theta_r + \phi_r = \phi_s \tag{8.1-6}$$

Finally, it is also useful to define a generic position as

$$\phi = P\phi_m/2 \tag{8.1-7}$$

The reason for the introduction of these electrical angles is that it will allow our analysis to be expressed so that all machines mathematically appear to be 2-pole machines, thereby providing considerable simplification.

Discrete Description of Distributed Windings

Rotating machinery generally has multiple sets of windings known as phases. Commonly, in a three-phase machine, the three-phase stator windings are referred to as the *a*-phase, *b*-phase, and *c*-phase, respectively, and denoted *as*, *bs*, and *cs*. In distributed winding machines, such as shown in Figure 8.1, these windings may be described using either a discrete or continuous formulation. The discrete description is based on the number of conductors in each slot; the continuous description is an abstraction based on an ideal distribution. A continuously distributed winding is desirable in order to achieve uniform torque. However, the conductors that make up the winding are not placed continuously around the stator but are rather placed into slots in the machine's stator and rotor structures, thereby leaving room for the stator and rotor teeth that are needed to conduct magnetic flux. Thus, a discrete winding distribution is used to approximate an ideal continuous winding. In reality, the situation is more nuanced. Since the slots and conductors have physical size, all distributions are continuous when viewed with sufficient resolution. Thus, the primary difference between these two descriptions is one of how we describe the winding mathematically. We will find that both descriptions have advantages in different situations, and so we will consider both.

Figure 8.4 illustrates the stator of a machine in which the stator windings are located in eight slots. The notation $N_{as,i}$ in Figure 8.4 indicates the number of *a*-phase stator conductors in the *i*th slot. The conductors are shown as open circles because conductors may be positive (coming out of the page or toward the front of the machine) or negative (going into the page or toward

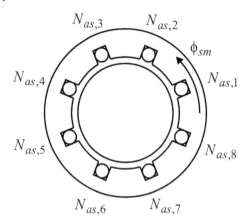

Figure 8.4 Slot structure.

the back of the machine). Generalizing this notation, $N_{x,i}$ is the number of conductors in the slot i of winding (or phase) x coming out of the page (or toward the front of the machine). In this example $x = as$. Often, a slot will contain conductors from multiple windings (phases). It is important to note that $N_{x,i}$ is a signed quantity—and that half the $N_{as,i}$ values will be negative since for every conductor that comes out of the page a conductor goes into the page.

The center of the ith slot and the ith tooth are located at

$$\phi_{ys,i} = \pi(2i-2)/S_y + \phi_{ys,1} \tag{8.1-8}$$

and

$$\phi_{yt,i} = \pi(2i-3)/S_y + \phi_{ys,1} \tag{8.1-9}$$

respectively, where S_y is the number of slots. In (8.1-8) and (8.1-9), $y = s$ for the stator, in which case $\phi_{ys,i}$ and $\phi_{yt,i}$ are relative to the stator. Similarly, $y = r$ when describing rotor windings in machines that have them. When describing rotor windings, $\phi_{ys,i}$ and $\phi_{yt,i}$ are relative to the rotor. The position $\phi_{ys,1}$ is the position of slot 1.

Since the number of conductors going into the page must be equal to the number of conductors out of the page (the conductor is formed into closed loops), we have that

$$\sum_{i=1}^{S_y} N_{x,i} = 0 \tag{8.1-10}$$

where x designates the winding (e.g., as). The total number of turns associated with the winding may be expressed

$$N_x = \sum_{i=1}^{S_y} N_{x,i}\, u(N_{x,i}) \tag{8.1-11}$$

where $u(\cdot)$ is the unit step function, which is one if its argument is greater or equal to zero, and zero otherwise.

In (8.1-11) and throughout this work we will use N_x to represent the total number of turns associated with winding x, $N_{x,i}$ to be the number of conductors in the ith slot, and $\mathbf{N}_x$ to be a vector whose elements correspond to the number of conductors in each slot. Additionally, if all the windings of the stator or rotor have the same number of turns, we will use the notation N_y to denote the

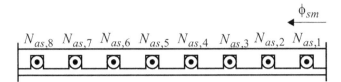

Figure 8.5 Developed diagram.

number of turns in the stator or rotor windings. For example, if $N_{as} = N_{bs} = N_{cs}$ (the usual case), then we will denote the number of turns in these windings as N_s.

It is sometimes convenient to illustrate features of a machine using a developed diagram. In the developed diagram, spatial features (such as the location of the conductors) are depicted against a linear axis. In essence, the machine becomes "unrolled." This process is best illustrated by an example; Figure 8.5 is the developed diagram corresponding to Figure 8.4. Note the independent axis is directed to the left rather than to the right. This is a convention that has been traditionally adopted in order to avoid the need to "flip" the diagram in three dimensions.

Continuous Description of Distributed Windings

The slot/tooth structure is used because the slot provides a region to place the winding conductors and the teeth provide a low reluctance path for magnetic flux between the stator and rotor. The use of a large number of slots allows the winding to be distributed, albeit in a discretized fashion. The continuous description of distributed winding represents the winding in terms of what it is desired to approximate—a truly distributed winding. The continuous description is based on conductor density, which is a measure of the number of conductors per radian, which varies as a function of position. We will describe the winding x of a machine with conductor density $n_x(\phi_m)$, where x again denotes winding (such as as). The conductor density may be positive or negative; positive conductors are considered here to be out of the page (toward the front of the machine).

The conductor density is often a sinusoidal function of position. A common choice for the a-phase stator conductor density in three-phase AC machinery is

$$n_{as}(\phi_{sm}) = N_{s1}\sin(P\phi_{sm}/2) - N_{s3}\sin(3P\phi_{sm}/2) \tag{8.1-12}$$

In this function, the first term represents the desired distribution; the second term allows for more effective slot utilization. The impact of the second term on slot utilization is explored in Problem 6 at the end of this chapter.

It will often be of interest to determine the total number of turns associated with a winding. This number is readily found by integrating the conductor density over all regions of positive conductors, so that the total number of turns may be expressed

$$N_x = \int_0^{2\pi} n_x(\phi_m) u(n_x(\phi_m)) d\phi_m \tag{8.1-13}$$

Symmetry Conditions on Conductor Distributions

Throughout this work, it is assumed that the conductor distribution obeys certain symmetry conditions. The first of these is that the distribution of conductors is periodic in a number of slots corresponding to the number of pole pairs. In particular, it is assumed that

$$N_{x,i+2S_y/P} = N_{x,i} \tag{8.1-14}$$

Second, it is assumed that the distribution of conductors is half-wave odd over a number of slots corresponding to one pole. This is to say

$$N_{x,i+S_y/P} = -N_{x,i} \tag{8.1-15}$$

While it is possible to construct an electric machine where these conditions are not met such as machines with a fractional number of slots per pole per phase, most electric machines satisfy these conditions.

In the case of the continuous winding distribution, the conditions corresponding to (8.1-14) and (8.1-15) may be expressed as

$$n_x(\phi_m + 4\pi/P) = n_x(\phi_m) \tag{8.1-16}$$
$$n_x(\phi_m + 2\pi/P) = -n_x(\phi_m) \tag{8.1-17}$$

Converting Between Discrete and Continuous Descriptions of Distributed Windings

Suppose that we have a discrete description of a winding consisting of the number of conductors of each phase in the slots. The conductor density could be expressed

$$n_x(\phi_m) = \sum_{i=1}^{S_y} N_{x,i} \delta(\phi_m - \phi_{ys,i}) \tag{8.1-18}$$

where $\delta(\cdot)$ is the unit impulse function and ϕ_m is relative to the stator or rotor reference axis for a stator or rotor winding, respectively.

Although (8.1-18) is in a sense a continuous description, normally we desire an idealized representation of the conductor distribution. To this end, we may represent the conductor distribution as a single-sided Fourier series of the form

$$n_x(\phi_m) = \sum_{j=1}^{J} a_j \cos(j\phi_m) + b_j \sin(j\phi_m) \tag{8.1-19}$$

where J is the number of terms used in the series and where

$$a_j = \frac{1}{\pi} \int_0^{2\pi} n_x(\phi_m) \cos(j\phi_m) d\phi_m \tag{8.1-20}$$

$$b_j = \frac{1}{\pi} \int_0^{2\pi} n_x(\phi_m) \sin(j\phi_m) d\phi_m \tag{8.1-21}$$

Substitution of (8.1-18) into (8.1-20)–(8.1-21) yields

$$a_j = \frac{1}{\pi} \sum_{i=1}^{S_y} N_{x,i} \cos(j\phi_{ys,i}) \tag{8.1-22}$$

$$b_j = \frac{1}{\pi} \sum_{i=1}^{S_y} N_{x,i} \sin(j\phi_{ys,i}) \tag{8.1-23}$$

Thus (8.1-19) along with (8.1-22)–(8.1-23) can be used to convert a discrete winding description to a continuous one.

It is also possible to translate a continuous winding description to a discrete winding description. To this end, one approach is to lump all conductors into the closest slot. This entails adding (or integrating, since we are dealing with a continuous function) all conductors within π/S_y of the center of the ith slot and considering them to be associated with the ith slot. This yields

$$N_{x,i} = \text{round}\left(\int_{\phi_{ys,i}-\pi/S_y}^{\phi_{ys,i}+\pi/S_y} n_x(\phi_m)d\phi_m\right) \quad (8.1\text{-}24)$$

where round(·) denotes a function that rounds the result to the next nearest integer.

End Conductors

Normally, our focus in describing a winding is on the conductors in the axial direction. However, for some purposes, such as calculating the winding resistance, it is important to be able to describe the number of conductors on the front and back ends of the machine connecting the slot conductors together. These conductors are referred to as end conductors.

Herein, we will focus our discussion on a discrete winding description. Consider Figure 8.6, which is a version of a developed diagram of the machine, except that instead of looking into the front of the machine, we are looking from the center of the machine outward in the radial direction. Therein, $N_{x,i}$ denotes the number of winding x conductors in the ith slot. Variables $L_{x,i}$ and $R_{x,i}$ denote the number of positive end conductors in front of the ith tooth directed to the left or right, respectively. These quantities are defined such as to be greater than or equal to zero. The net number of conductors directed in the counterclockwise direction when viewed from the front of the ith tooth is denoted $M_{x,i}$. In particular,

$$M_{x,i} = L_{x,i} - R_{x,i} \quad (8.1\text{-}25)$$

Unlike $L_{x,i}$ and $R_{x,i}$, $M_{x,i}$ can be positive or negative. Canceled conductors refer to the situation in which the same winding has conductors going to the left and the right over the end of the same

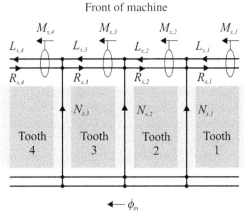

Figure 8.6 End conductors.

tooth. The number of canceled conductors in front of the ith tooth is denoted $C_{x,i}$. This quantity is defined as

$$C_{x,i} = \min(L_{x,i}, R_{x,i}) \tag{8.1-26}$$

Canceled conductors are undesirable in that they add to losses; however, some winding arrangements use them for manufacturing reasons.

It is possible to relate $M_{x,i}$ to the number of conductors in the slots. From Figure 8.6 it is apparent that

$$M_{x,i} = M_{x,i-1} + N_{x,i-1} \tag{8.1-27}$$

where the index operations are ring mapped, which is to say an index of $S_y + 1$ maps to one, and an index of 0 maps to S_y. In general, a ring mapped index may be expressed as $\mathrm{mod}(i-1, S_y) + 1$ where i is the original index and where $\mathrm{mod}(a, b)$ returns a modulus b. Since $M_{x,i}$ is the net number of conductors and $C_{x,i}$ is the number of canceled conductors, the total number of (unsigned) end conductors between slots $i-1$ and i is

$$E_{x,i} = |M_{x,i}| + 2C_{x,i} \tag{8.1-28}$$

The total number of end conductors is defined as

$$E_x = \sum_{i=1}^{S_y} E_{x,i} \tag{8.1-29}$$

Common Winding Arrangements

Before proceeding, it is convenient to consider some common machine winding schemes. Consider the 4-pole, 3.7 kW, 1800 rpm induction machine shown in Figure 8.1. As can be seen, the stator has 36 slots, which correspond to three slots per pole per phase. Figure 8.7 illustrates a common winding pattern for such a machine. Therein, each conductor symbol represents N conductors, going in or coming out as indicated. This is a double-layer winding, with each slot containing two groups of conductors. Both single- and double-layer winding arrangements are common in electric machinery. The number of a-phase conductors for the first 18 slots may be expressed as

$$\mathbf{N}_{as}|_{1-18} = N[0\ 0\ 0\ 1\ 2\ 2\ 1\ 0\ 0\ 0\ 0\ -1\ -2\ -2\ -1\ 0\ 0] \tag{8.1-30}$$

From (8.1-27) we have

$$\mathbf{M}_{as}|_{1-18} = M_{as,36} + N[0\ 0\ 0\ 0\ 1\ 3\ 5\ 6\ 6\ 6\ 6\ 5\ 3\ 1\ 0\ 0] \tag{8.1-31}$$

To proceed further, more details on the winding arrangement are needed.

Figure 8.8 illustrates some possible winding arrangements. In each case, the figure depicts the stator of a machine in an "unrolled" fashion similar to a developed diagram. However, the vantage point is that of an observer looking at the teeth from the center of the machine. Thus, each shaded area represents a tooth of the machine.

Figure 8.8(a) depicts a concentric winding arrangement wherein the a-phase conductors are organized in 12 coils, with 3 coils per set. Each coil is centered over a magnetic axis or pole associated with that phase. For this arrangement, $M_{as,36} = -3N$, $C_{as,i} = 0\ \forall\ i$, and $E_{as} = 88N$.

In Figure 8.8(b) a consequent pole winding arrangement is shown. In this arrangement, the windings are only wrapped around every other pole. From Figure 8.8(b), we have $M_{as,36} = 0$,

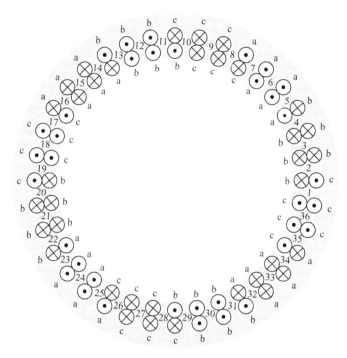

Figure 8.7 Stator winding for a 4-pole 36-slot machine.

$C_{as,i} = 0 \, \forall \, i$, and $E_{as} = 108N$. The increase in E_{as} will cause this arrangement to have a higher stator resistance than the concentric pole winding.

A lap winding is shown in Figure 8.8(c). Each coil of this winding is identical. For this arrangement $M_{as,36} = -3N$, as in the case of the concentric winding. However, in this winding $C_{as,6} = C_{as,15} = C_{as,24} = C_{as,34} = N$ and all other $C_{as,i} = 0$. The total number of end-turn segments is $96N$, which is better than the consequent pole winding, but not as good as the concentric winding.

Figure 8.8(d) depicts a wave winding, in which the winding is comprised of six coil groups. For this case, $M_{as,36} = -6N$, $C_{as,i} = 0$, and $E_{as} = 108N$. Like the consequent pole winding, a relatively high stator resistance is expected; however, the reduced number of coil groups (and the use of identical groups) offers a certain manufacturing benefit.

8.2 Winding Functions

Our first goal for this chapter was to set forth methods to describe distributed windings. Our next goal is to begin to analyze distributed winding devices. To this end, a valuable concept is that of the winding function [1]. The winding function has three important uses. First, it will be useful in determining the MMF caused by distributed windings. Second, it will be used to determine how much flux links a winding. Third, the winding function will be instrumental in calculating winding inductances.

The winding function is a description of how many times a winding links flux density at any given position. It may be viewed as the number of turns associated with a distributed winding. However, unlike the number of turns in a simple coil, we will find that the number of turns associated with a

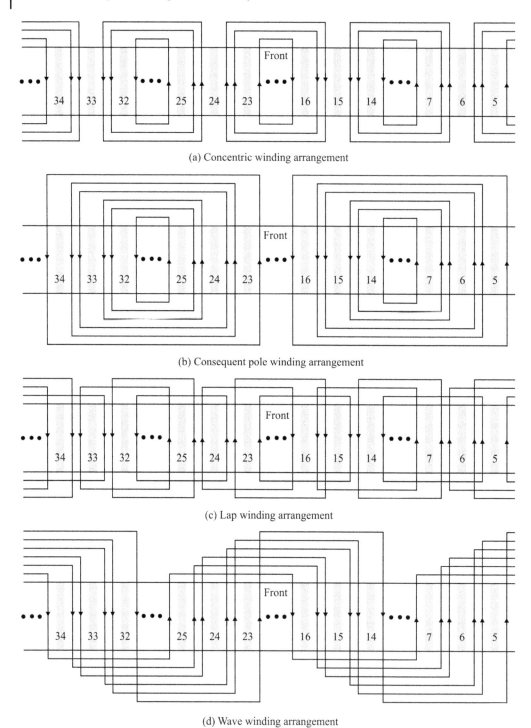

Figure 8.8 Winding arrangements.

Figure 8.9 Calculation of the winding function.

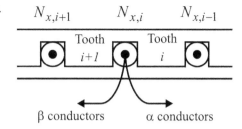

distributed winding is a function of position. Using this notion will allow us to formulate the mathematical definition of the winding function.

Let us now consider the discrete description of the winding function. Figure 8.9 illustrates a portion of the developed diagram of a machine, wherein it is arbitrarily assumed that the winding of interest is on the stator. Let $W_{x,i}$ denotes the number of times winding x links flux traveling through the ith tooth, where the direction for positive flux and flux density is taken to be from the rotor to the stator.

Now, let us assume that we know $W_{x,i}$ for some i. It can be shown that

$$W_{x,i+1} = W_{x,i} - N_{x,i} \tag{8.2-1}$$

To understand (8.2-1), suppose $N_{x,i}$ is positive. Of the $N_{x,i}$ conductors, suppose α of these conductors go to the right (where they turn back into other slots), and β of these conductors go to the left (where again they turn back into other slots). The α conductors form turns that link flux in tooth i but not flux in tooth $i + 1$ since they close the loop to the right. The β conductors form turns which are directed toward the left before closing the loop, and so do not link tooth i, but do link tooth $i + 1$, albeit in the negative direction (which can be seen using the right-hand rule and recalling that flux is considered positive from the rotor to the stator). Thus we have

$$W_{x,i+1} = W_{x,i} - \alpha - \beta \tag{8.2-2}$$

Since $N_{x,i} = \alpha + \beta$, (8.2-2) reduces to (8.2-1).

Manipulation of (8.2-1) yields an expression for the winding function. In particular,

$$W_{x,i} = W_{x,1} - \sum_{j=1}^{i-1} N_{x,j} \tag{8.2-3}$$

In order to determine $W_{x,1}$, we will require that the winding function possess the symmetry conditions on the conductor distribution as stated in (8.1-14) and (8.1-15). In particular, we require

$$W_{x,i+S_y/P} = -W_{x,i} \tag{8.2-4}$$

where the indexing operations are ring mapped with a modulus of S_y. Note that this requirement does not follow from (8.2-3); rather it is part of our definition of the winding function. Manipulating (8.2-3) with $i = 1 + S_y/P$ and using (8.2-4) yields

$$W_{x,1} = \frac{1}{2} \sum_{j=1}^{S_y/P} N_{x,j} \tag{8.2-5}$$

Using (8.2-5) and (8.2-1) the winding function can be computed for each tooth. It should be noted that it is assumed that S_y/P is an integer for the desired symmetry conditions to be met. In addition, for a 3-phase machine to have electrically identical phases while ensuring symmetry of each winding, it is further required that $S_y/(3P)$ be an integer.

Let us now consider the calculation of the winding function using a continuous description of the winding. In this case, instead of being a function of the tooth number, the winding function is a continuous function of position, which can be position relative to the stator ($\phi_m = \phi_{sm}$) for stator windings or position relative to the rotor ($\phi_m = \phi_{rm}$) for rotor windings. Let us assume that we know the value of the winding function at position ϕ_m, and desire to calculate the value of the winding function at position $\phi_m + \Delta\phi_m$. The number of conductors between these two positions is $n_x(\phi_m)\Delta\phi_m$, assuming $\Delta\phi_m$ is small. Using arguments identical to the derivation of (8.2-1) we have that

$$w_x(\phi_m + \Delta\phi_m) = w_x(\phi_m) - n_x(\phi_m)\Delta\phi_m \tag{8.2-6}$$

Taking the limit as $\Delta\phi_m \to 0$, we obtain

$$\frac{dw_x(\phi_m)}{d\phi_m} = -n_x(\phi_m) \tag{8.2-7}$$

Thus the winding function may be calculated as

$$w_x(\phi_m) = -\int_0^{\phi_m} n_x(\phi_m)d\phi_m + w_x(0) \tag{8.2-8}$$

In order to utilize (8.2-8), we must establish $w_x(0)$. As in the discrete case for computing $W_{x,1}$, we require that the winding function obey the same symmetry conditions as the conductor distribution, namely (8.1-17). Thus

$$w_x(\phi_m + 2\pi/P) = -w_x(\phi_m) \tag{8.2-9}$$

Note that (8.2-9) does not follow from (8.2-8); rather it is an additional part of the definition. Manipulating (8.2-8) with $\phi_m = 2\pi/P$ and using (8.2-9), we obtain

$$w_x(0) = \frac{1}{2}\int_0^{2\pi/P} n_x(\phi_m)\,d\phi_m \tag{8.2-10}$$

Substitution of (8.2-10) into (8.2-8) yields

$$w_x(\phi_m) = \frac{1}{2}\int_0^{2\pi/P} n_x(\phi_m)\,d\phi_m - \int_0^{\phi_m} n_x(\phi_m)\,d\phi_m \tag{8.2-11}$$

In summary, (8.2-1) along with (8.2-5) provides a means to calculate the winding function for a discrete winding description while (8.2-11) can be used for a continuous winding description. The winding function is a physical measure of the number of times a winding links the flux in a particular tooth (discrete winding description) or a particular position (continuous winding description). It may also be described as the number of turns going around a given tooth (discrete description) or a given position (continuous description).

Example 8.2A We will now consider the winding function for the machine shown in Figure 8.7. Recall that for this machine, $P = 4$ and $S_s = 36$. From Figure 8.7, observe that the first slot is at

$\phi_{sm} = 0$, hence $\phi_{ss,1} = 0$. The conductor distribution is given by (8.1-28) where N is an integer. Applying (8.2-5), we have $W_{x,1} = 3N$. Using (8.2-1), we obtain

$$W_{as}|_{1-18} = N[3\ 3\ 3\ 2\ 0\ -2\ -3\ -3\ -3\ -3\ -3\ -3\ -2\ 0\ 2\ 3\ 3\ 3] \qquad (8.2A-1)$$

The winding function is only given for the first 18 slots since the pattern is repetitive.

In order to obtain the continuous winding function, let us apply (8.1-22) and (8.1-23) where the slot positions are given by (8.1-8). Truncating the series (8.1-19) after the first two nonzero harmonics yields

$$n_{as} = N(7.221 \sin(2\phi_{sm}) - 4.4106 \sin(6\phi_{sm})) \qquad (8.2A-2)$$

Comparing (8.2A-2) to (8.1-12), we see that $N_{s1} = 7.221N$ and $N_{s3} = 4.4106N$. From (8.2-11), we obtain

$$w_{as} = N\left(\frac{7.221}{2} \cos(2\phi_{sm}) - \frac{4.4106}{6} \cos(6\phi_{sm})\right) \qquad (8.2A-3)$$

Figure 8.10 depicts the conductor distributions and winding function for the winding. The discrete description of the winding function is shown as a series of arrows suggesting a delta function representation. The corresponding continuous distribution (which is divided by 4) can be seen to have a relatively high peak. It is somewhat difficult to compare the discrete winding description to the continuous winding description since it is difficult to compare a delta function to a continuous function. The discrete representation of the winding function is shown as a set of horizontal lines spanning one tooth and one slot and centered on the tooth. These lines are connected to form a contiguous trace. The continuous representation of the winding function can be seen to be consistent with the discrete representation at the tooth locations. Had we chosen to include the next two harmonics, the error between the continuous winding function representation and the discrete winding function representation at the tooth centers would be further reduced.

The next step of this development will be the calculation of the MMF associated with a winding. As it turns out, this calculation is very straightforward using the winding function.

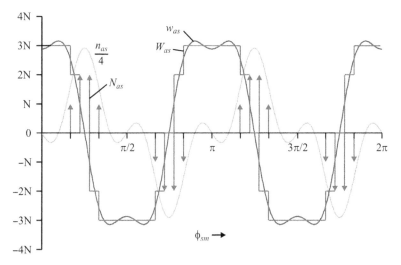

Figure 8.10 Conductor distribution and winding functions.

8.3 Air-Gap Magneto Motive Force

In this section, we consider the air-gap MMF, as well as the relationship of this MMF to the machine currents. We will find that the winding function is instrumental in establishing this relationship. In doing this, we will concentrate our efforts on the continuous winding description.

Let us begin by applying Ampere's law to the path shown in Figure 8.11. In particular, we have

$$\oint_{abcd} \mathbf{H} \cdot d\mathbf{l} = i_{enc}(\phi_m) \tag{8.3-1}$$

where $i_{enc}(\phi_m)$ is a function that describes the amount of current enclosed by the path. Expanding (8.3-1) we may write

$$\int_a^b \mathbf{H} \cdot d\mathbf{l} + \int_b^c \mathbf{H} \cdot d\mathbf{l} + \int_c^d \mathbf{H} \cdot d\mathbf{l} + \int_d^a \mathbf{H} \cdot d\mathbf{l} = i_{enc}(\phi_m) \tag{8.3-2}$$

The MMF across the air gap is defined as

$$F_g(\phi_m) \triangleq \int_{rotor}^{stator} \mathbf{H}(\phi_m) \cdot d\mathbf{l} \tag{8.3-3}$$

where the path of integration is directed in the radial direction. Because of this, we may rewrite (8.3-3) as

$$F_g(\phi_m) = \int_{rotor}^{stator} H_r(\phi_m) dl \tag{8.3-4}$$

where $H_r(\phi_m)$ is the outwardly directed radial component of the air-gap field intensity. It is further convenient to define the MMFs in the stator and rotor back iron as

$$F_{sb}(\phi_m) \triangleq \int_b^c \mathbf{H} \cdot d\mathbf{l} \tag{8.3-5}$$

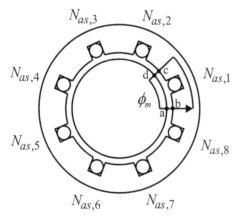

Figure 8.11 Path of integration.

$$F_{rb}(\phi_m) \triangleq \int_a^d \mathbf{H} \cdot d\mathbf{l} \tag{8.3-6}$$

As defined, these MMFs include a radial component in the teeth, and they are both defined in a predominately counterclockwise direction. Later, it may be convenient to break the radial component out as a separate MMF drop, but the given definitions are adequate for present purposes.

Substituting the definitions (8.3-3)–(8.3-6) into (8.3-2) yields

$$F_g(0) + F_{sb}(\phi_m) - F_g(\phi_m) - F_{rb}(\phi_m) = i_{enc}(\phi_m) \tag{8.3-7}$$

To proceed further, we must develop an expression for the current enclosed by the path. This current may be expressed as

$$i_{enc}(\phi_m) = \sum_{x \in X} \int_0^{\phi_m} n_x(\phi_m) i_x d\phi_m \tag{8.3-8}$$

where X denotes the set of all windings. Rearranging (8.3-8), we obtain

$$i_{enc}(\phi_m) = \sum_{x \in X} \left(\int_0^{\phi_m} n_x(\phi_m) d\phi_m \right) i_x \tag{8.3-9}$$

From (8.2-8), it can be shown that

$$\int_0^{\phi_m} n_x(\phi_m) d\phi_m = w_x(0) - w_x(\phi_m) \tag{8.3-10}$$

Combining (8.3-7), (8.3-9), and (8.3-10), we get

$$F_g(0) + F_{sb}(\phi_m) - F_g(\phi_m) - F_{rb}(\phi_m) = \sum_{x \in X} (w_x(0) - w_x(\phi_m)) i_x \tag{8.3-11}$$

Replacing ϕ_m by $\phi_m + 2\pi/P$ in (8.3-11) yields

$$F_g(0) + F_{sb}(\phi_m + 2\pi/P) - F_g(\phi_m + 2\pi/P) - F_{rb}(\phi_m + 2\pi/P)$$
$$= \sum_{x \in X} (w_x(0) - w_x(\phi_m + 2\pi/P)) i_x \tag{8.3-12}$$

Next, from the symmetry of the machine and the assumptions on the winding distribution, it will be assumed that all MMF terms are half-wave odd over a displacement of $2\pi/P$. Since the winding function has this property by definition, (8.3-12) may be written as

$$F_g(0) - F_{sb}(\phi_m) + F_g(\phi_m) + F_{rb}(\phi_m) = \sum_{x \in X} (w_x(0) + w_x(\phi_m)) i_x \tag{8.3-13}$$

Subtracting (8.3-11) from (8.3-13) and manipulating yields

$$-F_{sb}(\phi_m) + F_g(\phi_m) + F_{rb}(\phi_m) = \sum_{x \in X} w_x(\phi_m) i_x \tag{8.3-14}$$

The MMF source associated with the sum of the windings may be expressed as

$$F_X(\phi_m) = \sum_{x \in X} w_x(\phi_m) i_x \qquad (8.3\text{-}15)$$

Substitution of (8.3-15) into (8.3-14) yields

$$-F_{sb}(\phi_m) + F_g(\phi_m) + F_{rb}(\phi_m) = F_X(\phi_m) \qquad (8.3\text{-}16)$$

Expression (8.3-16) is important in that it relates the sum of the back iron and air-gap MMFs to the net MMF source provided by the windings.

At times, it will be convenient to define the stator and rotor source MMFs as

$$F_S(\phi_m) = \sum_{x \in X_S} w_x(\phi_m) i_x \qquad (8.3\text{-}17)$$

$$F_R(\phi_m) = \sum_{x \in X_R} w_x(\phi_m) i_x \qquad (8.3\text{-}18)$$

where X_S is the set of stator windings and (e.g., X_S = {'as', 'bs', 'cs'}) and X_R is the set of rotor windings (e.g., X_R = {'ar', 'br', 'cr' }). Thus

$$F_X(\phi_m) = F_S(\phi_m) + F_R(\phi_m) \qquad (8.3\text{-}19)$$

It is often the case that the back iron MMF drops are neglected. This approximation stems from the observation that if the flux density is finite and the permeability is high, then the field intensity must be small relative to its value in the air gap, and so $F_{rb}(\phi_m)$ and $F_{sb}(\phi_m)$ are small. In this case, from (8.3-16) and (8.3-19) the air-gap MMF may be readily expressed as

$$F_g(\phi_m) = F_S(\phi_m) + F_R(\phi_m) \qquad (8.3\text{-}20)$$

From the air-gap MMF drop, we may readily calculate the fields in the air gap. From (8.3-4), and assuming that the radial component of the field intensity is constant between the rotor and the stator, we have that

$$F_g(\phi_m) = H_g(\phi_m) g(\phi_m) \qquad (8.3\text{-}21)$$

whereupon the field intensity in the air gap may be expressed

$$H_g(\phi_m) = \frac{F_g(\phi_m)}{g(\phi_m)} \qquad (8.3\text{-}22)$$

Since $B = \mu_0 H$ in the air gap, the flux density in the air gap may be expressed

$$B_g(\phi_m) = \frac{\mu_0 F_g(\phi_m)}{g(\phi_m)} \qquad (8.3\text{-}23)$$

8.4 Rotating MMF

A goal of this chapter is to establish methods that can be used to determine electrical circuit models of electromechanical systems. To this end, we have just discussed how to calculate the MMF due to a distributed winding. In the next section, we will use this information in the calculation of the inductances of distributed windings. However, before we proceed into that discussion, it will be interesting to pause for a moment and consider our results thus far in regarding the operation of a machine.

8.4 Rotating MMF

Let us consider a three-phase stator winding. We will assume that the conductor distribution for the stator windings may be expressed as

$$n_{as}(\phi_{sm}) = N_{s1} \sin(P\phi_{sm}/2) - N_{s3} \sin(3P\phi_{sm}/2) \tag{8.4-1}$$

$$n_{bs}(\phi_{sm}) = N_{s1} \sin(P\phi_{sm}/2 - 2\pi/3) - N_{s3} \sin(3P\phi_{sm}/2) \tag{8.4-2}$$

$$n_{cs}(\phi_{sm}) = N_{s1} \sin(P\phi_{sm}/2 + 2\pi/3) - N_{s3} \sin(3P\phi_{sm}/2) \tag{8.4-3}$$

In (8.4-1)–(8.4-3), the second term to the right of the equal sign (the third harmonic term) is useful in achieving a more uniform slot fill.

In order to calculate the stator MMF for this system, we must first find the winding function. Using the methods of Section 8.3, the winding functions may be expressed

$$w_{as}(\phi_{sm}) = \frac{2N_{s1}}{P} \cos(P\phi_{sm}/2) - \frac{2N_{s3}}{3P} \cos(3P\phi_{sm}/2) \tag{8.4-4}$$

$$w_{bs}(\phi_{sm}) = \frac{2N_{s1}}{P} \cos(P\phi_{sm}/2 - 2\pi/3) - \frac{2N_{s3}}{3P} \cos(3P\phi_{sm}/2) \tag{8.4-5}$$

$$w_{cs}(\phi_{sm}) = \frac{2N_{s1}}{P} \cos(P\phi_{sm}/2 + 2\pi/3) - \frac{2N_{s3}}{3P} \cos(3P\phi_{sm}/2) \tag{8.4-6}$$

From (8.3-17), the total stator MMF may be expressed as

$$F_S(\phi_{sm}) = w_{as}(\phi_{sm})i_{as} + w_{bs}(\phi_{sm})i_{bs} + w_{cs}(\phi_{sm})i_{cs} \tag{8.4-7}$$

To proceed further, we need to inject currents into the system. Let us consider a balanced three-phase set of currents of the form

$$i_{as} = \sqrt{2}I_s \cos(\omega_e t + \phi_i) \tag{8.4-8}$$

$$i_{bs} = \sqrt{2}I_s \cos(\omega_e t + \phi_i - 2\pi/3) \tag{8.4-9}$$

$$i_{cs} = \sqrt{2}I_s \cos(\omega_e t + \phi_i + 2\pi/3) \tag{8.4-10}$$

where I_s is the rms magnitude of each phase current, ω_e is the ac electrical frequency, and ϕ_i is the phase of the a-phase current. Substitution of (8.4-4)–(8.4-6) and (8.4-8)–(8.4-10) into (8.4-7) and simplifying yields

$$F_S = \frac{3\sqrt{2}N_{s1}I_s}{P} \cos(P\phi_{sm}/2 - \omega_e t - \phi_i) \tag{8.4-11}$$

The result (8.4-11) is an important one. From (8.4-11), we can see that given a sinusoidal turns distribution and the appropriate currents we arrive at a MMF that is sinusoidal in space and time. It represents a wave equation. In other words, the resulting MMF is a traveling wave. To see this, consider the peak of the wave, wherein the argument to the cosine term in (8.4-11) is zero. In particular, at the peak of the wave we have

$$P\phi_{sm}/2 - \omega_e t - \phi_i = 0 \tag{8.4-12}$$

From (8.4-12) we obtain

$$\phi_{sm} = \frac{2}{P}\omega_e t + \frac{2}{P}\phi_i \tag{8.4-13}$$

Thus, the peak of the wave is moving at a speed of $2\omega_e/P$.

In synchronous machines, the rotor speed will be equal to the speed of the stator MMF wave in the steady-state, so it is clear that the speed will vary with ω_e. It can also be seen that as the number of poles is increased, the speed of the MMF wave (and hence the rotor) will decrease.

The creation of an MMF wave that travels in a single direction requires at least two currents and two windings. A single winding will produce an MMF with both forward and reverse traveling waves, which significantly reduces the efficiency of the machine. Polyphase machines with unbalanced excitation also yield forward and reverse traveling waves. An example of this is explored in Problem 9.

8.5 Flux Linkage and Inductance

We will now focus our attention on the calculation of the inductance and resistance of rotating electrical machines. To begin, it is convenient to view the flux linking any winding (winding x) in terms of leakage flux linkage λ_{xl} and magnetizing flux linkage λ_{xm}. The total flux linkage of a winding is the sum of these two components. Thus,

$$\lambda_x = \lambda_{xl} + \lambda_{xm} \tag{8.5-1}$$

The distinction between leakage flux linkage and magnetizing flux linkage is not always precise. However, leakage flux is associated with flux that does not travel across the air gap or couple both the rotor and stator windings. Magnetizing flux linkage λ_{xm} is associated with radial flux flow across the air gap and links both the stator and rotor windings.

Associated with the concept of leakage and magnetizing flux are the concepts of leakage and magnetizing inductance which relate their respective flux linkage components to current. In order to state these concepts mathematically, let x and y denote two windings (and will take on values of as, bs, cs, ar, br, and cr, etc.). Then the leakage and magnetizing inductance between two windings may be expressed

$$L_{xyl} = \frac{\lambda_{xl}|_{\text{due to } i_y}}{i_y} \tag{8.5-2}$$

$$L_{xym} = \frac{\lambda_{xm}|_{\text{due to } i_y}}{i_y} \tag{8.5-3}$$

From our definition, the mutual leakage inductance between a stator winding and a rotor winding will be zero. However, there will be leakage inductances between different stator windings and between different rotor windings.

The total inductance of a winding is defined as

$$L_{xy} = \frac{\lambda_x|_{\text{due to } i_y}}{i_y} \tag{8.5-4}$$

From (8.5-1)–(8.5-4) it is clear that the total inductance is the sum of the leakage and magnetizing inductance, hence

$$L_{xy} = L_{xyl} + L_{xym} \tag{8.5-5}$$

The leakage inductance of a winding can be viewed as parasitic, and it is a strong function of the details of the winding. We will consider the calculation of the leakage inductance in Section 8.7. At present, we will concentrate on the calculation of the magnetizing inductance between two

Figure 8.12 Calculation of flux linkage.

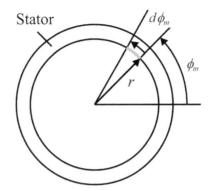

distributed windings. To this end, consider Figure 8.12, which shows a stator. Consider the incremental area along the inner surface of the stator, as shown. The radius to the inner stator surface is r and its position is ϕ_m. Note that by this drawing, ϕ_m is relative to the stator and could be designated ϕ_{sm}; however, as an identical argument could be made using any reference point, ϕ_m is used to denote position. The length of the edge segment is $rd\phi_m$. If the axial length of the machine is l, it follows that the incremental area is $lrd\phi_m$. For small $d\phi_m$, the incremental flux through this incremental area may be written as

$$d\Phi(\phi_m) = B(\phi_m)lrd\phi_m \tag{8.5-6}$$

Now recall that the winding function $w_x(\phi_m)$ describes how many times a winding x links the flux at a position ϕ_m. Thus, the contribution of the flux linking winding x through this incremental area may be expressed $w_x(\phi_m)d\Phi_m$ or $w_x(\phi_m)B(\phi_m)lrd\phi_m$. Adding up all the incremental areas along the stator, we have

$$\lambda_{xm} = \int_0^{2\pi} B(\phi_m)w_x(\phi_m)lrd\phi_m \tag{8.5-7}$$

Now that we have a means to calculate the flux linkage given the flux density and winding function, recall from (8.3-23) (and using (8.3-17), (8.3-18), and (8.3-20)) that the air gap flux density due to winding y may be expressed

$$B(\phi_m) = \frac{\mu_0}{g(\phi_m)} w_y(\phi_m) i_y \tag{8.5-8}$$

Substitution of (8.5-8) into (8.5-7) and manipulating, the flux in winding x due to the current in winding y is given by

$$\lambda_{xym} = \left(\mu_0 rl \int_0^{2\pi} \frac{w_x(\phi_m)w_y(\phi_m)}{g(\phi_m)} d\phi_m \right) i_y \tag{8.5-9}$$

From (8.5-9) it can be seen that the inductance between two windings may be expressed as

$$\frac{\lambda_{xym}}{i_y} = L_{xym} = \mu_0 rl \int_0^{2\pi} \frac{w_x(\phi_m)w_y(\phi_m)}{g(\phi_m)} d\phi_m \tag{8.5-10}$$

This relationship is valid for both self- and mutual inductance. When using (8.5-10) to calculate self-inductance $y = x$.

The expressions in (8.5-9) and (8.5-10) involve several approximations. First, they depend on an assumption made in (8.5-8) that the field is uniform across the air gap. In machines with large air gaps, or effectively large air gaps, it may be necessary to address the radial variation in the flux density. This is described in the context of a permanent magnet ac machine in Chapter 9. A second assumption associated with (8.5-9) and (8.5-10) is that they neglect the rather appreciable effect that slots may have on the magnetizing inductance. This may be accounted for using Carter's method, which is described in the next section.

Before concluding this section, consider multi-pole machines wherein all position-dependent quantities are periodic with a period of $4\pi/P$. Making use of the periodicity of field distribution, it is readily shown that (8.5-7) and (8.5-10) become

$$\lambda_{xm} = \int_0^{2\pi} B(\phi) w_x(\phi) lr d\phi \tag{8.5-11}$$

and

$$\frac{\lambda_{xym}}{i_y} = L_{xym} = \mu_0 rl \int_0^{2\pi} \frac{w_x(\phi) w_y(\phi)}{g(\phi)} d\phi \tag{8.5-12}$$

where ϕ is an electrical position defined by (8.1-7) and where the functions have been suitably modified so that their argument is electrical rather than mechanical position. Note that while (8.5-11) and (8.5-12) appear identical to (8.5-7) and (8.5-10), in the former all variables are periodic in 2π, which makes the integral easier to evaluate in the presence of discontinuities such as those arising from permanent magnets.

8.6 Slot Effects and Carter's Coefficient

In our previous analysis, we neglected the effects of slots on the stator and rotor. As it turns out, the effects of slots can be readily incorporated into the analysis by replacing the air gap g with a modified air gap g'. In particular, for the case of the stator slots, the modified air gap is calculated as

$$g' = g c_s \tag{8.6-1}$$

where c_s is the stator Carter's coefficient. We will now derive this result as well as a value for c_s.

The derivation of (8.6-1) begins with the consideration of Figure 8.13. This figure depicts the developed diagram over a small range of position w corresponding to one-half of a stator slot width plus one-half of a stator tooth width. Thus

$$w = \frac{1}{2} w_{ss} + \frac{1}{2} w_{st} \tag{8.6-2}$$

where w_{ss} is the stator slot width and w_{st} is the stator tooth width, both measured at the stator/air-gap interface.

Let us first consider the situation if we ignore the slot. In this case, it can be shown that the flux flowing across the air gap in the interval w may be expressed as

Figure 8.13 Carter's coefficient.

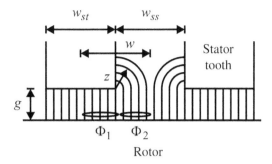

$$\Phi = \frac{\mu_0 Fl}{2g}(w_{ss} + w_{st}) \tag{8.6-3}$$

where l is the length of the machine and F is the MMF drop between the stator and rotor at that point. Because the slot is unaccounted for in (8.6-3), this expression is in error, because part of the flux Φ_2 will have to travel further. Our goal will be to establish a value g' such that

$$\Phi = \frac{\mu_0 Fl}{2g'}(w_{ss} + w_{st}) \tag{8.6-4}$$

is correct, or is at least a good approximation.

To this end, let us calculate the flux accounting for the fact that Φ_2 has to travel further than Φ_1. The total flux may be expressed

$$\Phi = \Phi_1 + \Phi_2 \tag{8.6-5}$$

The first term is readily expressed

$$\Phi_1 = \frac{\mu_0 F w_{st} l}{2g} \tag{8.6-6}$$

The second term is more involved. At a position z (see Figure 8.13) the distance from the rotor to the stator along the indicated path is $g + \pi z/2$. Thus, the field intensity along this path may be estimated as

$$H = \frac{F}{g + \pi z/2} \tag{8.6-7}$$

The flux Φ_2 may be expressed as

$$\Phi_2 = \int_{z=0}^{w_{ss}/2} Bl\,dz \tag{8.6-8}$$

Substitution of (8.6-7) into (8.6-8) and noting that the fields are in air yields

$$\Phi_2 = \frac{2\mu_0 Fl}{\pi} \ln\left(1 + \frac{\pi\, w_{ss}}{4g}\right) \tag{8.6-9}$$

The final step is to add (8.6-6) and (8.6-9) and to equate the result to (8.6-4). The result is (8.6-1) where

$$c_s = \frac{w_{ss} + w_{st}}{w_{st} + \frac{4g}{\pi}\ln\left(1 + \frac{\pi w_{ss}}{4g}\right)} \tag{8.6-10}$$

Observe that g, g', and c_s can all be functions of the position depending upon the rotor shape (as measured from the stator or the rotor), but this functional dependence is not explicitly shown.

The use of (8.6-1) and (8.6-10) is straightforward and very useful, because it allows us, with a simple substitution of g' for g, to account, albeit approximately, for the effects of the stator slots on magnetizing inductance calculations as well as flux linkage due to permanent magnets which is affected by the length of the air gap.

For machines with both stator and rotor slots, the concept of Carter's coefficient can still be used; however, in this case

$$g' = g c_s c_r \tag{8.6-11}$$

where

$$c_r = \frac{w_{rs} + w_{rt}}{w_{rt} + \frac{4 g c_s}{\pi} \ln\left(1 + \frac{\pi w_{rs}}{4 g c_s}\right)} \tag{8.6-12}$$

and where w_{rs} and w_{rt} are the width of the rotor slot and rotor tooth where it meets the air gap.

Before concluding, it should be noted that (8.6-10) and (8.6-12) are based on a geometry in which the tooth is rectangular (though arced at the air gap). In cases where the tooth geometry is more involved, the same methods can be used to find an alternate expression for Carter's coefficient for the given geometry. This is the topic of Problem 13.

8.7 Leakage Inductance

In this section, we consider the problem of the computation of leakage inductance of a stator winding. The method discussed herein is based on [2] which uses some techniques presented in [3]. As discussed in Chapter 2, the leakage flux linkage is associated with flux that does not cross the air gap. However, we will find that even without flux crossing the air gap, the windings of the machine will have mutual inductance. As a result, we will find that a machine has both self- and mutual leakage inductance.

In our development, we will assume that the leakage inductance of every winding is the same, and that the mutual leakage inductance between any two windings is the same. We will also assume that the leakage inductance paths are magnetically linear. With these assumptions the stator leakage flux linkage may be expressed in terms of the stator currents as

$$\lambda_{abcs,l} = \begin{bmatrix} L_{lp} & L_{lm} & L_{lm} \\ L_{lm} & L_{lp} & L_{lm} \\ L_{lm} & L_{lm} & L_{lp} \end{bmatrix} i_{abcs} \tag{8.7-1}$$

where L_{lp} is the self-leakage inductance of a phase, and L_{lm} is the mutual leakage inductance between phases.

Our goal is to derive expressions for L_{lp} and L_{lm}. To this end, let us consider Figure 8.14, which illustrates the ith slot of the machine, and a group of conductors within the slot, which includes all phases. Superimposed on the slot and conductor region is a simple MEC of the slot. Therein, $N_{as,i}$, $N_{bs,i}$, and $N_{cs,i}$ are the number of conductors of each phase within the slot, and P_{sl} denotes the permeance of the flux path within the slot. Consideration of Figure 8.14 yields that the contributions of the leakage inductance by the ith slot may be expressed as

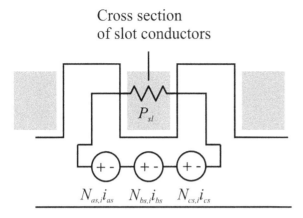

Figure 8.14 Slot leakage inductance.

$$L_{lpsl,i} = P_{sl} N_{as,i}^2 \tag{8.7-2}$$

$$L_{lmsl,i} = P_{sl} N_{as,i} N_{bs,i} \tag{8.7-3}$$

where the added subscript *sl* denotes the components of L_{lp} and L_{lm} due to slot leakage.

As it turns out, there is another component to the leakage inductance. This second component arises from the end conductors. Consider Figure 8.15, which depicts part of a longitudinal cross-section of the machine, including a cross-section of the end conductors across the *i*th back iron segment. Superimposed on this is a simple MEC in which $M_{as,i}$, $M_{bs,i}$, and $M_{cs,i}$ are the *a*-, *b*-, and *c*-phase end conductors across this segment, and P_{el} is the permeance associated with the flux path associated with this segment. Considering this simple MEC yields the contribution of the *i*th end conductor segment to the leakage inductance

$$L_{lpel,i} = P_{el} M_{as,i}^2 \tag{8.7-4}$$

$$L_{lmel,i} = P_{el} M_{as,i} M_{bs,i} \tag{8.7-5}$$

where the *el* in the subscript denotes end leakage.

Figure 8.15 End leakage inductance.

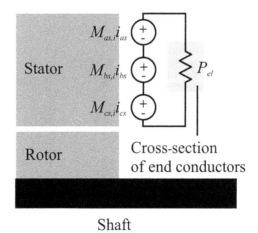

Adding the slot leakage and end leakage terms together, along with adding the contribution over all slot and end segments, yields

$$L_{lp} = P_{sl} \sum_{i=1}^{N_s} N_{as,i}^2 + P_{el} \sum_{i=1}^{N_s} M_{as,i}^2 \tag{8.7-6}$$

$$L_{lm} = P_{sl} \sum_{i=1}^{N_s} N_{as,i} N_{bs,i} + P_{el} \sum_{i=1}^{N_s} M_{as,i} M_{bs,i} \tag{8.7-7}$$

The next step is the calculation of the permeances P_{sl} and P_{el}. We will begin this endeavor with consideration of P_{sl}. This calculation will be based on a rectangular slot approximation that will include a provision for tooth tips that can be used to reduce the effective air gap, reduce the fields in the windings (and thereby loss), and provide additional structural support for the windings. It is assumed that the steel is infinitely permeable. We will break this permeance into seven terms corresponding to the seven flux paths depicted in Figures 8.16 and 8.17. Figure 8.16 and the lower portion of Figure 8.17 depict a cross-section of the teeth from viewed from the end of the machine along the axis of the machine. In Figure 8.17, the upper part of the figure is a view of the teeth looking radially outward from the center of the machine with the rotor removed. Both of these views are developed in the sense that the stator curvature has not been shown.

The permeances associated with path i will be denoted $P_{sl,i}$. The first three permeances correspond to paths that fully surround the slot conductors. This yields

$$P_{sl,1} = \frac{\mu_0 l}{\pi} \ln\left(1 + \frac{\pi \min(w_{tt}/2, g)}{w_{st}}\right) \tag{8.7-8}$$

$$P_{sl,2} = \frac{l d_{tt} \mu_0}{w_{st}} \tag{8.7-9}$$

$$P_{sl,3} = \frac{l(d_{si} - d_w)\mu_0}{w_{si}} \tag{8.7-10}$$

In (8.7-8), l is the length of the machine and g is the air gap used for slot leakage. In a synchronous or induction machine, this would be the actual air gap and this term will be very small. In a surface-mounted PM machine, g will include the actual air gap and the magnet depth, since many magnet materials have a permeability similar to that of air.

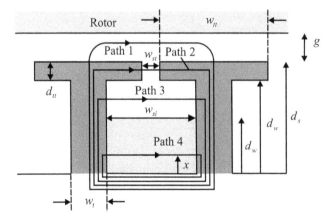

Figure 8.16 Slot leakage permeance due to paths 1–4. (Modified from Cassimere and Sudhoff [2].)

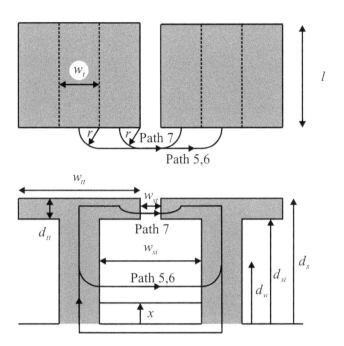

Figure 8.17 Slot leakage permeance due to paths 5–7. (Modified from Cassimere and Sudhoff [2].)

The permeance of Path 4 is calculated using the energy-based approach for finding permeance. This yields

$$P_{sl,4} = \frac{\mu_0 l d_w}{3 w_{si}} \qquad (8.7\text{-}11)$$

The final three paths considered are illustrated in Figure 8.17 and include an axial component. Again using an energy approach, we obtain

$$P_{sl,5} = \frac{\mu_0 d_w}{3\pi} \ln\left(1 + \frac{\pi w_t}{w_{si}}\right) \qquad (8.7\text{-}12)$$

$$P_{sl,6} = \frac{\mu_0}{\pi} \ln\left(1 + \frac{\pi w_t}{w_{si}}\right)(d_{si} - d_w) \qquad (8.7\text{-}13)$$

$$P_{sl,7} = \frac{\mu_0}{\pi} \ln\left(1 + \frac{\pi w_{tt}}{w_{st}}\right) d_{tt} \qquad (8.7\text{-}14)$$

Path 5 and Path 6 differ in that Path 5 is within the depth of the winding whereas Path 6 is above the winding. Path 7 corresponds to the region of the tooth tip. Adding all components together yields

$$P_{sl} = P_{sl,1} + P_{sl,2} + P_{sl,3} + P_{sl,4} + 2P_{sl,5} + 2P_{sl,6} + 2P_{sl,7} \qquad (8.7\text{-}15)$$

The coefficient of two appears on end permeances that occur on the front and the back of the machine.

The final step in our effort is the calculation of the end leakage permeance. This has two components: One is associated with flux within the winding bundle, and the other with flux on the exterior of the winding bundle. Using the same procedure as in Section 2.7 concerning exterior isolated leakage flux, where a cross-section of the end winding is the conductive region as shown in Figure 2.24, we obtain

$$P_{el,1} = \frac{\mu_0 L_{seg}}{d_r^2 l_r^2} \left[\begin{array}{l} \frac{d_r^4}{32} + \frac{d_r^3}{16}(l_r - d_r) + \frac{d_r^2}{64}(l_r - d_r)^2 \\ -\frac{d_r}{64}(l_r - d_r)^3 + \frac{1}{128}(l_r - d_r)^4 \ln\left(1 + \frac{2d_r}{l_r - d_r}\right) \end{array} \right] \quad (8.7\text{-}16)$$

where L_{seg} is the length of the segment of the end winding in the tangential direction. In particular,

$$L_{seg} = \frac{2\pi}{N_s}\left(r_s + d_s - \frac{d_w}{2}\right) \quad (8.7\text{-}17)$$

and

$$l_r = \max(l_{ew}, d_w) \quad (8.7\text{-}17)$$
$$d_r = \min(l_{ew}, d_w) \quad (8.7\text{-}18)$$

In (8.7-17) and (8.7-18) l_{ew} and d_w are the dimensions of the end winding segment as shown in Figure 8.18. If $l_r = d_r$, the expression of the permeance becomes

$$P_{el,1} = \frac{\mu_0 L_{seg}}{32} \quad (8.7\text{-}19)$$

Note that while this approach is rather crude, this term provides only a minor contribution to the overall leakage inductance and so the sensitivity of the leakage inductance to the error in (8.7-16) is small.

The second contribution of the end leakage permeance is associated with flux flow around the end winding segment, as shown in Figure 8.18. This figure depicts a cross-section of a machine that has been cut in half lengthwise through its rotational axis; the figure depicts the upper front region of the bisected machine. Dashed rectangles denote regions that will be treated by reluctance elements to compute the permeance.

Using the mean path approximation with the dashed rectangular elements shown in Figure 8.18, the reluctance of this path may be expressed

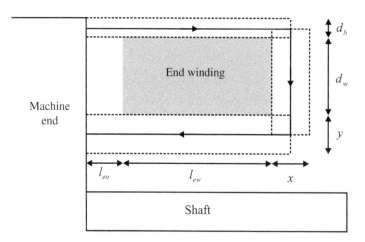

Figure 8.18 End winding permeance—exterior path. (Cassimere and Sudhoff [2] / with permission of IEEE.)

$$R = \frac{(l_{eo} + l_{ew} + x)}{d_b L_{seg} \mu_0} + \frac{(d_w + \frac{d_b}{2} + y)}{2x L_{seg} \mu_0} + \frac{(l_{eo} + l_{ew} + x)}{2y L_{seg} \mu_0} \qquad (8.7\text{-}20)$$

In order to determine the dimensions x and y, it is assumed that these are such that the reluctance of the path is minimized:

$$P_{el,2} = \frac{1}{R_{min}} \qquad (8.7\text{-}21)$$

where R_{min} is the minimum value of R for all x and y in (8.7-20).

At this point, the total end leakage permeance may be calculated as

$$P_{el} = 2P_{el,1} + 2P_{el,2} \qquad (8.7\text{-}22)$$

where the factor of two arises because of the fact that end permeances occur on the front and the back of the machine.

8.8 Resistance

Besides inductance, another electrical characteristic of a coil is its resistance. In this section, we consider the problem of finding the resistance of a distributed winding. Our approach will utilize the discrete winding description and will begin by considering the spatial volume of the winding.

The volume of a winding can be broken down into two parts: (a) the volume located in the slots and (b) the volume located in the end-turns. The portion of the volume of winding x located in the slots is denoted V_{xs} and is equal to the sum of the absolute value of the number of conductors in each slot times the volume of each conductor (which is, in turn, the length times the cross-sectional area). Thus,

$$V_{xs} = (l + 2e)a_c \sum_{i=1}^{S_y} |N_{x,i}| \qquad (8.8\text{-}1)$$

where a_c and e are the cross-sectional conductor area and axial distance from the end of the machine laminations to the center of the end-turn bundle, respectively.

The volume of conductor associated with the end-turn region is denoted V_{xe} and may be expressed as

$$V_{xe} = \frac{2\pi}{S_y} \bar{r}_x a_c E_x \qquad (8.8\text{-}2)$$

In (8.8-2), $2\pi/S_y$ is the angle of an end conductor sector and $\bar{r}_x$ is the mean radius (from the center of the machine) to the end conductor bundle. The product of these two factors is the length of an end conductor sector. Multiplying the end conductor sector length times E_x and the cross-sectional area of the conductor, a_c, yields the end conductor volume for the winding. This volume is that of one (front or back) end.

Since there are two end-turn regions, the total conductor volume, V_{xt}, associated with the winding may be expressed as

$$V_{xt} = V_{xs} + 2V_{xe} \qquad (8.8\text{-}3)$$

The length of the conductor associated with the winding is then given by

$$l_x = \frac{V_{xt}}{a_c} \tag{8.8-4}$$

whereupon the phase resistance may be calculated as

$$r_x = \frac{l_x}{a_c \sigma_c} \tag{8.8-5}$$

where σ_c is the conductivity of the conductor used for winding x.

8.9 Introduction to Reference Frame Theory

When analyzing electric machinery, one will find that there are generally three sets of equations: voltage equations that relate the applied voltage to the resistive drop and time rate of change of flux linkages, flux linkage equations that express the flux linkages in terms of rotor position and currents, and a torque equation that expresses the electromagnetic torque in terms of rotor position and current. In their original form, these equations can be difficult to use both because the input voltages are not constant (they are ac quantities) and because the flux linkage and torque equations are a function of rotor position. Reference frame theory provides a means by which the ac excitation of a machine can be transformed into something which, from a mathematical perspective, looks like dc excitation, and in which the rotor position dependence of the flux linkage and torque equations is eliminated, thereby producing considerable simplification.

Reference frame theory has had a long history, perhaps beginning with the work of Blondel [4]. Park [5] introduced the first explicit transformation, which is essentially the transformation we will use herein. This transformation, known as Park's transformation and also referred to as the transformation to the rotor reference frame, is a special case of a transformation to an arbitrary reference frame, an idea set forth in Krause and Thomas [6]. A thorough discussion of reference frame theory is set forth in Krause et al. [7]. Herein, we will focus our attention on one particular transformation, namely, the transformation to the rotor reference frame.

Park's Transformation

Let us consider a three-phase quantity in abc stator variables of the form

$$\mathbf{f}_{abcs} = [\, f_{as} \;\; f_{bs} \;\; f_{cs} \,]^T \tag{8.9-1}$$

where f can be a voltage, v, flux linkage, λ, or current, i. We will consider a mathematical transformation to an alternate representation of the form

$$\mathbf{f}^r_{qd0s} = \left[\, f^r_{qs} \;\; f^r_{ds} \;\; f_{0s} \,\right]^T \tag{8.9-2}$$

wherein the subscript qs denotes a q-axis stator quantity, the subscript ds denote a d-axis stator quantity, and the $0s$ subscript denotes a zero-sequence quantity. The superscript r is used to denote the rotor reference frame. The zero-sequence quantity does not carry the r superscript because it is reference frame independent. The transformation between abc variables and $qd0$ variables may be expressed

$$\mathbf{f}^r_{qd0s} = \mathbf{K}^r_s \mathbf{f}_{abcs} \tag{8.9-3}$$

where $\mathbf{K}^r_s$ is a transformation to the rotor reference frame which is given by

Figure 8.19 Geometric interpretation of Park's transformation.

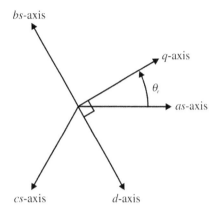

$$\mathbf{K}_s^r = \frac{2}{3} \begin{bmatrix} \cos\theta_r & \cos(\theta_r - 2\pi/3) & \cos(\theta_r + 2\pi/3) \\ \sin\theta_r & \sin(\theta_r - 2\pi/3) & \sin(\theta_r + 2\pi/3) \\ 1/2 & 1/2 & 1/2 \end{bmatrix} \quad (8.9\text{-}4)$$

In (8.9-4), θ_r denotes the electrical rotor position. It can be shown that

$$\mathbf{K}_s^{r^{-1}} = \begin{bmatrix} \cos\theta_r & \sin\theta_r & 1 \\ \cos(\theta_r - 2\pi/3) & \sin(\theta_r - 2\pi/3) & 1 \\ \cos(\theta_r + 2\pi/3) & \sin(\theta_r + 2\pi/3) & 1 \end{bmatrix} \quad (8.9\text{-}5)$$

A geometric interpretation of the transformation is depicted in Figure 8.19. Therein, the *as*-, *bs*-, and *cs*- axis of an electric machine are shown. The *q*- and *d*-axis can be viewed as the axis of a rotating coordinate system that is attached to the rotor of the electric machine. The *d*-axis, also known as the direct axis, is associated with the path taken by the magnetizing flux. The *q*-axis is in quadrature with the *d*-axis. The zero sequence is the mean value of the phase variables. Zero sequence quantities are often zero.

Transformation of a Balanced Set

A synchronous machine is a device in which the rotor moves at the same speed as stator MMF. This class of devices includes wound-rotor synchronous machines, permanent magnet synchronous machines, and reluctance machines. In any of these devices in the steady-state, ac quantities take on the form

$$\mathbf{f}_{abcs} = \sqrt{2}F_s \begin{bmatrix} \cos(\theta_r + \phi_f) \\ \cos(\theta_r + \phi_f - 2\pi/3) \\ \cos(\theta_r + \phi_f + 2\pi/3) \end{bmatrix} \quad (8.9\text{-}6)$$

where F_s is the *rms* value of waveform and ϕ_f is a-phase angle. This situation in which the three-phase quantities have identical amplitudes and a-phase separation of 120° is described as a three-phase balanced set and is a very desirable condition. Observe that the *abc* variables will vary with time unless the rotor is stationary.

Applying the transformation (8.9-3) to the three-phase balanced set (8.9-6) one obtains

$$\mathbf{f}^r_{qd0s} = \sqrt{2} F_s \begin{bmatrix} \cos(\phi_f) \\ -\sin(\phi_f) \\ 0 \end{bmatrix} \quad (8.9\text{-}7)$$

In making this transformation, the trigonometric identities of Appendix F are very useful. Note that the q- and d- axis variables are now constants and that the zero-sequence quantity is zero.

From (8.9-7) one can readily obtain the relationships

$$\sqrt{2} F_s = \sqrt{\left(f^r_{qs}\right)^2 + \left(f^r_{ds}\right)^2} \quad (8.9\text{-}8)$$

and

$$\phi_f = \text{angle}\left(f^r_{qs} - jf^r_{ds}\right) \quad (8.9\text{-}9)$$

where $j = \sqrt{-1}$.

Transformation of Voltage Equations

The stator voltage equations for balanced three-phase electric machinery can be expressed as

$$v_{as} = r_s i_{as} + \frac{d\lambda_{as}}{dt} \quad (8.9\text{-}10)$$

$$v_{bs} = r_s i_{bs} + \frac{d\lambda_{bs}}{dt} \quad (8.9\text{-}11)$$

$$v_{cs} = r_s i_{cs} + \frac{d\lambda_{cs}}{dt} \quad (8.9\text{-}12)$$

In matrix-vector form, we have

$$\mathbf{v}_{abcs} = r_s \mathbf{i}_{abcs} + \frac{d\lambda_{abcs}}{dt} \quad (8.9\text{-}13)$$

Writing each *abc* quantity in term of its *qd0* counterpart, we obtain

$$\mathbf{K}_s^{r^{-1}} \mathbf{v}^r_{qd0s} = r_s \mathbf{K}_s^{r^{-1}} \mathbf{i}^r_{qd0s} + \frac{d}{dt}\left\{\mathbf{K}_s^{r^{-1}} \lambda^r_{qd0s}\right\} \quad (8.9\text{-}14)$$

Premultiplying (8.9-14) by $\mathbf{K}_s^r$ and expanding the term in braces {} using the chain rule

$$\mathbf{v}^r_{qd0s} = r_s \mathbf{i}^r_{qd0s} + \mathbf{K}_s^r \left\{ \frac{d\mathbf{K}_s^{r^{-1}}}{dt} \lambda^r_{qd0s} + \mathbf{K}_s^{r^{-1}} \frac{d\lambda^r_{qd0s}}{dt} \right\} \quad (8.9\text{-}15)$$

Manipulating (8.9-15), one obtains

$$\mathbf{v}^r_{qd0s} = r_s \mathbf{i}^r_{qd0s} + \omega_r \begin{bmatrix} 0 & 1 & 0 \\ -1 & 0 & 0 \\ 0 & 0 & 0 \end{bmatrix} \lambda^r_{qd0s} + \frac{d\lambda^r_{qd0s}}{dt} \quad (8.9\text{-}16)$$

where ω_r is the electrical rotor speed, that is,

$$\omega_r = \frac{d\theta_r}{dt} \tag{8.9-17}$$

In scalar form (8.9-16) can be broken into qd0 components as

$$v_{qs}^r = r_s i_{qs}^r + \omega_r \lambda_{ds}^r + \frac{d\lambda_{qs}^r}{dt} \tag{8.9-18}$$

$$v_{ds}^r = r_s i_{ds}^r - \omega_r \lambda_{qs}^r + \frac{d\lambda_{ds}^r}{dt} \tag{8.9-19}$$

$$v_{0s} = r_s i_{0s} + \frac{d\lambda_{0s}}{dt} \tag{8.9-20}$$

We will use (8.9-18) and (8.9-19) in our consideration of the analysis and design of permanent magnet synchronous machines in Chapter 9.

Transformation of Flux Linkage Equations

From (8.9-3) the qd0 flux linkages are related to the abc flux linkages by

$$\lambda_{qd0s}^r = \mathbf{K}_s^r \lambda_{abcs} \tag{8.9-21}$$

To proceed further, one needs to specify the *abc* variable flux linkage equations, which are a function of the specific type of machine. Let us assume that these flux linkage equations are of the form of a permanent magnet synchronous machine. In particular,

$$\lambda_{abcs} = \mathbf{L}_s(\theta_r)\mathbf{i}_{abcs} + \lambda_{pm}(\theta_r) \tag{8.9-22}$$

In (8.9-22) $\mathbf{L}_s(\theta_r)$ is a rotor-position-dependent inductance matrix, and $\lambda_{pm}(\theta_r)$ is a vector of flux-linkage due to the permanent magnet. Substituting (8.9-22) into (8.9-21), and expressing the *abc* currents in terms of their qd0 counterparts, we obtain

$$\lambda_{qd0s}^r = \mathbf{K}_s^r \mathbf{L}_s(\theta_r) \mathbf{K}_s^{r^{-1}} \mathbf{i}_{qd0s}^r + \mathbf{K}_s^r \lambda_{pm}(\theta_r) \tag{8.9-23}$$

For a salient permanent magnet machine, it can be shown that

$$\mathbf{L_s}(\theta_{rm}) = L_{ls}\mathbf{I}_3 + L_A \begin{bmatrix} 1 & -1/2 & -1/2 \\ -1/2 & 1 & -1/2 \\ -1/2 & -1/2 & 1 \end{bmatrix} - L_B \begin{bmatrix} \cos 2\theta_r & \cos(2\theta_r - 2\pi/3) & \cos(2\theta_r + 2\pi/3) \\ \cos(2\theta_r - 2\pi/3) & \cos(2\theta_r + 2\pi/3) & \cos(2\theta_r) \\ \cos(2\theta_r + 2\pi/3) & \cos(2\theta_r) & \cos(2\theta_r - 2\pi/3) \end{bmatrix} \tag{8.9-24}$$

and

$$\lambda_{pm}(\theta_r) = \lambda_m \begin{bmatrix} \sin\theta_r \\ \sin(\theta_r - 2\pi/3) \\ \sin(\theta_r + 2\pi/3) \end{bmatrix} \tag{8.9-25}$$

where L_{ls}, L_A, L_B, and λ_m are constants and $\mathbf{I}_3$ is a 3 by 3 identity matrix.

Substitution of (8.9-24) and (8.9-25) into (8.9-23) yields

$$\lambda^r_{qd0s} = \begin{bmatrix} L_q & 0 & 0 \\ 0 & L_d & 0 \\ 0 & 0 & L_{ls} \end{bmatrix} \mathbf{i}^r_{qd0s} + \lambda_m \begin{bmatrix} 0 \\ 1 \\ 0 \end{bmatrix} \quad (8.9\text{-}26)$$

where

$$L_q = L_{ls} + \frac{3}{2}(L_A - L_B) \quad (8.9\text{-}27)$$

and

$$L_d = L_{ls} + \frac{3}{2}(L_A + L_B) \quad (8.9\text{-}28)$$

The drastic simplification of (8.9-22), (8.9-24), and (8.9-25) to (8.9-26) illustrates the usefulness of reference frame theory.

Transformation of Power

The instantaneous power into the stator of a three-phase machine may be expressed

$$P = v_{as}i_{as} + v_{bs}i_{bs} + v_{cs}i_{cs} \quad (8.9\text{-}29)$$

In vector form,

$$P = \mathbf{v}^T_{abcs}\mathbf{i}_{abcs} \quad (8.9\text{-}30)$$

Expressing (8.9-30) in terms of $qd0$ variables, one obtains

$$P = \left(\mathbf{K}^{r^{-1}}_s \mathbf{v}^r_{qd0s}\right)^T \mathbf{K}^{r^{-1}}_s \mathbf{i}^r_{qd0s} \quad (8.9\text{-}31)$$

which reduces to

$$P = \frac{3}{2}\left(v^r_{qs}i^r_{qs} + v^r_{ds}i^r_{ds} + 2v_{0s}i_{0s}\right) \quad (8.9\text{-}32)$$

8.10 Expressions for Torque

In Chapter 4 of this work, we developed some expressions for the computation of force and torque. These expressions are applicable to rotating electric machinery. However, for machinery with distributed windings in which the flux linkage equations can be expressed in $qd0$ variables without rotor position dependence, there is a general expression for the torque that is applicable across this entire class of machines. This expression can be derived using either an energy-based approach [8] as we used in Chapter 4 or a fields-based approach [9]. We will consider both approaches herein.

An Energy-Based Approach to Calculating Torque

Recall from Chapter 4, that the relationship between the energy being transferred to the coupling field is the sum of the energy contributed by the electrical system and that contributed by the mechanical system. In particular,

8.10 Expressions for Torque

$$W_f = W_e + W_m \tag{8.10-1}$$

Taking the partial derivative with respect to mechanical rotor position θ_{rm}, we have

$$\frac{\partial W_f}{\partial \theta_{rm}} = \frac{\partial W_e}{\partial \theta_{rm}} + \frac{\partial W_m}{\partial \theta_{rm}} \tag{8.10-2}$$

Our strategy will be to establish expressions for the required partial derivatives and substitute these into (8.10-2). The result will yield an expression for torque.

Our next step is to find an expression of the electrical input energy. Letting t_f be a dummy variable that represents the time at the instant of interest and t_0 be an initial time wherein the system is de-energized, we have that

$$W_e = \int_{t_0}^{t_f} P_e \, dt \tag{8.10-3}$$

where P_e is the electric power input that may be expressed as

$$P_e = e_{as}i_{as} + e_{bs}i_{bs} + e_{cs}i_{cs} + \sum_{k=1}^{K} e_k i_k \tag{8.10-4}$$

In (8.10-4), i_{as}, i_{bs}, and i_{cs} are the currents into the stator a-, b-, and c-phases, and e_{as}, e_{bs}, and e_{cs} are the time rate of change of the stator phase flux linkages (in other words, the stator phase voltages less the resistive drop). The variables i_k and e_k are the current into and voltage across (less resistive drops) circuits attached to the rotor of the electromechanical device. Denoting the a-, b-, and c-phase flux linkages as λ_{as}, λ_{bs}, and λ_{cs} and the flux linking the kth rotor circuit as λ_k, from Faraday's law and (5.1-83) we have

$$P_e = \frac{d\lambda_{as}}{dt}i_{as} + \frac{d\lambda_{bs}}{dt}i_{bs} + \frac{d\lambda_{cs}}{dt}i_{cs} + \sum_{k=1}^{K} \frac{d\lambda_k}{dt}i_k \tag{8.10-5}$$

Transforming the stator quantities to the rotor reference frame, we obtain

$$P_e = P_{e1} + P_{e2} \tag{8.10-6}$$

where

$$P_{e1} = \frac{3}{2}\left(\lambda_{ds}^r i_{qs}^r - \lambda_{qs}^r i_{ds}^r\right)\omega_r \tag{8.10-7}$$

and

$$P_{e2} = \frac{3}{2}\left(i_{qs}^r \frac{d\lambda_{qs}^r}{dt} + i_{ds}^r \frac{d\lambda_{ds}^r}{dt} + 2i_{0s}\frac{d\lambda_{0s}}{dt}\right) + \sum_{k=1}^{K}\frac{d\lambda_k}{dt}i_k \tag{8.10-8}$$

Thus

$$W_e = W_{e1} + W_{e2} \tag{8.10-9}$$

where

$$W_{e1} = \int_{t_0}^{t_f} \frac{3}{2}\left(\lambda_{ds}^r i_{qs}^r - \lambda_{qs}^r i_{ds}^r\right)\omega_r \, dt \tag{8.10-10}$$

and

$$W_{e2} = \int_{t_0}^{t_f} \left(\frac{3}{2}\left(i_{qs}^r \frac{d\lambda_{qs}^r}{dt} + i_{ds}^r \frac{d\lambda_{ds}^r}{dt} + 2i_{0s} \frac{d\lambda_{0s}}{dt} \right) + \sum_{k=1}^{K} \frac{d\lambda_k}{dt} i_k \right) dt \qquad (8.10\text{-}11)$$

Let us turn our attention to W_{e1}. Noting

$$\omega_r = \frac{P}{2} \frac{d\theta_{rm}}{dt} \qquad (8.10\text{-}12)$$

we have that

$$W_{e1} = \frac{3}{2} \frac{P}{2} \int_{t_0}^{t_f} \left(\lambda_{ds}^r i_{qs}^r - \lambda_{qs}^r i_{ds}^r \right) \frac{d\theta_{rm}}{dt} dt \qquad (8.10\text{-}13)$$

which may be expressed as

$$W_{e1} = \frac{3}{2} \frac{P}{2} \int_{\theta_{rm,0}}^{\theta_{rm,f}} \left(\lambda_{ds}^r i_{qs}^r - \lambda_{qs}^r i_{ds}^r \right) d\theta_{rm} \qquad (8.10\text{-}14)$$

where $\theta_{rm,f}$ is the rotor position at time t_f and $\theta_{rm,0}$ is θ_{rm} at time t_0.

Next, let us consider W_{e2}. In the rotor reference frame the flux-linkage equations are rotor position invariant, so W_{e2} is only a function of flux, and will henceforth be denoted $W_{e2}(\lambda)$, where the flux linkage vector λ is defined as

$$\lambda = [\lambda_{qs} \quad \lambda_{ds} \quad \lambda_{0s} \quad \lambda_1 \quad \cdots \quad \lambda_K]^T \qquad (8.10\text{-}15)$$

Our next step is to take the partial derivative of W_e with respect to mechanical rotor position. From (8.10-9) we have

$$\frac{\partial W_e}{\partial \theta_{rm}} = \frac{\partial W_{e1}}{\partial \theta_{rm}} + \frac{\partial W_{e2}}{\partial \theta_{rm}} \qquad (8.10\text{-}16)$$

Noting that the second term is zero (since W_{e2} is not a function of rotor position), substitution of (8.10-14) into (8.10-16) yields

$$\frac{\partial W_e}{\partial \theta_{rm}} = \frac{3}{2} \frac{P}{2} \left(\lambda_{ds}^r i_{qs}^r - \lambda_{qs}^r i_{ds}^r \right) \qquad (8.10\text{-}17)$$

We will next address the calculation of the field energy. In Chapter 4, we found that we could calculate the field energy by finding W_e with the mechanical position fixed. Thus

$$W_f = W_e|_{\text{Mechanical Position Fixed}} \qquad (8.10\text{-}18)$$

Observe that if the mechanical rotor position is fixed, W_{e1} is zero. Thus we have

$$W_f = W_{e2} \qquad (8.10\text{-}19)$$

However, recall that W_{e2} is not a function of position, and thus neither is W_f. It follows that

$$\frac{\partial W_f(\lambda)}{\partial \theta_{rm}} = 0 \qquad (8.10\text{-}20)$$

Our last energy to consider is mechanical energy W_m. From (4.1-11) (with T_e replacing f_e and θ_{rm} replacing x, as discussed in Section 4.8), we have that

$$\frac{\partial W_m}{\partial \theta_{rm}} = -T_e \tag{8.10-21}$$

Substitution of (8.10-17), (8.10-20), and (8.10-21) into (8.10-2) we have that

$$T_e = \frac{3}{2}\frac{P}{2}\left(\lambda_{ds}^r i_{qs}^r - \lambda_{qs}^r i_{ds}^r\right) \tag{8.10-22}$$

Although we have only considered the rotor reference frame herein, it is worthwhile noting that this result also holds in the arbitrary reference frame.

Equation (8.10-22) is an extremely useful and important result. It gives us an easy-to-use expression for electromagnetic torque for any three-phase machine in which the flux-linkage equations can be made rotor position invariant by transformation to the rotor reference frame. The expression is commonly applied to synchronous machines (including certain classes of permanent magnet machines), a variety of reluctance machines, and induction machines. It is highly useful in that it saves the effort of formally computing either the field or co-energy. A similar procedure can also be carried out for two-phase machines; in this case the corresponding result is

$$T_e = \frac{P}{2}\left(\lambda_{ds}^r i_{qs}^r - \lambda_{qs}^r i_{ds}^r\right) \tag{8.10-23}$$

A Field Approach to Calculating Torque

Equation (8.10-22) is an extremely useful result. However, the derivation assumed that the coupling field is conservative, which means, strictly speaking, that the method breaks down in materials in which exhibit magnetic hysteresis. In this section, an alternate method of deriving an expression for electromagnetic torque is set forth. It is valid for any distributed winding machine and is valid in the presence of hysteresis. However, there is a disadvantage in that one additional assumption will be made, which is that the slot/winding structure used for practical construction of the stator windings will produce the same amount of torque as the truly continuous distributed winding that it attempts to approximate. Clearly, slot-induced torque ripple will not be captured by this method; however, the same can be said of the previous derivation.

In this section, three-phase electric machinery with distributed windings will be considered. The radius of the location of the stator windings and length of the machine will be denoted r and l, respectively. An expression for torque production may be obtained starting with the Lorenz force equation, which states that the force acting on a single conductor in a machine may be expressed

$$F_c = il\mathrm{B}_r(\phi_{sm}) \tag{8.10-24}$$

where i is the current out of the page, $\mathrm{B}_r(\phi_{sm})$ is the air-gap flux density from the rotor to the stator, and F_c is the force on the conductor which will be at right angles to the conductor, and right angles to the radial flux density, and will be in the counter-clockwise direction relative to a line drawn from the center of the machine to the conductor. The torque on a conductor in the counterclockwise direction can be expressed as

$$T_c = ril\mathrm{B}_r(\phi_{sm}) \tag{8.10-25}$$

From (8.10-25), the total torque on the stator in the clockwise direction may be expressed as

$$T_{es} = rl \int_0^{2\pi} B_r(\phi_s) \mathbf{n}_{abcs}(\phi_{sm})^T \mathbf{i}_{abcs} d\phi_{sm} \tag{8.10-26}$$

where

$$\mathbf{n}_{abcs}(\phi_{sm}) = [n_{as}(\phi_{sm}) \quad n_{bs}(\phi_{sm}) \quad n_{cs}(\phi_{sm})]^T \tag{8.10-27}$$

and

$$\mathbf{i}_{abcs} = [i_{as} \quad i_{bs} \quad i_{cs}]^T \tag{8.10-28}$$

The electromagnetic torque on the rotor is related to the electromagnetic torque on the stator by

$$T_e = -T_{es} \tag{8.10-29}$$

For certain winding distributions, including sinusoidal, the turns density may be expressed as a linear function of the winding function as

$$\mathbf{n}_{abcs}(\phi_{sm}) = \frac{P}{2}\left[\mathbf{A}\mathbf{w}_{abcs}(\phi_{sm}) + [1 \quad 1 \quad 1]^T F_3(\phi_{sm})\right] \tag{8.10-30}$$

where P is the number of poles, $\mathbf{A}$ is a constant matrix,

$$\mathbf{w}_{abcs}(\phi_{sm}) = [w_{as}(\phi_{sm}) \quad w_{bs}(\phi_{sm}) \quad w_{cs}(\phi_{sm})]^T, \tag{8.10-31}$$

and $F_3(\phi_{sm})$ is an arbitrary scalar function of stator position. The purpose of introducing $F_3(\phi_{sm})$ into (8.10-30) is to address the case wherein the machine turns density contains a sinusoidal fundamental component plus a triple N harmonic content. Substitution of (8.10-30)–(8.10-31) into (8.10-26) yields

$$T_e = -rL\frac{P}{2}\int_0^{2\pi} B_r(\phi_{sm})\left(\mathbf{A}\mathbf{w}_{abcs}(\phi_{sm}) + [1 \quad 1 \quad 1]^T F_3(\phi_{sm})\right)^T \mathbf{i}_{abcs} d\phi_{sm} \tag{8.10-32}$$

Assuming that the sum of the phase currents is zero, the zero-sequence current is zero, and therefore the contribution of the $F_3(\phi_{sm})$ term in (8.10-32) is zero. The sum of the phase currents may be forced to zero by connecting the negative terminal of each phase together (to a point referred to as the neutral) and then leaving the neutral point floating. Such an arrangement is referred to as a wye-connection. Next, comparing (8.10-32) to (8.5-7), the torque may be expressed

$$T_e = -\frac{P}{2}\boldsymbol{\lambda}_{abcm}^T \mathbf{A}^T \mathbf{i}_{abcs} \tag{8.10-33}$$

where $\boldsymbol{\lambda}_{abcm}$ is a vector of phase magnetizing flux linkages. This is an interesting result in that it holds not only in the presence of magnetic saturation (as did (8.10-22)) but also in the presence of magnetic hysteresis.

As a special case, it is useful to consider a quasi-sinusoidally distributed machine given by (8.4-1)–(8.4-3). Comparing (8.4-1)–(8.4-3) to (8.4-4)–(8.4-6), and (8.10-30) it is apparent that for this winding distribution, $\mathbf{A}$ is given by

$$\mathbf{A} = \frac{\sqrt{3}}{6}\begin{bmatrix} 0 & 2 & -2 \\ -2 & 0 & 2 \\ 2 & -2 & 0 \end{bmatrix} \tag{8.10-34}$$

In this type of machine, it is often convenient to work in terms of *qd0* variables. Transforming (8.10-33) to the rotor reference using the transformation defined by (8.9-3) and (8.9-4), we obtain

$$T_e = -\frac{P}{2}\lambda_{qd0m}^{rT}\left[\mathbf{K}_s^{r-1}\right]^T \mathbf{A}^T \mathbf{K}_s^{r-1}\mathbf{i}_{qd0s}^r \qquad (8.10\text{-}35)$$

which reduces to

$$T_e = \frac{3}{2}\frac{P}{2}\left(\lambda_{dm}^r i_{qs}^r - \lambda_{qm}^r i_{ds}^r\right) \qquad (8.10\text{-}36)$$

which is very similar to (8.10-22) except that it is in terms of the magnetizing flux linkages. Like (8.10-22), (8.10-36) can be shown to be valid in any reference frame. In the common case wherein

$$\lambda_{qs} = \lambda_{qm} + L_{ls}()i_{qs} \qquad (8.10\text{-}37)$$
$$\lambda_{ds} = \lambda_{dm} + L_{ls}()i_{ds} \qquad (8.10\text{-}38)$$

and where $L_{ls}()$ denotes the leakage inductance which may be a function of stator current or magnetizing flux magnitude but is identical for the *q*- and *d*-axis, (8.10-22) and (8.10-36) are equivalent.

References

1. N. L. Shmitz and D. W. Novotny, *Introductory Electromechanics*. New York: Roland, 1965.
2. B. N. Cassimere and S. D. Sudhoff, Analytical design model for a surface mounted permanent magnet synchronous machine, *IEEE Transactions on Energy Conversion,* vol. **24**, no. 2, pp. 338–346, 2009.
3. D. C. Hanselman, *Brushless Permanent-Magnet Motor Design*. New York: McGrall-Hill, 1994.
4. A. Blondel, *Synchronous Motor and Converters, Part* **III**. New York: McGraw-Hill, 1912.
5. R. H. Park, Two-reaction theory of synchronous machines—Part I, *AIEE Transactions,* vol. **48**, no. 2, pp. 716–730, 1929.
6. P. C. Krause and C. H. Thomas, Simulation of symmetrial induction machinery, *IEEE Transactions on Power Apparatus and Systems,* vol. **84**, pp. 1038–1052, 1965.
7. P. C. Krause, O. Wasynczuk and S. D. Sudhoff, *Analysis of Electric Machinery and Drive Systems,* 2nd edition. Piscataway, NJ: IEEE Press/Wiley-Interscience, 2002.
8. D. C. Aliprantis, *Advances in Electric Machine Modeling and Evolutionary Parameter Identification (Ph. D. Thesis)*. West Lafayette, IN: Purdue University, 2003.
9. S. D. Sudhoff, B. T. Kuhn, P. L. Chapman and D. Aliprantis, *An advanced induction machine model for prediction of inverter-machine interaction,* in Power Electronics Specialists Conference, Vancouver, Canada, 2001.

Problems

1. The number of conductors in each slot of the *a*-phase of the stator of the machine are as follows:

$$N_{as} = 10.[1 \quad 2 \quad 2 \quad 1 \quad -1 \quad -2 \quad -2 \quad -1 \quad 1 \quad 2 \quad 2 \quad 1 \quad -1 \quad -2 \quad -2 \quad -1]^T$$

Compute and graph the winding function associated with this winding versus tooth number. Suppose the flux traveling from the rotor to the stator in each tooth is given by

$$\Phi_t = 10^{-3}.[-2 \ -1 \ 1 \ 2 \ 2 \ 1 \ -1 \ -2 \ -2 \ -1 \ 1 \ 2 \ 2 \ 1 \ -1 \ -2]^T$$

How much flux is linking the winding?

2. Suppose the turns (conductor) density of a winding function is given by

$$n_{br} = 262 \cos(8\phi_{rm} - 2\pi/3)$$

Compute the rotor b-phase winding function in terms of position measured from the rotor.

3. Consider (8.1-22) and (8.1-23) of the text, which are used to convert a discrete winding description to a continuous one. Instead of treating the discrete winding description as a series of delta functions when converting it to a continuous one, derive analogous expressions if it is assumed that the conductors within each slot are uniformly distributed over the angle spanned by the slot (call this angle Δ).

4. The conductor distribution for the first 18 slots of a 36 slot machine is given by

$$N_{as}|_{1-18} = N[0 \ 0 \ 0 \ 1 \ 2 \ 2 \ 1 \ 0 \ 0 \ 0 \ 0 \ -1 \ -2 \ -2 \ -1 \ 0 \ 0]^T$$

Using the impulse function approach (8.1-22) and (8.1-23), plot the spectrum of the resulting winding function over the first 50 harmonics of the fundamental (the fundamental being at $2\phi_{sm}$). Use a semi-log plot. On top of this, superimpose the results obtained using the results of Problem 3.

5. Consider the conductor density given by (8.1-12). Define the total conductor density as

$$n_t = |n_{as}| + |n_{bs}| + |n_{cs}|$$

Defining $\alpha_3 = N_{s3}/N_{s1}$, compute the value of α_3 which minimizes the peak value of $n_t(\phi_{sm})$. A numerical answer is acceptable.

6. Consider Problem 5. Derive the value of α_3 which minimizes the total number of conductors. A numerical answer is acceptable.

7. The winding function of the a- and b-phase stator windings of a machine are given by $w_{as} = 100 \cos(4\phi_{sm})$ and $w_{bs} = 250 \sin(4\phi_{sm})$. The a-phase current of the machine is given by $i_{as} = 10 \cos(\omega_e t + \pi/8)$. Determine the b-phase current and the air gap needed to achieve an air-gap flux density of $B(\phi_{sm}, t) = 1.2 \cos(\omega_e t + \pi/8 - 4\phi_{sm})$.

8. The conductor turns density of a 2-phase machine is given by

$$n_{as} = 100 \cos 2\phi_{sm}$$
$$n_{bs} = -100 \sin 2\phi_{sm}$$

The a- and b-phase currents are given by

$$i_{as} = 5 \cos(400t)$$
$$i_{bs} = 5 \sin(400t)$$

Express the total stator MMF. What is the speed and direction of the MMF?

9 The conductor turns density of a two-phase machine is given by

$$n_{as} = 100 \cos 2\phi_{sm}$$
$$n_{bs} = -100 \sin 2\phi_{sm}$$

The a- and b-phase currents are given by

$$i_{as} = 5 \cos(400t)$$
$$i_{bs} = 4 \sin(400t)$$

Express the total stator MMF as the sum of a forward and reverse traveling wave. This represents the situation in single-phase machines that have an auxiliary winding in series with a capacitor often used in residential applications.

10 Suppose the winding function of a-phase and b-phase of a stator is given by

$$w_{as} = \frac{2}{P} N_s \cos\left(\frac{P}{2}\phi_{sm}\right)$$
$$w_{bs} = \frac{2}{P} N_s \sin\left(\frac{P}{2}\phi_{sm}\right)$$

Express the mutual magnetizing inductance between the a-phase and b-phase stator windings in terms of N_s, N_r, P, the stator radius r, the machine length L, the air gap g, and the permeability of free space μ_0.

11 A wound-rotor induction machine has windings on the stator and the rotor. Suppose the winding function of a-phase of the stator is given by

$$w_{as} = \frac{2}{P} N_s \cos\left(\frac{P}{2}\phi_{sm}\right)$$

and the winding function of the b-phase of the rotor is given by

$$w_{br} = \frac{2}{P} N_r \sin\left(\frac{P}{2}\phi_{rm}\right)$$

Express the mutual inductance between the a-phase stator and b-phase rotor windings in terms of N_s, N_r, P, the rotor radius r, the rotor length L, the air gap g, the permeability of free space μ_0 and electrical rotor position θ_r.

12 Using (8.6-7) and (8.6-8) show (8.6-9).

13 Derive an expression for the stator Carter's coefficient analogous to (8.6-10) for a geometry with a tooth tip as shown in Figure 8.17. Assume that $w_{st} > 2d_{tt}$, and that flux doesn't flow into the top of the tooth or the side of the tooth follows that path it would have followed if the tooth tip was not present.

14 Express (8.7-1) in terms of $qd0$ variables.

15 Derive (8.7-12).

16 Derive (8.7-13).

17 Derive (8.7-14).

18 A permanent magnet synchronous machine is operating with a q-axis current of 5 A and a d-axis current of -2 A. Express the b-phase current as a function of rotor position.

19 The voltages applied to a permanent magnet synchronous machine are given by

$$v_{as} = 100 \cos(\theta_r + 0.1)$$
$$v_{bs} = 100 \cos(\theta_r - 2\pi/3 + 0.1)$$
$$v_{as} = 100 \cos(\theta_r + 2\pi/3 + 0.1)$$

What are the q- and d-axis voltages?

20 The voltages applied to a permanent magnet synchronous machine are given by

$$v_{as} = 102 \cos(\theta_r + 0.1)$$
$$v_{bs} = 100 \cos(\theta_r - 2\pi/3 + 0.1)$$
$$v_{as} = 100 \cos(\theta_r + 2\pi/3 + 0.1)$$

what are the q- and d-axis voltages?

21 Show that (8.9-23)–(8.9-25) reduce to (8.9-26)–(8.9-28)

9

Introduction to Permanent Magnet AC Machine Design

In the previous chapter, we set the stage for the analysis and design of distributed-winding-based rotating machinery. In this chapter, that material will be employed in the design of a permanent magnet ac (PMAC) machine. In keeping with the philosophy of this text, the prevalent design approach based on design rules coupled with detailed numerical analysis and manual design iteration is not used. Instead, the machine design problem is posed in a rigorous way as a formal mathematical optimization problem, as in references [1–3], which is similar to the approach we have taken for designing the inductor, the electromagnet, and the transformer.

As with our other design problems, the reader is forewarned that the approach has been simplified. Structural issues, thermal issues, and several loss mechanisms are neglected, and infinitely permeable magnetic steel is assumed, though saturation is considered. Even so, the design approach presented is nontrivial. It provides an organized and systematic approach to machine design, which is readily extended to include the aforementioned considerations. Indeed, we will consider such factors in Chapters 10 and 11.

9.1 Permanent Magnet Synchronous Machines

Figure 9.1 illustrates a two-pole surface-mounted radial flux permanent magnet synchronous machine (we will consider a P-pole design). The phase magnetic axes are shown, as well as the q- and d-axes. The stator is broken into two regions, the stator backiron and the slot/tooth region. The rotor includes a shaft, a magnetically inert region (which could be steel but need not be), a rotor backiron region, and permanent magnets. Arrows within the permanent magnet region indicate the direction of magnetization. Also shown in Figure 9.1 is the electrical rotor position, θ_r, position measured from the stator, ϕ_s, and position measured relative to the rotor ϕ_r. Since a two-pole machine is shown, these angles are identical to their mechanical counterparts, θ_{rm}, ϕ_{sm}, and ϕ_{rm}, for the device shown.

In the machine shown in Figure 9.1, the permanent magnets are located on the surface of the rotor structure, hence the description surface-mounted. Machines with interior magnets are also widely used. Figure 9.2 illustrates two possible interior magnetic arrangements, one with spoked magnets and another a more "standard" buried magnetic arrangement. A V-shaped magnetic arrangement, which concentrates magnet flux, is also commonly used.

Power Magnetic Devices: A Multi-Objective Design Approach, Second Edition. Scott D. Sudhoff.
© 2022 The Institute of Electrical and Electronics Engineers, Inc. Published 2022 by John Wiley & Sons, Inc.
Companion website: www.wiley.com/go/sudhoff/Powermagneticdevices

9 Introduction to Permanent Magnet AC Machine Design

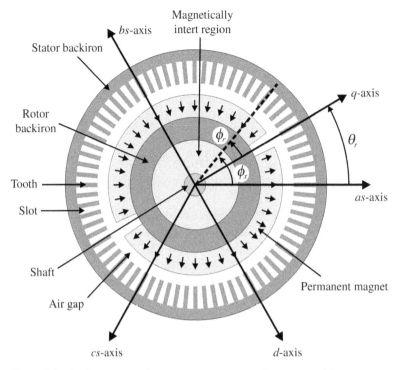

Figure 9.1 Surface-mounted permanent magnet synchronous machine.

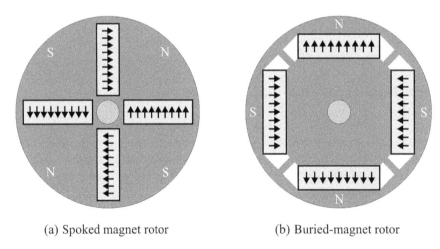

(a) Spoked magnet rotor (b) Buried-magnet rotor

Figure 9.2 Two interior magnetic arrangements.

The machines depicted in Figures 9.1 and 9.2 are considered radial since the flux flow in the air gap is in the radial direction. Axial machines in which the air-gap flux is directed axially have also been considered. The reader is referred to references [4–6] for a broad discussion of possible machine configurations; herein, we will focus our attention on the radial surface-mounted permanent magnet machine shown in Figure 9.1.

9.2 Operating Characteristics of PMAC Machines

Before considering the design of a PMAC machine, it is appropriate to briefly review its operating characteristics. One of the most important of these characteristics is that this class of machinery requires a power electronics device known as an inverter to operate. The inverter converts power from dc to three-phase ac with an angle determined by the instantaneous rotor position. In this section, the power converter will be discussed at a level sufficient for design purposes. For a much more involved discussion, the reader is referred to Krause et al. [7].

Machine Model in QD Variables

Let us begin our discussion of operating characteristics by considering the machine model. From our work in Section 8.9, the voltage equations for a PMAC may be expressed in the rotor reference frame as

$$v_{qs}^r = R_s i_{qs}^r + \omega_r \lambda_{ds}^r + \frac{d\lambda_{qs}^r}{dt} \tag{9.2-1}$$

$$v_{ds}^r = R_s i_{ds}^r - \omega_r \lambda_{qs}^r + \frac{d\lambda_{ds}^r}{dt} \tag{9.2-2}$$

$$v_{0s} = R_s i_{0s} + \frac{d\lambda_{0s}}{dt} \tag{9.2-3}$$

In (9.2-1)–(9.2-3), an upper case R for resistance is used in order to distinguish this symbol from that used for geometrical radius.

Most three-phase electric machinery is either wye- or delta-connected, as shown in Figure 9.3. In the case of a wye-connected machine, it can be shown that the zero-sequence current is zero since the sum of the phase currents must be zero. In the case of a delta-connected machine, the sum of the phase voltages must be zero, and so the zero-sequence voltage is zero. In an ideal machine, the connection would make little difference. However, in actual machines, the delta connection leads to circulating currents driven by harmonics in the permanent-magnet-induced back-emf, which will result in loss. Henceforth, we will restrict our attention to wye-connected machines, at which point the zero-sequence quantities are of little interest (since the zero-sequence current is zero) and will henceforth not be considered. The performance possibilities of machines with independent phases are explored in Chapman [8,9].

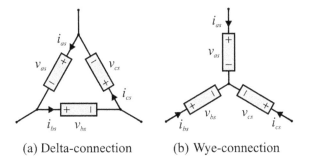

Figure 9.3 Wye and delta connections.

Next, we will consider flux linkage equations for the machine. We have not derived these yet but we will do so in Section 9.8. We will find that in terms of qd variables in the rotor reference frame, the flux linkage equations may be expressed as

$$\lambda_{qs}^r = L_q i_{qs}^r \tag{9.2-4}$$

$$\lambda_{ds}^r = L_d i_{ds}^r + \lambda_m \tag{9.2-5}$$

From our work in Section 8.10, the electromagnetic torque is given by

$$T_e = \frac{3}{2}\frac{P}{2}\left(\lambda_{ds}^r i_{qs}^r - \lambda_{qs}^r i_{ds}^r\right) \tag{9.2-6}$$

Although the machine model formed by (9.2-1)–(9.2-6) is not terribly complex, it can be simplified. For steady-state conditions, qd variables are constant. Substitution of the flux linkage equations (9.2-4) and (9.2-5) into the voltage equations (9.2-1) and (9.2-2) and setting the derivative terms equal to zero yields

$$v_{qs}^r = R_s i_{qs}^r + \omega_r L_d i_{ds}^r + \omega_r \lambda_m \tag{9.2-7}$$

$$v_{ds}^r = R_s i_{ds}^r - \omega_r L_q i_{qs}^r \tag{9.2-8}$$

Substitution of the flux linkage equations (9.2-4) and (9.2-5) into the torque equation (9.2-6) results in

$$T_e = \frac{3}{2}\frac{P}{2}\left(\lambda_m i_{qs}^r + \left(L_d - L_q\right)i_{qs}^r i_{ds}^r\right) \tag{9.2-9}$$

In (9.2-9), the torque contribution from the term with the factor λ_m is due to the permanent magnet, while the torque contribution from the term with the $(L_d - L_q)$ is due to saliency or the difference between the q- and d-axis inductances. We will use (9.2-7)–(9.2-9) heavily as we consider the operation and design of the machine.

Three-Phase Bridge Inverter

PMAC machines cannot be operated directly from a fixed voltage supply. They require a controllable voltage or current source, which normally takes the form of a power electronic converter known as an inverter, as shown in Figure 9.4. Therein, the voltage v_{dc} is a dc voltage that may

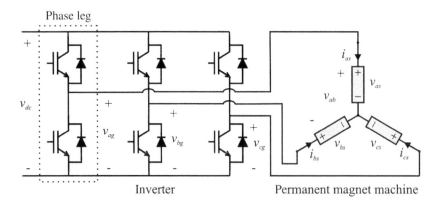

Figure 9.4 Three-phase bridge inverter and machine.

be from a battery or from a rectified ac source. The inverter consists of three-phase legs, each associated with one phase of the motor and each phase leg is made up of an upper and lower transistor and diode. The outputs of the inverter phase legs are connected to three phases of the machine.

At any point in time, a given transistor is either gated on so heavily as to be saturated or not gated on at all so as to be in the cutoff region of the operating curve. When a given transistor is on, the voltage drop is very low and so the power dissipated by the transistor is small. In the case where a transistor is off, there is a large voltage drop across the device but no current through the device, and so the power dissipation is zero. Except for transitions between the on- and off-states, typically either (i) the upper transistor of a given phase leg is turned on and the lower transistor is turned off or (ii) the upper transistor of a given phase leg is turned off and the lower transistor is turned on. Thus, neglecting the voltage drop across semiconductors in the on-state, the line-to-ground voltages v_{ag}, v_{bg}, and v_{cg} are either 0 or v_{dc}. It can be shown that the line-to-neutral voltages v_{as}, v_{bs}, and v_{cs} have instantaneous values of $-2v_{dc}/3$, $-v_{dc}/3$, 0, $v_{dc}/3$, or $2v_{dc}/3$.

The transistors of the inverter are turned on and off very rapidly, with switching frequencies ranging from a few kilohertz to hundreds of kilohertz, depending upon the ratings of the device (with larger ratings using lower switching frequencies). This rapid switching between transistor on and off states causes the machine to "see" a continuously controllable applied voltage.

In this work, our attention will be focused on the design of the machine. For a detailed explanation of the operation of the inverter, the reader is referred to Krause et al. [7]. We will simply view the inverter as a controllable voltage or current source. However, there are some constraints on the machine design that are imposed by the inverter, and so we will need to consider these.

One of the most important constraints on the inverter is the voltage limitation. The largest line-to-line voltage that can be produced by the inverter is given by

$$v_{llmx} = v_{dc} - 2v_{fs} \qquad (9.2\text{-}10)$$

where v_{fs} is the forward semiconductor drop across the transistor or diode. Clearly, the transistor and diode forward (on state) voltage drop will be different for each device and also a function of current, but we will simply represent both drops as a single constant.

To obtain (9.2-10), consider the voltage v_{ab} in Figure 9.4. Suppose operation at a given instant requires that the a-to-b phase line-to-line voltage v_{ab} be as large as possible. To do this, we will turn on the upper transistor of the a-phase leg and the lower transistor of the b-phase leg. The voltages v_{ag} and v_{bg} will be approximately v_{dc} and 0 but will deviate slightly from these values because of semiconductor voltage drops. If we assume the worst case for the direction of current flow with respect to the line-to-line voltage, the a-phase current is positive, so v_{ag} is $v_{dc} - v_{fs}$, and that the b-phase current is negative so that this current is flowing through the lower diode and v_{bg} is v_{fs}. For this condition, the a-to-b phase voltage is $v_{dc} - 2v_{fs}$, as given by (9.2-10). The inverter cannot produce a voltage waveform whose fundamental component requires a larger line-to-line voltage without introducing significant low-frequency harmonics. Of course, high-frequency voltage ripple is always present because of the inverter switching, but this is normally filtered by the machine inductance.

In addition to the voltage constraint, a current constraint will also be imposed. This is because the inverter transistor and diodes must carry the peak current seen by the machine, and it is important not to exceed the rated current for these devices.

Another factor that we will need to address is the inverter power loss. The inverter has two loss mechanisms: (i) conduction loss that is due to the voltage drop across conducting semiconductors and (ii) switching loss that is associated with turning on or off of the inverter transistors. While both forms of losses can impact the machine design, the conduction loss is particularly important. In order to get an approximate expression for conduction losses, let us suppose that the *a*-phase current in the machine may be expressed as

$$i_{as} = \sqrt{2}I_s \cos(\omega t + \phi) \tag{9.2-11}$$

where ω and ϕ are constants depending upon operating conditions. The absolute value of this current will be flowing through one of the phase leg semiconductors at any point of time. If the forward semiconductor drop is v_{fs}, the instantaneous power loss by the *a*-phase inverter leg is given by

$$p_{as} = \sqrt{2}v_{fs}I_s |\cos(\omega t + \phi)| \tag{9.2-12}$$

Taking the average of (9.2-12) and multiplying by the number of phases yield an average semiconductor power loss of

$$P_s = \frac{6\sqrt{2}}{\pi} v_{fs} I_s \tag{9.2-13}$$

Including semiconductor power loss in the machine design tends to result in a reduction of the required current and increase in the required voltage until a voltage limit is encountered. More detailed representations of semiconductor loss and inverter loss have been found to have limited impact on the machine design [10].

Voltage Source Operation

Inverters are typically controlled to act as a controllable voltage source or as a controllable current source. We will begin our investigation of the operating behavior of the machine by considering voltage source operation. In this operating mode, the inverter is controlled so that the fundamental components of the applied voltages are given by

$$v_{as} = \sqrt{2}V_s \cos(\theta_r + \phi_v) \tag{9.2-14}$$

$$v_{bs} = \sqrt{2}V_s \cos(\theta_r + \phi_v - 2\pi/3) \tag{9.2-15}$$

$$v_{cs} = \sqrt{2}V_s \cos(\theta_r + \phi_v + 2\pi/3) \tag{9.2-16}$$

These voltages form a three-phase balanced set with equal amplitudes and a phase displacement of $2\pi/3$ radians between phases. Observe that the voltages are a function of electrical rotor position. This is why an inverter is needed. At any given point of time, the electrical rotor position is sensed (or calculated using an estimator), and the desired voltages are determined. The inverter is then switched so that the desired voltage is obtained. Clearly, this process requires a controllable source and sensors (or estimators). A computational engine (microcontroller, digital signal processor, or the like) is also often used. In any case, from Section 8.9, we have

$$v_{qs}^r = \sqrt{2}V_s \cos\phi_v \tag{9.2-17}$$

$$v_{ds}^r = -\sqrt{2}V_s \sin\phi_v \tag{9.2-18}$$

which are constants.

Given the *q*- and *d*-axis voltages, we may solve (9.2-7) and (9.2-8) for the machine currents. This yields

$$i_{qs}^r = \frac{R_s\left(v_{qs}^r - \omega_r\lambda_m\right) - \omega_r L_d v_{ds}^r}{R_s^2 + \omega_r^2 L_d L_q} \tag{9.2-19}$$

$$i_{ds}^r = \frac{\omega_r L_q \left(v_{qs}^r - \omega_r \lambda_m\right) + R_s v_{ds}^r}{R_s^2 + \omega_r^2 L_d L_q} \qquad (9.2\text{-}20)$$

Once the q- and d-axis currents are found, we may compute the torque using (9.2-9).

Example 9.2A Let us consider the operating characteristics of a machine with the following parameters: $P = 4$, $R_s = 240$ mΩ, $L_q = L_d = 1.65$ mH, and $\lambda_m = 115$ mVs. We will assume that the machine is controlled in a voltage source-operating mode with $V_s = 10.2$ V, which corresponds to the maximum voltage that can be obtained from a 25 V dc voltage without the presence of low-frequency harmonics. We will take $\phi_v = 0$, which can be shown to maximize the stall torque. A discussion on the possibilities made available by manipulating ϕ_v is set forth in Krause et al. [7]. For our analysis, at any given speed, we will calculate the q- and d-axis voltages from (9.2-17) and (9.2-18), the q- and d-axis currents from (9.2-19) and (9.2-20), the rms current from (8.9-8), the electromagnetic torque from (9.2-9), the input power from (8.9-32), the output power as the product electromechanical torque and mechanical rotor speed (in rad/s), and the efficiency as the ratio of the output power to input power. The resulting machine characteristics versus speed are shown in Figure 9.5. Therein, the variables are normalized to the indicated maxima, which are the values at zero speed. In particular, $I_{smx} = 40.8$ A, $P_{inmx} = 1.25$ kW, and $T_{emx} = 19.9$ Nm. The rated current for the machine is 15.6 A. Thus, operating conditions with I_s/I_{smx} of greater than 0.38 cannot be maintained indefinitely. In a nonsalient machine in which $\phi_v = 0$, it can be shown that the normalized input power, torque, and q-axis current are all identical. Both the negated d-axis current and output power start at zero, reach a peak value, and then decrease to zero. Note that the efficiency is poor until the upper end of the speed range.

This study illustrates some important operating characteristics: operating the machine in a voltage source configuration requires the inverter to supply very large currents, and the machine will have poor efficiency over much of its speed range. We will next consider current source operation.

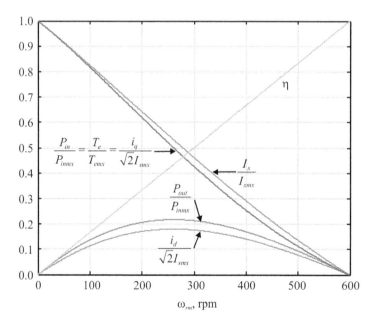

Figure 9.5 Voltage source fed PMAC machine characteristics.

Current Source Operation

An alternative to voltage source operation is current source operation. In this operating mode, the inverter is controlled so as to act as a current source. One way to accomplish this is analogous to voltage source operation, with the inverter currents regulated so that

$$i_{as} = \sqrt{2} I_s \cos(\theta_r + \phi_i) \tag{9.2-21}$$

$$i_{bs} = \sqrt{2} I_s \cos(\theta_r + \phi_i - 2\pi/3) \tag{9.2-22}$$

$$i_{cs} = \sqrt{2} I_s \cos(\theta_r + \phi_i + 2\pi/3) \tag{9.2-23}$$

In terms of qd variables, this yields

$$i_{qs}^r = \sqrt{2} I_s \cos \phi_i \tag{9.2-24}$$

$$i_{ds}^r = -\sqrt{2} I_s \sin \phi_i \tag{9.2-25}$$

Normally, however, when using this strategy, we start with a desired q- and d- axis current and then use the inverse transformation to determine the desired abc variable currents. In order to determine the desired q- and d-axis currents, let us consider a nonsalient machine in which $L_q = L_d = L_s$, where L_s will be considered to be the machine inductance since it is the same for the q- and d-axes. Suppose we wish to obtain a desired torque T_e^*. From (9.2-9), we conclude that the desired q-axis current is given by

$$i_{qs}^{r*} = \frac{4 T_e^*}{3 P \lambda_m} \tag{9.2-26}$$

Next, we need to establish an expression for the d-axis current. To this end, note that the power out of the machine may be expressed as

$$P_{in} = \frac{3}{2}\left(v_{qs}^r i_{qs}^r + v_{ds}^r i_{ds}^r\right) \tag{9.2-27}$$

Substitution of (9.2-7) and (9.2-8) into (9.2-27) yields

$$P_{in} = \frac{3}{2}\left(R_s i_{qs}^{r2} + \omega_r \lambda_m i_{qs}^r + R_s i_{ds}^{r2}\right) \tag{9.2-28}$$

Now, the terms involving i_{qs}^r are fixed since in order to achieve the desired torque, the desired q-axis current is already determined by (9.2-26). However, by inspection, we can minimize the input power by setting

$$i_{ds}^{r*} = 0 \tag{9.2-29}$$

Occasionally, there is a reason to inject d-axis current. From (8.9-8), the rms value of the fundamental component of the applied voltage may be expressed as

$$V_s = \frac{1}{\sqrt{2}} \sqrt{v_{qs}^{r2} + v_{ds}^{r2}} \tag{9.2-30}$$

Substitution of (9.2-7) and (9.2-8) into (9.2-30) yields

$$V_s = \frac{1}{\sqrt{2}} \sqrt{\left(R_s i_{qs}^r + \omega_r L_d i_{ds}^r + \omega_r \lambda_m\right)^2 + \left(R_s i_{ds}^r - \omega_r L_q i_{qs}^r\right)^2} \tag{9.2-31}$$

It can be readily shown that the peak value of the line-to-line voltage is $\sqrt{6}V_s$. Thus, the peak line-to-line voltage may be expressed as

$$v_{ll,pk} = \sqrt{3}\sqrt{\left(R_s i_{qs}^r + \omega_r L_d i_{ds}^r + \omega_r \lambda_m\right)^2 + \left(R_s i_{ds}^r - \omega_r L_q i_{qs}^r\right)^2} \quad (9.2\text{-}32)$$

which, for high speeds, may be approximated as

$$v_{ll,pk} = \sqrt{3}|\omega_r|\sqrt{\left(L_d i_{ds}^r + \lambda_m\right)^2 + \left(L_q i_{qs}^r\right)^2} \quad (9.2\text{-}33)$$

Clearly, as the speed increases, the required line-to-line voltage will increase and so eventually, the voltage limit (9.2-10) comes into play. However, by injecting a negative d-axis current, the line-to-line voltage requirement can be reduced, allowing an increase in the speed range.

It is possible to solve (9.2-31) for the precise amount of d-axis current needed to bring the voltage requirement to the available level [7]. However, there are penalties associated with doing this. First, there is a loss of efficiency since the d-axis current will cause resistive losses. Secondly, excessive d-axis current can demagnetize the magnets. In our design code, we will consider the q- and d-axis current commands to be part of the design space.

For salient machines, the problem of selecting the q- and d-axis currents is more involved since both components of current contribute to torque. However, here again, one can take the approach of making the q- and d-axis current commands part of the design space.

Example 9.2B Let us again consider the machine whose parameters are given in Figure 9.5, but now, consider current source operations. Our goal will be to determine the maximum amount of torque that can be obtained at any given speed subject to a maximum current constraint of $I_s = 15.6$ A and a maximum dc link voltage of 200 V.

We can pose this problem as an optimization problem. Let us make our parameter vector

$$\boldsymbol{\theta} = [\phi_i \ \omega_{rm}]^T \quad (9.2\text{B-}1)$$

From I_s (specified) and ϕ_i, we can calculate the q- and d-axis currents from (9.2-24) and (9.2-25). From the mechanical rotor speed, we can readily calculate the electrical rotor speed (8.1-5) and then the q- and d-axis voltages using (9.2-7) and (9.2-8). From these, we can compute the rms voltage as (9.2-30), the line-to-line voltage as $\sqrt{6}V_s$, and the torque from (9.2-9). We will then calculate a constraint variable

$$c = \text{lte}(v_{ll}, v_{dc}) \quad (9.2\text{B-}2)$$

and fitness as

$$\mathbf{f} = \begin{cases} \varepsilon(c-1)[1 \ 1]^T & c < 1 \\ \left[\omega_{rm} \ \dfrac{T_e}{1 + 10^{-6}\omega_{rm}}\right]^T & c = 1 \end{cases} \quad (9.2\text{B-}3)$$

In (9.2B-3), the first objective is speed; the second is basically torque. The reason for the 10^{-6} factor is so that at low speeds there is a trade-off between the first and second objectives so that we will obtain solutions in this region. Note that this does not taint our solution since for a given value of the first objective, the second objective is clearly maximized by maximizing torque. Maximizing $\mathbf{f}$ in (9.2B-3) yields a set of solutions, which map to the Pareto-front between torque and speed. The results are shown in Figure 9.6.

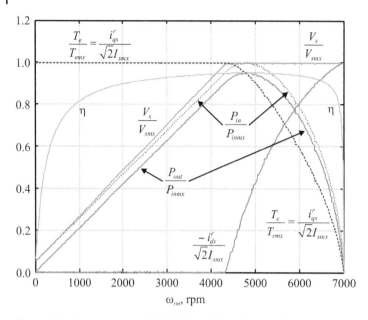

Figure 9.6 Current source fed PMAC machine characteristics.

As can be seen, for speeds below 4300 rpm, the maximum torque is constant, the d-axis current is zero, and the voltage and output power increase linearly with speed. At approximately 4300 rpm, the voltage constraint is encountered. At this point, negative d-axis current is used in order to mitigate the voltage constraint. Because the total current is limited, this forces a reduction in the q-axis current, which reduces torque. The output power remains approximately constant for a short interval after the voltage limit is encountered and then decreases, as the decrease in torque is more rapid than the increase in speed. Eventually, the torque and output power fall to zero. Observe that the efficiency is quite good over a very broad speed range.

At this point, we have completed our brief review of PMAC machine characteristics. We will now start setting forth the relationships we will need for machine design. We will begin by considering the machine geometry.

9.3 Machine Geometry

Figure 9.7 illustrates a cross section of the PMAC machine. As can be seen, the machine is divided into regions. Proceeding from the exterior of the machine to the interior, the outermost region of the machine is the stator backiron that extends from a radius of r_{sb} to r_{ss} from the center of the machine. In this region, the flux enters and leaves from the teeth and predominantly travels in the tangential direction. The next region is the slot/tooth region that contains the stator slots and teeth and stator conductors, as discussed in Chapter 8. The slot/tooth region extends from r_{st} to r_{sb}. The next region is the air gap that includes radii from r_{rg} to r_{st}. Proceeding inward, the permanent magnet region includes points with radii between r_{rb} and r_{rg} and consists of one of two types of material, either (i) a permanent magnet that will produce radial flux or (ii) a magnetically inert spacer that may be air (as shown). The rotor backiron extends from r_{ri} to r_{rb}. Flux enters and leaves the rotor backiron

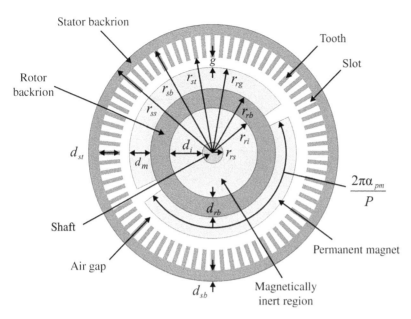

Figure 9.7 Surface-mounted permanent magnet synchronous machine.

predominantly in the radial direction, but the majority of the flux flow through the rotor backiron will be tangential. It serves a purpose similar to the stator backiron. The inert region (radii from r_{rs} to r_{ri}) mechanically transfers torque from the rotor backiron to the shaft. It is often just a continuation of the rotor backiron (possibly with areas removed to reduce mass) or could be a lightweight composite material. The material in this region does not serve a magnetic purpose, even if it is a magnetic material.

Variables depicted in Figure 9.7 include d_{sb}, the stator backiron depth; d_{st}, the stator tooth depth; g, the air-gap depth; d_m, the permanent magnet depth; d_{rb}, the rotor backiron depth; d_i, the magnetically inert region depth; and r_{rs}, the rotor shaft radius. The active length of the machine (the depth of the magnetic steel into the page) is denoted l. The quantity α_{pm} is the angular fraction of a magnetic pole occupied by the permanent magnet. All of these variables, with the exception of the radius of the rotor shaft, r_{rs}, which is assumed to be known, will be determined as part of the design process.

In terms of the parameters identified in the previous paragraph, the following may be readily calculated: r_{ri}, the rotor inert region radius; r_{rb}, the rotor backiron radius; r_{rg}, the rotor air-gap radius; r_{st}, the stator tooth inner radius; r_{sb}, the stator backiron inner radius; and r_{ss}, the stator shell radius. A stator shell, if present, is used for protection, mechanical strength, and thermal transfer. It will not be considered in our design.

Figure 9.8 depicts a portion of the stator consisting of one tooth and one slot (with half of a slot on either side of the tooth). Variables depicted therein, which have not been previously defined, include S_s, the number of stator slots; θ_{tt}, the angle spanned by the tooth tip at radius r_{st}; θ_{st}, the angle spanned by the slot at radius r_{st}; r_{si}, the radius to the inside tooth tip; θ_{ti}, the angle spanned by the tooth at radius r_{si}; θ_{tb}, the angle spanned by the tooth at radius r_{sb}; w_{tb}, the width of the tooth base; d_{tb}, the depth of the tooth base; d_{tte}, the depth of the tooth tip edge; and d_{ttc}, the depth of the tooth tip center.

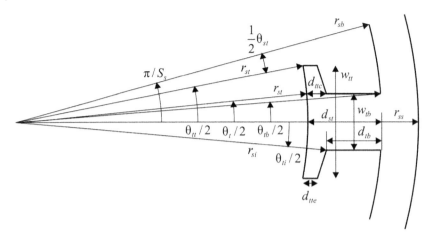

Figure 9.8 Slot and tooth dimensions.

For the purposes of design, it will be convenient to introduce the tooth fraction α_t and tooth tip fraction α_{tt}. The tooth fraction is defined as the angular fraction of the slot/tooth region occupied by the tooth at radius r_{st}. Hence,

$$\alpha_t = \frac{S_s \theta_t}{2\pi} \tag{9.3-1}$$

The tooth tip fraction is herein defined as the angular fraction of the slot/tooth region occupied by the tooth tip at radius r_{st}. It is defined as

$$\alpha_{tt} = \frac{S_s \theta_{tt}}{2\pi} \tag{9.3-2}$$

As previously noted, not all the variables in Figures 9.7 and 9.8 are independent. One choice of variables sufficient to define the geometry is given by

$$\mathbf{G}_I = \begin{bmatrix} r_{rs} & d_i & d_{rb} & d_m & g & d_{tb} & d_{ttc} & d_{tte} & \alpha_t & \alpha_{tt} & d_{sb} & \alpha_{pm} & l & P & S_s & \phi_{ss1} \end{bmatrix} \tag{9.3-3}$$

where a $\mathbf{G}$ is used to denote geometry and the subscript I serves as a reminder that these variables are considered independent. Note we have not discussed the last element of $\mathbf{G}_I$, namely, ϕ_{ss1}, in this chapter; it is the center location of the first slot as discussed in Chapter 8. Given $\mathbf{G}_I$, the locations of the slots and teeth may be calculated using (8.1-8) and (8.1-9); next, the remaining quantities in Figures 9.7 and 9.8 can be readily calculated as

$$r_{ri} = r_{rs} + d_i \tag{9.3-4}$$

$$r_{rb} = r_{ri} + d_{rb} \tag{9.3-5}$$

$$r_{rg} = r_{rb} + d_m \tag{9.3-6}$$

$$r_{st} = r_{rg} + g \tag{9.3-7}$$

$$r_{si} = r_{st} + d_{ttc} \tag{9.3-8}$$

$$\theta_t = 2\pi \alpha_t / S_s \tag{9.3-9}$$

$$\theta_{tt} = 2\pi \alpha_{tt} / S_s \tag{9.3-10}$$

9.3 Machine Geometry

$$\theta_{st} = \frac{2\pi}{S_s} - \theta_{tt} \tag{9.3-11}$$

$$w_{tb} = 2r_{st}\sin\left(\frac{\theta_t}{2}\right) \tag{9.3-12}$$

$$w_{tt} = 2r_{st}\sin\left(\frac{\theta_{tt}}{2}\right) \tag{9.3-13}$$

$$r_{sb} = \sqrt{(w_{tb}/2)^2 + (r_{st}\cos(\theta_t/2) + d_{tb} + d_{ttc})^2} \tag{9.3-14}$$

$$\theta_{tb} = 2\mathrm{asin}\left(\frac{w_{tb}}{2r_{sb}}\right) \tag{9.3-15}$$

$$\theta_{ti} = 2\mathrm{asin}\left(\frac{w_{tb}}{2r_{si}}\right) \tag{9.3-16}$$

$$d_{st} = r_{sb} - r_{st} \tag{9.3-17}$$

$$r_{ss} = r_{sb} + d_{sb} \tag{9.3-18}$$

Another geometrical variable of interest, although not shown in Figure 9.8, is the slot opening, that is, the distance between teeth. This is readily expressed as

$$w_{so} = 2r_{st}\sin\left(\frac{\theta_{st}}{2}\right) \tag{9.3-19}$$

Besides computing the dependent geometrical variables, there are several other quantities of interest that will prove useful in the design of the machine. The first of these is the area of a tooth base, which is the portion of the tooth that falls within $r_{si} \leq r \leq r_{sb}$ and is given by

$$a_{tb} = w_{tb}d_{tb} + \frac{r_{sb}}{2}\left(r_{sb}\theta_{tb} - w_{tb}\cos\left(\frac{\theta_{tb}}{2}\right)\right) - \frac{r_{si}}{2}\left(r_{si}\theta_{ti} - w_{tb}\cos\left(\frac{\theta_{ti}}{2}\right)\right) \tag{9.3-20}$$

The area of a tooth tip, which is the material at a radius $r_{st} \leq r \leq r_{si}$ from the center of the machine, may be expressed as

$$a_{tt} = \frac{1}{2}\begin{pmatrix} 2w_{tt}d_{tte} + (w_{tb} + w_{tt})(r_{si}\cos(\theta_{ti}/2) - r_{st}\cos(\theta_t/2) - d_{tte}) \\ -r_{st}^2\theta_{tt} + r_{st}w_{tt}\cos(\theta_{tt}/2) + r_{si}^2\theta_{ti} - r_{si}w_{tb}\cos(\theta_{ti}/2) \end{pmatrix} \tag{9.3-21}$$

The slot area is defined as the cross-sectional area of the slot between radii r_{sb} and r_{si}. This area is calculated as

$$a_{slt} = \frac{\pi}{S_s}\left(r_{sb}^2 - r_{si}^2\right) - a_{tb} \tag{9.3-22}$$

The total volume of all stator teeth, v_{st}, the back iron, v_{sb}, and the stator laminations, v_{sl}, may be formulated as

$$v_{st} = S_s(a_{tt} + a_{tb})l \tag{9.3-23}$$

$$v_{sb} = \pi\left(r_{ss}^2 - r_{sb}^2\right)l \tag{9.3-24}$$

$$v_{sl} = v_{st} + v_{sb} \tag{9.3-25}$$

The total volume of the rotor backiron, denoted v_{rb}, rotor inert region, v_{ri}, and permanent magnet, v_{pm}, is readily found from

$$v_{rb} = \pi\left(r_{rb}^2 - r_{ri}^2\right)l \tag{9.3-26}$$

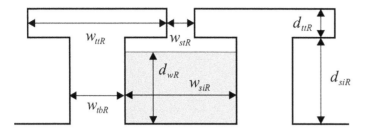

Figure 9.9 Rectangular slot approximation.

$$v_{ri} = \pi(r_{ri}^2 - r_{rs}^2)l \qquad (9.3\text{-}27)$$

$$v_{pm} = \pi(r_{rg}^2 - r_{rb}^2)\alpha_{pm}l \qquad (9.3\text{-}28)$$

For purposes of leakage inductance calculations, it is convenient to approximate the slot geometry as being rectangular, as depicted in Figure 9.9. There are many ways that such an approximation can be accomplished. One approach is as follows. First, the width of the tooth tip is approximated as the circumferential length of the actual tooth tip

$$w_{ttR} = r_{st}\theta_{tt} \qquad (9.3\text{-}29)$$

The depth of the rectangular approximation to the tooth tip is set so that the tooth tip has the same cross-sectional area. In particular,

$$d_{ttR} = \frac{a_{tt}}{w_{ttR}} \qquad (9.3\text{-}30)$$

Next, the width of the slot between the stator tooth tips is approximated by the circumferential distance between the tooth. Thus,

$$w_{stR} = r_{st}\theta_{st} \qquad (9.3\text{-}31)$$

The width of the slot between the base of the tips is taken as the average of the distance of the chord length across the slot at the top of the slot (but under the tooth tip) and the chord distance across the slot at the bottom of the slot. This yields

$$w_{siR} = r_{si}\sin\left(\frac{\pi}{S_s} - \frac{\theta_{ti}}{2}\right) + r_{sb}\sin\left(\frac{\pi}{S_s} - \frac{\theta_{tb}}{2}\right) \qquad (9.3\text{-}32)$$

Maintaining the area of the slot and the area of the tooth base, the depth of the slot (exclusive of the tooth tip) and the width of the tooth base are set in accordance with

$$d_{siR} = \frac{a_{slt}}{w_{siR}} \qquad (9.3\text{-}33)$$

$$w_{tbR} = \frac{a_{tb}}{d_{siR}} \qquad (9.3\text{-}34)$$

Note that this approach is not consistent in that it does not require $w_{siR} + w_{tbR} = w_{ttR} + w_{stR}$. However, this does not matter in the primary use of the model—the calculation of the slot leakage permeance. The final parameter shown in Figure 9.9 is the depth of the winding within the slot, d_{wR}. This parameter will not be considered a part of the stator geometry but rather as part of the winding.

Before concluding this section, it is appropriate to organize our calculations in order to support our design efforts. In (9.3-3), we defined a list of independent variables, which define the machine geometry, and organized them into a vector $\mathbf{G}_I$. Based on this, we found a host of related variables, which will also be of use. It is convenient to define these dependent variables as a vector,

$$\mathbf{G}_D = \begin{bmatrix} r_{ri} & r_{rb} & r_{rg} & r_{st} & r_{ri} & \theta_t & \theta_{tt} & w_{tb} & w_{tt} & r_{sb} & \theta_{tb} & \theta_{ti} & d_{st} & r_{ss} & w_{so} & \cdots \\ a_{slt} & a_{tt} & a_{tb} & v_{st} & v_{sb} & v_{sl} & v_{rb} & v_{ri} & v_{pm} & w_{ttR} & d_{ttR} & w_{stR} & w_{siR} & d_{siR} & w_{tbR} \end{bmatrix} \tag{9.3-35}$$

where again $\mathbf{G}$ denotes geometry and the subscript D indicates dependent variables. We may summarize our calculations (9.3-4)–(9.3-34) as a vector-valued function F_G such that

$$\mathbf{G}_D = F_G(\mathbf{G}_I) \tag{9.3-36}$$

This view of our geometrical calculations will be useful as we develop computer codes to support machine design, directly suggesting the inputs and outputs of a subroutine/function calls to make geometrical calculations. Finally, other calculations required to perform will require knowledge of both $\mathbf{G}_I$ and $\mathbf{G}_D$; it will therefore be convenient to define

$$\mathbf{G} = \begin{bmatrix} \mathbf{G}_I & \mathbf{G}_D \end{bmatrix} \tag{9.3-37}$$

9.4 Stator Winding

It is assumed that the conductor distribution is sinusoidal with the addition of a third harmonic term as discussed in Section 8.1 and given, in a slightly different but equivalent form, by (8.1-12). In particular, the assumed conductor density is given by

$$n_{as}(\phi_{sm}) = N_{s1}^* \left(\sin\left(\frac{P}{2}\phi_{sm}\right) - \alpha_3^* \sin\left(3\frac{P}{2}\phi_{sm}\right) \right) \tag{9.4-1}$$

$$n_{bs}(\phi_{sm}) = N_{s1}^* \left(\sin\left(\frac{P}{2}\phi_{sm} - \frac{2\pi}{3}\right) - \alpha_3^* \sin\left(3\frac{P}{2}\phi_{sm}\right) \right) \tag{9.4-2}$$

$$n_{cs}(\phi_{sm}) = N_{s1}^* \left(\sin\left(\frac{P}{2}\phi_{sm} + \frac{2\pi}{3}\right) - \alpha_3^* \sin\left(3\frac{P}{2}\phi_{sm}\right) \right) \tag{9.4-3}$$

where N_{s1}^* is the desired fundamental amplitude of the conductor density and α_3^* is the ratio between the third harmonic component and the fundamental component.

In order to construct the machine, we need to specify the specific number of conductors of each phase to be placed in each slot. To this end, we can use the results from Section 8.1. Using (8.1-24) in conjunction with (9.4-1)–(9.4-3) yields

$$N_{as,i} = \text{round}\left(\frac{4N_{s1}^*}{P} \left(\sin\left(\frac{P}{2}\phi_{ss,i}\right) \sin\left(\frac{\pi P}{2S_s}\right) - \frac{\alpha_3^*}{3} \sin\left(\frac{3P}{2}\phi_{ss,i}\right) \sin\left(\frac{3P\pi}{2S_s}\right) \right) \right) \tag{9.4-4}$$

$$N_{bs,i} = \text{round}\left(\frac{4N_{s1}^*}{P} \left(\sin\left(\frac{P}{2}\phi_{ss,i} - \frac{2\pi}{3}\right) \sin\left(\frac{\pi P}{2S_s}\right) - \frac{\alpha_3^*}{3} \sin\left(\frac{3P}{2}\phi_{ss,i}\right) \sin\left(\frac{3P\pi}{2S_s}\right) \right) \right) \tag{9.4-5}$$

$$N_{cs,i} = \text{round}\left(\frac{4N_{s1}^*}{P}\left(\sin\left(\frac{P}{2}\phi_{ss,i} + \frac{2\pi}{3}\right)\sin\left(\frac{\pi P}{2S_s}\right) - \frac{\alpha_3^*}{3}\sin\left(\frac{3P}{2}\phi_{ss,i}\right)\sin\left(\frac{3P\pi}{2S_s}\right)\right)\right) \tag{9.4-6}$$

where $N_{as,i}$, $N_{bs,i}$, and $N_{cs,i}$ are the number of conductors of the respective phases in the ith slot and where $\phi_{ss,i}$ denotes the mechanical location of the center of the ith stator slot, which is given by (8.1-8) in terms of $\phi_{ss,1}$, which is the location of the center of the first slot. This angle takes on a value of 0 if the a-phase magnetic axis is aligned with the first slot or π/S_s if it is desired to align the a-phase magnetic axis with the first tooth.

The total number of conductors in the ith slot is given by

$$N_{s,i} = |N_{as,i}| + |N_{bs,i}| + |N_{cs,i}| \tag{9.4-7}$$

For some of our magnetic analysis, we will use the continuous rather than discrete description of the winding. Once the number of conductors in each slot are computed using (9.4-4)–(9.4-6), then from (8.1-23) with $j = P/2$ and (9.4-1), the effective values of N_{s1} and α_3 are given by

$$N_{s1} = \frac{1}{\pi}\sum_{i=1}^{S_s} N_{as,i}\cos\left(\frac{P}{2}\phi_{ss,i}\right) \tag{9.4-8}$$

$$\alpha_3 = -\frac{1}{\pi N_{s1}}\sum_{i=1}^{S_s} N_{as,i}\cos\left(3\frac{P}{2}\phi_{ss,i}\right) \tag{9.4-9}$$

It is also necessary to establish an expression to describe the end conductor distribution. The end conductor distribution for each winding may be calculated in terms of the slot conductor distribution using the methods of Section 8.1. In particular, repeating (8.2-27) for convenience, the net end conductor distribution for winding x is expressed as

$$M_{x,i} = M_{x,i-1} + N_{x,i-1} \tag{9.4-10}$$

Using (9.4-10) requires knowledge of the net number of end conductors $M_{x,1}$ on the end of tooth 1. This, and the number of canceled conductors in each slot, $C_{x,i}$ (see Section 8.1), determines the type of winding (lap, wave, concentric). For the purposes of this chapter, let us take the number of canceled conductors to be zero and require the end winding conductor arrangement to be symmetric in the sense that for any end conductor count over tooth i, the end conductor count over the diametrically opposed tooth (in an electrical sense) has the opposite value. Mathematically,

$$M_{x,i} = -M_{x,S_s/P+i} \tag{9.4-11}$$

From (9.4-10) and (9.4-11), it can be shown that

$$M_{x,1} = -\frac{1}{2}\sum_{i=1}^{S_s/P} N_{x,i} \tag{9.4-12}$$

which is the subject of problem 3. Thus, once the slot conductor distribution is known, (9.4-12) and (9.4-10) can be used to find an end conductor distribution. The distribution chosen here corresponds to a concentric winding. Note that (9.4-12) can yield a noninteger result. In this case, minor alterations to the end conductor arrangement can be used to provide proper connectivity with an integer number of conductors.

In addition to the conductor distribution, it is also necessary to compute the wire cross-sectional area. To this end, the concept of packing factor is useful. The packing factor is defined as the maximum (over all slots) of the ratio of the total conductor cross-sectional area within the slot to the total slot area and will be denoted k_{pf}. Typical packing factors for a round wire range from 0.3 to 0.5. Assuming that it is advantageous not to waste the slot area, the conductor cross-sectional area and diameter may be expressed as

$$a_c = \frac{a_{slt} k_{pf}}{\|\mathbf{N}_s\|_{\max}} \tag{9.4-13}$$

$$d_c = \sqrt{\frac{4a_c}{\pi}} \tag{9.4-14}$$

where $\|\mathbf{N}_s\|_{\max}$ denotes the maximum element of the vector $\mathbf{N}_s$, whose elements are defined by (9.4-7). If desired, a_c and d_c can be adjusted to match a standard wire gauge. In this case, the gauge selected should be the one with the largest conductor area, which is smaller than that calculated using (9.4-13).

It will be necessary to compute the depth of the winding within the slot for the rectangular slot approximation. This may be readily expressed as

$$d_{wR} = \frac{\|\mathbf{N}_s\|_{\max} a_c}{k_{pf} w_{siR}} \tag{9.4-15}$$

Also of interest is the dimension of the end winding bundle in the direction parallel to the rotor shaft. Assuming the same depth as calculated by (9.4-15), this dimension may be approximated as

$$l_{ew} = \frac{\||M_{as}| + |M_{bs}| + |M_{cs}|\|_{\max} a_c}{k_{pf} d_{wR}} \tag{9.4-16}$$

An expression for the total volume of stator conductor per phase, v_{cd}, will also be required to conduct the design. From (8.8-1)–(8.8-3), we obtain

$$v_{cd} = (l + 2l_{eo}) a_c \sum_{i=1}^{S_s} |N_{as,i}| + \frac{2\pi}{S_s}(r_{st} + r_{sb}) a_c \sum_{i=1}^{S_s} |M_{as,i}| \tag{9.4-17}$$

where l_{eo} is the end winding offset, which is the amount of overhang of the end winding between the end of the stator stack and the end winding bundle. The end winding offset is a function of the manufacturing process. In general, it is desirable to make this as small as possible, though extremely small values may increase leakage inductance and core loss somewhat.

As in the case of the stator geometry, it is convenient to organize the variables discussed into independent and dependent variables, which will show the relationship of the variables from a programming point of view. To this end, it is convenient to organize the independent variables of the winding description as

$$\mathbf{W}_I = \begin{bmatrix} N_{s1}^* & \alpha_3^* & k_{pf} & l_{eo} \end{bmatrix} \tag{9.4-18}$$

The output of our winding calculations is encapsulated by the vector

$$\mathbf{W}_D = \begin{bmatrix} N_{s1} & \alpha_3 & \mathbf{N}_{as} & \mathbf{N}_{bs} & \mathbf{N}_{cs} & \mathbf{N}_s & \mathbf{M}_{as} & \mathbf{M}_{bs} & \mathbf{M}_{cs} & a_c & d_{wR} & l_{ew} & v_{cd} \end{bmatrix} \tag{9.4-19}$$

Functionally, we have

$$\mathbf{W}_D = \mathbf{F}_W(\mathbf{W}_I, \mathbf{G}) \tag{9.4-20}$$

It will also prove convenient to define

$$\mathbf{W} = \begin{bmatrix} \mathbf{W}_D & \mathbf{W}_I \end{bmatrix} \tag{9.4-21}$$

9.5 Material Parameters

As part of the design process, we will also need to select materials for the stator steel, the rotor steel, the conductor, and the permanent magnet. We will use s_t, r_t, c_t, and m_t as integer variables denoting the stator steel type, the rotor steel type, the conductor type, and the permanent magnet type. Based on these variables, the material parameters can be established using tabulated functions in accordance with

$$\mathbf{S} = \mathrm{F}_{sc}(s_t) \tag{9.5-1}$$
$$\mathbf{R} = \mathrm{F}_{sc}(r_t) \tag{9.5-2}$$
$$\mathbf{C} = \mathrm{F}_{cc}(c_t) \tag{9.5-3}$$
$$\mathbf{M} = \mathrm{F}_{mc}(m_t) \tag{9.5-4}$$

where the subscripts sc, cc, and mc denote steel catalog, conductor catalog, and magnet catalog, respectively and where $\mathbf{S}$, $\mathbf{R}$, $\mathbf{C}$, and $\mathbf{M}$ are vectors of material parameters for the stator steel, rotor steel, conductor, and magnet, respectively, and may be expressed as

$$\mathbf{S} = [\rho_s \;\; B_{s,lim}] \tag{9.5-5}$$
$$\mathbf{R} = [\rho_r \;\; B_{r,lim}] \tag{9.5-6}$$
$$\mathbf{C} = [\rho_c \;\; \sigma_c \;\; J_{lim}] \tag{9.5-7}$$
$$\mathbf{M} = [\rho_m \;\; B_r \;\; \chi_m \;\; H_{lim}] \tag{9.5-8}$$

In (9.5-5)–(9.5-8), ρ denotes volumetric mass density, $B_{s,lim}$ and $B_{r,lim}$ denote the flux density limits on the stator and rotor steel so as to avoid saturation, σ_c is the conductor conductivity, J_{lim} is the recommended maximum allowed current density, and B_r, χ_m, and H_{lim} are the permanent magnet parameters.

Permanent magnet material is generally operated in the second quadrant if it is positively magnetized or fourth quadrant if it is negatively magnetized. It is important to make sure that

$$H \geq H_{lim} \quad \text{(positively magnetized)} \tag{9.5-9}$$
$$H \leq -H_{lim} \quad \text{(negatively magnetized)} \tag{9.5-10}$$

in order to avoid demagnetization, where H_{lim} is a minimum allowed field intensity to avoid demagnetization, which is a negative number whose magnitude is less than that of the intrinsic coercive force, H_{ci}, which is discussed in Section 2.9.

Conductor, steel, and permanent magnet data are listed in Appendices A, C, and D, respectively. It should be remembered that many material properties are temperature-dependent. This is particularly important in the case of permanent magnet materials. While we will not consider temperature variation in this chapter, we will develop the background to do so in Chapter 10.

9.6 Stator Currents and Control Philosophy

In our design, we will consider a machine connected to a current-regulated inverter and that, through the action of the inverter controls, the machine currents are regulated to be equal to the commanded q- and d-axis currents, i_{qs}^{r*} and i_{ds}^{r*}. It is assumed that the currents are equal to the commanded values. This is reasonable, assuming the use of a synchronous current regulator [7] or similar technique.

From our work in Chapter 8, the rms value and phase angle of the current are readily expressed as

$$I_s = \frac{1}{\sqrt{2}} \sqrt{\left(i_{qs}^{r*}\right)^2 + \left(i_{ds}^{r*}\right)^2} \tag{9.6-1}$$

$$\phi_i = \text{angle}\left(i_{qs}^r - ji_{ds}^r\right) \tag{9.6-2}$$

Although the calculations of this section are straightforward, for the sake of consistency, they will be organized as in previous sections. We will define an input vector, an output vector, and a functional relationship as

$$\mathbf{I}_I = \begin{bmatrix} i_{qs}^{r*} & i_{ds}^{r*} \end{bmatrix} \tag{9.6-3}$$

$$\mathbf{I}_D = \begin{bmatrix} I_s & \phi_i \end{bmatrix} \tag{9.6-4}$$

and

$$\mathbf{I}_D = \mathbf{F}_I(\mathbf{I}_I) \tag{9.6-5}$$

respectively. The amalgamation of variables associated with the current is

$$\mathbf{I} = \begin{bmatrix} \mathbf{I}_I & \mathbf{I}_D \end{bmatrix} \tag{9.6-6}$$

9.7 Radial Field Analysis

The objective of this section is a magnetic analysis of the machine. In this work, thus far, an MEC approach has generally been used and could be used here. However, we will utilize a different approach in this chapter—an analytical field solution. Part of the reason for this is to illustrate a variety of techniques. Another reason is that the large effective air gaps (the air gap plus permanent magnet) of PM machines make an analytical solution reasonably accurate. A key assumption in developing an analytical field solution is that the MMF drop across the steel portions of the machine is negligible. Unless the steel becomes highly saturated, this is a reasonable assumption because the relative permeability of most permanent magnet materials is very low compared to that of steel, and so the MMF drop across the permanent magnet and air gap dominates that of the steel.

In performing our analysis, we will take the rotor position to be fixed. This may strike the reader as overly restrictive. However, as stated in the introduction, it will be our objective to produce a design that yields a constant torque T_e^*. In such a machine, except for the perturbation caused by slot effects, the flux and current densities are traveling waves that rotate but do not change in magnitude under steady-state conditions. Thus, ideally, it is only necessary to consider a single position, which could, for example, be taken to be zero. In practice, slot effects are a factor, so we will consider several fixed rotor positions, although all positions will be within one slot/tooth pitch of zero.

In order to analyze the field in the machine, we note that from Section 8.3 in the absence of rotor currents, the air-gap MMF drop is equal to the stator MMF. In particular, from (8.3-20),

$$F_g(\phi_{sm}) = F_s(\phi_{sm}) \tag{9.7-1}$$

In the definition of air-gap MMF used in defining (8.3-20), it is important to recall that the definition of air-gap MMF drop given by (8.3-3) extended from the rotor steel to the stator steel. In particular, in terms of the dimensions of Figure 9.7,

$$F_g(\phi_{sm}) = \int_{r_{rb}}^{r_{st}} H(r,\phi_{sm})dr \qquad (9.7\text{-}2)$$

where $H(r,\phi_{sm})$ denotes the radial component of field intensity.

For the purposes at hand, it will be convenient to define

$$F_{pm}(\phi_{sm}) = \int_{r_{rb}}^{r_{rg}} H(r,\phi_{sm})dr \qquad (9.7\text{-}3)$$

$$F_a(\phi_{sm}) = \int_{r_{rg}}^{r_{st}} H(r,\phi_{sm})dr \qquad (9.7\text{-}4)$$

which describe (i) the MMF drop across the range of radii spanned by the permanent magnet and (ii) the MMF drop across the air gap, respectively. Comparing (9.7-1)–(9.7-4), it is clear that

$$F_a(\phi_{sm}) + F_{pm}(\phi_{sm}) = F_s(\phi_{sm}) \qquad (9.7\text{-}5)$$

In the following subsections, we will establish the expression for each term in (9.7-5) and then use these to establish an expression for the radial flux density in the machine.

Stator MMF

The first step in determining the stator MMF is to determine the winding functions. The conductor density distribution is given by (9.4-1)–(9.4-3) with the replacement of N_{s1}^* by N_{s1} and α_3^* by α_3. In particular, for the a-phase, we have

$$n_{as}(\phi_{sm}) = N_{s1}\left(\sin\left(\frac{P}{2}\phi_{sm}\right) - \alpha_3 \sin\left(3\frac{P}{2}\phi_{sm}\right)\right) \qquad (9.7\text{-}6)$$

Applying (8.2-11)–(9.7-6) yields the a-phase winding function,

$$w_{as}(\phi_{sm}) = \frac{2N_{s1}}{P}\left(\cos\left(\frac{P\phi_{sm}}{2}\right) - \frac{\alpha_3}{3}\cos\left(3\frac{P\phi_{sm}}{2}\right)\right) \qquad (9.7\text{-}7)$$

Expressions of the b- and c-phases are similarly derived.

From (8.4-7), the stator MMF may be expressed as

$$F_s = w_{as}i_{as} + w_{bs}i_{bs} + w_{cs}i_{cs} \qquad (9.7\text{-}8)$$

Substitution of the winding functions and the expressions for currents (9.2-21)–(9.2-23) into (9.7-8) and simplifying yields the expression for stator MMF, namely,

$$F_s(\phi_{sm}) = \frac{3\sqrt{2}N_{s1}I_s}{P}\cos\left(\frac{P}{2}\phi_{sm} - \theta_r - \phi_i\right) \qquad (9.7\text{-}9)$$

Alternately, in terms of qd variables, we obtain

$$F_s(\phi_{sm}) = \frac{3N_{s1}}{P}\left(\cos\left(\frac{P}{2}\phi_{sm} - \theta_r\right)i_{qs}^r - \sin\left(\frac{P}{2}\phi_{sm} - \theta_r\right)i_{ds}^r\right) \qquad (9.7\text{-}10)$$

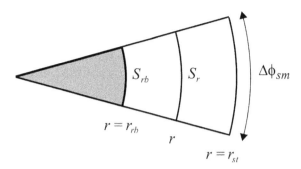

Figure 9.10 Thin sector of machine.

Radial Field Variation

Before establishing expressions for the permanent magnet and air-gap MMF drop, it is necessary to describe how the radial component of flux density varies with the radius from the center of the machine. In Chapter 8, we considered the flux density to be constant. However, in this case, the distance from the rotor steel to the stator steel is much larger than in a typical induction or synchronous machine, and so it is appropriate to take the radial variation of the flux density into account when determining MMF components.

In order to establish the radial variation in the field, consider Figure 9.10. Therein, a cross section of an angular slice of the machine is shown. Assuming that the flux density is entirely radial for radii between the rotor backiron and the stator teeth, then the flux through surface S_{rb} at the rotor backiron radius must, by Gauss's law, be equal to the flux through the surface S_r at an arbitrary radius, whereupon it follows that the flux density of an arbitrary radius is given by

$$B(r,\phi_{sm}) = \frac{r_{rb}}{r} B_{rb}(\phi_{sm}) \quad r_{rb} \leq r \leq r_{st} \tag{9.7-11}$$

where B_{rb} is the radial flux density at the rotor backiron radius.

Air-Gap MMF Drop

In the air gap, the field intensity and flux density are related by

$$B(r,\phi_{sm}) = \mu_0 H(r,\phi_{sm}) \quad r_{rg} \leq r \leq r_{st} \tag{9.7-12}$$

Manipulating (9.7-4), (9.7-11), and (9.7-12), one obtains

$$F_a(\phi_{sm}) = B_{rb}(\phi_{sm}) R_g \tag{9.7-13}$$

where R_g is a quasi-reluctance that may be expressed as

$$R_g = \frac{r_{rb}}{\mu_0} \ln\left(1 + \frac{g}{r_{rb} + d_m}\right) \tag{9.7-14}$$

Note that the accuracy of (9.7-14) can be improved by replacing g by g_{eff}, which is the effective air gap determined using Carter's coefficient, which compensates for the effect of the missing steel in the stator slots on the air-gap MMF. This is discussed in Section 8.6.

Permanent Magnet MMF

The next step in our development is to calculate the MMF across the permanent magnet region. Assuming that the knee of the magnetization curve is avoided (which will be a design constraint), the relationship between flux density and field intensity for $r_{rb} \leq r \leq r_{rg}$ may be expressed as

$$B = \begin{cases} \mu_0(1+\chi_m)H + B_r & \text{positively magnetized} \\ \mu_0(1+\chi_m)H - B_r & \text{negatively magnetized} \\ \mu_0 H & \text{inert region between magnets} \end{cases} \quad (9.7\text{-}15)$$

It is convenient to represent (9.7-15) as

$$B = \mu_0 \mu_{rm}(\phi_{rm}) H + B_m(\phi_{rm}) \quad (9.7\text{-}16)$$

where $B_m(\phi_{rm})$ is due to the residual flux density in the permanent magnet and $\mu_{rm}(\phi_{rm})$ is the relative permeability of the permanent magnet region (including the inert material), and ϕ_{rm} denotes the position as measured from the q-axis of the rotor, and both B and H refer to their respective radial components from the rotor to the stator. These functions are illustrated in Figure 9.11 in a developed diagram form.

The spatial dependence of $B_m(\phi_{rm})$ and $\mu_{rm}(\phi_{rm})$ is illustrated in the second and third traces of Figure 9.11, respectively. From this figure, we may express $\mu_{rm}(\phi_{rm})$ and $B_m(\phi_{rm})$ as

$$B_m(\phi_{rm}) = -\text{sqw}_s\left(\frac{P}{2}\phi_{rm}, \alpha_{pm}\right) B_r \quad (9.7\text{-}17)$$

$$\mu_{rm}(\phi_{rm}) = 1 + \left|\text{sqw}_s\left(\frac{P}{2}\phi_{rm}, \alpha_{pm}\right)\right| \chi_m \quad (9.7\text{-}18)$$

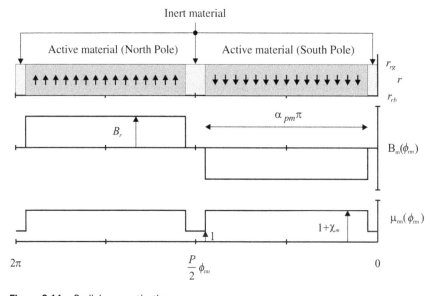

Figure 9.11 Radial magnetization.

where $\text{sqw}_s(\cdot)$ is the square wave function with sine symmetry defined as

$$\text{sqw}_s(\theta, \alpha) = \begin{cases} 1 & \sin(\theta) \geq \sin(\pi(1-\alpha)/2) \\ -1 & \sin(\theta) \leq -\sin(\pi(1-\alpha)/2) \\ 0 & |\sin(\theta)| < \sin(\pi(1-\alpha)/2) \end{cases} \quad (9.7\text{-}19)$$

Manipulating (9.7-3), (9.7-11), and (9.7-16), one obtains the expression for the MMF drop across the permanent magnet region, in particular,

$$F_{pm}(\phi_{sm}) = R_m(\phi_{sm}) B_{rb}(\phi_{sm}) - F_m(\phi_{sm}) \quad (9.7\text{-}20)$$

where

$$R_m(\phi_{sm}) = \frac{r_{rb}}{\mu_0 \mu_{rm}(\phi_{sm} - \theta_{rm})} \ln\left(1 + \frac{d_m}{r_{rb}}\right) \quad (9.7\text{-}21)$$

and

$$F_m(\phi_{sm}) = \frac{d_m}{\mu_0 \mu_{rm}(\phi_{sm} - \theta_{rm})} B_m(\phi_{sm} - \theta_{rm}) \quad (9.7\text{-}22)$$

It should be observed that $R_m(\phi_{rm})$ is not a reluctance; however, it plays a similar role. It takes on two values, depending upon stator and rotor positions. For positions under the permanent magnet, we have

$$R_m(\phi_{sm}) = R_{pm} = \frac{r_{rb}}{\mu_0(1 + \chi_m)} \ln\left(1 + \frac{d_m}{r_{rb}}\right) \quad (9.7\text{-}23)$$

and for positions under the inert region, we have

$$R_m(\phi_{sm}) = R_i = \frac{r_{rb}}{\mu_0} \ln\left(1 + \frac{d_m}{r_{rb}}\right) \quad (9.7\text{-}24)$$

The function $F_m(\phi_{rm})$ can be thought of as an MMF source resulting from the permanent magnet.

Solution for Radial Flux Density

At this point, it is possible to solve for the radial flux density. Manipulating (9.7-5), (9.7-11), (9.7-13), and (9.7-20), one obtains

$$B(r, \phi_{sm}) = \frac{r_{rb}}{r} \frac{F_m(\phi_{sm}) + F_s(\phi_{sm})}{R_m(\phi_{sm}) + R_g} \quad r_{rb} \leq r \leq r_{st} \quad (9.7\text{-}25)$$

We will use this result extensively in the sections to follow. It should be noted that θ_{rm} is an implicit argument of $F_m(\phi_{sm})$, $F_s(\phi_{sm})$, and $R_m(\phi_{sm})$. Because of this, we will sometimes denote the radial flux density given by (9.7-25) as $B(r, \phi_{sm}, \theta_{rm})$ when it is important to remember this functional dependence.

The analysis leading to (9.7-25) neglects the MMF drop in the steel. It should be mentioned that it is rather straightforward to modify the approach leading to (9.7-25) to include the MMF drop across the teeth and consider magnetic saturation. In this case, the result analogous to (9.7-25) must be solved numerically but is still computationally efficient. The reader is referred to Alsawalhi and Sudhoff [11].

9.8 Lumped Parameters

The goal of this section is to set forth a procedure to calculate the parameters of the lumped parameter model of the machine—namely, the stator resistance R_s, the q- and d-axis inductances L_q and L_d, and the flux linkage due to the permanent magnet λ_m. We will use these parameters in our calculation of electromagnetic torque, as well as to compute the required inverter voltage.

We will begin with the parameters associated with the flux linkage equations. The abc variable flux linkage equation machine may be expressed as

$$\lambda_{abcs} = \lambda_{abcl} + \lambda_{abcm} \tag{9.8-1}$$

where λ_{abcl} is the leakage flux linkage and λ_{abcm} is the magnetizing flux linkage. The leakage flux linkage is due to stator flux that does not cross the air gap; the magnetizing flux linkage terms are associated with flux that crosses the air gap.

As discussed in Section 8.7, the leakage flux linkage term is similar for many different forms of rotating electric machinery. Therein, the leakage flux was expressed as

$$\lambda_{abcl} = \begin{bmatrix} L_{lp} & L_{lm} & L_{lm} \\ L_{lm} & L_{lp} & L_{lm} \\ L_{lm} & L_{lm} & L_{lp} \end{bmatrix} \mathbf{i}_{abcs} \tag{9.8-2}$$

A method to calculate L_{lp} and L_{lm} in terms of the machine geometry was set forth in Section 8.7.

From our work in Section 8.5, using (8.5-7), the magnetizing flux linkages of the three phases may be expressed as

$$\lambda_{abcm} = lr_{st} \int_0^{2\pi} \mathbf{w}_{abcs}(\phi_{sm}) \mathrm{B}(r_{st}, \phi_{sm}) d\phi_{sm} \tag{9.8-3}$$

It is desired to find the flux linkage equations in $qd0$ variables. Using Park's transformation, (9.8-1) becomes

$$\lambda^r_{qd0s} = \lambda^r_{qd0l} + \lambda^r_{qd0m} \tag{9.8-4}$$

In order to find the $qd0$ leakage flux, we transform (9.8-2) to the rotor reference frame. Discarding the zero sequence,

$$\lambda^r_{qdl} = L_{ls} \mathbf{i}^r_{qds} \tag{9.8-5}$$

where

$$L_{ls} = L_{lp} - L_{lm} \tag{9.8-6}$$

In terms of q- and d-axis variables, (9.8-3) becomes

$$\lambda_{qdm} = lr_{st} \int_0^{2\pi} \mathbf{K}^r_s(\theta_r)\big|_{\mathrm{utr}} \mathbf{w}_{abcs}(\phi_{sm}) \mathrm{B}(r_{st}, \phi_{sm}) d\phi_{sm} \tag{9.8-7}$$

where utr denotes the upper two rows. Evaluating (9.8-7) using Park's transformation, the winding functions of the form (9.7-7) and the expression for flux density (9.7-25) along with its constituent relationships (9.7-10) and (9.7-22)–(9.7-24) yield

$$\lambda_{qm} = L_{qm} i^r_{qs} \tag{9.8-8}$$

$$\lambda_{dm} = L_{dm}i^r_{ds} + \lambda_m \tag{9.8-9}$$

where

$$L_{qm} = \frac{6lr_{rb}N^2_{s1}}{P^2}\left[\frac{\pi(1-\alpha_{pm}) + \sin(\pi\alpha_{pm})}{R_i + R_g} + \frac{\pi\alpha_{pm} - \sin(\pi\alpha_{pm})}{R_{pm} + R_g}\right] \tag{9.8-10}$$

$$L_{dm} = \frac{6lr_{rb}N^2_{s1}}{P^2}\left[\frac{\pi(1-\alpha_{pm}) - \sin(\pi\alpha_{pm})}{R_i + R_g} + \frac{\pi\alpha_{pm} + \sin(\pi\alpha_{pm})}{R_{pm} + R_g}\right] \tag{9.8-11}$$

$$\lambda_m = \frac{8r_{rb}lN_{s1}}{P(R_{pm} + R_g)}\frac{d_m}{\mu_0(1+\chi_m)}B_r\sin\left(\frac{\pi\alpha_{pm}}{2}\right) \tag{9.8-12}$$

The accuracy of (9.8-10) and (9.8-11) may be improved by replacing the length of the machine, l, with the effective length l_{eff}, as described in Cassimere and Sudhoff [2]. This accounts for end-flux paths in the machine. This improvement does not apply to the flux linkage due to the permanent magnet (9.8-12).

Combining (9.8-5), (9.8-8), and (9.8-9), we have

$$\lambda^r_{qs} = L_q i^r_{qs} \tag{9.8-13}$$

$$\lambda^r_{ds} = L_d i^r_{ds} + \lambda_m \tag{9.8-14}$$

where

$$L_q = L_{ls} + L_{qm} \tag{9.8-15}$$

$$L_d = L_{ls} + L_{dm} \tag{9.8-16}$$

At this point, the procedure to establish the parameters of the flux linkage equations has been established. The stator resistance, R_s, is readily calculated using the procedure set forth in Section 8.8.

In order to organize our calculations for design, it is useful to give the calculations functional dependence. The calculation of the lumped parameter models does not require any inputs over those already defined. The outputs of this analysis are encapsulated into a vector of electrical parameters **E** as

$$\mathbf{E} = \begin{bmatrix} R_s & L_q & L_d & \lambda_m \end{bmatrix} \tag{9.8-17}$$

Functionally, the outputs of this section may be described in the form

$$\mathbf{E} = F_E(\mathbf{M}, \mathbf{C}, \mathbf{G}, \mathbf{W}) \tag{9.8-18}$$

9.9 Ferromagnetic Field Analysis

In this section, the problem of determining the flux density waveforms in the ferromagnetic portions of the machine is considered. This is desirable for several reasons. First, we will place a limit on the maximum value of flux density so that we do not overly magnetically saturate the steel. Second, the flux density waveforms in the stator backiron and teeth can be used to determine hysteresis and eddy current losses. Finally, the problem of computing the minimum field intensity in the positively magnetized portion of the permanent magnet is addressed. This is necessary in order to check for demagnetization.

Stator Tooth Flux

We begin the development by computing the flux density in a stator tooth as a function of rotor position. To this end, we will assume that all the air-gap radial flux within $\pm \pi/S_s$ of the center of the tooth cumulates in the tooth. Thus, the flux in the ith tooth is expressed as

$$\Phi_{t,i}(\theta_{rm}) = r_{st} l \int_{\phi_{st,i} - \pi/S_s}^{\phi_{st,i} + \pi/S_s} B(r_{st}, \phi_{sm}, \theta_{rm}) d\phi_{sm} \qquad (9.9\text{-}1)$$

Because of the slot and tooth structure of the machine, the field varies as the rotor moves over the angle occupied by a slot and tooth in a more involved way than a simple rotation. Thus, we will consider a number of discrete mechanical rotor positions as the rotor position varies over a sector of the machine consisting of one slot and one tooth. To this end, let $\boldsymbol{\theta}_{rms}$ denote the vector of mechanical rotor position values over a sector. In particular, the elements of $\boldsymbol{\theta}_{rms}$ are given by

$$\theta_{rms,j} = -\frac{\pi}{S_s} + \frac{2\pi}{S_s} \frac{j-1}{J} \quad j \in [1 \cdots J] \qquad (9.9\text{-}2)$$

where j is a rotor position index variable and J is the number or positions considered.

As a next step, let $\boldsymbol{\Phi}_{ts}$ denote a matrix of tooth flux values as the mechanical rotor position varies over the slot/tooth sector. In particular, define the elements of $\boldsymbol{\Phi}_{ts}$ as

$$\Phi_{ts:i,j} = \Phi_{t,i}(\theta_{rms,j}) \quad i \in [1 \cdots S_s 2/P] \quad j \in [1 \cdots J] \qquad (9.9\text{-}3)$$

In (9.9-3), i is the tooth index and j is the rotor position index.

The operation of a PMAC machine is such that the flux in a tooth at a given rotor position is equal to the flux in the next tooth at that position plus a mechanical displacement of one slot plus one tooth. Thus,

$$\Phi_{t,i}(\theta_{rm}) = \Phi_{t,i+1}(\theta_{rm} + 2\pi/S_s) \qquad (9.9\text{-}4)$$

Using (9.9-3) and (9.9-4), it is possible to synthesize a vector of tooth flux values for tooth 1 over a rotational cycle of rotor positions. In particular, this vector is denoted $\boldsymbol{\Phi}_{t1c}$ and has elements that may be formulated as

$$\Phi_{t1c,j + J(i-1)} = \Phi_{ts:\mathrm{mod}(2S_s/P - i + 1, 2S_s/P) + 1, j} \qquad (9.9\text{-}5)$$

The elements in $\boldsymbol{\Phi}_{t1c}$ correspond to mechanical rotor positions over a cycle given by

$$\theta_{rmc,j + J(i-1)} = \theta_{rms,j} + \frac{2\pi}{S_s}(i-1) \quad i \in [1 \cdots 2S_s/P] \quad j \in [1 \cdots J] \qquad (9.9\text{-}6)$$

We will use $\boldsymbol{\theta}_{rmc}$ to denote the corresponding electrical rotor positions. From $\boldsymbol{\Phi}_{t1c}$, a vector of flux density values in the first tooth as the rotor position varies over one cycle is readily expressed as

$$\mathbf{B}_{t1c} = \frac{\boldsymbol{\Phi}_{t1c}}{w_{tb} l} \qquad (9.9\text{-}7)$$

and the maximum tooth flux density as

$$B_{tmx} = \|\mathbf{B}_{t1c}\|_{\max} \qquad (9.9\text{-}8)$$

where $\|\cdot\|_{\max}$ returns the absolute value of the element of its matrix or vector argument with the greatest absolute value.

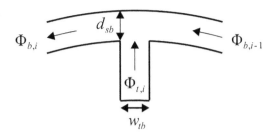

Figure 9.12 Backiron flux calculation.

Stator Backiron Flux

The flux density in the stator backiron is also of interest. Let $\Phi_{b,i}$ denote the flux in the backiron segment i. In order to calculate the flux density in the stator backiron, consider Figure 9.12. Clearly, the backiron flux in segment i of the machine is related to the flux in tooth i and the flux in segment i-1 by

$$\Phi_{b,i}(\theta_{rm}) = \Phi_{b,i-1}(\theta_{rm}) + \Phi_{t,i}(\theta_{rm}) \quad (9.9\text{-}9)$$

where the segment index operations are modulus S_s, so that $\Phi_{b,1-1} = \Phi_{b,S_s}$. Assuming that the fields in the machine are half-wave odd, it follows that the flux in the backiron segment S_s may be expressed as

$$\Phi_{b,S_s}(\theta_{rm}) = -\frac{1}{2}\sum_{n=1}^{S_s/P}\Phi_{t,n}(\theta_{rm}) \quad (9.9\text{-}10)$$

Thus, for a given rotor position, (9.9-9) and (9.9-10) may be used to determine the backiron fluxes from the tooth fluxes.

Using (9.9-9)–(9.9-10), a matrix of backiron segment fluxes $\mathbf{\Phi}_{bs}$ is created. Here, the rows correspond to segment number and the columns to rotor position. Thus, the elements are assigned as

$$\Phi_{bs:i,j} = \Phi_{b,i}(\theta_{rms,j}) \quad (9.9\text{-}11)$$

Using $\mathbf{\Phi}_{bs}$, it is possible to determine the flux in backiron segment 1 over a cycle of rotor position using an approach identical to that used in calculating the flux in tooth 1. In particular,

$$\Phi_{b1c,j+J(i-1)} = \Phi_{bs:\text{mod}(S_s-i+1,S_s)+1,j} \quad i \in [1\cdots 2S_s/P] \quad j \in [1\cdots J] \quad (9.9\text{-}12)$$

whereupon the flux density in backiron segment 1 is calculated as

$$\mathbf{B}_{b1c} = \frac{\mathbf{\Phi}_{b1c}}{d_{sb}l} \quad (9.9\text{-}13)$$

$$\mathbf{B}_{sbmx} = \|\mathbf{B}_{b1c}\|_{\max} \quad (9.9\text{-}14)$$

Stator Core Loss

A method of establishing the flux density waveform for tooth 1 and backiron segment 1 has just been established. These waveforms describe flux density as a function of mechanical rotor position. In order to obtain waveforms in the time domain, we can calculate a time vector as

$$t = \frac{\theta_{rmc}}{\omega_{rm}} \quad (9.9\text{-}15)$$

so that we effectively have time domain waveforms $B_{t1c}(t)$ and $B_{b1c}(t)$, represented by vectors of points. From these waveforms, we may use any of the methods described in Chapter 6 to calculate the power loss density in the teeth and in the backiron, which we will denote as p_{ct} and p_{cb}. The resulting core loss may then be calculated as

$$P_c = p_{ct}v_{st} + p_{cb}v_{sb} \qquad (9.9\text{-}16)$$

Rotor Flux

At this point, a means of calculating the stator flux density waveforms in tooth 1 and backiron segment 1 has been set forth. It is unnecessary to calculate these waveforms in other teeth, as they will simply be phase-shifted from the waveform in tooth 1 and backiron segment 1. For the next step in our development, let us consider the problem of calculating the rotor fields. Since the rotor field is essentially dc (viewed from the rotor), our focus will be on computing the extrema in the fields so that we can avoid heavy saturation and demagnetization.

Let us first consider the rotor backiron flux density since it is closely related to the stator backiron flux density. Consider Figure 9.13. Therein, a portion of the stator and rotor is shown in the developed diagram form. The key point in this figure is that because the backiron rotor flux is governed by a relationship analogous to that governed by the stator backiron flux, the rotor flux at the indicated positions and directions will be equal to the stator flux at the corresponding segments. This conclusion can also be reached by consideration of Gausses' law. Considering all segments and all rotor positions, an estimate of the peak tangential flux density in the rotor is given by

$$B_{rbt,mx} = \frac{1}{d_{rb}l}\|\Phi_{bs}\|_{\max} \qquad (9.9\text{-}17)$$

Now, let us consider the peak radial flux density in the rotor. From (9.7-25), with $r = r_{rb}$, we have

$$B(r_{rb},\phi_{sm}) = \frac{F_m(\phi_{sm}) + F_s(\phi_{sm})}{R_p(\phi_{sm}) + R_g} \qquad (9.9\text{-}18)$$

In (9.9-18), $F_m(\phi_{sm})$ and $R_p(\phi_{sm})$ are constant except for points of discontinuity. The maximum radial flux density must either be at an extrema of $F_s(\phi_{sm})$ or at one of the points of discontinuity.

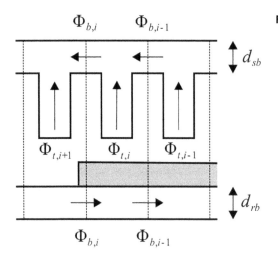

Figure 9.13 Backiron flux.

Because of symmetry, it is sufficient to consider the maximum of $F_s(\phi_{sm})$ and the two points on the edges of positively magnetized permanent magnet regions. This yields

$$B_{rbr,mx} = \max \begin{pmatrix} \left|\dfrac{F_{st1}}{R_i + R_g}\right|, \left|\dfrac{F_{m,pk} + F_{st1}}{R_{pm} + R_g}\right|, \left|\dfrac{F_{m,pk} + F_{st2}}{R_{pm} + R_g}\right|, \\ \left|\dfrac{F_{st2}}{R_i + R_g}\right|, \left|\dfrac{F_m(2\phi_i/P) + F_{s,pk}}{R_p(2\phi_i/P) + R_g}\right| \end{pmatrix} \qquad (9.9\text{-}19)$$

where F_{st1} and F_{st2} are the stator MMF at the edges of the positively magnetized permanent magnet regions (with the rotor position at zero), which are given by

$$F_{st1} = F_s\left(\dfrac{(3-\alpha_{pm})\pi}{P}\right) \qquad (9.9\text{-}20)$$

$$F_{st2} = F_s\left(\dfrac{(3+\alpha_{pm})\pi}{P}\right) \qquad (9.9\text{-}21)$$

and where $F_{m,pk}$ and $F_{s,pk}$ are the peak values of the permanent magnet and stator MMF, which are given by

$$F_{m,pk} = \dfrac{d_m B_r}{\mu_0(1+\chi_m)} \qquad (9.9\text{-}22)$$

$$F_{s,pk} = \dfrac{3\sqrt{2}N_{s1}I_s}{P} \qquad (9.9\text{-}23)$$

At this point, we have developed expressions for the peak tangential flux density from (9.9-17) in the rotor backiron as well as the peak radial flux density entering the outer edge of the backiron given by (9.9-19). The question arises about the interaction of these two field components and whether this interaction could result in magnetic saturation even if the individual components are bounded. As it turns out, this is not the case. First, it should be remembered that the peak tangential flux density in the backiron and the peak radial flux density do not occur at the same spatial location. Indeed, they are spatially separated by 90 electrical degrees. Second, it should be remembered that the radial component of the flux density is only the radial component of the flux density at the outer surface of the rotor backiron. Consider Figure 9.14. Therein, the backiron region of the machine is shown. Consider a radial component of the flux density B_r entering a region of cross-sectional area S_r. Let B_{t1} and B_{t2} be the flux density through surfaces S_{t1} and S_{t2}, respectively. Now, suppose that $|B_r| < B_{sat}, |B_{t1}| < B_{sat}$, and $|B_{t2}| < B_{sat}$, where B_{sat} is the flux density level considered to be saturated. Now, consider the flux density B_i, which we will take to be uniform across an

Figure 9.14 Flux density in rotor backiron.

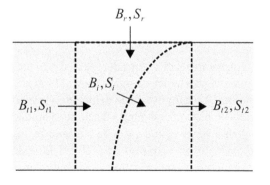

intermediate surface S_i. Given our limits on the input flux densities, along with the fact that $S_i \geq \min(S_r + S_{t1}, S_{t2})$, it follows that $|B_i| < B_{sat}$; in other words, the flux could distribute itself so as to avoid saturation.

Permanent Magnetic Field Intensity

Now, let us consider the field intensity in the permanent magnet. Our interest is computing the minimum (most negative) field intensity in the positively magnetized region. Combining (9.7-16) with (9.7-25) in a region of positive magnetization yields

$$H = \frac{\frac{r_r}{r}\frac{F_{m,pk} + F_s}{R_{pm} + R_g} - B_r}{\mu_0(1 + \chi_m)} \tag{9.9-24}$$

where, from (9.7-22), we obtain

$$F_{m,pk} = \frac{d_m B_r}{\mu_0(1 + \chi_m)} \tag{9.9-25}$$

Assuming that $F_{m,pk} + F_s > 0$, it follows that the minimum field intensity occurs, where F_s is a minimum and where the radius is a maximum. Thus,

$$H_{mn} = \frac{\frac{r_r}{r_{rg}}\frac{F_{m,pk} + F_{s,mn}}{R_{pm} + R_g} - B_r}{\mu_0(1 + \chi_m)} \tag{9.9-26}$$

where the minimum stator MMF over the positively magnetized region of the permanent magnet can be shown to be

$$F_{s,mn} = \begin{cases} -F_{s,pk} & 3 - \alpha_{pm} \leq \mod(2 + 2\phi_i/\pi, 4) \leq 3 + \alpha_{pm} \\ \min(F_{st1}, F_{st2}) & \text{otherwise} \end{cases} \tag{9.9-27}$$

At this point, it is convenient to define the functional dependence of our field calculations. These calculations require no inputs beyond those previously defined. The output of the analysis of this section is

$$\mathbf{F} = [\theta_{rm} \ B_{t1c} \ B_{b1s} \ B_{rbt,mx} \ B_{rbr,mx} \ H_{mn} \ P_c] \tag{9.9-28}$$

The calculations of this section may thus be organized as

$$\mathbf{F} = \mathbf{F}_F(\mathbf{M}, \mathbf{G}, \mathbf{W}, \mathbf{I}, \omega_{rm}) \tag{9.9-29}$$

9.10 Formulation of Design Problem

At this point, we are ready to consider the design of the machine. There are many ways to formulate the design problem. Herein, an approach will be given, which is readily tailored to a specific application and can be readily expanded to include a large variety of considerations. We will specifically consider the problem of designing a machine to produce a positive torque T_e^* at a mechanical speed ω_{rm}^* using an inverter with a given dc bus voltage, v_{dc}. Our approach will be to (i) formulate the design problem as an optimization problem in which we attempt to minimize mass and loss and (ii) then to use an optimization algorithm to arrive at the best choice of parameters.

Design Space

An important early step in the design process is to identify the parameters we are free to choose. We will use $\boldsymbol{\theta}$ to denote the vector of free design parameters. One possible choice of a parameter vector is

$$\boldsymbol{\theta} = \begin{bmatrix} S_t & r_t & c_t & m_t & \mathbf{G}_I & \mathbf{W}_I & \mathbf{I}_I \end{bmatrix} \tag{9.10-1}$$

Using this choice, every other quantity of the machine can be calculated. However, it is often the case that some of these parameters are fixed prior to the initiation of the design process. For our purposes, it is assumed that the shaft radius r_{rs} is known. Other variables that we will consider to be effectively fixed include the number of slots S_s, which we will compute based on the number of poles. The number of slots should be an integer multiple of the number of phases times the number of poles so that the machine can be constructed in a symmetric fashion. In particular, for a three-phase machine, we will take the number of slots to be

$$S_s = 3n_{spp}P \tag{9.10-2}$$

where n_{spp} is the number of slots per pole per phase.

We will also assume a packing factor k_{pf}, end winding offset l_{eo}, and the location of stator slot 1, ϕ_{ss1}, to be fixed. In order to simplify the design problem further, we will assume that we will not have tooth tips. Thus, $d_{ttc} = d_{tte} = 0$ and $\theta_{tt} = \theta_t$. This further reduces the design space. In our design process, we will also require the mass to be less than M_{mxa} and have a power less than P_{lmxa}. We will base our magnetic analysis on J rotor positions.

Let us denote the vector of fixed design parameters as $\mathbf{D}$, where

$$\mathbf{D} = \begin{bmatrix} v_{dc} & T_e^* & \omega_{rm}^* & n_{spp} & k_m & k_{pf} & l_{e0} & r_{sh} & v_{fs} & J & \phi_{ss1} & m_{mxa} & P_{lmxa} & \alpha_{tar} & \alpha_{so} \end{bmatrix} \tag{9.10-3}$$

In (9.10-3), k_m, α_{tar}, and α_{so} are previously undefined and are related to forthcoming constraints on demagnetization, tooth aspect ratio, and slot opening, respectively. They will be formally defined in the discussion on constraints. With these variables associated with $\mathbf{D}$ set a priori, and in the absence of a tooth tip, the parameter vector becomes

$$\boldsymbol{\theta} = \begin{bmatrix} S_t & r_t & c_t & m_t & \hat{\mathbf{G}}_I & \hat{\mathbf{W}}_I & \mathbf{I}_I \end{bmatrix} \tag{9.10-4}$$

where

$$\hat{\mathbf{G}}_I = \begin{bmatrix} d_i & d_{rb} & d_m & g & d_{tb} & \alpha_t & d_{sb} & \alpha_{pm} & l & P_p \end{bmatrix} \tag{9.10-5}$$

and

$$\hat{\mathbf{W}}_I = \begin{bmatrix} N_{s1}^* & \alpha_3^* \end{bmatrix} \tag{9.10-6}$$

In (9.10-5), P_p is the number of pole pairs, which is one-half the number of poles. This is so that P_p is in the set of integers rather than having a variable P that is in the set of even integers. Even with the fixed parameters, our design is left with 5 discrete variables (the material types and number of pole pairs) and 13 continuous variables for a total of 18 degrees of freedom.

Design Metrics

While there are many possible design metrics of interest, herein, we will focus on two—mass and loss. The mass may be readily expressed as

$$M = v_{sl}\rho_s + v_{rb}\rho_r + v_{pm}\rho_m + 3v_{cd}\rho_c \tag{9.10-7}$$

We will consider three components of the loss. First, we will consider the resistive loss in the machine, which is readily expressed as

$$P_r = 3R_s I_s^2 \qquad (9.10\text{-}8)$$

Semiconductor loss, denoted P_s, is given by (9.2-13). Core loss, P_c, is given by (9.9-16). The sum of these loss components is given by

$$P_l = P_s + P_r + P_c \qquad (9.10\text{-}9)$$

We will attempt to select the parameter of the machine to minimize the total mass M and the power loss P_l, subject to operating constraints. These constraints are our next topic.

Design Constraints

In order to ensure proper operation of the machine, we will enforce a number of constraints on the design. Our first two constraints will be related to the machine geometry and structural issues. First, it is desired to keep the length of the teeth reasonable compared to their width. Thus, we will require that

$$c_1 = \text{lte}(d_{st}, \alpha_{tar} w_{tb}) \qquad (9.10\text{-}10)$$

where α_{tar} is the allowed tooth aspect ratio (depth to width).

In order to make the machine easier to wind, we will next require that the wire diameter multiplied by a slot opening factor α_{so} (which is >1) be less than the width of the slot opening w_{so}. This leads to

$$c_2 = \text{lte}(d_c \alpha_{so}, w_{so}) \qquad (9.10\text{-}11)$$

Another constraint is that the current density within the wire does not exceed an acceptable value. Thus, we choose

$$c_3 = \text{lte}(I_s/a_c, J_{lim}) \qquad (9.10\text{-}12)$$

where J_{lim} is the maximum allowable rms current density for the material, m_c. In essence, this is a thermal limit, which, if rigorously treated, should be a function of how well the machine conducts heat away from the winding. However, the allowed current density will be treated as a constant herein.

In order to limit the search space, we place a limit on the maximum mass. In particular,

$$c_4 = \text{lte}(M, M_{mxa}) \qquad (9.10\text{-}13)$$

The next constraint we will consider is designed to ensure that there is adequate dc link voltage. The peak line-to-line voltage may be calculated using (9.2-30). This voltage must be less than the dc link voltage if we are to obtain the desired currents without introducing low-frequency harmonics. This leads to the constraint

$$c_5 = \text{lte}(v_{ll,pk}, v_{dc} - 2v_{fs}) \qquad (9.10\text{-}14)$$

In (9.10-14), observe that twice the semiconductor forward voltage drop has been subtracted from the dc link voltage in order to account for switch drops.

The next constraints are related to the field analysis and, in particular, to ensuring that we do neither overly saturate the material nor demagnetize the permanent magnet. While our design is focused on a single operating point, it seems desirable that the fields be limited under unexcited conditions (no current) as well as the loaded conditions. Therefore, we will apply our field analysis twice: first ensuring the fields are appropriate when the machine has zero current and then also constraining the fields under loaded conditions.

Let us denote $\mathbf{B}_{t1cnc}$, $\mathbf{B}_{bs1snc}$, $B_{rbtnc,\,mx}$, $B_{rbrnc,mx}$, and H_{ncmn} as the flux density in tooth 1 versus position, the flux density in backiron segment 1 versus rotor position, the maximum tangential flux density in the rotor backiron, the maximum radial flux density in the rotor backiron, and the minimum field intensity in the permanent magnet, respectively, for no current conditions. We impose the constraints

$$c_6 = \text{lte}(\|B_{t1cnc}\|_{\max}, B_{s,\,lim}) \tag{9.10-15}$$

$$c_7 = \text{lte}(\|B_{b1cnc}\|_{\max}, B_{s,\,lim}) \tag{9.10-16}$$

$$c_8 = \text{lte}(B_{rbtnc,mx}, B_{r,\,lim}) \tag{9.10-17}$$

$$c_9 = \text{lte}(B_{rbrnc,mx}, B_{r,\,lim}) \tag{9.10-18}$$

$$c_{10} = \text{gte}(H_{mnnc}, H_{lim}) \tag{9.10-19}$$

In (9.10-15)–(9.10-16), $B_{s,\,lim}$ is a maximum allowed flux density for the stator steel, which is chosen so as to avoid heavy saturation. Similarly, in (9.10-17)–(9.10-18), $B_{r,\,lim}$ is the maximum allowed flux density in the rotor steel. In (9.10-19), H_{lim} is the maximum allowed field intensity in the permanent magnet. This will be based on the intrinsic coercive force. In particular, we will take the maximum allowed field intensity as

$$H_{lim} = k_m H_{ci} \tag{9.10-20}$$

where H_{ci} is the intrinsic coercive force of the material in the temperature range of interest and k_m is normally between 0.5 and 0.75 and relates the field intensity limit to the intrinsic coercive force.

Next, we impose constraints on the field under the designated operating point. This leads to the constraints

$$c_{11} = \text{lte}(\|B_{t1c}\|_{\max}, B_{s,\,lim}) \tag{9.10-21}$$

$$c_{12} = \text{lte}(\|B_{b1c}\|_{\max}, B_{s,\,lim}) \tag{9.10-22}$$

$$c_{13} = \text{lte}(B_{rbt,mx}, B_{r,\,lim}) \tag{9.10-23}$$

$$c_{14} = \text{lte}(B_{rbr,mx}, B_{r,\,lim}) \tag{9.10-24}$$

$$c_{15} = \text{gte}(H_{mn}, H_{lim}) \tag{9.10-25}$$

Our next constraint will be that the desired torque is obtained. The torque is readily calculated from the lumped parameter model using (9.2-9). However, it is appropriate to adjust the torque to address the core loss. The machine equations in the lumped parameter model represent stator-resistive losses but not core losses. One method of incorporating core loss in the lumped parameter machine model is to adjust the torque by an amount sufficient to account for core loss. The corrected torque may be formulated as

$$T_{ec} = \begin{cases} T_e & \omega_{rm} = 0 \\ T_e - \dfrac{P_c}{\omega_{rm}} & \omega_{rm} \neq 0 \end{cases} \tag{9.10-26}$$

Assuming motor operation at positive speed, the torque constraint may be expressed as

$$c_{16} = \text{gte}(T_{ec}, T_e^*) \tag{9.10-27}$$

Our final constraint will be designed to focus our attention to reasonable loss levels. Thus,

$$c_{17} = \text{lte}(P_l, P_{lmxa}) \tag{9.10-28}$$

where P_{lmxa} is the maximum allowed loss.

Design Fitness

Our fitness function will be very similar to work on other devices. In particular, our fitness function will be taken as

$$\mathbf{f} = \begin{cases} \varepsilon[1 \ 1]^T \left(\dfrac{C_S - N_C}{N_C}\right) & C_S < C_I \\ \left[\dfrac{1}{M} \ \dfrac{1}{P_l}\right]^T & C_S = N_C \end{cases} \tag{9.10-29}$$

In (9.10-24), N_C, C_S, and C_I represent the number of constraints, the number of constraints satisfied, and the number of constraints imposed to a given point in the evaluation of the fitness function, and ε is a small positive number (e.g., 10^{-10}). As constraints are evaluated, the number of constraints satisfied and imposed are compared at appropriate points, and if the number of constraints satisfied fall below the number imposed, the execution of the fitness function terminates early. If all constraints are satisfied, then the fitness function elements are the reciprocals of mass and power loss.

The pseudo-code used to determine design fitness is listed in Tables 9.1 and 9.2. Table 9.1 lists the main code for fitness functional evaluation; Table 9.2 lists a code sequence executed whenever the number of constraints satisfied is compared to that imposed. Note that this is not done after checking each constraint; rather, it is done after computing $c_1 - c_4$ (as these involve low computational intensity), again after c_5, which involves the analysis to require the lumped parameter model, and then after evaluating $c_6 - c_{12}$, which all require the same field analysis.

Our next step will be to employ this algorithm and approach to design a machine. This is the topic of the next section.

9.11 Case Study

In this section, we will undertake a case study in machine design. The design requirements and fixed parameters of this design are enumerated in Table 9.3. Therein, $\phi_{ss,1}$ is the location of the first stator slot, which is given in terms of the number of stator slots, S_s (calculated using (9.10-2)), in such a way that the first tooth is centered at $\phi_{sm} = 0$. Our goal is to determine a set of designs that define the performance possibilities in the trade-off between loss and mass.

The ranges and types of all parameters used in this example are listed in Table 9.4. The ranges are determined by experience and engineering judgment. Fortunately, only crude estimates of the ranges are needed and we can check the appropriateness of the selected range by inspecting the parameter distribution of the final generation.

The final step needed to perform the design using a genetic algorithm is to select a population size and the number of generations, both of which are set to 3000 in this study. Figure 9.15 depicts the

9.11 Case Study

Table 9.1 Pseudo-Code for Calculation of the Fitness Function

1) Initialization and material selection Initialize number of constraints to $N_C = 17$
   Assign field of material vectors (structures) **S**, **R**, **C**, and **M** based on θ
2) Calculate machine geometry Assign independent fields of **G** based on θ and **D**
   Calculate dependent fields of **G** using (9.3-36)
   Evaluate c_1 using (9.10-10)
3) Perform winding calculations Assign independent fields of **W** based on θ and **D**
   Calculate dependent fields of **W** using (9.4-20)
   Evaluate c_2 using (9.10-11)
4) Current calculations Assign independent fields of **I** based on θ
   Calculate dependent fields of **I** using (9.6-5)
   Determine constraint c_3 using (9.10-12)
5) Mass calculations Compute total mass using (9.10-7)
   Determine constraint c_4 using (9.10-13)
   Test constraints (Table 9.2)
6) Analyze electrical performance Determine electrical parameter **E** from , **C**, **G**, and **W** using (9.8-18)
   Determine line-to-line voltage using (9.2-30)
   Determine torque using (9.2-9)
   Determine c_5 using (9.10-14)
   Test constraints (Table 9.2)
7) Perform field analysis under no current conditions Assign fields of **I** to zero
   Determine **F** from **M**, **G**, **W**, **I**, and ω_{rm} using (9.9-29)
   Compute c_6-c_{10} using (9.10-15)-(9.10-19)
   Test constraints (Table 9.2)
8) Perform field analysis under operational conditions Assign fields of **I** as in Section 9.6
   Determine **F** from **M**, **G**, **W**, **I**, and ω_{rm} using (9.9-29)
   Compute c_{11}-c_{15} using (9.10-21)-(9.10-25)
   Calculate corrected torque using (9.10-26)
   Compute c_{16} using (9.10-27)
9) Compute losses Compute semiconductor loss P_s using (9.2-13)
   Compute resistive loss P_r using (9.10-8)
   Compute total loss P_l using (9.10-9)
   Compute constraint c_{17} using (9.10-28)
   Test constraints (Table 9.2)
10) Compute fitness Compute fitness using (9.10-29)
    Return

Table 9.2 Pseudo-Code for Check of Constraints Satisfied Against Imposed

```
update C_s
update C_I
if (C_s < C_I)
```
$$\mathbf{f} = \varepsilon \left(\frac{C_S - N_C}{N_C} \right) \begin{bmatrix} 1 \\ 1 \end{bmatrix}$$
```
    return
end
```

Table 9.3 Design Specifications

Parameter	Value	Parameter	Value	Parameter	Value
v_{dc}	400 V	k_{pf}	0.5	ϕ_{ss1}	π/S_s
T_e^*	20 Nm	l_{eo}	1 cm	m_{mxa}	10 kg
ω_{rm}^*	5000 RPM	r_{sh}	2 cm	P_{mxa}	1000 W
n_{spp}	2	v_{fs}	2 V	α_{tar}	10
k_m	0.75	J	3	α_{so}	1.5

Table 9.4 Parameter Ranges

Parameter	Description	Type	Min	Max	Gene
s_t	Stator steel type	int	1	4	1
r_t	Rotor steel type	int	1	4	2
c_t	Conductor type	int	1	2	3
m_t	Magnet type	int	1	7	4
P_p	Pole pairs	int	4	6	5
d_i	Depth of inert region (m)	lin	0	10^{-1}	6
d_{rb}	Depth of rotor backiron (m)	log	10^{-3}	5×10^{-2}	7
d_m	Magnetic depth (m)	log	10^{-3}	5×10^{-2}	8
g	Air gap (m)	lin	5×10^{-4}	2×10^{-3}	9
d_{tb}	Depth of tooth base (m)	log	10^{-3}	5×10^{-2}	10
α_t	Tooth fraction	lin	0.05	0.95	11
d_{sb}	Depth of stator backiron (m)	log	10^{-3}	5×10^{-2}	12
α_{pm}	Magnetic fraction	lin	0.05	0.95	13
l	Length (m)	log	10^{-2}	0.5	14
N_{s1}^*	Peak fundamental conductor density (cond/rad)	log	10	10^3	15
α_3^*	Coefficient of third harmonic conductor density	lin	0.1	0.7	16
i_{qs}^{r*}	Q-axis current (A)	log	0.1	50	17
i_{ds}^{r*}	D-axis current (A)	lin	−50	0	18

results of the study in terms of the trade-off between electromagnetic mass and loss. Note that the electromagnetic mass includes the stator and rotor laminations, the windings, and the magnet but does not include the machine housing or shaft, and the losses do not include windage loss (the frictional loss associated with the rotor moving through air). However, the basic approach may be readily extended to include these effects. As can be seen, as the mass varies from 3.5 to 10 kg, the power loss drops from 750 W to just below 400 W. Design 38 (the 38th design in terms of increasing mass) is highlighted in Figure 9.15. We will consider this design in detail after we consider some of the general trends in parameters as a function of mass.

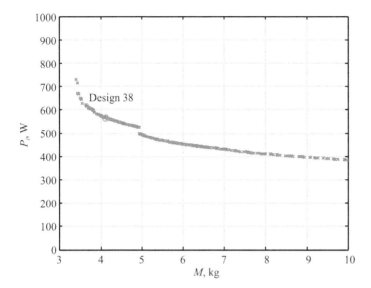

Figure 9.15 Pareto-optimal front.

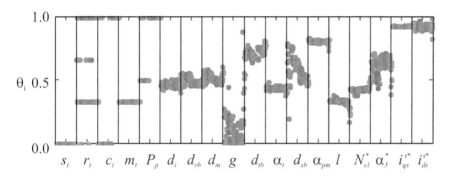

Figure 9.16 Parameter distribution.

The parameter distribution of the final population is given in Figure 9.16. As a reminder, each point in this plot represents a normalized value of a parameter of a member of the population, with parameters of higher mass individuals on the left side of each window and parameters of lower mass individuals on the right side of each window. From Figure 9.16, we see that almost all designs use stator steel corresponding to $s_t = 1$ (M19), which corresponds to 0 in terms of a normalized value. The rotor steel type r_t can be seen to take on all of the allowed values. It appears that many designs use a conductor type $c_t = 1$ (normalized value 0), which corresponds to copper, though some of the least massive designs use $c_t = 2$ (normalized value 1), which corresponds to aluminum. The magnet type is designated m_t; almost all designs use $m_t = 3$ (normalized value 0.33), which corresponds to a plastic-encapsulated NdFeB.

Also from Figure 9.16, observe that the maximum number of pole pairs P_p is usually chosen in this study. The depth of the stator backiron d_{sb} is highly correlated with mass. The remaining design variables show less correlation with mass. Observe that the air-gap values g in this case favor the minimum allowed value. The tooth fraction α_t, permanent fraction α_{pm}, stack length l, fundamental

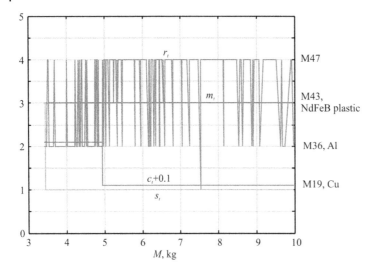

Figure 9.17 Material selection versus electromagnetic mass.

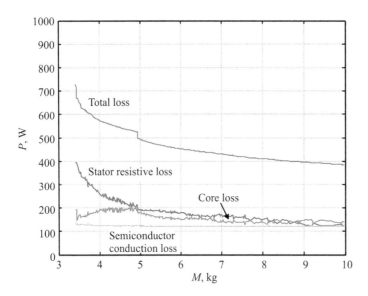

Figure 9.18 Power loss components versus electromagnetic mass.

component of turns density N_{s1}^*, and q- and d-axis currents i_{qs}^{r*} and i_{ds}^{r*} all have tight distributions except for very low mass designs.

In Figures 9.17–9.21, a variety of machine parameters are plotted versus machine mass. In each case, each point plotted represents the appropriate parameter and mass for a given design, and points for each nondominated design are plotted. The designs are sorted in terms of increasing mass, and the points are connected through a line in order to form a continuous trace, thereby giving the viewer a sense of how a given parameter varies with mass.

Figure 9.17 illustrates the material selections as a function of electromagnetic mass. Let us begin our discussion with the conductor type c_t. There, it can be seen that the least massive designs use

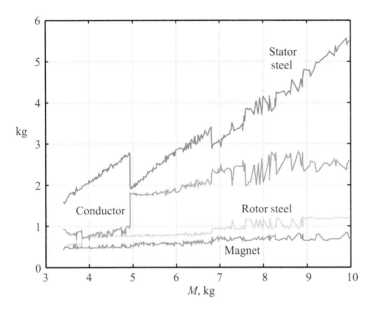

Figure 9.19 Component mass versus electromagnetic mass.

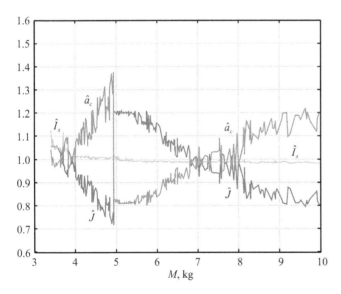

Figure 9.20 Current-related parameters versus electromagnetic mass.

aluminum though the majority of the designs use copper. Almost all machines on the front can be seen to use M19 for the stator steel. The rotor steel is quite varied, though M36 and M47 appear to be the most common choices. M36 is the least dense of the steels considered; M47 has the highest saturation flux density. Note that in our study, Hiperco50 was not an allowed choice because it is a fairly high-cost material. Note that in the stator steel, the fields are alternating and causing core loss, while in the rotor steel, the fields are predominantly dc. Thus, core loss characteristics are much

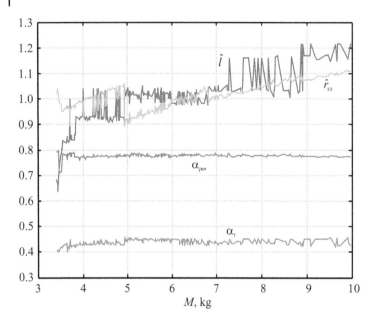

Figure 9.21 Machine parameters versus electromagnetic mass.

more important in the stator than in the rotor. Finally, we see that all designs used plastic-encapsulated NdFeB magnet material.

Figure 9.18 illustrates loss components versus mass for the nondominated designs. Core loss and semiconductor loss are relatively constant as mass varies from 3.5 to 10 kg. However, the stator resistance loss drops significantly with mass. This indicates that the biggest driving force in the trade-off between loss and mass is the stator-resistive loss. Observe that while the total loss is a relatively smooth function of mass, the component losses have a degree of irregularity. Much of this is driven by the discrete nature of some design choices.

The breakdown of component mass versus total mass is depicted in Figure 9.19. Therein, the magnet and rotor steel mass can be seen to be relatively constant. The conductor mass increases as the total mass increases. The increase in conductor mass enables a reduction in stator resistance. The stator mass increases markedly with increasing electromagnetic mass, as the stator must become larger in order to accommodate a lower current density of lower resistance stator windings.

Figure 9.20 illustrates current-related quantities as a function of electromagnetic mass. Herein, all quantities are shown relative to the mean values over all nondominated designs. Relative values are indicated with "^". The mean values for the stator current I_s, current density J, and conductor area a_c are 22.7 A, 6.28 A/mm^2, and 3.69 mm^2, respectively. As can be seen, the stator current is approximately constant after an initial decline. Observe that the current density initially falls, increases in a stepwise fashion, and then falls again. The reason for the step increase is that the conductor material changes from aluminum to copper.

Some additional machine parameters are shown versus electromagnet mass in Figure 9.21. In particular, the tooth fraction α_t, permanent magnet fraction α_{pm}, normalized stator shell radius $\hat{r}_{ss}$, and normalized length $\hat{l}$ are shown. The mean shell radius and length are 10.9 and 3.65 cm, respectively. As can be seen, the permanent magnet fraction, tooth faction, and length do not

Table 9.5 Machine Design 38 from Pareto-Optimal Front

```
Design Data --------------------       Electrical Model -------------------
Outside Diameter: 21.9 cm              Number of Poles: 12
Total Length: 7.98 cm                  Nominal Stator Resistance: 156 mΩ
Active Length: 3.34 cm                 Q-Axis Inductance: 0.666 mH
Number of Poles: 12                    D-Axis Inductance: 0.701 mH
Number of Slots: 72                    Flux Linkage Due to PM: 70.5 mVs
Stator Material Type: M19
Rotor Material Type: M36               Operating Point Performance Data
Conductor Type: Aluminum               --------
Permanent Magnet Type: NdFeB Plastic   Speed: 5000 RPM
Permanent Magnet Fraction: 77.4%       Frequency: 500Hz
Permanent Magnet Depth: 0.616 cm       Q-Axis Voltage: 217 V
Shaft Radius: 2 cm                     D-Axis Voltage: -68 V
Inert Radius: 7.12 cm                  Peak Line-to-Line Voltage: 394 V
Rotor Iron Radius: 7.8 cm              Q-Axis Current: 32.2 A
Air Gap: 0.747 mm                      D-Axis Current: -4.37 A
Slot Depth: 1.8 cm                     Peak Line Current: 32.5 A
Tooth Fraction: 43 %                   Current Density: 5.59 A rms/mm^2
Stator Backiron Depth: 0.649 cm        Torque: 20.4 Nm
Rotor Backiron Depth: 0.679 cm         Corrected torque: 20 Nm
Fund. Conductor Density: 77.2 cond/    Semiconductor Conduction Loss: 124 W
rad                                    Machine-Resistive Losses: 246 W
Third Harmonic Conductor Density: 49%  Machine Core Loss: 196 W
Conductor Diameter: 2.29 mm            Total Loss: 566 W
Stator Iron Mass: 2.09 kg              Machine Efficiency: 95.9%
Rotor Iron Mass: 0.745 kg              Inverter Efficiency: 98.9%
Conductor Mass: 0.805 kg               Machine/Inverter Efficiency: 94.9%
Magnet Mass: 0.463 kg                  Stator Tooth B/Limit: 99.9%
Mass: 4.11 kg                          Stator Backiron B/Limit: 99.7%
A-Phase Winding Pattern (first Pole):  Rotor Peak Tangential B/Limit: 99.4%
0 2 9 9 2 0                            Rotor Peak Radial B/Limit: 50.9%
Minimum Conductors Per Slot: 11        PM Demagnetization/Limit: 38.8%
Maximum Conductors Per Slot: 11
Packing Factor: 50%
```

change markedly with mass. The stator shell radius increases significantly with mass, though there is a sharp step decrease at just under 5 kg where the machine transitions from an aluminum to copper winding.

The parameters for design 38 on the Pareto-optimal front are listed in Table 9.5. In the operating point performance data, it is interesting that a small d-axis current is being used, suggesting a slight amount of flux weakening. This is compatible with the fact that the peak line-to-line voltage is close to the maximum value allowed of 394 V. It is also interesting to observe that the stator tooth and rotor tangential flux densities are all at the maximum value allowed. The stator backiron flux density is slightly lower than this, which may result in reduced backiron core loss. This stator backiron is in a sense a less critical region since there are no conflicting pressures to conduct flux and current. Observe that the permanent magnet is not close to its demagnetization limit. Hence, the coercive force of the magnet chosen was well in excess of what was needed.

A cross section of the machine is depicted in Figure 9.22. Therein, the darkest region is machine shaft. The light area adjacent to the shaft is a magnetically inert region, which must have some structure but only serves mechanical purposes. Proceeding outward in the radial direction, the next region

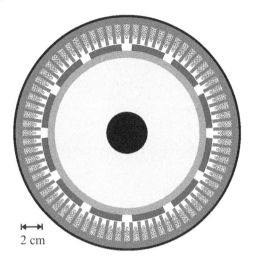

Figure 9.22 Design 38 cross section.

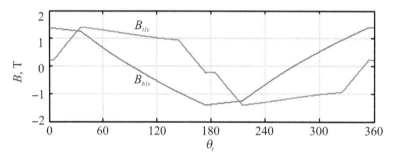

Figure 9.23 Design 38 flux density versus rotor position.

is the rotor backiron followed by the permanent magnets. Figure 9.23 depicts the flux density in a tooth and in a backiron segment. It is interesting that, although the machine will have a sinusoidal back-emf waveform, neither of the flux density waveforms is sinusoidal. The tooth flux density waveform can be viewed as a trapezoidal component due to the permanent magnet added to a sinusoidal component due to stator current. Nevertheless, the approximately sinusoidal conductor distribution yields a machine with a sinusoidal back-emf waveform (at least in terms of line-to-line measurement).

9.12 Extensions

The objective of this chapter was to introduce the design of PMAC machines. Although the approach considered was fairly detailed, several refinements are desirable before using this approach. For high-speed and large machine, structural analysis is important. Rather than limiting current density, it is more appropriate to include a thermal analysis and to limit temperature. This is the subject of the next chapter. In addition to these basic extensions, there are many other enhancements and variations that can be considered. The use of an MEC rather than an analytical solution for the field avoids the need for hard limits on the flux density [12]. While we considered semiconductor conduction losses, the consideration of switching losses can be incorporated into the machine design [13]. An alternate ac permanent magnet machine with a pronounced saliency is considered in Krizan and Sudhoff [14]. The effect of a permanent magnet type on the design of the machine is considered in Krizan and Sudhoff [15]. The connection of a permanent magnet machine to a passive rectifier rather than an inverter for use as a dc generator is considered in Bash et al. [16]. In Alsawalhi and Sudhoff, the idea of making the machine rotationally asymmetrical to increase torque density in one direction by sacrificing it in the other is explored [11]. In Kasha and Sudhoff, the use of a pole composed of multiple magnets is explored [17]. The design of an internal permanent magnet machine using this approach is set forth in Lin et al. [18].

Another straightforward extension is the consideration of multiple operating points. This may be readily incorporated by requiring that all operational constraints be satisfied at every desired operating point and computing a weighted loss metric in terms of the individual operating points. This is the subject of problem 14.

References

1. S. D. Sudhoff, J. Cale, B. Cassimere and M. Swinney, "Genetic algorithm based design of a permanent magnet synchronous machine," in *International Electric Machines and Drives Conference*, San Antonio, 2005.
2. B. N. Cassimere and S. D. Sudhoff, "Analytical design model for a surface mounted permanent magnet synchronous machine," *IEEE Transactions on Energy Conversion*, vol. **24**, no. 2, pp. 338–346, 2009.
3. B. N. Cassimere and S. D. Sudhoff, "Population based design of permanent magnet synchronous machines," *IEEE Transactions on Energy Conversion*, vol. **24**, no. 2, pp. 347–357, 2009.
4. D. C. Hanselman, *Brushless Permanent-Magnet Motor Design*, New York, NY: McGrall-Hill, 1994.
5. D. Hanselman, *Brushless Motors Magnetic Design, Performance, and Control*, New York, NY: E-Man Press, 2012.
6. S. A. Nasar, I. Boldea and L. E. Unnewehr, *Permanent Magnet, Releuctance, and Self-Synchronous Motors*, Boca Raton: CRC Press, 1993.
7. P. C. Krause, O. Wasynczuk and S. D. Sudhoff, *Analysis of Electric Machinery and Drive Systems, 2nd Edition*, Piscataway, NJ: IEEE Press/Wiley-Interscience, 2002.
8. P. L. Chapman, "Optimal current control strategies for non-sinusoidal permanet-magnet synchronous machine drives," *IEEE Transactions on Energy Conversion*, vol. **14**, no. 3, pp. 1043–1050, 1999.
9. P. L. Chapman, "A multiple reference frame synchronous estimator/regulator," *IEEE Transactions On Energy Conversion*, vol. **15**, no. 2, pp. 192–202, 2000.
10. J. Krizan and S. D. Sudhoff, "Modeling semiconductor losses for population based electric machinery design," in *Applied Power Electronics Conference and Exposition*, Oralando, 2012.
11. J. Y. Alsawalhi and S. D. Sudhoff, "Design optimization of asymmetric salient permanent magnet synchronous machines," *IEEE Transactions on Energy Conversion*, vol. **31**, no. 4, pp. 1315–1324, 2016.
12. M. Bash and S. Pekarek, "Modeling of salient-pole wound-rotor synchronous machines for population-based design," *IEEE Transactions on Energy Conversion*, vol. **26**, no. 2, pp. 381–392, 2011.
13. J. Krizan and S. D. Sudhoff, "Modeling semiconductor losses for population based electric machinery design," in *Applied Power Electronics Conference and Exposition*, Orlando, FL, 2012.
14. J. A. Krizan and S. D. Sudhoff, "A design model for salient permanent-magnet machines with investigation of saliency and wide-speed-range performance," *IEEEE Transactions on Energy Conversion*, 2012; **28**(1): 95–105.

15 J. A. Krizan and S. D. Sudhoff, "Theoretical performance boundaries for permanent magnet machines as a function of magnet type," in *Power and Energy Society General Meeting*, San Diego, CA, 2012.
16 M. Bash, S. Pekarek and S. Sudhoff, "A comparison of permanent magnet and wound rotor synchronous machines for portable power generation," in *Power and Energy Conference at Illinois*, Urbana-Chaimpaign, IL, 2010.
17 E. A. Kasha and S. D. Sudhoff, "Multi-objective design optiization of a surface-mounted modular permanent-magnet pole machine," in *2016 IEEE Power and Energy Conference at Illinois*, Urbana, IL, 2016, 2016.
18 R. Lin, S. Sudhoff and V. C. d. Nascimento, "A multi-physics design method for V-shape interior permanent-magnet machines based on multi-objective optimization," *IEEE Transactions on Energy Conversion*, vol. 35, no. 2, pp. 651–661, 2020.
19 S. D. Sudhoff, "*MATLAB codes for Power Magnetic Devices: A Multi-Objective Design Approach*, 2nd Edition," Piscataway, NJ: Wiley-IEEE Press [Online]. Available: http://booksupport.wiley.com.

Problems

1 Derive (9.3-12)–(9.3-16).

2 Derive (9.3-20) and (9.3-21).

3 From (9.4-10) and (9.4-11), derive (9.4-12).

4 Suppose round conductors are arranged in a grid such that the center points are aligned with the vertices of the grid. What is the highest packing factor that can be obtained?

5 Suppose that round conductors could be placed arbitrarily. What is the theoretical limit on packing factor?

6 Starting with (9.7-4), derive expressions analogous to (9.7-13) and (9.7-14) with the radial field variation neglected.

7 Starting with (9.7-3), derive an expressions analogous to (9.7-20)–(9.7-22) with the radial field variation neglected.

8 Derive (9.8-8) and (9.8-10).

9 Derive (9.8-9), (9.8-11), and (9.8-12).

10 Assuming that the machine current is sinusoidal, derive (9.2-13).

11 Would the lower term in (9.10-21) be valid for negative speed operation?

12 A slot liner is often used to prevent shorts of the stator windings to the stator laminations. Investigate the impact of placing a slot liner of 0.25 mm thickness on the case study. The software used for the case study is available at reference [19].

13 PMAC machines are often designed to operate over a range of speeds and torques. Modify the code used in the case study (available at reference [19]) to design for multiple operating points. Assume that all constraints must be met at all operating points. For the loss objective, define the loss as a weighted sum of the losses at each operating point. Determine the Pareto optimal front for a machine that must operate at the following operating points:

Operating Point 1:	1000 rpm, 20 Nm, 40% loss weight
Operating Point 2:	5000 rpm, 20 Nm, 40% loss weight
Operating Point 3:	6000 rpm, 15.6 Nm, 20% loss weight

Thus, for this case study, the aggregate loss is $0.4P_{l,op1} + 0.4P_{l,op2} + 0.2P_{l,op3}$, where $P_{l,op1}$, $P_{l,op2}$, and $P_{l,op3}$ are the losses at operating points 1, 2, and 3, respectively.

10

Introduction to Thermal Equivalent Circuits

In the previous chapters, our design process considered thermal limitations indirectly through a constraint on current density. It has often been mentioned that a better approach would be to perform a thermal analysis and limit temperature. Such a thermal analysis is the topic of this chapter. Like magnetic analysis, there are many approaches to performing thermal analysis, including thermal equivalent circuits (TECs), which are very similar to Magnetic Equivalent Circuit (MECs). Like magnetic systems, finite element analysis can also be used. In this chapter, we will focus on the former approach. Thermal analysis can be quite involved, and this chapter will only serve as a workable introduction. However, the approach will prove sufficient for many situations, particularly since uncertainties in thermal properties may negate any advantage of a more detailed analysis.

10.1 Heat Energy, Heat Flow, and the Heat Equation

The thermal energy density of a material may be expressed as

$$e = c\rho T \tag{10.1-1}$$

where e has units of J/m³, c is the specific heat capacity in J/kg·K, and T is temperature in K. The thermal energy in material sample Ω of volume V_Ω may be expressed as

$$E_\Omega = \int_{V_\Omega} e\, dV \tag{10.1-2}$$

To proceed, it will be convenient to define the spatial average of a quantity ζ over a volume V_Ω as

$$\langle \zeta \rangle = \frac{1}{V_\Omega} \int_{V_\Omega} \zeta\, dV \tag{10.1-3}$$

where the $\langle \rangle$ will be used to indicate a spatial average (as opposed to a temporal average). The mean temperature of a material over a region Ω may thus be expressed as

$$\langle T_\Omega \rangle = \frac{1}{V_\Omega} \int_{V_\Omega} T_\Omega\, dV \tag{10.1-4}$$

Manipulating (10.1-1)–(10.1-4), we obtain

$$E_\Omega = C_\Omega \langle T_\Omega \rangle \tag{10.1-5}$$

where C_Ω is thermal capacitance with units of J/K and which may be expressed as

$$C_\Omega = c\rho V_\Omega \tag{10.1-6}$$

Fourier's Law governs conductive heat flow in a material. In particular, it states that heat flux is proportional to the gradient of the temperature. Thus, heat flux may be expressed

$$\mathbf{Q}'' = -\left(k_x \frac{\partial T}{\partial x} \mathbf{a}_x + k_y \frac{\partial T}{\partial y} \mathbf{a}_y + k_z \frac{\partial T}{\partial z} \mathbf{a}_z \right) \tag{10.1-7}$$

where $\mathbf{Q}''$ is a vector, which denotes the heat flux in W/m², k_x, k_y, and k_z are thermal conductivities in W/m·K and $\mathbf{a}_x$, $\mathbf{a}_y$, and $\mathbf{a}_z$ are unit vectors in the x-, y-, and z-directions.

In a point that may be confusing to electrical engineers, heat flux may be thought of as a flow per unit area, as opposed to magnetic flux, which is flow. Thus heat flux is more analogous to magnetic flux density than magnetic flux. Heat flux is also commonly denoted $\mathbf{q}''$ though we will use $\mathbf{Q}''$ herein.

Heat transfer rate is somewhat analogous to magnetic flux or electric current. The heat transfer rate through a surface Γ is defined as

$$\dot{Q}_\Gamma = \int_{S_\Gamma} \mathbf{Q}'' \cdot d\mathbf{S} \tag{10.1-8}$$

and has units of W.

Net heat transfer through a given surface from a time t_1 to a time t_2 may be expressed as

$$Q_\Gamma = \int_{t_1}^{t_2} \dot{Q}_\Gamma dt \tag{10.1-9}$$

The heat equation governs how temperature varies within a material as a function of space and time and will be the starting point for our thermal analysis. In order to derive this equation, consider a small incremental cuboid of material as shown in Figure 10.1. Therein, x, y, and z denote coordinates; and Δx, Δy, and Δz are small perturbations to those coordinates. The quantity $\dot{Q}_x(x)$ denotes the thermal heat transfer in the direction of the x-axis at location x.

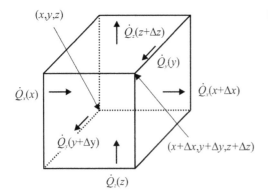

Figure 10.1 An elemental cuboid.

The time derivative of the thermal energy stored in the cube is equal to the sum of the heat transfer rates flowing into the cube plus the power being dissipated within the cube. Thus

$$\frac{dE}{dt} = P + \dot{Q}_x(x) - \dot{Q}_x(x + \Delta x) \\ + \dot{Q}_y(y) - \dot{Q}_y(y + \Delta y) + \dot{Q}_z(z) - \dot{Q}_z(z + \Delta z) \quad (10.1\text{-}10)$$

where P is the total power converted to heat in the cube. For our purposes, P will normally be the electrical power dissipated in the cube from either resistive or core losses.

From (10.1-10), for sufficiently small Δx, Δy, and Δz we have that

$$\Delta x \Delta y \Delta z \frac{de}{dt} = \Delta x \Delta y \Delta z p + \\ \dot{Q}_x(x) - \dot{Q}_x(x + \Delta x) + \\ \dot{Q}_y(y) - \dot{Q}_y(y + \Delta y) + \\ \dot{Q}_z(z) - \dot{Q}_z(z + \Delta z) \quad (10.1\text{-}11)$$

where e is thermal energy density (i.e., $E/(\Delta x \Delta y \Delta z)$, and p is power dissipation density, i.e., $P/(\Delta x \Delta y \Delta z)$).

From our definition of heat transfer rate (10.1-8) and the expression for heat flux (10.1-7) we have that

$$\dot{Q}_x(x) - \dot{Q}_x(x + \Delta x) = -\int_{S_x} k_x \frac{\partial T(x,y,z)}{\partial x} dS + \int_{S_x} k_x \frac{\partial T(x + \Delta x, y, z)}{\partial x} dS \quad (10.1\text{-}12)$$

which may be rearranged as

$$\dot{Q}_x(x) - \dot{Q}_x(x + \Delta x) = -k_x \int_{S_x} \left[\frac{\partial T(x + \Delta x, y, z)}{\partial x} - \frac{\partial T(x,y,z)}{\partial x} \right] dS \quad (10.1\text{-}13)$$

Again, for small Δx, (10.1-13) may be written as

$$\dot{Q}_x(x) - \dot{Q}_x(x + \Delta x) = k_x \int_{S_x} \Delta x \frac{\partial^2 T(x,y,z)}{\partial x^2} dS \quad (10.1\text{-}14)$$

which for small Δx, Δy, and Δz becomes

$$\dot{Q}_x(x) - \dot{Q}_x(x + \Delta x) = k_x \Delta x \Delta y \Delta z \frac{\partial^2 T(x,y,z)}{\partial x^2} \quad (10.1\text{-}15)$$

Substitution of (10.1-15) and analogous expressions for the y- and z-axis into (10.1-11) yields the heat equation

$$\frac{de}{dt} = p + k_x \frac{\partial^2 T}{\partial x^2} + k_y \frac{\partial^2 T}{\partial y^2} + k_z \frac{\partial^2 T}{\partial z^2} \quad (10.1\text{-}16)$$

We will use (10.1-16) extensively in our development of TECs.

10.2 Thermal Equivalent Circuit of One-Dimensional Heat Flow

We will now use the heat equation to derive a TEC of a region with one-dimensional heat flow. This circuit will be exact in the steady state and will be capable of predicting transient behavior, albeit only approximately. We will consider three-dimensional heat flow in Section 10.3.

To begin this development, consider a region Ω of material shown in Figure 10.2. The temperature of the region on the plane at $x = 0$ is $T_{\Omega 0x}$ and the temperature on the plane at $x = l_{\Omega x}$ is $T_{\Omega lx}$, where $l_{\Omega x}$ is the length of the cuboidal region in the x-direction. We will assume that heat can only enter and leave from the two aforementioned planes, and that the heat transfer rate entering the plane at $x = 0$ is $\dot{Q}_{\Omega 0x}$ and that the heat transfer rate leaving the region at the plane at $x = l_{\Omega lx}$ is $\dot{Q}_{\Omega lx}$. The temperature at location x is denoted $T_\Omega(x)$, which is assumed to be independent of y and z positions.

Manipulating (10.1-16) for steady-state conditions with temperature variation limited to the x-coordinate, we have

$$\frac{d^2 T_\Omega(x)}{dx^2} = -\frac{p_\Omega}{k_{\Omega x}} \qquad (10.2\text{-}1)$$

Integrating (10.2-1) with respect to x twice, the temperature distribution may be expressed as

$$T_\Omega(x) = -\frac{1}{2}\frac{p_\Omega}{k_{\Omega x}}x^2 + c_1 x + c_0 \qquad (10.2\text{-}2)$$

where c_1 and c_0 are constants. We have the boundary conditions

$$T_\Omega(0) = T_{\Omega 0x} \qquad (10.2\text{-}3)$$

$$T_\Omega(l_{\Omega x}) = T_{\Omega lx} \qquad (10.2\text{-}4)$$

which allow c_0 and c_1 to be found. Obtaining expressions for c_0 and c_1 and substituting these quantities back into (10.2-2) one obtains the steady-state temperature distribution within the sample. In particular,

$$T_\Omega(x) = -\frac{1}{2}\frac{p_\Omega}{k_{\Omega x}}x^2 + \left(T_{\Omega lx} - T_{\Omega 0x} + \frac{1}{2}\frac{p_\Omega}{k_{\Omega x}}l_{\Omega x}^2\right)\frac{x}{l_{\Omega x}} + T_{\Omega 0x} \qquad (10.2\text{-}5)$$

From (10.1-4) and (10.2-5), the mean temperature of the region for steady-state conditions can be found. In particular,

$$\langle T_\Omega \rangle = \frac{1}{12}\frac{p_\Omega}{k_{\Omega x}}l_{\Omega x}^2 + \frac{1}{2}(T_{\Omega lx} + T_{\Omega 0x}) \qquad (10.2\text{-}6)$$

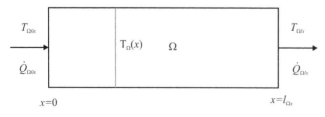

Figure 10.2 One-dimensional heat flow example.

The expression (10.2-5) can also be used in conjunction with (10.1-7) to find the heat flux within the sample. In particular,

$$Q''_\Omega(x) = p_\Omega x - \frac{k_{\Omega x}}{l_{\Omega x}}\left(T_{\Omega l} - T_{\Omega 0} + \frac{1}{2}\frac{p_\Omega}{k_{\Omega x}}l_{\Omega x}^2\right) \tag{10.2-7}$$

Using the definition of heat transfer rate (10.1-8) with (10.2-7) yields

$$\dot{Q}_{\Omega 0x} = -\frac{k_{\Omega x}S_{\Omega x}}{l_{\Omega x}}\left(T_{\Omega lx} - T_{\Omega 0x} + \frac{1}{2}\frac{p_\Omega}{k_{\Omega x}}l_{\Omega x}^2\right) \tag{10.2-8}$$

and

$$\dot{Q}_{\Omega lx} = -\frac{k_{\Omega x}S_{\Omega x}}{l_{\Omega x}}\left(T_{\Omega lx} - T_{\Omega 0x} - \frac{1}{2}\frac{p_\Omega}{k_{\Omega x}}l_{\Omega x}^2\right) \tag{10.2-9}$$

It is convenient to define the thermal resistance of the region in the x-direction as

$$R_{\Omega x} = \frac{l_{\Omega x}}{2k_{\Omega x}S_{\Omega x}} \tag{10.2-10}$$

where $S_{\Omega x}$ is the surface area in the x-direction (edge on in Figure 10.2). Using (10.2-10), and noting that the total power dissipation in the region may be expressed

$$P_\Omega = V_\Omega p_\Omega \tag{10.2-11}$$

(10.2-8) and (10.2-9) may be written as

$$-2R_{\Omega x}\dot{Q}_{\Omega 0x} = T_{\Omega lx} - T_{\Omega 0x} + P_\Omega R_{\Omega x} \tag{10.2-12}$$

and

$$-2R_{\Omega x}\dot{Q}_{\Omega lx} = T_{\Omega lx} - T_{\Omega 0x} - P_\Omega R_{\Omega x} \tag{10.2-13}$$

Next, from (10.2-6), (10.2-10), and (10.2-11) we may write

$$P_\Omega = \frac{6}{R_{\Omega x}}\left(\langle T_\Omega\rangle - \frac{1}{2}(T_{\Omega lx} + T_{\Omega 0x})\right) \tag{10.2-14}$$

Manipulating (10.2-12)–(10.2-14) yields

$$T_{\Omega 0x} = T_{\Omega cx} + R_{\Omega x}\dot{Q}_{\Omega 0x} \tag{10.2-15}$$

and

$$T_{\Omega lx} = T_{\Omega cx} - R_{\Omega x}\dot{Q}_{\Omega lx} \tag{10.2-16}$$

where

$$T_{\Omega cx} = \langle T_\Omega\rangle - \frac{1}{3}R_{\Omega x}\left(\dot{Q}_{\Omega 0x} - \dot{Q}_{\Omega lx}\right) \tag{10.2-17}$$

The quantity $T_{\Omega cx}$ will represent the temperature of an internal node in our circuit.

Equations (10.2-15)–(10.2-17) form the basis of the steady-state portion of the equivalent circuit. At this point, we could use (10.2-6) to find $\langle T_\Omega\rangle$ whereupon (10.2-15)–(10.2-17) could be solved for the heat transfer rates. Instead, however, let us attempt to develop a TEC that will approximate the thermal transients. We will accomplish this by assuming that (10.2-15)–(10.2-17) are valid during transient conditions, but that the mean temperature of the sample is not given by (10.2-6) but is instead a state variable.

In order to determine $\langle T_\Omega \rangle$, let us take the spatial average of (10.1-16) with the y- and z-axis terms neglected. We have

$$\frac{d\langle e(x) \rangle}{dt} = \langle p(x) \rangle + k_{\Omega x} \left\langle \frac{\partial^2 T(x)}{\partial x^2} \right\rangle \tag{10.2-18}$$

Multiplying (10.2-18) by the volume of the sample

$$\frac{dE_\Omega}{dt} = P_\Omega + V_\Omega k_x \left\langle \frac{\partial^2 T_\Omega(x)}{\partial x^2} \right\rangle \tag{10.2-19}$$

The spatial average of the second derivative of temperature may be expressed

$$\left\langle \frac{\partial^2 T}{\partial x^2} \right\rangle = \frac{1}{V_\Omega} \int_0^{l_{\Omega y}} \int_0^{l_{\Omega z}} \int_0^{l_{\Omega x}} \frac{\partial^2 T_\Omega}{\partial x^2} \, dx \, dy \, dz \tag{10.2-20}$$

from which we obtain

$$\left\langle \frac{\partial^2 T_\Omega}{\partial x^2} \right\rangle = \frac{1}{V_\Omega} \int_0^{l_{\Omega y}} \int_0^{l_{\Omega z}} \left. \frac{\partial T_\Omega}{\partial x} \right|_{x = l_{\Omega x}} - \left. \frac{\partial T_\Omega}{\partial x} \right|_{x = 0} dz \, dy \tag{10.2-21}$$

Using (10.1-7) and (10.1-8) in conjunction with (10.2-21),

$$\left\langle \frac{\partial^2 T_\Omega}{\partial x^2} \right\rangle = \frac{1}{V_\Omega k_{\Omega x}} (\dot{Q}_{\Omega 0 x} - \dot{Q}_{\Omega l x}) \tag{10.2-22}$$

Next we substitute (10.2-22) into (10.2-19). This yields

$$\frac{dE_\Omega}{dt} = P_\Omega + \dot{Q}_{\Omega 0 x} - \dot{Q}_{\Omega l x} \tag{10.2-23}$$

From (10.1-5) and (10.2-23), we obtain the final result.

$$\frac{d\langle T_\Omega \rangle}{dt} = \frac{1}{C_\Omega} (P_\Omega + \dot{Q}_{\Omega 0 x} - \dot{Q}_{\Omega l x}) \tag{10.2-24}$$

Together (10.2-15)–(10.2-17) and (10.2-24) correspond to the TEC shown in Figure 10.2. Note that for steady-state conditions, the result exactly corresponds to the original partial differential equation (PDE). The dynamic part of the circuit given by (10.2-23) is also an exact result in the sense that no additional approximations are made. Even so, the circuit shown in Figure 10.3 is only approximate in its prediction of thermal transients, because the steady-state temperature distribution was used in deriving (10.2-15)–(10.2-17). We will explore this in Example 10.2A.

Example 10.2A In this example, let us compare the transient predictions of the TEC shown in Figure 10.3 to the numerical solution of the PDE (10.2-1). To this end, we will consider a bar of material 10 cm in length and 2 cm by 2 cm in cross section. The specific heat capacity, mass density, and thermal conductivity of the bar are assumed to be 469 J/kg·K, 7500 kg/m³, and 16.7 W/m·K, respectively. We will assume that both ends of the bar are held at 26°C, and that the sides of the bar

Figure 10.3 Thermal equivalent circuit for one-dimensional heat flow.

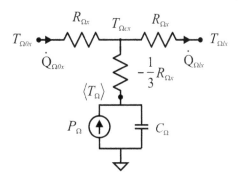

are insulated so that no heat flow occurs. In this example, the electrical power dissipation within the bar is 10 W.

Figure 10.4 illustrates the mean temperature versus time as computed using the TEC shown in Figure 10.3 as well as by a numerical solution of the original PDE represented by (10.2-1). The PDE was solved by discretizing the length into 400 segments. Both the TEC and the PDE used a times step of 0.5 ms with a forward Euler integration algorithm, which, while having poor numerical performance in general, is very straightforward and adequate for this simple problem. In Figure 10.4, the solution of the PDE is actually a temperature profile versus time. Herein, the mean temperature is computed so that it may be compared to the predictions of the TEC. Observe the predictions are generally consistent, but certainly not identical. The error that is present results from the fact that the TEC is derived based on the steady-state temperature profile.

The difference between the PDE solution and the TEC solution is further illustrated in Figure 10.5. Therein, the heat transfer rate out of the end, denoted $\dot{Q}_{\Omega lx}$ is shown as calculated using the two methods. Here again, the two methods are generally consistent, but not identical, particularly in the initial part of the study. This discrepancy is caused by the fact that the TEC does not correctly account for the spatial variation in temperature during transient conditions.

Before concluding this example, it is interesting to consider Figure 10.6, which shows the spatial temperature variation as a function of time. Therein, each trace represents T_Ω versus x at a particular instant of time; the traces are separated in time by 30s. Observe that at $t = 30$ s in particular, the temperature profile has a large central flat region that is decidedly not parabolic as in (10.2-2), which was used in the derivation of the TEC. It is this difference in shape that explains the error in the TEC. Observe that as time goes on, the spatial variation in temperature becomes increasingly parabolic as assumed by the TEC.

When conducting a thermal analysis, it is often the case that the peak temperature will be of interest. If $p_\Omega \neq 0$, then the location of the temperature extremum may be found by differentiation of (10.2-5) with respect to position and solving for the position where the derivative is equal to zero. This yields the position x_{ex} where the extremum point occurs. In particular,

$$x_{ex} = \left(\frac{T_{\Omega lx} - T_{\Omega 0x}}{l_{\Omega x}} + \frac{p_\Omega l_{\Omega x}}{2k_{\Omega x}} \right) \frac{k_{\Omega x}}{p_\Omega} \tag{10.2-25}$$

The corresponding extremum point is obtained by substitution of (10.2-25) into (10.2-5), which yields

$$T_{\Omega,ex} = \frac{k_{\Omega x}}{2p_\Omega} \left(\frac{T_{\Omega lx} - T_{\Omega 0x}}{l_{\Omega x}} + \frac{p_\Omega l_{\Omega x}}{2k_{\Omega x}} \right)^2 + T_{\Omega 0x} \tag{10.2-26}$$

Of course, (10.2-26) only yields the peak temperature if $0 \leq x_{ex} \leq l_{ex}$. To account for the possibility that this might not be the case, as well as to address the case where $p_\Omega = 0$, the peak temperature may be calculated as

$$T_{\Omega,pk} = \begin{cases} \max(T_{\Omega 0x}, T_{\Omega lx}) & p_\Omega = 0 \\ \max(T_{\Omega 0x}, T_{\Omega lx}) & p_\Omega \neq 0, x_{ex} < 0 \text{ or } x_{ex} > l_{\Omega x} \\ \max(T_{\Omega 0x}, T_{\Omega lx}, T_{\Omega,ex}) & p_\Omega \neq 0, 0 \leq x_{ex} \leq l_{\Omega x} \end{cases} \quad (10.2\text{-}27)$$

Example 10.2B Let us reconsider Example 10.2A for steady-state conditions, and in particular investigate the peak temperature within the bar. Using (10.2-10), we first obtain $R_{\Omega x} = 7.49$ K/W. Next, solving the TEC in Figure 10.3 for steady-state conditions, one finds that $T_{\Omega cx}$ and $\langle T_\Omega \rangle$ are 63.4°C and 38.5°C, respectively. Observe that the value of mean temperature matches the steady-state value of temperature predicted in Figure 10.4, as would be expected. Finally, using (10.2-25)–(10.2-27) one obtains a peak temperature, $T_{\Omega,pk}$ of 44.7°C, which is consistent with Figure 10.6 for large values of time.

While all of these calculations are straightforward, there is an important point in this example. Note that the temperature of the center node, $T_{\Omega cx}$, is significantly higher than the peak temperature in the sample. This is a result of the fact that the temperature distribution is parabolic. Consider that the ends are cooler than the center. If we linearize the temperature distribution on the two ends, the intersection of the two lines will correspond to a temperature $T_{\Omega cx}$. However, this point will be well above the peak of the parabola. It follows that when searching for the peak temperature in a device, one cannot take the peak temperature of all the node values. Rather, one must determine the peak temperature in each region, and also consider the temperatures of the nodes on the physical boundaries. The temperatures at center point nodes, such as $T_{\Omega cx}$, are artificial and do not have direct physical meeting.

Before concluding this section, it should be observed that there are many cases in which we use the model of a region with one-dimensional heat flow to represent a lossless and low-density medium such as air or insulating paper. For these cases, $P_\Omega = 0$ since the material is lossless and C_Ω can be neglected since the region has little thermal capacitance. In such a case, our

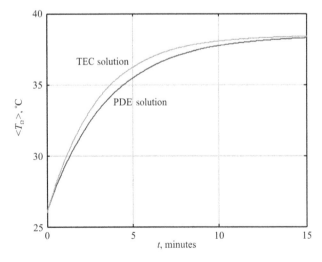

Figure 10.4 Mean temperature versus time.

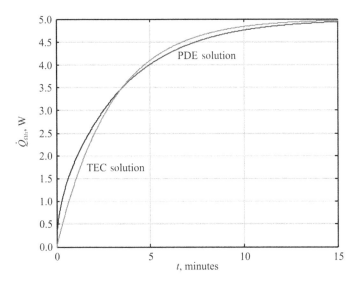

Figure 10.5 Heat transfer rate versus time.

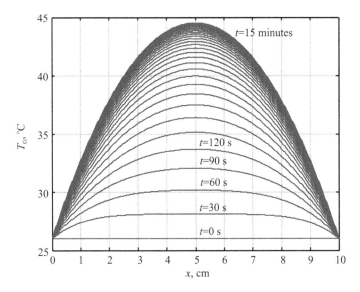

Figure 10.6 Temperature profile versus time.

equivalent circuit in Figure 10.3 reduces to that shown in Figure 10.7. In this case, the expression for $R_{\Omega x}$ given by (10.2-10) is replaced by

$$R_{\Omega x} = \frac{l_{\Omega x}}{k_{\Omega x} S_{\Omega x}} \tag{10.2-28}$$

so the single thermal resistance in Figure 10.7 is equivalent to the series connection of two thermal resistances in Figure 10.3.

Figure 10.7 Special case for one-dimensional heat flow.

$T_{\Omega 0x} \;\text{—}\!\!\!\bigwedge\!\!\!\text{—}\; T_{\Omega lx}$ with $R_{\Omega x}$

10.3 Thermal Equivalent Circuit of a Cuboidal Region

In this section, we will develop a TEC for a cuboidal region. We begin our development with the heat equation applied to a cuboidal region Ω, which occupies the volume given by $0 \leq x \leq l_{\Omega x}$, $0 \leq y \leq l_{\Omega y}$, and $0 \leq z \leq l_{\Omega z}$. From (10.1-16), we have

$$\frac{de_\Omega}{dt} = p_\Omega + k_{\Omega x}\frac{\partial^2 T_\Omega}{\partial x^2} + k_{\Omega y}\frac{\partial^2 T_\Omega}{\partial y^2} + k_{\Omega z}\frac{\partial^2 T_\Omega}{\partial z^2} \tag{10.3-1}$$

where e_Ω and p_Ω are the thermal energy and power loss densities in the region, respectively, and where $k_{\Omega x}$, $k_{\Omega y}$, and $k_{\Omega z}$ are the thermal conductivities in the three axes.

In order to solve (10.3-1), we will assume that heat flow is independent in each axis. Thus our solution will be similar to the one-dimensional case, but will have a component in each axis. In particular, our assumed temperature distribution will be

$$T_\Omega = c_{2x}x^2 + c_{1x}x + c_{2y}y^2 + c_{1y}y + c_{2z}z^2 + c_{1z}z + c_0 \tag{10.3-2}$$

where the subscripted c variables are spatial coefficients that are constants for a given set of inputs and boundary conditions. Applying (10.1-3)–(10.3-2), we have

$$\langle T_\Omega \rangle = \frac{1}{3}c_{2x}l_{\Omega x}^2 + \frac{1}{2}c_{1x}l_{\Omega x} + \frac{1}{3}c_{2y}l_{\Omega y}^2 + \frac{1}{2}c_{1y}l_{\Omega y} + \frac{1}{3}c_{2z}l_{\Omega z}^2 + \frac{1}{2}c_{1z}l_{\Omega z} + c_0 \tag{10.3-3}$$

We will now focus our attention on heat flow in the x-axis; the development for the other axes is analogous. From (10.3-2), we can obtain that the mean temperature of the region on the plane at $x = 0$ is

$$\langle T_{\Omega 0x} \rangle = \frac{1}{3}c_{2y}l_{\Omega y}^2 + \frac{1}{2}c_{1y}l_{\Omega y} + \frac{1}{3}c_{2z}l_{\Omega z}^2 + \frac{1}{2}c_{1z}l_{\Omega z} + c_0 \tag{10.3-4}$$

Likewise, the spatial mean temperature of the region on the plane at $x = l_{\Omega x}$ is given by

$$\langle T_{\Omega lx} \rangle = c_{2x}l_{\Omega x}^2 + c_{1x}l_{\Omega x} + \frac{1}{3}c_{2y}l_{\Omega y}^2 + \frac{1}{2}c_{1y}l_{\Omega y} + \frac{1}{3}c_{2z}l_{\Omega z}^2 + \frac{1}{2}c_{1z}l_{\Omega z} + c_0 \tag{10.3-5}$$

It is important to realize that the assumptions on the boundary condition made here are different than in the case of the one-dimensional heat flow considered in Section 10.2. Therein, the temperature boundary conditions on the plane $x = 0$ and $x = l_{\Omega x}$ were assumed to be constants. Here, they are not taken to be constant, but rather described by (10.3-3) with $x = 0$ on one side and $x = l_{\Omega x}$ on the other. The relationships (10.3-4) and (10.3-5) are for the mean values over these respective planes marking the ends of the sample in the x-direction.

Comparing (10.3-3)–(10.3-4) and (10.3-5), we can show

$$\langle T_{\Omega 0x} \rangle = \langle T_\Omega \rangle - \frac{1}{3}c_{2x}l_{\Omega x}^2 - \frac{1}{2}c_{1x}l_{\Omega x} \tag{10.3-6}$$

and

$$\langle T_{\Omega lx} \rangle = \langle T_\Omega \rangle + \frac{2}{3}c_{2x}l_{\Omega x}^2 + \frac{1}{2}c_{1x}l_{\Omega x} \tag{10.3-7}$$

The next step in our development will be to calculate the heat transfer rates at the planes $x = 0$ and $x = l_{\Omega x}$. From (10.3-2),

$$\frac{\partial T_\Omega}{\partial x} = 2c_{2x}x + c_{1x} \tag{10.3-8}$$

Applying (10.1-7) and (10.1-8)–(10.3-8) yields

$$\dot{Q}_{\Omega 0x} = -c_{1x}k_{\Omega x}l_{\Omega y}l_{\Omega z} \tag{10.3-9}$$

$$\dot{Q}_{\Omega lx} = -(2c_{2x}l_{\Omega x} + c_{1x})k_{\Omega x}l_{\Omega y}l_{\Omega z} \tag{10.3-10}$$

Rearranging (10.3-9) and (10.3-10), we have

$$c_{1x} = -\frac{\dot{Q}_{\Omega 0x}}{k_{\Omega x}l_{\Omega y}l_{\Omega z}} \tag{10.3-11}$$

$$c_{2x} = \frac{\dot{Q}_{\Omega 0x} - \dot{Q}_{\Omega lx}}{2k_{\Omega x}l_{\Omega y}l_{\Omega z}l_{\Omega x}} \tag{10.3-12}$$

Substitution of the values of c_{2x} and c_{1x} into (10.3-6) and (10.3-7) yields

$$\langle T_{\Omega 0x} \rangle = T_{\Omega cx} + R_{\Omega x}\dot{Q}_{\Omega 0x} \tag{10.3-13}$$

and

$$\langle T_{\Omega lx} \rangle = T_{\Omega cx} - R_{\Omega x}\dot{Q}_{\Omega lx} \tag{10.3-14}$$

where

$$T_{\Omega cx} = \langle T_\Omega \rangle - \frac{1}{3}R_{\Omega x}(\dot{Q}_{\Omega 0x} - \dot{Q}_{\Omega lx}) \tag{10.3-15}$$

and where $R_{\Omega x}$ is given by (10.2-10). The relationships (10.3-13)–(10.3-15) will be the basis for the steady-state portion of our TEC. Repeating the process for the y- and z-axis heat flows, we obtain analogous results.

The expressions (10.3-13)–(10.3-15) were obtained based on the steady-state temperature distribution within the region of interest. As in the case of one-dimensional heat flow, we could find $\langle T_\Omega \rangle$ using (10.3-3). Instead of taking this approach, however, we will instead come up with a dynamic formulation of $\langle T_\Omega \rangle$ so that we may approximately predict transient behavior.

In order to obtain the dynamic portion of the TEC, we begin by taking the spatial average of (10.3-1):

$$\left\langle \frac{de_\Omega}{dt} \right\rangle = \langle p_\Omega \rangle + k_{\Omega x}\left\langle \frac{\partial^2 T_\Omega}{\partial x^2} \right\rangle + k_{\Omega y}\left\langle \frac{\partial^2 T_\Omega}{\partial y^2} \right\rangle + k_{\Omega z}\left\langle \frac{\partial^2 T_\Omega}{\partial z^2} \right\rangle \tag{10.3-16}$$

Multiplying (10.3-16) by the volume of the sample, we have

$$\frac{dE_\Omega}{dt} = P_\Omega + V_\Omega k_{\Omega x}\left\langle \frac{\partial^2 T_\Omega}{\partial x^2} \right\rangle + V_\Omega k_{\Omega y}\left\langle \frac{\partial^2 T_\Omega}{\partial y^2} \right\rangle + V_\Omega k_{\Omega z}\left\langle \frac{\partial^2 T_\Omega}{\partial z^2} \right\rangle \tag{10.3-17}$$

where the volume V_Ω is given by

$$V_\Omega = l_{\Omega x}l_{\Omega y}l_{\Omega z} \tag{10.3-18}$$

Using (10.2-22) for the x-axis, and analogous results for the y- and z-axes, (10.3-17) becomes

$$\frac{dE_\Omega}{dt} = P_\Omega + \dot{Q}_{\Omega 0x} - \dot{Q}_{\Omega lx} + \dot{Q}_{\Omega 0y} - \dot{Q}_{\Omega ly} + \dot{Q}_{\Omega 0z} - \dot{Q}_{\Omega lz} \qquad (10.3\text{-}19)$$

Finally, manipulating (10.1-5) and (10.3-19), we obtain

$$\frac{d\langle T_\Omega \rangle}{dt} = \frac{1}{C_\Omega}\left(P_\Omega + \dot{Q}_{\Omega 0x} - \dot{Q}_{\Omega lx} + \dot{Q}_{\Omega 0y} - \dot{Q}_{\Omega ly} + \dot{Q}_{\Omega 0z} - \dot{Q}_{\Omega lz}\right) \qquad (10.3\text{-}20)$$

Together (10.3-13)–(10.3-15), along with analogous results for the y- and z-axis, and (10.3-20) suggest the TEC for a cuboidal element shown in Figure 10.8. For an alternate (and earlier) derivation of this circuit, the reader is referred to Wrobel and Mellor [1].

It will often be of interest to compute the peak temperature within the cuboid. To this end, let us define the x-axis temperature component as

$$T_x = c_{2x}x^2 + c_{1x}x \qquad (10.3\text{-}21)$$

It will also prove convenient to define T_{lx} as T_x at $x = l_{\Omega x}$. Thus

$$T_{lx} = c_{2x}l_{\Omega x}^2 + c_{1x}l_{\Omega x} \qquad (10.3\text{-}22)$$

Note that T_{lx} is not, in general, equal to $T_{\Omega lx}$. Analogous quantities for the y- and z-axis components are analogously defined. Comparing (10.3-21) to (10.3-2), it is clear that

$$T_\Omega = T_x + T_y + T_z + c_0 \qquad (10.3\text{-}23)$$

Let us temporarily assume that $c_{2x} \neq 0$. The value of x at which the extremum point of T_x occurs is given by

$$x_e = -\frac{c_{1x}}{2c_{2x}} \qquad (10.3\text{-}24)$$

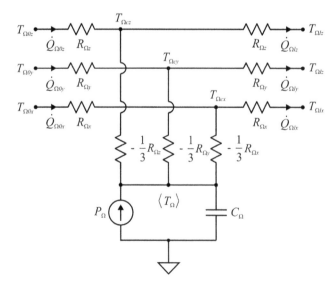

Figure 10.8 Thermal equivalent circuit of a cuboidal region. (Based on [1].)

Substitution of (10.3-24) into (10.3-21) yields the extremum value of T_x, denoted T_{ex}, given by

$$T_{ex} = -\frac{c_{1x}^2}{4c_{2x}} \tag{10.3-25}$$

Accounting for the possibilities that c_{2x} may be zero, or that the extremum point x_e does not fall in the interval $[0, l_{\Omega x}]$, or that c_{2x} may be positive, the peak value of T_{ax} may be expressed

$$T_{x,pk} = \begin{cases} \max(0, T_{lx}) & c_{2x} = 0 \text{ or } (c_{2x} \neq 0 \text{ and } (x_e < 0 \text{ or } x_e > l_{\Omega x})) \\ \max(0, T_{ex}, T_{lx}) & c_{2x} \neq 0 \text{ and } 0 \leq x_e \leq l_{\Omega x} \end{cases} \tag{10.3-26}$$

Carrying out this process for all three axes, the peak temperature in the cuboid is given by

$$T_{\Omega,pk} = T_{x,pk} + T_{y,pk} + T_{z,pk} + c_0 \tag{10.3-27}$$

where from (10.3-3), we obtain

$$c_0 = \langle T_\Omega \rangle - \left(\frac{1}{3}c_{2x}l_{\Omega x}^2 + \frac{1}{2}c_{1x}l_{\Omega x} + \frac{1}{3}c_{2y}l_{\Omega y}^2 + \frac{1}{2}c_{1y}l_{\Omega y} + \frac{1}{3}c_{2z}l_{\Omega z}^2 + \frac{1}{2}c_{1z}l_{\Omega z}\right) \tag{10.3-28}$$

In (10.3-28), c_{1x} and c_{2x} are calculated using (10.3-11) and (10.3-12) in terms of x-axis heat transfer rate components that are determined using the TEC. Quantities associated with the y- and z-axis are likewise determined.

It has been observed that this approach tends to overestimate the peak temperature because this TEC treats the heat flow of the three axes somewhat independently. An alternate approach to estimating the peak temperature would be to simply take

$$T_{\Omega,pk} = \max\left(T_{\Omega 0x}, T_{\Omega 0y}, T_{\Omega 0z}, T_{\Omega lx}, T_{\Omega ly}, T_{\Omega lz}, \langle T_\Omega \rangle\right) \tag{10.3-29}$$

10.4 Thermal Equivalent Circuit of a Cylindrical Region

In this section, we will consider the TEC of a cylindrical region. The development will prove similar, but not identical, to that of the cuboidal region we considered in the previous section.

Let us consider the geometry shown in Figure 10.9. Therein, the area between circles of radius $r_{\Omega i}$ and $r_{\Omega o}$ is considered. Between these radii, we will consider a region that spans an arc of θ_Ω. The z-axis is not shown, but is in the direction out of the page. The height of our cylindrical region is $l_{\Omega z}$ in the z-axis. Normally, we will use the results of this section when considering an object that is a full cylinder. However, it is often the case that we will only want to represent a portion of the cylinder. For example, when analyzing a motor, the angle θ_Ω may be chosen so as to represent the span of one-half a slot plus one-half of a tooth. In any case, the volume of the region of interest is expressed as

$$V_\Omega = \frac{1}{2}\theta_\Omega(r_{\Omega o}^2 - r_{\Omega i}^2)l_{\Omega z} \tag{10.4-1}$$

From (10.1-16), the heat flow equation for the region of interest may be expressed as

$$\frac{de_\Omega}{dt} = p_\Omega + k_{\Omega r}\frac{\partial^2 T_\Omega}{\partial x^2} + k_{\Omega r}\frac{\partial^2 T_\Omega}{\partial y^2} + k_{\Omega z}\frac{\partial^2 T_\Omega}{\partial z^2} \tag{10.4-2}$$

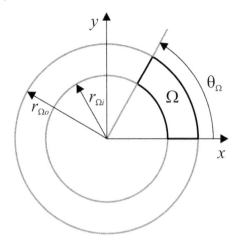

Figure 10.9 Cylindrical region.

In (10.4-2), we have assumed that the thermal conductivity in the x- and y-axes are the same, and equal to $k_{\Omega r}$. The thermal conductivity in the z-axis is $k_{\Omega z}$.

Because of symmetry, it is convenient to use a cylindrical coordinate system (r, ϕ, z) where the radius r and angle ϕ are related to the positions x and y as

$$x = r \cos \phi \tag{10.4-3}$$

$$y = r \sin \phi \tag{10.4-4}$$

From (10.4-3) and (10.4-4), we can relate differential changes in radius and angle to differential changes in x- and y-axis position. In particular,

$$\begin{bmatrix} \partial r \\ \partial \phi \end{bmatrix} = \begin{bmatrix} \cos \phi & \sin \phi \\ -\dfrac{\sin \phi}{r} & \dfrac{\cos \phi}{r} \end{bmatrix} \begin{bmatrix} \partial x \\ \partial y \end{bmatrix} \tag{10.4-5}$$

Manipulating (10.4-2)–(10.4-5), we obtain the heat flow equation in cylindrical coordinates. This relationship is given by

$$\frac{de_\Omega}{dt} = p_\Omega + k_{\Omega r}\left(\frac{\partial^2 T_\Omega}{\partial r^2} + \frac{1}{r}\frac{\partial T_\Omega}{\partial r} + \frac{1}{r^2}\frac{\partial^2 T_\Omega}{\partial \phi^2}\right) + k_{\Omega z}\frac{\partial^2 T_\Omega}{\partial z^2} \tag{10.4-6}$$

We will assume that temperature is not a function of ϕ (i.e., there is no tangential heat flow), whereupon (10.4-6) reduces to

$$\frac{de_\Omega}{dt} = p_\Omega + k_{\Omega r}\left(\frac{\partial^2 T_\Omega}{\partial r^2} + \frac{1}{r}\frac{\partial T_\Omega}{\partial r}\right) + k_{\Omega z}\frac{\partial^2 T_\Omega}{\partial z^2} \tag{10.4-7}$$

In order to solve (10.4-7), we will assume the same form of temperature variation as we assumed in the case of a cuboidal element, with the addition of a $\ln(r)$ term. We will see the role of this term in solving (10.4-7) momentarily. With the $\ln(r)$ term, the assumed temperature profile is

$$T_\Omega = c_{2r}r^2 + c_{1r}r + c_{lr}\ln(r) + c_{2z}z^2 + c_{1z}z + c_0 \tag{10.4-8}$$

From (10.4-8), we can readily show that

$$\frac{\partial^2 T_\Omega}{\partial r^2} + \frac{1}{r}\frac{\partial T_\Omega}{\partial r} = 4c_{2r} + \frac{c_{1r}}{r} \tag{10.4-9}$$

Note from the right-hand side of (10.4-9) that the $\ln(r)$ term disappears. It is a part of the natural response of the differential equation. Also, by comparing the right-hand side of (10.4-9)–(10.4-7), if (10.4-8) is to be a solution of (10.4-7) we must have that

$$c_{1r} = 0 \tag{10.4-10}$$

which is to say there is no term proportional to radius.

For our region of interest, the spatial average may be readily expressed

$$\langle \zeta \rangle = \frac{2}{l_{\Omega z}(r_{\Omega o}^2 - r_{\Omega i}^2)} \int_0^{l_{\Omega z}} \int_{r_{\Omega i}}^{r_{\Omega o}} \zeta\, r\, dr\, dz \tag{10.4-11}$$

In (10.4-4), it is assumed that ζ is only a function of radius and z-coordinate. Applying (10.4-11)–(10.4-8), we obtain

$$\langle T_\Omega \rangle = \frac{1}{2}c_{2r}(r_{\Omega o}^2 + r_{\Omega i}^2) + c_{lr}\left(\frac{r_{\Omega o}^2 \ln(r_{\Omega o}) - r_{\Omega i}^2 \ln(r_{\Omega i})}{r_{\Omega o}^2 - r_{\Omega i}^2} - \frac{1}{2}\right) + \frac{1}{3}c_{2z}l_{\Omega z}^2 + \frac{1}{2}c_{1z}l_{\Omega z} + c_0 \tag{10.4-12}$$

In developing our TEC, we will examine heat flow along the axial (z-axis) and radial directions.

Axial Heat Flow

We will first consider heat flow in the axial direction. To this end, it is convenient to find the spatial averages of the temperatures of the two ends of our region, namely, the end at $z = 0$ and the end at $z = l_{\Omega z}$. These are spatial averages, but averages over surfaces (i.e., the ends) rather than averages over the volume as defined by (10.4-11). From (10.4-8), we obtain

$$\langle T_{\Omega 0z} \rangle = \frac{1}{2}c_{2r}(r_{\Omega o}^2 + r_{\Omega i}^2) + c_{lr}\left(\frac{r_{\Omega o}^2 \ln(r_{\Omega o}) - r_{\Omega i}^2 \ln(r_{\Omega i})}{r_{\Omega o}^2 - r_{\Omega i}^2} - \frac{1}{2}\right) + c_0 \tag{10.4-13}$$

$$\langle T_{\Omega lz} \rangle = \frac{1}{2}c_{2r}(r_{\Omega o}^2 + r_{\Omega i}^2) + c_{lr}\left(\frac{r_{\Omega o}^2 \ln(r_{\Omega o}) - r_{\Omega i}^2 \ln(r_{\Omega i})}{r_{\Omega o}^2 - r_{\Omega i}^2} - \frac{1}{2}\right) + c_{2z}l_{\Omega z}^2 + c_{1z}l_{\Omega z} + c_0 \tag{10.4-14}$$

Comparing (10.4-13) and (10.4-14) to (10.4-12) it can be seen that

$$\langle T_{\Omega 0z} \rangle = \langle T_\Omega \rangle - \frac{1}{3}c_{2z}l_{\Omega z}^2 - \frac{1}{2}c_{1z}l_{\Omega z} \tag{10.4-15}$$

and

$$\langle T_{\Omega lz} \rangle = \langle T_\Omega \rangle + \frac{2}{3}c_{2z}l_{\Omega z}^2 + \frac{1}{2}c_{1z}l_{\Omega z} \tag{10.4-16}$$

Next, we note that from (10.4-8) we have

$$\frac{\partial T_\Omega}{\partial z} = 2c_{2z}z + c_{1z} \tag{10.4-17}$$

Applying (10.1-7) and (10.1-8)–(10.4-17) with $z = 0$ and $z = l_{\Omega z}$,

$$\dot{Q}_{\Omega 0z} = -\frac{1}{2}\theta_\Omega k_{\Omega z} c_{1z}(r_{\Omega o}^2 - r_{\Omega i}^2) \tag{10.4-18}$$

$$\dot{Q}_{\Omega lz} = -\frac{1}{2}\theta_\Omega k_{\Omega z}(2c_{2z}l_{\Omega z} + c_{1z})(r_{\Omega o}^2 - r_{\Omega i}^2) \tag{10.4-19}$$

Manipulating (10.4-15)–(10.4-16) and (10.4-18)–(10.4-19), we obtain

$$\langle T_{\Omega 0z}\rangle = T_{\Omega cz} + R_{\Omega z}\dot{Q}_{0z} \tag{10.4-20}$$

$$\langle T_{\Omega lz}\rangle = T_{\Omega cz} - R_{\Omega z}\dot{Q}_{lz} \tag{10.4-21}$$

where

$$T_{\Omega cz} = \langle T_\Omega\rangle - \frac{1}{3}R_{\Omega z}(\dot{Q}_{0z} - \dot{Q}_{lz}) \tag{10.4-22}$$

and

$$R_{\Omega z} = \frac{l_{\Omega z}}{k_{\Omega z}(r_{\Omega o}^2 - r_{\Omega i}^2)\theta_\Omega} \tag{10.4-23}$$

This is essentially the same process and results as for the cuboidal element of Section 10.3. We will find, however, that our results will vary when considering heat flow in the radial direction.

Radial Heat Flow

We will begin our consideration of heat flow in the radial direction by looking at the boundary conditions. From (10.4-8) and (10.4-10), the spatial averages of temperature over the surfaces formed by the arcs at radii $r_{\Omega i}$ and $r_{\Omega o}$, which extend from $z = 0$ to $z = l_{\Omega z}$ may be expressed as

$$\langle T_{\Omega ir}\rangle = c_{2r}r_{\Omega i}^2 + c_{lr}\ln(r_{\Omega i}) + \frac{1}{3}c_{2z}l_{\Omega z}^2 + \frac{1}{2}c_{1z}l_{\Omega z} + c_0 \tag{10.4-24}$$

$$\langle T_{\Omega or}\rangle = c_{2r}r_{\Omega o}^2 + c_{lr}\ln(r_{\Omega o}) + \frac{1}{3}c_{2z}l_{\Omega z}^2 + \frac{1}{2}c_{1z}l_{\Omega z} + c_0 \tag{10.4-25}$$

Comparing (10.4-12), (10.4-24), and (10.4-25), we obtain

$$\langle T_{\Omega ir}\rangle = \langle T_\Omega\rangle + c_{2r}\left(\frac{1}{2}r_{\Omega i}^2 - \frac{1}{2}r_{\Omega o}^2\right) - c_{lr}\left(\frac{r_{\Omega o}^2}{r_{\Omega o}^2 - r_{\Omega i}^2}\ln\left(\frac{r_{\Omega o}}{r_{\Omega i}}\right) - \frac{1}{2}\right) \tag{10.4-26}$$

$$\langle T_{\Omega or}\rangle = \langle T_\Omega\rangle + c_{2r}\left(\frac{1}{2}r_{\Omega o}^2 - \frac{1}{2}r_{\Omega i}^2\right) - c_{lr}\left(\frac{r_{\Omega i}^2}{r_{\Omega o}^2 - r_{\Omega i}^2}\ln\left(\frac{r_{\Omega o}}{r_{\Omega i}}\right) - \frac{1}{2}\right) \tag{10.4-27}$$

From (10.4-8) and (10.4-10), we have

$$\frac{\partial T_\Omega}{\partial r} = 2c_{2r}r + \frac{c_{lr}}{r} \tag{10.4-28}$$

Using (10.4-28) in conjunction with (10.1-7) and (10.1-8), we obtain

$$\dot{Q}_{\Omega ir} = -\theta_\Omega k_{\Omega r}(2c_{2r}r_{\Omega i}^2 + c_{lr})l_{\Omega z} \tag{10.4-29}$$

$$\dot{Q}_{\Omega or} = -\theta_\Omega k_{\Omega r}(2c_{2r}r_{\Omega o}^2 + c_{lr})l_{\Omega z} \tag{10.4-30}$$

Rearranging (10.4-29) and (10.4-30), it can be shown that

$$c_{2r} = \frac{\dot{Q}_{\Omega ir} - \dot{Q}_{\Omega or}}{2k_{\Omega r}\theta_\Omega l_{\Omega z}(r_{\Omega o}^2 - r_{\Omega i}^2)} \tag{10.4-31}$$

$$c_{lr} = \frac{\dot{Q}_{\Omega or} r_{\Omega i}^2 - \dot{Q}_{\Omega ir} r_{\Omega o}^2}{k_{\Omega r} \theta_\Omega l_{\Omega z} (r_{\Omega o}^2 - r_{\Omega i}^2)} \tag{10.4-32}$$

Substitution of (10.4-31) and (10.4-32) into (10.4-26) yields

$$\langle T_{\Omega ir} \rangle = T_{\Omega cr} + R_{\Omega ir} \dot{Q}_{\Omega ir} \tag{10.4-33}$$

where

$$T_{\Omega cr} = \langle T_\Omega \rangle + R_{\Omega tr} (\dot{Q}_{\Omega ir} - \dot{Q}_{\Omega or}) \tag{10.4-34}$$

In (10.4-33) and (10.4-34), the thermal resistances are given by

$$R_{\Omega tr} = -\frac{1}{4 k_{\Omega r} \theta_\Omega (r_{\Omega o}^2 - r_{\Omega i}^2) l_{\Omega z}} \left(r_{\Omega i}^2 + r_{\Omega o}^2 - \frac{4 r_{\Omega i}^2 r_{\Omega o}^2}{r_{\Omega o}^2 - r_{\Omega i}^2} \ln\left(\frac{r_{\Omega o}}{r_{\Omega i}}\right) \right) \tag{10.4-35}$$

and

$$R_{\Omega ir} = \frac{1}{2 k_{\Omega r} \theta_\Omega l_{\Omega z}} \left(\frac{2 r_{\Omega o}^2}{r_{\Omega o}^2 - r_{\Omega i}^2} \ln\left(\frac{r_{\Omega o}}{r_{\Omega i}}\right) - 1 \right) \tag{10.4-36}$$

In (10.4-35), the t in the subscript is to suggest the bottom leg of a t-shaped equivalent circuit, which is the form this portion of the equivalent circuit will take.

Likewise, substitution of (10.4-31) and (10.4-32) into (10.4-27) produces

$$\langle T_{\Omega ro} \rangle = \langle T_\Omega \rangle + R_{\Omega tr} (\dot{Q}_{\Omega ir} - \dot{Q}_{\Omega or}) - R_{\Omega or} \dot{Q}_{\Omega or} \tag{10.4-37}$$

where

$$R_{\Omega or} = \frac{1}{2 k_{\Omega r} \theta_\Omega l_{\Omega z}} \left(1 - \frac{2 r_{\Omega i}^2}{r_{\Omega o}^2 - r_{\Omega i}^2} \ln\left(\frac{r_{\Omega o}}{r_{\Omega i}}\right) \right) \tag{10.4-38}$$

Observe that in the case of the cylindrical element, the thermal resistances of the two surfaces to the temperature node $T_{\Omega cr}$ are not identical as they are in the case of a cuboidal element.

To finish our equivalent circuit, let us take the spatial average of (10.4-7) and multiply by the volume. This yields

$$V_\Omega \left\langle \frac{de_\Omega}{dt} \right\rangle = V_\Omega \langle p_\Omega \rangle + k_{\Omega r} V_\Omega \left\langle \frac{\partial^2 T_\Omega}{\partial r^2} + \frac{1}{r} \frac{\partial T_\Omega}{\partial r} \right\rangle + k_{\Omega z} V_\Omega \left\langle \frac{\partial^2 T_\Omega}{\partial z^2} \right\rangle \tag{10.4-39}$$

From (10.4-8), (10.4-10), and (10.4-11), we have

$$\left\langle \frac{\partial^2 T_\Omega}{\partial r^2} + \frac{1}{r} \frac{\partial T_\Omega}{\partial r} \right\rangle = 4 c_{2r} \tag{10.4-40}$$

Using (10.4-40) in conjunction with (10.4-39), (10.4-31), and (10.4-1), we obtain

$$\left\langle \frac{\partial^2 T_\Omega}{\partial r^2} + \frac{1}{r} \frac{\partial T_\Omega}{\partial r} \right\rangle = \frac{(\dot{Q}_{\Omega ir} - \dot{Q}_{\Omega or})}{V_\Omega k_{\Omega r}} \tag{10.4-41}$$

Next, from (10.4-8), (10.4-18), (10.4-19), and (10.4-1), we have

$$\left\langle \frac{\partial^2 T_\Omega}{\partial z^2} \right\rangle = \frac{\dot{Q}_{\Omega oz} - \dot{Q}_{\Omega lz}}{k_{\Omega z} V_\Omega} \tag{10.4-42}$$

Combining (10.4-39), (10.4-41), and (10.4-42),

$$\frac{dE_\Omega}{dt} = P_\Omega + \dot{Q}_{\Omega ir} - \dot{Q}_{\Omega or} + \dot{Q}_{\Omega 0z} - \dot{Q}_{\Omega lz} \tag{10.4-43}$$

which, using (10.1-5) and (10.1-6) may be expressed as

$$\frac{d\langle T_\Omega \rangle}{dt} = \frac{1}{C_\Omega} \left(P_\Omega + \dot{Q}_{\Omega ir} - \dot{Q}_{\Omega or} + \dot{Q}_{\Omega 0z} - \dot{Q}_{\Omega lz} \right) \tag{10.4-44}$$

where

$$C_\Omega = \frac{1}{2} c_\Omega \rho_\Omega l_{\Omega z} \theta_\Omega \left(r_{\Omega o}^2 - r_{\Omega i}^2 \right) \tag{10.4-45}$$

Together (10.4-20)–(10.4-22), (10.4-33)–(10.4-34), (10.4-37), and (10.4-44) comprise the equivalent circuit shown in Figure 10.10. Therein, the spatial average of the exterior node temperatures is not explicitly indicated.

Before concluding this section, let us consider the calculation of the peak temperature within the region. To this end, let us define

$$T_r = c_{2r} r^2 + c_{lr} \ln(r) \tag{10.4-46}$$

as well as

$$T_{ri} = c_{2r} r_{Ai}^2 + c_{lr} \ln(r_{\Omega i}) \tag{10.4-47}$$

and

$$T_{ro} = c_{2r} r_{\Omega o}^2 + c_{lr} \ln(r_{\Omega o}) \tag{10.4-48}$$

Note that T_{ri} and T_{ro} are the boundary values of T_r; they are not in general equal to $T_{\Omega ri}$ and $T_{\Omega ro}$. Next, we define T_z as in the cuboidal element case of Section 10.3. From the definitions of T_r and T_z, (10.4-8), and recalling that $c_{1r} = 0$, we have

$$T_\Omega = T_r + T_z + c_0 \tag{10.4-49}$$

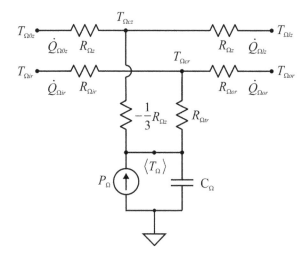

Figure 10.10 Thermal equivalent circuit of cylindrical region.

We will establish the peak value of our sample by adding the peak value of T_r to the peak value of T_z. To establish the peak value of T_r, we differentiate (10.4-46) with respect to radius and set the resulting expression equal to zero. This yields

$$r_{er} = \sqrt{-\frac{c_{lr}}{2c_{2r}}} \qquad (10.4\text{-}50)$$

Clearly, (10.4-50) only holds if $c_{2r} \neq 0$, $c_{lr}/c_{2r} \leq 0$. Substitution of (10.4-50) into (10.4-46) yields

$$T_{er} = \frac{1}{2}c_{lr}\left(\ln\left(-\frac{c_{lr}}{2c_{2r}}\right) - 1\right) \qquad (10.4\text{-}51)$$

Taking into account for the various possibilities for whether or not T_{er} is the extremum point, the peak value of T_r can be calculated as

$$T_{r,pk} = \begin{cases} \max(T_{ri}, T_{ro}) & c_{2r} = 0 \text{ or } (c_{2r} \neq 0 \text{ and } c_{lr}/c_{2r} > 0) \text{ or} \\ & (c_{2r} \neq 0 \text{ and } c_{lr}/c_{2r} < 0 \text{ and } (r_{er} < r_{\Omega i} \text{ or } r_{er} > r_{\Omega o})) \\ \max(T_{ri}, T_{er}, T_{ro}) & c_{2r} \neq 0 \text{ and } c_{lr}/c_{2r} < 0 \text{ and } r_{\Omega i} \leq r_{er} \leq r_{\Omega o} \end{cases}$$
$$(10.4\text{-}52)$$

The peak value of T_z, which is denoted $T_{z,pk}$, may be computed as in the cuboidal case. Finally, the peak value with the element may be calculated as

$$T_{\Omega,pk} = T_{r,pk} + T_{z,pk} + c_0 \qquad (10.4\text{-}53)$$

In (10.4-53), the constant c_0 may be calculated as

$$c_0 = \langle T_\Omega \rangle - \left(\frac{1}{2}c_{2r}(r_{\Omega o}^2 + r_{\Omega i}^2)\right) + c_{lr}\left(\frac{r_{\Omega o}^2 \ln(r_{\Omega o}) - r_{\Omega i}^2 \ln(r_{\Omega i})}{r_{\Omega o}^2 - r_{\Omega i}^2} - \frac{1}{2}\right) + \frac{1}{3}c_{2z}l_{\Omega z}^2 + \frac{1}{2}c_{1z}l_{\Omega z}\right) \qquad (10.4\text{-}54)$$

which may be derived using (10.4-12). Therein constants c_{2r} and c_{lr} are obtained using (10.4-31) and (10.4-32), where the heat transfer rates are computed using the TEC. The variables c_{2z} and c_{1z} are computed as in the cuboidal element case.

As in the case with the cuboidal element, an alternate approach to estimating the peak temperature is simply to take

$$T_{\Omega,pk} = \max(T_{\Omega ir}, T_{\Omega 0z}, T_{\Omega or}, T_{\Omega lz}, \langle T_\Omega \rangle) \qquad (10.4\text{-}55)$$

keeping in mind that the node temperatures represent averages over surfaces.

10.5 Inhomogeneous Regions

When modeling heat flow within a power magnetic device, there are regions such as the windings where it is difficult to represent each different material individually. Consider Figure 10.11(a) that shows the cross section of a winding bundle. Therein, the bundle can be seen to include the individual conductors, the insulation surrounding the conductors, as well as the air or potting material (a nonmagnetic, nonconductive material that adds additional electrical insulation, prevents abrasion, and improves heat transfer). Representing every part of the winding bundle would be rather burdensome. As an alternative, one approach is to homogenize such an inhomogeneous region by representing it as an effective material, though perhaps an anisotropic one (i.e., having different properties in different directions).

One strategy to do this is as suggested in Hettegger et al. [2]. To illustrate this approach, consider Figure 10.11, which depicts a winding bundle. Three materials are assumed to be present: the conductor material, the insulation material, and air or potting material. The thermal conductivities of these three materials will be denoted k_c, k_i, and k_a, respectively. Likewise, the mass densities will be denoted ρ_c, ρ_i, and ρ_a. Finally, the specific heat capacities are symbolized c_c, c_i, and c_a.

We will assume the conductor bundle is of cross section w by d. The radius of each of the N conductors in the region will be denoted r_c, and the thickness of the insulation on the conductor will be designated t_i. In order to define an effective material, we begin by computing the total cross-sectional of area of each material. The total cross section of the conductor, insulation, or air/potting material may be expressed as

$$a_c = N\pi r_c^2 \tag{10.5-1}$$

$$a_i = N\pi\left((r_c + t_i)^2 - r_c^2\right) \tag{10.5-2}$$

$$a_a = wd - a_c - a_i \tag{10.5-3}$$

Let us define the aspect ratio of the winding bundle as

$$\zeta = \frac{w}{d} \tag{10.5-4}$$

The basic strategy to determine the homogenized material is shown in Figure 10.11(b). In this approach, we *temporarily* rearrange the material into three bulk regions corresponding to conductor, to insulator, and to air (or potting material). This temporary arrangement is not how we will model the region—it is only for derivational purposes. Our end result will be a homogenous effective material that will fill the entire region as shown in Figure 10.11(c).

The dimension of the conductor region is a rectangle of width w_c by depth d_c. The insulator region is a region is a backward "L" shape. The width of the upright part of our "L" is denoted w_i, and the depth of the bottom part of the "L" is denoted d_i. The air/potting material region is also shaped as a backward "L" with width w_a and depth d_a as depicted in Figure 10.11(b). For each material, the dimension will be determined so that the appropriate area is maintained and so that ratio of width

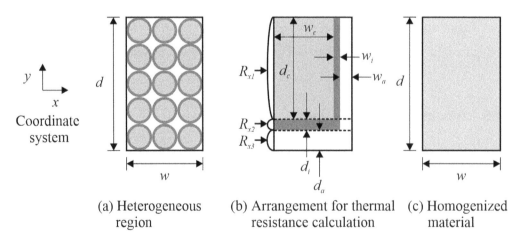

(a) Heterogeneous region (b) Arrangement for thermal resistance calculation (c) Homogenized material

Figure 10.11 Representation of homogenized region. (Based on [2].)

over depth is in all cases equal to ζ. We will find that this choice causes the homogenized thermal conductivities to be the same in the x- and y-axes.

Let us first consider the conductor region. We require

$$w_c d_c = a_c \tag{10.5-5}$$

$$\frac{w_c}{d_c} = \zeta \tag{10.5-6}$$

Solving (10.5-5) and (10.5-6) yields

$$w_c = \sqrt{a_c \zeta} \tag{10.5-7}$$

$$d_c = \sqrt{\frac{a_c}{\zeta}} \tag{10.5-8}$$

Next, we consider the insulator region. Imposing our requirements on the total area and aspect ratio, we have

$$(w_c + w_i)(d_c + d_i) - w_c d_c = a_i \tag{10.5-9}$$

$$\frac{w_i}{d_i} = \zeta \tag{10.5-10}$$

Solving (10.5-9) and (10.5-10) yields

$$d_i = \sqrt{d_c^2 + \frac{a_i}{\zeta}} - d_c \tag{10.5-11}$$

$$w_i = d_i \zeta \tag{10.5-12}$$

Finally, from Figure 10.11(b), it can be seen that we must have

$$w_a = w - w_c - w_i \tag{10.5-13}$$

$$d_a = d - d_c - d_i \tag{10.5-14}$$

It can be readily shown that this choice satisfies our two requirements: (a) the area is as desired, that is,

$$wd - (w_c + w_i)(d_c + d_i) = a_i \tag{10.5-15}$$

and (b) we have the desired aspect ratio

$$\frac{w_a}{d_a} = \zeta \tag{10.5-16}$$

Our goal is to determine effective material parameters that will allow us to represent the region with a homogeneous (but perhaps anisotropic) effective material. To this end, let us compute the thermal resistance across the width of the sample. This thermal resistance can be expressed in terms of three parallel paths in the x-direction represented by R_{x1}, R_{x2}, and R_{x3} as indicated in Figure 10.11 (b). Using our results from the one-dimensional heat flow case (temporarily disregarding the source) (10.2-28), and assuming that thermal resistances can be added in series like electrical resistances, the three thermal resistances may be expressed

$$R_{x1} = \frac{w_c}{k_c d_c l} + \frac{w_i}{k_i d_c l} + \frac{w_a}{k_a d_c l} \tag{10.5-17}$$

$$R_{x2} = \frac{w_c + w_i}{k_i d_i l} + \frac{w_a}{k_a d_i l} \qquad (10.5\text{-}18)$$

$$R_{x3} = \frac{w}{k_a d_a l} \qquad (10.5\text{-}19)$$

The net thermal resistance is the parallel combination of the thermal resistances of the three paths. Assuming that parallel thermal resistances can be treated like electrical resistances, the net thermal resistance across the width in the x-direction may be expressed as

$$R_x = \frac{1}{\frac{1}{R_{x1}} + \frac{1}{R_{x2}} + \frac{1}{R_{x3}}} \qquad (10.5\text{-}20)$$

Our assumptions on the treatment of series and parallel connections of thermal resistances will be verified in Section 10.7. For the present, the reader is asked to accept that thermal resistances may be treated as electrical resistances.

Now, if we had an effective material with thermal conductivity of k_{xye} in the x-direction, the thermal resistance in the x-direction would be expressed as

$$R_x = \frac{w}{k_{xye} l d} \qquad (10.5\text{-}21)$$

Substitution of (10.5-17)–(10.5-19) into (10.5-20) and equating the result with (10.5-21) yields an expression for the effective thermal conductivity in the x-direction as

$$k_{xye} = \frac{1}{\frac{1}{k_c} + \frac{d_i}{k_i d_c} + \frac{d_a}{k_a d_c}} + \frac{1}{\frac{d_c + d_i}{k_i d_i} + \frac{d_a}{k_a d_i}} + \frac{1}{\frac{d}{k_a d_a}} \qquad (10.5\text{-}22)$$

Now, if we were to repeat the process, but instead of considering the thermal resistance across the width, we consider the thermal resistance across the depth in the y-direction, we would obtain an effective value of thermal resistance in the y-direction as

$$k_{xye} = \frac{1}{\frac{1}{k_c} + \frac{w_i}{k_i w_c} + \frac{w_a}{k_a w_c}} + \frac{1}{\frac{w_c + w_i}{k_i w_i} + \frac{w_a}{k_a w_i}} + \frac{1}{\frac{w}{k_a w_a}} \qquad (10.5\text{-}23)$$

As a result of the aspect ratio requirements (10.5-6), (10.5-10), and (10.5-16), the results (10.5-22) and (10.5-23) are equivalent. This is a desirable result because from a physical standpoint, it seems reasonable that the thermal conductivity in the x- and y-directions is the same. Achieving this result was the reason the requirement was imposed. Hence, we have established a value for the effective thermal conductivity in the x- and y- direction, denoted k_{xye}. Although this value will hold in the x- and y- directions, as alluded to earlier, our effective material will be anisotropic. This is because the effective thermal conductivity will be different in the z-axis.

In the x- and y-axis, observe that the region is, from a macroscopic viewpoint, essentially isotropic. However, in the z-axis (coming out of the page in Figure 10.11(b)), the situation is different. This is because the conductors, which typically have good thermal conductivity, run contiguously in this axis whereas they do not form a contiguous path in the x- and y-axes.

Returning to Figure 10.11(b), let us consider the thermal resistance in the z-direction. In this case, we will have three parallel paths, one for the conductor region, one for the insulation region, and one for the air/potting material region. The net thermal resistance will be that of the three regions in parallel. Thus

$$R_z = \frac{1}{\frac{a_c k_c}{l} + \frac{a_i k_i}{l} + \frac{a_a k_a}{l}} \qquad (10.5\text{-}24)$$

where l is the length of the material in the z-direction. The thermal resistance of the effective material may be expressed as

$$R_z = \frac{l}{k_{ze}wd} \qquad (10.5\text{-}25)$$

Equating (10.5-24) and (10.5-25), we obtain the effective thermal resistance in the z-direction. In particular,

$$k_{ze} = \frac{a_c}{wd}k_c + \frac{a_i}{wd}k_i + \frac{a_a}{wd}k_a \qquad (10.5\text{-}26)$$

Clearly, the effective thermal conductivity in the z-direction is not the same as in the x- and y- direction and so we can see that our effective material is anisotropic.

We have now established the thermal conductivity of the effective material. We need to also to define the effective value of density and specific heat capacity. To this end, the mass of the region may be expressed as

$$M = a_c l \rho_c + a_i l \rho_i + a_a l \rho_a \qquad (10.5\text{-}27)$$

In terms of the effective density, the mass may be expressed as

$$M = wdl\rho_e \qquad (10.5\text{-}28)$$

Comparing (10.5-27) and (10.5-28), the effective density is given by

$$\rho_e = \frac{a_c \rho_c + a_i \rho_i + a_a \rho_a}{wd} \qquad (10.5\text{-}29)$$

Next, let us consider the effective specific heat capacity. The heat capacity of the region may be expressed as

$$C = a_c l \rho_c c_c + a_i l \rho_i c_i + a_a l \rho_a c_a \qquad (10.5\text{-}30)$$

In terms of the effective material, we have

$$C = wdl\rho_e c_e \qquad (10.5\text{-}31)$$

Equating (10.5-30) and (10.5-31), we obtain the effective value of effective heat capacity. In particular,

$$c_e = \frac{a_c \rho_c c_c + a_i \rho_i c_i + a_a \rho_a c_a}{wd\rho_e} \qquad (10.5\text{-}32)$$

At this point, we have established the parameters of an effective anisotropic material with which to represent to represent a combination of material. Although it has been noted already, it is worth again observing that Figure 10.11(b) does not reflect our strategy for modeling the region; rather our strategy is represented by Figure 10.11(c) wherein we replace the region with a uniform material. Although the properties of the effective material shown in Figure 10.11(c) are derived from Figure 10.11(b) there is a significant difference. In Figure 10.11(c), we are only representing a single material, not breaking the material into separate regions. As a result, the peak temperature in the conductor that we would calculate using Figure 10.11(c) will be significantly different than if we left the situation as in Figure 10.11(b)—for in this later case the temperature node corresponding to the left wall in Figure 10.11(b) would effectively dictate the temperature of the entire winding because of the high thermal conductivity of the conductor and the contiguous contact of the entire conductive region to the left wall. Of course, this does not represent the original situation in which the conductor is broken apart into subregions by insulation and air. In our single material equivalent in Figure 10.11(c) this effect is represented by use of the effective thermal conductivities. Note that we can calculate the peak temperature in the conductor using the method of calculating the peak temperature in a uniform cuboidal element described in Section 10.3. This procedure will be illustrated in the following example.

Example 10.5A Let us consider a winding in a slot and attempt to calculate the peak temperature within the slot. Let us suppose that the slot is 2.4 cm wide and 2.8 cm deep, and 7.1 cm long. The slot contains 50 conductors of 11-gauge copper wire that has a radius of 1.15 mm and has insulation of 34.3 μm. Assume the thermal conductivity of the copper, the insulation, and air are 400, 0.175, and 0.024 W/K·m, respectively. The operating conditions are such that the current density in the conductor is 7.5 A/mm². Finally, suppose the walls of the slot are at a temperature of 50°C, that the ends of the slot are at 52°C, and that the top of the slot is at a temperature of 53°C.

We will solve this problem by first computing the power dissipation within the slot. Based on the current density and the conductor radius, the conductor current is 31.2 A. The resistance of each conductor is 287 mΩ. This yields a total power loss of 13.9 W for the entire slot. Note that while electrical resistivity increases with temperature, we will neglect this effect for the present.

Our next step is to determine a homogenized representation of the slot material. Using (10.5-22) and (10.5-26), we obtain a thermal conductivity in the x-direction (across the width of the slot) and the y-direction (across the depth of the slot) of 42.1 mW/K·m and obtain a thermal conductivity in the z-direction (along the length of the slot) of 123.7 W/K·m. The large difference in the thermal conductivity arises from the fact that the conductor is an excellent thermal conductor that runs contiguously in the z-direction (along the length), but that is thermally insulated by the electrical insulation, and especially air, in the x- and y-directions. The use of a thermally conductive potting material in the slot can improve the situation.

The next step is to apply the TEC of a cuboidal element to determine the mean and peak temperature. The first step is to compute the thermal resistances. From (10.2-10), one obtains $R_x = 143$ K/W (across half the width of slot), $R_y = 195$ K/W (across half the depth of the slot), and $R_z = 0.437$ K/W (across half the length of the slot). From the TEC, one can derive

$$\langle T_A \rangle = \frac{\frac{1}{3}P_A + \frac{T_{A0x} + T_{Alx}}{R_{Ax}} + \frac{T_{A0y} + T_{Aly}}{R_{Ay}} + \frac{T_{A0z} + T_{Alz}}{R_{Az}}}{2\left(\frac{1}{R_{Ax}} + \frac{1}{R_{Ay}} + \frac{1}{R_{Az}}\right)} \quad (10.5\text{A-1})$$

which yields $\langle T_A \rangle = 53.0°$C. Also from the equivalent circuit, it can be shown that

$$T_{Acx} = 3\langle T_A \rangle - T_{A0x} - T_{Alx} \quad (10.5\text{A-2})$$

Similar results are readily obtained for the other axis. This yields $T_{Acx} = 58.9°$C, $T_{Acy} = 55.9°$C, and $T_{Acz} = 54.9°$C.

Next, again using the equivalent circuit with known temperatures at all nodes, the heat transfer rates are readily established. In particular, we obtain $\dot{Q}_{A0x} = -62.3$ mW, $\dot{Q}_{Alx} = 62.3$ mW, $\dot{Q}_{A0y} = -30.4$ mW, $\dot{Q}_{Aly} = 15.0$ mW, $\dot{Q}_{A0z} = -6.88$ W, and $\dot{Q}_{A0z} = 6.88$ W. As we can see, although heat is traveling from all surfaces of the region, most of the heat is leaving through the ends of the slot because of the high thermal conductivity. This is a somewhat unusual case in that the ends of the slot are at fixed temperature. Under other conditions, we would see the heat flow into the sides of the slot.

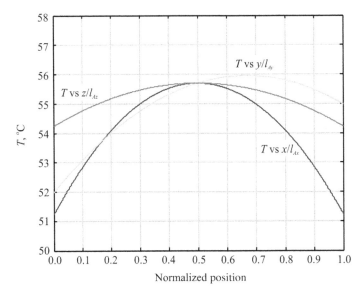

Figure 10.12 Spatial temperature dependence.

With the heat transfer rates known, we may use (10.3-11) to obtain $c_{1x} = 745$ K/m, $c_{1y} = 424$ K/m, and $c_{1z} = 82.7$ K/m and (10.3-12) to find $c_{2x} = -31.0$ kK/m², $c_{2y} = -11.3$ kK/m², and $c_{2z} = -1.17$ kK/m². Next, from (10.3-28), $c_0 = 319$ K.

As a final step in the process, from (10.3-24)–(10.3-26), one obtains $T_{ex} = 4.47$ K, $T_{ey} = 3.97$ K, and $T_{ez} = 1.47$ K, whereupon the peak temperature may be found from (10.3-27) which yields $T_{A,pk} = 56.0°C$, which is 3°C higher than the mean temperature.

Before concluding this example, it is interesting to plot the temperature through the sample. This is shown in Figure 10.12. Therein, the trace labeled T versus y/l_{Ay} is the temperature at $x = l_{Ax}/2$ and $z = l_{Az}/2$ as y is varied. Similarly, T versus z/l_{Az} is the temperature at $x = l_{Ax}/2$ and $y = l_{Ay}/2$ as z is varied. Finally, the trace labeled T versus x/l_{Ax} is the temperature at $y = l_{Ay}/2$ and $z = l_{Az}/2$ as x is varied. In the case of the y-axis, which is directed along the depth of the slot, the temperature distribution is asymmetrical. This is because the bottom of the slot was taken to be at a lower temperature than the top of the slot.

Finally, note that the methodology for homogenizing the material may be indicative of a worst-case analysis. This is because in reality, we would expect the copper conductors to intermittently contact each other down the slot. This would effectively improve the thermal conductivity in the x- and y-axes.

10.6 Material Boundaries

Thus far in this chapter, our attention has been focused on heat flow within a region. In this section, we begin to consider what happens when we have multiple regions of materials with different properties. This includes the contact between different solid regions, as well as the interface between solids and gas or liquids.

Contact Resistance

The heat transfer rate across a boundary between two materials may be modeled as

$$\dot{Q}_{ct} = \frac{\Delta T_{ct}}{R_{ct}} \qquad (10.6\text{-}1)$$

where ΔT_{ct} is the temperature difference across the contact and R_{ct} is the contact resistance. The contact resistance may be expressed as

$$R_{ct} = \frac{1}{A_{ct} h_{ct}} \qquad (10.6\text{-}2)$$

where A_{ct} is the contact area and h_{ct} is a contact heat transfer coefficient.

The reason for the contact resistance is closely related to surface roughness. As the two surfaces come together, air or some other fluid (such as thermal grease) fill the voids in the contact. The contact heat transfer coefficient is a function of material, of surface roughness, of the contact pressure, and of the fluid used to fill the void (vacuum, air, thermal grease, etc).

In Mellor et al. [3], data are given for an iron–iron contact between stator laminations and a stator shell. For this interface, with air as the fluid, the value of h_{ct} is about 1 kW/K·m² at a contact pressure of 5 MPa. Note that while some theoretical results are available, the best results are obtained experimentally. This is often an area of significant uncertainty.

Convective Heat Transfer

Convective heat transfer refers to the transfer of heat from a solid object to a moving fluid. The movement of the fluid may be natural (for example, heating causing a fluid to rise) or forced wherein the fluid such as air is moved by a fan.

The heat transfer rate from the solid to the fluid may be modeled as

$$\dot{Q}_{cv} = \frac{\Delta T_{cv}}{R_{cv}} \qquad (10.6\text{-}3)$$

where ΔT_{cv} is the difference in temperature between the solid and the fluid and R_{cv} is the effective thermal resistance. The thermal resistance may be expressed

$$R_{cv} = \frac{1}{h_{cv} A_{cv}} \qquad (10.6\text{-}4)$$

where h_{cv} is the convective heat transfer coefficient, and A_{cv} is the contact area between the solid and the fluid.

The relationships embodied by (10.6-3) and (10.6-4) may give the impression that the calculation of convective heat transfer is a simple process. It is not, the principal difficulty being the calculation of the convective heat transfer coefficient h_{cv}, which is a strong function of the fluid parameters and fluid velocity. Part of the difficulty is that the fluid temperature will vary with distance from the surface, as will the fluid velocity. To make matters more difficult, the fluid flow will often be laminar (i.e., smoothly flowing in layers) over some regions in the fluid and turbulent in the others. There is no bypassing that predicting h_{cv} is an involved problem of which a thorough explanation would require something like the length of this book. To this end, the reader is referred to Bergman et al. [4]. However, some tractable and useful results relevant to machine design are given in references [3,5]. For natural convection to air, values of h_{cv} range from 2 to 10 W/K·m². For water, the value would increase to roughly 200 W/K·m². Forced convection using a fan for increased air flow is often used in machine to increase h_{cv}.

Radiation

All objects with temperatures above absolute zero radiate heat. The heat transfer rate leaving the surface of a solid object may be expressed as

$$\dot{Q}_{r,out} = \sigma \varepsilon_s A_s T_s^4 \tag{10.6-5}$$

In (10.6-5), A_s and T_s are the area and temperature of the solid; ε_s is a material (and surface) property called the emissivity, which describes how efficient a radiator the material/surface is. The emissivity is between 0 and 1. The term σ is the Stefan–Boltzmann constant, which is equal to 56.7 nW/K⁴·m².

Objects also absorb radiation from their surroundings. For a small object surrounded by an isothermal surface of temperature T_{sur}, the heat transfer rate into the small object is the same as that object, at the temperature of the surroundings, would have radiated [4]. That is,

$$\dot{Q}_{r,in} = \sigma \varepsilon_s A_s T_{sur}^4 \tag{10.6-6}$$

The net heat transfer rate due to radiation may thus be expressed as

$$\dot{Q}_r = \sigma \varepsilon_s A_s \left(T_s^4 - T_{sur}^4 \right) \tag{10.6-7}$$

The use of appropriate units is particularly important in radiation calculations, wherein Kelvin must be used. Recall that Kelvin and Celsius units are essentially equivalent except for an offset. Because of the nonlinearity of radiation with temperature, this offset would render (10.6-5)–(10.6-7) invalid.

Clearly, heat flow due to radiation is nonlinear with respect to temperature. It is sometimes convenient to linearize (10.6-7). This yields

$$\dot{Q}_r = \frac{T_s}{R_r} + \dot{Q}_{r0} \tag{10.6-8}$$

where

$$R_r = \frac{1}{\sigma \varepsilon_s A_s 4 T_{s0}^3} \tag{10.6-9}$$

and

$$\dot{Q}_{r0} = -\sigma \varepsilon_s A_s \left(3 T_{s0}^4 + T_{sur}^4 \right) \tag{10.6-10}$$

In (10.6-9) and (10.6-10), T_{s0} is the temperature about which (10.6-7) is linearized. This temperature may be selected, for example, as the ambient temperature; or it can be computed iteratively as part of a numerical solution of a TEC. In terms of a TEC, R_r is a resistance to ground and $\dot{Q}_{r0}$ is a heat transfer rate to ground.

For a given temperature T_{s0}, (10.6-8) may be used with the remainder of the circuit, which can then be solved for T_s. This value of T_s is then used for T_{s0} in the subsequent iteration. The process repeats until the temperature estimate converges.

Emissivity of a variety of materials is listed in [4]. Highly polished metals often have an emissivity of <0.1. Oxidized metals usually have an emissivity greater than 0.5. Radiation is often ignored in the analysis of power magnetic devices, but can become important for devices that operate at temperatures significantly higher than those of the surroundings.

10.7 Thermal Equivalent Circuit Networks

In this section, we consider the interconnection of TECs of individual regions of a device in order to form and solve a TEC for an entire device. To this end, this section will begin with a discussion of the TEC laws. Next, just as in the case of a magnetic equivalent circuit, we will formulate a standard branch to facilitate a systematic analysis. The solution of the TEC using nodal analysis will then be discussed. Finally, some notation for TECs will be introduced in order to reduce our effort in describing complex thermal circuits.

Thermal Equivalent Circuit Laws

In electrical circuits, the interconnection of circuit components is based on Kirchoff's voltage and current laws, which state that the sum of the voltage drops across a closed path is zero, and that the sum of the currents going into a node is zero. We also observed that similar relationships applied in the case of magnetic equivalent circuits. In particular, we had that the sum of MMF drops was equal to the sum of the MMF sources and that the sum of the magnetic fluxes into a magnetic node are zero.

In the case of TECs, the same analogous laws apply. First, the sum of the changes in temperature around a closed path must be zero. Second, the sum of the heat transfer rates into a node must be zero.

With regard to the first point, consider the temperature at three points in the system, denoted T_A, T_B, and T_C. Let $T_{xy} = T_x - T_y$ where x and y may be A, B, or C. Clearly $T_{AB} + T_{BC} + T_{CA} = 0$. Extending this idea to the general case, one rapidly concludes that the sum of the temperature differences around any closed loop is zero.

The situation regarding the heat transfer rates is somewhat more subtle. In general, our TEC nodes will be of two types. First, there are nodes that are internal to the TEC of an individual element, such as a region of one-directional heat flow, a cuboid, or a cylindrical region. Second, there are nodes that correspond to the boundaries between different elements.

Let us first consider those nodes that are constructed mathematically, which correspond to internal nodes within elements. For example, in the case of the equivalent circuit for one-dimensional heat flow shown in Figure 10.3, the nodes whose temperatures are designated $T_{\Omega cx}$ and $\langle T_\Omega \rangle$ fall into this category. The heat transfer rates going into or out of these nodes sum to zero because the circuits were constructed so that this property would hold. A similar situation exists of $T_{\Omega cx}$, $T_{\Omega cy}$, $T_{\Omega cz}$, and $\langle T_\Omega \rangle$ in Figure 10.8, and $T_{\Omega cz}$, $T_{\Omega cr}$, and $\langle T_\Omega \rangle$ in Figure 10.10.

The second category of nodes is those corresponding to the boundary between elements. Let us consider the case where the boundary between elements is a flat rectangle and is surrounded by cuboid as shown in Figure 10.1. Let us make this cuboid infinitesimally thin so that, while encompassing the boundary, it has no volume. With no volume, there is no power dissipation within the element, as well as no energy stored. It follows from (10.1-10) that the sum of the heat transfer rates into our thin cuboid, which will represent the boundary between two elements, is zero. This argument can be extended for a more general interface, but we will not do so here.

10.7 Thermal Equivalent Circuit Networks

The conclusion is that our TECs can be treated much like electrical or magnetic equivalent circuits. Thus, concepts of voltage (temperature) division, current (heat transfer rate) division, the series or parallel connection of resistances, and nodal and mesh analysis still apply to thermal as well as electrical and magnetic circuits.

Standard Branch

In order to facilitate the systematic formulation of the equations describing a TEC, it is convenient to define a standard branch. One possible choice in branch is depicted in Figure 10.13. Therein, i denotes the branch number and $n_{p,i}$ and $n_{n,i}$ denote the positive and negative node numbers. The temperature across the branch and heat transfer rate into the branch are denoted $T_{b,i}$ and $\dot{Q}_{b,i}$ respectfully. The term $G_{b,i}$ will be the thermal conductance, which is the inverse of the thermal resistance. In our branch equations, it will be assumed that the conductance can go to zero, but resistance must be nonzero; therefore, a conductance is specified rather than a resistance. There are three source terms in the branch, an independent power source $P_{s,i}$, a dependent power source $P_{d,i}$, and a temperature source $T_{s,i}$. It is normally the case that when a temperature source is present, the negative node is the ground node.

The standard branch shown in Figure 10.13 is similar to the standard branch for the nodal form of our magnetic equivalent circuit shown in Figure 2.26. The primary differences are that the orientation of the heat (power) source has been altered for convenience of the branch in normal usage, along with the introduction of the dependent source.

The purpose for the dependent source is to accommodate the situation where resistance (and hence the power dissipation) is affected by temperature. Let us consider a region of a thermal circuit corresponding to N conductors with a length l_c, cross section a_c, resistivity ρ, and carrying a current i_c. The power dissipation may be expressed as

$$P = \frac{N l_c i_c^2 \rho}{a_c} \qquad (10.7\text{-}1)$$

Figure 10.13 Standard branch.

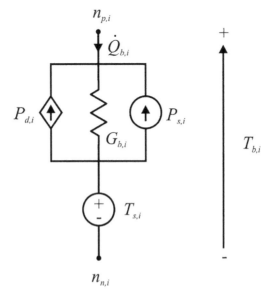

The resistivity varies with temperature. The dependence has been traditionally described by

$$\rho = \rho_0\big(1 + \alpha_\rho(T - T_0)\big) \tag{10.7-2}$$

where ρ_0 is the resistivity at T_0, and α_ρ is a coefficient of temperature dependence.

Substitution of (10.7-2) into (10.7-1) yields

$$P = \underbrace{\frac{Nli_c^2\rho_0(1-\alpha_\rho T_0)}{a_c}}_{P_{s,i}} + \underbrace{\frac{Nli_c^2\rho_0\alpha_\rho}{a_c}T}_{P_{d,i}} \tag{10.7-3}$$

The latter term in (10.7-3) has the form

$$P_{d,i} = G_{d,i}T_{N,d(i)} \tag{10.7-4}$$

where $d(i)$ returns the index of the governing temperature. This form motivates the incorporation of the dependent source in the standard branch shown in Figure 10.13.

Observe that, mathematically, the heat transfer rate into the standard branch may be expressed as

$$\dot{Q}_{b,i} = G_{b,i}(T_{b,i} - T_{s,i}) - P_{s,i} - P_{d,i} \tag{10.7-5}$$

Nodal Network Analysis

Once we determine a TEC for a component (such as an inductor or a motor), we will of course need to solve the TEC. As in the case of electrical or magnetic circuits, two methods to do this are nodal analysis and mesh analysis. In the case of magnetic circuits, we explored the question and concluded that mesh analysis was advantageous. However, the thermal problem is different in that the nonlinearities are predominantly in source terms rather than conductance (analogous to permeance) terms as in the magnetic equivalent circuit. In addition, the number of nodes in a TEC tends to be quite large, making the enumeration of the meshes rather more difficult than the compilation of node equations. Because of this, we will utilize a nodal approach.

Using nodal analysis, we will formulate a system of equations of the form

$$G_N T_N = P_N \tag{10.7-6}$$

where $T_N \in \mathbb{R}^{N_n}$ is the vector of nodal temperature, $G_N \in \mathbb{R}^{N_n \times N_n}$ is a network thermal conductance matrix, and $P_N \in \mathbb{R}^{N_n}$ is the network thermal power vector, and where N_n is the number of nodes in the TEC.

In order to obtain this form, first we note that the temperature across branch i may be expressed as

$$T_{b,i} = T_{N,n_{p,i}} - T_{N,n_{n,i}} \tag{10.7-7}$$

where $T_{N,n_{p,i}}$ and $T_{N,n_{n,i}}$ are the temperatures of the positive and negative nodes of branch i, respectively. Substitution of (10.7-4) and (10.7-7) into (10.7-5) yields

$$\dot{Q}_{b,i} = G_{b,i}\big(T_{N,n_{p,i}} - T_{N,n_{n,i}} - T_{s,i}\big) - P_{s,i} - G_{d,i}T_{N,d(i)} \tag{10.7-8}$$

Our next step is to sum the heat transfer rates into node j, which must be zero. This yields

$$\sum_{i \in L_{p,j}} \dot{Q}_{b,i} - \sum_{i \in L_{n,j}} \dot{Q}_{b,i} = 0 \tag{10.7-9}$$

In (10.7-9), $L_{p,j}$ is the set of branches that have node j as the positive node, and $L_{n,j}$ is the set of branches that have node j as the negative node.

Now, let us suppose branch k is in $L_{p,j}$. Using (10.7-8) and (10.7-9), we may write

$$G_{b,k}\left(T_{N,n_{p,k}} - T_{N,n_{n,k}} - T_{s,k}\right) - P_{s,k} - G_{d,k}T_{N,d(k)} + \sum_{i \in L_{p,j}/k} \dot{Q}_{b,i} - \sum_{i \in L_{n,j}} \dot{Q}_{b,i} = 0 \quad (10.7\text{-}10)$$

where $L_{p,j}/k$ denotes the list of branches which have node j ($=n_{p,k}$) as the positive node with the kth branch removed from the set. Manipulating (10.7-10), we have

$$G_{b,k}\left(T_{N,n_{p,k}} - T_{N,n_{n,k}}\right) - G_{d,k}T_{N,d(k)} + \sum_{i \in L_{p,j}/k} \dot{Q}_{b,i} - \sum_{i \in L_{n,j}} \dot{Q}_{b,i} = G_{b,k}T_{s,k} + P_{s,k} \quad (10.7\text{-}11)$$

The expression (10.7-11) suggests how the elements of the standard branch will manifest themselves in the G_N matrix and P_N vector.

Now let us suppose that branch k is in $L_{n,j}$. In this case, we have

$$-G_{b,k}\left(T_{N,n_{p,k}} - T_{N,n_{n,k}} - T_{s,k}\right) + P_{s,k} + G_{d,k}T_{N,d(k)} + \sum_{i \in L_{p,j}} \dot{Q}_{b,i} - \sum_{i \in L_{n,j}/k} \dot{Q}_{b,i} = 0 \quad (10.7\text{-}12)$$

Manipulating (10.7-12) slightly, we have

$$G_{b,k}\left(T_{N,n_{n,k}} - T_{N,n_{p,k}}\right) + G_{d,k}T_{N,d(k)} + \sum_{i \in L_{p,j}} \dot{Q}_{b,i} - \sum_{i \in L_{n,j}/k} \dot{Q}_{b,i} = -G_{b,k}T_{s,k} - P_{s,k} \quad (10.7\text{-}13)$$

Observe that in this case $j = n_{n,k}$.

Together (10.7-11) and (10.7-13) suggest the Thermal Nodal Analysis Formulation Algorithm (TNAFA) described in pseudo-code form in Table 10.1. We will use this algorithm when we explore the use of TECs.

Graphical Element Representation

We are now prepared to perform a thermal analysis of devices using TECs. Before doing this, however, it should be observed that the TEC for our devices can become fairly involved, even for rather simple devices. In order to make our TEC diagrams more concise, some graphical shorthand notation will prove convenient. To this end, some standard circuit symbols and combinations thereof, and their shorthand notation, are shown in Figure 10.14. These symbols and combinations of symbols tend to occur very often and so the shorthand notation will save considerable space.

In addition to the shorthand of Figure 10.14, note that our equivalent circuits for the one-dimensional heat flow element (Figure 10.3), the cuboidal element (Figure 10.8), and cylindrical element (Figure 10.10) are each comprised of a small network. We will represent these three networks as shown in Figure 10.15. To this end, let us consider the use of the one-dimensional element. In Figure 10.15, Ω denotes the name of the element, N_{cx} is replaced by the name of the central node in the T-equivalent circuit (the node whose temperature is denoted $T_{\Omega cx}$ in Figure 10.3), and N_Ω is the number of the node whose temperature is that of the mean temperature of the one-dimensional element. The partial dot marked with a $0x$ is the node corresponding to $T_{\Omega 0x}$ in Figure 10.3; likewise the partial dot marked lx is the node corresponding to $T_{\Omega lx}$ in Figure 10.3. The terminology will become clearer when we use it in the case study of the next section.

Table 10.1 Thermal Nodal Analysis Formulation Algorithm

```
P_N = 0
G_N = 0
for i = 0 to i = N_b
   if n_p, i > 0
      G_{N:n_{p,i},n_{p,i}} = G_{N:n_{p,i},n_{p,i}} + G_{b,i}
      G_{N:n_{p,i},d(i)} = G_{N:n_{p,i},d(i)} - G_{d,i}
      P_{N,n_{p,i}} = P_{N,n_{p,i}} + G_{b,i}T_{s,i} + P_{s,i}
   end
   if n_n, i > 0
      G_{N:n_{n,i},n_{n,i}} = G_{N:n_{n,i},n_{n,i}} + G_{b,i}
      G_{N:n_{n,i},d(i)} = G_{N:n_{n,i},d(i)} + G_{d,i}
      P_{N,n_{p,i}} = P_{N,n_{p,i}} - G_{b,i}T_{s,i} - P_{s,i}
   end
   if n_p, i > 0 and n_n, i > 0
      G_{N:n_{p,i},n_{n,i}} = G_{N:n_{p,i},n_{n,i}} - G_{b,i}
      G_{N:n_{n,i},n_{p,i}} = G_{N:n_{n,i},n_{p,i}} - G_{b,i}
   end
end
```

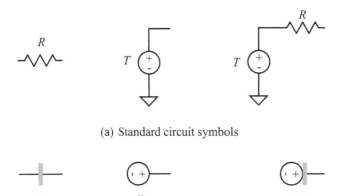

(a) Standard circuit symbols

(b) Concise circuit symbols

Figure 10.14 Concise circuit symbols.

10.8 Case Study: Thermal Model of Electromagnet

In this section, we will derive and use a TEC for the electromagnet we first considered in Chapter 5. Recall that in this application we used EI-core arrangement, and that the coil was attached to a 12 V voltage source. The device was designed to provide an attractive force of 2500 N at distance of 1 mm. In this section, we will develop the thermal model and use it in two examples. In the first example, we will analyze the design we developed in the case study of Section 5.4. In the second example, we

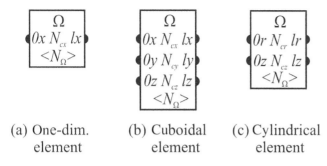

(a) One-dim. element (b) Cuboidal element (c) Cylindrical element

Figure 10.15 Thermal equivalent circuit elements.

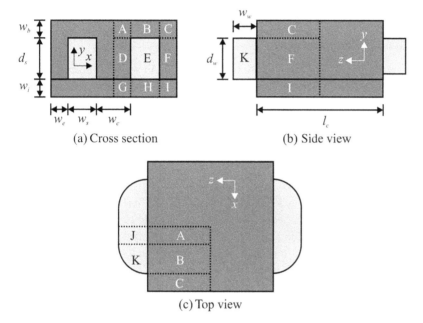

Figure 10.16 Cuboids of EI core electromagnet arrangement.

will incorporate the TEC model into the design process and recalculate the Pareto-Optimal front between mass and loss, but now including a thermal analysis.

Figure 10.16 illustrates a cross section, side view, and front view of the EI-core arrangement. As indicated, we will break the device into 11 cuboidal elements labeled A through I. Cuboids J, K, and E represent the winding bundle; the remainder represents the core.

Assuming thermal symmetry, we will only analyze one-quarter of the device. Note that whether the device is thermally symmetric or not is a function of orientation because heat transfer coefficients are different for vertical versus horizontal surfaces. However, in our initial formulation of the TEC, we will neglect this effect.

Before setting forth the TEC, we will need to address several issues. One of these is the treatment of a rounded corner (element K). We will also need to address the issue of the thermal resistance between the winding and the core. Finally, we will have to consider the thermal resistance across the air gap.

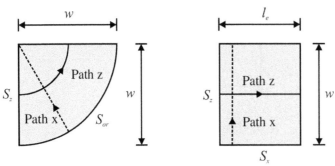

(a) Anisotropic corner element (b) Effective corner element

Figure 10.17 Corner element.

Thermal Representation of a Rounded Corner

Let us begin our deliberations with the consideration of cuboid K. As we have discussed, this cuboid is different from the others in that it is not a rectangular prism. One approach would be to represent this region using a cylindrical element. However, the cylindrical element model derived in Section 10.4 was for an axis symmetric situation in which there is no heat flow in the tangential direction. This is not the case here. Another approach would be to derive a new model starting with the heat flow equation expressed in cylindrical coordinates given by (10.4-6). A third, and simpler, approach is to represent the region as an effective cuboid. This is the approach we will take herein.

The construction of the effective cuboid is depicted in Figure 10.17. Therein, we start with the winding corner element of Figure 10.17(a), which will have anisotropic properties since it will be thermally more conductive in the direction of the conductors than in other directions. The conductors are assumed to be in the direction of the z-path, which will eventually become our z-axis in our effective cuboid. The y-axis is assumed to be out of the page.

The process of obtaining our effective corner element will be to "bend" it into a cuboid as shown in Figure 10.17(b). Therein, the principal parameter to be determined is the effective length, l_e. Choosing

$$l_e = \frac{\pi}{4} w \qquad (10.8\text{-}1)$$

will keep the volume of the two regions the same, keep the area S_z, which is defined to be looking into the cross section along the z-path, to be the same, and keep S_y, the cross section looking into the elements from the y-axis, the same.

The area looking into the direction of path x will not be the same, however. It is readily shown that

$$S_{or} = 2S_x \qquad (10.8\text{-}2)$$

The relationship (10.8-2) makes physical sense in that the cuboid has two surfaces of area S_x while the corner element has a single surface of area S_{or}. This will affect how we connect the element to the other elements and ambient sources. In particular, since S_{or} is in contact with the ambient, it follows that both S_x surfaces (at $x = 0$ and $x = l_x$) should be connected to the ambient.

Winding to Core Resistance

Another issue we will need to address in constructing the TEC is the thermal resistance associated with the core-winding interface. This is depicted in Figure 10.18. Therein, Figure 10.18(a) depicts a core-winding interface, which involves several regions. First, there is often a slot liner that serves to provide additional electrical insulation between the winding and the core, and it prevents the core from abrading the winding. Next, there may be a small air gap between the slot liner and the winding. Such an air gap is considered to have a thickness g_{wc}. Proceeding to the right, the next object is the winding bundle. The air gap and space between conductors may be air, or may be a potting material that helps hold the winding in place and also can improve the thermal conductivity.

Our goal at this point is to find a simple way to represent this interface. The way in which we will do this is to derive an expression for an effective thermal heat transfer coefficient between the core and the winding, denoted h_{cw}. In terms of this coefficient, the thermal resistance between the core and the winding bundle may be expressed as

$$R_{cw} = \frac{1}{S_{cw} h_{cw}} \quad (10.8\text{-}3)$$

where S_{cw} is the contact area from the core to the winding and h_{cw} is the effective heat transfer coefficient.

In order to derive this value, we will utilize the homogenized situation shown in Figure 10.18(b). Therein, the gap between the slot liner and winding is denoted g_{wce}. This effective gap is designed to incorporate g_{wc} if present, but also takes into account the nonuniform surface posed by the conductors. In order to estimate g_{wce} we will require that the area of the air/potting region be the same for Figure 10.18(a) and 10.18(b). This yields

$$g_{wce} = g_{wc} + \frac{4-\pi}{4} r_c \quad (10.8\text{-}4)$$

where r_c is the conductor radius.

The thermal resistance of the homogenized interface may then be expressed as the thermal resistance of the slot liner plus the thermal resistance of the air/potting region. This yields

$$R_{cw} = \frac{1}{S_{cw} h_{sl}} + \frac{g_{wce}}{S_{cw} k_p} \quad (10.8\text{-}5)$$

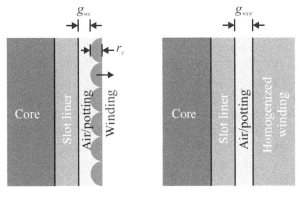

(a) Core-winding interface   (b) Homogenized interface

Figure 10.18 Core-winding interface.

In (10.8-5), the first term is the resistance of the slot liner, and h_{sl} is heat transfer coefficient of the slot liner. The second term is the thermal resistance of the air/potting material, which is assumed to have a thermal conductivity of k_p. Manipulating (10.8-3), (10.8-4), and (10.8-5), we arrive at the expression for effective heat transfer coefficient in (10.8-3). In particular,

$$h_{cw} = \frac{4 h_{sl} k_p}{4 k_p + h_{sl}(4 g_{wc} + (4-\pi) r_c)} \tag{10.8-6}$$

Air Gap Thermal Resistance

One final issue we will need to address is the contact point between the E-core and the I-core. If the gap g is nonzero, we can use a one-dimensional element to compute the thermal resistance of the air between the E-core and U-core. If the E and I core are in contact, then we will use a heat transfer coefficient. In order to make sure that this interface resistance is a continuous function of air gap, one approach is to take the thermal resistance as

$$R_g = \max\left(\frac{g}{k_a S_g}, \frac{1}{h_{cc} S_g}\right) \tag{10.8-7}$$

In (10.8-7), S_g is the thermal cross section of the air gap. Finally k_a and h_{cc} denote the thermal conductivity of air and the core-to-core heat transfer coefficient. This latter quantity varies with pressure, but we will approximate it as a constant.

Thermal Equivalent Circuit Architecture

At this point, we are ready to detail the construction of the TEC. Using the nomenclature of Figures 10.14 and 10.15, the equivalent circuit is depicted in Figure 10.19. Therein, the letters A through I indicate the cuboidal elements corresponding to those in Figure 10.16. The numbers listed are node numbers, and an O denotes an open-circuit connection to one of the element's nodes.

In general (but with exceptions), the x-axis is to the right and the y-axis is up. However, elements J and K, which are in front of elements D and E, are shown below those elements. The temperature T_a denotes ambient temperature.

Let us first consider elements A, B, and C. Observe that the $0x$ node of element A is considered to be open since no heat flow will cross the central plane of the electromagnet by thermal symmetry. The lx node of A is linked to the $0x$ node of element B, the lx node of element B is linked to the $0x$ node of element C, and the lx-node of the C element is connected to the ambient. The ly and $0z$ nodes of these three elements, which correspond to the top faces in Figure 10.16(c) and front faces shown in Figure 10.16(a), respectively, are also connected to ambient. The lz nodes are open circuited because heat will not flow across the corresponding boundary by symmetry. Consideration of the surface elements of these elements yields the thermal resistance values shown in Table 10.2. Therein, h_{ca} denotes the heat transfer coefficient from the core to air.

Next let us consider the row of elements D, E, and F. In this case, elements D and F are core material, while element E represents the winding bundle. This element is based on a homogenized representation of the winding, insulator, and air using the approach of Section 10.5. In applying that

10.8 Case Study: Thermal Model of Electromagnet

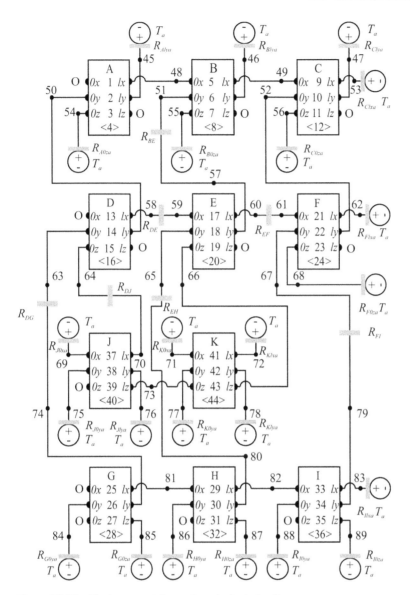

Figure 10.19 Electromagnet thermal equivalent circuit.

approach, w and d in (10.5-1)–(10.5-32) are taken to be w_w and d_w, respectively. The x- and y-direction within the elements are taken to be orthogonal to the direction of the conductors, which are parallel to the z-direction. This matches the treatment of the rest of the elements in the case of element E, but this will not be the case when we consider elements J and K.

The lx node of D is connected to the $0x$ node of E through thermal resistance R_{DE} and the lx node of E is connected to the $0x$ node of F through R_{EF}. These thermal resistances are present because of the differences between materials and are calculated using (10.8-3)–(10.8-6). When calculating R_{DE}, we take $g_{wc} = 0$ since the winding is wound around this part of the core and $g_{wc} = w_s - w_w$ when calculating R_{EF}. In both cases $S_{cw} = d_w l_c / 2$.

Table 10.2 Thermal Resistances to Ambient

Elements A–C:

$$R_{A l y a} = \frac{4}{h_{ca} w_c l_c} \qquad R_{B l y a} = \frac{2}{h_{ca} w_s l_c} \qquad R_{C l y a} = \frac{2}{h_{ca} w_e l_c} \qquad R_{C l x a} = \frac{2}{h_{ca} w_b l_c}$$

$$R_{A 0 z a} = \frac{2}{h_{ca} w_c w_b} \qquad R_{B 0 z a} = \frac{1}{h_{ca} w_s w_b} \qquad R_{C 0 z a} = \frac{1}{h_{ca} w_b w_e}$$

Elements D–F:

$$R_{F l x a} = \frac{2}{h_{ca} d_s l_c} \qquad R_{F 0 z a} = \frac{1}{h_{ca} d_s w_e}$$

Elements J–K:

$$R_{J 0 x a} = \frac{2}{h_{ca} w_c d_w} \qquad R_{K 0 x a} = \frac{4}{h_{ca} \pi w_w d_w} \qquad R_{K l x a} = \frac{4}{h_{ca} \pi w_w d_w}$$

$$R_{J 0 y a} = \frac{2}{h_{ca} w_c w_w} \qquad R_{J l y a} = \frac{2}{h_{ca} w_c w_w} \qquad R_{K 0 y a} = \frac{4}{h_{ca} \pi w_w^2} \qquad R_{K l y a} = \frac{4}{h_{ca} \pi w_w^2}$$

Elements G–I:

$$R_{G 0 y a} = \frac{4}{h_{ca} l_c w_c} \qquad R_{H 0 y a} = \frac{2}{h_{ca} l_c w_s} \qquad R_{I l x a} = \frac{2}{h_{ca} l_c w_i} \qquad R_{I 0 z a} = \frac{1}{h_{ca} w_i w_e}$$

$$R_{G 0 z a} = \frac{2}{h_{ca} w_c w_i} \qquad R_{H 0 z a} = \frac{1}{h_{ca} w_s w_i} \qquad R_{I 0 y a} = \frac{2}{h_{ca} l_c w_e}$$

Observe that, in general, the *ly* nodes of D–F are connected to the *0y* nodes of nodes A–C, and that the *0y* nodes of D–F are connected to *ly* nodes of G–H. Recall elements J and K are spatially in front of elements D and E. Element E is connected to the elements above and below, B and H, by resistances R_{BE} and R_{EH}. The first of these is calculated using (10.8-3)–(10.8-6) with $g_{wc} = 0$ and $S_{cw} = w_w l_c/2$. The reluctance R_{HE} is calculated as a one-dimensional element through air using (10.2-28) with $l_{Ax} = d_s - d_w + g$, $S_{Ax} = w_w l_c/2$, and $k_{Ax} = k_a$.

Elements D and F are connected to elements G and I in the *y*-direction by thermal resistances R_{DG} and R_{FI}. Note that the respective elements may or may not be in direct contact, depending upon the air gap *g*. Their thermal resistances are calculated using (10.8-7) with $S_g = l_c w_c/4$ when calculating R_{DG} and $S_g = l_c w_e/2$ in the case of R_{FI}.

The next row in our TEC are elements J and K, which both represent the winding, and are in front of elements D and E in Figure 10.16(a). Because these elements are anisotropic, their orientation is important. Recall that when homogenizing the winding, the direction of the conductors is taken (in a local sense) to be in the *z*-axis. In the case of element J, our local *x*- and *z*-axes are in the direction of the electromagnet *z*- and *x*-axes, respectively. In the case of element K, the situation is more subtle, because the conductors bend. At the *0x*, *0y*, and *0z* nodes of element K, the alignment of the axes is as in element J. However, for the *lx*, *ly*, and *lz* nodes of element K, the local coordinate system axes have bent to match those of the system.

All *y*-nodes of elements J and K have a thermal resistance to the ambient. The *0x* node of element J and the *0x* and *lx* nodes of element K also have a thermal resistance to ambient. All of these resistances are itemized in Table 10.2. In the case of element K, some explanation is in order. Recall that this element is bent as illustrated in Figure 10.17. In order to achieve the same surface area to ambient in the *x*-direction in Figure 10.17(b) as is presented by the area S_{or} in Figure 10.17(a), both of the *x*-nodes have been connected to the ambient temperature.

A final point that merits comment is that the *lx* node of element J is connected to the *0z* node of element D through R_{DJ}. Again, it is important to remember that the coordinate system of element J has been rotated. This thermal resistance may be calculated using (10.8-3) where $S_{cw} = w_c d_w/2$ and

h_{cw} is given by (10.8-3)–(10.8-6) with $g_{wc} = 0$ and h_{sl} and k_p being the heat transfer coefficient of the slot liner and thermal conductivity of air or the potting material, respectively.

The final row of elements is that including G, H, and I. As can be seen, all $0y$ and lz nodes are connected to the ambient, as is the lx node of element I. The open-circuited nodes have no heat transfer by way of symmetry. The elements G, H, and I are connected in the y-direction to D, E, and F, respectively, through thermal resistances R_{DG}, R_{EH}, and R_{FI}, respectively, as discussed previously in this section.

At this point, the description of the TEC is complete. Observe that even though the electromagnet is in a simple device, its TEC, with 89 nodes, is much more involved than the magnetic equivalent circuit for the same device.

Electro-Thermal Analysis

The TEC of an electromagnet was set forth in the previous section. However, one aspect of the circuit not discussed was the calculation of the losses in the various elements.

Because the excitation of the electromagnet is dc, losses will only occur in elements J, K, and E, and may be expressed as

$$P_\Omega = R_\Omega i^2 \tag{10.8-8}$$

where P_Ω and R_Ω are the power loss and the portion of the coil resistance associated with element Ω, and where for this example Ω is J, K, or E. The resistance R_Ω may be expressed as

$$R_\Omega = \frac{l_{c\Omega}}{a_c \sigma_\Omega} \tag{10.8-9}$$

where $l_{c\Omega}$ is the total length of conductor in element Ω, a_c is the cross section of the conductor material (which is assumed to be constant in all elements), and σ_Ω is the conductivity of the conductor in element Ω.

The length of the conductor in element Ω may be expressed as

$$l_{c\Omega} = \frac{V_{c\Omega}}{a_c} \tag{10.8-10}$$

where $V_{c\Omega}$ is the volume of conductor within the element. Combining (10.8-8) through (10.8-10), we obtain

$$P_\Omega = \frac{V_{c\Omega}}{\sigma_\Omega} J^2 \tag{10.8-11}$$

where J is the current density given by

$$J = \frac{i}{a_c} \tag{10.8-12}$$

The same conductor is assumed to be used in all elements since the coil is continuous. However, the resistivity of each element can be different since the resistivity will be a function of temperature. Treating the temperature within the element as the mean temperature, the resistivity may be approximated as

$$\sigma_\Omega = \frac{\sigma_{0c}}{1 + \alpha_c(\langle T_\Omega \rangle - T_{0c})} \tag{10.8-13}$$

where σ_{0c} is the conductivity of the conductor at a temperature T_{0c} and α_c is a temperature coefficient for the conductor.

The total electrical resistance of the coil may be calculated as

$$R = \frac{4(V_J + V_K + V_E)}{a_c^2 \sigma_{cl}} \tag{10.8-14}$$

where σ_{cl} is the mean conductivity of the coil given by (10.8-13) with Ω taken as cl and where $\langle T_{cl} \rangle$ is the mean temperature of the coil given by

$$\langle T_{cl} \rangle = \frac{\langle T_J \rangle V_J + \langle T_K \rangle V_K + \langle T_E \rangle V_E}{V_J + V_K + V_E} \tag{10.8-15}$$

Using the total resistance, the coil current may be expressed as

$$i = \frac{V}{R} \tag{10.8-16}$$

In considering (10.8-8)–(10.8-16), the coupled nature of the electro-thermal problem is evident. The element temperatures are a function of loss, the loss is a function of temperature and current, the current is a function of applied voltage and electrical resistance, and the electrical resistance is a function of temperature.

Table 10.3 lists pseudo-code for an algorithm to conduct the electro-thermal analysis. Therein, lines 1–10 involve initialization, while the remaining code carries out the iterative analysis. In line 1, it is first assumed that the power dissipation in elements J, K, and E are zero. With this assumption, in line 2 $\mathbf{P}_N$ and $\mathbf{G}_N$ are calculated using the TNAFA, which is listed in Table 10.1. In line 3, copies of these matrices are formulated for later use. In line 4, present and previous temperature vectors, whose elements are the temperature at each node, are initialized to the ambient temperature. The mean coil temperature is also initialized as the ambient temperature.

Based on the assumed mean coil temperature, the conductivity, resistance, current, and current density are found sequentially as indicated in lines 5–8. Note that these are initial estimates.

The iteration counter and error are initialized in lines 9 and 10; after this, the main body of the electro-thermal iteration occurs. The first step in doing this is to assign $\mathbf{P}_N$ to be equal to the initial value. Next, the conductivity and power dissipation with the winding elements J, K, and E are updated in lines 13–18, and the vector of node powers $\mathbf{P}_N$ is updated in lines 19–21 in accordance with adding a power source using the algorithm of Table 10.1. The updated nodal temperature vector is then computed in line 22.

Lines 23–27 focus on a revised electrical analysis by first finding the mean coil temperature, then computing the coil conductivity, resistance, current, and current density. After this has been done, the error is computed by comparing the present temperature vector estimate to the previous. The error is defined to be that of the element with the maximum absolute value of difference. Finally, the previous temperature estimated is updated to be that of the present estimate, and the iteration counter is updated.

Although not shown in the pseudo-code, post-processing steps, such as computing the maximum temperature over surface elements, can be readily carried out by inspecting the appropriate elements of the temperature vector. Likewise, the peak temperature within each element can be readily found within each element as discussed in Sections 10.2-10.4.

Table 10.3 Electrothermal Analysis Algorithm

```
01. set P_J = 0, P_K = 0, P_E = 0
02. calculate P_N and G_N Using Thermal Nodal Analysis
    Formulation Algorithm (Table 10.1)
03. G_N0 = G_N, P_N0 = P_N
04. set T = T_A, T_old = T_A, and ⟨T_cl⟩ = T_A
05. calculate σ_cl using ⟨T_cl⟩ and (10.8-13)
06. calculate R using (10.8-14)
07. calculate i using (10.8-16)
08. calculate J using (10.8-12)
09. k = 1
10. e = 0
11. while (k = 1) or ((k ≤ k_max) and (e > e_max))
12.     P_N = P_N0
13.     calculate σ_J using (10.8-13) and ⟨T_J⟩ = T_20
14.     calculate P_J using ρ_J and V_cJ using (10.8-11)
15.     calculate σ_K using (10.8-13) and ⟨T_K⟩ = T_40
16.     calculate P_K using ρ_K and V_cK using (10.8-11)
17.     calculate σ_E using (10.8-13) and ⟨T_E⟩ = T_44
18.     calculate P_E using ρ_E and V_cE using (10.8-11)
19.     P_N,20 = P_N,20 + P_J
20.     P_N,40 = P_N,40 + P_K
21.     P_N,44 = P_N,44 + P_E
22.     solve G_N T = P_N
23.     calculate ⟨T_cl⟩ using (10.8-15)
24.     calculate σ_cl using ⟨T_cl⟩ and (10.8-13)
25.     calculate R using (10.8-14)
26.     calculate i using (10.8-16)
27.     calculate J using (10.8-12)
28.     e = max (|T_old - T|)
29.     T_old = T
30.     k = k + 1
31. end
```

Examples

We will use the electromagnet TEC in two examples. In Example 10.8A, we will conduct a thermal analysis of the electromagnet design considered in Chapter 5. Recall that this design was conducted without a thermal analysis. We will find that as a result of thermal heating, the resistance of the coil will be higher than anticipated in Chapter 5, resulting in less current and less force. In Example 10.8B, we will incorporate our thermal analysis into the design process and recalculate the Pareto-optimal front.

Example 10.8A In this example, we will perform a thermal analysis of the electromagnet design considered in Section 5.4. In particular, we will consider the design shown in Figure 5.16, whose parameters are listed in Table 5.4. This design featured a modest current density of 3.03 A/mm². Performing a thermal analysis will require some additional information, which is listed in Table 10.4. Therein, t_i is the thickness of the wire insulation, and g_{wc} is the gap between the winding and slot liner in the inner post of the device, which is assumed to be zero. The terms k_{max} and e_{max} are the maximum number of iterations and allowed error in the thermal analysis, respectively. The second column lists assumed heat transfer coefficients. The terms h_{cc}, h_{sl}, h_{ca}, and h_{wa} are the heat transfer coefficient from core-to-core, of the slot liner, from the core to air, and from the winding to air, respectively. The heat transfer coefficient from winding to air is assumed to be higher than that from core to air since the wire will present a larger surface area. Note that surfaces would ideally be given different heat transfer coefficients depending upon their orientation with respect to gravity since the air flow will vary depending upon orientation, but that refinement is not used herein. The last column lists assumed values of thermal conductivity. In particular, k_c, k_{H50}, k_a, and k_i are the thermal conductivity of copper, Hyperco 50, air, and wire insulation, respectively. Another possible refinement is adjusting the core thermal conductivity to be different in the direction across the laminations than in the plane of the lamination, since the laminations insulation will decrease the thermal conductivity in that direction. That refinement is not included herein.

The first step in performing the analysis is computing the homogenized parameters of the winding region. Following the procedure of Section 10.5, we obtain $k_{wx} = k_{wy} = 0.188$ W/m·K and $k_{wz} = 269$ W/m·K.

At this point, we may perform our analysis using the algorithm shown in pseudo-code form in Table 10.3. It is interesting to view the convergence of the error. From $k = 1$ to $k = 7$, we obtain $e = 150$ K, 37.1 K, 6.42 K, 1.16 K, 0.208 K, 37.5 mK, and 6.7 mK. While the rate of convergence could be improved by using a Newton–Raphson algorithm, the number of iterations required does not seem excessive.

The maximum temperature recorded at any node is 417 K or 144°C; however, it must be remembered that the maximum temperature over the set of nodes does not correspond to the maximum temperature in the device because the temperature profile within each element is parabolic. The maximum temperature within the device is actually 379 K or 106°C. This temperature occurs in element K which corresponds to the end-turns of the coil. The peak temperature of adjacent element J, also part of the end-turns, is nearly identical; the peak temperature of the element E, which corresponds to the conductor within the slot, is 369 K or 96°C.

The peak surface temperature of the device is important as a safety consideration. This is found by inspecting the temperature at all nodes connected to ambient. The highest surface temperature is 338 K or 65°C, which would be unpleasant.

Table 10.4 Electromagnet Thermal Parameters

$t_i = 27$ μm	$h_{cc} = 500$ W/m²·K	$k_c = 385$ W/m·K
$g_{wc} = 0$ m	$h_{sl} = 560$ W/m²·K	$k_{H50} = 29.8$ W/m·K
$k_{max} = 100$	$h_{ca} = 15$ W/m²·K	$k_a = 26.3$ mW/m·K
$e_{max} = 0.01$ K	$h_{wa} = 25$ W/m²·K	$k_i = 400$ mW/m·K

As a result of the increase in coil temperature, the resistance of the coil increases from 4.82 to 5.99 Ω; and as a result the current will fall from 2.49 to 2.00 A, which will reduce the attractive force associated with the electromagnet, at least after it has reached thermal equilibrium. In order to create a design in which this is taken into account, let us reconsider the design process and include thermal analysis is the design. This is the topic of the next example.

Example 10.8B In this example, we will reconsider the design example of the electromagnet carried out in Section 5.3. We will use the same specifications as are given in Table 5.2, and use the same design space as specified in Table 5.3, with the one exception, which is that the maximum allowed number of turns will be increased from 10^3 to 10^4. The design problem for the electromagnet features six design constraints, and these will also remain unchanged, except for constraint 3. Originally, this constraint was imposed as a limit on current density. Because we are performing a thermal analysis, it is not necessary to limit current density. However, we will limit the peak winding temperature. Thus, our third constraint will become

$$c_3 = \text{lte}(T_{wpk}, T_{wmxa}) \tag{10.8B-1}$$

where T_{wpk} is the peak winding temperature (evaluated based on elements J, K, and E) and T_{wmxa} will be a new design specification which is the maximum allowed winding temperature. We will assume an ambient temperature of 298 K or 25°C.

The pseudo-code for the calculation of the fitness function is listed in Table 10.5. Except for the first three lines, which now incorporate a thermal analysis, this code is identical to the pseudo-code which was given in Table 5.1 in which we first considered the electromagnet.

Thermal data needed to support the electro-thermal analysis is listed in Table 10.6. In addition to that data, we will need thermal conductivity for all the possible core and conductor materials. This is listed in Table 10.6. Therein k_c and k_a are the thermal conductivity of copper and aluminum; k_{M19}, k_{M36}, k_{M43}, k_{M47}, and k_{H50} are the thermal conductivities of samples of the steels which fall into the respective classes.

The resulting Pareto-optimal front of this study is shown in Figure 10.20. Therein, several fronts are shown. The front labeled "Designs w/o Thermal Analysis" is the front obtained in Chapter 5, in which no thermal analysis was carried out. The front labeled $T_{wpk} < 180°C$ was obtained by setting $T_{wmxa} = 180°C$. Pareto-optimal fronts obtained with $T_{wmxa} = 155°C$, $T_{wmxa} = 130°C$, and $T_{wmxa} = 105°C$ are also shown. These temperature breakpoints correspond to different NEMA (National Electrical Manufacturers Association) insulation classes.

There are several points of interest in Figure 10.20. First, note that the designs conducted with the thermal analysis appear to be worse than those that do not. However, this is misleading: The designs that included thermal analysis are based on a more realistic model. Recall from Example 10.8A that the designs for Chapter 5 are not likely to meet specifications once the design reaches thermal equilibrium. Because conductivity goes down with increasing temperature, the designs that included thermal analysis take into account the higher resistance that would result from the increase in conductor temperature.

Another point of interest in Figure 10.20 is that as the peak allowed temperature is increased, the volume of the design can be reduced. Observe that all the Pareto-optimal fronts that include thermal analysis coincide; the difference is how low a volume can be obtained.

In the remainder of this example, we will consider the Pareto-optimal front in which $T_{wmxa} = 180°C$, so that we see the broadest range of objective function. The final parameter

Table 10.5 Calculation of Fitness Function

```
calculate V_E
conduct electro-thermal analysis using TEC
calculate P_E
calculate constraints c₁ through c₆
C_S = c₁ + c₂ + c₃ + c₄ + c₅ + c₆
C_I = 6
if (C_S < C_I)
```
$$\mathbf{f} = \varepsilon\left(\frac{C_s - C}{C}\right)\begin{bmatrix}1\\1\end{bmatrix}$$
```
    return
end
construct MEC and calculate c₇ and c₈
C_s = C_s + c₇ + c₈
C_I = C_I + 2
if (C_S < C_I)
```
$$\mathbf{f} = \varepsilon\left(\frac{C_s - C}{C}\right)\begin{bmatrix}1\\1\end{bmatrix}$$
```
    return
end
```
$$f = \begin{bmatrix}\frac{1}{V_E}\\ \frac{1}{P_E}\end{bmatrix}$$
```
return
```

Table 10.6 Thermal Conductivities

k_c = 385 W/m·K	k_{M19} = 16.7 W/m·K	k_{M36} = 18.8 W/m·K
k_a = 205 W/m·K	k_{M43} = 20.9 W/m·K	k_{M47} = 37.7 W/m·K
		k_{H50} = 29.8 W/m·K

distribution for the study, which used a population size of 2000 over 4000 generations, is shown in Figure 10.21. Like the study of Chapter 5, the core material chosen and conductor material chosen consistently had normalized values of 1 and 0, which correspond to Hyperco50 and copper, respectively.

The current density versus volume is shown in Figure 10.22. Compared to the designs of Chapter 5, the current density is somewhat higher for a given volume. The large increase in current density at a volume of 0.93 L is due to a change in conductor size.

In Figure 10.23, the conductor counts are shown as a function of volume. Variables depicted include N_w, the width of the winding in terms of conductors, N_d, the depth of the winding in terms of conductors, and sqrt(N), the square root of the number of turns. The variable N_d increases

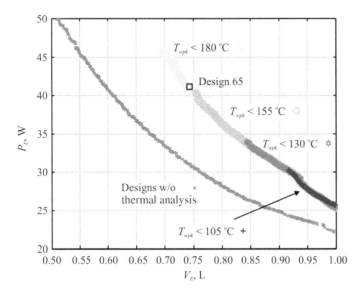

Figure 10.20 Pareto-optimal fronts.

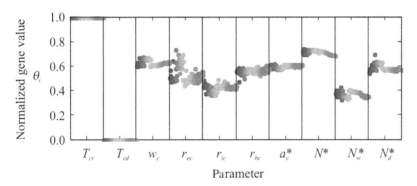

Figure 10.21 Gene distribution.

gradually with volume but undergoes a step decrease every time N_w increases. In general, the number of turns increases with volume. The discontinuity at a volume of 0.93 L is caused by a change in conductor size.

Electromagnet width dimensions w_i, w_c, w_e, and w_b are shown in Figure 10.24. These widths are fairly consistent with volume, except at the point where the conductor size changes.

Figure 10.25 shows additional electromagnet dimensions versus volume. Variables depicted include the core length l_c, the slot depth d_s, 100 times the conductor radius $100r_c$, the square root of the slot area a_s, the width of the core w_c, and the width of the slot w_s. The correlation of N_d with l_c and d_s is evident. The change in conductor radius at a volume of 0.93 L can also be seen.

Figure 10.26 illustrates the peak winding temperature versus volume. In general, temperature goes down with increasing volume, though not perfectly monotonically due to the discrete nature of some of the design variables. The change in conductor size causes a particularly noticeable irregularity.

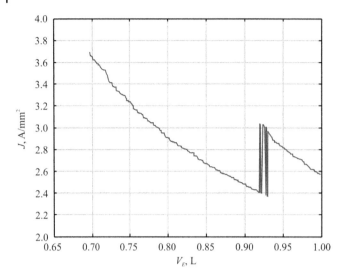

Figure 10.22 Current density versus volume.

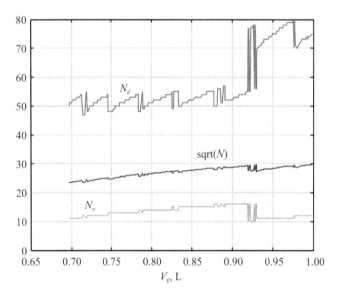

Figure 10.23 Conductor counts versus volume.

Table 10.7 lists design information for Design 65 from the Pareto-optimal front. This design has essentially the same volume as Design 250 from the Pareto-optimal front discussed in Chapter 5. Comparing the parameters in Table 10.7 to those in Table 5.4, one can see that although the volume is the same, the power dissipation of the Design 65 is higher than that of Design 250, despite using fewer turns of larger diameter wire. Design 65 is also somewhat taller and narrower than Design 250.

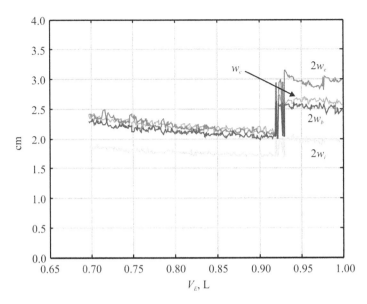

Figure 10.24 Widths versus volume.

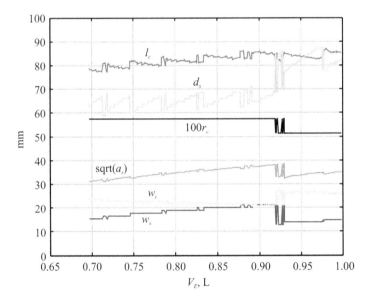

Figure 10.25 Assorted dimensions versus volume.

At this point, it is natural to consider what strategies may be used to improve the performance of the device. One possibility is to embed the winding in thermally conductivity potting material to improve the heat transfer from the winding to the core. An important refinement to the design process would be to add a constraint limiting the surface temperature. Both of these refinements are straightforward.

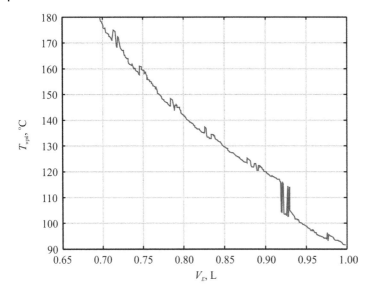

Figure 10.26 Peak winding temperature versus volume.

Table 10.7 Design 65 Data

El Core Data	Winding Data	Metrics
Material = Hiperco50	Material = Copper	V_E = 742 mL
w_c = 2.31 cm	AWG = 17	P_E = 41.1 W
w_e = 1.14 cm	a_c = 1.04 mm^2	i_E = 3.43 A
w_i = 9.00 mm	N = 624	J_E = 3.30 A/mm^2
w_b = 1.09 cm	N_w = 12	R_E = 3.50 Ω
w_s =1.65 cm	N_d = 53	h_E = 8.70 cm
d_s = 6.60 cm	d_w = 6.40 cm	w_E = 7.90 cm
l_c = 7.90 cm	w_w = 1.45 cm	l_E = 10.8 cm
		T_{wpk} = 159°C

References

1 R. Wrobel and P. H. Mellor, A general cuboidal element for three-dimensional thermal modeling, *IEEE Transactions on Magnetics*, vol. **46**, no. 8, pp. 3197–3200, 2010.
2 M. Hettegger, O. Biro, A. Stermecki and G. Ofner, Temperature rise determination of an induction motor under blocked rotor conditions, in *5th IET International Conference on Power Electronics, Machines, and Drives (PEMD 2010)*, 2010.
3 P. H. Mellor, D. Roberts and D. R. Turner, Lumped parameter thermal model for electrical machines of TEFC design, *IEE Proceedings-B*, vol. **138**, no. 5, pp. 205–218, 1991.
4 T. L. Bergman, A. S. Lavine, F. P. Incropera and D. P. Dewitt, *Fundamentals of Heat and Mass Transfer*, 7th edition, Hoboken: John Wiley & Sons, Inc., 2011.

5 D. Staton, A. Boglietti and A. Cavagnino, Solving the more difficult aspects of electric motor thermal analysis in small and medium size industrial induction motors, *IEEE Transactions on Energy Conversion*, vol. **20**, no. 3, pp. 620–628, 2005.

6 S. D. Sudhoff, *MATLAB Codes For Power Magnetic Devices: A Multi-Objective Design Approach*, [online], 2nd edition. Available: http://booksupport.wiley.com.

Problems

1 Using (10.2-8)–(10.2-11) show that we obtain (10.2-12)–(10.2-13).

2 Using (10.2-6),(10.2-10), and (10.2-11) show that we obtain (10.2-14).

3 Using (10.2-12)–(10.2-14) show that we obtain (10.2-15)–(10.2-17).

4 Suppose the temperature within a sample of material is given by

$$T = 200 + 7x - 2x^2 + 10y - y^2 + 10z - 20z^2$$

where $0 \leq x \leq 1$, $0 \leq y \leq 2$, $0 \leq z \leq 0.5$, and the temperature is given in Kelvin. Find the mean temperature of the sample and the peak temperature of the sample.

5 Consider Problem 4. Determine $\langle T_{0x} \rangle$ and $\langle T_{lx} \rangle$.

6 Rework Example 10.2B if the length of the bar is reduced to 5 cm, and all other parameters are the same.

7 Consider Example 10.2B. How much power can be dissipated within the bar if the peak temperature in the bar is not to exceed 100°C?

8 From (10.3-2) obtain (10.3-4).

9 From (10.3-2) obtain (10.3-5).

10 Using (10.3-3) and (10.3-4) show (10.3-6).

11 Using (10.3-3) and (10.3-5) show (10.3-7).

12 Using (10.1-7), (10.1-8), (10.3-8), and (10.4-12) derive (10.3-9) and (10.3-10).

13 Using (10.2-10), (10.3-6), (10.3-7), (10.3-11), and (10.3-12) show that we obtain (10.3-13)–(10.3-15).

14 Using (10.4-8) obtain (10.4-9).

15 Using (10.4-8), (10.4-10), and (10.4-11) show (10.4-12).

16 From (10.4-8) to (10.4-10) show (10.4-13) and (10.4-14).

17 Using (10.4-12)–(10.4-14) and (10.4-18)–(10.4-19) obtain (10.4-20)–(10.4-23).

18 Using (10.4-8) and (10.4-10) show (10.4-24) and (10.4-25).

19 Comparing (10.4-12) to (10.4-24) and (10.4-25) obtain (10.4-26) and (10.4-27).

20 Using (10.4-26), (10.4-31), and (10.4-32) obtain (10.4-33)–(10.4-36).

21 Rework Example 10.5A if a potting material with the same thermal conductivity as the insulation is used in place of air in the slot.

22 Suppose a core material is at 150°C, and is in a room with a 25°C ambient temperature. Assume the heat transfer coefficient from the core to air is 10 W/m²·K, and the emissivity of the core material is 0.5. Compare the heat flow due to convection to that due to radiation.

23 Derive (10.8-4).

24 Using (10.8-3)–(10.8-5) derive (10.8-6).

25 Consider Example10.8B. Rework this example assuming a potting material with the same thermal conductivity as the insulation is used in place of the air in the slot. Assume the density of potting material is 4000 kg/m³. You may make use of the code set forth in [6].

11

Alternating Current Conductor Losses

In this chapter, we consider ac conductor losses. Two principal loss mechanisms are considered. The first is skin effect, which is the tendency of an ac current to exhibit a current density within a conductor that is higher toward the outside of the conductor than toward the interior of the conductor. We will consider this effect in the context of both strip conductors and then round conductors. Next, we will focus our attention on proximity effect. In essence, when a conductor is exposed to a time-varying field, an eddy current is induced in the conductor that results in loss. The term proximity effect is employed because the source for this external field is often the surrounding conductors. A general formulation to calculate proximity effect is given, followed by some results for particular geometries. The chapter concludes by considering the ac conductor losses in rotating electric machinery and in a UI-core inductor.

11.1 Skin Effect in Strip Conductors

Skin effect is a phenomenon in which current density becomes concentrated at the outside of a conductor rather than being uniformly distributed. The high concentration of current density at the outside of the conductor leads to higher ohmic losses than would otherwise occur. Skin effect is associated with AC currents and becomes increasingly pronounced as frequency increases. In this section, we will begin our work by studying strip conductors. This is for two reasons. First, in some applications, strip conductors can be used instead of round conductors, and they will have lower skin-effect losses for the same conductor area. Second, skin effect in strip conductors is easier to analyze and thus makes a good introduction to the topic.

Consider a thin strip of conductor as presented in Figure 11.1(a). Therein, V_{+-} is the voltage drop across the conductor and I is the current into the conductor, which flows to the right. The conductor has length l and a thickness $2R$, where R is used because it will be suggestive of radius when we consider cylindrical conductors in the next section. The cross section of the conductor is presented in Figure 11.1(b). As can be seen, the cross section of the conductor is width w by thickness $2R$. We will assume that the strip conductor dimensions are such that the length is much greater than the width, which is, in turn, much greater than the thickness, that is, $l \gg w \gg 2R$.

In Figure 11.1, r is the distance above the center plane of the conductor measured in the z-axis as indicated. The variables J_y and E_y will be used to denote the current density and electric field in the assumed direction of the y-axis. The quantities H_x and B_x will denote the field intensity and flux density in the x-axis. These variables will all be a function of time and of position r.

Power Magnetic Devices: A Multi-Objective Design Approach, Second Edition. Scott D. Sudhoff.
© 2022 The Institute of Electrical and Electronics Engineers, Inc. Published 2022 by John Wiley & Sons, Inc.
Companion website: www.wiley.com/go/sudhoff/Powermagneticdevices

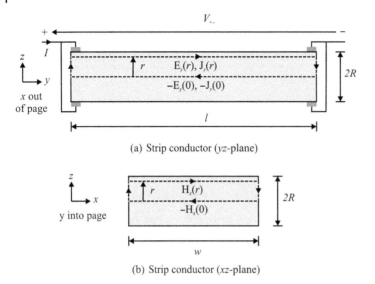

(a) Strip conductor (yz-plane)

(b) Strip conductor (xz-plane)

Figure 11.1 Strip conductor.

Let us consider the loop formed by the dashed line in Figure 11.1(a). From Maxwell's equations in integral form, we have

$$\oint \mathbf{E} \cdot d\mathbf{l} = -\int_S \frac{\partial \mathbf{B}}{\partial t} \cdot d\mathbf{S} \tag{11.1-1}$$

Applying (11.1-1) to the dashed path in Figure 11.1(a), while ignoring end effects (the contribution of the z-component of the fields at the $y = 0$ and $y = l$ conductor end portions of the path integral), we have

$$E_y(r)l - E_y(0)l = -\int_0^r -\frac{\partial B_x}{\partial t} l\, dr \tag{11.1-2}$$

The minus sign in (11.1-2) arises from the fact that the path in Figure 11.1(a) defines the flux density as being into the page, whereas the direction of the x-axis is out of the page. Simplifying (11.1-2), we have

$$E_y(r) - E_y(0) = \int_0^r \frac{\partial B_x}{\partial t} dr \tag{11.1-3}$$

In the frequency domain, (11-3) becomes

$$\widetilde{E}_y(r) - \widetilde{E}_y(0) = j\omega \int_0^r \widetilde{B}_x dr \tag{11.1-4}$$

Differentiating (11.1-4) with respect to r, we have

$$\frac{d\widetilde{E}_y}{dr} = j\omega \widetilde{B}_x \tag{11.1-5}$$

11.1 Skin Effect in Strip Conductors

The expression (11.1-5) is a fundamental relationship we will need to analyze skin effect. Let us next consider applying Ampere's law to the dashed path in Figure 11.1(b). Ampere's law states the following:

$$\oint H \cdot dl = i_{enc} \tag{11.1-6}$$

Ignoring the edges of the path (in essence, assuming $w \gg 2R$) and assuming H only varies in the z-axis, we have

$$w(H_x(r) - H_x(0)) = \int_0^r J_y(r) w \, dr \tag{11.1-7}$$

In (11.1-7), the field intensity and current density are written as functions to emphasize their dependence on position. Taking the derivative of (11.1-7) with respect to position, we have

$$\frac{\partial H_x}{\partial r} = J_y \tag{11.1-8}$$

For sinusoidal steady-state conditions, (11.1-8) may be expressed as

$$\frac{d\tilde{H}_x}{dr} = \tilde{J}_y \tag{11.1-9}$$

where the partial derivative is replaced by the total derivative operator since (11.1-9) is not a function of time.

The (11.1-5) and (11.1-9) are both key points in our development. To tie these together, we will utilize the material relationships

$$J_y = \sigma E_y \tag{11.1-10}$$

and

$$B_x = \mu H_x \tag{11.1-11}$$

where σ and μ are the conductivity and permeability, respectively, of the conductor (which is assumed to be magnetically linear). The (11.1-10) and (11.1-11) also apply to phasor representations.

Manipulating (11.1-5), (11.1-10), and (11.11), the field intensity may be expressed as

$$\tilde{H}_x = \frac{1}{j\omega\mu\sigma} \frac{d\tilde{J}_y}{dr} \tag{11.1-12}$$

Substitution of (11.1-12) into (11.1-9) yields

$$\frac{d^2 \tilde{J}_y}{dr^2} = j\omega\mu\sigma \tilde{J}_y \tag{11.1-13}$$

The expression (11.1-13) governs the current distribution with the conductor.

In order to solve (11.1-13), we will use D to denote differentiation with respect to r, whereupon the characteristic polynomial of (11.1-13) may be expressed as

$$D^2 - j\omega\mu\sigma = 0 \tag{11.1-14}$$

which has roots at

$$D = \pm \sqrt{j\omega\sigma\mu} \tag{11.1-15}$$

Defining the skin depth δ as

$$\delta = \sqrt{\frac{2}{\omega\sigma\mu}} \qquad (11.1\text{-}16)$$

and noting that

$$\sqrt{j} = \frac{1+j}{\sqrt{2}} \qquad (11.1\text{-}17)$$

the roots of the characteristic polynomial may be written as

$$D = \pm d \qquad (11.1\text{-}18)$$

where

$$d = \frac{1+j}{\delta} \qquad (11.1\text{-}19)$$

The solution of (11.1-13) may now be written as

$$\widetilde{J}_y(r) = J_1 e^{dr} + J_2 e^{-dr} \qquad (11.1\text{-}20)$$

where J_1 and J_2 are complex valued constants.

In order to solve for J_1 and J_2, note that from (11.1-20), we have

$$\frac{d\widetilde{J}_y(r)}{dr} = d\left(J_1 e^{dr} - J_2 e^{-dr}\right) \qquad (11.1\text{-}21)$$

From (11.1-21), we have

$$\frac{d\widetilde{J}_y(0)}{dr} = d(J_1 - J_2) \qquad (11.1\text{-}22)$$

and

$$\frac{d\widetilde{J}_y(R)}{dr} = d\left(J_1 e^{dR} - J_2 e^{-dR}\right) \qquad (11.1\text{-}23)$$

We will need (11.1-22) and (11.1-23) to compute J_1 and J_2. However, before we can do this, we will also need some additional expressions. Rearranging (11.1-12), we obtain

$$\frac{d\widetilde{J}_y(r)}{dr} = j\omega\mu\sigma\widetilde{H}_x(r) \qquad (11.1\text{-}24)$$

By symmetry, at $r = 0$, there is no x-axis component of the field intensity. From (11.1-24), it follows that

$$\frac{d\widetilde{J}_y(0)}{dr} = 0 \qquad (11.1\text{-}25)$$

From (11.1-22) and (11.1-23), we have

$$J_2 = J_1 \qquad (11.1\text{-}26)$$

Next, since that path enclosed by the dashed loop in Figure 11.1(b) encloses half the current, from Ampere's law

$$\tilde{H}_x(R) = \frac{I}{2w} \qquad (11.1\text{-}27)$$

From (11.1-16), (11.1-24), and (11.1-27), we arrive at

$$\frac{d\tilde{J}_y(R)}{dr} = \frac{jI}{\delta^2 w} \qquad (11.1\text{-}28)$$

Using (11.1-19), (11.1-23), and (11.1-28), one obtains the following:

$$J_1 = \frac{I}{\delta w(1-j)} \frac{1}{e^{dR} - e^{-dR}} \qquad (11.1\text{-}29)$$

At this point, the solution for the current density is nearly complete. Using (11.1-20), (11.1-26), and (11.1-29), we have

$$\tilde{J}_y(r) = \frac{I}{\delta w(1-j)} \frac{e^{dr} + e^{-dr}}{e^{dR} - e^{-dR}} \qquad (11.1\text{-}30)$$

Example 11.1A Let us consider the current distribution within a rectangular piece of copper conductor. The piece of material is 51 mm wide and 3.21 mm thick. The permeability is that of free space, and the resistivity is 59.6 MS/m. We will plot $|\tilde{J}_y|/J_{dc}$, $\text{real}(\tilde{J}_y)/J_{dc}$, and $\text{imag}(\tilde{J}_y)/J_{dc}$ as a function of the distance from the center of the conductor, where J_{dc} is the uniform current density that would be obtained for direct current (DC) excitation. We will assume an excitation frequency of 5 kHz.

The problem is straightforward to solve using (11.1-30). The results are presented in Figure 11.2. The property that the current distribution is greatest toward the outside of the conductor is clear. It is interesting to observe that at the center of the conductor, the real part of the current density is negative – in essence, the current in this region is actually flowing opposite to the intended direction. There is also a significant imaginary component to the current density. As a result, the actual power loss in conducting the intended current is significantly more than if the current density were uniformly distributed.

Using (11.1-30), we can readily find the impedance associated with the strip conductor. From (11.1-2), the voltage from the positive to negative node in Figure 11.1(a) may be expressed as

$$\tilde{V}_{+-} = -\int_{-}^{+} -\tilde{E}_y(R) dl \qquad (11.1\text{-}31)$$

In (11.1-31), the extra minus sign arises from the assumed direction of the electric field relative to the positive and negative nodes in Figure 11.1. This expression reduces to

$$\tilde{V}_{+-} = \tilde{E}_y(R) l \qquad (11.1\text{-}32)$$

Using (11.1-10), the electric field may be written in terms of the current density so that (11.1-32) becomes

$$\tilde{V}_{+-} = \frac{\tilde{J}_y(R) l}{\sigma} \qquad (11.1\text{-}33)$$

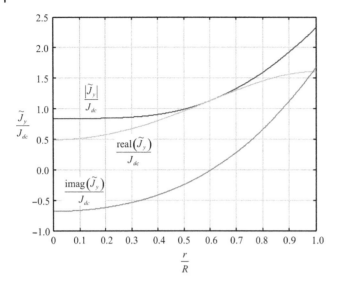

Figure 11.2 Current density distribution in a strip conductor.

Substitution of (11.1-30) into (11.1-33) gives

$$\frac{\tilde{V}_{+-}}{\tilde{I}} = \frac{l}{\delta\sigma w(1-j)} \frac{e^{dR} + e^{-dR}}{e^{dR} - e^{-dR}} \tag{11.1-34}$$

Thus, the impedance of the conductor may be expressed as

$$Z = \frac{l\coth(dR)}{\delta\sigma w(1-j)} \tag{11.1-35}$$

It should be remembered that the impedance expressed by (11.1-35) is an internal impedance, resulting from the fields interior to the conductor; the fields exterior to the conductor will add an inductive reactance.

Example 11.1B In this example, we will use (11.1-35) to look at the impedance presented by the strip conductor considered in Figure 11.1A. Using (11.1-35), the impedance will be calculated and normalized to the DC impedance. The normalized magnitude, real part, and imaginary part of the impedance will then be plotted as frequency is varied from 10 Hz to 10 kHz. The results are presented in Figure 11.3. As can be seen, at low frequency, the impedance is real and is equal to its DC value. As frequency increases, both the real and reactive components of the impedance increase markedly with frequency. Also presented in Figure 11.3 is the real part of the measured impedance of the bar, denoted as Z_{ms}, relative to the measured DC impedance of the bar, $Z_{dc,ms}$. The impedance was measured based on a voltage measured over an approximately 3.03 m of a 4.88-m segment strip conductor while a current was passed through it. Two return conductors with identical series impedances of 0.1 Ω were placed at 22.9 cm to either side of the segment under test, so that their fields would essentially cancel in the region of the test conductor. By measuring the voltage drop across the segment, and the current into the segment, the impedance was readily determined. The reactive component is not shown since this includes an external as well as internal component, and the external component is not represented by (11.1-35). As can be seen, the real part of the measured impedance is in rough agreement with the predicted value. The dip in the real part of

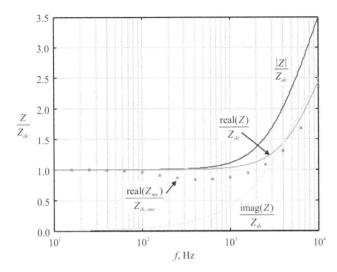

Figure 11.3 Impedance characteristics of a strip conductor.

the measured impedance is attributed to measurement error and to the influence of the return path of the conductors.

11.2 Skin Effect in Cylindrical Conductors

Let us now consider skin effect in cylindrical conductors. To this end, let us consider Figure 11.4, wherein a length of the cylindrical conductor is presented. As in the previous section, the conductor is assumed to be oriented in the direction of the y-axis, and the electric field and current density in the y-axis are denoted as E_y and J_y, respectively. The field intensity will be in the tangential direction. Since in the plane of the voltage loop, the field is in the x-axis direction, we will designate the magnetic field intensity and flux density as H_x and B_x, respectively, in order to keep the same nomenclature as the previous section. It is understood that off of the indicated plane, these field components are tangential.

In order to analyze this situation, we will begin by considering the voltage loop. As it turns out, the analysis in Section 11.1, which begins with (11.1-1) and culminates in (11.1-5), still applies and thus does not need to be repeated. We will next apply Ampere's law to the circular path of radius r in the xz-plane. This yields

$$2\pi r H_x(r) = \int_0^r J_y(r) 2\pi r \, dr \tag{11.2-1}$$

Differentiating (11.2-1) with respect to radius yields:

$$r\frac{\partial H_x(r)}{\partial r} + H_x(r) = J_y(r) r \tag{11.2-2}$$

For sinusoidal steady-state conditions, (11.2-2) becomes

$$r\frac{d\widetilde{H}_x(r)}{dr} + \widetilde{H}_x(r) = \widetilde{J}_y(r) r \tag{11.2-3}$$

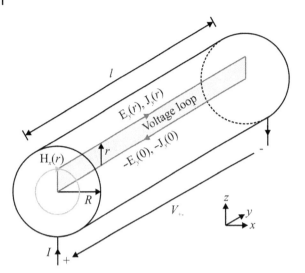

Figure 11.4 Cylindrical conductor.

Combining (11.1-5) with the material relationships expressed in (11.1-10) and (11.1-11), and (11.2-3), we obtain the differential equation which governs the current distribution. In particular, we have

$$r\frac{d^2\widetilde{J}_y}{dr^2} + \frac{d\widetilde{J}_y}{dr} - rj\omega\sigma\mu\widetilde{J}_y = 0 \tag{11.2-4}$$

Note that this is a different form as compared to (11.1), and so the solution will be fundamentally different.

In order to solve (11.2-4), we will need to use Bessel functions. The Bessel may be expressed as follows:

$$x^2\frac{d^2y}{dx^2} + x\frac{dy}{dx} + (x^2 - v^2)y = 0 \tag{11.2-5}$$

where v is a constant that is related to the order of the Bessel. For our purposes, we will be interested in the Bessel of order zero, wherein $v = 0$.

The solution to the Bessel of zero order may be expressed as

$$y = c_1 J_B(x) + c_2 Y_B(x) \tag{11.2-6}$$

where $J_B(x)$ is the Bessel function of the first kind of order zero, and $Y_B(x)$ is the Bessel function of the second kind of order zero, and where c_1 and c_2 are constants whose values are determined by boundary conditions. Note that it is more common to denote the Bessel functions of order zero as $J_0(x)$ and $Y_0(x)$; however, we will diverge from this notation in order to avoid confusion with the current density terms.

The Bessel functions of order zero may be expressed as

$$J_B(x) = 1 + \sum_{m=1}^{\infty} \frac{(-1)^m x^{2m}}{2^{2m}(m!)^2} \quad x \geq 0 \tag{11.2-7}$$

and

$$Y_B(x) = \frac{2}{\pi}\left[\left(\gamma + \ln\left(\frac{2}{x}\right)\right)J_B(x) + \sum_{m=1}^{\infty}\frac{(-1)^{m+1}H_m}{2^{2m}(m!)^2}x^{2m}\right] \quad x > 0 \qquad (11.2\text{-}8)$$

In (11.2-8),

$$H_m = \frac{1}{m} + \frac{1}{m-1} + \cdots \frac{1}{2} + 1 \qquad (11.2\text{-}9)$$

and

$$\gamma = \lim_{m\to\infty}(H_m - \ln m) \cong 0.5772 \qquad (11.2\text{-}10)$$

It is important for our purposes to note that $J_B(0)$ is finite, but $Y_B(0)$ is not. A thorough discussion of Bessel and their solutions is set forth in Boyce and Diprima [1]. As a final note in our discussion of Bessel equation's, it will be convenient to denote

$$J_B'(x) = \frac{d}{dx}J_B(x) \qquad (11.2\text{-}11)$$

In order to use Bessel functions as a solution to (11.2-4), it is necessary to manipulate the form of (11.2-4) to match (11.2-5). To this end, we will define

$$r = \kappa \hat{r} \qquad (11.2\text{-}12)$$

where

$$\kappa = \sqrt{\frac{j}{\omega\sigma\mu}} \qquad (11.2\text{-}13)$$

Replacing r with $\kappa\hat{r}$ in (11.2-4) and manipulating yields

$$\hat{r}^2\frac{d^2\tilde{J}_y}{d\hat{r}^2} + \hat{r}\frac{d\tilde{J}_y}{d\hat{r}} + \hat{r}^2\tilde{J}_y = 0 \qquad (11.2\text{-}14)$$

which is in the form of a Bessel of order zero.

The solution for current density may thus be expressed as

$$J_y = J_1 J_B(\hat{r}) + J_2 Y_B(\hat{r}) \qquad (11.2\text{-}15)$$

where J_1 and J_2 are constants determined by boundary conditions. In order to determine these constants, let us first consider the condition at $r = 0$ at which $\hat{r} = 0$. If we assume that the current density is finite, and noting that $Y_B(0)$ is infinite, we immediately conclude that J_2 is zero. Thus, our expression for current density reduces to

$$J_y = J_1 J_B(r/\kappa) \qquad (11.2\text{-}16)$$

To find J_1, we note that

$$\frac{dJ_y(R)}{dr} = \frac{J_1}{\kappa}J_B'(R/\kappa) \qquad (11.2\text{-}17)$$

We also note that

$$\tilde{H}_x(R) = \frac{\tilde{I}}{2\pi R} \qquad (11.2\text{-}18)$$

From (11.1-8), (11.1-13), (11.1-14), and (11.2-18), we have

$$\frac{d\tilde{J}_y(R)}{dr} = \frac{j\omega\mu\sigma\tilde{I}}{2\pi R} \qquad (11.2\text{-}19)$$

Equating the right-hand sides of (11.2-17) and (11.2-19), we obtain

$$J_1 = -\frac{\tilde{I}}{2\pi R\kappa J_B'(R/\kappa)} \qquad (11.2\text{-}20)$$

Substituting (11.2-20) into (11.2-15), we may express the current density as

$$\tilde{J}_y = -\frac{\tilde{I} J_B(r/\kappa)}{2\pi R\kappa J_B'(R/\kappa)} \qquad (11.2\text{-}21)$$

In evaluating (11.2-21), the derivative term associated with $J_B'(R/\kappa)$ is computed numerically.

Our original goal was to find the internal impedance of a conductor. Using the definition of impedance in (11.1-33) and (11.2-21), the internal impedance may be expressed as

$$Z = -\frac{l J_B(R/\kappa)}{2\pi R\kappa\sigma J_B'(R/\kappa)} \qquad (11.2\text{-}22)$$

Example 11.2A Let us use (11.2-22) to predict the AC resistance of a segment of #8 American wire gauge (AWG) wire. This wire has a nominal diameter of 3.23 mm. Figure 11.5 illustrates the real part of the impedance calculated by (11.2-22) normalized to its DC value. Also presented in Figure 11.5 is the real part of the measured impedance for the segment of wire normalized to its DC value, denoted by $\text{real}(Z_{ms})/Z_{dc,ms}$. The impedance was measured based on a voltage measured over an approximately 3.03 m of a 4.88 m segment of wire while a current was passed through it. Two return conductors with identical series impedances of 0.1 Ω were placed at 22.9 cm to either side of the segment under test, so that their fields would essentially cancel in the region of the test

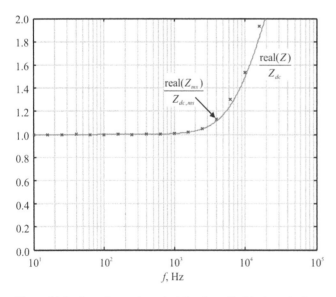

Figure 11.5 Impedance characteristic of a cylindrical conductor.

11.3 Proximity Effect in a Single Conductor

conductor. By measuring the voltage drop across the segment and the current into the segment, the impedance was readily determined. As can be seen, the predicted result is in good agreement with the observed behavior.

Proximity effect is a loss mechanism in which a time-varying field from nearby conductors induces eddy currents in a given conductor. In order to estimate this effect, let us consider Figure 11.6, wherein a rectangular conductor of width w_c, depth d_c, and length l_c (into the page) is presented. Let us suppose that the conductor is exposed to a time-varying flux density B_x from an external source and directed along the x-axis.

The time-varying magnetic field will induce currents into the conductor, which results in loss. Our goal is to estimate this loss. To this end, let us consider the current in a band of width dy, which flows out of the page at a distance y above the centerline of the conductor and has a return path in a band of width dy at a distance y below the centerline of the conductor. These current bands are shaded in Figure 11.6. We will assume that the conductor is long relative to the other dimensions so that we can neglect end effects.

The voltage for the loop consisting of the upper band and lower band may be expressed as

$$v = ri + \frac{d\lambda}{dt} \tag{11.3-1}$$

where the resistance of the loop may be expressed as

$$r = \frac{2l_c}{\sigma w_c dy} \tag{11.3-2}$$

and the flux linking the loop caused by the external flux source is given by

$$\lambda = 2yl_c B_x \tag{11.3-3}$$

By using (11.3-1), (11.3-2), and (11.3-3), the current in the shaded band may be expressed as

$$i = -yw_c \sigma \frac{dB_x}{dt} dy \tag{11.3-4}$$

Our goal is to calculate the power loss associated with this induced current. The instantaneous differential power loss associated with our band may be expressed as follows:

$$dp = i^2 r \tag{11.3-5}$$

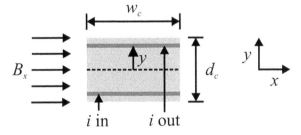

Figure 11.6 Rectangular conductor exposed to a magnetic field.

Substitution of (11.3-2) and (11.3-4) into (11.3-5) gives

$$dp = 2\sigma w_c l_c \left(\frac{dB_x}{dt}\right)^2 y^2 dy \qquad (11.3\text{-}6)$$

so that the total instantaneous power dissipated by the conductor may be expressed as

$$p = \int_0^{d_c/2} 2\sigma w_c l_c \left(\frac{dB_x}{dt}\right)^2 y^2 dy \qquad (11.3\text{-}7)$$

which yields

$$p = \frac{1}{12} d_c^3 w_c l_c \sigma \left(\frac{dB_x}{dt}\right)^2 \qquad (11.3\text{-}8)$$

Taking a time average of (11.3-8), we obtain

$$\bar{p} = \frac{1}{12} d_c^3 w_c l_c \sigma \frac{1}{T} \int_0^T \left(\frac{dB_x}{dt}\right)^2 dt \qquad (11.3\text{-}9)$$

Had the field been oriented along the y-axis, we would obtain

$$\bar{p} = \frac{1}{12} w_c^3 d_c l_c \sigma \frac{1}{T} \int_0^T \left(\frac{dB_y}{dt}\right)^2 dt \qquad (11.3\text{-}10)$$

Now, let us consider the case of a round conductor, as presented in Figure 11.7. In this case, the development is very similar to the rectangular case, and it can be shown that (11.3-7) still applies, except that w_c is a function of y and so is denoted as $w_c(y)$, and $d_c/2$ is replaced by r_c in the upper limit of integration. Using geometrical arguments, one can readily show that

$$w_c(y) = 2\sqrt{r_c^2 - y^2} \qquad (11.3\text{-}11)$$

Substituting (11.3-11) into (11.3-7), simplifying, and then taking the time average, we obtain the following:

$$\bar{p} = \frac{\pi}{4} \sigma r_c^4 l_c \frac{1}{T} \int_0^T \left(\frac{dB_x}{dt}\right)^2 dt, \qquad (11.3\text{-}12)$$

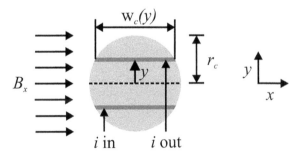

Figure 11.7 Round conductor exposed to a magnetic field.

which is equivalent to the result given, for example, in R. S. Charles [2]. Clearly, (11.3-12) could be applied to a field excitation in the y-axis by replacing B_x with B_y. If ϕ denotes the orientation of a resultant field such that

$$B_x = B_p \cos \phi \qquad (11.3\text{-}13)$$

and

$$B_y = B_p \sin \phi \qquad (11.3\text{-}14)$$

then

$$\bar{p} = \frac{\pi}{4} \sigma r_c^4 l_c \frac{1}{T} \int_0^T \left(\frac{dB_p}{dt}\right)^2 dt \qquad (11.3\text{-}15)$$

It should be noted that (11.3-9), (11.3-10), and (11.3-15) are approximate, in that they neglect the effect of the induced eddy current on the field. This will cause an overestimation of the predicted loss. We saw this same effect (for much the same reason) when studying eddy current core loss in Section 6.1.

11.4 Independence of Skin and Proximity Effects

In this chapter, we first considered skin effect in Sections 11.1 and 11.2. Therein, we neglected fields due to proximity effect. We then began to consider proximity effect in Section 11.3, wherein we did not consider skin effect. A rather unpleasant possibility is that the two loss mechanisms may interact. As it turns out, this is not the case [3]. The objective of this section is to demonstrate why the two loss mechanisms may be treated separately.

Consider Figure 11.8, which shows a conductor of indeterminate shape (though it as shown as an ellipse) that is centered at the origin. Let J(x, y) denote the current density as some arbitrary position. The loss associated with a region of dimension dx by dy may be expressed as

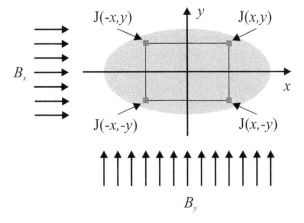

Figure 11.8 Current density in a symmetric conductor.

$$dp = \underbrace{\frac{l_c}{dxdy\sigma}}_{r}\underbrace{(J(x,y)dxdy)^2}_{i^2} \tag{11.4-1}$$

where l_c is the length into the page, σ is the conductivity, r is the resistance of the differential element, and i is the current through the differential element. Simplifying (11.4-1), we obtain

$$dp = \frac{l_c}{\sigma}J^2(x,y)dxdy \tag{11.4-2}$$

Integrating the differential power loss over the cross section of the conductor,

$$p = \frac{l_c}{\sigma}\iint_S J^2(x,y)dxdy \tag{11.4-3}$$

where S denotes the cross section of the conductor.

Let us use $\langle\rangle_S$ to denote a spatial average over a region S. In particular,

$$\langle x \rangle_S = \frac{1}{|S|}\int_S x\,dS \tag{11.4-4}$$

This definition will be applied in one, two, or three dimensions as appropriate; $|S|$ will denote the length, area, or volume accordingly.

In terms of a spatial average, (11.4-3) is written as

$$p = \frac{l_c|S|}{\sigma}\langle J^2(x,y)\rangle_S \tag{11.4-5}$$

The current density may be expressed as

$$J(x,y) = J_s(x,y) + J_{Bx}(x,y) + J_{By}(x,y) \tag{11.4-6}$$

where $J_s(x,y)$ is the current density due to the current flow in the conductor, $J_{Bx}(x,y)$ is the current density due to the imposed flux density in the x-axis, and $J_{By}(x,y)$ is the current density due to the imposed flux density in the y-axis. Substituting (11.4-6) into (11.4-4) and expanding, we have

$$p = \frac{l_c|S|}{\sigma}\begin{pmatrix} \langle J_s^2(x,y)\rangle_S + \langle J_{Bx}^2(x,y)\rangle_S + \langle J_{By}^2(x,y)\rangle_S + \\ 2\langle J_s(x,y)J_{Bx}(x,y)\rangle_S + 2\langle J_s(x,y)J_{By}(x,y)\rangle_S + 2\langle J_{Bx}(x,y)J_{By}(x,y)\rangle_S \end{pmatrix}. \tag{11.4-7}$$

Since the conductors are magnetically linear materials, superposition applies, and the current densities can be computed separately. Even so, the final three terms in (11.4-7) suggest that there is ample opportunity for the current density components to interact to impact losses.

The reason the components do not interact is symmetry. The current density due to the net current flow in the conductor is symmetric about both axes, thus,

$$J_s(x,y) = J_s(x,-y) = J_s(-x,y) = J_s(-x,-y). \tag{11.4-8}$$

The current densities due to the externally sourced flux density in the x- and y-axes are symmetric about the x- and y- axes and have odd symmetry about the y- and x-axes, since one half of the conductor forms the return path for the other half. Therefore,

$$J_{Bx}(x,y) = J_{Bx}(-x,y) = -J_{Bx}(x,-y) = -J_{Bx}(-x,-y) \tag{11.4-9}$$

and

$$J_{By}(x,y) = -J_{By}(-x,y) = J_{By}(x,-y) = -J_{By}(-x,-y). \tag{11.4-10}$$

Now, let us consider the term $\langle J_s(x,y)J_{Bx}(x,y)\rangle_S$. Observe that in evaluating this average, the contribution to the spatial average at point (x,y) is given by $J_s(x,y)J_{Bx}(x,y)dxdy$. The contribution to the average of point $(x,-y)$ is $J_s(x,-y)J_{Bx}(x,-y)dxdy$, which, from (11.4-8) and (11.4-9), is equal to $-J_s(x,y)J_{Bx}(x,y)dxdy$. Thus, the contributions cancel out, and it follows that

$$\langle J_s(x,y)J_{Bx}(x,y)\rangle_S = 0. \tag{11.4-11}$$

Following similar lines of reasoning, we can show that

$$\langle J_s(x,y)J_{By}(x,y)\rangle_S = 0 \tag{11.4-12}$$

and

$$\langle J_{Bx}(x,y)J_{By}(x,y)\rangle_S = 0. \tag{11.4-13}$$

Incorporating (11.4-11) to (11.4-13) into (11.4-7), we have

$$p = \frac{l_c|S|}{\sigma}\left(\langle J_s^2(x,y)\rangle_S + \langle J_{Bx}^2(x,y)\rangle_S + \langle J_{By}^2(x,y)\rangle_S\right) \tag{11.4-14}$$

From (11.4-14), we can see that the AC power loss in the conductor is the superposition of the skin effect loss, the proximity-effect loss due to x-axis field, and the proximity-effect loss due to y-axis field. Note that in non-round conductors, the use of this conclusion requires orienting the references axes so that the conductor is symmetric about the axis as shown in Figure 11.8.

11.5 Proximity Effect in a Group of Conductors

In Section 11.3, we found expressions for the proximity-effect loss in rectangular and round conductors. In Section 11.4, we found that skin-effect losses and proximity-effect losses act independently, and we also determined that the proximity-effect loss caused by a field in one axis is independent of the proximity-effect loss in the orthogonal axis, at least in the case of symmetric conductors. We will now consider the calculation of the proximity-effect loss for an entire winding.

Proximity-Effect Loss in Terms of Flux Density

Using the (11.3-9), (11.3-10), and (11.3-12), we note that in all cases, the power loss in a conductor may be expressed as follows:

$$\overline{P}_{wr} = c_{xwr}\overline{\left(\frac{dB_x}{dt}\right)^2} + c_{ywr}\overline{\left(\frac{dB_y}{dt}\right)^2} \tag{11.5-1}$$

where

$$c_{xwr} = \frac{1}{12}\sigma d_{cw}^3 w_{cw} l_{cwr} \tag{11.5-2}$$

and

$$c_{ywr} = \frac{1}{12}\sigma w_{cw}^3 d_{cw} l_{cwr} \tag{11.5-3}$$

for rectangular conductors and

$$c_{xwr} = c_{ywr} = \frac{\pi}{4}\sigma r_{cw}^4 l_{cr} \tag{11.5-4}$$

for round conductors. In (11.5-1) to (11.5-4), w denotes a winding (for example, as, bs, and cs), and the subscript r denotes a region, (for example, slt for slot).

Our ultimate goal is not to compute the proximity-effect loss for a segment of a single conductor, but rather an entire winding. To achieve this goal, we will assume that a winding is uniformly distributed throughout a region, whereupon the proximity effect-loss of a winding region may be expressed as

$$\overline{P}_{pwr} = |N_w|_r c_{xwr} \left\langle \left(\frac{dB_x}{dt}\right)^2 \right\rangle_r + |N_w|_r c_{ywr} \left\langle \left(\frac{dB_y}{dt}\right)^2 \right\rangle_r \tag{11.5-5}$$

where $|N_w|_r$ is the number of conductors of winding w in region r.

Defining the loss constants for the x- and y-axes as

$$\gamma_{xwr} = |N_w|_r c_{xwr} \tag{11.5-6}$$

and

$$\gamma_{ywr} = |N_w|_r c_{ywr} \tag{11.5-7}$$

we may express (11.5-5) as

$$\overline{P}_{pwr} = \gamma_{xwr} \left\langle \left(\frac{dB_x}{dt}\right)^2 \right\rangle_r + \gamma_{ywr} \left\langle \left(\frac{dB_y}{dt}\right)^2 \right\rangle_r \tag{11.5-8}$$

Round Conductors

In the case of round conductors, the loss constants in the x- and y-directions are the same, so we define

$$\gamma_{wr} = \gamma_{xwr} = \gamma_{ywr} \tag{11.5-9}$$

Let B_p denote flux density along a flux line. It may be positive or negative; it is not the magnitude of the flux. At any point, we may write

$$B_x = B_p \cos(\phi(x,y)) \tag{11.5-10}$$

and

$$B_y = B_p \sin(\phi(x,y)) \tag{11.5-11}$$

where $\phi(x,y)$ is the angle of the flux line relative to the x-axis. Manipulating (11.5-8) to (11.5-11), we obtain the simpler result

$$\overline{P}_{pwr} = \gamma_{wr} \left\langle \left(\frac{dB_p}{dt}\right)^2 \right\rangle_r \tag{11.5-12}$$

Dynamic Resistance and Multi-winding Systems

Now, let us consider a multi-winding system. Although the methodology could be extended to any number of windings, let us assume there are two windings, which we will designate a and b. The flux density may be expressed as

$$B_x = B_{xa} + B_{xb} \tag{11.5-13}$$

and

$$B_y = B_{ya} + B_{yb} \tag{11.5-14}$$

In (11.5-13) and (11.5-14), B_{xa} is the flux density caused by the a winding, and B_{xb} is the flux density associated with the b winding. The flux density in the y-axis is similarly partitioned. The ability to associate field components with individual sources is contingent upon magnetic linearity. Substitution of (11.5-13) and (11.5-14) into (11.5-8) yields

$$\overline{P}_{pwr} = \gamma_{xwr} \left\langle \left(\frac{dB_{xa}}{dt}\right)^2 + 2\overline{\left(\frac{dB_{xa}}{dt}\frac{dB_{xb}}{dt}\right)} + \left(\frac{dB_{xb}}{dt}\right)^2 \right\rangle_r + \gamma_{ywr} \left\langle \left(\frac{dB_{ya}}{dt}\right)^2 + 2\overline{\left(\frac{dB_{ya}}{dt}\frac{dB_{yb}}{dt}\right)} + \left(\frac{dB_{yb}}{dt}\right)^2 \right\rangle_r \tag{11.5-15}$$

where w is a or b.

To proceed further, let us indicate current normalized flux density with a "^" and define it as

$$\hat{B} = \frac{B|_{due\ to\ i_w}}{i_w} \tag{11.5-16}$$

In terms of current normalized flux density, (11.5-16) becomes

$$\overline{P}_{pwr} = \gamma_{xwr} \left\langle \hat{B}_{xa}^2 \overline{\left(\frac{di_a}{dt}\right)^2} + 2\hat{B}_{xa}\hat{B}_{xb} \overline{\left(\frac{di_a}{dt}\frac{di_b}{dt}\right)} + \hat{B}_{xb}^2 \overline{\left(\frac{di_b}{dt}\right)^2} \right\rangle_r + \gamma_{ywr} \left\langle \hat{B}_{ya}^2 \overline{\left(\frac{di_a}{dt}\right)^2} + 2\hat{B}_{ya}\hat{B}_{yb} \overline{\left(\frac{di_a}{dt}\frac{di_b}{dt}\right)} + \hat{B}_{yb}^2 \overline{\left(\frac{di_b}{dt}\right)^2} \right\rangle_r \tag{11.5-17}$$

The sum of $\overline{P}_{par}$ and $\overline{P}_{pbr}$ yields the proximity-effect loss for a region, which may be obtained by applying (11.5-17) to the a and b windings and summing the results. The sum may be written as

$$\overline{P}_{pr} = \overline{\frac{d\mathbf{i}^T}{dt} \mathbf{D}_r \frac{d\mathbf{i}}{dt}} \tag{11.5-18}$$

where $\mathbf{i}$ is a current vector defined as

$$\mathbf{i} = [i_a\ i_b]^T \tag{11.5-19}$$

and $\mathbf{D}_r$ is the dynamic resistance matrix for the region, which may be expressed as

$$\mathbf{D}_r = \underbrace{(\gamma_{xar} + \gamma_{xbr})\left\langle \begin{matrix} \hat{B}_{xa}^2 & \hat{B}_{xa}\hat{B}_{xb} \\ \hat{B}_{xa}\hat{B}_{xb} & \hat{B}_{xb}^2 \end{matrix} \right\rangle}_{\mathbf{D}_{xr}} + \underbrace{(\gamma_{yar} + \gamma_{ybr})\left\langle \begin{matrix} \hat{B}_{ya}^2 & \hat{B}_{ya}\hat{B}_{yb} \\ \hat{B}_{ya}\hat{B}_{yb} & \hat{B}_{yb}^2 \end{matrix} \right\rangle}_{\mathbf{D}_{yr}} \tag{11.5-20}$$

The separate computation of $\mathbf{D}_{xr}$ and $\mathbf{D}_{yr}$ components is convenient when these are computed analytically.

In the case of round conductors, we may use (11.5-12) instead of (11.5-8) as a beginning point, which yields the alternate expression

$$\mathbf{D}_r = (\gamma_{ar} + \gamma_{br}) \left\langle \begin{matrix} \hat{B}_{pa}^2 & \hat{B}_{pa}\hat{B}_{pb} \\ \hat{B}_{pa}\hat{B}_{pb} & \hat{B}_{pb}^2 \end{matrix} \right\rangle \tag{11.5-21}$$

The total loss due to proximity effect may be computed by summing up the losses in all regions. In particular,

$$\overline{P}_p = \sum_{r \in R} \overline{P}_{pr} \tag{11.5-22}$$

where R denotes a set of regions, which might include the slot region and end region, for example. We may also express the total proximity-effect loss as

$$\overline{P}_p = \overline{\frac{d\mathbf{i}^T}{dt} \mathbf{D} \frac{d\mathbf{i}}{dt}} \tag{11.5-23}$$

where

$$\mathbf{D} = \sum_{r \in R} \mathbf{D}_r \tag{11.5-24}$$

In the abovementioned range from (11.5-23) to (11.5-24), $\mathbf{D}$ is referred to as the dynamic resistance matrix for the entire component. Clearly, for a single winding system, (11.5-23) reduces to

$$\overline{P}_p = \mathbf{D} \overline{\left(\frac{di}{dt}\right)^2} \tag{11.5-25}$$

As mentioned previously, this method of calculating proximity-effect losses is a close variant of that set forth in Sullivan [2]. One disadvantage of the approach is that it does not account for the fact that the eddy current affects the field distribution. There are many other approaches in the literature. In one of these approaches, a matrix of conductors is represented as a set of foil (strip) conductors, a problem with an analytical solution (P. L. Dowell) [4]. This method tends to work well for closely packed windings. Another approach, which works well on loose windings is based on a detailed analysis of a round conductor in a uniform field (J. A. Ferreira) [5]. A discussion of the two methods, and a way to combine them, is set forth in Nan and Sullivan [6,7].

11.6 Relating Mean-Squared Field and Leakage Permeance

In order to use the results of the previous section in order to calculate the proximity-effect loss, it is necessary to calculate the mean-squared field and the dynamic resistance matrix. It is often the case that the region of interest is associated with a leakage permeance, since we have considered many leakage paths that pass through the winding. Let us consider the case where (a) we have round conductors, and (b) the region for which we found the leakage permeance corresponds to a conductor region in which we wish to find the proximity effect.

Let us consider the approach we often used to find leakage permeance for such a region. Recall that such derivations began by finding the energy associated with the region, which was calculated as

$$E = \frac{1}{2} \int_{V_r} B_p H_p dV \tag{11.6-1}$$

In (11.6-1), B_p and H_p are, respectively, the flux density and field intensity over some path, and V_r is the volume of the region.

Next, we note that for the magnetically linear conditions (since the leakage path was through a low permeability conductor), we have

$$E = \frac{1}{2} N^2 i^2 P \tag{11.6-2}$$

Equating (11.6-1) and (11.6-2),

$$P = \frac{\int_{V_r} B_p H_p dV}{N^2 i^2} \tag{11.6-3}$$

which may be expressed as

$$P = \frac{\int_{V_r} \hat{B}_p^2 dV}{\mu_0 N^2} \tag{11.6-4}$$

Comparing (11.6-4) to our definition of the spatial average, we conclude

$$\left\langle \hat{B}_p^2 \right\rangle_r = \frac{\mu_0 N^2 P}{V_r} \tag{11.6-5}$$

Finally, combining (11.6-5) with (11.5-4), (11.5-6) with $|N_w|_r = N_{pr}N$, and (11.5-21) (for a one-winding case), we have

$$D_r = \frac{\mu_0 \pi N^3 N_{pr} \sigma r_c^4 l_c P}{4 V_r} \tag{11.6-6}$$

where r_c is the conductor radius, N is the number of conductors, and l_c is the length of one strand of conductor in the region of interest. In deriving (11.6-6), note that (11.5-9) applies since we are considering round conductors, and $|N_r|_w = N_{pr}N$, where N_{pr} is the number of parallel strands that make up a conductor.

The results obtained in (11.6-5) and (11.6-6) are quite useful, in that the effort we placed into computing a leakage permeance also yields the mean-squared field and dynamic resistance as well. The result must be used with some care, however, since it only applies to the situation in which the region of the leakage permeance is an exact match to the region of interest for the proximity effect.

11.7 Mean-Squared Field for Select Geometries

In this section, we will use the results of Section 11.6 to find the mean-squared field and dynamic resistance matrix for a number of common geometries.

Exterior Adjacent Conductors

Exterior adjacent conductors refer to the situation in which a conductor bundle resides on the outside of a core, as presented in Figure 2.23. This is a situation in which we calculated a leakage permeance of region exactly corresponding to a region where we will wish to calculate the proximity-effect loss. From (11.6-5) and (11.6-6), we have

$$\left\langle \hat{B}_p^2 \right\rangle_{ea} = \frac{\mu_0 N^2 P_{eali}}{d_w w_w l} \tag{11.7-1}$$

and

$$D_{ea} = \frac{\mu_0 \pi N^3 N_{pr} \sigma r_c^4 P_{eali}}{4 d_w w_w} \tag{11.7-2}$$

where the subscript ea denotes exterior adjacent and where P_{eali} is the interior portion of the exterior adjacent leakage inductance as given by (2.7-41).

Exterior Isolated and Non-gapped Closed-Slot Conductors

Exterior isolated conductors are conductors in which the winding is far from the core material, as presented in Figure 2.24. Therein, we will consider the interior flux path. A similar flux path can be present in a situation in which a winding is surrounded by a non-gapped closed slot, like path P5 in Figure 7.16. Referring to Figure 2.24, this is another situation where a region over which we calculated leakage permeance corresponds to a region where we will wish to calculate proximity effect. From (11.6-5) and (11.6-6), we have

$$\left\langle \hat{B}_p^2 \right\rangle_{ei} = \frac{\mu_0 N^2 P_{eili}}{d_w w_w l_c} \tag{11.7-3}$$

and

$$D_{ei} = \frac{\mu_0 \pi N^3 N_{pr} \sigma r_c^4 P_{eili}}{4 d_w w_w} \tag{11.7-4}$$

In (11.7-3) and (11.7-4), P_{eili} is the interior portion of the exterior isolated leakage permeance as given by (2.7-52).

Open-Slot Conductors

Another situation of interest, particularly in rotating machinery, is when we have conductors in an open slot. To this end, let us consider the region associated with Path 4 in Figure 8.16. Applying (11.6-5), we obtain

$$\left\langle \hat{B}_p^2 \right\rangle_{os} = \frac{\mu_0 N^2 P_{sl,4}}{d_w w_{si} l} \tag{11.7-5}$$

where $P_{sl,4}$ is given by (8.7-11). In (11.7-5), N is the number of conductors in the slot, which are assumed to all be associated with one winding or phase. The case where multiple phases are in one slot is considered in Section 11.8.

Substitution of (8.7-11) into (11.7-5) yields

$$\left\langle \hat{B}_p^2 \right\rangle_{os} = \frac{1}{3}\left(\frac{\mu_0 N}{w_{si}}\right)^2 \tag{11.7-6}$$

and

$$D_{os} = \frac{\mu_0^2 \pi \sigma N^3 N_{pr} r_c^4 l}{12 w_{si}^2} \tag{11.7-7}$$

Gapped Closed-Slot Conductors

Next, consider a winding in a gapped closed slot, like we have in UI- and EI-core inductors. Unfortunately, in this situation, we will not be able to apply our results for the leakage permeances. This is primarily due to the interaction between the leakage paths.

Consider the situation presented in Figure 11.9, which depicts a winding bundle in a gapped closed slot. Therein, the point O denotes the origin of the coordinate system (the direction of the axis is shown to the left of the core). The geometrical parameters are the same as we have used before, though we have introduced c_v and c_h as clearances between the winding bundle to the core in the horizontal (x) and vertical (y) directions, respectively.

The winding bundle in Figure 11.9 has dimensions w_w by d_w as indicated. Within the winding bundle, four subregions, numbered 1–4, are indicated by different shading levels. Each region will be treated separately.

We will estimate the dynamic resistance of the region of the winding by considering three field components. The first of these is a horizontal slot leakage flux, denoted as B_h. The second is the vertical slot leakage flux denoted as B_v. The third is the component of the fringing flux, which intersects the windings, denoted as B_f. Based on our work in Section 2.7, we have components B_h and B_v that are given by

$$B_h = \frac{\mu_0 N i y}{d_w w_s} \tag{11.7-8}$$

and

$$B_v = \frac{\mu_0 N i x}{w_w (d_s + g)} \tag{11.7-9}$$

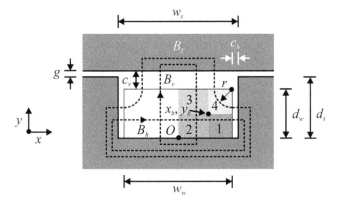

Figure 11.9 Gapped closed slot conductors.

In (11.7-8) and (11.7-9), x and y are relative to the origin O in Figure 11.9.

We previously considered fringing flux in Section 2.6. In that development, we assumed a path similar, but not identical, to the one presented in Figure 11.9, with the difference being that in Section 2.6, we took the center of curvature of the winding to be at the upper inner corner of the U-core, which is to say a point at a distance c_h to the right and c_v above the center of curvature used in Figure 11.9. Since these clearances are normally small, the differences are modest, though the approach used in Section 2.6 is more representative. However, the path chosen herein will prove much more convenient for calculating proximity-effect loss.

Applying Ampere's law to the path shown for the fringing flux, ignoring the magnetomotive force (MMF) drop in the core, and also ignoring the fact that the path shown may not include all conductors, the fringing field may be expressed as

$$B_f = \frac{\mu_0 Ni}{2c_w + 2c_h + \pi r} \tag{11.7-10}$$

When we studied the UI-core inductor fringing flux in Section 2.6 and leakage flux in Section 2.7, we noted, in passing, that we were neglecting the interaction of these two components. Let us try a different approach here. The basis of this approach will be dividing the winding into four regions as presented in Figure 2.19. These regions occupy half of the winding; we will not concern ourselves with the other half because of symmetry. In region 1, we will assume that only horizontal leakage flux exists. In region 2, we will assume that we have horizontal and vertical leakage flux. In region 3, we will assume that we only have vertical leakage flux. Finally, in region 4, we will assume we only have fringing flux. In essence, we are assuming that the fringing path and leakage paths do not coexist, but rather that there is a preferred path in each region. The point (x_b, y_b) in Figure 2.19 denotes a boundary point between the regions.

We will compute the location of the boundary point by finding the point at radius r at which the fringing flux density is equal to the leakage flux density. This point will be denoted as r_{mx}. This will establish the size of region 4.

We will begin by comparing the fringing field expression to the horizontal leakage field. Equating (11.7-8) and (11.7-10) and noting that

$$y = d_w - r \tag{11.7-11}$$

we obtain

$$r_{mxh} = \begin{cases} \max\left(\frac{1}{2}\left(-b_h - \sqrt{b_h^2 - 4c_h}\right), d_w\right) & b_h^2 \geq 4c_h \\ d_w & b_h^2 < 4c_h \end{cases} \tag{11.7-12}$$

where

$$b_h = \frac{2}{\pi}(c_v + c_h) - d_w \tag{11.7-13}$$

$$c_h = \frac{1}{\pi} d_w (w_s - 2c_v - 2c_h) \tag{11.7-14}$$

In (11.7-12), r_{mxh} is r_{mx}, based on a comparison between the fringing field and the horizontal leakage field. This value is limited to d_w because of the geometry.

Next, we will compare the fringing field to the vertical leakage field. Equating (11.7-9) and (11.7-10) with

$$x = \frac{w_w}{2} - r \tag{11.7-15}$$

yields

$$r_{mxv} = \begin{cases} \max\left(\frac{1}{2}\left(-b_v - \sqrt{b_v^2 - 4c_v}\right), w_w/2\right) & b_v^2 \geq 4c_v \\ w_w/2 & b_v^2 < 4c_v \end{cases} \quad (11.7\text{-}16)$$

as an estimate for r_{mx} based on the vertical field component. This value is limited to $w_w/2$ because of the geometry. In (11.7-16),

$$b_h = \frac{2}{\pi}(c_v + c_h) - \frac{1}{2}w_w \quad (11.7\text{-}17)$$

and

$$c_h = \frac{1}{\pi}w_w(d_s + g - c_v - c_h) \quad (11.7\text{-}18)$$

Once the two estimates for r_{mx} are found, we take the minimum of the two. Thus,

$$r_{mx} = \min(r_{mxv}, r_{mxh}) \quad (11.7\text{-}19)$$

In terms of r_{mx}, the boundary point defining the four regions is given by

$$x_b = \frac{1}{2}w_w - r_{mx} \quad (11.7\text{-}20)$$

and

$$y_b = d_w - r_{mx} \quad (11.7\text{-}21)$$

With the boundaries of the regions defined, and the flux densities given by ranging from (11.7-8) to (11.7-10), we have

$$\langle \hat{B}_x^2 \rangle_1 = \langle \hat{B}_x^2 \rangle_2 = \frac{1}{3}\left(\frac{\mu_0 N y_b}{d_w w_s}\right)^2 \quad (11.7\text{-}22)$$

$$\langle \hat{B}_y^2 \rangle_2 = \langle \hat{B}_y^2 \rangle_3 = \frac{1}{3}\left(\frac{\mu_0 N x_b}{w_w(d_s + g)}\right)^2 \quad (11.7\text{-}23)$$

$$\langle \hat{B}_p^2 \rangle_4 = \frac{1}{2\pi}\left(\frac{\mu_0 N}{r_{mx}}\right)^2 \left[\ln\left(1 + \frac{\pi r_{mx}}{2c_h + 2c_v}\right) - \frac{\pi r_{mx}}{2c_h + 2c_v + \pi r_{mx}}\right] \quad (11.7\text{-}24)$$

The next step is to compute the loss constants. From (11.5-6) and (11.5-7), and noting that each region contains only a fraction of the conductors, we have

$$\gamma_1 = NN_{pr}\frac{2y_b r_{mx}}{w_w d_w}c \quad (11.7\text{-}25)$$

$$\gamma_2 = NN_{pr}\frac{2x_b y_b}{w_w d_w}c \quad (11.7\text{-}26)$$

$$\gamma_3 = NN_{pr}\frac{2x_b r_{mx}}{w_w d_w}c \quad (11.7\text{-}27)$$

$$\gamma_4 = NN_{pr}\frac{2r_{mx}^2}{w_w d_w}c \quad (11.7\text{-}28)$$

where

$$c = \frac{\pi}{4}\sigma r_c^4 l_c \qquad (11.7\text{-}29)$$

In ranging from (11.7-25) to (11.7-29), the number of conductors includes both sides of the winding so that we obtain the total loss (recall regions 1–4 only included half the winding). Note that it has been assumed that round conductors are used.

From (11.5-20), (11.5-21), and (11.5-24), we obtain

$$D_{gcs} = \gamma_1 \left\langle \hat{B}_x^2 \right\rangle_1 + \gamma_2 \left(\left\langle \hat{B}_x^2 \right\rangle_2 + \left\langle \hat{B}_y^2 \right\rangle_2 \right) + \gamma_3 \left\langle \hat{B}_y^2 \right\rangle_3 + \gamma_4 \left\langle \hat{B}_p^2 \right\rangle_4 \qquad (11.7\text{-}30)$$

Incorporating ranging from (11.7-25) to (11.7-19) into (11.7-30) yield our desired result, namely that the dynamic resistance for the gapped closed slot is given by

$$D_{gcs} = \frac{NN_{pr}\pi\sigma r_c^4 l_c}{2w_w d_w} \left(y_b r_{mx} \left\langle \hat{B}_x^2 \right\rangle_1 + x_b y_b \left(\left\langle \hat{B}_x^2 \right\rangle_2 + \left\langle \hat{B}_y^2 \right\rangle_2 \right) + x_b r_{mx} \left\langle \hat{B}_y^2 \right\rangle_2 + r_{mx}^2 \left\langle \hat{B}_p^2 \right\rangle_4 \right) \qquad (11.7\text{-}31)$$

where the mean-squared field terms are given by ranging from (11.7-22) to (11.7-24).

We will use this expression in Section 11.9 to determine proximity-effect losses in a UI-core inductor. Before we embark on this investigation, however, let us consider a simpler situation in which we compute the ac losses in rotating electric machine.

11.8 Conductor Losses in Rotating Machinery

Thus far, in this chapter, we have established a basis for the calculation of skin-effect and proximity-effect losses. In this section, we will apply our analysis to study AC slot losses in rotating machinery. In our discussion of rotating machinery in Chapters 8 and 9, we only considered core loss and dc resistive loss. In the case of high-speed machines, wherein the electrical frequency is high, the skin effect and proximity effect could become important. We will focus our attention on the losses within the slots of the machine because this is where the conductors would normally experience the highest fields.

Figure 11.10 depicts a stator's slot. In order to estimate the proximity-effect losses, we will use the rectangular slot approximation discussed in Chapters 8 and 9. In the figure, the darker area is the stator steel and the lighter shaded area represents the conductor region. While we obtained a result

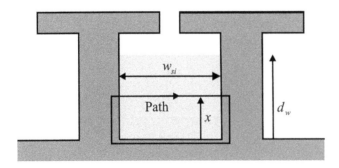

Figure 11.10 Slot geometry.

11.8 Conductor Losses in Rotating Machinery

for the dynamic resistance of the slot in Section 11.7, that particular result was for a slot with only one winding; we will derive a more general result here. Using Ampere's law, assuming that the permeability of the conductor is close to that of free space, and employing our definition of current normalized flux density in (11.5-16), we obtain

$$\hat{B}_{xas} = \frac{\mu_0 N_a x}{w_{si} d_w} \tag{11.8-1}$$

and

$$\hat{B}_{xbs} = \frac{\mu_0 N_b x}{w_{si} d_w} \tag{11.8-2}$$

where N_a and N_b are the number of a- and b-phase conductors directed into the page in the slot of interest. Using (11.8-1) and (11.8-2) and taking the appropriate spatial averages over the region of the winding within the slot yields

$$\left\langle \hat{B}_{xas}^2 \right\rangle_{slt} = \frac{1}{3} \left(\frac{\mu_0 N_a}{w_{si}} \right)^2 \tag{11.8-3}$$

$$\left\langle \hat{B}_{xbs}^2 \right\rangle_{slt} = \frac{1}{3} \left(\frac{\mu_0 N_b}{w_{si}} \right)^2 \tag{11.8-4}$$

$$\left\langle \hat{B}_{xas} \hat{B}_{xbs} \right\rangle_{slt} = \frac{1}{3} \left(\frac{\mu_0}{w_{si}} \right)^2 N_a N_b \tag{11.8-5}$$

Note that (11.8-3) and (11.8-4) match the result given in (11.7-6). From (11.5-4), (11.5-6), and (11.5-7), we have

$$\gamma_{xasslt} + \gamma_{xbsslt} = \frac{\pi}{4} \sigma r_c^4 l_c (|N_a| + |N_b|) N_{pr} \tag{11.8-6}$$

where l_c is the axial length of the slot, and σ and r_c are the conductivity and radius of the conductors, which are assumed to be the same for both phases. Substituting ranging from (11.6-5) to (11.6-6) into (11.5-20) and neglecting any y-component of the field, we obtain

$$\mathbf{D}_{slt} = \frac{\pi \sigma l_c r_c^4 \mu_0^2 N_{pr}}{12 w_{si}^2} (|N_a| + |N_b|) \begin{bmatrix} N_a^2 & N_a N_b \\ N_a N_b & N_b^2 \end{bmatrix} \tag{11.8-7}$$

In order to determine the proximity-effect loss, we must specify the currents. We will assume that the machine is a three-phase device, so that in steady state, the currents are given by

$$i_{as} = \sqrt{2} I_s \cos(\omega_e t + \phi) \tag{11.8-8}$$

and

$$i_{bs} = \sqrt{2} I_s \cos(\omega_e t + \phi - 2\pi/3) \tag{11.8-9}$$

Note that since it would be highly unusual for a three-phase machine to have all three phases in a single slot, the results for a two-winding system are adequate. Taking the temporal derivative of the current waveforms of (11.8-8) and (11.8-9) and substituting the result into (11.5-23) yields

$$\overline{P}_{slt} = \frac{\pi \sigma l_c r_c^4 \mu_0^2 I_s^2 \omega_e^2 N_{pr}}{12 w_{si}^2} (|N_a| + |N_b|)(N_a^2 - N_a N_b + N_b^2) \tag{11.8-10}$$

The result from (11.8-10) yields the proximity-effect losses in one slot, conducting only a- and b-phase conductors. If we assume that the proximity-effect loss is the same in all slots (which it normally is), the total proximity-effect loss in all slots is given by

$$\overline{P}_{tslt} = \frac{\pi \sigma l_c r_c^4 \mu_0^2 I_s^2 \omega_e^2 N_{pr} S_s}{12 w_{si}^2} (|N_a| + |N_b|)(N_a^2 - N_a N_b + N_b^2) \tag{11.8-11}$$

where S_s is the number of slots, and N_a and N_b are the number of (signed) conductors in a single slot containing only a- and b-phase conductors. Recall N_{pr} is the number of parallel strands in a conductor.

Example 11.8A In this example, we will examine how well (11.8-11) predicts proximity-effect losses in practice. In particular, a slot is created in the shape of an elongated oval running track. The mean length of the oval (and hence the length of the slot) is 78 cm, and the width of the slot is 6.93 mm. The slot is occupied by $N = 24$ turns of #8 AWG conductor ($r_c = 0.51$ mm). The conductor consists of a single strand. The conductivity of copper is 59.6 MS. We will attempt to compute the apparent resistance of the coil versus frequency and compare this to what is measured using a four-wire LCR meter from 10 Hz to 10 kHz.

We begin our calculation with the dc resistance, which may be expressed as

$$r_{dc} = \frac{l_c N \sigma}{\pi r_c^2} \tag{11.8A-1}$$

The power loss due to the DC coil resistance is, thus,

$$P_{dc} = r_{dc} I_s^2 \tag{11.8A-2}$$

where I_s is the rms current.

Next, using (11.8-11), with $S_s = 1$, $N_{pr} = 1$, $N_a = N$, and $N_b = 0$ we observe that the total loss due to proximity effect is given by

$$\overline{P}_{tslt} = \frac{\pi \sigma l_c r_c^4 \mu_0^2 I_s^2 \omega_e^2 N^3}{12 w_{si}^2} \tag{11.8A-3}$$

Thus, the total loss from the dc resistance plus proximity effect is given by

$$P_t = \left(r_{dc} + \frac{\pi \sigma l_c r_c^4 \mu_0^2 \omega_e^2 N^3}{12 w_{si}^2} \right) I_s^2 \tag{11.8A-4}$$

Defining the proximity effect resistance as

$$r_{pe} = \frac{\pi \sigma l_c r_c^4 \mu_0^2 \omega_e^2 N^3}{12 w_{si}^2} \tag{11.8A-5}$$

we have

$$P_t = (r_{dc} + r_{pe}) I_s^2 \tag{11.8A-6}$$

Thus, the apparent resistance of the coil is given by

$$r = r_{dc} + r_{pe} \tag{11.8A-7}$$

In this development, we have neglected both skin-effect and the magnetic core loss. However, in general, proximity effect is much more significant than skin effect. With regard to core loss, the

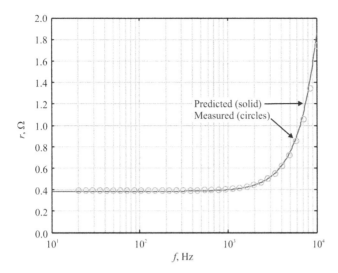

Figure 11.11 Predicted and measured resistance of a coil in an open slot.

magnetic core is a ferrite, and so core loss is very modest in the frequency range considered. The results from evaluating (11.8A-1), (11.8A-5), and (11.8A-7) are shown as a function of frequency in Figure 11.11. Also shown is the measured coil resistance versus frequency. As can be seen, the results are in reasonable agreement.

Example 11.8B In this example, we consider the slot losses in a permanent magnet AC (PMAC) machine. In particular, let us compute the slot losses for Design 38 from the case study presented in Section 9.11. The relevant parameters for this machine are as follows: $P = 12$, $S_s = 72$, $N_{pr} = 1$, $l = 3.34$ cm, $w_{si} = 5$ mm, $d_w = 1.80$ cm, and $r_c = 1.15$ mm, the conductor chosen was copper with $\sigma = 59.6$ MS. We will consider slot 3, for which $N_a = 9$ and $N_b = -2$. The sum of the absolute value of a-phase conductors over all the slots is 264. It is desired to find the losses at an output torque of 20 Nm and a speed of 5000 rpm. At this operating point, the phase current is 23.0 Arms.

Let us find the power that would be lost based on the dc resistance. We will find only that component of the resistance due to the conductors in the slot. The portion of the dc stator resistance may be expressed as

$$r_{dc} = \frac{l \sum_{i=1}^{S_s} |N_{as,i}|}{\pi r_c^2 \sigma} \tag{11.8B-1}$$

The loss due to the resistance may be expressed as

$$P_{dc} = 3 r_{dc} I_s^2 \tag{11.8B-2}$$

For the parameters of this machine at the cited operating point, $r_{dc} = 35.927$ mΩ yielding $P_{dc} = 57.016$ W.

Next, let us compute the ac resistance due to skin effect. Based on the number of poles, we find that $\omega_r = 3140$ rad/s, which corresponds to 500 Hz.

From (11.2-22), the ac resistance may be calculated as

$$r_{ac} = -\text{real}\left(\frac{l\sum_{i=1}^{S_s}|N_{as,i}|J_0(r_c/\kappa)}{2\pi r_c\kappa\sigma J_0'(r_c/\kappa)}\right) \quad (11.8\text{B-}3)$$

where κ is given by (11.2-13). The ohmic losses, including skin effect, may then be expressed as

$$P_{ac} = 3r_{ac}I_s^2 \quad (11.8\text{B-}4)$$

Evaluating (11.8B-3), we have $r_{ac} = 35.944$ mΩ, yielding $P_{ac} = 57.044$ W. Note this ac loss includes the dc loss. The difference between the calculations, defined by

$$\Delta P_{ac} = P_{ac} - P_{dc} \quad (11.8\text{B-}5)$$

is on the order of 28 mW. Clearly, skin effect is irrelevant to this design.

Let us now evaluate (11.8-11) to find the proximity-effect loss. This yields $\overline{P}_{tslt} = 24.1$ W. This loss is significant relative to the dc resistive loss of 57.0 W (note that this is only the portion of the resistive loss associated with the slots). It would have increased the machine loss predictions from 566 W (see Table 9.5) to 590 W, an increase of 4%.

11.9 Conductor Losses in a UI-Core Inductor

Let us now consider the dc, skin-effect, and proximity-effect losses in a UI-core inductor, as presented in Figure 11.12. Therein, the inductor dimensions are labeled just as they were in Chapter 3. As we have discussed previously, the horizontal leakage flux density denoted as B_h, vertical leakage flux density denoted as B_v, and fringing flux component denoted as B_f will all cause proximity-effect losses and contribute to the dynamic resistance denoted as D_{gcs}. The exterior

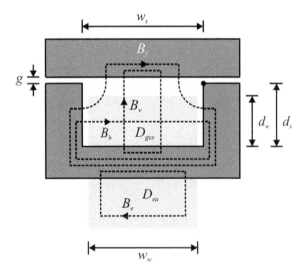

Figure 11.12 UI-core inductor.

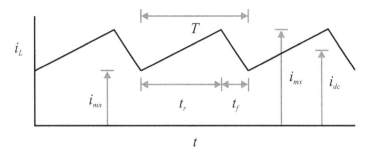

Figure 11.13 Assumed current waveform.

adjacent leakage flux density denoted as B_{ea} also results in a dynamic resistance component denoted as D_{ea}.

The conductor losses will be a function of the current waveform we inject, and so the application is very important. We will assume that the inductor is in a dc to dc converter. While the focus of this book is not on power electronic converters, the waveform presented in Figure 11.13 would be representative of a number of converters. Therein, the inductor current i_L has a sawtooth shape, rising from a minimum value of i_{mn} to a maximum value of i_{mx} over a rise time of t_r, and then falls back to i_{mn} over a fall time of t_f. For a variety of converter circuits, these quantities are straightforward to calculate in terms of input voltage, output voltage, inductance, duty cycle, and frequency.

The rise and fall are approximately linear in time. The period of the waveform is T. Clearly,

$$T = t_r + t_f \tag{11.9-1}$$

The frequency of the waveform will be denoted as f and is the reciprocal of T; the radian frequency ω is $2\pi f$.

One type of power electronic converter that could produce such a waveform is the buck converter presented in Figure 11.14. The function of this converter is to produce an output voltage that is equal to some fraction of the input voltage in an efficient way. Therein, the transistor is turned on during the period t_r. During this period, the transistor carries the rising inductor current. The transistor is then turned off during the interval t_f. During this interval, the diode conducts the falling inductor current.

Let us first consider the DC losses. Assuming a linear rise and fall in the currents, the steady-state DC component of the current is readily expressed as follows:

$$i_{dc} = \frac{1}{2}(i_{mx} + i_{mn}) \tag{11.9-2}$$

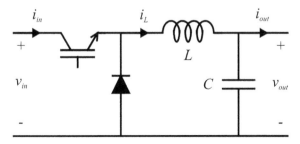

Figure 11.14 Buck converter.

Using the approach of Section 3.2, the total wire length may be expressed as

$$l_w = \frac{w_w d_w (2l_c + 2w_b + \pi d_w) k_{pf}}{\pi r_c^2} \tag{11.9-3}$$

where r_c is the conductor radius. The dc resistance is expressed as

$$r_{dc} = \frac{l_w}{\pi r_c^2 \sigma} \tag{11.9-4}$$

and the dc power loss is given by

$$P_{dcr} = r_{dc} i_{dc}^2 \tag{11.9-5}$$

Let us, next, compute the losses due to the AC component of the current. To this end, it is convenient to express the current waveform in terms of a Fourier series (McGillem and Cooper) [8]. Recall that an arbitrary periodic signal denoted as $x(t)$ may be approximated as

$$x(t) \approx \frac{a_0}{2} + \sum_{n=1}^{\infty} a_n \cos(n\omega t) + b_n \sin(n\omega t) \tag{11.9-6}$$

where ω is the frequency of the fundamental component and

$$a_n = \frac{2}{T} \int_0^T x(t) \cos(n\omega t) dt \tag{11.9-7}$$

and

$$b_n = \frac{2}{T} \int_0^T x(t) \sin(n\omega t) dt \tag{11.9-8}$$

Thus, the current waveform may be expressed as

$$i_L = i_{dc} + \sum_{n=1}^{\infty} a_n \cos(n\omega t) + b_n \sin(n\omega t) \tag{11.9-9}$$

Using (11.9-7) and (11.9-8) in conjunction with the waveform presented in Figure 11.13, we obtain

$$a_n = \frac{2}{n^2 \phi_r \phi_f} (i_{mx} - i_{mn})(\cos(n\phi_r) - 1) \tag{11.9-10}$$

and

$$b_n = \frac{2}{n^2 \phi_r \phi_f} (i_{mx} - i_{mn}) \sin(n\phi_r) \tag{11.9-11}$$

where

$$\phi_r = 2\pi \frac{t_r}{T} \tag{11.9-12}$$

and

$$\phi_f = 2\pi \frac{t_f}{T} \tag{11.9-13}$$

11.9 Conductor Losses in a UI-Core Inductor

From (11.9-10) and (11.9-11), we may readily compute the rms value of the nth harmonic component of the current as

$$i_{rms,n} = \frac{1}{\sqrt{2}}\sqrt{a_n^2 + b_n^2} \qquad (11.9\text{-}14)$$

In order to compute the ac resistive losses, from (11.2-13), we may calculate κ for the nth harmonic, which is at a frequency $n\omega$, as

$$\kappa_n = \sqrt{\frac{j}{n\omega\sigma\mu}} \qquad (11.9\text{-}15)$$

Likewise, using (11.2-12), the resistance of the nth harmonic may be expressed as

$$r_n = -\text{Re}\left(\frac{l_w J_0(r_c/\kappa_n)}{2\pi r_c \kappa_n \sigma J_0'(r_c/\kappa_n)}\right) \qquad (11.9\text{-}16)$$

where l_w is given by (11.9-3).

The ac resistive loss associated with the nth harmonic is given by

$$P_{acr,n} = r_n i_{rms,n}^2 \qquad (11.9\text{-}17)$$

The total ac resistive loss may be expressed as

$$P_{acr} = \sum_{n=1}^{\infty} P_{acr,n} \qquad (11.9\text{-}18)$$

Now, let us turn our attention to proximity-effect losses. The dynamic resistance will have two terms, one for the region inside the core and one for the region outside the core. The dynamic resistance may thus be expressed as

$$D = D_{ea} + D_{gcs} \qquad (11.9\text{-}19)$$

where D_{ea} is given by (11.7-2), and D_{gcs} is given by (11.7-31). Note that the expression D_{ea} given by (11.7-2) is in terms of leakage permeance P_{eali}, which is, in turn, given by (2.7-41). When evaluating this permeance, the length l should be calculated as

$$l = l_c + 2w_b \qquad (11.9\text{-}20)$$

so that the proximity-effect loss calculation includes not only the exterior winding beneath the core but also the end winding in the front and rear faces.

From Figure 11.13, we may readily show that

$$\overline{\left(\frac{di}{dt}\right)^2} = \frac{(i_{mx} - i_{mn})^2}{t_r t_f} \qquad (11.9\text{-}21)$$

Combining (11.5-23) and (11.9-21) for this single-input system,

$$P_p = D\frac{(i_{mx} - i_{mn})^2}{t_r t_f}. \qquad (11.9\text{-}22)$$

The total conductor loss may be expressed as

$$P_{cdl} = P_{dcr} + P_{acr} + P_p \qquad (11.9\text{-}23)$$

Example 11.9A Let us consider the losses in a UI-core inductor. In particular, let us consider Design 50 of the UI-core inductor we first considered in Section 3.4, and whose parameters are listed in Table 3.4. Let us consider a current waveform of the form presented in Figure 11.13, with $i_{mn} = 9.5$ A and $i_{mx} = 10.5$ A. Let us assume that the rise time is 90% of the period, and the fall time is 10% of the period. For the buck converter presented in Figure 11.14, the ratio of rise time to period is the duty cycle, which in this case is 0.9. We will examine the losses versus frequency. In this study, it is important to realize that if one keeps the ripple current constant and increases the frequency, then the input and output voltages must rise linearly with frequency. Thus, in this study, as we increase frequency, we also increase output power.

First, using ranging from (11.9-2) to (11.9-5), we determine that $r_{dc} = 155\ m\Omega$, $i_{dc} = 10$ A, and $P_{dcr} = 15.5$ W. This loss is frequency independent.

Next, let us consider the AC resistive losses. This requires breaking the current into a spectrum using (11.9-10), (11.9-11), and (11.9-14). The rms value of the fundamental component of the waveform is calculated to be 0.246 A, which implies that ac resistive losses might be relatively small since the ac component of the current is quite modest. From the current spectrum, the ac resistive losses are calculated using ranging from (11.9-15) to (11.9-18). At 20 kHz, the loss is only 17 mW (based on the first 10 harmonics), and thus it is negligible. This is despite the fact that the ac resistance at 20 kHz is 1.2 times that of the dc resistance. The small amplitude of the ac current explains why this term is relatively unimportant. The ac resistive loss is plotted versus switching frequency in Figure 11.15. Therein, the dc resistive loss P_{dcr}, ac resistive loss P_{acr}, proximity-effect loss P_p, and the total conductor loss P_{cdl} are presented. As can be seen, the ac resistive loss is negligible in this study. The situation would be different in an application without a dc component.

Next, let us consider proximity-effect loss. The first step is to calculate the dynamic resistance. Using (11.7-2) and (11.7-31), we obtain $D_{ea} = 4.99 \cdot 10^{-10}$ Hs and $D_{gcs} = 2.76 \cdot 10^{-9}$ Hs, for a total $D = 3.26 \cdot 10^{-9}$ Hs. Observe that the dynamic resistance was mostly associated with the slot, where the fields are stronger. The proximity-effect loss is then calculated using (11.9-22). Since t_r and t_f are both inversely proportional to frequency, we find that proximity-effect loss goes up with frequency squared, at least initially. We see this in Figure 11.15, where proximity-effect losses rapidly become

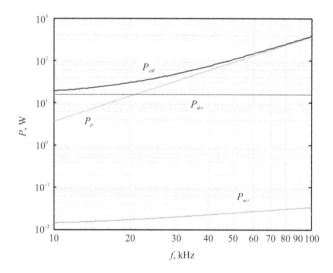

Figure 11.15 Conductor losses.

the dominant loss mechanism. At 20 kHz, proximity-effect losses are 14.5 W, which approaches the dc loss.

Before concluding this study, it is interesting to note that if we had posed the study differently and kept the input and output voltage the same and instead calculated the current ripple as a function of frequency, then the proximity-effect losses would have been approximately independent of frequency because current ripple would be inversely proportional to frequency.

11.10 Closing Remarks

In this chapter, we have examined methods to calculate skin-effect and proximity-effect losses in a general way. We also considered these losses for particular cases, including rotating machinery and a UI-core inductor. In both of these cases, the analysis presented could be readily introduced into our design framework. This would be very straightforward in the case of rotating machinery, wherein we would replace the machine's DC resistance with the AC resistance and include proximity-effect loss given by (11.8-11). We could also take a similar approach with regard to the inductor. However, in that case, it is more appropriate to design the inductor with the power converter, so that the semiconductor conduction and switching losses, voltage ripple current specifications, switching frequency, and capacitor selection could simultaneously be considered. This will be the topic of Chapter 13. Additional examples of this type of analysis are set forth by Shane and Sudhoff and Wang [9,10].

References

1 W. E. Boyce and R. C. Diprima, *Elementary Differential and Boundary Value Problems*, 4th edition, New York: John Wiley & Sons, 1986.
2 R. S. Charles, Computationally efficient winding loss calculation with multiple windings, arbitrary waveforms, and two-dimensional or three-dimensional field geometry, *IEEE Transactions on Power Electronics*, vol. **16**, no. 1, pp. 142–150, 2001.
3 J. A. Ferreira, *Electromagnetic modleing of power electronic conversters*, Boston: Kluwer Acadmic, 1989.
4 P. L. Dowell, Effects of eddy currents in transformer windings, *Proceedings of the IEEE*, vol. **113**, no. 8, pp. 1387–1394, 1966.
5 J. A. Ferreira, "Improved analytical modleing of conductive losses in magnetic components," *IEEE Transactions on Power Electronics*, vol. **9**, no. 1, pp. 127–131, 1994.
6 X. Nan and C. R. Sullivan, "An improved calculation of proximity-effect loss in high-frequency windings of round condutors," In *IEEE Power Electronics Specialists Conference*, 2003.
7 X. Nan and C. R. Sullivan, "Simplified high-accuracy calculation of Eeddy-current loss in round-wire windings," In *IEEE Power Electronics Specialists Conference*, Aachen, 2004.
8 C. D. McGillem and G. R. Cooper, *Continuous and discrete signal and system analysis*, 2nd edition, New York: CBS College Publishing, 1984.
9 G. Shane and S. D. Sudhoff, "Design paradigm for permanent magnet inductor-based power converters," *IEEE Transactions on Energy Conversion,* vol. **28**, no. 4, pp. 880–893, 2013.
10 S. Wang, Automated design approaches for power electronics converters, Doctoral Disseration, Purdue University, West Lafayette, IN, 2010.

Problems

1. From (11.1-5), (11.1-10), (11.1-11), and (11.2-3) show that we obtain (11.2-4).

2. Using (11.2-12) and (11.2-13), show that (11.2-4) becomes (11.2-14).

3. Consider a round conductor with a radius of 2 mm. Plot the magnitude of the current density versus radius at 1, 10, and 100 kHz. Normalize the current density so that it is plotted relative to the current density we would have for the dc (uniform distribution) case.

4. Using (11.3-7) and (11.3-11), obtain (11.3-12).

5. Show that using (11.3-12), which should also be applied to the y-axis, (11.3-13), and (11.3-14), we obtain (11.3-15).

6. Starting with (11.7-8), derive (11.7-22).

7. Starting with (11.7-9), derive (11.7-23).

8. Starting with (11.7-10), derive (11.7-24).

9. Using (11.7-8), (11.7-10), and (11.7-11), derive ranging from (11.7-12) to (11.7-14).

10. Using (11.7-9), (11.7-11), and (11.7-15), derive ranging from (11.7-16) to (11.7-18).

11. Using (11.8-1), derive (11.8-3).

12. Using (11.8-1) and (11.8-2), derive (11.8-5).

13. Using (11.5-23), (11.8-7), (11.8-8), and (11.8-9), obtain (11.8-10).

14. Consider Example 11.8B. Plot proximity-effect loss versus speed, as speed is varied from 1000 to 10 000 rpm.

15. Using Figure 11.13 and ranging from (11.9-7) to (11.9-8), obtain ranging from (11.9-10) to (11.9-13).

16. Consider Example 11.9A. Prove that if the voltages had been constant and that as a result the current ripple $(i_{mx} - i_{mn})$ was inversely proportional to frequency, then the proximity-effect loss would be independent of frequency.

12

Parasitic Capacitance

The focus of this book is magnetics, and so the reader may wonder why there is a chapter on capacitance. Unfortunately, all magnetic components include parasitic capacitance, which limits their performance, and therefore, it is important to be able to address capacitance in the design process. In this chapter, several capacitance mechanisms will be discussed, including turn-to-turn capacitance, layer-to-core capacitance, and layer-to-layer capacitance. Associated with these capacitances are both static and dynamic aspects.

12.1 Modeling Approach

Before considering the calculation of the capacitance associated with a magnetic component, it is appropriate to consider what we will do with this knowledge. To answer this question, let us suppose we are considering an inductor design. Thus far, our equivalent circuit model of an inductor would be the portion of Figure 12.1 enclosed by the dotted line. Therein, our inductance is denoted L, the winding resistance, which may include skin and proximity effects, is denoted R_s, and R_p may optionally be used to represent core loss. The frequency range of the applicability of the portion of the circuit in the dotted line is quite limited but can be improved by including the capacitive element in Figure 12.1. Observe that at low frequencies, we observe that the capacitive impedance is very large, and so our low-frequency model is unperturbed. At high frequencies, the inductive impedance becomes very large and so the capacitance dominates the behavior. Even so, this model is only valid over a limited range of frequencies; at some point past the first resonance, the frequency response of the inductor will deviate substantially from that predicted by Figure 12.1. Nevertheless, our goal will be to predict C, which will enable us to predict the first resonant frequency, thereby bounding the useful frequency range of the device. This first resonant frequency can then be the basis of a constraint in our design fitness function. Indeed, in Chapter 13, we will design a dc to dc converter in which we place just such a constraint on the inductor design.

To proceed to predict the capacitance, it must be realized that it can be viewed as having several components, including turn-to-turn capacitance denoted C_{tt}, coil-to-core capacitance C_{clcr}, and layer-to-layer capacitance C_{ll}. Thus, in Figure 12.1, we have

$$C = C_{tt} + C_{clcr} + C_{ll} \qquad (12.1\text{-}1)$$

In the next section, we will review electrostatics and capacitance of some simple geometries. In subsequent sections, we will discuss the calculation of C_{tt}, C_{clcr}, and C_{ll}.

Power Magnetic Devices: A Multi-Objective Design Approach, Second Edition. Scott D. Sudhoff.
© 2022 The Institute of Electrical and Electronics Engineers, Inc. Published 2022 by John Wiley & Sons, Inc.
Companion website: www.wiley.com/go/sudhoff/Powermagneticdevices

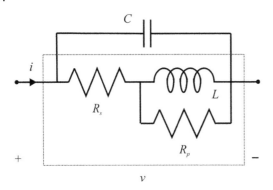

Figure 12.1 Inductor equivalent circuit.

12.2 Review of Electrostatics

The goal of this section is to review the relationships used to calculate capacitance. Recall that for a linear capacitor, the charge on each plate q is related to the capacitance C and capacitor voltage v by

$$q = Cv \qquad (12.2\text{-}1)$$

from which it follows that the current into the capacitor may be expressed as

$$i = C \frac{dv}{dt} \qquad (12.2\text{-}2)$$

In order to calculate the capacitance of a simple geometry, we need three relationships. The first of these is the definition of voltage between nodes a and b in terms of the electric field between the nodes, which is given by

$$v_{ab} = \int_a^b \mathbf{E} \cdot d\mathbf{l} \qquad (12.2\text{-}3)$$

The second of these is the relationship between electric flux density and the electric field in a given medium with a permittivity ε, a relationship that may be expressed as

$$\mathbf{D} = \varepsilon \mathbf{E} \qquad (12.2\text{-}4)$$

Permittivity is often expressed as the product of relative permittivity and the permittivity of free space, that is,

$$\varepsilon = \varepsilon_r \varepsilon_0 \qquad (12.2\text{-}5)$$

where the permeability of free space $\varepsilon_0 \approx 8.854187818 \cdot 10^{-12}$ F/m.

Finally, we have the relationship between the charge enclosed by a surface and the electric flux density on that surface. This is Gauss's law for electrostatics, which may be expressed as

$$q_{enc} = \oint_S \mathbf{D} \cdot d\mathbf{S} \qquad (12.2\text{-}6)$$

Figure 12.2 Electric field at a conductor's surface.

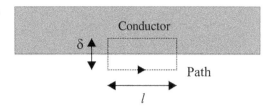

In order to gain insight into the behavior of electric fields, one property that is useful is that the electric field is orthogonal to the surface of a perfect conductor. To see this, from Maxwell's equations in an integral form, we have that

$$\oint \mathbf{E} \cdot d\mathbf{l} = -\int_S \frac{\partial \mathbf{B}}{\partial t} \cdot d\mathbf{S} \tag{12.2-7}$$

which states that the integral of the electric field over a closed path is equal to the integral of the time rate of the change of flux density over the surface bounded by the path. Let us suppose we are in a region wherein the magnetic flux density is zero and thus the integral of the electric field around a path is zero. Such a path is shown in Figure 12.2. Therein, we have a rectangular path with dimensions δ by l, where the dimension l is parallel to the conductor's surface. As we make l small and δ very small, and note that within the conductor the electric field is small (zero in a perfect conductor), and recalling the assumption that we are in region where the flux density is negligible, (12.2-7) reduces to

$$E_t l = 0 \tag{12.2-8}$$

where E_t is the tangential component of **E** in the direction of the horizontal portion of the path outside of the conductor. Hence, the electric field is normal to a conductor's surface.

Another property of electric fields that will be mentally useful comes from Gauss's law for electrostatics (12.2-6). From this relationship, in a charge-free region, we have that

$$\oint_S \mathbf{D} \cdot d\mathbf{S} = 0 \tag{12.2-9}$$

If we define electric flux as

$$\Phi = \int_S \mathbf{D} \cdot d\mathbf{S} \tag{12.2-10}$$

Then, the sum of the electric flux into any closed surface is equal to zero, provided that the closed surface does not contain a net charge.

Parallel Plate Capacitance

Let us now derive the capacitance of a simple parallel plate capacitor. Such a capacitor is shown in Figure 12.3. The capacitor consists of two conductive plates, each of length w and width l and separated by a distance g with a dielectric having a permittivity of ε. We assume that the electric field lines are directed from the top plate to the lower plate and the electric field is constant. As can be seen, the electric field is assumed to emanate and terminate at the conductors in the orthogonal direction, and the assumption of a uniform field satisfies (12.2-9).

436 | *12 Parasitic Capacitance*

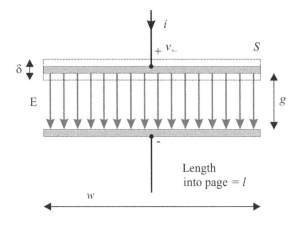

Figure 12.3 Parallel plate capacitor.

From (12.2-3), the voltage across the plates may be expressed as

$$v_{+-} = Eg \qquad (12.2\text{-}11)$$

where E is the component of **E** in the direction of the shown field lines.

Next, let us consider the application of (12.2-6) to Figure 12.3. Consider the closed surface that forms a narrow cuboid around the upper plate whose cross section is shown as a dotted line. Allowing δ to approach zero, so that the integral of $\mathbf{D} \cdot d\mathbf{S}$ over the sides is zero, and assuming that the electric field and hence electric flux density on the top surface is zero and that the electric field and electric flux density are uniform over the lower surface, we have that

$$q = wlD \qquad (12.2\text{-}12)$$

where D is the electric flux density in the direction of the field lines. Combining (12.2-4), (12.2-11), and (12.2-12), we have

$$q = \frac{wl\varepsilon}{g} v_{+-} \qquad (12.2\text{-}13)$$

Comparing (12.2-1) to (12.2-13), we obtain the well-known result for the capacitance of a parallel plate capacitor

$$C = \frac{wl\varepsilon}{g} \qquad (12.2\text{-}14)$$

Next, let us consider a related case wherein we have two layers of dielectric medium: one of thickness g_1 and permittivity ε_1 and the second with thickness g_2 and permittivity ε_2. We may treat this situation as two capacitors in series, which yields a net capacitance of

$$C = \frac{wl\varepsilon_{\mathit{eff}}}{g_1 + g_2} \qquad (12.2\text{-}15)$$

where $\varepsilon_{\mathit{eff}}$ is the effective permittivity of a metamaterial composed of the described layers. The effective permittivity may be expressed as

$$\varepsilon_{\mathit{eff}} = \frac{\varepsilon_1 \varepsilon_2 (g_1 + g_2)}{\varepsilon_1 g_2 + \varepsilon_2 g_1} \qquad (12.2\text{-}16)$$

Figure 12.4 Cylindrical plate capacitor.

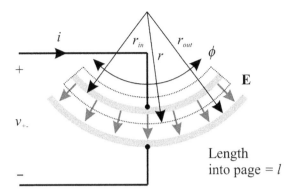

This treatment is useful because it allows us to treat multiple layers of dielectrics as a single material in the subsequent development.

Curved Plate Capacitance

Now, let us consider the case of the capacitance of two circularly curved plates as shown in Figure 12.4. Here, we have an inner plate with radius r_{in}, an outer plate with radius r_{out}, both plates spanning an angle ϕ, and extending into the page with a distance l. If ϕ is set to 2π, this represents a cylindrical capacitor. The case where $\phi < 2\pi$ is unusual as a capacitor but will prove to be useful in considering the capacitance between adjacent conductors of a winding.

To analyze this case, begin by considering the surface formed by the dotted line. On the upper part of the dotted line, it is assumed that the electric field is zero. From (12.2-6) and assuming that all electric flux exits through the lower curved surface, we have that

$$q = D_r r l \phi \tag{12.2-17}$$

where D_r is the magnitude of the electric flux density, which, from Figure 12.4, is assumed to be radially directed. From (12.2-4) and (12.2-17), we can solve for the magnitude of the radially directed electric field. In particular, we have

$$E_r = \frac{q}{l r \phi \varepsilon} \tag{12.2-18}$$

Next, from our definition of voltage (12.2-3) and taking any radial path from the inner plate to the outer plate,

$$v_{+-} = \frac{q}{l \phi \varepsilon} \ln\left(\frac{r_{out}}{r_{in}}\right) \tag{12.2-19}$$

Thus,

$$q = \frac{l \phi \varepsilon}{\ln\left(\frac{r_{out}}{r_{in}}\right)} v_{+-} \tag{12.2-20}$$

Comparing (12.2-1) to (12.2-20),

$$C = \frac{l \phi \varepsilon}{\ln\left(\frac{r_{out}}{r_{in}}\right)} \tag{12.2-21}$$

Before concluding this example, from (12.2-21), we conclude that the differential capacitance per unit angle is given by

$$\frac{dC}{d\theta} = \frac{l\varepsilon}{\ln\left(\frac{r_{out}}{r_{in}}\right)} \quad (12.2\text{-}22)$$

where upon capacitance may be calculated as the integral of the parallel differential capacitances as

$$C = \int_{-\phi/2}^{\phi/2} \frac{dC}{d\theta} d\theta \quad (12.2\text{-}23)$$

In (12.2-22) and (12.2-23), we have used θ as a dummy variable of integration so that ϕ can be used to express the limits of integration. Such an approach is circular in this example (on multiple levels), but we will find it to be extremely useful in our next example.

Insulated Conductor Capacitance

It will prove to be useful to consider a cylindrical conductor over a planar conductor termed a quasi-conductor in Figure 12.5 for reasons that will become evident. This arrangement will prove to be useful in the study of turn-to-turn capacitance, coil-to-core capacitance, and finally layer-to-layer capacitance. The cylindrical conductor in Figure 12.5 represents an insulated wire. The conductor and wire radius are r_c and r_w, respectively, and the thickness and relative permittivity of the wire insulation are denoted t_{wi} and ε_{rwi}, respectively. The auxiliary insulation could represent a layer of insulating paper between the layers of a winding or between the winding and a magnetic core. Its thickness and relative permittivity are denoted t_{ai} and ε_{rai}, respectively. The medium between the wire and the auxiliary insulation will have a relative permittivity of ε_{rm}. This medium could be free space whereupon $\varepsilon_{rm} = 1$ or could represent the potting material.

In Figure 12.5, the electric field lines are assumed to roughly follow the line of incremental capacitors labeled dC_1, dC_2, and dC_3, which correspond to the capacitance associated with the wire

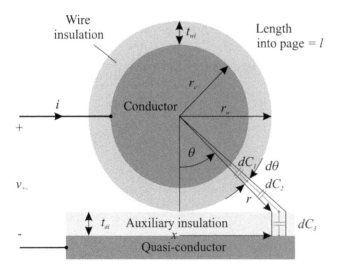

Figure 12.5 Conductor and quasi-conductor.

insulation, the medium between conductors, and the auxiliary insulation regions, respectively. From (12.2-22), we have

$$dC_1 = \frac{l\varepsilon_{rwi}\varepsilon_0}{\ln\left(\frac{r_w}{r_c}\right)} d\theta \qquad (12.2\text{-}24)$$

and

$$dC_2 = \frac{l\varepsilon_{rm}\varepsilon_0}{\ln\left(\frac{r}{r_w}\right)} d\theta \qquad (12.2\text{-}25)$$

In (12.2-25), note that from geometrical considerations,

$$r = \frac{r_w}{\cos\theta} \qquad (12.2\text{-}26)$$

Next, from (12.2-14), we have that

$$dC_3 = \frac{l\varepsilon_{rai}\varepsilon_0}{t_{ai}} dx \qquad (12.2\text{-}27)$$

From geometrical considerations,

$$x = r\sin\theta \qquad (12.2\text{-}28)$$

Using (12.2-26) and (12.2-28), we can show that

$$dx = \frac{r_w}{\cos^2\theta} d\theta \qquad (12.2\text{-}29)$$

whereupon (12.2-27) becomes

$$dC_3 = \frac{l\varepsilon_{rai}\varepsilon_0 r_w}{t_{ai}\cos^2\theta} d\theta \qquad (12.2\text{-}30)$$

The total differential capacitance is the three capacitors in series; thus,

$$dC = \frac{1}{\frac{1}{dC_1} + \frac{1}{dC_2} + \frac{1}{dC_3}} \qquad (12.2\text{-}31)$$

Substitution of (12.2-24), (12.2-25), and (12.2-30) into (12.2-32) and using (12.2-26), we have that

$$C_{hp}(\theta_{mx}) = \int_0^{\theta_{mx}} \frac{\varepsilon_0 l r_w}{\frac{r_w}{\varepsilon_{rwi}}\ln\left(\frac{r_w}{r_c}\right) + \frac{r_w}{\varepsilon_{rm}}\ln\left(\frac{1}{\cos\theta}\right) + \frac{t_{ai}}{\varepsilon_{rai}}\cos^2\theta} d\theta \qquad (12.2\text{-}32)$$

In (12.2-32), the function $C_{hp}(\theta_{mx})$ denotes the capacitance of one-half of a conductor to a plane. The variable θ_{mx} is chosen based on the application of this geometry. To the best of the author's knowledge, an analytical solution of (12.2-32) does not exist. However, the simple one-variable integration is readily and efficiently evaluated numerically. We will use (12.2-32) extensively in the next section.

It should be noted that alternate expressions for this capacitance can be obtained by modifying the assumptions on the path; for the case without auxiliary insulation, the reader is referred to Massarini [1].

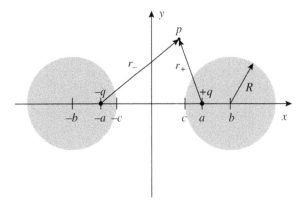

Figure 12.6 Capacitance of conductive cylinders.

Capacitance between Isolated Conductive Cylinders

Let us now consider the capacitance between two isolated conductive cylinders. This may sound similar to the work that we have already done, as round conductors are conductive cylinders. The difference is that in this case, the two cylinders will be isolated from other conductors so that we will need to include the impact of the electric field in all directions. The situation is shown in Figure 12.6. Therein, we have two conductive cylinders each of radius R and length l whose centers are separated by a distance $2b$. We will assume that the cylinders are long so that we may neglect field components in the length direction. We will find this situation useful when we discuss capacitance associated with structures. The process of finding the capacitance will be a bit different from our prior analysis, and we will be able to obtain an exact solution using the approach covered by Hause [2].

To begin, consider the situation in which there are no conductive cylinders present, only a line charge $\lambda = q/l$ at a position $(a,0)$, where l is the length, which is assumed to be long relative to the other dimensions so that we may utilize a 2D analysis. Using (12.2-4) and (12.2-6), we find that the electric field caused charge q, which, at a distance r_+ from this charge and radially directed away from the charge, is given by

$$E_+ = \frac{q}{2\pi r_+ \varepsilon l} \tag{12.2-33}$$

Based on (12.2-33) and (12.2-3), the voltage from point p relative to infinity is given by

$$v_{p\infty+} = \frac{q}{2\pi\varepsilon l} \ln\left(\frac{r_\infty}{r_+}\right) \tag{12.2-34}$$

where r_∞ denotes a distance very far from the charge.

As a next step, let us introduce a negative charge $-q$ at a position $(-a,0)$. The voltage at point p due to this negative charge is given by

$$v_{p\infty-} = \frac{-q}{2\pi\varepsilon l} \ln\left(\frac{r_\infty}{r_-}\right) \tag{12.2-35}$$

where r_- is the distance between point p and the negative charge. The potential at point p due to both charges is thus

$$v_{p\infty} = \frac{q}{2\pi\varepsilon l} \ln\left(\frac{r_-}{r_+}\right) \tag{12.2-36}$$

Expanding r_+ and r_- in (12.2-36), we have

$$v_{p\infty} = \frac{q}{2\pi\varepsilon l} \ln\left(\frac{\sqrt{(x+a)^2 + y^2}}{\sqrt{(x-a)^2 + y^2}}\right) \tag{12.2-37}$$

Inspecting (12.2-37), we can see that an equipotential surface is formed at the point for the set of points wherein the argument to the natural log function is constant. Let the argument to the natural log function be denoted k, which may be expressed as

$$k = e^{\frac{2\pi\varepsilon l v_{p\infty}}{q}} \tag{12.2-38}$$

and as

$$k = \frac{\sqrt{(x+a)^2 + y^2}}{\sqrt{(x-a)^2 + y^2}} \tag{12.2-39}$$

Evaluating (12.2-38) for a given desired equipotential $v_{p\infty}$ and charge q, thus determining k in (12.2-39), the equipotential surface is a set of points described by

$$y^2 + (x-b)^2 = R^2 \tag{12.2-40}$$

where

$$b = a\frac{k^2 + 1}{k^2 - 1} \tag{12.2-41}$$

and

$$R = \sqrt{b^2 - a^2} \tag{12.2-42}$$

Thus, we see that one equipotential surface (with the positive charge) is a circle, which is not centered about the charge but rather about the point $(b,0)$. There will be a corresponding negative equipotential around the negative charge located at $(-b,0)$. In this case, the charge takes on a negative value with a magnitude of that of the positive case and a constant analogous to k will take on the reciprocal of the value it takes in the corresponding positive case.

At this point, we can introduce the conducing cylinders into our analysis. In particular, if we place conducting cylinders of radius R centered at points $(-b,0)$ and $(b,0)$, inside the cylinders, the charge will move to the surface of the cylinders. Outside of our cylinders, nothing changes as we have not disturbed the field.

We can now find the capacitance between the cylinders. We know that the charge on the cylinders is $+q$ and $-q$. If we find voltage between the cylinders, we can find the capacitance. The voltage between the cylinders is denoted v_{cc} and is equal to the voltage at point $(c,0)$, which is less than the voltage at point $(-c,0)$, where

$$c = b - R \tag{12.2-43}$$

Using (12.2-36),

$$v_{cc} = \frac{q}{2\pi\varepsilon l}\left[\ln\left(\frac{a+c}{a-c}\right) - \ln\left(\frac{a-c}{a+c}\right)\right] \tag{12.2-44}$$

Manipulating (12.2-42)–(12.2-44) yields

$$v_{cc} = \frac{q}{\pi\varepsilon l}\ln\left(\frac{b}{R} + \sqrt{\left(\frac{b}{R}\right)^2 - 1}\right) \tag{12.2-45}$$

Thus, the capacitance between cylinders or radius R whose centers are separated by a distance $2b$ is given by

$$C = \frac{q}{v_{cc}} = \frac{\pi\varepsilon l}{\ln\left(\frac{b}{R} + \sqrt{\left(\frac{b}{R}\right)^2 - 1}\right)} \tag{12.2-46}$$

12.3 Turn-to-Turn Capacitance

We will now begin our study of parasitic capacitance in earnest. To this end, there are many types of parasitic capacitances, including turn-to-turn, coil-to-core, and layer-to-layer. We will begin our study with a turn-to-turn capacitance of three geometries: a simple coil, an orthogonally wound core, and an orthocyclicly wound core.

Simple Coil

Let us begin by considering the turn-to-turn capacitance of a simple coil, as shown in Figure 12.7. Therein, two cases are shown, (i) a simple coil in free space and (ii) a simple coil wound around an insulating paper surrounding a core.

In computing the capacitance between a turn and the next turn, we note that we can, in fact, use (12.2-32), which is the capacitance between the right side of a cylindrical conductor and the right side of a plane, as shown in Figure 12.5. If we have two cylindrical conductors in space (one above

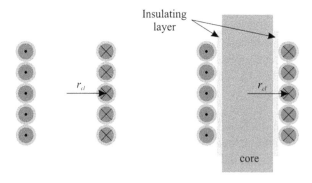

(a) Simple coil in free space   (b) Simple coil around core

Figure 12.7 Turn-to-turn capacitance of a simple coil.

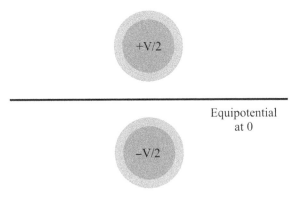

Figure 12.8 Symmetry conditions for capacitance.

and one below), the plane between them forms an equipotential surface and so, for analysis purposes, that plane, even though it has no physical instantiation, may be considered to be a conductor, as shown in Figure 12.8. We must realize, however, that the capacitance value given by (12.2-32) is only the capacitance from the upper conductor to the plane; we will also, in series with this, have the capacitance from the plane to the lower conductor. Adding these capacitors in series, we see that the capacitance from conductor to conductor will be half of that from conductor to plane. We must also consider that the value of capacitance in (12.2-32) is the contribution from the right-half side of the diagram in Figure 12.5. We need to account for the left side. This will result in a doubling of the capacitance. Thus, the static turn-to-turn capacitance may be expressed as

$$C_{tt0} = C_{hp}\left(\frac{\pi}{2}\right) \qquad (12.3\text{-}1)$$

where, in (12.2-32), since there are no physical limits on θ_{mx}, it is taken to be $\pi/2$, the auxiliary insulation thickness $t_{ai} = 0$ as there is none, and the length is taken as the mean length of a turn, which is accomplished by setting $l = 2\pi r_{cl}$.

In (12.3-1), we have used the term "static" and added a "0" subscript. The reason for this terminology is that it is based on an electrostatic analysis. Later, we will consider dynamic capacitance in which we assume a voltage profile, which is brought about by time-varying excitation (dynamic capacitance). However, for the moment, keep in mind that these capacitance values do not represent the entire coil but rather only the capacitance between one turn and an adjacent one.

In the case of the simple winding around the core, note that on the side of conductors with a core, θ_{mx} is constrained to $\pi/4$ (as an unsigned angle); going further would cause the assumed flux line to enter the insulating layer and core region. On the side of the conductors opposite to the core, θ_{mx} is still constrained to $\pi/2$. Keeping in mind the series connection of capacitance between the upper conductor to our imaginary plane and then that plane to the lower conductor, we have

$$C_{tt0} = \frac{1}{2}\left(C_{hp}\left(\frac{\pi}{4}\right) + C_{hp}\left(\frac{\pi}{2}\right)\right) \qquad (12.3\text{-}2)$$

Turn-to-Turn Capacitance of an Orthogonally Wound Coil

Now, let us consider the case of the turn-to-turn capacitance in a multilayer winding such as that shown in Figure 12.9. We will first consider the case of an orthogonal winding arrangement such as

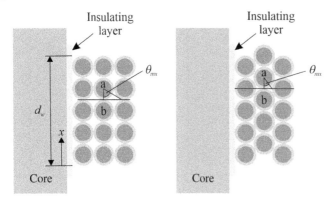

(a) Orthogonal winding   (b) Orthocyclic winding

Figure 12.9 Turn-to-turn capacitance in multilayer windings.

that shown in Figure 12.9(a). In particular, let us consider the turn-to-turn capacitance between conductors "a" and "b". The critical aspect of determining the capacitance is to establish the value of θ_{mx} to be used. Based on the assumed electric field path shown in Figure 12.4, and limiting θ_{mx} so that the field lines do not run into an adjacent conductor, and assuming that the insulating wire is much thinner than the conductor, we obtain $\theta_{mx} = \pi/3$. Recalling the fact that (12.2-32) is the capacitance on one side of the conductor centerline and that we have a series connection of capacitances, we have

$$C_{tt0} = C_{hp}\left(\frac{\pi}{3}\right) \tag{12.3-3}$$

where the length l used in (12.2-32) is based on the mean length of the turn, which varies with the layer. Note that this choice of $\theta_{mx} = \pi/3$ ignores the impact of the field lines of adjacent conductors. We could avoid this difficulty by taking $\theta_{mx} = \pi/4$. However, given that our approach tends to underestimate capacitance, the larger value of $\theta_{mx} = \pi/3$ is recommended. Actually, the question is not as critical as one might suppose because most of the capacitance comes from the region where the two conductors are in close physical proximity.

It should be observed that (12.3-3) is for an interior layer, which is to say that there is a layer on either side. On the innermost layer of a multilayer winding, we have

$$C_{tt0} = \frac{1}{2}\left(C_{hp}\left(\frac{\pi}{4}\right) + C_{hp}\left(\frac{\pi}{3}\right)\right) \tag{12.3-4}$$

On the outermost layer of a multilayer winding,

$$C_{tt0} = \frac{1}{2}\left(C_{hp}\left(\frac{\pi}{3}\right) + C_{hp}\left(\frac{\pi}{2}\right)\Big|\right) \tag{12.3-5}$$

Observe that both (12.3-4) and (12.3-5) are evaluated with $t_{ai} = 0$ in (12.2-32).

Turn-to-Turn Capacitance in an Orthocyclicly Wound Coil

Returning to Figure 12.9, the orthocyclicly wound coil exhibits a better packing factor. We will again use (12.2-32) to determine the turn-to-turn capacitance. Using geometry and again assuming thin insulation, we can show that the value of θ_{mx}, which avoids the assumed field path from overlapping an adjacent conductor, is

$$\theta_{mx,oc} = \mathrm{acot}\left(2\cos\left(\frac{\pi}{6}\right) - 1\right) \approx 0.9389 \quad (53.8°) \tag{12.3-6}$$

Thus,

$$C_{tt0} = C_{hp}(\theta_{mx,oc}) \tag{12.3-7}$$

Another approach to choose θ_{mx} would be such that the assumed path does not overlap with the assumed path for other conductors; this yields $\theta_{mx} = \pi/6$. Again, the value of l used in (12.2-32) is the mean length of a turn, which will vary by layer.

As in the case of the orthogonally wound core, there are two special cases. In the case of the innermost layer, we have

$$C_{tt0} = \frac{1}{2}\left(C_{hp}\left(\frac{\pi}{4}\right) + C_{hp}(\theta_{mx,oc})\right) \tag{12.3-8}$$

and in the case of the outermost layer,

$$C_{tt0} = \frac{1}{2}\left(C_{hp}(\theta_{mx,oc}) + C_{hp}\left(\frac{\pi}{2}\right)\right) \tag{12.3-9}$$

Note that both (12.3-8) and (12.3-9) are evaluated with $t_{ai} = 0$ in (12.2-32).

Dynamic Turn-to-Turn Capacitance

Thus far, we have found an expression for the capacitance between two adjacent turns of a given layer. We will henceforth denote this capacitance $C_{tt0,\,i}$, where i denotes the layer number. The turn-to-turn capacitance varies with the layer because the mean length of the turn varies and because of the special cases of the innermost and outermost layers. We will denote the number of turns on the ith layer as $N_{tl,i}$. Our goal is to calculate the value of the dynamic turn-to-turn capacitance C_{tt} to be used in (12.1-1) with Figure 12.1. We use the term dynamic capacitance because we assume that we have time-varying (i.e., dynamic) excitation such that there is an equal voltage associated with every turn.

With the assumption that there is an equal voltage associated with every turn, the voltage across a given turn may be expressed as

$$v_t = \frac{v}{N} \tag{12.3-10}$$

which is to say that the voltage across the device is uniformly distributed among the turns. The energy stored across by the net turn-to-turn (i.e., dynamic) capacitance C_{tt} may be expressed as

$$E = \frac{1}{2}C_{tt}v^2 \tag{12.3-11}$$

Using (12.3-10), the energy stored between two turns in the i'th layer is given by

$$E_{tt,i} = \frac{1}{2}C_{tt0,i}\left(\frac{v}{N}\right)^2 \tag{12.3-12}$$

Noting that a layer with $N_{tl,i}$ turns has $N_{tl,i} - 1$ turn-to-turn capacitances, the total energy storage due to turn-to-turn capacitance may be expressed as

$$E = \frac{1}{2} \sum_{i=1}^{N_l} C_{tt0,i} \left(\frac{v}{N}\right)^2 (N_{tl,i} - 1) \qquad (12.3\text{-}13)$$

where N_l is the number of layers. Equating (12.3-11) and (12.3-13),

$$C_{tt} = \frac{1}{N^2} \sum_{i=1}^{N_l} C_{tt0,i} (N_{tl,i} - 1) \qquad (12.3\text{-}14)$$

Observe that from (12.3-14), the turn-to-turn capacitance decreases with the number of turns. For this reason, the turn-to-turn capacitance is normally only a modest contributor to the total capacitance, except in the case of a single-layer winding, particularly in an inductor without a core.

Example 12.3A As an example, let us compute the capacitance associated with a coil in free space. The core itself consists of 30 turns of 8 AWG wire. The nominal wire diameter is 3.264 mm, and the thickness of the wire insulation is 44.5 μm. The insulation consists of a base coat of polyester glass (with a permittivity ranging from 4 to 4.5) with a top coat of polyamide (with a permittivity ranging from 2 to 2.5). The mean radius of the coil is 17.8 mm. Let us compute the coil capacitance. As a first step, we will compute the effective permittivity of the insulation system. Assuming, with little justification, that the permittivities of the base coat and top coats are 4.25 and 2.25 and that the thicknesses of the base and top coats are equal, we use (12.2-16) to find an effective relative permittivity of 2.94. From the radius of the coil, we obtain a mean turn length of 11.2 cm. Next, from (12.3-1), we find a C_{tt0} of 21.0 pF. Finally, we use (12.3-14) to find the dynamic turn-to-turn capacitance of 0.676 pF. Measuring the capacitance of the coil, we obtain 0.591 pF, an error of 14%. Although this is appreciable, there is considerable uncertainty in the insulation system. Indeed, depending upon the assumptions made within the stated range of uncertainties in the insulation system, capacitance values of 0.548–0.849 pF could be obtained. Also, there is uncertainty in the measurement. The measured value was based on frequency response data from 10 kHz to 10 MHz, which ends before the first resonance of the LC circuit is observed and from this perspective is not ideal. Finally, another source of error is that for a coil in free space, the assumed equal distribution of the voltage among the turns is questionable. As an interesting alternative method, the reader is referred to Grandi [3].

12.4 Coil-to-Core Capacitance

Another source of capacitance is that from coil to core. In the case of a magnetic steel or nanocrystalline core, the core is conductive, at least in certain directions. Even in the case of a ferrite core, the conductivity, while not high, is still much greater than that of an insulator. Thus, for the purposes of computing capacitance, we will treat the core as a conducting surface.

For this discussion, let as again refer to Figure 12.9. From our work in Section 12.2 and taking x to be in the upward direction in this figure, we have that the capacitance per unit length in the x-direction between the core and the first layer of the coil may be expressed as

$$\hat{C}_{clcr} = \frac{2 C_{hp}\left(\frac{\pi}{3}\right)}{2 r_w} \qquad (12.4\text{-}1)$$

where the hat is used to denote per unit length in the x-direction. With the connection shown, note that the difference in voltage between the beginning of the first turn and the end of the last turn of the first layer may be expressed as

$$\Delta v_1 = \frac{N_{tl,1}}{N} v \tag{12.4-2}$$

To proceed further, the connection of the core must be considered. This will impact the capacitance. If we assume that the core is connected to the negative terminal, then the voltage between the first layer and the core may be expressed as

$$v(x) = \frac{x}{d_w} \Delta v_1 \tag{12.4-3}$$

where d_w is the depth of the winding (and, in particular, the first layer), which is given by

$$d_w = 2 r_w N_{tl,1} N_{pr} \tag{12.4-4}$$

where N_{pr} is the number of parallel strands.

Our next step is to determine the coil-to-core capacitance C_{clcr} such that the energy stored in a parallel capacitance is equal to that stored along the first layer. In particular,

$$\frac{1}{2} C_{clcr} v^2 = \frac{1}{2} \hat{C}_{clcr} \int_0^{d_w} v^2(x) dx \tag{12.4-5}$$

Manipulation of (12.4-1) through (12.4-5) yields

$$C_{clcr} = \frac{2}{3} C_{hp} \left(\frac{\pi}{3}\right) N_{pr} \frac{N_{tl,1}^3}{N^2} \tag{12.4-6}$$

Next, let us consider what would happen if there were no intentional electrical connection between the coil and the core. In this case, there is still a parasitic conductive and capacitive connection. In order to have no net current going into the core, the average electrical potential between the coil and the core should be zero. Therefore, we have

$$v(x) = \Delta v_1 \left(\frac{x}{d_w} - \frac{1}{2} \right) \tag{12.4-7}$$

Repeating our previous derivation with (12.4-7) replacing (12.4-3), we obtain

$$C_{clcr} = \frac{1}{6} C_{hp} \left(\frac{\pi}{3}\right) N_{pr} \frac{N_{tl,1}^3}{N^2} \tag{12.4-8}$$

which is one-fourth the value of the previous case. Clearly, the electrical connection has a profound effect on the coil-to-core capacitance.

Finally, let us now consider a case where the core is grounded and where the center point of the applied voltage is also grounded, which is to say that the potentials of the negative and positive terminals are at $-v/2$ and $v/2$, respectively. This could be achieved by connecting the two terminals to a balanced high-impedance grounding system. In this case, the potential between the core and the negative terminal is $-v/2$ and so the potential between the core and the first layer is given by

$$v(x) = -\frac{1}{2} v + \Delta v_1 \frac{x}{d_w} \tag{12.4-9}$$

Using (12.4-9) in lieu of (12.4-3) yields a coil-to-core capacitance of

$$C_{clcr} = \frac{1}{6} C_{hp}\left(\frac{\pi}{3}\right) N_{pr} N_{tl,1} \left(3 - 6\frac{N_{tl,1}}{N} + 4\frac{N_{tl,1}^2}{N^2}\right) \quad (12.4\text{-}10)$$

Example 12.4A As a second example, let us consider a single-layer coil wound around an I-core. In particular, we have an I-core with a magnetic cross section of 25.07 by 25.46 mm, wrapped with an insulating paper with a thickness of $t_p = 0.254$ mm and a relative permittivity of $\varepsilon_{rp} = 2.6$. Fifty turns of 18 AWG wire are wrapped around the paper. The wire diameter is 1.204 mm, and the wire insulation has a thickness of 33.02 μm and a relative permittivity of 3.52. The outside cross section of the assembly (core, paper, wire) is 28.62 mm by 28.76 cm, when measured near the center of the sides. Our goal is to compute the capacitance.

We begin our analysis with the calculation of the turn-to-turn capacitance. The length of a turn may be approximated as

$$l = p_I + 8(t_p + r_w) \quad (12.4\text{A-}1)$$

where p_I is the perimeter of the I-core about which the winding is wound. Using (12.4A-1), we obtain $l = 10.91$ cm. We may use (12.3-2), which, in turn, utilizes (12.2-32) to find the static turn-to-turn capacitance. When performing this evaluation, note that $t_{ul} = 0$ since there is no auxiliary insulation between turns. Using (12.3-2), we obtain a static turn-to-turn capacitance of 13.6 pF, and from (12.3-14), we obtain a dynamic turn-to-turn capacitance of 0.266 pF. As it turns out, this capacitance will prove inconsequential.

Next, let us compute the coil-to-core capacitance. However, before doing this, we must resolve a discrepancy. Based on the size of the I-core, the radius of the wire, and the thickness of the insulating paper, the cross section of the outside of the winding should be 27.62 by 28.01 mm, which is smaller than what is measured. The reason for the difference is the bowing of the conductor, which occurs between the corners. Thus, while the conductor is very close to the paper at the corners, an air gap develops toward the middle of each side. Based on the dimensions, the air gap in the middle of each side is $t_g = 0.4345$ mm. This may seem like a small amount, but it has a profound effect on capacitance.

In order to address this, recall that (12.2-16) we derived an expression for the relative permittivity of a metamaterial that represented the series connection of two materials placed in series. We will take a similar approach here. Suppose we have an insulating material of 1 with a constant thickness of t_{ai1} and a relative permittivity of ε_{rai1}. In series with this insulating layer, we have a second insulating layer (nominally free space or potting material), which varies in thickness from 0 (at the corners) to t_{ai2} (at the center of each side) and has a relative permittivity of ε_{rai2}. Assuming that the slope is low, we obtain an "almost" parallel plate capacitor. We will assume that the trajectory of the wire over the insulator is a parabola so that at the center, the wire is parallel to the core and that the wire touches the core at both ends. Taking the thickness of the homogenized material to be the mean thickness of this trajectory plus the thickness of the paper, it can be shown that the effective thickness is given by

$$t_{ai} = t_{ai1} + \frac{2}{3} t_{ai2} \quad (12.4\text{A-}2)$$

and effective relative permittivity is given by

$$\varepsilon_{ai} = \frac{\varepsilon_{a1}\varepsilon_{a2}t_{ae}}{2\sqrt{t_{a2}\varepsilon_{a1}(t_{a2}\varepsilon_{a1} + t_{a1}\varepsilon_{a2})}} \left[\ln\left(\frac{t_{a2}\varepsilon_{a1} + t_{a1}\varepsilon_{a2} + \sqrt{t_{a2}\varepsilon_{a1}(t_{a2}\varepsilon_{a1} + t_{a1}\varepsilon_{a2})}}{t_{a2}\varepsilon_{a1} + t_{a1}\varepsilon_{a2} - \sqrt{t_{a2}\varepsilon_{a1}(t_{a2}\varepsilon_{a1} + t_{a1}\varepsilon_{a2})}}\right)\right]$$

$$(12.4\text{A-}3)$$

The derivations of (12.4A-2) and (12.4A-3) are left as an exercise. For our example, (12.4A-2) and (12.4A-3) with $t_{ai1} = t_p$, $\varepsilon_{rai1} = \varepsilon_{rp}$, $t_{ai2} = t_g$, and $\varepsilon_{rai2} = 1$ yield an effective value of an auxiliary insulation thickness of $t_{ai} = 543.7$ μm and $\varepsilon_{rai} = 1.6861$.

We can now evaluate the coil-to-core capacitance. For a single-layer winding, both (12.4-8) and (12.4-10) are applicable and equivalent. Taking the length of the coil-to-core interface of a turn to be

$$l = p_I + 8t_{ai} \tag{12.4A-4}$$

one calculates the coil-to-core capacitance to be 12.21 pF, so the total capacitance is 12.47 pF. Clearly, in this example, the turn-to-turn capacitance is inconsequential. Based on the frequency response between 600 kHz and 6 MHz, a range that captures the first resonance of the frequency response, one obtains a measured capacitance of 12.13 pF. Thus, the error in our measurement is less than 3%. Capacitance calculations normally do not work out so well. As a final note, had the air region between the insulating paper and the coil been ignored, the capacitance would have been calculated to be 25.91 pF. Clearly, a small gap can lead to a large change in capacitance.

12.5 Layer-to-Layer Capacitance

We will next consider layer–layer capacitance, which is often the dominant source of capacitance, as observed by Biela [4]. In this case, we will see that the connection of the coils of a winding will have an impact on the observed value of capacitance. Consider Figure 12.10, wherein one side of an orthogonal winding is shown. In this case, we have included the provision for layer-to-layer insulation, which can be used to reduce the capacitance or reduce the electric field between layers. For the orthogonal winding, two methods of connecting the coils are shown. In Figure 12.10(a), a standard winding arrangement is shown. This arrangement is physically the easiest to implement. A flyback winding arrangement is shown in Figure 12.10(b). The flyback arrangement offers reduced capacitance. From a magnetic point of view, the two arrangements are identical.

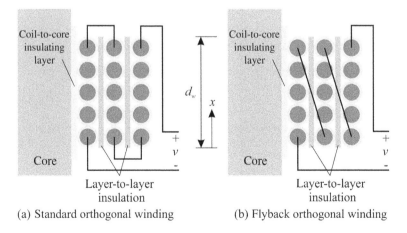

Figure 12.10 Calculation of layer-to-layer capacitance.

We will begin our analysis by calculating the static turn-to-turn capacitance between two turns on different layers. In particular, the static turn-to-turn capacitance associated with layer i (which is between layer i and layer $i + 1$) may be expressed as

$$C_{ltt0,i} = \begin{cases} C_{hp}\left(\dfrac{\pi}{3}\right) & \text{orthogonal} \\ C_{hp}(\theta_{mx,oc}) & \text{orthocyclic} \end{cases} \tag{12.5-1}$$

where the length of the turn is taken as

$$l = \frac{1}{2}(l_{t,i} + l_{t,i+1}) \tag{12.5-2}$$

and the thickness of the auxiliary insulation is taken to be half that of the paper used, and the permittivity of the auxiliary insulation is taken to be that of the insulating paper. This is because of the assumed symmetry conditions in Figure 12.5 and the derivation of $C_{hp}(\theta_{mx})$. In essence, each conductor is associated with half the thickness of the insulating paper. It should be observed that it is often the case that the layers do not line up perfectly. The error because of this effect is modest and explored to some extent in reference [5].

Next, the capacitance per unit length in the x-direction as shown in the center of Figure 12.10 is given by

$$\hat{C}_{ll,i} = \frac{C_{ltt0,i}}{2r_w} \tag{12.5-3}$$

The total voltage associated with the two layers may be expressed as

$$\Delta v_i = \frac{N_{tl,i} + N_{tl,i+1}}{N} v \tag{12.5-4}$$

In the case of the standard winding, the voltage between the layers may be expressed as

$$v_{ll}(x) = \left(1 - \frac{x}{d_{w,i}}\right)\Delta v_i \tag{12.5-5}$$

In order to determine the capacitance between layer i and layer $i + 1$, an energy argument is used. In particular, denoting the capacitance between these layers as $C_{ll,i}$, we have that

$$\frac{1}{2}C_{ll,i}v^2 = \frac{1}{2}\int_0^{d_{w,i}} \hat{C}_{ll,i}v_{ll}^2(x)dx \tag{12.5-6}$$

where $d_{w,i}$ is the depth of the interface between the two layers and may be expressed as

$$d_{w,i} = 2r_w \min(N_{tl,i}, N_{tl,i+1})N_{pr} \tag{12.5-7}$$

In (12.5-7), N_{pr} is the number of parallel conductors used to implement a winding, as described in Chapter 8 in the case of a transformer. The parallel conductor associated with a turn is assumed to be stacked vertically in Figures 12.9 and 12.10.

Manipulating (12.5-3) through (12.5-7), the layer-to-layer capacitance between layer i and $i + 1$ is given by

$$C_{ll,i} = \frac{1}{3}C_{ltt0,i}N_{pr}\min(N_{tl,i}, N_{tl,i+1})\left(\frac{N_{tl,i} + N_{tl,i+1}}{N}\right)^2 \tag{12.5-8}$$

whereupon the total layer-to-layer capacitance is given by

$$C_{ll} = \sum_{i=1}^{N_l-1} C_{ll,i} \qquad (12.5\text{-}9)$$

Now, let us consider the situation for the flyback winding. The analysis is very similar, except the voltage between the layers does not vary with x. In particular,

$$v_{ll}(x) = \frac{1}{2}\Delta v_i \qquad (12.5\text{-}10)$$

Repeating our analysis using (12.5-10) in lieu of (12.5-5), one obtains

$$C_{ll,i} = \frac{1}{4} C_{ltt0,i} N_{pr} \min(N_{tl,i}, N_{tl,i+1}) \left(\frac{N_{tl,i} + N_{tl,i+1}}{N}\right)^2 \qquad (12.5\text{-}11)$$

Example 12.5A In this example, we will consider the capacitance calculation of a two-layer winding, which includes a layer of insulating paper between the core and the first layer and a layer of insulating paper between the two winding layers.

In particular, we have an I-core with a magnetic cross section of 25.5 by 25.6 mm. The insulating paper used has a thickness of $t_p = 0.254$ mm and a relative permittivity of $\varepsilon_{rp} = 2.6$. Fifty turns of 18 AWG wire are wrapped around the paper. The wire diameter is 1.204 mm, and the wire insulation has a thickness of 33.02 μm and a relative permittivity of 3.52. Then, another layer of insulating paper is wrapped around the first layer, and a second winding layer of 50 turns is wrapped, so that the total number of turns is 100.

As in the case of Example 12.4A, we will have to address the fact that coils, paper layers, and core are not in constant contact. In particular, measuring the outside dimensions of the coil, one measures the cross section to be 32.9 by 32.2 mm, where the measurement is taken at the center of the sides. This indicates an unintended gap thickness of $t_g = 0.9440$ mm at the center of the coils. This gap includes regions on each side of both layers of the insulating paper. Taking the same approach as in example 12.4A, we can homogenize the paper/unintended air gap region. This yields a total thickness of the two layers of insulating paper and the gap to be 1.137 mm with an effective relative permittivity of 1.677. This value of thickness includes both layers of insulating paper. Thus, going forward, we will treat the effective thickness of each insulation layer to be $t_{eil} = 0.5687$ mm with $\varepsilon_{eil} = 1.677$. Clearly, we do not know that the gap is equally distributed between the coil–core region and the coil-to-coil region, but it seems a reasonable assumption in the lack of any other knowledge.

We will now find the turn-to-turn capacitance. The length of the first and second layers may be expressed as

$$l_1 = p_I + 8(t_{ai} + r_w) \qquad (12.5\text{A-}1)$$
$$l_2 = p_I + 8(2t_{ai} + 3r_w) \qquad (12.5\text{A-}2)$$

Using (12.3-4) and (12.3-5), we obtain the static capacitances of the two layers to be 13.50 and 15.56 pF, respectively. Note that when evaluating (12.3-4) and (12.3-5), we use the lengths given by (12.5A-1) and (12.5A-2). It is important that in this calculation, t_{ai} is set to 0 and ε_{ai} is set to any positive number in (12.2-32). From (12.3-31), we obtain $C_{tt} = 0.1424$ pF, which, like in the case of example 12.4A, will prove inconsequential.

Next, we will compute the core-to-coil capacitance. For this calculation, we will estimate the turn length of

$$l = p_I + 8t_{ai} \qquad (12.5\text{A-}3)$$

Using (12.4-10) with $t_{ai} = t_{eil}$ and $\varepsilon_{ai} = \varepsilon_{eil}$, we obtain $C_{clcr} = 11.9$ pF. This will prove significant but not dominant.

Finally, we compute the layer-to-layer capacitance. From (12.5-10), with the mean value of (12.5A-1) and (12.5A-2) as the length, and using an auxiliary insulation thickness of $t_{ai} = t_{eil}/2$ (because each winding layer has half the full thickness) in (12.2-32), we obtain $C_{ll} = 41.57$ pF for a total capacitance of 53.62 pF. Based on the frequency response between 100 kHz and 1 MHz, a region that captures the first resonance of the system, the capacitance is measured to be 51.50 pF, so that the error of the calculated value is less than 5%.

12.6 Capacitance in Multi-Winding Systems

In the case of multi-winding systems, such as the core-type transformer with parallel connected winding such as that shown in Figure 12.11, there are additional capacitive components, particularly winding-to-winding capacitance. In the case of multi-winding systems, the exact nature of the connections, while not important magnetically, can make a significant difference in terms of analysis, modeling, and the amount of parasitic capacitance. The connection scheme shown in Figure 12.11 is attractive in that it is symmetric and the differential-mode and common-mode equivalent electrical circuits are decoupled. Observe that in this arrangement, the two coils of the primary (and secondary) winding are connected in parallel.

Figure 12.12 depicts an equivalent circuit of the transformer shown in Figure 12.11. Therein, v_{pd} is the differential primary voltage defined as

$$v_{pd} = v_{p+} - v_{p-} \tag{12.6-1}$$

where v_{p+} and v_{p-} are the node potentials relative to node P in Figures 12.11 and 12.12. Likewise, v_{sd} is the differential secondary voltage defined as

$$v_{sd} = v_{s+} - v_{s-} \tag{12.6-2}$$

The differential primary and secondary voltages were simply referred to as the primary and secondary voltages in Chapter 7.

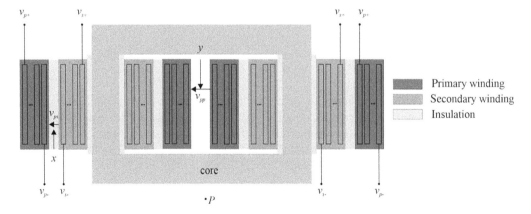

Figure 12.11 Core-type transformer.

Figure 12.12 Transformer equivalent circuit.

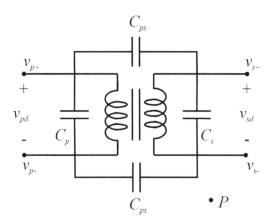

The circuit shown in Figure 12.12 cannot be applied to an arbitrary transformer. As we will see, the connections and details of the winding are very important. However, with the structure shown in Figure 12.11, this circuit does apply. For other windings and connections, the equivalent circuit shown in Figure 12.12 may exhibit the capacitance between the positive nodes of the primary and secondary, which are different from those connecting to negative nodes; in addition, cross capacitance between the positive nodes of the primary winding and the negative nodes of the secondary comes into play.

In Figure 12.12, the capacitance C_s is readily calculated as

$$C_s = 2(C_{tts} + C_{clcrs} + C_{lls}) \qquad (12.6\text{-}3)$$

where C_{tts}, C_{clcrs}, and C_{lls} are the turn-to-turn, coil-to-core, and layer-to-layer capacitances of the secondary winding, respectively, which may be found using methods already described. The factor of 2 in (12.6-3) arises because there are two coils in parallel. Finally, when computing C_{clcrs} with this connection strategy (12.4-9) applies since the core voltage will be the mean of the positive and negative secondary node potentials.

Likewise, the primary capacitance may be expressed as

$$C_p = 2C_{ttp} + 2C_{llp} + C_{clclpp} + C_{clclps} \qquad (12.6\text{-}4)$$

In (12.6-4), C_{ttp} and C_{llp} are the turn-to-turn and layer-to-layer capacitances, which are calculated in the same way as for an isolated winding. The term C_{clclpp} is due to the interaction of the primary coils, and the term C_{clclps} is due to the interaction of the primary coils with the secondary coils and is related to C_{ps}. Note that alternatively, it would have been possible to have added an alternate value of capacitance to the secondary winding. In (12.6-4), the stray capacitive interaction between the primary winding and core is neglected.

Let us consider the calculation of C_{clclpp}. To this end, consider our work with the transformer in Chapter 7 and, in particular, Figure 7.9. Therein, neglecting the dimension w_{cec}, the shape of the outside of the primary winding when viewed from above is that of a cylinder with the two halves separated by a distance l_{cc}. Because of the symmetry involved with one primary coil and the opposite primary coil (considering the interface inside of the core), it seems reasonable to estimate the capacitance as that of two conductive cylinders of radius r_{po}, length d_{pw}, and whose centers are separated by $2r_{po} + c_{pp}$ in parallel with a parallel plate capacitor with an area of $l_{cc}d_{pw}$ and a plate separation of c_{pp}. Thus, from (12.2-14) and (12.2-46), the capacitance per unit length (in this case in the direction of d_{pw}) may be expressed as

$$\hat{C}_{clclpp} = \frac{l_{cc}\varepsilon}{c_{pp}} + \frac{\pi\varepsilon}{\ln\left(1 + \frac{c_{pp}}{2r_{po}} + \sqrt{\left(1 + \frac{c_{pp}}{2r_{po}}\right)^2 - 1}\right)} \quad (12.6\text{-}5)$$

We must now consider the voltage profile between the cores. Neglecting partial turns, from Figure 12.11, the voltage between the two coils of the primary winding may be expressed as

$$v_{pp} = v_{pd}\left(1 - \frac{N_{pud}}{N_p}\right) \quad (12.6\text{-}6)$$

In (12.6-6), it is assumed that the layers are full so that the number of turns per layer is N_{pud} (see Figure 7.10). Computing the energy stored using (12.6-6), one obtains

$$E = \frac{1}{2}v_{pd}^2 \hat{C}_{clclpp} d_w \left(\frac{N_{pud}}{N_p} - 1\right)^2 \quad (12.6\text{-}7)$$

Thus,

$$C_{clclpp} = \hat{C}_{clclpp} d_w \left(\frac{N_{pud}}{N_p} - 1\right)^2 \quad (12.6\text{-}8)$$

Our next step is to calculate C_{clclps}. The first step is to calculate the capacitance per unit length. The capacitances per unit length of the primary and secondary conductors are $C_{hp}(\pi/3)|_p / r_{pw}$ and $C_{hp}(\pi/3)|_s / r_{sw}$, respectively, where r_{pw} and r_{sw} are the primary and secondary wire radii, respectively. In evaluating these quantities, the value of t_{ai} in (12.2-32) is one-half the layer-to-layer thickness. Adding these capacitances per unit length in series,

$$\hat{C}_{clclps0} = \frac{C_{hp}\left(\frac{\pi}{3}\right)|_p C_{hp}\left(\frac{\pi}{3}\right)|_s}{r_{sw} C_{hp}\left(\frac{\pi}{3}\right)|_p + r_{pw} C_{hp}\left(\frac{\pi}{3}\right)|_s} \quad (12.6\text{-}9)$$

Next, referring to Figure 12.11, and assuming full layers, the voltage on the primary and secondary coils at a distance x above the bottom of the coils relative to node P may be expressed as

$$v_p = v_{p-} + \frac{x}{d_w} v_{pd} \frac{N_{pud}}{N_p} \quad (12.6\text{-}10)$$

$$v_s = v_{s-} + \frac{x}{d_w} v_{sd} \frac{N_{sud}}{N_s} \quad (12.6\text{-}11)$$

The primary coil to secondary voltage is given by

$$v_{ps} = v_p - v_s \quad (12.6\text{-}12)$$

The energy stored in the region between the primary and secondary coils (on the left side) is thus given by

$$E_l = \frac{1}{2}\hat{C}_{clclps0} \int_0^{d_w} v_{ps}^2 dx \quad (12.6\text{-}13)$$

Combining (12.6-10)–(12.6-13), we obtain

$$E_l = \frac{1}{2}\hat{C}_{clclps0}d_w\left[(v_{p-} - v_{s-})^2 + (v_{p-} - v_{s-})m + \frac{1}{3}m^2\right] \quad (12.6\text{-}14)$$

where

$$m = \frac{N_{pud}}{N_p}v_{pd} - \frac{N_{sud}}{N_s}v_{sd} \quad (12.6\text{-}15)$$

If we repeat the analysis on the right side, we obtain

$$E_r = \frac{1}{2}\hat{C}_{clclps0}d_w\left[(v_{p+} - v_{s+})^2 - (v_{p+} - v_{s+})m + \frac{1}{3}m^2\right] \quad (12.6\text{-}16)$$

Adding the energy stored on the left and right sides and making the approximation

$$v_{sd} \approx v_{pd}\frac{N_s}{N_p} \quad (12.6\text{-}17)$$

we obtain

$$E = \frac{1}{2}C_{ps}(v_{p+} - v_{s+})^2 + \frac{1}{2}C_{ps}(v_{p-} - v_{s-})^2 + \frac{1}{2}C_{clclps}v_{pd}^2 \quad (12.6\text{-}18)$$

where

$$C_{ps} = \hat{C}_{clclps0}d_w \quad (12.6\text{-}19)$$

and

$$C_{clclps} = \hat{C}_{clclps0}d_w\frac{2(N_{pud} - N_{sud})^2 + 3(N_s - N_p)(N_{pud} - N_{sud})}{3N_p^2} \quad (12.6\text{-}20)$$

12.7 Measuring Capacitance

Let us now consider the problem of measuring capacitance. This can be readily accomplished by measuring the device frequency response and, in particular, the impedance versus frequency.

Let us consider the case of a single-winding system. We may use Figure 12.1 as an equivalent circuit for the winding system. Therein, C represents the effective value of the total dynamic capacitance, L represents the inductance of the winding, R_s represents the series resistance of the winding, and R_p is a resistance used to represent core loss. As suggested in reference [5], the resistances are frequency-dependent and may be represented in the forms

$$R_s(\omega) = R_{so}\left(\frac{\omega}{\omega_b}\right)^{n_s} \quad (12.7\text{-}1)$$

and

$$R_p(\omega) = R_{po}\left(\frac{\omega}{\omega_b}\right)^{-n_p} \quad (12.7\text{-}2)$$

where R_{so}, n_s, R_{po}, and n_p are positive constants, and ω_b is a base frequency, which is typically taken to be 1 rad/s. Note that because of their respective positions in the circuit, the losses associated with both of these terms increase with frequency. From Figure 12.1, the impedance of the equivalent circuit is readily expressed as

$$Z(\mathbf{x}, \omega) = \frac{j\omega L(R_s(\omega) + R_p(\omega)) + R_s(\omega)R_p(\omega)}{-\omega^2 LC(R_s(\omega) + R_p(\omega)) + j\omega(CR_s(\omega)R_p(\omega) + L) + R_p(\omega)} \quad (12.7\text{-}3)$$

where $\mathbf{x}$ is the vector of circuit parameters, that is,

$$\mathbf{x} = \begin{bmatrix} L & R_p & n_p & R_s & n_s & C \end{bmatrix}^T \quad (12.7\text{-}4)$$

Next, the impedance of the winding is measured at a number of points (tens to hundreds). We will denote the radian frequency and measured impedance at the ith point as ω_i and Z_i, respectively. Defining the error in fitting the ith data point as

$$e_i(\mathbf{x}) = \left| \frac{Z_i - Z(\mathbf{x}, \omega_i)}{Z_i} \right| \quad (12.7\text{-}5)$$

we may define a fitness function

$$f(\mathbf{x}) = \frac{N}{\sum_i e_i^2(\mathbf{x})} \quad (12.7\text{-}6)$$

where N is the number of points. Our parameter vector is thus obtained as

$$\mathbf{x} = \operatorname{argmax}(f(\mathbf{x})) \quad (12.7\text{-}7)$$

and we obtain $\mathbf{x}$, and therefore C, as the solution of a single-objective optimization problem that is readily solved using, for example, a genetic algorithm.

In obtaining the frequency response, it is important to select the correct frequency range. Ideally, this should include the first resonance and must avoid further resonances because the equivalent circuit loses validity between the first and subsequent resonances. As an additional complication, when using the method for conductive cores (silicon steel or nanocrystalline), it may be necessary to make the inductance frequency-dependent as well as the resistances. We will now consider an example to illustrate the method.

Example 12.7A In order to illustrate the procedure for measuring capacitance, let us consider the measurement of the capacitance of the inductor described in Example 12.5A. A photograph of this inductor appears in Figure 12.13, along with the inductors used in Examples 12.3A and 12.4A. As a first step, the impedance of the inductor is measured from 100 Hz to 10 MHz with a Keysight E4990A 20 Hz to 120 MHz impedance analyzer. When conducting this test, the inductor is suspended in free space with a string in order to minimize stray capacitance contributions. The resulting impedance frequency response is shown in Figure 12.14.

As can be seen in Figure 12.14, the inductor, like all distributed parameter systems, exhibits multiple resonances, which our simple model (Figure 12.1) cannot capture. Therefore, we truncate the impedance frequency response data to between 100 kHz and 1 MHz, a region that captures the first

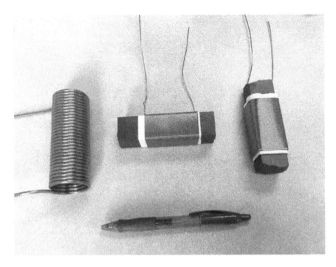

Figure 12.13 Photograph of inductors for Examples 12.3A (leftmost), 12.4 A (center), and 12.5A (rightmost).

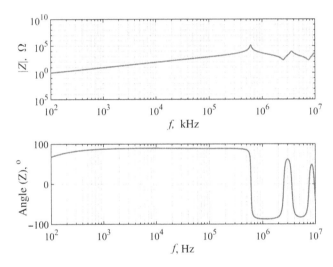

Figure 12.14 Full impedance frequency response of a two-layer inductor.

resonance. This data (along with the resulting fitted impedance) is shown in Figure 12.15. Based on the truncated data range, we will solve (12.7-7) using a genetic algorithm. The domain of the six parameters is shown in Table 12.1.

Performing the optimization, one obtains $L = 1.352\,\text{mH}$, $R_p = 994.6\,\text{M}\Omega$, $n_p = 0.5757$, $R_s = 22.33\,\Omega$, $n_s = 0$, and $C = 51.50$ pF. In this particular problem, it would seem that the parallel resistance and frequency dependence of the resistances were probably not necessary. The frequency response based on (12.7-3) with these parameters is seen to match the measured data very well in Figure 12.15.

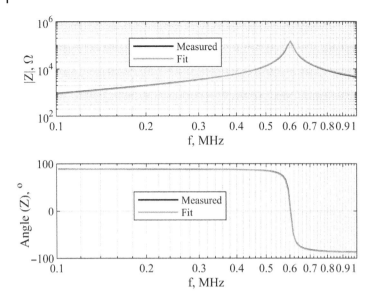

Figure 12.15 Measured and fitted impedance frequency responses.

Table 12.1 Parameter Domain

Parameter	L	R_p	n_p	R_s	n_s	C
Minimum	10^{-6}	10^{3}	0	10^{-5}	0	10^{-15}
Maximum	10^{-1}	10^{9}	3	10^{12}	5	10^{-9}
Encoding	log	log	lin	log	lin	log
Chromosome	1	1	1	1	1	1

References

1 M. K. A. Massarini, "Self-capacitance of inductors," *IEEE Transactions on Power Electronics*, vol. **12**, no. 4, pp. 671–676, 1997.
2 J. M. H.A. Hause, "*Electromagnetic Fields and Energy*," Cambridge, MA: MIT; 2020 [Online]. Available from: web.mit.edu/6.013_book/www/chapter4/4.6.html. [Accessed 27 January 2020].
3 M. K. U. R. G. Grandi, "Stray capacitances of single-layer solenoid air-core inductors," *IEEE Transactions on Industry Applications*, vol. **35**, no. 5, pp. 1162–1168, 1999.
4 J. K. J. Biela, "Using transformer parasitics for resonant converters—a review of the calculation of the stray capacitance of transformers," *IEEE Transactions on Industry Applications*, vol. **44**, no. 1, pp. 223–233, 2008.
5 S. D. Sudhoff, H. Singh, V. S. Duppalli and R. R. Swanson, "Capacitance of UR-core and C-core common mode inductors," *IEEE Power and Energy Technology Systems Journal*, vol. **6**, no. 2, pp. 113–121, 2019.
6 S. D. Sudhoff, "*MATLAB Codes for Power Magnetic Devices: A Multi-Objective Design Approach*, 2nd ed. Piscataway, NJ: Wiley-IEEE Press; 2014 [Online]. Available from: http://booksupport.wiley.com.

Problems

1. Starting with (12.2-14), derive (12.2-15) and (12.2-16).

   Using (12.2-3) and (12.2-18), derive (12.2-20).

3. Using (12.2-26) and (12.2-28), show (12.2-29).

4. Using (12.2-24), (12.2-25), (12.2-26), (12.2-30), and (12.2-31), obtain (12.2-32).

5. Using (12.2-3) and (12.3-33), obtain (12.2-34).

6. From (12.2-39), we obtain (12.2-40)–(12.2-42).

7. From (12.2-42) through (12.2-44), derive (12.2-45).

8. Using Figure 12.9(b), derive (12.3-6).

9. Consider Example 12.3A. Using the assumed value of the effective relative permittivity, plot $C_{hp}(\theta)$ assuming a turn length of 1 m.

10. Using (12.4-1), (12.4-2), (12.4-4), (12.4-5), and the voltage profile of (12.4-3), derive (12.4-6).

11. Using (12.4-1), (12.4-2), (12.4-4), (12.4-5), and the voltage profile of (12.4-7), derive (12.4-8).

12. Using (12.4-1), (12.4-2), (12.4-4), (12.4-5), and the voltage profile of (12.4-9), derive (12.4-10).

13. From geometry, derive (12.4A-2) and (12.4A-3).

14. Repeat Example 12.4A if a potting material with a relative permittivity of 2.5 is used.

15. From (12.5-3) through (12.5-7), derive (12.5-8).

16. Computing energy stored from (12.6-6), derive (12.6-7) and (12.6-8).

17. From (12.6-10) through (12.6-13), derive (12.6-14) through (12.6-16).

18. From (12.6-14) through (12.6-17), derive (12.6-18) through (12.6-20).

13

Buck Converter Design

While the focus of this book is magnetics, not power electronics, a chief power magnetic application is in power electronics, and so it makes sense not to design power magnetic components in isolation but rather to design them together with the power electronic converter in which they are used. In this chapter, such an approach is illustrated for a buck converter, which efficiently reduces dc voltage from one level to another.

13.1 Buck Converter Analysis

A buck converter is depicted in Figure 13.1. Therein, dashed rectangles encapsulate single components such as inductors and capacitors that are represented with multiple circuit elements. For example, an inductor is represented as an ideal inductor in series with a parasitic resistance. Elements of this circuit include an input inductor with inductance L_{in} and resistance r_{in}, an input capacitor with capacitance C_{in} and effective series resistance r_{Cin}, a power transistor, a power diode, an output inductor with inductance L_{out} and resistance r_{out}, and an output capacitor of capacitance C_{out} and effective series resistance r_{Cout}. The power electronic switch could be a variety of devices but is commonly an insulated-gate bipolar transistor (IGBT) or a metal–oxide–semiconductor field-effect transistor (MOSFET). The power diode may be PN or Schottky diode. Variables of interest include the input voltage and current v_{in} and i_{in}, the transistor and diode currents i_t and i_d, the switch point voltage v_s, the output inductor current i_l, the input and output capacitor currents i_{Cin} and i_{Cout}, and finally the output voltage and current v_{out} and i_{out}, respectively.

Operation

The operation of the buck converter can be in continuous mode or discontinuous mode. Typical waveforms for continuous mode are depicted in Figure 13.2. In this mode, the output inductor current is always greater than zero. This is the operating mode considered herein.

In operating the buck converter, the transistor in Figure 13.1 is operated as a switch. It is either completely on (saturated) or completely off (cutoff). The lightly shaded regions of Figure 13.2 correspond to those times when the transistor is on. The switching period is the length of time that transpires between turning the transistor on, going through a cycle, and then turning it on again. It is denoted T_{sw}. The switching frequency f_{sw} is the reciprocal of this value. The duty cycle d is the fraction of the time the transistor is on; thus, the transistor is on for a duration dT_{sw} and off for a duration $(1-d)T_{sw}$ of each cycle.

Power Magnetic Devices: A Multi-Objective Design Approach, Second Edition. Scott D. Sudhoff.
© 2022 The Institute of Electrical and Electronics Engineers, Inc. Published 2022 by John Wiley & Sons, Inc.
Companion website: www.wiley.com/go/sudhoff/Powermagneticdevices

462 | *13 Buck Converter Design*

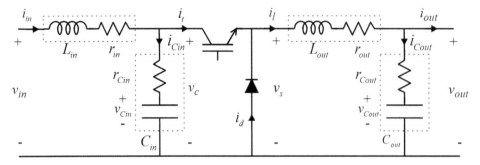

Figure 13.1 Buck converter.

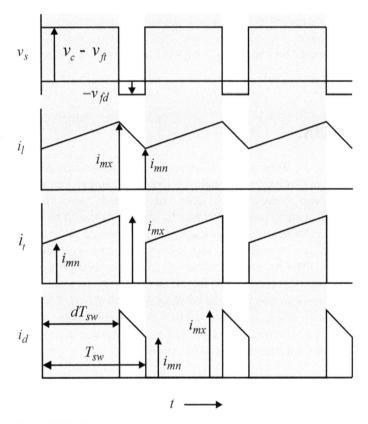

Figure 13.2 Buck converter operation.

We will denote the forward voltage across the switch when it is on as v_{ft}, where the "ft" subscript stands for forward transistor (voltage drop). Likewise, we will denote the forward diode voltage drop as v_{fd}. Temporarily, we will consider these small voltage drops to be constants even though we realize that they are functions of current.

When the transistor is on, the voltage drop across the transistor is v_{ft}, and so the switching point voltage v_s is the input capacitor voltage less the forward transistor voltage drop, that is, $v_c - v_{ft}$. This voltage is greater than the output voltage, and so during this time, the output inductor and transistor currents i_l and i_t typically rise almost linearly from a value i_{mn} to a value i_{mx}. The diode current is zero. When the transistor is turned off, the inductor current commutes from the transistor to the diode. Since the diode is now conducting, the voltage across the diode becomes v_{fd} and so the switching point voltage goes to $-v_{fd}$. The voltage across the output inductor is $-v_{fd} - v_{out}$ during this interval, and so the inductor and diode currents fall from i_{mx} back to i_{mn}, at which point the transistor is turned back on.

Time-Domain Analysis

We will begin our examination of the buck converter by setting forth a state model for the device, such as that used in a time-domain simulation. The assumed inputs to the model will be the input voltage v_{in} and the output current i_{out}. For this model, our states (a minimum set of variables that, together with model inputs, can be used to calculate everything else in a model) will be taken to be the input inductor current i_{in}, the internal input capacitor voltage v_{Cin}, the output inductor current i_l, and the internal output capacitor voltage v_{Cout}. We will represent capacitors as a resistance in series with an ideal capacitor; the voltage across the ideal capacitor is referred to as the "internal" voltage. The model outputs will be the input current i_{in}, which is also a state, and the output voltage v_{out}.

The function of a state model is to compute the time derivatives of the state variables in terms of state variables and inputs. Given this ability, a time-domain simulation algorithm such as the Runge–Kutta method can be used to perform a time-domain simulation. We will use the model to derive the operating characteristics of the converter.

The first step in our model is to calculate the transistor current, which may be expressed as

$$i_t = \begin{cases} i_l & \text{transistor on} \\ 0 & \text{transistor off} \end{cases} \tag{13.1-1}$$

Note that since i_l is a state, it can be considered a known quantity at any instant of time. Once the transistor current is calculated and since the input current i_{in} is a state, the current into the input capacitor may be expressed as

$$i_{Cin} = i_{in} - i_t \tag{13.1-2}$$

The time derivative of the voltage across this ideal capacitor (another state) may be expressed as

$$\frac{dv_{Cin}}{dt} = \frac{1}{C_{in}} i_{Cin} \tag{13.1-3}$$

The terminal voltage across the physical input capacitor is given by

$$v_c = v_{Cin} + r_{Cin} i_{Cin} \tag{13.1-4}$$

Once the input capacitor voltage is known and since the input voltage is known as a model input, the time derivative of the input current may be found. In particular,

$$\frac{di_{in}}{dt} = \frac{v_{in} - v_c - r_{in} i_{in}}{L_{in}} \tag{13.1-5}$$

Also, at this point, the switching point voltage may be calculated. In particular,

$$v_s = \begin{cases} v_c - v_{ft} & \text{transistor on} \\ -v_{fd} & \text{transistor off} \end{cases} \tag{13.1-6}$$

The next step is focused on the output capacitor. Since we have assumed that the output current is known (a model input), the output capacitor current and voltage may be expressed as

$$i_{Cout} = i_l - i_{out} \tag{13.1-7}$$

$$v_{out} = r_{Cout} i_{Cout} + v_{Cout} \tag{13.1-8}$$

where v_{Cout} is the voltage across the "ideal" capacitor used in the capacitor model. At this point, the time derivative of this "ideal" capacitor voltage and output inductor current states may be found as

$$\frac{dv_{Cout}}{dt} = \frac{1}{C_{out}} i_{Cout} \tag{13.1-9}$$

$$\frac{di_l}{dt} = \frac{v_s - v_{out} - r_{out} i_l}{L_{out}} \tag{13.1-10}$$

In this development, it has been assumed that the input voltage v_{in} and output current i_{out} are known values, which are model inputs. If this is not the case, the model must be adjusted. For example, if the output of the converter were connected to a resistive load, then

$$i_{out} = \frac{v_{out}}{R_{ld}} \tag{13.1-11}$$

where R_{ld} is a load resistance. Manipulating (13.1-7), (13.1-8), and (13.1-11), we can readily show that

$$v_{out} = \frac{R_{ld}(r_{Cout} i_l + v_{Cout})}{R_{ld} + r_{Cout}} \tag{13.1-12}$$

In this case, the calculation sequence of (13.1-7) and then (13.1-8) is replaced by the sequences (13.1-12), (13.1-11), and then (13.1-7).

Average-Value Analysis

On analyzing the operating of the buck converter, we will focus our work on the steady-state performance. To this end, it will be useful to think in terms of average values. Let $\mathbf{T}_{ss}$ denote an interval of time of duration T_{sw} such that the system is in the periodic steady state. The average value of a variable is defined as

$$\bar{x} = \frac{1}{T_{sw}} \int_{t \in \mathbf{T}_{ss}} x(t) dt \tag{13.1-13}$$

It will also be useful to remind ourselves of the definition of the root mean square of a variable,

$$x_{rms} = \sqrt{\frac{1}{T_{sw}} \int_{t \in \mathbf{T}_{ss}} x^2(t) dt} \tag{13.1-14}$$

Now, let us apply the average-value analysis to the system of equations (13.1-1)–(13.1-10). In doing so, there are two ideas that will be important. First, since we are considering steady-state performance, the average value of the time derivative of any variable is zero. Otherwise, we are not in the steady state. This does not mean that the instantaneous value of the time derivative is zero. Secondly, it should be observed that the resistances r_{Cin}, r_{Cout}, r_{in}, and r_{out} are small in that voltage drops associated with these resistances are small relative to the input and output voltages.

13.1 Buck Converter Analysis

This results in the output inductor current waveform being nearly triangular in shape. Third, for normal operation, we will assume that the input capacitor voltage ripple and output capacitor voltage ripple are small relative to their average values. We will enforce design constraints to this effect.

Let us first consider the input inductor. Applying (13.1-13) to (13.1-5) and requiring that the time average value of the time derivative is zero,

$$\bar{v}_{in} = r_{in}\bar{i}_{in} + \bar{v}_c \tag{13.1-15}$$

For the input capacitor, applying (13.1-13) to (13.1-3) and setting the average time derivative to zero,

$$\bar{i}_{Cin} = 0 \tag{13.1-16}$$

whereupon from (13.1-2) and (13.1-4), we have

$$\bar{i}_{in} = \bar{i}_t \tag{13.1-17}$$

$$\bar{v}_c = \bar{v}_{Cin} \tag{13.1-18}$$

Next, from the uppermost trace of Figure 13.2 and (13.1-6), and using our small ripple approximation, the average switching point voltage is given by

$$\bar{v}_s = d(\bar{v}_c - v_{ft}) - v_{fd}(1-d) \tag{13.1-19}$$

where d is the duty cycle as previously discussed.

Now, let us consider the switching point current and its relationship to the output inductor current. Combining (13.1-6) and (13.1-10), the output inductor current is governed by

$$L_{out}\frac{di_l}{dt} = \begin{cases} v_c - v_{ft} - v_{out} - r_{out}i_l & \text{transistor on} \\ -v_{fd} - v_{out} - r_{out}i_l & \text{transistor off} \end{cases} \tag{13.1-20}$$

If the input capacitor voltage ripple and output capacitor voltage ripple are small, and r_{out} is small so that the voltage drop $r_{out}i_l$ is small relative to the other terms, then (13.1-20) is well approximated by

$$L_{out}\frac{di_l}{dt} = \begin{cases} \bar{v}_c - v_{ft} - \bar{v}_{out} - r_{out}\bar{i}_l & \text{transistor on} \\ -v_{fd} - \bar{v}_{out} - r_{out}\bar{i}_l & \text{transistor off} \end{cases} \tag{13.1-21}$$

Observe that in (13.1-21), the time derivative of the inductor current is a constant value while the transistor is on and a different constant value while the transistor is off, so that we would expect the inductor current to take on the saw-tooth wave shape shown in Figure 13.2.

Applying (13.1-13) to the second and third traces of Figure 13.2, with the assumption of a saw-tooth wave shape, we can readily show

$$\bar{i}_l = \frac{1}{2}(i_{mx} + i_{mn}) \tag{13.1-22}$$

and that the average input current, which is equal to the average transistor current (13.1-17), is given by

$$\bar{i}_{in} = d\bar{i}_l \tag{13.1-23}$$

Averaging (13.1-10) and assuming steady state, we have

$$\bar{v}_s = r_{out}\bar{i}_{out} + \bar{v}_{out} \tag{13.1-24}$$

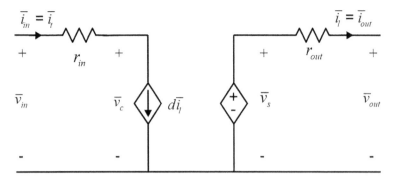

Figure 13.3 Buck converter average-value model.

As in the case of the input capacitor, the average output capacitor current must be zero whereupon

$$\bar{i}_l = \bar{i}_{out} \tag{13.1-25}$$

$$\bar{v}_{out} = \bar{v}_{Cout} \tag{13.1-26}$$

Combining (13.1-15), (13.1-17), and (13.1-23)–(13.1-25), one obtains the average-value model shown in Figure 13.3. Therein, the expression for $\bar{v}_s$ is given by (13.1-19).

Now that we have an average-value model, our next step will be to derive expressions for the ripple in the input and output voltages and currents. These quantities are important in order to address power quality constraints placed upon the converter. We will begin by considering the input filter.

Input Filter Ripple

The ripple component of a signal is the non-dc portion. In particular, we may express a variable x as

$$x(t) = \bar{x} + \tilde{x}(t) \quad t \in \mathbf{T}_{ss} \tag{13.1-27}$$

where $\bar{x}$ and $\tilde{x}(t)$ denote the steady-state average-value and ripple components, respectively. It will also be convenient to define peak-to-peak ripple as

$$\Delta x = \max_{t \in \mathbf{T}_{ss}} x(t) - \min_{t \in \mathbf{T}_{ss}} x(t) \tag{13.1-28}$$

Next, let us consider the ripple in the input capacitor voltage and input inductor current. To this end, consider Figure 13.4, which illustrates input filter waveforms. As in Figure 13.2, the shaded regions depict intervals wherein the transistor is on. The first trace represents the transistor current. In Figure 13.2, the transistor current exhibits a waveform similar to the dashed line in Figure 13.4. In order to simplify our analysis, and in keeping with the small ripple approximation, we will instead assume that the switch current waveform is that of the solid line in the first trace.

The average input current is equal to the average transistor current $\bar{i}_{in} = d\bar{i}_l$. Neglecting the ripple in the input inductor current yields the input capacitor current waveform shown in the second trace of Figure 13.4, from which we obtain

$$i_{Cin,rms} = \bar{i}_l \sqrt{d(1-d)} \tag{13.1-29}$$

and, using (13.1-3), that

$$\Delta v_{Cin} = \frac{1}{f_{sw} C_{in}} d(1-d) \bar{i}_l \tag{13.1-30}$$

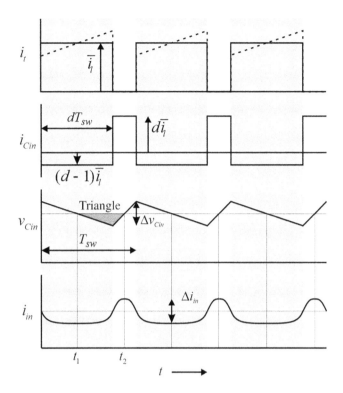

Figure 13.4 Input filter waveforms.

We will now consider the input inductor current ripple. Previously, we ignored the input current ripple when estimating the input capacitor ripple. However, now that we have an estimate for the input capacitor voltage ripple, we can use that estimate to approximate the input inductor current ripple. To this end, consider the interval $t_1 \leq t \leq t_2$ in Figure 13.4. Approximating $r_{in}i_{in} = r_{in}\bar{i}_{in}$ and $v_c = v_{Cin}$, and noting that $\bar{v}_{Cin} = v_{in} - r_{in}\bar{i}_{in}$ from (13.1-4), we may write

$$\Delta i_{in} = \frac{1}{L_{in}} \int_{t_1}^{t_2} \bar{v}_{Cin} - v_{Cin} dt \tag{13.1-31}$$

Noting $t_2 = t_1 + T_{sw}/2$ and that the integral in (13.1-31) is the absolute value of the area of the triangle in Figure 13.4, we obtain

$$\Delta i_{in} = \frac{1}{8C_{in}L_{in}f_{sw}^2} d(1-d)\bar{i}_l \tag{13.1-32}$$

Output Filter Ripple

Now, we will consider the output filter waveforms. Let us begin our analysis by computing the minimum and maximum inductor current values, i_{mn} and i_{mx}, as shown in Figure 13.2. To this end, consider (13.1-21). During the on period wherein the transistor conducts for a duration dT_{sw},

$$\frac{di_l}{dt} = \frac{i_{mx} - i_{mn}}{dT_{sw}} = \frac{\bar{v}_{Cin} - v_{ft} - \bar{v}_{out} - r_{out}\bar{i}_l}{L_{out}} \tag{13.1-33}$$

During the off period and noting that the diode conducts for a duration $T_{sw} - dT_{sw} = (1-d)T_{sw}$, we have

$$\frac{di_l}{dt} = \frac{i_{mn} - i_{mx}}{(1-d)T_{sw}} = \frac{-v_{fd} - \bar{v}_{out} - r_{out}\bar{i}_l}{L_{out}} \tag{13.1-34}$$

Manipulating (13.1-33) and (13.1-34), we arrive at the peak-to-peak inductor current ripple. In particular,

$$\Delta i_l = i_{mx} - i_{mn} = \frac{1}{f_{sw}L_{out}}(\bar{v}_{Cin} - v_{ft} + v_{fd})d(1-d) \tag{13.1-35}$$

Observe that from (13.1-22) and (13.1-35), we may find

$$i_{mn} = \bar{i}_l - \frac{1}{2}\Delta i_l \tag{13.1-36}$$

$$i_{mx} = \bar{i}_l + \frac{1}{2}\Delta i_l \tag{13.1-37}$$

From the inductor current waveform of Figure 13.2 and (13.1-14), it can also be shown that

$$i_{l,rms} = \sqrt{\bar{i}_l^2 + \frac{1}{12}\Delta i_l^2} \tag{13.1-38}$$

It will also be useful to describe the inductor current waveform as a Fourier series. From the waveform shown in Figure 13.2, the Fourier series representation of the inductor current waveform is given by

$$i_l = \bar{i}_l + \sum_{n=1}^{\infty} a_{l,n}\cos(n2\pi f_{sw}t) + b_{l,n}\sin(n2\pi f_{sw}t) \tag{13.1-39}$$

where

$$a_{l,n} = \frac{1}{2\pi^2 n^2 d(1-d)}\Delta i_l(\cos(n2\pi d) - 1) \tag{13.1-40}$$

$$b_{l,n} = \frac{1}{2\pi^2 n^2 d(1-d)}\Delta i_l \sin(n2\pi d) \tag{13.1-41}$$

A final quantity of interest with regard to the inductor current will be the mean value of the square of the time derivative of the inductor current. Again, from Figure 13.2, we have that

$$\overline{\left(\frac{di_l}{dt}\right)^2} = \frac{\Delta i_l^2 f_{sw}^2}{d(1-d)} \tag{13.1-42}$$

Our next goal will be to estimate the output voltage ripple. To this end, we will assume $\Delta v_{out} = \Delta v_{Cout}$ (thus ignoring the capacitor ESR) and, as a worst-case scenario, assume that all of the output inductor current ripple goes into the output capacitor. Thus, the capacitor current waveform is as depicted in the second trace of Figure 13.5.

The capacitor voltage will rise from its minimum value to its maximum value from time t_1 to time t_2. From (13.1-9),

$$\Delta v_{Cout} = \frac{1}{C_{out}}\int_{t_1}^{t_2} i_{Cout}\,dt \tag{13.1-43}$$

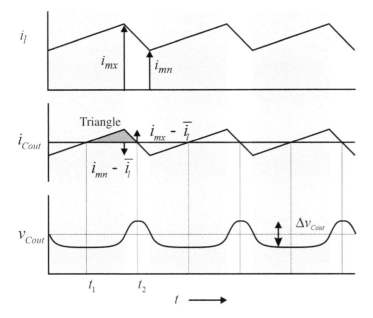

Figure 13.5 Output filter waveforms.

Noting that the integral in (13.1-43) is the area of the triangle in the second trace of Figure 13.5 and noting that $i_{mx} - \bar{i}_l = \Delta i_l/2$ and that $t_2 - t_1 = T_{sw}/2$, we have that

$$\Delta v_{Cout} = \frac{\Delta i_l}{8 f_{sw} C_{out}} \tag{13.1-44}$$

It is also readily shown that

$$i_{Cout,rms} = \frac{\Delta i_l}{2\sqrt{3}} \tag{13.1-45}$$

13.2 Semiconductors

In a power electronic circuit, the semiconductors are one of the most significant sources of loss. For a given device, this loss has two components: the conduction loss associated with the forward voltage drop, which occurs as a part of conducting current, and the switching loss, which occurs when a device turns on or off.

Conduction Loss

We will assume the existences of functions $p_{tc}(i)$ and $p_{dc}(i)$, which yield the instantaneous power drop associated with the forward voltage drop of the transistor and diode as a function of current. More generally, these drops will be a function of junction temperature as well, but here, we will forgo that level of detail. These functions may either be interpolated functions to measured data or fitted functions. One possible fitted form is

$$p_{xc}(i) = \sum_{k=1}^{K} \alpha_{tc,k} \left(\frac{i}{i_{xb}} \right)^{\beta_{xc,k}} \tag{13.2-1}$$

where x is t for the transistor or d for the diode, i_{xb} is the base current for the device of interest, and $\alpha_{xc,k}$ and $\beta_{xc,k}$ are constants. It is important to distinguish the base voltage from the rated voltage for the device. Note that this form for the power drop is more general than what results from the simple forward voltage drop considered in the previous section. We will resolve this discrepancy before the end of this section.

The average conduction loss in the switch or diode may be expressed as

$$P_{xc} = \frac{1}{T_{sw}} \int_{t \in T_{ss}} p_{xc}(i_x(t)) dt \qquad (13.2\text{-}2)$$

Assuming the inductor current waveform of Figure 13.2 and a conduction loss function of the form (13.2-1), (13.2-2) yields

$$P_{tc} = \frac{d}{\Delta i_l} \sum_{k=1}^{K} \frac{\alpha_{tc,k}}{i_{tb}^{\beta_{tc,k}}(\beta_{tc,k}+1)} \left(i_{mx}^{\beta_{tc,k}+1} - i_{mn}^{\beta_{tc,k}+1} \right) \qquad (13.2\text{-}3)$$

and

$$P_{dc} = \frac{1-d}{\Delta i_l} \sum_{k=1}^{K} \frac{\alpha_{dc,k}}{i_{db}^{\beta_{dc,k}}(\beta_{dc,k}+1)} \left(i_{mx}^{\beta_{dc,k}+1} - i_{mn}^{\beta_{dc,k}+1} \right) \qquad (13.2\text{-}4)$$

Switching Loss

When a semiconductor (transistor or diode) is fully on, its instantaneous power distribution is very low because the forward voltage drop is small. When it is fully off, the instantaneous power drop is essentially zero because the current through the device is essentially zero. However, when transitioning between on and off states, the instantaneous power loss can be very high because the device simultaneously experiences a large forward voltage and forward current at the same time. The net impact of this is an energy loss associated with every switching event. Thus, transistors experience turn-on loss and turn-off losses. PN junction diode losses are normally dominated by the turn-off loss.

Switching losses are proportional to the voltage that the semiconductor must block in the off state and the value of current that was or will be conducted following the change in switching state. We will denote the energy lost during a switching event as $E_{xy}(v, i)$, where x is again t for transistor or d for diode, and y is on or off to denote energy lost in turning on or turning off. In this form, v is the voltage being blocked in the off state, and i is the current being conducted just before turning off or just after turning on. The function $E_{xy}(v, i)$ could either be interpolated from measured data or be a mathematical expression based on fitting to measured data. For the latter case, one possible form is

$$E_{xy}(v, i) = \sum_{k=1}^{K} \alpha_{xy,k} \left(\frac{i}{i_{xb}} \right)^{\beta_{xy,k}} \left(\frac{v}{v_{xb}} \right)^{\gamma_{xy,k}} \qquad (13.2\text{-}5)$$

where v_{bx} is the base-blocking voltage for the switch and $\alpha_{xy,k}$ and $\beta_{xy,k}$ are constants.

Noting that $E_{xy}(v_b, i)$ describes the energy lost per event, the average switching loss in the transistor and diode may be expressed as

$$P_{tsw} = f_{sw} \left(E_{ton}(v, i_{mn}) + E_{toff}(v, i_{mx}) \right) \qquad (13.2\text{-}6)$$

$$P_{dsw} = f_{sw} \left(E_{don}(v, i_{mx}) + E_{doff}(v, i_{mn}) \right) \qquad (13.2\text{-}7)$$

The total power losses in the transistor and diode are given by summing the conduction and switching loss in each device. Thus,

$$P_{xtl} = P_{xc} + P_{xsw} \qquad (13.2\text{-}8)$$

where xtl denotes the transistor or diode total loss.

Sample data loosely based on a 1200 V, ~20 A silicon carbide MOSFET and a silicon carbide Schottky diode is listed in Table 11.1. Note that the current rating of both parts is a function of temperature. This data is taken from [1] and the datasheets [2,3] and has been simplified. The data listed here is only meant to be illustrative and to be used in the problems of this chapter and for the case study. The diode switching losses are extremely low; herein, they have been neglected. Also, note that some data related to thermal performance has also been included; this will be used in the following section.

Effective Forward Voltage Drops

The conduction loss calculation of (13.2-2)–(13.2-4) is not fully compatible with the simple constant forward voltage drop model used in Figure 13.1 and that model does not include switching losses at all. However, we can reconcile these models. To this end, based on the waveforms of Figure 13.1, and the simple forward drop model used, the transistor and diode losses may be expressed as

$$P_{ttl} = v_{ft} d \bar{i}_l \qquad (13.2\text{-}9)$$

$$P_{dtl} = v_{fd}(1-d)\bar{i}_l \qquad (13.2\text{-}10)$$

Thus, we may reconcile the two models (the models of this section and that of a constant voltage drop), if we define the forward transistor and diode drops as

$$v_{ft} = \frac{P_{ttl}}{d\bar{i}} \qquad (13.2\text{-}11)$$

$$v_{fd} = \frac{P_{dtl}}{(1-d)\bar{i}_l} \qquad (13.2\text{-}12)$$

Doing this requires an iterative approach, in that we must assume an initial estimate for v_{ft} and v_{fd} and then find the inductor current ripple, whereupon the total losses are calculated using (13.2-1)–(13.2-8), which are then used to update a revised estimate of v_{ft} and v_{fd} using (13.2-11)–(13.2-12). However, this process converges very quickly. The details of this process will be considered in Section 13.7.

Encapsulation

For the purposes of creating a structured design code, it is convenient to encapsulate the calculations of this section. Our goal is to calculate the effective forward voltage drops and power losses in each semiconductor. In particular, we wish to calculate the variables in the structure $\mathbf{S}$ defined as

$$\mathbf{S} = \begin{bmatrix} v_{ft} & P_{ttl} & P_{tsw} & P_{tc} & v_{fd} & P_{dtl} & P_{dsw} & P_{dc} \end{bmatrix} \qquad (13.2\text{-}13)$$

These variables will be calculated based on the blocking voltages v, i_{mn}, i_{mx}, f_{sw}, d and the semiconductor parameters, which will be stored in structures $\mathbf{D}_{St}$ and $\mathbf{D}_{Sd}$ (example data for which is given in Table 11.1). In particular, we may write

$$\mathbf{S} = F_{Sbk}(v, i_{mn}, i_{mx}, f_{sw}, d, \mathbf{D}_{St}, \mathbf{D}_{Sd}) \qquad (13.2\text{-}14)$$

where $F_{Sbk}()$ is a function that encapsulates the semiconductor calculations.

13.3 Heat Sink

Because of the heat generated in the power semiconductors, it is necessary to use a heat sink to keep the junction temperature below a maximum allowed value. There are many technologies available—such as passive air, forced air, liquid-cooled, and a variety of advanced strategies. Herein, we will consider a passive air-cooled scenario in which the transistor and diode are each mounted on their own heat sink.

Figure 13.6 depicts a simple thermal-equivalent circuit for a semiconductor device on a heat sink. Therein, P_{xtl} represent the total device dissipation as calculated by (13.2-1)–(13.2-8), T_{xj} is the junction temperature of the device, R_{xjc} is the thermal resistance from the semiconductor junction to its case, R_{xch} is the thermal resistance from the case to the heat sink, R_{xha} is the thermal resistance between the heat sink and ambient, and T_a is the ambient temperature, usually taken as a worst case. As in the previous sections, x may be t for transistor or d for diode. The thermal resistance R_{xjc} is a function of the device and can be obtained from the device data sheet. The thermal resistance R_{xch} is a function of how the device is connected to the heat sink and depends on the thermal grease used between the device and heat sink and how well (in terms of correct thickness and uniformity) it is applied. Herein, the thermal resistance R_{xch} will be assumed to be a known parameter based on experience. The thermal resistance R_{xha} is that of the heat sink to air.

From Figure 13.6, we can readily show that if we desire the maximum allowed junction temperature to be T_{xjmxa} at a worst-case ambient temperature of T_a, then the maximum allowed thermal resistance of the heat sink is given by

$$R_{xha} = \frac{T_{xjmxa} - T_a}{P_{xtl}} - R_{xjc} - R_{xch} \qquad (13.3\text{-}1)$$

In the event that (13.3-1) calls for a negative heat-sink resistance, there is no feasible solution.

In [4], a heat-sink model is set forth to predict the mass of an aluminum plate fin passively cooled heat sink based on the required thermal resistance. In particular, the heat-sink mass is expressed as

$$M_{xh} = \frac{a_{h1}}{(R_{xha}/R_{bh})^{n_{h1}}} + \frac{a_{h1}}{(R_{xha}/R_{bh})^{n_{h2}}} \qquad (13.3\text{-}2)$$

where $a_{h1} = 0.1516$ kg, $a_{h2} = 7.5568 \cdot 10^{-5}$ kg, $n_{h1} = 1.1688$, and $n_{h2} = 5.5445$, and $R_{bh} = 1°C/W$. We will include these parameters in a structure D_{PAH}.

Thus, our design approach will be to calculate the losses in each semiconductor, calculate the required heat-sink thermal resistance, impose a constraint that the required resistance be greater than zero, and then finally calculate the mass of each of the heat sinks, which will be added to the total mass. We encapsulate this sequence with the functional form

$$[M_{xh} \ c_{xh}] = F_{PAH}(P_{xtl}, T_A, R_{xch}, D_{Sx}, D_{PAH}) \qquad (13.3\text{-}3)$$

where $c_{xh} = 1$ indicates a valid solution (with a positive heat sink to ambient resistance) and $c_{xh} = 0$ indicates an invalid solution. The subscript "PAH" denotes a passive air-cooled heat sink. Recall

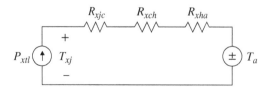

Figure 13.6 Semiconductor/heat-sink thermal-equivalent circuit.

Table 13.1 Sample Semiconductor Parameters

Symbol (Units)	Value	Symbol (Units)	Value
v_{bt} (A)	800 V	v_{bd}	n/a
i_{bt} (A)	1	i_{bd}	1 A
α_{tc} (W)	[1.303 6.4 × 10^{-3}]	α_{dc} (W)	[0.9784 2.39 × 10^{-2}]
β_{tc}	[1 2.774]	β_{dc}	[1 2.067]
α_{ton} (J)	[274.1 3.752 5.852] × 10^{-7}	α_{don} (J)	n/a
β_{ton}	[0 1 2]	β_{don}	n/a
γ_{ton}	[1 1 1]	γ_{don}	n/a
α_{toff} (J)	[574.8 −9.938 2.454] × 10^{-7}	α_{doff} (J)	n/a
β_{toff}	[0 1 2]	β_{doff}	n/a
γ_{toff}	[1 1 1]	γ_{doff}	n/a
R_{tjc} (°C/W)	0.65	R_{djc} (°C/W)	0.6
$T_{tj,\,mxa}$ (°C)	150	$T_{dj,\,mxa}$ (°C)	175

that the semiconductor parameter structure D_{sx} includes the maximum allowed junction temperature T_{xjmxa} and maximum junction to case thermal resistance R_{xjc}.

Example 13.3A In this example, let us consider both semiconductor loss and heat-sink requirements of a buck converter as a function of current ripple. In particular, let us assume the transistor and diode whose parameters are described in Table 13.1. Let us further assume a blocking voltage $\bar{v}_c = 750$ V and $\bar{i}_l = 25$ A. We will assume a switching frequency of 40 kHz and a duty cycle of $d = 0.5\overline{3}$, which should yield an output voltage on the order of 400 V and an output power of 10 kW. Finally, we will assume an ambient temperature of 50°C and junction-to-case resistances of 0.6°C/W for both the transistor and the diode. We will observe the semiconductor losses as Δi_l is varied from 0 to 50 A, which is the range of the continuous mode of operation (i.e., at any point of time, either the transistor or diode is conducting).

Figure 13.7 depicts the total, conduction, and switching losses in the transistor as well as the total diode loss. As can be seen, the transistor loss greatly exceeds the diode loss. This is because the diode has a lower conduction loss due to the nature of the semiconductor junctions, coupled with the characteristic Schottky diodes acting nearly ideally and having inconsequential switching losses. Further, both the diode loss and transistor switching loss are seen to be relatively constant with respect to the current ripple. However, after a gentle decrease, the transistor conduction loss is seen to significantly increase for high values of ripple. The impact of this increase in losses can be seen in Figure 13.8, wherein the required head sink masses are shown. As can be seen, the diode heat-sink mass is inconsequential. However, once the output inductor current ripple exceeds 30 A, the transistor heat-sink mass increases sharply. After a current ripple of approximately 48.5 A, there is no passive air-cooled solution.

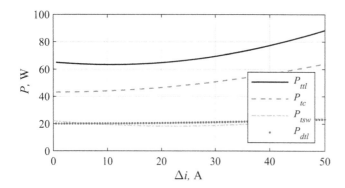

Figure 13.7 Semiconductor loss versus current ripple.

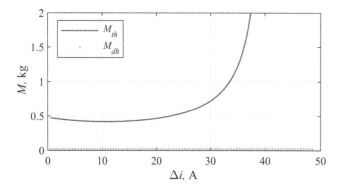

Figure 13.8 Required heat-sink mass versus current ripple.

13.4 Capacitors

As part of the design process, we will need to select both input and output capacitors. There are several types of capacitors that can be used, including ceramic, electrolytic, and film capacitors. In ceramic capacitors, a ceramic material is used for the dielectric. In an electrolytic capacitor, the positive plate or anode is metal and the negative plate or cathode is formed by a solid, liquid, or gel electrolyte. An oxide forms over the metal to form the dielectric. In a film capacitor, a plastic film forms the dielectric.

Of these capacitor types, electrolytic capacitors tend to be the most energy dense. However, their capacitance falls off markedly with frequency and their lifetime is limited. Thus, herein, we will focus our attention on polypropylene film capacitors, which are very commonly used in power electronics applications.

Although polypropylene capacitors are generally more reliable than electrolytic capacitors, we will find that lifetime is still an issue. The expected lifetime is a function of the operating voltage relative to the nominal voltage, as well as the temperature. Thus, we will have constraints to this effect. Also, we will want to know the capacitor's mass so that it can be included in the system mass.

Our basic approach to capacitor modeling will be as follows. First, we will a priori select a capacitor family from a vendor for the input capacitor and a potentially second family based on the output capacitor. Then, we will pick a capacitor from the family to be our capacitance. In doing this, we will assume we may use N_c capacitors from a given family in parallel. Series combinations are of course possible, but we will forego this situation.

The capacitor family is selected based on the operating voltage and the nominal voltage of the capacitor family. Let $\mathbf{C}_{FCx}$ be a vector of possible capacitance values within a family of capacitor to be used for $x \in [in, out]$ and $\mathbf{M}_{FCx}$, $\mathbf{R}_{FCx}$, $\mathbf{G}_{FCx}$, and $\mathbf{T}_{FCx}$ be vectors of the corresponding mass, effective series resistance, thermal conductivity, and temperature limit, respectively. Let j be an index over the set of capacitors within the set. Now suppose we wish to obtain a desired capacitance C_x^*. For each capacitance value in the catalog, we need at least

$$N_{Cx,j} = \text{ceil}\left(\frac{C_x^*}{C_{Fx,j}}\right) \tag{13.4-1}$$

capacitors in parallel to achieve the desired capacitance. We will choose j in order to obtain the required capacitance with minimum mass. In particular, we select

$$j = j^* = \text{argmin}(M_{FCx,j} N_{Cx,j}) \tag{13.4-2}$$

The actual capacitance, mass, effective series resistance, thermal conductance, and allowed temperature are given by

$$C_x = C_{FCx,j^*} N_{Cx,j^*} \tag{13.4-3}$$

$$M_{Cx} = M_{FCx,j^*} N_{Cx,j^*} \tag{13.4-4}$$

$$R_{Cx} = R_{FCx,j^*} / N_{Cx,j^*} \tag{13.4-5}$$

$$G_{Cx} = G_{FCx,j^*} N_{Cx,j^*} \tag{13.4-6}$$

$$T_{Cx,mxa} = T_{FC,j^*} \tag{13.4-7}$$

This sequence of calculations is represented by the function call

$$\mathbf{C}_x = \mathbf{F}_C(C_x^*, \mathbf{D}_{Cx}) \tag{13.4-8}$$

where

$$\mathbf{C}_x = [C_x \ M_{Cx} \ R_{Cx} \ G_{Cx} \ T_{Cx,mxa}] \tag{13.4-9}$$

and

$$\mathbf{D}_{Cx} = [\mathbf{C}_{FCx} \ \mathbf{M}_{FCx} \ \mathbf{R}_{FCx} \ \mathbf{G}_{FCx} \ \mathbf{T}_{FCx}] \tag{13.4-10}$$

In (13.4-9), recall that the italic nonbold C_x is the capacitor capacitance, whereas the nonitalic bold $\mathbf{C}_x$ is a structure of capacitor data.

Once a capacitor is selected, using the rms capacitor current given by (13.1-29) for the input capacitor and (13.1-45) for the output capacitor, the capacitor power loss and temperature are readily calculated as

$$P_{Cx} = R_{Cx} i_{Cx,rms}^2 \tag{13.4-11}$$

$$T_{Cx} = P_{Cx}/G_{Cx} + T_A \tag{13.4-12}$$

Clearly, this is a simplistic behavioral model. However, it is one that is commonly supplied by manufacturers with the capacitor data sheets. It is limited, however, in that it does not, for example, consider the impact of air flow.

13.5 UI-Core Input Inductor

The buck converter includes both an input inductor and an output inductor. The output inductor is the chief concern in this effort. It is the dominant inductor as it sees both a larger current and much larger voltage swing. While it is possible to design both inductors simultaneously, here we will take a different approach. In many cases, when we wish to design a system, if we simultaneously design every component of the system, the design space gets very large. One way to combat this problem is to use simplified design models for less critical parts of the system. In this case, we leverage this approach by using a simplified model of the input inductor. This simplified model will give us the input inductor mass and loss and represent the trade-off between these metrics. However, it will not include the details of the input inductor design. This will have to be done after the system-level optimization.

Thus, for our converter design, we will use a metamodel to represent the input inductor. A metamodel is a model of a more detailed model. From the perspective of our work in dc/dc converter design, the metamodel is very easy to use. In essence, for a given inductance, it utilizes the current density to parameterize the trade-off between mass and loss. While the details in deriving such a metamodel are set forth in [5], it should be noted that the formation is very much based on our work in Chapter 3.

The use of the metamodel [5] is as follows. First, we define an energy metric E_{Lin}^* as

$$E_{Lin} = \frac{1}{2} L_{in}^* \tilde{i}_{in}^2 \tag{13.5-1}$$

where L_{in}^* will be the desired input inductance. Next, the trade-off between mass M_{Lin} and loss P_{Lin} may be expressed as

$$M_{Lin} = c_M E_{Lin}^* \prod_{k=1}^{K_m} \left(J_{Lin} \left(E_{Lin}^* \right)^{1/3} + b_{M,k} \right)^{n_{M,k}} \tag{13.5-2}$$

$$P_{Lin} = c_P \left(E_{Lin}^* \right)^{1/3} \prod_{k=1}^{K_P} \left(J_{Lin} \left(E_{Lin}^* \right)^{1/3} + b_{P,k} \right)^{n_{P,k}} \tag{13.5-3}$$

where J_{Lin} is the input inductor current density, and all remaining variables (other than the energy metric) are model constants stored in a structure $\mathbf{D}_{Lin}$. We can assume to know J_{Lin} because it will be part of our design space once we define it in Section 13.8. Values for the constants based on a ferrite core UI-core inductor are set forth in Table 13.2, taken from [5].

By evaluating (13.5-2) and (13.5-3) at a desired current density, the mass and loss are known, and the series resistance may be expressed as

$$r_{Lin} = P_{Lin}/\tilde{i}_{in}^2 \tag{13.5-4}$$

The inductor metamodel may be readily calculated by the function call

$$\begin{bmatrix} M_{Lin} & P_{Lin} \end{bmatrix}^T = \mathbf{F}_{UILdcmm}(E_{Lin}, J_{Lin}, \mathbf{D}_{Lin}) \tag{13.5-5}$$

where the "UILdcmm" indicates a UI-core inductor, dc applications, metamodel.

It is insightful to briefly consider the predictions of the metamodel. Note that from the form of (13.5-2) and (13.5-3), the mass and loss are only a function of the energy metric and not a function of inductance or current independently. This makes intuitive sense. The form of (13.5-2) and (13.5-3) also indicates that the mass and loss of an inductor versus inductance are not linear. We will explore this in Example 13.5A.

Table 13.2 Inductor Metamodel Parameters (UI-Core, P-Ferrite)

$c_M = 5.9851$		$c_P = 1.1021 \cdot 10^{-5}$	$K_M = 7$	$K_P = 7$
k	C	$n_{M,k}$	$b_{P,k}$	$n_{P,k}$
1	0	0.24700	0	0.54482
2	100.05	0.24673	1.1658×10^3	0.25254
3	100.05	-1.3215	1.1658×10^3	0.17114
4	3.4677×10^6	-1.2423	1.1659×10^3	0.24906
5	8.2537×10^6	2.4809	5.1412×10^4	-0.15241
6	7.3079×10^7	-2.0633	4.4344×10^5	0.52755
7	1.0430×10^8	1.5530	1.2330×10^6	-0.59614

(Source: reproduced with permission from Ref. [5]. © IEEE.)

Example 13.5A In this example, we examine the predictions of the metamodel in terms of mass and loss of an input inductor as a function of required inductance. In particular, let us assume an input current of 13.33 A (which corresponds to the input current of example 13.3A) and a current density of 4 A/mm². Our goal is to use the metamodel to predict mass and loss as a function of inductance. Using (13.5-2) and (13.5-3) with the parameters of Table 11.2, one obtains the results shown in Figures 13.9 and 13.10. The interesting aspect of this study is that mass and loss increase sublinearly with inductance. An implication of this is that one inductor can be made to be smaller than the series combination of two smaller inductors whose inductances sum to the same value.

13.6 UI-Core Output Inductor

We will now turn our attention to the output inductor. We will use a UI-core inductor so that we can take advantage of our work in Chapter 3. The geometry of the inductor is shown in Figure 13.11. Therein, we recognize all the variables we previously considered in Chapter 3. As a refinement to our work in Chapter 3, we will take a more detailed view of the winding. In particular, we include the impact of the bending radius of the wire on the design and also include a provision for coil-to-core and layer-to-layer insulation.

The modeling of the UI-core inductor will have five parts. First, we will consider the geometry, which is an extension of Chapter 3. Next, we will include the magnetic analysis based on our work from Chapters 2 and 3, followed by a loss anaysis, which comes from our work in Chapters 6 and 11. We will then consider a thermal model using the development of Chapter 10, followed by a capacitance analysis using our work from Chapter 12.

Geometry

The focus of this section is to define a set of independent variables for the UI-core inductor and to, from that set, derive an expression for all remaining variables. To this end, one possible structure for geometric variables is

$$\mathbf{G}_{UII} = \begin{bmatrix} g & w_{bc} & l_c & r_{ib} & r_{eb} & N_{tpl}^* & N_l^* & N_{pr}^* & a_{wcpr}^* & c_w & c_d & t_{clc}^* & t_{ll}^* & \mathbf{D}_{UI} \end{bmatrix} \qquad (13.6\text{-}1)$$

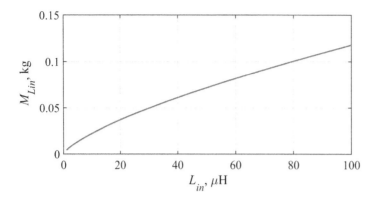

Figure 13.9 Input inductor mass versus inductance.

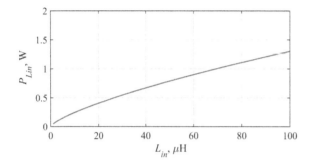

Figure 13.10 Input inductor loss versus inductance.

In (13.6-1), g is the air gap and w_{bc} is the width of the U-core base center, which is smaller than the width of the base and marks the length of the vertical travel of the conductors. The vertical travel w_{bc} and making two bends with bend radius r_b cause the first layer of conductors to fit around the U-core and coil-to-core insulation. The variable l_c is the core length as defined in Chapter 3 and defined in Figure 13.11. The ratios r_{ib} and r_{eb} are the width of the I-core and U-core ends relative to the U-core base, respectively. The variables N^*_{tpl}, N^*_l, and N^*_{pr} are the desired number of turns per layer, number of layers, and number of parallel conductors, as real numbers, respectively, which are similar to N^*_w and N^*_d in Chapter 3. In particular, since these parameters behave in a quasi-continuous fashion, it makes since to treat them as real numbers and then round them to integers. The use of a parallel conductor facilitates the use of a smaller conductor diameter, which is advantageous in terms of the bending radius and proximity-effect loss. The total (over all parallel strands) desired cross-sectional area of the wire conductor is a^*_{wcpr}. The clearances c_w and c_d are the same as in Chapter 3 and as shown in Figure 13.11. Variables t^*_{clc} and t^*_{ll} are the desired thickness of the coil-to-core and layer-to-layer insulation.

The last field of $\mathbf{G}_{UI}$ is a substructure of the UI-core inductor design data, $\mathbf{D}_{UI}$, which is of the form

$$\mathbf{D}_{UI} = \begin{bmatrix} k_{bd} & k_b & h_{wa} & \varepsilon_{rp} & k_p & T_{wpk,mxa} & N_h & r_{\Phi rq} & \mathbf{D}_{CD} & \mathbf{D}_{CR} \end{bmatrix} \tag{13.6-2}$$

where the first seven fields are the bending radius constant (allowed bending radius/wire diameter), the build factor, the winding-to-air heat transfer coefficient, the relative permittivity of the potting

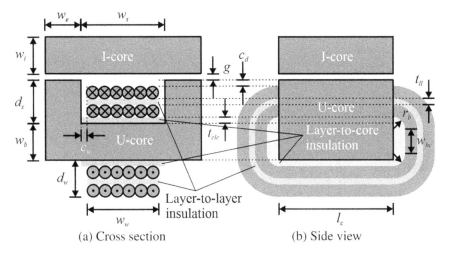

Figure 13.11 UI-core inductor.

material, the thermal conductivity of the potting material, the maximum allowed winding temperature, the number of harmonics used in the ac analysis, and the required flux factor, respectively. Each of these terms will be discussed in more detail as we proceed.

The last two fields of $\mathbf{D}_{UI}$ are $\mathbf{D}_{CD}$ and $\mathbf{D}_{CR}$, which are substructures of conductor and core data, respectively. In particular,

$$\mathbf{D}_{CD} = [t_{wi} \; \varepsilon_{rwi} \; k_{wi} \; \rho_{cd} \; k_{cd} \; \sigma_{0cd} \; \alpha_{cd} \; T_{0cd}] \tag{13.6-3}$$

where the fields are the wire insulation thickness, the relative permittivity of the wire insulation, the thermal conductivity of the wire insulation, the conductor mass density, the thermal conductivity, nominal conductivity, temperature coefficient of conductivity, and the nominal temperature, respectively.

The substructure $\mathbf{D}_{CR}$ includes data on the core and includes the mass density of the core p_{cr} and the magnetic properties of the core, which include the parameters of the anhysteretic B–H characteristic as specified, for example, by (6.5-33) and the parameters of the core loss model as specified, for example, by (6.6-8).

Herein, it is assumed that an insulating paper is used for the coil-to-core insulation and for the layer-to-layer insulation. As a result, the coil-to-core insulation thickness comes in discreet thickness depending upon the number of layers used and the thickness of those layers. For simplicity, we will calculate the actual coil-to-core and layer-to-layer insulation thicknesses as

$$[t_{clc} \; \varepsilon_{rclc} \; k_{clc}] = \mathrm{F}_{IP}(t_{clc}^*) \tag{13.6-4}$$

$$[t_{ll} \; \varepsilon_{rll} \; k_{ll}] = \mathrm{F}_{IP}(t_{ll}^*) \tag{13.6-5}$$

where $\mathrm{F}_{IP}()$ is a function that rounds the result to the closest available thickness of insulating paper. Because the relative permittivity and thermal conductivity are a function of thickness, these parameters are also computed by the thickness function. From a geometrical point of view, we need neither permittivity nor thermal conductivity; however, these fields are important in computing capacitance and in thermal modeling.

The advantage of a large value of t_{clc} and t_{ll} is that it will decrease capacitance; the disadvantage is that it will increase resistance by virtue of longer turns. The impact of layer-to-layer insulation is more nuanced. As in the case of the coil-to-core insulation, it will increase resistance but even more significantly reduce capacitance. However, in addition, the layer-to-layer insulation reduces the thermal conductivity of the coil.

The remainder of the dependent geometrical variables is calculated as follows. First, we calculate the number of turns per layer, number of layers, and number of parallel conductors as

$$N_{tpl} = \text{round}\left(N_{tpl}^*\right) \tag{13.6-6}$$

$$N_l = \text{round}\left(N_l^*\right) \tag{13.6-7}$$

$$N_{pr} = \text{round}\left(N_{pr}^*\right) \tag{13.6-8}$$

In Chapter 3, N_w and N_d were used for N_{tpl} and N_l, respectively. Requiring all winding layers to be full, the number of turns is given by

$$N = N_{tpl} N_l \tag{13.6-9}$$

Next, we will perform calculations related to the wire size. First, the actual wire conductor area (of one of the parallel strands) is computed by finding the closest achievable area using a standard wire gauge as

$$a_{wc} = F_{SWG}\left(\frac{a_{wcpr}^*}{N_{pr}}\right) \tag{13.6-10}$$

where $F_{SWG}()$ is essentially a lookup function. In (13.6-10), a_{wcpr}^* is the total conductor cross-sectional area desired (including all parallel conductors), while a_{wc} will be the cross section of a single conductor. After this step, the radius of the wire conductor and radius of the wire can be found as

$$r_{wc} = \sqrt{\frac{a_{wc}}{\pi}} \tag{13.6-11}$$

$$r_w = r_{wc} + t_{wi} \tag{13.6-12}$$

The width and depth of the winding bundle and slot are then found as

$$w_w = 2r_w(N_{tpl} + 1)N_{pr}k_b \tag{13.6-13}$$

$$d_w = N_l 2r_w k_b + t_{clc} + (N_l - 1)t_{ll} \tag{13.6-14}$$

$$w_s = w_w + 2c_w \tag{13.6-15}$$

$$d_s = d_w + c_d \tag{13.6-16}$$

In (13.6-13), observe that $(N_{tpl} + 1)$ is used rather than N_{tpl} since a conductor must overlap laterally in completing a turn. Similarly, in (13.6-14), $(N_l - 1)$ is used rather than N_l since the layer-to-layer insulation is only placed between layers.

The next step is the calculation of the bending radius,

$$r_b = \max(t_{clc}, 2k_{bd}r_w) \tag{13.6-17}$$

wherein the max() operator is used to ensure that $w_b \geq w_{bc} \geq 0$. Note that for the common case that $r_b > t_{clc}$, a filler will be required between the core and the insulating paper on the ends of the core.

Referring back to Figure 13.11, the width of the U-core base is then found as

$$w_b = w_{bc} + 2r_b - 2t_{clc} \tag{13.6-18}$$

13.6 UI-Core Output Inductor

The remaining core widths and core mass are then found using

$$w_i = w_b r_{ib} \tag{13.6-19}$$

$$w_e = w_b r_{eb} \tag{13.6-20}$$

$$M_{cr} = \rho_{cr}((w_b + w_i)(w_s + 2w_e) + 2d_s w_e)l_c \tag{13.6-21}$$

In order to compute capacitance, it is necessary to establish a vector $\mathbf{l}_t$ that includes the length of the conductors in each layer. The elements of this vector can be expressed as

$$l_{t,i} = \sqrt{(2w_{bc} + 2l_c + 2\pi(r_b + r_w k_b(2i-1) + t_{ll}(i-1)))^2 + (2r_w k_b N_{pr})^2} \tag{13.6-22}$$

which includes the impact of the skew of the winding (i.e., the conductors must move laterally in order to complete a turn). The total length of the coil of parallel wires may be found as

$$l_{wc} = N_{tpl} \sum_{i=1}^{N_l} l_{t,i} \tag{13.6-23}$$

At this point, the mass of the wire and the complete inductor (neglecting the minor contribution of the insulation) can be calculated from

$$M_{wc} = a_{wc} l_{wc} \rho_{cd} N_{pr} \tag{13.6-24}$$

$$M_L = M_{cr} + M_{wc} \tag{13.6-25}$$

The last step for our geometrical calculations will be to find the area of the magnetic cross section, the magnetic mean path length, and the magnetic volume of each region of the magnetic equivalent circuit—that is, the I-core, U-core base, and U-core end (legs). In particular, we have

$$a_{mcI} = w_i l_c \tag{13.6-26}$$

$$l_{mpI} = w_s + w_e \tag{13.6-27}$$

$$V_{mI} = a_{mcI} l_{mpI} \tag{13.6-28}$$

$$a_{mcUb} = w_b l_c \tag{13.6-29}$$

$$l_{mpUb} = w_s + w_e \tag{13.6-30}$$

$$V_{mUb} = a_{mcUb} l_{mpUb} \tag{13.6-31}$$

$$a_{mcUe} = w_e l_c \tag{13.6-32}$$

$$l_{mpUe} = d_s + \frac{1}{2} w_b \tag{13.6-33}$$

$$V_{mpUb} = a_{mcUb} l_{mpUb} \tag{13.6-34}$$

We will summarize the calculations of this section with a function call

$$\mathbf{G}_{UID} = \mathrm{F}_{GUI}(\mathbf{G}_{UII}, \mathbf{D}_{UI}) \tag{13.6-35}$$

where $\mathbf{G}_{UII}$ and $\mathbf{D}_{UI}$ have already been defined, and

$$\mathbf{G}_{UID} = [t_{clc} \; \varepsilon_{rclc} \; k_{clc} \; t_{ll} \; \varepsilon_{rll} \; k_{ll} \; N_{tpl} \; N_l \; N_{tpl} \; N_{pr} \; N \; a_{wc} \; r_{wc} \; r_w \; w_w \; d_w \; w_s \; d_s \; r_b \; w_b \; w_i \; w_e \cdots \\ M_{cr} \; \mathbf{1}_t^T \; l_{wc} \; M_{wc} \; M_L \; a_{mcI} \; l_{mpI} \; V_{mI} \; a_{mcUb} \; l_{mpUb} \; V_{mUb} \; a_{mcUe} \; l_{mpUe} \; V_{mUe}] \tag{13.6-36}$$

Finally, as we have done elsewhere, we can amalgamate the dependent and independent geometrical variables into a single structure

$$\mathbf{G}_{UI} = [\mathbf{G}_{UII} \quad \mathbf{G}_{UID}] \tag{13.6-37}$$

Magnetic Analysis

Once the geometric analysis is complete, a magnetic analysis can be conducted. To this end, our first step will be to calculate the incremental inductance of the inductor. In particular, we will calculate the incremental inductance over the range of assumed ripple current

$$L_{inc} = \frac{\lambda(i_{mx}) - \lambda(i_{mn})}{\Delta i_l} \tag{13.6-38}$$

where the flux linkage is calculated using the UI-core magnetic equivalent circuit (MEC) from Chapter 2. As part of (13.6-38), we will also define c_m as a logical variable, which is 1 if the MEC convergences and 0 if not. Further, we will evaluate r_Φ as defined by (3.3-36) in order to make sure the intended magnetic path is used.

We will compute core loss by assuming that the flux density waveforms are saw-tooth like the inductor current. Again using the MEC, we will find the change flux density between the minimum and maximum inductor currents in the I-core, U-core base, and U-core ends (legs) as

$$\Delta B_I = B_I(i_{mx}) - B_I(i_{mn}) \tag{13.6-39}$$

$$\Delta B_{Ub} = B_{Ub}(i_{mx}) - B_{Ub}(i_{mn}) \tag{13.6-40}$$

$$\Delta B_{Ue} = B_{Ue}(i_{mx}) - B_{Ue}(i_{mn}) \tag{13.6-41}$$

Referring to Figure 2.25, B_I may be obtained from mesh flux Φ_1, B_{Ub} from the difference of mesh fluxes $\Phi_3 - \Phi_2$, and B_{Ue} from mesh flux Φ_2. The flux densities are readily determined by dividing the appropriate flux (or difference thereof) by the corresponding magnetic cross section. From these results, we will be able to compute the core loss. However, we will postpone that discussion until the next subsection on losses. Note that for the purposes of simplification, we will compute the core loss on the ends of U-core based on the flux through the end closer to the base, that is, Φ_2, rather than breaking each end of the U-core into lower and upper halves, separated by nodes 2 and 7 in Figure 2.25, as is done in the MEC. The result of this will be a very slightly higher estimation of loss.

The dynamic resistance of the UI-core inductor can be found using (11.9-19), where it is expressed in terms of the dynamic resistance in the exterior region D_{ea} (11.7-2) and the dynamic resistance of a gapped closed-slot region D_{gcs} (11.7-31). Note that rather than finding the dynamic resistances as cited, we will find the ratio of dynamic resistances over the conductivity, that is,

$$D_{o\sigma} = \frac{D}{\sigma} \tag{13.6-42}$$

in order to eliminate conductivity from the basic expressions. This is useful when incorporating the change of conductivity with temperature.

We may encapsulate these calculations with the function call

$$\mathbf{M}_{UI} = F_{MUI}(\mathbf{G}_{UI}, \bar{i}_l, \Delta i_l) \tag{13.6-43}$$

where

$$\mathbf{M}_{UI} = [c_m \quad L_{inc} \quad \Delta B_I \quad \Delta B_{Ub} \quad \Delta B_{Ue} \quad r_\Phi \quad D_{o\sigma}] \tag{13.6-44}$$

It is important to observe that none of the magnetic calculations, as posed, are a function of temperature.

UI-Core Inductor Losses

We will now consider the losses in the UI-core output inductor. First, we have the dc-resistive loss, which is readily calculated with the sequence

$$R_L = \frac{l_{wc}}{a_{wc}\sigma N_{pr}} \tag{13.6-45}$$

$$P_{dcr} = R_L \bar{i}_l^2 \tag{13.6-46}$$

Observe that in (13.6-45), the conductivity σ is in general a function of temperature.

Next, we will compute the ac losses. We begin by establishing a vector of frequencies

$$\mathbf{f} = f_{sw}[1 \quad 2 \quad \cdots \quad N_h] \tag{13.6-47}$$

where N_h is the number of harmonics we wish to include. Assuming that the output inductor current has a saw-tooth waveshape, the spectrum of the output inductor current is given by (13.1-40)–(13.1-41), whereupon the square of the zero-to-peak magnitude of the nth harmonic may be expressed as

$$m_{l,n}^2 = a_{l,n}^2 + b_{l,n}^2 = 2\left(\frac{\Delta i_l}{2\pi^2 n^2 d(1-d)}\right)^2 (1 - \cos(n2\pi d)) \tag{13.6-48}$$

Calculating the ac resistance at each frequency in (13.6-47) using the real part of the impedance given by (11.2-22) divided by N_{pr}, here called $r_{ac}(f)$, which is a function of frequency, the ac power loss is then calculated as

$$P_{acr} = \frac{1}{2} \sum_{n=1}^{N_h} r_{ac}(f_n) m_{l,n}^2 \tag{13.6-49}$$

It should be noted that in this application, the ac-resistive losses will probably be minor; they will be dominated by the proximity-effect loss, which, from (11.5-25) and (13.1-42), may be expressed as

$$P_p = D\frac{\Delta i_l^2 f_{sw}^2}{d(1-d)} \tag{13.6-50}$$

At this point, the total winding loss can be expressed as

$$P_w = P_{dcr} + P_{acr} + P_p \tag{13.6-51}$$

The next step is the calculation of the core loss. To compute the core loss, we will assume that the flux density waveforms are saw-tooth in shape like the inductor current. Based on this waveform and the MSE loss model (6.3-16), the power loss density in region x is given by

$$p_x = \sum_{k=1}^{K} k_{h,k}\left(\frac{2}{\pi^2}\frac{1}{d(1-d)}\right)^{\alpha_k - 1}\left(\frac{f_{sw}}{f_b}\right)^{\alpha_k}\left(\frac{\Delta B}{2B_b}\right)^{\beta_k} + k_e\frac{(\Delta B_x)^2 f_{sw}^2}{d(1-d)} \tag{13.6-52}$$

In (13.6-52), $x \in [I, Ub, Ue]$, for the I-core, base of U-core, and ends (legs) of U-core, respectively. After the power loss densities are found, the core loss is computed with

$$P_{cr} = p_I V_{mI} + p_{Ub} V_{mUb} + 2p_{Ue} V_{mUe} \tag{13.6-53}$$

It should be noted that the MSE, while very convenient, is somewhat wanting in applications involving a dc offset [6]. However, in the case of ferrite materials, the core loss should be relatively low and so a large percentage error would make little difference. In the case of, for example, nanocrystalline materials, the k_e eddy-current term in (13.6-51) can be much larger; the accuracy of this term is not impacted by dc offsets or nonsinusoidal waveforms.

Finally, the total loss may be calculated as

$$P_t = P_w + P_{cr} \tag{13.6-54}$$

Related to the total loss, it will prove convenient to define an effective resistance such that

$$r_{Leff} = \frac{P_t}{\bar{i}_l^2} \tag{13.6-55}$$

As in other aspects of the work, it will prove convenient to encapsulate the loss calculations as

$$\mathbf{P}_{UI} = F_{PUI}(\mathbf{G}_{UI}, \bar{i}_l, \Delta i_l, f_{sw}, \mathbf{M}_{OP}, \sigma, d) \tag{13.6-56}$$

where

$$\mathbf{P}_{UI} = \begin{bmatrix} P_t & P_w & P_{dcr} & P_{acr} & P_p & P_{cr} & r_{Leff} \end{bmatrix} \tag{13.6-57}$$

Thermal Model

We must also consider the thermal performance of the inductor. We know from our work in Chapter 10 that the layer-to-layer insulation will reduce the winding capacitance. On the other hand, it is detrimental to the thermal performance because the layer-to-layer electrical insulation also acts as thermal insulation.

Herein, we will utilize a simplified thermal model that focuses on the winding bundle, as shown in Figure 13.12. In this representation, we will assume that no heat crosses between the winding and the core, an approximation that is facilitated by the coil-to-core insulating layer. Thus, winding heat is assumed to only leave through the sides or top of the winding bundle. Observe that Figure 13.12 also includes a coordinate system, which will be used; the z-axis is out of the page. However, it will be assumed that the winding temperature is isothermal in the z-axis. The length of the winding bundle will be taken to be that obtained by "unrolling" our racetrack-shaped coil into a line; taking the length into the z-axis to be the mean turn length, we have

$$l_z = \frac{1}{N_l} \sum_{i=1}^{N_l} l_{t,i} \tag{13.6-58}$$

Also, observe that for thermal analysis purposes, we will exclude the coil-to-core insulation (which is assumed to provide an adiabatic boundary); thus, the winding depth is taken to be

$$d'_w = d_w - t_{clc} \tag{13.6-59}$$

Our basic analysis strategy will be to homogenize the winding region into a single cuboidal thermal-equivalent circuit element. This will be done in two steps. First, using the procedure set forth in Section 10.5 and illustrated in Figure 10.11, we will homogenize the winding layers. A homogenized winding layer represents the conductors, their insulation, and whatever medium (air or potting material) that exists between the conductors. The layer-to-layer insulation is not included. Next, we will use a second homogenization process to combine the homogenized winding layers of

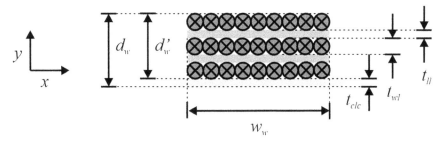

Figure 13.12 Cross-section of the winding bundle.

our first step with the insulating layers, thus creating a homogenized model of the entire winding bundle. The reason for the use of the second step is that the homogenization process set forth in Section 10.5 assumes the presence of only three materials, not four, and, more significantly, assumes that the homogenized material should be isotropic in the x- and y-directions, which will not be the case for the second step.

Proceeding with the first step, we observe that each winding layer is one layer of conductors, the wire insulation, and the potting material (if any). The thickness of a winding layer is given by

$$t_{wl} = 2r_w k_b \tag{13.6-60}$$

To homogenize the layer, we begin by calculating its aspect ratio using (10.5-4)

$$\zeta = \frac{w_w}{t_{wl}} \tag{13.6-61}$$

The total cross-sectional areas of the conductor and insulator within a winding are given by

$$a_{cl} = a_{wc} N_{tpl} N_{pr} \tag{13.6-62}$$

$$a_{il} = \pi \left(r_w^2 - r_{wc}^2 \right) N_{tpl} N_{pr} \tag{13.6-63}$$

Using our work in Section 10.5 and applying our specific parameter names in this chapter with the more general nomenclature of Section 10.5, that is, setting $a_c = a_{cl}$, $a_i = a_{il}$, $w = w_w$, $d = t_{wl}$, $k_i = k_{wi}$ (thermal conductivity of wire insulation), and $k_a = k_p$ (thermal conductivity of potting), we can calculate the effective thermal conductivity of the layer in the x- and y-directions using the sequences (10.5-7), (10.5-11), (10.5-14), and finally (10.5-22), whereupon we denote the effective thermal conductivity of the layer as $k_{wl} = k_{xye}$.

Our next step is another exercise in homogenization. In our view of the winding bundle, we now have alternating layers of winding (with thickness t_{wl} and thermal conductivity k_{wl}) and layer-to-layer insulation (with thickness t_{ll} and thermal conductivity k_{ll}). Based on this representation, if we find the total thermal resistance across the winding bundle in the x- and then y-directions and equate that to the thermal resistance of a homogeneous material in the x- and y-directions, respectively, we obtain expressions for the thermal conductivity of the homogenized winding bundle in the x- and y-directions. In particular, we obtain

$$k_{xe} = \frac{k_{wl} t_{wl} N_l + k_{ll} t_{ll} (N_l - 1)}{d'_w} \tag{13.6-64}$$

$$k_{ye} = \frac{d'_w k_{wl} k_{ll}}{t_{wl} N_l k_{ll} + t_{ll} (N_l - 1) k_{wl}} \tag{13.6-65}$$

At this point, we can create a thermal-equivalent circuit of the winding bundle based on a cuboidal element as set forth in Section 10.3. In particular, this yields the equivalent circuit shown in Figure 13.13. Therein, observe that there are no connections to the z-axis nodes because we assumed no heat flow in the z-axis. Also, there is no connection to the $0y$ node; as we assumed no heat flow between the winding coil and the core. From our work in Section 10.5, the thermal resistances x- and y-central nodes to the corresponding exterior nodes may be expressed as

$$R_{wx} = \frac{w_w}{2k_{xe}d'_w l_z} \tag{13.6-66}$$

$$R_{wy} = \frac{d'_w}{2k_{ye}w_w l_z} \tag{13.6-67}$$

In addition to R_{wx} and R_{yx}, we must represent the contact thermal resistance between the winding bundle and the ambient air. From Section 10.6

$$R_{ax} = \frac{1}{N_l t_{wl} l_z h_{wa}} \tag{13.6-68}$$

$$R_{ay} = \frac{1}{w_w l_z h_{wa}} \tag{13.6-69}$$

where h_{wa} is the winding-to-air heat transfer coefficient. Note that in (13.6-68), it is assumed that heat transfer to air does not occur from the layer-to-layer insulation, and so it is only the conductor area that determines R_{ax}. One could argue that the actual surface area of the conductor–ambient contact is $\pi t_{wl} N_l l_z / 2$ rather than $N_l t_{wl} l_z$ in (13.6-68); however, given that much of the excess area is poorly oriented for heat flow, the more conservative value is taken.

Clearly, the TEC shown in Figure 13.13 can be simplified. Eliminating the open-circuited branches, removing the thermal capacitance (to form a static model), and performing circuit reductions, the simplified TEC shown in Figure 13.14 is obtained. Therein,

$$R_{tx} = \frac{1}{6}R_{wx} + \frac{1}{2}R_{ax} \tag{13.6-70}$$

$$R_{ty} = \frac{2}{3}R_{wy} + R_{ay} \tag{13.6-71}$$

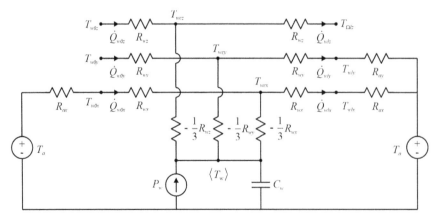

Figure 13.13 Winding bundle thermal-equivalent circuit.

Figure 13.14 Simplified winding bundle thermal-equivalent circuit.

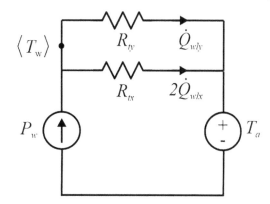

Combining the x- and y-branches, the net combination of these thermal resistances is given by

$$R_n = \frac{R_{tx}R_{ty}}{R_{tx} + R_{ty}} \tag{13.6-72}$$

Now that the thermal-equivalent circuit has been identified, it is, for later algorithm development purposes, appropriate to encapsulate the sequence of calculations as

$$\mathbf{T}_{UI} = \mathrm{F}_{TRWBp}(\mathbf{G}_{UI}) \tag{13.6-73}$$

where

$$\mathbf{T}_{UI} = \begin{bmatrix} l_z & w_w & d'_w & k_{ex} & k_{ey} & R_{tx} & R_{ty} & R_n \end{bmatrix} \tag{13.6-74}$$

The "TRWBp" subscript denotes a "thermal rectangular winding bundle parameters" since the model is not specific to a UI-core inductor.

Note that the width of the winding w_w is not really a calculation associated with the thermal model, but it will be convenient to copy this field from $\mathbf{G}_{UI}$ to simplify the parameter passing later on. Also, note that at this point, we have not actually used the thermal model. We will consider its use next; however, it is convenient to separate those aspects of the thermal model, which can be calculated from the geometry from those that must be calculated within an electrothermal analysis.

Within the electrothermal analysis, the first step will be to calculate the mean winding temperature from the thermal-equivalent circuit as

$$\langle T_w \rangle = R_n P_w + T_a \tag{13.6-75}$$

We will use the mean winding temperature to calculate the winding conductivity, which impacts the winding losses.

For constraint purposes, it will be desirable to calculate the peak winding temperature. To this end, the first step is to calculate the heat flows. In particular,

$$\dot{Q}_{wlx} = -\dot{Q}_{w0x} = \frac{1}{2} \frac{R_{ty}}{R_{tx} + R_{ty}} P_w \tag{13.6-76}$$

$$\dot{Q}_{wly} = \frac{R_{tx}}{R_{tx} + R_{ty}} P_w \tag{13.6-77}$$

Recall that $\dot{Q}_{w0y}$, $\dot{Q}_{w0z}$, and $\dot{Q}_{wlz}$ are all zero.

The spatial dependence within the cuboid representing the homogenized winding bundle is given by (10.3-2). From Section 10.3 and the conditions imposed on the equivalent circuit ($\dot{Q}_{wlx} = -\dot{Q}_{w0x}$ and $\dot{Q}_{w0y} = 0$), we can show that the coefficients in (10.3-2) are given by

$$c_{1x} = \frac{\dot{Q}_{wlx}}{k_{ex} d'_w l_z} \tag{13.6-78}$$

$$c_{2x} = \frac{-\dot{Q}_{wlx}}{k_{ex} w_w d'_w l_z} \tag{13.6-79}$$

$$c_{2y} = \frac{-\dot{Q}_{wly}}{2 k_{ey} w_w d'_w l_z} \tag{13.6-80}$$

and that $c_{1y} = c_{1z} = c_{2z} = 0$. From (10.3-28),

$$c_0 = \langle T_w \rangle - \left(\frac{1}{3} c_{2x} w_w^2 + \frac{1}{2} c_{1x} w_w + \frac{1}{3} c_{2y} d'^2_w \right) \tag{13.6-81}$$

Noting that $c_{2y} < 0$, it can be shown that the peak temperature occurs at $x = w_w/2$ and $y = 0$, whereupon from (10.3-2), the peak temperature is given by

$$T_{wpk} = \frac{1}{4} c_{2x} w_w^2 + \frac{1}{2} c_{1x} w_w + c_0 \tag{13.6-82}$$

At this point, we can encapsulate this set of calculations as

$$[\langle T_w \rangle \quad T_{wpk}] = F_{TRWBm}(\mathbf{T}_{UI}, T_a, P_w) \tag{13.6-83}$$

Capacitance

The final computation with respect to the UI-core output inductor will be its capacitance. This will be very important because we will need to ensure that the first resonant frequency of the inductor is well above the switching frequency, typically by an order of magnitude. Using the methods of Chapter 12, the dynamic turn-to-turn, dynamic coil-to-core, and dynamic layer-to-layer capacitances denoted C_{tt}, C_{clcr}, and C_{ll}, respectively, may be calculated. The total inductor capacitance may be expressed as

$$C_t = C_{tt} + C_{clcr} + C_{ll} \tag{13.6-84}$$

The calculations of Chapter 12 will be encapsulated as

$$\mathbf{C}_{UI} = F_{CRWB}(\mathbf{G}_{UI}) \tag{13.6-85}$$

where

$$\mathbf{C}_{UI} = [C_t \quad C_{tt} \quad C_{clcr} \quad C_{ll}] \tag{13.6-86}$$

13.7 Operating Point Analysis

In this section, we will consider the electrothermal analysis of an operating point. We will assume we know all parameters of the system, the average input voltage $\bar{v}_{in}$, the average output voltage $\bar{v}_{out}$, the average output current $\bar{i}_{out}$, and the desired output inductor current ripple Δi_l. Note that we assume the existence of a closed-loop control system such that the desired output voltage is achieved. As part of this analysis, we will need to compute the required duty cycle d. Note that, because of the way we have formulated the design problem, Δi_l will assume to be known. We will find it convenient to make Δi_l a design variable rather than f_{sw} even though that choice may seem more natural. The advantage is that with such a choice, not all of the magnetic analysis of the UI-core output inductor needs to be part of the iterative steady-state operating point analysis.

We begin our development by considering the buck converter average-value model shown in Figure 13.3 and (13.1-19), from which

$$v_{out} = (\bar{v}_{in} - r_{in}d\bar{i}_{out} - v_{ft})d - v_{fd}(1-d) - r_{out}\bar{i}_{out} \tag{13.7-1}$$

This form is a quadratic in duty cycle, and so the duty cycle may be calculated using the sequence

$$\alpha = r_{in}\bar{i}_{out} \tag{13.7-2}$$

$$\beta = -(v_{in} - v_{ft} + v_{fd}) \tag{13.7-3}$$

$$\gamma = v_{out} + v_{fd} + r_{out}\bar{i}_{out} \tag{13.7-4}$$

$$\delta = \beta^2 - 4\alpha\gamma \tag{13.7-5}$$

$$d = \frac{-\beta - \sqrt{\delta}}{2\alpha} \quad \delta \geq 0 \tag{13.7-6}$$

Note that if $\delta < 0$, there is no solution. The choice of root in (13.7-6) is so that the duty cycle calculated is the one close to the ratio of the output to the input voltage. This is explored in Problem 14. Unfortunately, many of the parameters in (13.7-2)–(13.7-6) such as the switch forward drops and the output inductor resistance (which is a function of temperature) are unknown, and so an iterative approach is required. The operating point analysis algorithm (OPAA) below is one approach to this end.

Operating Point Analysis Algorithm

Step 1: Initialize output inductor current. First, we find the average output inductor current using (13.1-25). Then, we determine the minimum and maximum values of the output inductor currents i_{mn} and i_{mx} using (13.1-36) and (13.1-37).

Step 2: Initialize mean winding temperature to the ambient temperature, that is, $\langle T_w \rangle = T_a$, and calculate the output inductor winding conductivity from (10.8-13) and then take the output resistance to be the output inductor resistance, that is,

$$r_{out} = \frac{l_{wc}}{N_{pr}a_{wc}\sigma} \tag{13.7-7}$$

Step 3: Initialize duty cycle. As an approximation, we initialize the duty cycle to

$$d = \frac{\bar{v}_{out}}{\bar{v}_{in}} \tag{13.7-8}$$

Step 4: Initialize input resistance to be equal to the input inductor resistance, that is, $r_{in} = r_{Lin}$, where r_{Lin} is calculated from (13.5-4) with $\bar{i}_{in} = d\bar{i}_{out}$, and P_{Lin} is found from (13.5-5), where both E_{Lin}^* and J_{Lin} can be considered to be known as they will be part of the search space.

Step 5: Initialize semiconductor losses. We next analyze the semiconductor losses. In particular, we call the semiconductor model (13.2-14) with its argument $v = \bar{v}_{in}$ and $f_{sw} = 0$. This gives us initial values for v_{ft} and v_{fd} though these estimates do not yet account for switching loss.

13 Buck Converter Design

Step 6: Initialize vector of unknowns. The final initialization step is to create a vector of unknown variables

$$\mathbf{x}_{old} = [d \quad f_{sw} \quad v_{ft} \quad v_{fd} \quad \langle T_w \rangle] \tag{13.7-9}$$

and an iteration count $k = 1$. We also create a convergence flag $c_{OP} = 1$, which is 1 if the analysis is valid and 0 otherwise.

Step 7: Duty cycle update. At this point, we have a crude estimate for the operating point of the converter. The procedure to generate an improved estimate begins with recalculating the duty cycle using (13.7-2)–(13.7-6). If $\delta < 0$, which indicates that the analysis is not valid, then we set $c_{OP} = 0$ and execution terminates. Otherwise, we use (13.1-23) to find the average input current $\bar{i}_{in}$.

Step 8: Input resistance update. Next, we compute an improved value for the input resistance. Formerly, we set the input resistance of the converter to be the resistance of the input inductor. However, we will adjust this value in order to represent the losses in the input capacitor (which does not appear in Figure 13.3). The rms input capacitor ripple is given by (13.1-29), whereupon the input capacitor loss may be expressed as

$$P_{Cin} = r_{Cin} i_{Cin,rms}^2 \tag{13.7-10}$$

The effective input resistance is taken to be

$$r_{in} = \frac{P_{Lin} + P_{Cin}}{\bar{i}_{in}^2} \tag{13.7-11}$$

where P_{Lin} is known from Step 4. In this way, the capacitor power loss is represented in our model, though not exactly in the correct spot. At this point, the input capacitor voltage is calculated as

$$\bar{v}_c = \bar{v}_{in} - r_{in}\bar{i}_{in} \tag{13.7-12}$$

Step 9: Switching frequency update. The switching frequency is then calculated. Rearranging (13.1-35)

$$f_{sw} = \frac{1}{\Delta i_l L_{out}} (\bar{v}_c - v_{ft} + v_{fd}) d(1-d) \tag{13.7-13}$$

Once we have an updated switching frequency, we can update the semiconductor loss information using (13.2-14) with the blocking voltage set to $\bar{v}_c$.

Step 10: Output inductor temperature update. The calculation of the UI-core output inductor temperature and loss is the next step. Based upon the current estimate of the mean winding temperature, $\langle T_w \rangle$, the conductivity is calculated using (10.8-13). Next, the UI-core output inductor losses are computed using (13.6-56). The UI-core thermal model (13.6-83) is then called, yielding updated values of $\langle T_w \rangle$ and T_{wpk}.

Step 11: Output resistance update. Our next step is to compute an updated estimate of the output resistance. First, the rms output capacitor current is found using (13.1-45). Then, the output capacitor loss is computed as

$$P_{Cout} = r_{Cout} i_{Cout,rms}^2 \tag{13.7-14}$$

The output resistance may then be computed as

$$r_{out} = r_{Leff} + \frac{P_{Cout}}{\bar{i}_l^2} \tag{13.7-15}$$

where r_{Leff} was calculated in Step 10 in evaluating (13.6-56). Observe that this value of resistance accounts for dc conductive, ac conductive, and core losses in the inductor as well as the resistive losses in the output capacitor. This is done to maintain power balance.

Step 12: Convergence check. We now determine whether another iteration is required. We first form a vector of updated solution variables

$$\mathbf{x} = \begin{bmatrix} d & f_{sw} & v_{ft} & v_{fd} & T_{wmn} \end{bmatrix} \tag{13.7-16}$$

We define the difference and average between our two estimates

$$\mathbf{x}_\Delta = |\mathbf{x}_{old} - \mathbf{x}| \tag{13.7-17}$$

$$\mathbf{x}_a = \frac{1}{2}|\mathbf{x}_{old} + \mathbf{x}| \tag{13.7-18}$$

Using these quantities, we define an error metric

$$e = \max_i \left(\mathbf{x}_{\Delta,i} - r_e \mathbf{x}_{a,i} - a_e \right) \tag{13.7-19}$$

In (13.7-19), r_e is a relative error constant and a_e is an absolute error constant. These constants are both small numbers, for example, on the order of 10^{-6}. For e to be negative, then the magnitude of the difference of a quantity between the current and previous estimates must either be very small in absolute terms or it must be small relative to the value of that quantity for all elements of $\mathbf{x}$.

The final aspect of this step is to replace the vector of old estimates with the new ones and increment the iteration count. This is accomplished with

$$\mathbf{x}_{old} = \mathbf{x} \tag{13.7-20}$$

$$k = k + 1 \tag{13.7-21}$$

At this point, a decision is made. If $c_{OP} = 1$ and $k < k_{mxit}$, where k_{mxit} is the maximum allowed number of iterations, and $e > 0$, we return to Step 7 and perform another update. If $c_{OP} = 1$, $k \leq k_{mxit}$, and $e \leq 0$, we are finished and have a valid solution and terminate the algorithm. If $c_{OP} = 0$ or $k > k_{mxit}$, we leave $c_{OP} = 0$ and terminate the algorithm, in this case, without a solution.

Step 13: Related calculations. Although not a formal part of the steady-state solution, this is a convenient time to compute the voltage and current ripple in the converter, as well as the capacitor temperatures. In particular, we can compute the input capacitor voltage ripple using (13.1-30), wherein the capacitance value can be determined from the design parameters, as will be discussed in Section 13.8. Next, from the input inductor energy metric and input current, the input inductor inductance can be calculated by rearranging (13.5-1). In particular,

$$L_{in} = \frac{2E^*_{Lin}}{\bar{i}^2_{in}} \tag{13.7-22}$$

whereupon the input inductor current ripple is calculated using (13.1-32). Finally, the output capacitor voltage ripple is found using (13.1-44). The capacitor temperatures can be found from (13.4-12).

Encapsulation

As with other components of our analysis, we will encapsulate this work as a function call. In particular,

$$\mathbf{B}_{OP} = \mathbf{F}_{BOP}(\mathbf{G}_{UI}, \mathbf{M}_{UI}, \mathbf{T}_{UI}, \mathbf{C}_{IN}, \mathbf{C}_{OUT}, E_{Lin}, P_{Lin}, v_{in}, v_{out}, i_{out}, \Delta i_l) \tag{13.7-23}$$

where

$$\mathbf{B}_{OP} = \begin{bmatrix} c_{OP} & f_{sw} & d & \mathbf{S} & P_{Cin} & T_{Cin} & P_{Cout} & T_{Cout} & L_{in} & r_{Lin} & \mathbf{P}_{UI} & T_{wpk} & \langle T_w \rangle & \Delta i_{Lin} & \Delta v_{Cin} & \Delta i_l & \Delta v_{Cout} \end{bmatrix} \quad (13.7\text{-}24)$$

In (13.7-24), all fields have been defined. Note that, for convenience, Δi_l is copied from the input arguments to $\mathbf{B}_{OP}$.

13.8 Design Paradigm

We will now use all of the analysis set forth in this chapter to develop a buck converter/UI-core output inductor design paradigm based on multiobjective optimization. Our goal will be to design a converter to reduce an input voltage $\bar{v}_{in}$ to an output voltage $\bar{v}_{out}$. We will assume the presence of a closed-loop control system, which will adjust the duty cycle with load so that the exact value of the output voltage is obtained in the steady state. This can be readily accomplished with a proportional plus integral control, among other strategies. We will design our converter based on a full-load output current $\bar{i}_{out}$. We begin by identifying the search space. In particular, one possibility is the search space indicated defined by (13.8-1)

$$\boldsymbol{\theta} = \begin{bmatrix} g & w_{bc} & l_c & r_{ib} & r_{eb} & N^*_{tpl} & N^*_l & N^*_{pr} & J^*_{Lout} & c_w & c_d & t^*_{clc} & t^*_{ll} & C^*_{in} & C^*_{out} & E^*_{Lin} & J_{Lin} & \Delta i_l \end{bmatrix} \quad (13.8\text{-}1)$$

All of these variables in the search space have already been discussed except for the desired full-load output current density, J^*_{Lout}. This variable facilitates specifying the desired cross-sectional area. In particular, we will take

$$a^*_{wcpr} = \frac{\bar{i}_{out}}{J^*_{Lout}} \quad (13.8\text{-}2)$$

Another point of interest in (13.8-1) is that a desired value of output inductance is not in the search space; this is because the geometrical information in (13.8-1) determines the output inductance. Also, observe that the switching frequency is not a design variable. While this would have been possible, from (13.1-35) and noting that the input capacitor voltage is essentially the input voltage and that the duty cycle is essentially fixed by the ratio of the output to input voltage, the specification of the switching frequency is accomplished indirectly by the output inductance and current ripple Δi_l.

Fixed design information will be captured in a structure $\mathbf{D}$ defined by

$$\mathbf{D} = \begin{bmatrix} \bar{v}_{in} & \bar{v}_{out} & \bar{i}_{out} & r_{\Phi rq} & T_{Cin,mxa} & T_{Cout,mxa} & T_{Twpk,mxa} & \Delta v_{Cin,mxa} & \Delta i_{Lin,mxa} & \Delta v_{out,mxa} & r_{ff} & f_{sw,mxa} \\ T_A & R_{tch} & R_{dch} & M_{mxa} \mathbf{D}_{Cin} & \mathbf{D}_{Cout} & \mathbf{D}_{UI} & \mathbf{D}_{Lin} & \mathbf{D}_{st} & \mathbf{D}_{sd} & \mathbf{D}_{PUI} & \mathbf{D}_{St} & \mathbf{D}_{Sd} \end{bmatrix} \quad (13.8\text{-}3)$$

Most of the variables and substructures in $\mathbf{D}$ have been previously discussed. However, there are some new variables including the resonant frequency factor r_{ff}, maximum allowed switching frequency $f_{sw,mxa}$, and maximum allowed mass M_{mxa}, which will be discussed in more detail as we set forth constraints.

As with our previous design work, we will pose the design problem as a multiobjective optimization problem. Our objectives will be to minimize loss and mass, subject to a variety of constraints. To this end, one possible procedure will now be set forth algorithmically, with the various constraints introduced in the order they would be addressed in the execution of the fitness function. In particular, the steps in the fitness function evaluation are as follows.

Step 1: Constraint initialization. As a first step, we initialize the number of constraints $C = 15$, the constraints satisfied $C_s = 0$, and the constraints imposed $C_i = 0$.

Step 2: Capacitance data. From (13.4-8), the capacitor data is determined as

$$\mathbf{C}_{in} = \mathbf{F}_C(\mathbf{C}^*_{in}, \mathbf{D}_{Cin}) \tag{13.8-4}$$

$$\mathbf{C}_{out} = \mathbf{F}_C(\mathbf{C}^*_{out}, \mathbf{D}_{Cout}) \tag{13.8-5}$$

Step 3: Geometrical calculations for the UI-core output inductor are then performed. From (13.6-35),

$$\mathbf{G}_{UID} = \mathbf{F}_{GUI}(\mathbf{G}_{UII}, \mathbf{D}_{UI}) \tag{13.8-6}$$

At this point, we compute our first constraint. In particular, the winding is easier to construct if its depth is less than its width; thus, we impose

$$c_1 = \text{lte}(d_w, w_w) \tag{13.8-7}$$

Next, returning to our work from Chapter 1, we execute the algorithm of Table 1.6.

Step 4: Output inductor magnetic analysis is carried out using (13.6-43) as

$$\mathbf{M}_{UI} = \mathbf{F}_{MUI}(\mathbf{G}_{UI}, \bar{i}_l, \Delta i_l) \tag{13.8-8}$$

Before proceeding, we check convergence

$$c_2 = c_m \tag{13.8-9}$$

and execute Table 1.6. If the algorithm has converged, we check the flux ratio constraint

$$c_3 = \text{gte}(r_\Phi, r_{\Phi rq}) \tag{13.8-10}$$

and again execute Table 1.6.

Step 5: Thermal analysis initialization is performed using (13.6-73). In particular,

$$\mathbf{T}_{UI} = \mathbf{F}_{TRWBp}(\mathbf{G}_{UI}) \tag{13.8-11}$$

Step 6: Output inductor capacitance. After computing the inductor capacitance from (13.6-85)

$$\mathbf{C}_{UI} = \mathbf{F}_{CRWB}(\mathbf{G}_{UI}) \tag{13.8-12}$$

we can compute the inductor resonance frequency

$$f_r = \frac{1}{2\pi\sqrt{C_t L_{inc}}} \tag{13.8-13}$$

Later, we will compare the resonant frequency with the switching frequency.

Step 7: Input inductor parameters. The next step is to compute the parameters of the input inductor. From (13.5-5),

$$\begin{bmatrix} M_{Lin} & P_{Lin} \end{bmatrix}^T = \mathbf{F}_{UILdcmm}(E_{Lin}, J_{Lin}, \mathbf{D}_{Lin}) \tag{13.8-14}$$

Step 8: Operating point analysis is carried out using the OPAA set forth in Section 13.7. In particular, from (13.7-24),

$$\mathbf{B}_{OP} = F_{BOP}(\mathbf{G}_{UI}, \mathbf{M}_{UI}, \mathbf{T}_{UI}, \mathbf{C}_{IN}, \mathbf{C}_{OUT}, E_{Lin}, P_{Lin}, v_{in}, v_{out}, i_{out}, \Delta i_l) \quad (13.8\text{-}15)$$

After computing the operating point, we first check for convergence using

$$c_4 = c_{op} \quad (13.8\text{-}16)$$

and then executing Table 1.6. Next, we constrain the capacitor temperatures as well as the voltage and current ripple with the constraints

$$c_5 = \text{gte}(T_{Cin}, T_{Cin,mxa}) \quad (13.8\text{-}17)$$
$$c_6 = \text{gte}(T_{Cout}, T_{Cout,mxa}) \quad (13.8\text{-}18)$$
$$c_7 = \text{gte}(T_{wpk}, T_{Twpk,mxa}) \quad (13.8\text{-}19)$$
$$c_8 = \text{gte}(\Delta v_{Cin}, \Delta v_{Cin,mxa}) \quad (13.8\text{-}20)$$
$$c_9 = \text{gte}(\Delta i_{Lin}, \Delta i_{Lin,mxa}) \quad (13.8\text{-}21)$$
$$c_{10} = \text{gte}(\Delta v_{out}, \Delta v_{out,mxa}) \quad (13.8\text{-}22)$$

For the output inductor to be effective, its resonant frequency must be much higher than the switching frequency. Therefore, we impose the constraint

$$c_{11} = \text{gte}(f_r, f_{sw} r_{ff}) \quad (13.8\text{-}23)$$

Also, it is appropriate to limit the switching frequency in order to address limitations imposed because of semiconductor edge rate concerns or limitations imposed by the gate drive circuitry. To this end, we add the constraint

$$c_{12} = \text{gte}(f_{sw}, f_{sw,mxa}) \quad (13.8\text{-}24)$$

After calculating the constraints, we again execute the algorithm of Table 1.6, with the exception that the number of constraints imposed is incremented by 8 and that the number of constraints satisfied is incremented with the sum of c_5 through c_{12}.

Step 9: Find heat-sink masses. As a next step, we address the transistor and diode heat sinks. From (13.3-3),

$$[M_{th} \ c_{13}] = F_{PAH}(P_{ttl}, T_A, R_{tch}, D_{St}, D_{PAH}) \quad (13.8\text{-}25)$$
$$[M_{dh} \ c_{14}] = F_{PAH}(P_{dtl}, T_A, R_{dch}, D_{Sd}, D_{PAH}) \quad (13.8\text{-}26)$$

At this point, we again execute Table 1.6, with suitable modification to the number of constraints satisfied and imposed.

Step 10: Compute total mass and loss from

$$M = M_{Cin} + M_{Cout} + M_{Lin} + M_{Lout} + M_{th} + M_{dh} \quad (13.8\text{-}27)$$
$$P_{loss} = P_{Cin} + P_{Cout} + P_{Lin} + P_{Lout,t} + P_{ttl} + P_{dtl} \quad (13.8\text{-}28)$$

Observe that the "total" mass is somewhat misleading as it does not include structure or control system. One can think of it as an "electromagnetic" mass, that is, the mass of the electrical, magnetic, and thermal components. At this point, we add one final constraint, in that we will limit the maximum mass of interest using

$$c_{15} = \text{lte}(M, M_{mxa}) \tag{13.8-29}$$

At this point, we once again execute Table 1.6.

Step 11: Compute fitness. If the fitness function execution reaches this point, all constraints are passed. We may compute the fitness as

$$f = \begin{bmatrix} \dfrac{1}{M} & \dfrac{1}{P_{loss}} \end{bmatrix}^T \tag{13.8-30}$$

This completes the description of our fitness function execution. In the next section, we will use this process to carry out a design example.

13.9 Case Study

In the previous section, we considered one possible fitness function for the multiobjective design optimization of a buck converter/UI-core output inductor. We will now use that fitness function in a design example. In particular, we will consider the design of a 10 kW, 750 to 400 V buck converter. In addition, several design choices will be made a priori. The principal specifications and a priori design choices for this converter are listed in Table 13.3.

Of the a priori design choices, first we will choose a solid magnet wire. For this, we will choose copper conductors corresponding to American wire gauges (AWGs) with insulation parameters listed in Table 13.3. Therein, the bending radius constant, build factor, and winding-to-air heat transfer coefficient are also listed. For layer-to-layer-insulation, a Nomex 410 paper catalog will be used (see [7]). To improve thermal performance, we will assume the use of a potting material with parameters listed in Table 13.3, which also include the peak allowed winding temperature, which is based on an insulation class. In order to simplify the design, we will also a priori choose an MN60-LL core material. Clearly, this could have also been part of the search space. Likewise, we will also choose the semiconductors a priori. In particular, we will use the silicon carbide MOSFET and a diode whose parameters are listed in Table 13.1. We will assume that the heat-sink parameters are as given in Section 13.3. The assumed values for the case-to-heat-sink thermal resistances are listed in Table 13.3. For the input inductor, we will utilize the metamodel parameters given by Table 13.2. We will use the capacitor data set forth in [7], selecting the 800-V nominal voltage for the input capacitor and the 500-V nominal voltage for the output capacitors. The assumed values of maximum allowed temperature listed in Table 13.3 are selected based on the desired lifetime. The remaining parameters listed in the table are algorithmic. These numbers were based on engineering judgment; however, there is great latitude in values, which yield acceptable results.

The search space is set forth in Table 13.4. Therein, many of the parameters are set via experimentation—making sure that the gene values do not become clustered on limits unless there is a physical limit. Others are based on what is deemed reasonable based on experience—for example, the insulating paper thickness and current density values. Here again, there is great latitude in many of the values listed in terms of obtaining favorable results.

One of the most critical values in Table 13.4 is the maximum value allowed for Δi_l. In fact, we will find that for the parameters listed, the maximum value allowed will be consistently chosen. To understand the motivation for the particular value, note that from (13.1-35), if we are expecting a relatively constant input voltage and a relatively constant desired output voltage, the current ripple Δi_l would not be expected to greatly change with load. For the continuous mode of operation, $\Delta i_l < 2\bar{i}_l$. Thus, for an output current of 25 A, we need Δi_l to be less than 50 A, which is much larger than the value indicated in Table 13.4. However, if we desire that the operation remain continuous

Table 13.3 Buck Converter Specifications and A Priori Design Choices

Symbol	Description	Value
$\bar{v}_{in}$	Input voltage	750 V
$\bar{v}_{out}$	Output voltage	400 V
$\bar{i}_{out}$	Output current	25 A
$\Delta v_{Cin,\,mxa}$	Input capacitor pk–pk voltage ripple, max allowed	37.5 V
$\Delta i_{Lin,\,mxa}$	Input inductor pk–pk current ripple, max allowed	133 mA
$\Delta v_{Cout,\,mxa}$	Output capacitor pk–pk voltage ripple, max allowed	4 V
$f_{sw,\,mxa}$	Switching frequency, max allowed	100 kHz
r_{ff}	Resonant frequency factor	20
M_{mxa}	Mass, max allowed	10 kg
T_a	Ambient temperature	40°C
t_{wi}	Thickness of wire insulation	33 µm
k_{wi}	Wire insulation thermal conductivity	0.2 W/(m·K)
ε_{rwi}	Wire insulation relative permittivity	3.52
k_{bd}	Wire bending radius constant	6
k_b	Winding build factor	1.05
h_{wa}	Winding-to-air heat transfer coefficient	23 W/(m²K)
k_p	Potting material thermal conductivity	1.48 W/(m·K)
ε_{rp}	Potting material relative permeability	6
$T_{wpk,\,mxa}$	Peak allowed winding temperature	160°C
N_h	Number of harmonics used in loss analysis	10
$r_{\Phi,\,rq}$	Minimum flux factor	0.9
R_{tch}	Assumed transistor to heat-sink thermal resistance	0.5°C/W
R_{dch}	Assumed diode to heat-sink thermal resistance	0.5°C/W
$T_{Cin,\,mxa}$	Maximum allowed input capacitor temperature	85°C
$T_{Cout,\,mxa}$	Maximum allowed output capacitor temperature	85°C
r_e	OPAA relative error constant (see Section 13.7)	10^{-6}
a_e	OPAA absolute error constant	10^{-6}
k_{mxit}	OPAA maximum allowed iterations	20

for low loads, then a reduced value of Δi_l is required. The 10 A listed in Table 10.4 will keep operation continuous down to approximately 20% of the rated load. Of course, there is a question of whether it is necessary to require continuous operation. Here, strictly speaking, the answer is no—but only provided that controller—which we have not addressed—can maintain suitable transient characteristics while in discontinuous mode.

Figure 13.15 illustrates the parameter distribution after a run. In this study, there were 4000 individuals and 4000 generations. The parameter numbers in Figure 13.15 correspond to the element number in (13.8-1). As can be seen, parameters 1, 3, 4–6 (with a bifurcation), 7, 10, 12, 13, and 18, which correspond to $g, l_c, r_{ib}, r_{eb}, N^*_{tpl}, N^*_l, c_w, t^*_{clc}, t^*_{ll}$, and Δi_l converge to more-or-less static values.

Table 13.4 Buck Converter Design Space

Par.	Description	Min.	Max.	Enc.	Gene
g	Out. ind. air gap (m)	10^{-4}	10^{-2}	log	1
w_{bc}	Out. ind. center base width (m)	10^{-3}	10^{-1}	log	2
l_c	Out. ind. core length (m)	10^{-3}	2×10^{-1}	log	3
r_{ib}	Out. ind. I/base width ratio	0.5	1.5	log	4
r_{eb}	Out. ind. end/base width ratio	0.5	1.5	log	5
N_{tl}^*	Out. ind. desired turns per layer	0.51	100.4	log	6
N_l^*	Out. ind. desired number of layers	0.51	10.49	log	7
N_{pr}^*	Out. ind. number of parallel cond.	0.51	4.49	lin	8
J_{Lout}^*	Out. ind. des. current density (A/m²)	10^5	10^7	log	9
c_w	Out. ind. horizontal clearance (m)	10^{-3}	5×10^{-2}	log	10
c_d	Out. ind. vertical clearance (m)	10^{-3}	5×10^{-2}	log	11
t_{clc}^*	Out. ind. des. cl-to-cr ins. thick. (m)	$5 \cdot 10^{-5}$	5×10^{-3}	lin	12
t_{ll}^*	Out. ind. des. lyr-to-lyr ins. thick. (m)	0	5×10^{-3}	lin	13
C_{in}^*	Desired inp. capacitor capacitance (F)	10^{-6}	5×10^{-3}	log	14
C_{out}^*	Desired out. capacitor capacitance (F)	10^{-6}	5×10^{-3}	log	15
E_{Lin}^*	Inp. ind. energy metric (J)	10^{-3}	10^{-1}	log	16
J_{Lin}	Inp. ind. current density (A/m²)	10^5	10^7	log	17
Δi_l	Out. ind. current ripple (A)	0.1	10	log	18

Parameter 8, N_{pr}^*, is loosely clustered at the upper end of its domain; this is because of the rounding operator coupled with a relatively small set of allowed values. Elements 2, 9, 11, and 14–17, which correspond to w_{bc}, J_{Lout}^*, c_d, C_{in}^*, C_{out}^*, E_{Lin}^*, and J_{Lin}, vary significantly as we move along the Pareto-optimal front, with the most massive designs on the left side of the parameter distribution window and the least massive design toward the right.

The Pareto-optimal front is shown in Figure 13.16. Therein, we can see that the loss varies from 180 to 80 W as the mass varies from just under 2 to 10 kg. Again, it should be emphasized that the mass does not include structure or auxiliary features. Design 100 is the 100th design in the order of increasing mass; we will examine this design in more detail momentarily. This data is also shown in an alternate form in Figure 13.17, where the design is viewed in terms of efficiency versus specific power density. Clearly, all designs along the front are fairly efficient. The inclusion of a thermal model leads to this; overly lossy designs become problematic from a thermal point of view.

While the output inductor current density increases as we move toward less massive designs (this can be seen from element 9, J_{Lout}^*, in Figure 13.15), this is not the only or even dominant change. A significant change in switching frequency occurs as we move along the front. This can be seen in Figure 13.18. For the lowest mass designs, the switching frequency is in excess of 50 kHz; for the most efficient and highest mass designs, the switching frequency drops to around 15 kHz.

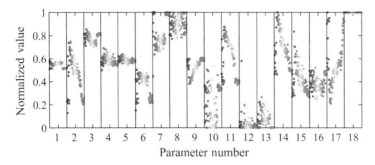

Figure 13.15 Buck converter parameter distribution.

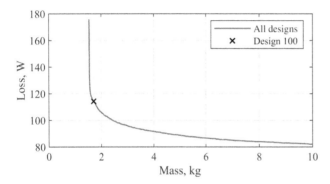

Figure 13.16 Buck converter Pareto-optimal front.

Details of design 100 are set forth in Table 13.5. Therein, the first item listed is the passive component values. Observe that the output inductor resonant frequency is 1.022 MHz; given the 49.65-kHz switching frequency and the resonant frequency factor of $r_{ff} = 20$, the minimum allowed resonant frequency was 0.993 MHz; thus, it can be seen that the resonant frequency constraint was active and that the output inductor capacitance was important to the design. It is interesting that the switching frequency is well below our limit of 100 kHz.

The UI-core output inductor winding utilized four parallel conductors; this reduced the skin-effect and proximity-effect losses. The use of Litz wire would have been an alternate approach. A cross section of the UI-core output inductor is given in Figure 13.19. Interestingly, observe that the winding does not occupy the entire slot. This reduces the proximity-effect loss by keeping the winding out of high-field regions associated with fringing flux. Here again, Litz wire could have potentially been put to good effect; however, the lower packing density may reduce the benefit.

The converter mass was dominated by the output inductor (68%), with the transistor heat sink (22.8%) and input capacitor (5.88%) being the next most massive elements.

The converter losses were dominated by the transistor (60%), diode (17%), and output inductor (21%). The output inductor losses were dominated by the dc-resistive loss (16% of total loss), followed by the proximity-effect loss (4% of total loss). Skin-effect and core losses were nearly negligible.

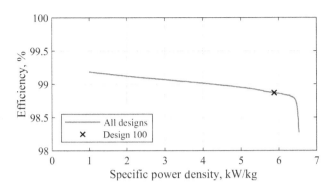

Figure 13.17 Efficiency versus specific power density.

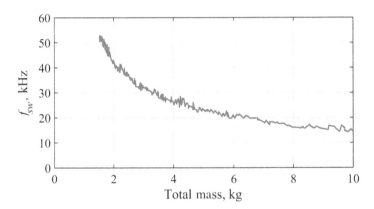

Figure 13.18 Switching frequency versus mass.

Table 13.5 Design 100

Component Data	Full-Load Operating Point Data
Input capacitance: 60 µF	Inp. voltage: 750 V
Output capacitance: 12 µF	Inp. current: 13.49 A
Input inductance: 39.89 µH	Inp. power: 1.011e + 04 W
Output inductance: 373.6 µH	Out. voltage: 400 V
Trn–Trn cap.: 1.644 pF	Out. current: 25 A
Lyr–Lyr cap.: 63.09 pF	Out. power: 1e + 04 W
Cl-to-Cr cap.: 0.2296 pF	Switching freq.: 49.65 kHz
Total cap.: 64.96 pF	Duty cycle: 0.5394
Resonant freq.: 1.022 MHz	Pk–Pk out. rip.: 2.098 V
UI-Core Output Inductor	Pk–Pk out. ind. rip.: 10 A
Core length: 6.987 cm	Pk–Pk inp. cap. rip.: 2.085 V
Slot width: 2.36 cm	Pk–Pk inp. cur. rip.: 0.1316 A
Slot depth: 1.814 cm	Full-Load Loss
Base width: 1.907 cm	Converter total: 114.3 W
End width: 1.801 cm	Tran. total: 68.58 W (60%)
I width: 1.79 cm	Tran. cond.: 44.49 W (38.92%)
Air gap: 1.065 mm	Tran. switch.: 24.09 W (21.08%)

(Continued)

Table 13.5 (Continued)

Total width: 5.962 cm	Diode total: 19.93 W (17.43%)
Total height: 6.903 cm	Diode cond.: 19.93 W (17.43%)
Total length: 9.558 cm	Diode switch.: 0 W (0%)
Number of layers: 10	Inp. ind.: 1.153 W (1.009%)
Turns per layer: 2	Out. ind. total: 24.39 W (21.34%)
Number of turns: 20	Out. ind. DC: 17.87 W (15.64%)
Number of parallel conductors: 4	Out. ind. AC: 0.5398 W (0.4723%)
Wire diameter: 1.217 mm	Out. ind. prox.: 4.9 W (4.287%)
Conductor length (each): 4.701 m	Out. ind. core: 1.075 W (0.9404%)
Cl–Cr ins.: 0.08 mm	Inp. cap.: 0.1863 W (0.163%)
Lyr-to-Lyr ins.: 0 mm	Out. cap.: 0.06666 W (0.05833%)
Mass--	Temperature---
Converter Total: 1.701 kg	Ambient temp.: 40°C
Tran. hs: 0.3875 kg (22.79%)	Inp. cap. temp.: 40.67°C
Diode. hs: 0.01993 kg (1.172%)	Out. cap. temp.: 41.52°C
Inp. ind.: 0.04233 kg (2.489%)	Out. ind. pk wind. temp.: 153.1°C
Out. ind.: 1.136 kg (66.79%)	Out. ind. mn wind. temp.: 149.2°C
Inp. cap.: 0.1 kg (5.88%)	
Out. cap.: 0.015 kg (0.882%)	

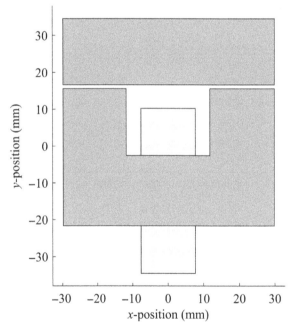

Figure 13.19 UI-core output inductor cross section.

In terms of temperature, observe that the temperature limit on the capacitors was not an issue; they barely rose above the ambient temperature. The inductor peak winding temperature is close to the limit of 160°C, however.

13.10 Extensions

Clearly, there are many improvements that could be made to the example. Some very straightforward changes would be to include the provision to choose semiconductors and the core material type. With a little more effort, the ability to design with Litz wire or film conductor could be added. Finally, the consideration of multiple operating points would be valuable, so that the converter is optimized over an expected range of loads rather than a single loading condition.

There have been a variety of papers in the literature that use this approach. In [4], the design of cascaded dc–dc converter is considered. The design of a permanent-magnet inductor-based buck converter is set forth in [8]; in this approach, the inductor size is greatly reduced by using a permanent magnet to remove the bias flux. The design of permanent magnet ac (PMAC) machine, rectifier, and buck converter, and control, is set forth in [9] and also described in [1,10]. The design of systems of power converters is set forth in [11].

References

1 B. Zhang and et al., "Prediction of Pareto-optimal performance improvements in a power conversion system using GaN devices," in *2017 IEEE 5th Workshop on Wide Bandgap Power Devices and Applications*, Albuquerque, NM, 2017.

2 C4020120A Datasheet. Durham, NC: Cree, Inc., 2016

3 C2M0080120D Datasheet. Durham, NC: Cree, Inc., 2019

4 S. Wang, *Automated Design Approaches for Power Electronics Converters*, West Lafayette, IN: PhD Dissertation, Purdue University, 2010.

5 S. Sudhoff, G. Shane and H. Suryanarayana, "Magnetic-equivalent-circuit-based scaling laws for low-frequency magnetic devices," *IEEE Transactions on Energy Conversion*, vol. **28**, no. 3, p. 9, 2013.

6 J. Cale, S. Sudhoff and R. Chan, "A field-extrema hysteresis loss model for high-frequency ferrimagnetic materials," *IEEE Transactions on Magnetics*, vol. **44**, no. 7, pp. 1728–1736, 2008.

7 S. D. Sudhoff, "*MATLAB Codes for Power Magnetic Devices: A Multi-Objective Design Approach*, 2nd Edition," [Online]. Available: http://booksupport.wiley.com.

8 G. Shane and S. Sudhoff, "Design paradigm for permanent-magnet-inductor-based power converters," *IEEE Transactions on Energy Conversion*, vol. **28**, no. 4, pp. 880–893, 2013.

9 B. Zhang, *A Design Paradigm for DC Generation System*, West Lafayette, IN: PhD Dissertation, Purdue University, 2019.

10 B. Zhang, S. Sudhoff, S. Pekarek and J. Neely, "Optimization of a wide bandgap based generation system," in *IEEE Electric Ship Technologies Symposium (ESTS)*, Arlington, VA, 2017.

11 H. Suryanarayana and S. Sudhoff, "Design paradigm for power electronics-based DC distribution systems," *IEEE Journal of Emerging and Selected Topics in Power Electronics*, vol. **5**, no. 1, pp. 51–63, 2017.

Problems

1 Using (13.1-7), (13.1-8), and (13.1-11), obtain (13.1-12).

2 From Figure 13.4, derive (13.1-29).

3. From Figure 13.4 and (13.1-3), show (13.1-30).

4. Deduce (13.1-32) from Figure 13.4 and (13.1-31).

5. From (13.1-33) and (13.1-34), arrive at (13.1-35).

6. Obtain (13.1-38) from Figure 13.2 and (13.1-14)

7. From Figure 13.2, express (13.1-40) and (13.1-41).

8. From Figure 13.2, show (13.1-42).

9. Deduce (13.1-44) and (13.1-45) from Figure 13.5.

10. Using (13.2-1) and (13.2-2), derive (13.2-3) and (13.2-4).

11. Using (13.3-2), plot the heat-sink mass versus the desired thermal resistance for a thermal resistance of 0.1–10 K/W. Use a log–log plot. Use the parameters appearing just after (13.3-2).

12. Assuming a triangular current waveform and the MSE loss model, derive (13.6-52).

13. Derive (13.6-64) and (13.6-65). Carefully justify the choice or root.

14. From (13.7-1), obtain (13.7-2)–(13.7-6).

15. Explore the impact of Δi_l on the Pareto-optimal front. Using the code obtained from [7], compare the Pareto-optimal front obtained with the maximum allowed value of Δi_e changed from 10 to 20A.

16. Modify the OPAA of Section 13.7 so that f_{sw} is an input and Δi_l is an output.

17. Modify the design algorithm set forth in Section 13.8 so that f_{sw} replaces Δi_l as part of the search space in (13.8-1). Assume that you have a modified OPAA such as that described in problem 16 above. You may need to modify constraints.

18. Starting from the code base set forth in [7], modify the code so that f_{sw} replaces Δi_l in the search space. Compare the PO front thus obtained to that in the text.

14

Three-Phase Inductor Design

In Chapter 13, we considered the design of an inductor in a buck converter. In this chapter, we will again consider the design of an inductor in a power electronic converter, in this case a three-phase inductor in a dc to ac converter. This represents a circuit which could either be used as part of an active rectifier, or as part of renewable power grid interface.

14.1 System Description

Figure 14.1 depicts the system we will consider. Therein we have a dc voltage v_{dc} at the dc terminals of a three-phase bridge. This voltage may be established by a battery, a capacitor, a photo-voltaic array, or some other means. On an appropriate time scale, this voltage may also be regulated by the control of the inverter, whose leg-to-bottom-rail voltages are denoted as v_{ar}, v_{br}, and v_{cr}, respectively. Neglecting forward voltage drops, these voltages switch between 0 and v_{dc} in such a manner that the desired currents are obtained. The currents flowing out of the converter and into a three-phase inductor are denoted as i_{al}, i_{bl}, and i_{cl}. These nominally sinusoidal ac currents will include a significant ripple component because of the switching of the power semiconductors. The purpose of the inductor is to limit the switching-induced current ripple.

The voltages across the inductive filter are v_{al}, v_{bl}, and v_{cl}. The output of the filter is assumed to be an ideal three-phase voltage source with phase-to-neutral voltages of v_{au}, v_{bu}, and v_{cu}. Denoting three-phase quantities as a vector of the form

$$\mathbf{f}_{abcx} = \begin{bmatrix} f_{ax} & f_{bx} & f_{cx} \end{bmatrix}^T \tag{14.1-1}$$

the ac side utility voltages may be expressed as

$$\mathbf{v}_{abcu} = \sqrt{2} V_u \begin{bmatrix} \cos \theta_e \\ \cos(\theta_e - 2\pi/3) \\ \cos(\theta_e + 2\pi/3) \end{bmatrix} \tag{14.1-2}$$

where

$$\theta_e = \omega_e t + \theta_{e0} \tag{14.1-3}$$

In (14.1-2), V_u is the rms value of the line-to-neutral (l–n) utility voltage, and in (14.1-3), ω_e is the radian ac side frequency, and θ_{e0} is the phase, which will be considered as zero.

Physically, the voltages may represent a utility connection, or the voltages across a capacitive filter, wherein the capacitance is large enough that harmonics in the voltages may be neglected.

Power Magnetic Devices: A Multi-Objective Design Approach, Second Edition. Scott D. Sudhoff.
© 2022 The Institute of Electrical and Electronics Engineers, Inc. Published 2022 by John Wiley & Sons, Inc.
Companion website: www.wiley.com/go/sudhoff/Powermagneticdevices

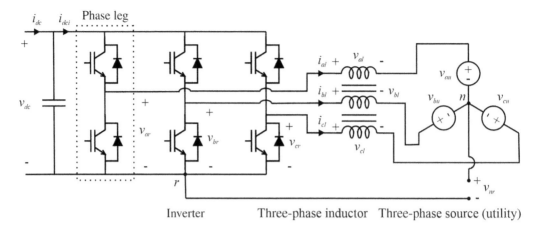

Figure 14.1 A dc–ac converter.

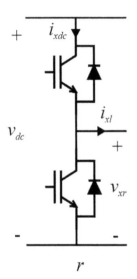

Figure 14.2 Phase leg of three-phase bridge converter.

Note that the neutral point of this three-phase source is not connected to the bottom rail of the inverter. As a result, the sum of the inductor phase current is zero. The neutral to bottom-rail voltage will be designated as v_{nr}.

Phase-Leg Operation

In order to understand the operation of the three-phase bridge converter, it is useful to first consider a phase leg as shown in Figure 14.2. Therein, $x \in [a, b, c]$.

The phase leg state is denoted as s_x. If $s_x = 1$, the upper transistor is on; if $s_x = 0$, the lower transistor is on. In an actual converter, there is also a deadtime in which neither transistor is on, in order to avoid shoot through (a condition in which both transistors turn on at the same time, and which can be fatal to the bridge), but we will neglect that brief state herein. Denoting the transistor forward on-state voltage drop as v_{ft} and diode forward on-state voltage drop as v_{fd}, from Figure 14.2, we can readily show that

$$v_{xr} = \begin{cases} v_{dc} - v_{ft} & s_x = 1, \; i_{xl} \geq 0 \\ v_{dc} + v_{fd} & s_x = 1, \; i_{xl} < 0 \\ -v_{fd} & s_x = 0, \; i_{xl} \geq 0 \\ v_{ft} & s_x = 0, \; i_{xl} < 0 \end{cases} \qquad (14.1\text{-}4)$$

and that

$$i_{xdc} = \begin{cases} i_{xl} & s_x = 1 \\ 0 & s_x = 0 \end{cases} \qquad (14.1\text{-}5)$$

Modulation

The determination of the switching signal s_x is a function of the modulation strategy. Here, we will consider two strategies, sine-triangle modulation and extended sine-triangle modulation. Other strategies are described in [1].

In sine-triangle modulation, we compare a low-frequency sinusoidal phase duty-cycles to a high-frequency triangular carrier wave c. In particular, the phase duty-cycles are taken to be of the form:

$$\mathbf{d}_{abc} = d \begin{bmatrix} \cos \theta_c \\ \cos(\theta_c - 2\pi/3) \\ \cos(\theta_c + 2\pi/3) \end{bmatrix} - d_3 \cos(3\theta_c) \begin{bmatrix} 1 \\ 1 \\ 1 \end{bmatrix} \qquad (14.1\text{-}6)$$

where θ_c is the converter angle that, in the steady-state, has a time rate of change of ω_e, and is related to the angle of our ac source by

$$\theta_c = \theta_e + \phi_c \qquad (14.1\text{-}7)$$

where ϕ_c is a constant in the steady state. The variable d is the duty cycle amplitude, which is also commonly referred to as the modulation index.

For pure sine-triangle modulation, $d_3 = 0$ and $d \leq 1$ if overmodulation is to be avoided. For extended sine-triangle modulation $d_3 = d/6$, and $d \leq 2/\sqrt{3}$ to avoid overmodulation. Overmodulation occurs when the peak absolute value of the phase duty cycle exceeds 1, and results in a low-frequency distortion [1]. Herein, this mode of operation will not be considered. Once the phase duty cycles are established, the switching state of each phase is found in accordance with

$$s_x = d_x > c \qquad (14.1\text{-}8)$$

This process is illustrated in Figure 14.3. Therein, the upper traces show the carrier signal c (high-frequency triangle wave) and a phase duty cycle d_x (low-frequency sinusoid). The lower trace shows the resulting switching signal, s_x. It should be noted that the separation between the switching frequency (that of the carrier wave) and the fundamental frequency (that of the phase duty cycle) is typically much greater than that depicted in Figure 14.3, but showing a typical case makes it difficult to observe the operation.

To go further, it is useful to introduce the idea of a fast average. In particular, the fast average of a signal is defined by:

$$\hat{x}(t) = \frac{1}{T_{sw}} \int_{t-T_{sw}}^{t} x(\tau) d\tau \qquad (14.1\text{-}9)$$

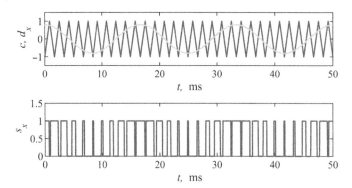

Figure 14.3 Sine-triangle modulation.

The fast-average of a signal can be thought of as a signal's low-frequency components. Assuming the arbitrary phase duty cycles $\mathbf{d}_{abc}$ vary slowly as compared to the fast-average period and neglecting the forward semiconductor voltage drops, it can be shown that

$$\hat{\mathbf{v}}_{abcr}(t) = \frac{1}{2} v_{dc} \left(\begin{bmatrix} 1 & 1 & 1 \end{bmatrix}^T + \mathbf{d}_{abc} \right) \tag{14.1-10}$$

In (14.1-10), it is assumed that the phase duty cycles vary slowly enough that there is no meaningful difference between the phase duty cycles $\mathbf{d}_{abc}$ and their fast average, and so a fast–average "^" is not used.

Assuming the phase duty signals, $\mathbf{d}_{abc}$ of (14.1-6) are used in (14.1-10), and neglecting the forward semiconductor voltage drops, one can show

$$\hat{\mathbf{v}}_{abcr}(t) = \frac{1}{2} v_{dc} \begin{bmatrix} 1 \\ 1 \\ 1 \end{bmatrix} + \frac{1}{2} dv_{dc} \begin{bmatrix} \cos(\theta_c) \\ \cos(\theta_c - 2\pi/3) \\ \cos(\theta_c + 2\pi/3) \end{bmatrix} - \frac{1}{2} d_3 v_{dc} \cos(3\theta_c) \begin{bmatrix} 1 \\ 1 \\ 1 \end{bmatrix} \tag{14.1-11}$$

Thus, we can see that the low-frequency component of the phase to bottom-rail voltage has a low-frequency sinusoidal component, albeit with a dc offset and a third-harmonic component. Ideally, neither of these latter two components will directly impact the ac portion of the system because they cancel in terms of line-to-line voltages.

Control

The control objectives for our converter depend on the application, and we will not consider the control system in any detail. Rather, we will assume the existence of a closed loop control system, which will determine the phase duty cycles such that the desired control objectives are achieved. This may be to achieve a desired output voltage, as for example in [2], or to regulate power flow as in [3]. For a discussion of several control strategies, as well as other modulation strategies, the reader is referred to [1].

QD Transformation

To proceed further, it will be useful to utilize reference frame theory, which was also used in the case of the permanent magnet ac machine. In the case of our converter, the qd transformation is useful for the three reasons. First, from Figure 14.1 and Kirchoff's current law, it is clear that

$i_{al} + i_{bl} + i_{cl} = 0$. The transformation to qd variables will facilitate enforcing this constraint. Second, the transformation to qd variables will also provide a path whereby the voltage v_{nr} and the $v_{dc}/2$ dc offset and third harmonic component of $\hat{v}_{abcr}$ disappear. Third, with appropriate choice of reference frame, we can magnetically decouple the phases (at least in the magnetically linear case), in that the q-axis and d-axis flux–linkage equations are independent of each other.

A thorough discussion of reference frame theory is set forth in [1]; here we will consider a brief review. Consider a three-phase set of phase variables $\mathbf{f}_{abcx}$ of the form given by (14.1-1). The transformation to the arbitrary reference frame may be expressed

$$\mathbf{f}_{qd0x} = \mathbf{K}_s \mathbf{f}_{abcx} \tag{14.1-12}$$

where the form of the $qd0$ variables is

$$\mathbf{f}_{qd0x} = \begin{bmatrix} f_{qx} & f_{dx} & f_{0x} \end{bmatrix}^T \tag{14.1-13}$$

and where

$$\mathbf{K}_s = \frac{2}{3} \begin{bmatrix} \cos\theta & \cos(\theta - 2\pi/3) & \cos(\theta + 2\pi/3) \\ \sin\theta & \sin(\theta - 2\pi/3) & \sin(\theta + 2\pi/3) \\ \frac{1}{2} & \frac{1}{2} & \frac{1}{2} \end{bmatrix} \tag{14.1-14}$$

The subscript "s" in the transformation is a holdover from the analysis of electric machines, wherein it designates "stator transformation," but for our purposes, we can think of it as designated "stationary circuit transformation." In (14.1-14), θ is the position of the reference frame. The "d" and "q" subscripts stand for the "direct" and "quadrature" axis from an electric machine theory; the "0" is for zero sequence. Zero-sequence variables are often zero, and from the transformation (14.1-14) it can be readily seen that the zero-sequence inductor current in our system must be zero.

The inverse transformation is given by

$$\mathbf{f}_{abcx} = \mathbf{K}_s^{-1} \mathbf{f}_{qd0x} \tag{14.1-15}$$

where

$$\mathbf{K}_s^{-1} = \begin{bmatrix} \cos\theta & \sin\theta & 1 \\ \cos(\theta - 2\pi/3) & \sin(\theta - 2\pi/3) & 1 \\ \cos(\theta + 2\pi/3) & \sin(\theta + 2\pi/3) & 1 \end{bmatrix} \tag{14.1-16}$$

Let us consider the transformation of a three-phase balanced set of phase variables of the form

$$\mathbf{f}_{abcx} = \sqrt{2} F_x \begin{bmatrix} \cos\theta_{fx} \\ \cos(\theta_{fx} - 2\pi/3) \\ \cos(\theta_{fx} + 2\pi/3) \end{bmatrix} \tag{14.1-17}$$

From (14.1-12), (14.1-14), and (14.1-17) we readily obtain

$$\mathbf{f}_{qd0x} = \sqrt{2} F_x \begin{bmatrix} \cos(\theta_{fx} - \theta) \\ -\sin(\theta_{fx} - \theta) \\ 0 \end{bmatrix} \tag{14.1-18}$$

Observe that the zero-sequence voltage is zero.

Using the nomenclature of [1], the reference frame used is designated with a superscript. For example, variables in the rotor reference frame of a permanent magnet ac machine, wherein $\theta = \theta_r$ were designated with a superscript "r" in Chapters 8 and 9. Herein, much of our work will be in a stationary reference frame, in which the q-axis and d-axis will carry an "s" superscript. In this reference frame, θ is a constant. Since the zero-sequence inductor current is zero, we will only retain the q-axis and d-axis variables which will be denoted

$$\mathbf{f}_{qdx}^s = \begin{bmatrix} f_{qx}^s \\ f_{dx}^s \end{bmatrix} \tag{14.1-19}$$

It is normally the case that in the stationary reference frame, θ is taken to be zero. That is not the case herein. For reasons which will become apparent, in our work, we will take the position of the stationary reference frame to be

$$\theta = \pi/6 \tag{14.1-20}$$

In this reference frame, it will be convenient to note that

$$\mathbf{K}_s^s \big|_{utr} = \frac{1}{3} \begin{bmatrix} \sqrt{3} & 0 & -\sqrt{3} \\ 1 & -2 & 1 \end{bmatrix} \tag{14.1-21}$$

and that

$$\left(\mathbf{K}_s^s\right)^{-1}\big|_{ltc} = \frac{1}{2} \begin{bmatrix} \sqrt{3} & 1 \\ 0 & -2 \\ -\sqrt{3} & 1 \end{bmatrix} \tag{14.1-22}$$

where "utr" and "ltc" denote the upper two rows and left two columns, respectively.

Circuit Analysis

We are now in a position to analyze the circuit shown in Figure 14.1. Using Kirchhoff's voltage law we obtain

$$-\mathbf{v}_{abcr} + \mathbf{v}_{abcl} + \mathbf{v}_{abcu} + \mathbf{v}_{nr}\begin{bmatrix} 1 & 1 & 1 \end{bmatrix}^T = 0 \tag{14.1-23}$$

where the inductor voltage may be expressed as

$$\mathbf{v}_{abcl} = r_l \mathbf{i}_{abcl} + p\boldsymbol{\lambda}_{abcl} \tag{14.1-24}$$

In (14.1-24), r_l is the resistance of an inductor coil, which is assumed to be the same for all phases.

In the magnetically linear case, the inductor flux linkage may be related to the inductor current by

$$\boldsymbol{\lambda}_{abcl} = \mathbf{L}\mathbf{i}_{abcl} \tag{14.1-25}$$

where

$$\mathbf{L} = \begin{bmatrix} L_e & M_{ec} & M_{ee} \\ M_{ec} & L_c & M_{ec} \\ M_{ee} & M_{ec} & L_e \end{bmatrix} \tag{14.1-26}$$

The form of (14.1-26) assumes of an inductor made from three legs, with one leg per phase. There are two outer legs that are essentially identical, and an inner leg. One such arrangement is shown in

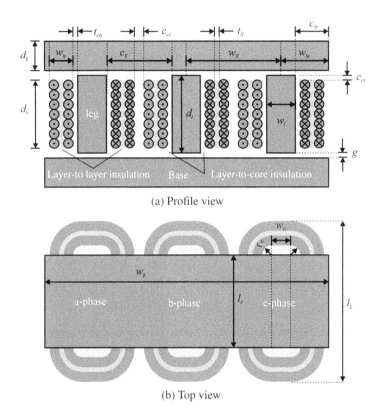

Figure 14.4 Inductor topology.

Figure 14.4. It is assumed that the a- and c-phases are located at the ends, and the b-phase is in the center. Note that this arrangement causes a degree of asymmetry. In (14.1-26), L_e and L_c denote the self-inductance of phases on the end leg and center leg, respectively. Likewise, M_{ec} and M_{ee} represent the mutual inductance between a phase located on an end leg with a phase located on the center leg, and between two phases that are both on end legs, respectively.

Substitution of (14.1-24) into (14.1-23), transforming to $qd0$ variables, rearranging, and discarding the zero-sequence component yields

$$\mathbf{v}_{qdr}^s = r_l \mathbf{i}_{qdl}^s + p\lambda_{qdl}^s + \mathbf{v}_{qdu}^s \qquad (14.1\text{-}27)$$

From (14.1-2) and (14.1-18) and retaining only the q-axis and d-axis variables we obtain

$$\mathbf{v}_{qdu}^s = \sqrt{2} V_u \begin{bmatrix} \cos(\theta_e - \pi/6) \\ -\sin(\theta_e - \pi/6) \end{bmatrix} \qquad (14.1\text{-}28)$$

Converting (14.1-25) to $qd0$ variables, we have

$$\lambda_{qd0l} = \mathbf{K}_s \mathbf{L} \mathbf{K}_s^{-1} \mathbf{i}_{qd0l} \qquad (14.1\text{-}29)$$

If $L_e = L_c$ and $M_{ee} = M_{ec}$, then the matrix $\mathbf{K}_s \mathbf{L} \mathbf{K}_s^{-1}$ is diagonal and independent of θ. However, even if this is not the case, choosing $\theta = \pi/6$ yields the diagonal form

$$\lambda_{qd}^s = \begin{bmatrix} L_q & 0 \\ 0 & L_d \end{bmatrix} \mathbf{i}_{qd}^s \qquad (14.1\text{-}30)$$

where

$$L_q = L_e - M_{ee} \tag{14.1-31}$$

and

$$L_d = \frac{1}{3}L_e + \frac{2}{3}L_c - \frac{4}{3}M_{ec} + \frac{1}{3}M_{ee} \tag{14.1-32}$$

Most other values of θ yields an inductance matrix wherein the q-axis and d-axis are coupled magnetically (unless all self-inductances are equal and all mutual-inductances are equal).

Steady-State Analysis

Let us now take a look at the steady-state analysis of the circuit of Figure 14.1. First, we will assume that the desired fast-average current is given by

$$\hat{\mathbf{i}}_{abcl} = \sqrt{2}I_l \begin{bmatrix} \cos(\theta_e + \phi_{il}) \\ \cos(\theta_e - 2\pi/3 + \phi_{il}) \\ \cos(\theta_e + 2\pi/3 + \phi_{il}) \end{bmatrix} \tag{14.1-33}$$

From (14.1-2) and (14.1-33), the power injected into the three-phase voltage source is given by

$$P = 3V_u I_l \cos(\phi_{il}) \tag{14.1-34}$$

Clearly, if our only objective is to inject or receive real power from the source, we would set ϕ_{il} to 0 or π so that this could be done with the minimum possible inductor current.

In any case, to proceed, from (14.1-28), the phasor representation of the three-phase voltage source may be expressed as

$$\tilde{\mathbf{v}}_{qdu}^s = V_u e^{-j\pi/6} \begin{bmatrix} 1 \\ j \end{bmatrix} \tag{14.1-35}$$

From (14.1-18) and (14.1-33), the desired inductor currents are expressed in the stationary reference frame as

$$\hat{\mathbf{i}}_{qdl}^s = \sqrt{2}I_l \begin{bmatrix} \cos(\theta_e + \phi_{il} - \pi/6) \\ -\sin(\theta_e + \phi_{il} - \pi/6) \end{bmatrix} \tag{14.1-36}$$

and as phasors as

$$\tilde{\mathbf{i}}_{qdl}^s = I_l e^{j(\phi_{il} - \pi/6)} \begin{bmatrix} 1 \\ j \end{bmatrix} \tag{14.1-37}$$

From (14.1-27) and (14.1-30) we have

$$\tilde{\mathbf{v}}_{qdr}^s = r_l \tilde{\mathbf{i}}_{qdl}^s + j\omega_e \begin{bmatrix} L_q & 0 \\ 0 & L_d \end{bmatrix} \tilde{\mathbf{i}}_{qdl}^s + \tilde{\mathbf{v}}_{qdu}^s \tag{14.1-38}$$

which allows us to determine the necessary inverter voltages to achieve a desired current. Transforming (14.1-10) to the stationary reference frame and extracting the qd components, we have

$$\hat{\mathbf{v}}_{qdr}^s = \frac{1}{2} v_{dc} \mathbf{d}_{qd}^s \tag{14.1-39}$$

where $\mathbf{d}_{qd}^s$ is the vector of qd components of the transformed phase duty cycles. Thus, once the voltages required are determined from (14.1-38), the phasor representation of the required duty cycles in the stationary reference frame can be obtained from

$$\tilde{\mathbf{d}}_{qd}^s = \frac{2}{v_{dc}} \tilde{\mathbf{v}}_{qdr}^s \qquad (14.1\text{-}40)$$

whereupon their phasor representation can be transformed back into phase variables as

$$\tilde{\mathbf{d}}_{abc} = \left(\mathbf{K}_s^s\right)^{-1}\Big|_{ltc} \tilde{\mathbf{d}}_{qd}^s \qquad (14.1\text{-}41)$$

and finally expressed in the time domain as

$$\mathbf{d}_{abc} = \sqrt{2} \begin{bmatrix} |\tilde{d}_a| \cos\left(\theta_e + \text{angle}\left(\tilde{d}_a\right)\right) \\ |\tilde{d}_b| \cos\left(\theta_e + \text{angle}\left(\tilde{d}_b\right)\right) \\ |\tilde{d}_c| \cos\left(\theta_e + \text{angle}\left(\tilde{d}_c\right)\right) \end{bmatrix} + d_{cm} \begin{bmatrix} 1 \\ 1 \\ 1 \end{bmatrix} \qquad (14.1\text{-}42)$$

In (14.1-42), d_{cm} is a scalar signal which is common to all the three duty cycles. Comparing (14.1-42) to (14.1-6), we can see that for that case $d_{cm} = -d_3 \cos(3\theta_c)$. However, in (14.1-6), balanced excitation was assumed, which have not been assumed here. This is because of the possibility that L_q is not equal L_d, a situation that may require a different strategy for selecting d_{cm}.

Time-Domain Simulation

Unlike the dc-to-dc converter of the previous chapter, there is no simple analytical description of the current waveform with sufficient fidelity to facilitate a design. Nonetheless, a detailed description of both the fundamental and ripple components of the current waveform are needed. One way to obtain this information is using time-domain simulation.

There are a plethora of circuit and state-variable-based time-domain simulation tools which can be used for this purpose, and the focus of this work is not on the art and science of time-domain simulation of power-electronics-based systems. Nevertheless, for the sake of completeness, an algorithm for a very simple but useful and self-contained simulation of the system as shown in Figure 14.1 will be set forth.

For our purposes, we will concentrate on the computation of the steady-state waveforms. To this end, we will assume that we have conducted a steady-state analysis, so that the phase duty cycles as expressed by (14.1-42) and are known a priori. We will also round the switching frequency so that it is an integer multiple of the fundamental frequency. This makes the solution periodic, and involves very little error assuming that the switching frequency is one or two orders of magnitude greater than the fundamental frequency.

The first step of the simulation will be to create a time vector, and at the same time create the switching signals which we will need. To this end, the simplest approach is to create a time vector where the kth time is given by

$$t_k = \frac{(k-1)}{K-1} \frac{1}{f_e} \qquad (14.1\text{-}43)$$

where K is the number of points, and where f_e is the fundamental frequency $\omega_e/(2\pi)$. Note that (14.1-43) yields a time step of

$$h = \frac{1}{K-1}\frac{1}{f_e} \tag{14.1-44}$$

For good results, it is required that $h < 1000/f_{sw}$ so that the transitions of the switching state are adequately captured. Once the time vector is established, we may take

$$\theta_{e,k} = \omega_e t_k \tag{14.1-45}$$

and then use (14.1-42) to determine the value of phase duty cycles at time t_k. Comparing these duty cycles to a triangular carrier wave of frequency f_{sw} allows the gate signals to be calculated at kth instant using (14.1-8).

Using the phase duty cycles, and neglecting forward semiconductor drops, we can then use (14.1-4) to predict the phase line-to-bottom rail voltages at every instant of time. We will denote these voltages at the kth instant as $\mathbf{v}_{abcr,k}$. Transforming the phase, line-to-bottom rail voltages to the stationary reference frames using (14.1-12) and (14.1-21) yields $\mathbf{v}^s_{qdr,k}$. Evaluating (14.1-28) at time instant k yields $\mathbf{v}^s_{qdu,k}$. Now, from (14.1-27) and (14.1-30), we have the decoupled (thanks to the qd transformation) ordinary differential equations

$$pi^s_{ql} = \frac{v^s_{qr} - v^s_{qu} - r_l i^s_{ql}}{L_q} \tag{14.1-46}$$

$$pi^s_{dl} = \frac{v^s_{dr} - v^s_{du} - r_l i^s_{dl}}{L_d} \tag{14.1-47}$$

In (14.1-46) and (14.1-47), it is assumed that the inductor is operating in the magnetically linear regime.

At this point, we must choose a numerical integration method. Here, we will utilize the second order backward trapezoidal method. This is a simple second order method which is numerically stable for all time steps. These yields update equations of

$$h = t_{k+1} - t_k \tag{14.1-48}$$

$$i^s_{ql,k+1} = \frac{(2L_q - r_l h)i^s_{ql,k} + h\left(v^s_{qr,k+1} - v^s_{qu,k+1} + v^s_{qr,k} - v^s_{qu,k}\right)}{(2L_q + r_l h)} \tag{14.1-49}$$

$$i^s_{dl,k+1} = \frac{(2L_d - r_l h)i^s_{dl,k} + h\left(v^s_{dr,k+1} - v^s_{du,k+1} + v^s_{dr,k} - v^s_{du,k}\right)}{(2L_d + r_l h)} \tag{14.1-50}$$

In order to utilize (14.1-48)–(14.1-50), we need the initial condition. Assuming we do not know this, one alternative is to assume $i^s_{ql,1} = i^s_{dl,1} = 0$. Then, we simulate one fundamental frequency cycle. If the simulation is in the steady state, the average value of the q-axis and d-axis currents should be zero. Therefore, if these averages are not both zero, we update the estimated initial conditions as

$$i^s_{ql,1} = i^s_{ql,K} - \alpha f_e \int_0^{1/f_e} i^s_{ql} dt \tag{14.1-51}$$

$$i^s_{dl,1} = i^s_{dl,K} - \alpha f_e \int_0^{1/f_e} i^s_{dl} dt \tag{14.1-52}$$

and repeat the one fundamental cycle simulation. In (14.1-51) and (14.1-52), α is a convergence factor (typically around 0.8) and the integration in these equations is done numerically based on the time vector and time discretized current. The process is repeated until it converges, which usually takes a handful of iterations. Here, the requirement for convergence will be the maximum allowed error between the end ($k = K$) and beginning ($k = 1$) values of both the q-axis and d-axis currents is less than $e_{alw}\sqrt{2}I_l$, where e_{alw} is an algorithm parameter. The advantage of this approach is that it requires less simulation time than simply allowing the simulation to reach steady state.

Once convergence is obtained, the currents at the kth instant of time are known. From these currents, we can determine the corresponding values of the time derivatives of the currents using (14.1-46) and (14.1-47). We will use the time derivatives of the currents in order to calculate proximity effect losses and to find the time derivative of the flux density waveforms that will be used to find core loss.

It is convenient to encapsulate the results of this section with the call

$$\mathbf{S}_D = F_S(\mathbf{S}_I) \tag{14.1-53}$$

where the independent variables associated with the time-domain simulation are

$$\mathbf{S}_I = \begin{bmatrix} v_{dc} & V_u & \omega_e & \theta_{e0} & I_l & \phi_{il} & L_q & L_d & r_l & f_{sw} & t^*_{mx} & h^* & d_{lm} & \alpha & e_{alw} \end{bmatrix} \tag{14.1-54}$$

and where d_{lm} is the limit on the peak value of the phase duty cycle (an error is flagged if this limit is exceeded), t^*_{mx} is the maximum time to simulate (typically set to one cycle), h^* is the desired time step used to determine K by solving (14.1-44) for K and rounding the result to an integer, and then reapplying (14.1-44) to determine the exact value of h. The remaining terms have been defined. The dependent variables are

$$\mathbf{S}_D = \begin{bmatrix} \mathbf{t} & \mathbf{i}^s_{ql} & \mathbf{i}^s_{dl} & \mathbf{pi}^s_{ql} & \mathbf{pi}^s_{dl} & i^s_{qlrms} & i^s_{dlrms} & \overline{\left(pi^s_{ql}\right)^2} & \overline{\left(pi^s_{dl}\right)^2} & \overline{\left(pi^s_{ql}pi^s_{dl}\right)^2} \end{bmatrix} \tag{14.1-55}$$

The vectors $\mathbf{t}$, $\mathbf{i}^s_{ql}$, $\mathbf{i}^s_{dl}$, $\mathbf{pi}^s_{ql}$, $\mathbf{pi}^s_{dl}$ are the waveforms associated with the system wherein each element represent a particular instant of time, and the remaining variables are metrics based on these waveforms. Therein, i^s_{qlrms} and i^s_{dlrms}, are the root-mean-square (rms) values of the q-axis and d-axis inductors currents, and the other terms are temporal means of the square and products of the time derivatives.

Amalgamating these structures, we have

$$\mathbf{S} = [\mathbf{S}_I \ \mathbf{S}_D] \tag{14.1-56}$$

Note that prior to the simulation, we will use (14.1-35)–(14.1-42) to calculate the duty cycles as part of (14.1-53). Also, each quantity in bold on the right-hand side of (14.3-55) is a vector, whereupon each value of a given vector corresponds to a particular instant of time.

Before concluding this sketch of the process of conducting a time-domain simulation, one important refinement should be mentioned. Using (14.1-43) yields a regularly spaced time vector. However, the disadvantage of this approach is that a very small time step is then required in order to accurately calculate the times of the power electronic switching events, which is important if a good representation of the waveforms is to be achieved. By adding a time point just before and after each switching event and just after each switching event, so that the time of the events are well represented in the simulation, a much larger time step can be used everywhere else (that is, everywhere not associated with a switching event). This drastically increases the speed of the simulation. This

14 Three-Phase Inductor Design

technique is used in the studies presented herein. In many simulation languages, this is referred to as scheduling or discrete event handling.

Now, let us turn our attention to an example in which we will exercise our steady-state calculations and a time-domain simulation.

Example 14.1A In this example, let us consider the performance of the system shown in Figure 14.1. We will assume that $\hat{v}_{dc} = 400$ V, $V_u = 230/\sqrt{3}$ V, $\omega_e = 2\pi 60$ rad/s, $L_d = L_q = 2$ mH, $r_l = 100$ mΩ, and $C_{dc} = 100$ μF. Furthermore, we will assume we wish to deliver 5 kW to the source. Selecting $\phi_{il} = 0$, from (14.1-34), one obtains $I_l = 12.55$ A. Using (14.1-35), (14.1-37)–(14.1-38), (14.1-40)–(14.1-42), and taking $d_{cm} = 0$ yields

$$d_{abc} = 0.9502 \begin{bmatrix} \cos(\theta_e + 0.07048) \\ \cos(\theta_e + 0.07048 - 2\pi/3) \\ \cos(\theta_e + 0.07048 + 2\pi/3) \end{bmatrix} \tag{141A1}$$

Applying this modulating signal to a sine-triangle modulation strategy, and a switching frequency of 12 kHz yields the waveforms as shown in Figure 14.5. As can be seen in the first trace, the inverter/inductor currents have a fundamental component equal to the desired value plus a ripple component resulting from the inverter switching. The second trace depicts the ripple in the dc inverter voltage. Therein, it is assumed that the dc capacitor completely supplies the dc inverter current ripple.

In order to provide a comparison to our next example, the rms phase current ripple is 533.9 mA for each phase. This is the rms value of the phase current less its desired fundamental component. The rms value of the dc current ripple is 6.821 A. Finally, the dc voltage ripple is 726.3 mV, rms.

Impact of Saliency

Before proceeding further, it is interesting to assess the impact of saliency, that is, the impact of imbalance arising from the difference between L_q and L_d. There are many potential impacts of saliency. These include, potentially, the need to operate at higher dc voltages and increased common-mode issues (we will discuss the common mode performance in the next chapter). However, one of the biggest impacts of saliency is on the dc bus voltage.

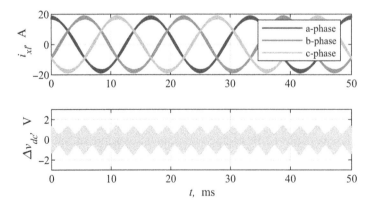

Figure 14.5 Time-domain waveforms with $L_d = L_q$.

To understand the impact of saliency on the bus voltage, from Section 4.7, the energy stored in the inductor may be expressed as

$$E_l = \frac{1}{2}\mathbf{i}_{abcl}^T \mathbf{L} \mathbf{i}_{abcl} \qquad (14.1\text{-}57)$$

or

$$E_l = \frac{1}{2}\mathbf{i}_{abcl}^T \boldsymbol{\lambda}_{abcl} \qquad (14.1\text{-}58)$$

In the terms of $qd0$ variables (14.1-58) may be rewritten as

$$E_l = \frac{3}{4}\left(i_{ql}\lambda_{ql} + i_{dl}\lambda_{dl} + 2i_{0l}\lambda_{0l}\right) \qquad (14.1\text{-}59)$$

In the absence of zero-sequence current, and using (14.1-30)

$$E_l = \frac{3}{4}\left(L_q i_{ql}^2 + L_d i_{dl}^2\right) \qquad (14.1\text{-}60)$$

Substituting (14.1-36) into (14.1-60), the energy stored in the inductor may be expressed as

$$E_l = \frac{3}{4}I_l^2\left(L_q + L_d + (L_q - L_d)\cos(2\theta_e + 2\phi_{il} - \pi/3)\right) \qquad (14.1\text{-}61)$$

If we assume that the dc input capacitor supplies the time-varying component of the inductor energy, the dc capacitor voltage ripple to due to inductor saliency is given by

$$\Delta v_{dcs} = -\frac{3}{4}\frac{I_l^2}{v_{dc0}}\frac{(L_d - L_q)}{C_{dc}}\cos(2\theta_e + 2\phi_{il} - \pi/3) \qquad (14.1\text{-}62)$$

Let us explore this effect with an example.

Example 14.1B In this example, let us repeat Example 14.1A, except that L_q is decreased to 1.8 mH, and L_d is increased to 2.2 mH. Observe that the average of these two inductances is the same as in Example 14.1A. In this manner, we can explore the impact of saliency. Repeating our calculations from last time, we can show that now the phase duty cycles are given by

$$d_{abc} = \begin{bmatrix} 0.9442\cos(\theta_e + 0.06738) \\ 0.9507\cos(\theta_e + 0.07750 - 2\pi/3) \\ 0.9558\cos(\theta_e + 0.06656 + 2\pi/3) \end{bmatrix} \qquad (14.1\text{B-}1)$$

Notice that the phase duty cycles no longer form a balanced three-phase set. The resulting waveforms are shown in Figure 14.6. On the casual inspection, the phase currents seem not to have changed much, though there are some differences. In Example 14.1A, the rms phase current ripples were all 533.9 mA. In this example, the phase current ripples are 569.6, 485.2, and 567.0 mA, for the a-, b-, and c-phases, respectively. The dc current ripple was previously 6.822 A; this value is unchanged from Example 14.1A.

While the phase currents have not changed much, the dc input voltage ripple has changed significantly, and is larger in this case. Previously, the dc voltage ripple was 726.3 mV, rms. Now, however, the dc voltage ripple has increased to 1.104 V. Superimposed on the voltage ripple is the ripple component due to saliency given by (14.1-62). Clearly, this effect has made an impact. It may strike the reader as curious that the change in dc bus voltage occurred without a significant increase in dc current. The reason for the change in the capacitor voltage is that the small change in dc current occurs at a very low frequency (twice the fundamental), where the capacitor impedance is large.

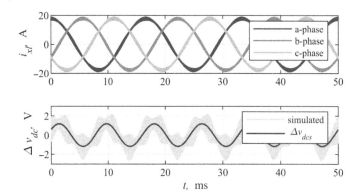

Figure 14.6 Time-domain waveforms with $L_d \neq L_q$.

Design Approach

Before proceeding any further, it is appropriate to briefly discuss the formulation of the design process. This of course can be done many ways, and with varying scope. For example, our goal might be a comprehensive design of the inverter including the input capacitor selection, switching frequency, and inductor design, analogous to what we did in Chapter 13. This is possible, but because the three-phase inductor design problem is more involved, we will take on less scope herein, and assume that a value of desired inductance is already known, and that it is our task to design that inductor.

To this end, our input will be a desired phase inductance L^*, which is the desired value of inductance if we had three ideal non-coupled inductors. Thus, one of our design objectives will be design the inductor so that L_d and L_q are both close to L^*. Another input to our design process will be the current waveforms. We will need these in order to calculate the proximity effect and core losses.

Finally, as we move forward, note that it is typical that in the design process, we could consider many operating points, ensure that design requirements are met at each operating point, and then used a weighted average of the losses to determine an overall efficiency. Here, for simplicity, we will focus on a single full-load operating point, the extension to multiple operating points being straightforward, though not trivial.

The remainder of this chapter will proceed as follows. In Section 14.2, we will consider a three-phase inductor geometry. Then, in Section 14.3, we will develop a magnetic equivalent circuit (MEC) which is the heart of our analysis required for the design. In Section 14.4, the use of the MEC to evaluate key design specifications, in particular, the determination of inductance and certain coefficients which facilitate loss calculations. The computation of losses is set forth in Section 14.5. Finally, a design paradigm and case study are set forth in Sections 14.6 and 14.7, respectively.

14.2 Inductor Geometry

The focus of this section is to set forth an inductor geometry, and an independent set of variables that define that geometry. Several possible inductor geometries are set forth in [4], including some inherently symmetrical topologies [5]. Herein, the asymmetrical I-core based 3-phase inductor geometry shown in Figure 14.4 will be considered. One possible choice of independent variable is

$$\mathbf{G}_I = \begin{bmatrix} g & w_{lc} & l_c & r_{bl} & N_{tpl}^* & N_l^* & N_{pr}^* & a_{wcpr}^* & t_{clc}^* & t_{ll}^* & c_{cc} & c_{cr} & c_{le} & \varepsilon_{rp} & k_p & \mathbf{D}_{CD} & \mathbf{D}_{CR} \end{bmatrix} \quad (14.2\text{-}1)$$

In (14.2-1), g, w_{lc}, and l_c are all defined in Figure 14.4. The variable r_{bl} is the ratio of the depth of the base to the width of the leg. The variables N_{tpl}^*, N_l^*, N_{pr}^*, a_{wcpr}^*, t_{clc}^*, and t_{ll}^* are all as in Section 13.6. The clearances c_{cc}, c_{cr}, and c_{le} are defined in Figure 14.4. As in Section 13.6, ε_{rp} and k_p are the permittivity and thermal conductivity of the potting material, if any is used. The substructure $\mathbf{D}_{CD}$ of conductor parameters is defined by (13.6-3); $\mathbf{D}_{CR}$ is a substructure of core parameters.

Given the elements of $\mathbf{G}_I$, we can calculate remaining geometrical variables as follows. First, we will utilize the sequence (13.6-4)–(13.6-12) and (13.6-17) to find parameters related to the winding bundle; this is essentially unchanged from Chapter 13. However, the orientation of the winding leg has changed and so (13.6-13)–(13.6-14) become

$$w_w = N_l 2 r_w k_b + (N_l - 1) t_{ll} \quad (14.2\text{-}2)$$

$$d_w = 2 r_w (N_{tpl} + 1) N_{pr} k_b \quad (14.2\text{-}3)$$

As noted in Chapter 13, in (14.2-3), we have $(N_{tpl} + 1)$ rather than N_{tpl} so that there is room to complete the last turn. The volume occupied by the winding may be expressed as

$$V_w = d_w \left[2(l_c + w_{lc}) w_w + \pi (r_b + w_w)^2 - \pi r_b^2 \right] \quad (14.2\text{-}4)$$

At this point, from geometrical considerations we have

$$w_l = w_{lc} + 2 r_b - 2 t_{clc} \quad (14.2\text{-}5)$$

$$d_l = d_w + 2 c_{cr} \quad (14.2\text{-}6)$$

$$c_{ll} = 2 w_w + 2 t_{clc} + c_{cc} \quad (14.2\text{-}7)$$

$$w_{ll} = w_l + c_{ll} \quad (14.2\text{-}8)$$

$$w_{le} = \frac{1}{2} w_l + c_{le} \quad (14.2\text{-}9)$$

The width and depth of the base I-cores are given by

$$w_b = 2 w_{ll} + 2 w_{le} \quad (14.2\text{-}10)$$

$$d_b = w_l r_{bl} \quad (14.2\text{-}11)$$

The mass of the leg I-core and the base I-core are found as

$$M_{lcr} = \rho_{cr} d_l w_l l_c \quad (14.2\text{-}12)$$

$$M_{bcr} = \rho_{cr} d_b w_b l_c \quad (14.2\text{-}13)$$

Modifying (13.6-22), the length of a turn on the ith layer is expressed as

$$l_{t,i} = \sqrt{(2 w_{lc} + 2 l_c + 2 \pi (r_b + r_w k_b (2i - 1) + t_{ll}(i-1)))^2 + (2 r_w k_b N_{pr})^2} \quad (14.2\text{-}14)$$

which include the impact of skew on the winding which is helical in nature. The total length of coil wire (per phase) l_{wc} may be expressed by (13.6-23) and the coil mass M_{wc} from (13.6-24).

The total inductor mass (core plus conductor, but foregoing the insulation paper and potting material) is then given by

$$M_L = 3 M_{lcr} + 2 M_{bcr} + 3 M_{wc} \quad (14.2\text{-}15)$$

The overall inductor length, depth, and width are given by

$$l_L = l_c + 2(r_b + w_w) \tag{14.2-16}$$

$$d_L = d_l + 2g + 2d_b \tag{14.2-17}$$

$$w_L = \max(w_b, 2w_{ll} + w_l + 2w_w + 2t_{clc}) \tag{14.2-18}$$

The inductor MEC will have three main components reluctances associated with the core. This will include the reluctance of a phase leg, R_p, the reluctance of a base section marking the path from one phase leg to the next, R_b, and finally the reluctance of the vertical portion within the base as the flux enters the base vertically and then bends in the horizontal direction. We will term this reluctance as R_{pb}. The area of the magnetic cross sections, lengths of the magnetic paths, and volumes (magnetic) of each of these reluctances are calculated as

$$a_{mcp} = w_l l_c \tag{14.2-19}$$

$$l_{mpp} = d_l \tag{14.2-20}$$

$$V_{mp} = d_l w_l l_c \tag{14.2-21}$$

$$a_{mcb} = d_b l_c \tag{14.2-22}$$

$$l_{mpb} = w_{ll} \tag{14.2-23}$$

$$V_{mb} = d_b w_{ll} l_c \tag{14.2-24}$$

$$a_{mcpb} = w_l l_c \tag{14.2-25}$$

$$l_{mppb} = \frac{1}{2} d_b \tag{14.2-26}$$

$$V_{mpb} = \frac{1}{2} d_b w_l l_c \tag{14.2-27}$$

As we have done earlier in work, it will be convenient to summarize the calculations of this section with a function call

$$\mathbf{G}_D = \mathrm{F}_G(\mathbf{G}_I, \mathbf{D}_I) \tag{14.2-28}$$

where $\mathbf{G}_I$ has already been defined, and

$$\mathbf{G}_D = [t_{clc} \ \varepsilon_{rclc} \ k_{clc} \ t_{ll} \ \varepsilon_{rll} \ k_{ll} \ N_{tpl} \ N_l \ N_{pr} \ N \ a_{wc} \ r_{wc} \ r_b \ r_w \ w_w \ d_w \ w_l \ d_l \ c_{ll} \ w_{ll} \ w_{le} \ w_b \ d_b \cdots$$
$$M_{lcr} \ M_{bcr} \ \mathbf{1}_t^T \ l_{wc} \ M_{wc} \ M_L \ l_L \ d_L \ w_L \ V_L \ a_{mcp} \ l_{mpp} \ V_{mp} \ a_{mcb} \ l_{mpb} \ V_{mb} \ a_{mcpb} \ l_{mppb} \ V_{mpb}] \tag{14.2-29}$$

Finally, as we have done elsewhere we can amalgamate the dependent and independent geometrical variables into a single structure

$$\mathbf{G} = [\mathbf{G}_I \ \mathbf{G}_D] \tag{14.2-30}$$

14.3 Magnetic Equivalent Circuit

In this section, we will focus on the development of a MEC of the three-phase inductor as shown in Figure 14.4, which we will use as a design model. Although this model will be fairly detailed, it is appropriate to begin by considering an extremely simple model. This will help lay the groundwork for our design model.

Figure 14.7 Elementary three-phase inductor MEC.

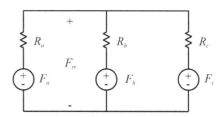

Elementary Analysis

Before developing a detailed MEC for using as a design model, let us first consider the highly simplified MEC as shown in Figure 14.7. Therein, we assume the inductor has three phase legs that are tied together with a core of very high permeability. The reluctances R_a, R_b, and R_c represent the air gap reluctances of the three phases, and F_a, F_b, and F_c represent the phase magnetomotive forces (MMFs). The phase MMF may be expressed

$$F_x = Ni_x \quad x \in [a,b,c] \tag{14.3-1}$$

where N is the number of turns, which is common to all the phases.

Using the elementary circuit analysis techniques, we can quickly solve for the MMF between the upper and lower "rails," which is given by

$$F_{rr} = \frac{F_a R_b R_c + F_b R_a R_c + F_c R_a R_b}{R_b R_c + R_a R_c + R_a R_b} \tag{14.3-2}$$

Now, let us consider the special but desired case that the reluctance of each phase is identical. In this case (14.3-2) reduces to

$$F_{rr} = \frac{1}{3}(F_a + F_b + F_c) \tag{14.3-3}$$

Now, considering an even more special, but nevertheless very common case that the sum of the phase MMFs is zero. From (14.3-3)

$$F_{rr} = 0 \tag{14.3-4}$$

Clearly, with $F_{rr} = 0$ the phases of the inductor do not interact, and the three-phase inductor performs similarly to three independent inductors. This is the desired situation. However, by sharing a common magnet core, the overall mass and loss of the inductor can be reduced.

Note that in this case, wherein $F_{rr} = 0$, the total MMF drop across the air gap is the phase MMF. In an arrangement in which there are two air gaps such as we are considering, the MMF drop across each air gap is approximately one-half the phase MMF.

Detailed Magnetic Equivalent Circuit Architecture

The equivalent circuit as shown in Figure 14.7 is very simple and its accuracy will be limited. Figure 14.8 depicts a more detailed MEC which we will use as a design model. Therein, $R_p(\Phi)$ and $R_b(\Phi)$ represent the reluctance of a phase I-core and half of a base I-core, respectively (leg-to-leg). The reluctance $R_{pb}(\Phi)$ represent the reluctance of flux entering the base I-cores in the vertical direction before bending into the horizontal direction. The reluctance across an end air gap (a- or c-phases), and the reluctance across the center air gap are denoted as R_{gc} and R_{ge},

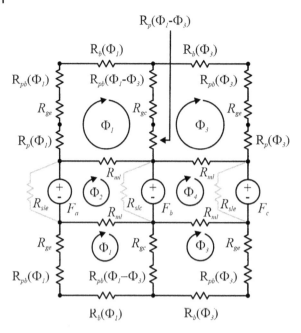

Figure 14.8 The three-phase inductor MEC.

respectively. Self-leakage reluctances of end and center phases are R_{sle} and R_{slc}. Finally, we will have a mutual leakage reluctance R_{ml}. As earlier in this work, the self-leakage inductances, being across source elements, will not be represented with the MEC. Also note that, because of the symmetry of the geometry only four rather than six mesh fluxes are used. Because of this, it will not be necessary to represent the lower ten branches.

We begin our development with expressions for the core reluctances. Using the methods of Chapter 2, we have

$$R_p(\Phi) = \frac{l_{mpp}}{a_{mcp}\mu_B(\Phi/a_{mcp})} \tag{14.3-5}$$

$$R_{pb}(\Phi) = \frac{l_{mppb}}{a_{mcpb}\mu_B(\Phi/a_{mcpb})} \tag{14.3-6}$$

$$R_b(\Phi) = \frac{l_{mpb}}{a_{mcb}\mu_B(\Phi/a_{mcb})} \tag{14.3-7}$$

Next, we will consider the calculation of fringing fluxes. To this end, we could use the approach of Chapter 2. That worked well, but the reader we will recall that the assumed flux path intersected the winding, and this was not considered in the derivation of the fringing permeance. Since, this flux path will be important in determining proximity effect losses, let us use this opportunity to refine our analysis.

Interior Fringing

Figure 14.9 depicts the interior fringing and leakage paths which will be considered. Observe that this path is slightly different than we took in, for example, Figure 2.17 in that it is a pure arc, rather than an arc with a short straight segment with a length of the air gap as in Figure 2.17. Therein, "m,"

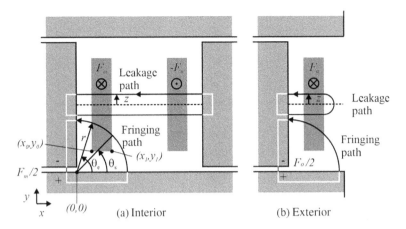

Figure 14.9 Interior and exterior fringing and leakage paths.

"n," and "o" are MMF sources that each could take on values of "a," "b," or "c," depending upon the situation. Note that from our earlier work, the MMF drop across each air gap will be approximately one half of the MMF produced by the coil associated with that air gap.

We will begin by deriving an expression for the interior fringing permeance. To this end, note the position of the origin of the x–y coordinate system. In this system, the inner lower corner of the winding has coordinates of

$$x_0 = t_{clc} \qquad (14.3\text{-}8)$$

$$y_0 = c_{cr} + g \qquad (14.3\text{-}9)$$

and a radius of

$$r_0 = \sqrt{x_0^2 + y_0^2} \qquad (14.3\text{-}10)$$

from the designated origin. The outer lower corner has coordinates

$$x_1 = x_0 + w_w \qquad (14.3\text{-}11)$$

$$y_1 = y_0 \qquad (14.3\text{-}12)$$

with a radius

$$r_1 = \sqrt{x_1^2 + y_1^2} \qquad (14.3\text{-}13)$$

from the origin. We will take the largest radius associated with the fringing path as

$$r_{mx} = \min\left(g + \frac{1}{2}d_l, \frac{1}{2}c_{ll}\right) \qquad (14.3\text{-}14)$$

Now, let us consider the fringing permeance for Region 0, which we will define as the region wherein $g \leq r \leq r_0$. Noting that the MMF drop across the air gap is approximately $F_m/2$, the magnitude of the field intensity along the arc of the path is given by

$$H = \frac{F_m}{\pi r} \qquad (14.3\text{-}15)$$

Using the methods of Chapter 2, this yields an inner corner Region 0 fringing permeance contribution of

$$P_{fic0} = \frac{\Phi_{fic0}}{F_m/2} = \frac{2\mu_0 l_c}{\pi} \ln\left(\frac{\min(r_0, r_{mx})}{g}\right) \quad (14.3\text{-}16)$$

where Φ_{fic0} is the fringing flux associated with this path and $F_m/2$ is the MMF across the path. Note that this is the permeance of a single corner.

Next, provided that $r_{mx} > r_0$ we define Region 1 to consist of paths satisfying $r_0 \leq r \leq \min(r_1, r_{mx})$. If $r_{mx} > r_1$, we define Region 2 to consist of paths satisfying $r_1 \leq r \leq r_{rmx}$. Within these regions, we define two angles, which are both a function of the radius of the path, and are also indicated in Figure 14.9. In particular,

$$\theta_s = \begin{cases} \operatorname{asin}\left(\dfrac{y_0}{r}\right) & \text{Region 1} \\ \operatorname{acos}\left(\dfrac{x_1}{r}\right) & \text{Region 2} \end{cases} \quad (14.3\text{-}17)$$

and

$$\theta_e = \operatorname{acos}\left(\frac{x_0}{r}\right) \quad (14.3\text{-}18)$$

In terms of the radius of the path, and the angles θ_s and θ_e, we can find the area of the winding enclosed by the path. Using geometry, we have

$$a = \begin{cases} \dfrac{1}{2}(r\cos\theta_s - x_0)(r\sin\theta_e - y_0) + \dfrac{1}{2}r^2(\theta_e - \theta_s - \sin(\theta_e - \theta_s)) & \text{Region 1} \\ w_w\left(\dfrac{1}{2}r(\sin\theta_e + \sin\theta_s) - y_0\right) + \dfrac{1}{2}r^2(\theta_e - \theta_s - \sin(\theta_e - \theta_s)) & \text{Region 2} \end{cases} \quad (14.3\text{-}19)$$

The next step is to compute the field intensity. Noting that the air gap MMF drop is approximately $F_m/2$, and that the path has enclosed a current of $aF_m/(w_w d_w)$, the magnitude of the field intensity along the path may be expressed as

$$H = \frac{(w_w d_w - 2a)F_m}{\pi w_w d_w r} \quad (14.3\text{-}20)$$

Using energy arguments from Chapter 2, the fringing permeance associated with Regions 1 and 2 of an inner corner may be expressed as

$$P_{fic12} = \frac{\Phi_{fic12}}{F_m/2} = \frac{2\mu_0 l_c}{\pi w_w^2 d_w^2} \int_{\min(r_0, r_{mx})}^{r_{mx}} \frac{(w_w d_w - 2a)^2}{r} dr \quad (14.3\text{-}21)$$

When evaluating (14.3-21), it is important to recall that a is a function of r, θ_s, and θ_e, and these latter two quantities are they themselves a function of r. For this reason, the integration must be carried out numerically. As in the case of P_{fic0}, P_{fic12} corresponds to a single inner corner.

In addition to computing permeances, we will also need to find dynamic resistance to facilitate the calculation of proximity effect losses, as we discussed in Chapter 11. We will assume round conductors, so that we only need to know the magnitude of flux density. Using (14.3-20), the contribution of a single inner corner fringing path (see Figure 14.9a) to the spatial mean of the square of the normalized flux density within a winding is given by

$$\left\langle \hat{B}_{p,m}^2 \right\rangle_{m,fic} = \frac{\mu_0^2 N^2 l_c}{V_w \pi^2 w_w^2 d_w^2} \int_{\min(r_0, r_{mx})}^{r_{mx}} \frac{(w_w d_w - 2a)^2}{r} (\theta_e - \theta_s) dr \qquad (14.3\text{-}22)$$

where $m \in [a, b, c]$ and V_w is the volume of the winding bundle. The subscript "p,m" in (14.3-22) indicates a direction of the flux density is along the path, and that it is due to winding "m." In our analysis, our paths will be chosen such that they do not overlap. The "m,fic" subscript indicates the region of winding "m" and that this term is the subregion of fringing in an inner corner. Observe that we have taken the spatial mean of the contribution over the entire volume of the winding, even though the loss occurs in a particular subregion of the coil. This will simplify our book keeping, though it will reduce the fidelity of any thermal model we might develop. Later, we will use this result to find the dynamic resistance. As in the case of P_{fic12}, observe that a, θ_s, and θ_e are all functions of radius.

Interior Leakage

We will next consider the leakage path shown in Figure 14.9(a). This path is characterized by z, and we will consider leakage in the region $0 \leq z \leq z_{mx}$, where

$$z_{mx} = \max(0, d_l/2 + g - r_{mx}) \qquad (14.3\text{-}23)$$

and r_{mx} is given by (14.3-14). The reason it is so bounded is that for values of z beyond this point, it is assumed that the field is dominated by the fringing component.

From Ampere's law, we have that within the core window we have

$$H = \frac{(F_m - F_n)z}{d_w c_{ll}} \qquad (14.3\text{-}24)$$

We will also assume the existence of a fringing path at the same vertical position as the leakage path in Figure 14.9(a), but which enters the core on the front (or back) faces of the core, follows a semi-circular arc out of the core, becomes parallel to the indicated path, and then follows a semi-circular arc back into the face of the core. Taking r to be the distance of a path in front of the core, which is also the radius of the semi-circles, we have that

$$H = \frac{(F_m - F_n)z}{d_w(\pi r + c_{ll})} \qquad (14.3\text{-}25)$$

Next, using energy arguments as we discussed in Section 2.7, and, noting from Figure 14.8 that the energy is stored in two permeances in series, we have that

$$P_{ml} = \frac{1}{R_{ml}} = \frac{4\mu_0 z_{mx}^3}{3 d_w} \left[\frac{l_c}{c_{ll}} + \frac{2}{\pi} \ln\left(1 + \frac{\pi w_l}{2 c_{ll}}\right) \right] \qquad (14.3\text{-}26)$$

In deriving (14.3-26), the first term corresponds to (14.3-24) and the second term corresponds to (14.3-25). With respect to this second term, the energy integral is taken over half the face of an I-core, thus from $r = 0$ to $r = w_l/2$; the other half is "reserved" for other leakage paths leading to the opposite direction. Another approach would be to integrate over the entire face and ignore interaction between leakage fields.

From (14.3-24) and (14.3-25), and our work in Chapter 11, we can also show that the spatial averages of the normalized flux density in the m-region (that is, the region of coil m) may be expressed as

$$\left\langle \hat{B}_{p,m}^2 \right\rangle_{m,ml} = \frac{2\mu_0^2 N^2 z_{mx}^3 w_w}{3 V_w d_w^2 c_{ll}} \left(\frac{l_c}{c_{ll}} + \frac{2w_l}{2c_{ll} + \pi w_l} \right) \tag{14.3-27}$$

$$\left\langle \hat{B}_{p,n}^2 \right\rangle_{m,ml} = \left\langle \hat{B}_{p,m}^2 \right\rangle_{m,ml} \tag{14.3-28}$$

$$\left\langle \hat{B}_{p,m} \hat{B}_{p,n} \right\rangle_{m,ml} = - \left\langle \hat{B}_{p,m}^2 \right\rangle_{m,ml} \tag{14.3-29}$$

where $m, n \in [a, b, c]$, "p,m" and "p,n" refer to flux density in the direction of the path due to winding "m" and "n," respectively, and "m,ml" refers to the region of winding "m," due to the mutual leakage path. Again, the spatial average is taken over the entirety of coil "m." As in the case of the leakage permeance, the region of integration for the term arising from (14.3-25) is calculated by varying $r = 0$ to $r = w_l/2$.

Exterior Fringing and Leakage

Our next step is to calculate the fringing and leakage flux associated with a winding on the end of the inductor, as shown in Figure 14.9(b). In this case, the process and results are nearly identical to the interior case. The only difference is that the maximum radius of the leakage flux is taken as

$$r_{mx} = \min\left(g + \frac{1}{2} d_l, c_{le} \right) \tag{14.3-30}$$

rather than (14.3-14).

With this change, we can compute the permeance associated with Region 0 of an outer corner, P_{foc0} using (14.3-16), the permeance associated with Regions 1 and 2 of the outer corner, P_{foc12} using (14.3-21), and the spatial mean of the normalized flux density $\left\langle \hat{B}_{p,m}^2 \right\rangle_{m,foc}$ using (14.3-22) where $m \in [\text{'}a\text{'}, \text{'}c\text{'}]$.

The exterior leakage path is different however. We will assume the exterior path shown in Figure 14.9(b). With this path, the magnitude of the field intensity along the path may be expressed as

$$H = \frac{2F_o z}{d_w (2t_{clc} + 2w_w + \pi z)} \tag{14.3-31}$$

Using energy arguments, we obtain

$$P_{el} = \frac{2\mu_0 l_c}{d_w^2 \pi} \left(z_{mx}^2 - \frac{4(t_{clc} + w_w)}{\pi} z_{mx} + \frac{8(t_{clc} + w_w)^2}{\pi^2} \ln\left(1 + \frac{\pi z_{mx}}{2(t_{clc} + w_w)} \right) \right) \tag{14.3-32}$$

where r_{mx} given by (14.3-30) and z_{mx} is given by (14.3-23).

Finally, from (14.3-31) the spatial mean of the square of the normalized flux density in the region of winding "o" may be expressed

$$\left\langle \hat{B}_{p,o}^2 \right\rangle_{o,el} = \frac{8\mu_0^2 N^2 w_w l_c}{\pi^2 V_w d_w^2} \left[z_{mx} + \frac{2(t_{clc} + w_w) z_{mx}}{2(t_{clc} + w_w) + \pi z_{mx}} - \frac{4(t_{clc} + w_w)}{\pi} \ln\left(1 + \frac{\pi z_{mx}}{2(t_{clc} + w_w)} \right) \right] \tag{14.3-33}$$

Face Fringing and Leakage

Now, let us consider the fringing and leakage paths associated with the front and back faces of the core. These paths are depicted in Figure 14.10.

Figure 14.10 Face fringing and leakage permeances.

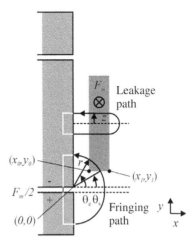

We will start with the fringing path. As can be seen, the geometry is the similar to Figure 14.9(a) except that we will have

$$x_0 = r_b \tag{14.3-34}$$

$$y_0 = c_{cr} \tag{14.3-35}$$

where r_b is the bending radius (see the lower portion of Figure 14.4), and

$$r_{mx} = \min\left(\frac{1}{2}d_l, d_b\right) \tag{14.3-36}$$

The expressions (14.3-10)–(14.3-13) and (14.3-17)–(14.3-19) still hold.

As in the case of interior and exterior fringing permeances, we will have three regions depending upon the radius, r. For Region 0, in which $0 \leq r \leq r_0$, we have that

$$H = \frac{F_m}{2(\pi r + g)} \tag{14.3-37}$$

from which we conclude the fringing associated with the front face of Region 0 is given by

$$P_{ff0} = \frac{\mu_0 w_l}{\pi} \ln\left(1 + \frac{\pi \min(r_0, r_{mx})}{g}\right) \tag{14.3-38}$$

In Region 1 ($r_0 \leq r \leq r_1$) and Region 2 ($r_1 \leq r \leq r_{mx}$) we have

$$H = \frac{(w_w d_w - 2a)F_m}{2w_w d_w (g + \pi r)} \tag{14.3-39}$$

Again using the results of Section 2.7, and in particular (2.7-9) and (2.7-19), we obtain the permeance associated with (one) face fringing in Regions 1 and 2 as

$$P_{ff12} = \frac{\mu_0 w_l}{w_w^2 d_w^2} \int_{\min(r_0, r_{mx})}^{r_{mx}} \frac{(w_w d_w - 2a)^2}{(g + \pi r)} dr \tag{14.3-40}$$

Again, it should be noted that (14.3-40) corresponds to either the upper or lower fringing flux on a single face.

The mean value of the normalized flux density in winding m due to current in winding m associated with one of these fringing paths is readily expressed as

$$\left\langle \hat{B}_{p,m}^2 \right\rangle_{m,ff} = \frac{\mu_0^2 N^2 w_l}{4 V_w w_w^2 d_w^2} \int_{\min(r_0, r_{mx})}^{r_{mx}} \frac{(w_w d_w - 2a)^2}{(g + \pi r)^2} r(\theta_e - \theta_s) dr \tag{14.3-41}$$

Recall that a given winding will see four such contributions—top and bottom, front and back.

Next, let us calculate the face leakage permeance. This is essentially identical to the calculation of the exterior leakage permeance, except that r_{mx} is calculated using (14.3-36), and r_b replaces t_{clc} and w_l replace l_c in our derivation. Thus, we have that

$$P_{fl} = \frac{2\mu_0 w_l}{d_w^2 \pi} \left(z_{mx}^2 - \frac{4(r_b + w_w)}{\pi} z_{mx} + \frac{8(r_b + w_w)^2}{\pi^2} \ln\left(1 + \frac{\pi z_{mx}}{2(r_b + w_w)}\right) \right) \tag{14.3-42}$$

and that

$$\left\langle \hat{B}_{p,o}^2 \right\rangle_{o,fl} = \frac{8\mu_0^2 N^2 w_w w_l}{\pi^2 V_w d_w^2} \left[z_{mx} + \frac{2(r_b + w_w) z_{mx}}{2(r_b + w_w) + \pi z_{mx}} - \frac{4(r_b + w_w)}{\pi} \ln\left(1 + \frac{\pi z_{mx}}{2(r_b + w_w)}\right) \right] \tag{14.3-43}$$

where $m \in [a, b, c]$. Each winding will see two such contributions (front and back).

Permeance Aggregation

As this point, we can aggregate the permeance contributions in order to calculate the remaining permeances and reluctances in Figure 14.8. In particular, we have

$$P_{ge} = \frac{1}{R_{ge}} = \frac{w_l l_c \mu_0}{g} + P_{fic0} + P_{fic12} + P_{foc0} + P_{foc12} + 2P_{ff0} + 2P_{ff12} \tag{14.3-44}$$

$$P_{gc} = \frac{1}{R_{gc}} = \frac{w_l l_c \mu_0}{g} + 2P_{fic0} + 2P_{fic12} + 2P_{ff0} + 2P_{ff12} \tag{14.3-45}$$

$$P_{sle} = \frac{1}{R_{sle}} = P_{el} + 2P_{fl} \tag{14.3-46}$$

$$P_{slc} = \frac{1}{R_{slc}} = 2P_{fl} \tag{14.3-47}$$

Recall that R_{ml} is set forth in (14.3-26), and that the core reluctances are defined by (14.3-5)–(14.3-7). At this point, all elements of the MEC depicted in Figure 14.8 are complete.

Dynamic Resistance

We are now also in a position to compute proximity effect losses. From (11.5-18), proximity effect losses for the three-phase inductor may be expressed as

$$P_p = \overline{\frac{d\mathbf{i}_{abcl}^T}{dt} \mathbf{D} \frac{d\mathbf{i}_{abcl}}{dt}} \qquad (14.3\text{-}48)$$

For our inductor, we have a degree of symmetry. The *a*-phase and *c*-phase are identical, and have identical interactions with the *b*-phase. This leads to a dynamic resistance matrix of the form

$$\mathbf{D} = \begin{bmatrix} D_{se} & D_m & 0 \\ D_m & D_{sc} & D_m \\ 0 & D_m & D_{se} \end{bmatrix} \qquad (14.3\text{-}49)$$

where

$$D_{se} = \gamma \left(2\langle \hat{B}_m^2 \rangle_{m,fic} + 2\langle \hat{B}_m^2 \rangle_{m,foc} + 4\langle \hat{B}_m^2 \rangle_{m,ff} + \langle \hat{B}_m^2 \rangle_{m,el} + \langle \hat{B}_m^2 \rangle_{m,ml} + 2\langle \hat{B}_m^2 \rangle_{m,fl} \right) \qquad (14.3\text{-}50)$$

$$D_{sc} = \gamma \left(4\langle \hat{B}_m^2 \rangle_{m,fic} + 4\langle \hat{B}_m^2 \rangle_{m,ff} + 2\langle \hat{B}_m^2 \rangle_{m,ml} + 2\langle \hat{B}_m^2 \rangle_{m,fl} \right) \qquad (14.3\text{-}51)$$

$$D_m = \gamma \langle \hat{B}_m \hat{B}_n \rangle_{m,ml} \qquad (14.3\text{-}52)$$

From our work in Section 11.5, and assuming round conductors, we have that

$$\gamma = \frac{\pi}{4} \sigma r_{wc}^4 l_{wc} N_{pr} \qquad (14.3\text{-}53)$$

It will sometimes be convenient to compute the dynamic resistances without incorporating the γ term. This is because this term is a function of the conductors and specifically the temperature through the conductivity. Therefore, we will use a prime to denote dynamic resistance normalized to γ, i.e.

$$D' = \frac{1}{\gamma} D \qquad (14.3\text{-}54)$$

The expression (14.3-48) can be simplified. If we assume that, following the geometrical symmetry,

$$\overline{\left(\frac{di_c}{dt}\right)^2} = \overline{\left(\frac{di_a}{dt}\right)^2} \qquad (14.3\text{-}55)$$

and that

$$\overline{\left(\frac{di_b}{dt}\frac{di_c}{dt}\right)} = \overline{\left(\frac{di_a}{dt}\frac{di_b}{dt}\right)} \qquad (14.3\text{-}56)$$

then (14.3-48) and (14.3-49) reduce to

$$P_p = 2D_{se}\overline{\left(\frac{di_a}{dt}\right)^2} + D_{sc}\overline{\left(\frac{di_b}{dt}\right)^2} + 4D_m \overline{\left(\frac{di_a}{dt}\frac{di_b}{dt}\right)} \qquad (14.3\text{-}57)$$

In terms of *qd* variables, (14.3-57) becomes

$$P_p = \frac{3}{2} D_{se} \overline{\left(\frac{di_{qs}^s}{dt}\right)^2} + \left(\frac{1}{2} D_{se} + D_{sc} - 2D_m\right) \overline{\left(\frac{di_{ds}^s}{dt}\right)^2} + \sqrt{3}(D_{se} - 2D_m) \overline{\left(\frac{di_{qs}^s}{dt}\frac{di_{ds}^s}{dt}\right)} \qquad (14.3\text{-}58)$$

Magnetic Equivalent Circuit Encapsulation

At this point, we have completely described the formulation of the MEC for the three-phase inductor, as well as quantities needed for the dynamic resistance matrix. We will let $\mathbf{D}_{\text{MEC}}$ denote a structure which includes all data required formulate the MEC. In addition to that data needed to formulate the MEC, this structure will also include related terms that are closely associated with the MEC analysis, but not formally a part of the MEC including the leakage permeances P_{sle} and P_{slc} as well as the gamma-normalized dynamic resistances D'_{se}, D'_{sc}, and D'_{m}. Without itemizing the remaining fields of $\mathbf{D}_{\text{MEC}}$ for brevity (this essentially includes all the terms described in this section), we encapsulate our analysis as

$$\mathbf{D}_{\text{MEC}} = \text{F}_{\text{MECsetup}}(\mathbf{G}) \tag{14.3-59}$$

where $\mathbf{G}$ is the geometry structure of Section 14.2. Observe that we have separated the formulation of the MEC from its execution. This is because we will execute the MEC multiple times, and because some of the data generated in the MEC formulation is used in the context of other analysis (for example, the dynamic resistance matrix).

Magnetic Equivalent Circuit Execution

Upon formulating the MEC, for a given set of phase currents we can solve the MEC for mesh fluxes. Let us assume that we are given vectors of phase currents $\mathbf{i}_{as}$, $\mathbf{i}_{bs}$, $\mathbf{i}_{cs}$, each of length J. For the jth element these vectors, we first compute the phase MMFs using (14.3-1), and solve the MEC for the mesh fluxes. The corresponding values of flux linkage are given by

$$\lambda_{as,j} = N\left(P_{sle}F_{a,j} + \Phi_2\right) \tag{14.3-60}$$

$$\lambda_{bs,j} = N\left(P_{slc}F_{b,j} + \Phi_4 - \Phi_2\right) \tag{14.3-61}$$

$$\lambda_{cs,j} = N\left(P_{sle}F_{c,j} - \Phi_4\right) \tag{14.3-62}$$

The fluxes does not carry a "j" subscript in (14.3-60)–(14.3-62) because we will recompute them for every element "j" of the phase currents. Rather, the flux subscripts in (14.3-60)–(14.3-62) refer to mesh numbers as defined by Figure 14.8.

In order to compute losses, we will also wish to compute the flux densities corresponding to the currents. In particular, we will compute the flux density in reluctances corresponding to R_p in the a-phase (an end leg), to R_{pb} in the a-phase (an end leg), to R_p in the b-phase (the center leg), to R_{pb} in the b-phase (the center leg), and R_b between the a- and b-phases. In particular,

$$B_{pe,j} = \Phi_1/a_{mcp} \tag{14.3-63}$$

$$B_{pbe,j} = \Phi_1/a_{mcpb} \tag{14.3-64}$$

$$B_{pc,j} = (\Phi_3 - \Phi_1)/a_{mcp} \tag{14.3-65}$$

$$B_{pbc,j} = (\Phi_3 - \Phi_1)/a_{mcpb} \tag{14.3-66}$$

$$B_b = \Phi_1/a_{mcb} \tag{14.3-67}$$

We will encapsulate this set of calculations with the function call

$$\left(\boldsymbol{\lambda}_{as}, \boldsymbol{\lambda}_{bs}, \boldsymbol{\lambda}_{cs}, \mathbf{B}_{pe}, \mathbf{B}_{pbe}, \mathbf{B}_{pc}, \mathbf{B}_{pbc}, \mathbf{B}_b, c\right) = \text{F}_{\text{MEC}}(\mathbf{D}_{\text{MEC}}, \mathbf{i}_{as}, \mathbf{i}_{bs}, \mathbf{i}_{cs}) \tag{14.3-68}$$

where the elements of the indicated vectors are defined by (14.3-60)–(14.3-67), and c is a logical flag which is true if the MEC converges successfully at all of the test points.

14.4 Magnetic Analysis

In this section, we will consider the magnetic analysis of an operating point. This will include assessing the incremental inductance, low-frequency core loss, high-frequency core loss, and proximity-effect loss. We will begin our work with the computation of incremental inductance.

Incremental Inductance

In the dc–ac converter, the purpose of the three-phase inductor is to filter (reduce) the high-frequency current ripple which exists as a perturbation to the desired low-frequency fundamental component. Thus, it makes sense that we are interested in incremental inductance.

Our analysis will begin by creating two sets of test points. We will use these points to establish the incremental inductance, and also to formulate a means to rapidly estimate the flux-density waveforms in terms of current waveforms.

To establish the first set, we consider a set of currents of equal amplitude and uniformly spaced in terms of angular separation in the qd plane. In particular, the currents associated with the kth test point of the first set may be established from

$$\theta_{et1,k} = \frac{2\pi(k-1)}{K} \tag{14.4-1}$$

$$i^s_{qst1,k} = I_{t1} \cos \theta_{et1,k} \tag{14.4-2}$$

$$i^s_{dst1,k} = -I_{t1} \sin \theta_{et1,k} \tag{14.4-3}$$

where K is the number of test points, and I_{t1} is the peak current associated with our operating point, i.e.

$$I_{t1} = \sqrt{2} I_l \tag{14.4-4}$$

These points form the innermost ring of points in Figure 14.11. Once these test points are established, we will convert the currents from qd to abc form using the inverse transformation (14.1-15) and (14.1-22), and then call the MEC to establish the phase flux linkages using (14.3-68). The phase flux linkages are then transformed back to qd variables using the forward transformation of (14.1-12) and (14.2-21). This process yields vectors of flux linkages λ^s_{qst1} and λ^s_{dst1}.

We will then create a second set of test points, which correspond to the outer ring of point in Figure 14.11. The location of the kth test point of the second set is given by

$$\theta_{et2,k} = \frac{2\pi(k-1/2)}{K} \tag{14.4-5}$$

$$i^s_{qst2,k} = I_{t2} \cos \theta_{et2,k} \tag{14.4-6}$$

$$i^s_{dst2,k} = -I_{t2} \sin \theta_{et2,k} \tag{14.4-7}$$

In (14.4-6) and (14.4-7), there is some degree-of-freedom in selecting I_{t2}. However, if we define

$$h = \frac{2 I_{t1} \tan(\pi/K)}{1 - \sqrt{3} \tan(\pi/K)} \tag{14.4-8}$$

and I_{t2} as

$$I_{t2} = \sqrt{I_{t1}^2 + \sqrt{3} h I_{t1} + h^2} \tag{14.4-9}$$

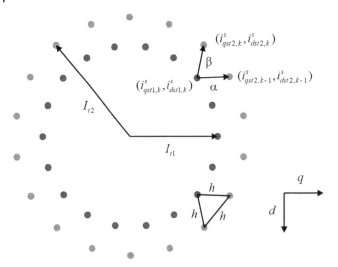

Figure 14.11 Test points.

then the Euclidean distance between any two of the points $\left(i^s_{qst1,k}, i^s_{dst1,k}\right)$, $\left(i^s_{qst2,k}, i^s_{dst2,k}\right)$, $\left(i^s_{qst,k-1}, i^s_{dst2,k-1}\right)$ is equal to h. This is not critical, but has a certain elegance.

Once the two sets of test currents are established, we will again convert the currents from qd to abc form using the inverse transformation (14.1-15) and (14.1-22), and then call the MEC to establish the phase flux linkages using (14.3-68). The phase flux linkages are then transformed back to qd variables using the forward transformation (14.1-12) and (14.1-21). This process yields vectors of flux linkages λ^s_{qst1}, λ^s_{dst1}, λ^s_{qst2}, and λ^s_{dst2}.

One of the goals of establishing the test points is to create a set of current perturbations in order to numerically determine incremental inductance. For each point k in the first set, we create an α perturbation wherein

$$\Delta i^s_{qs\alpha,k} = i^s_{qst2,k} - i^s_{qst1,k} \tag{14.4-10}$$

$$\Delta i^s_{ds\alpha,k} = i^s_{dst2,k} - i^s_{dst1,k} \tag{14.4-11}$$

$$\Delta \lambda^s_{qs\alpha,k} = \lambda^s_{qst2,k} - \lambda^s_{qst1,k} \tag{14.4-12}$$

$$\Delta \lambda^s_{ds\alpha,k} = \lambda^s_{dst2,k} - \lambda^s_{dst1,k} \tag{14.4-13}$$

and a β perturbation for which

$$\Delta i^s_{qs\beta,k} = i^s_{qst2,k-1} - i^s_{qst1,k} \tag{14.4-14}$$

$$\Delta i^s_{ds\beta,k} = i^s_{dst2,k-1} - i^s_{dst1,k} \tag{14.4-15}$$

$$\Delta \lambda^s_{qs\beta,k} = \lambda^s_{qst2,k-1} - \lambda^s_{qst1,k} \tag{14.4-16}$$

$$\Delta \lambda^s_{ds\beta,k} = \lambda^s_{dst2,k-1} - \lambda^s_{dst1,k} \tag{14.4-17}$$

From these two perturbations, and noting that the linearized inductance matrix should be symmetric, we have the overdetermined system

14.4 Magnetic Analysis

$$\begin{bmatrix} \Delta \lambda_{q\alpha,k}^s \\ \Delta \lambda_{d\alpha,k}^s \\ \Delta \lambda_{q\beta,k}^s \\ \Delta \lambda_{d\beta,k}^s \end{bmatrix} = \begin{bmatrix} \Delta i_{q\alpha,k}^s & 0 & \Delta i_{d\alpha,k}^s \\ 0 & \Delta i_{d\alpha,k}^s & \Delta i_{q\alpha,k}^s \\ \Delta i_{q\beta,k}^s & 0 & \Delta i_{d\beta,k}^s \\ 0 & \Delta i_{d\beta,k}^s & \Delta i_{q\beta,k}^s \end{bmatrix} \begin{bmatrix} L_{qq,k} \\ L_{dd,k} \\ L_{qd,k} \end{bmatrix} \quad (14.4\text{-}18)$$

Solving (14.4-18) in a least-squares sense yields the incremental inductances $L_{qq,k}$, $L_{dd,k}$, and $L_{qd,k}$ associated with the kth test point (of the first set of test points).

In the ideal case, all $L_{qq,k}$ and $L_{dd,k}$ would be equal to the desired phase inductance, L^*, and all $L_{qd,k}$ would be equal to zero. To assess our closeness to this desired case, we define

$$L_{s,mn} = \min\left(\mathbf{L}_{qq} \cup \mathbf{L}_{dd}\right) \quad (14.4\text{-}19)$$

$$L_{s,mx} = \max\left(\mathbf{L}_{qq} \cup \mathbf{L}_{dd}\right) \quad (14.4\text{-}20)$$

$$L_{m,mx} = \max\left(|\mathbf{L}_{qd}|\right) \quad (14.4\text{-}21)$$

In our next section, we will place constraints on $L_{s,mn}$, $L_{s,mx}$, and $L_{m,mx}$ as part of the design process.

Low-Frequency Core Loss

Core loss is a difficult to evaluate in the presence of both low- and high-frequency components, if one is to proceed with complete rigor. Here, we will take a decoupled approach. In particular, referring to the combined MSE model, we will apply the hysteresis portion of the model to the low-frequency component of the flux density waveforms, and the eddy portion of the model to the high-frequency component of the flux density waveforms. To this end, for each point k in the first set of test points, we will record the flux density in region, i.e. $B_{x,k}$ where "x" denotes region, thus, $x \in [pe, pbe, pc, pbc, b]$ which correspond to an end post, the post-base transition region of an end leg a center post, the post-base transition region of the center leg, and a base region, respectively. We will also establish a corresponding time within a cycle, given by

$$t_k = \frac{\theta_{et1,k}}{\omega_e} \quad (14.4\text{-}22)$$

We now have time histories of the flux density waveforms and can apply the sMSE model to compute core loss density (computing the time derivatives of the flux densities numerically) using the sMSE model of (6.3-16), whereupon the low-frequency core loss, which we will associate with hysteresis, may be expressed as

$$P_h = 2p_{h,pe}V_{mp} + p_{h,pc}V_{mp} + 4p_{h,pbe}V_{mpb} + 2p_{h,pbc}V_{mpb} + 4p_{h,b}V_{mb} \quad (14.4\text{-}23)$$

High-Frequency Core Loss

In computing the low-frequency losses, we consider a number of points K, which is on the order of 10^2, as we are representing the fundamental component of the waveform. To represent the full waveform, with all the switching frequency components, would require many more points. For example, if we represent one cycle of a 60 Hz waveform with a 1 μs time step, we would arrive at the need to compute the flux density at 10^4 points, which is modestly computationally demanding.

To circumvent this difficulty, recall that our goal will be such as to try to make the inductor behave somewhat ideally, i.e., to act as if it were linear with decoupled phases. If we accomplish this goal, the magnetic material should behave approximately linearly. Thus, we should be able to reasonably relate the flux density in the various regions to the currents as

$$\mathbf{B}_x = c_{Bxq}\mathbf{i}_{qs}^s + c_{Bxd}\mathbf{i}_{ds}^s \qquad (14.4\text{-}24)$$

The calculation of currents at on the order of 10^4–10^5 points is trivial, because it can be done a priori to the design optimization. From (14.4-24), we can readily calculate the flux density corresponding to the currents after we have established numerical values for c_{Bxq} and c_{Bxd}.

In order to obtain values for c_{Bxq} and c_{Bxd}, from the form of (14.4-24), and using the flux densities from the first set of test points, we may solve for the coefficients c_{Bxq} and c_{Bxd} by solving the over-determined system

$$\begin{bmatrix} B_{x,1} \\ B_{x,2} \\ \vdots \\ B_{x,K} \end{bmatrix} = \begin{bmatrix} i_{qst1,1}^s & i_{dst1,1}^s \\ i_{qst1,2}^s & i_{dst1,2}^s \\ \vdots & \vdots \\ i_{qst1,K}^s & i_{dst1,K}^s \end{bmatrix} \begin{bmatrix} c_{Bxq} \\ c_{Bxd} \end{bmatrix} \qquad (14.4\text{-}25)$$

This will allow us to very rapidly compute flux density waveforms in terms of the currents.

Once we have the coefficients, we can differentiate (14.4-24) to yield

$$p\mathbf{B}_x = c_{Bxq}p\mathbf{i}_{qs}^s + c_{Bxd}p\mathbf{i}_{ds}^s \qquad (14.4\text{-}26)$$

where the time derivatives of the currents come from the time domain simulation that was discussed in Section 14.1. Then, from (6.3-14) we can compute the eddy current loss density in each region, whereupon we can compute the eddy current loss as

$$P_e = 2p_{e,pe}V_{mp} + p_{e,pc}V_{mp} + 4p_{e,pbe}V_{mpb} + 2p_{e,bc}V_{mpb} + 4p_{e,b}V_{mb} \qquad (14.4\text{-}27)$$

Proximity Effect Losses

Proximity effect losses can be calculated using (14.3-58). However, in order for this calculation to not be a function of temperature (to facilitate future thermal analysis), we will define

$$P'_p = \frac{P_p}{\gamma} \qquad (14.4\text{-}28)$$

where γ is defined by (14.3-53). From (14.3-54), (14.3-58), and (14.4-28), the gamma-normalized proximity effect lost may be expressed as

$$P'_p = \frac{3}{2}D'_{se}\overline{\left(\frac{di_{qs}^s}{dt}\right)^2} + \left(\frac{1}{2}D'_{se} + D'_{sc} - 2D'_m\right)\overline{\left(\frac{di_{ds}^s}{dt}\right)^2} + \sqrt{3}(D'_{se} - 2D'_m)\overline{\left(\frac{di_{qs}^s}{dt}\frac{di_{ds}^s}{dt}\right)}$$

$$(14.4\text{-}29)$$

Encapsulation

We will encapsulate the results of the section call by the function call

$$\mathbf{M} = F_{MA}(\mathbf{G}, \mathbf{D}_{MEC}, \mathbf{S}, K) \qquad (14.4\text{-}30)$$

where

$$\mathbf{M} = \begin{bmatrix} c_{MA} & L_{s,mn} & L_{s,mx} & L_{m,mn} & P_h & P_e & P'_p \end{bmatrix} \quad (14.4\text{-}31)$$

and where c_{MA} is a logical variable which is true if all magnetic analysis used has converged and false otherwise. It is true if and only if the value of c returned from (14.3-68) is true for both the sets of test points.

14.5 Inductor Design Paradigm

In this section, we will develop a design paradigm for the three-phase inductor. Unlike our work in the previous chapter, we will not consider the inclusion of either a thermal model, or capacitance. While these aspects of the design are very important, this extension is left to the reader in the interest of brevity. We will begin our discussion with the design space.

Design Space

A first step in the design process is to identify the degree-of-freedom, which in this case come primarily from the independent variables of $\mathbf{G}_I$ as specified by (14.2-1). However, certain fields have been deleted. In particular, since we are not considering thermal analysis or capacitance, we will eliminate the thickness of the coil-to-core insulation t^*_{clc} and layer-to-layer insulation t^*_{ll}, and make these variables design specific. Furthermore, there is no need to retain ε_{rp} and k_p, the permittivity and thermal conductivity of the potting material. Note that while we could certainly make the conductor and core material part of the design space, here, we will select it a priori. With these allowances, our design space becomes

$$\boldsymbol{\theta} = \begin{bmatrix} g & w_{lc} & l_c & r_{bl} & N^*_{tpl} & N^*_l & N^*_{pr} & a^*_{wc} & c_{cc} & c_{cr} & c_{le} \end{bmatrix} \quad (14.5\text{-}1)$$

Associated with our design will be a structure $\mathbf{D}$ of design information. In particular, we have

$$\mathbf{D} = \begin{bmatrix} \mathbf{S} & t^*_{clc} & t^*_{ll} & k_b & k_{bd} & \mathbf{D}_{CD} & \mathbf{D}_{CR} & T_w & L^* & k_R & k_S & k_M & K \end{bmatrix} \quad (14.5\text{-}2)$$

Recall that $\mathbf{S}$ is a vector of data from our time domain simulation, k_b is the winding build factor, k_{bd} is the conductor bending ratio (bending radius/strand diameter), and $\mathbf{D}_{CD}$ and $\mathbf{D}_{CR}$ are structures of conductor and core data, respectively. The assumed temperature of the winding is T_w. As previously defined, L^* is the desired value of inductance. We yet have to define k_R, k_S, and k_M, but these are all constants related to the ideality of the inductance. Finally, recall that K is the number of points used to evaluate the inductance. From a coding point of view, $\mathbf{D}$ may also contain other data such as wire gauge and insulating paper data.

One point of special note in (14.5-2) is the use of simulation data $\mathbf{S}$ as a part of the design specification. This simulation requires input, as specified by (14.1-54). While most of this data are known (for example, the dc and utility voltage), it is observed that lumped inductor parameters L_q, L_d, and r_l are also required. In conducting this simulation, we will set $L_q = L_d = L^*$, and, in our fitness function, place constraints such that the desired incremental inductance is obtained. However, there is still the question of what value to use for r_l since it is not known a priori to the design. To this end, any reasonable estimate can be used, because the voltage drop across this resistance should be much less than the voltage drop associated with the inductance. Which is to say, in practice, the waveform data specified in (14.1-55), and in particular i^s_{qlrms}, i^s_{dlrms}, $\overline{\left(pi^s_{ql}\right)^2}$, $\overline{\left(pi^s_{dl}\right)^2}$, and $\overline{\left(pi^s_{ql}pi^s_{dl}\right)^2}$ that are used in the design process are rather insensitive to r_l. This will be the subject of an exercise.

Metrics

In our previous work, we have often used loss as one metric, and mass or volume as the other. Clearly, there is often a desire to minimize mass, and volume, and while they are often highly correlated, there is a difference. While we could certainly perform an optimization in three metrics, for the purposes of illustration, let us consider another approach. In this approach, we first calculate the total inductor mass using (14.2-15), and then compute the circumscribing volume as

$$V_L = w_L d_L l_L \tag{14.5-3}$$

Next, we will define the geometric size as the geometric mean of mass and volume. In particular, we define

$$S_L = \sqrt{M_L V_L} \tag{14.5-4}$$

which we will use as our first metric to minimize.

Our second metric to minimize will be loss. Given the temperature, we may calculate the conductor conductivity σ from the reciprocal of (10.7-2), which is in terms of resistivity. Next, we can calculate the phase resistance using (13.6-45). It can be shown that in the absence of zero sequence voltage or current, instantaneous power in qd variables may be expressed as

$$P = \frac{3}{2}\left(v_{qs}i_{qs} + v_{ds}i_{ds}\right) \tag{14.5-5}$$

which holds in any reference frame. From (14.5-5), the average resistive power loss may be expressed as

$$P_r = \frac{3}{2}r_l\left(\left(i^s_{qs,rms}\right)^2 + \left(i^s_{ds,rms}\right)^2\right) \tag{14.5-6}$$

where $i^s_{qs,rms}$ and $i^s_{ds,rms}$ are the rms values of the q-axis and d-axis current, which we can obtain by processing the results of the time domain simulation.

In computing the resistive losses, we will neglect the skin effect losses, though including them would not be particularly difficult. The reason is that the skin effect losses will be dominated by proximity effect losses. To calculate the proximity effect lost, we first calculate γ using (14.3-53), whereupon the proximity effect loss may be expressed as

$$P_p = P'_p \gamma \tag{14.5-7}$$

where the gamma-normalized proximity effect loss P'_p is calculated as a part of (14.4-30). The eddy current and hysteresis components of core loss, P_e and P_h, are also known from the magnetic analysis represented by (14.4-30).

At this point, the total inductor loss may be expressed as

$$P_L = P_r + P_p + P_e + P_h \tag{14.5-8}$$

Constraints

Now that we have set forth metrics, let us consider constraints. The first of these will be related to the leakage reluctances within the MEC. It is desired that flux flow through the intended magnetic path through the core dominates the flux flow through magnetic paths associated with the leakage inductance.

To this end, let us consider the MEC as shown in Figure 14.8. For the purposes of formulating constraints, let us take all the reluctances associated with the core to be zero. In this case, looking at the top left side of the equivalent circuit, we see that the reluctance $R_{ge} + R_{gc}$ is in parallel with R_{ml}. Since we wish the flux to flow through the former reluctance rather than the latter, we will require

$$c_1 = \text{lte}(R_{ge} + R_{gc}, R_{ml}/k_R) \tag{14.5-9}$$

where k_R is a reluctance constant, a design specification. This will be a relatively large number, a typical value being 20.

Likewise, keeping in mind that under idealistic conditions, the MMF between the top and bottom nodes of the equivalent circuit is zero, we will require that the sum of the air gap reluctances in phase leg be much less than the leakage reluctance. This leads to the constraints

$$c_2 = \text{lte}(2R_{ge}, R_{sle}/k_R) \tag{14.5-10}$$

$$c_3 = \text{lte}(2R_{gc}, R_{slc}/k_R) \tag{14.5-11}$$

where we use the same reluctance constant as before.

The next constraint will be the convergence of the MEC. In particular, we take

$$c_4 = c_{MA} \tag{14.5-12}$$

We will also require that the minimum incremental inductance be greater than the target inductance with the constraint

$$c_5 = \text{gte}(L_{s,mn}, L^*) \tag{14.5-13}$$

Note that it is not necessary to bound the inductance from above. In order both to reduce size and reduce loss, the inductance obtained will always approach the minimum allowed value.

In addition, we will limit the ratio of the maximum to minimum incremental inductance. This is essentially a requirement on the ideality of the inductor, addressing both saliency, and to some extent magnetic linearity. In particular, we impose

$$c_6 = \text{lte}\left(\frac{L_{s,mx}}{L_{s,mn}}, k_S\right) \tag{14.5-14}$$

where k_S is the saliency specification. A typical number is 1.05.

A related constraint will be on the mutual inductance between the q-axis and d-axis. The constraint imposed is

$$c_7 = \text{lte}(L_{m,mx}, k_M L^*) \tag{14.5-15}$$

where k_M is a mutual inductance specification, where a typical value is 0.05.

Calculation of Fitness

In calculating the fitness, we will take a slightly different approach as compared to what we have before. After evaluate constraint i we will update the number of constraints imposed c_I and the number of constraints satisfied as c_S as

$$c_S = c_S + c_i \quad (14.5\text{-}16)$$

$$c_I = c_I + 1 \quad (14.5\text{-}17)$$

Then, at strategic points in the algorithm execution, we will compare the number of constraints satisfied to the number of imposed. If the number of constraints satisfied fall below the number of constraints imposed, we will calculate the fitness as

$$f = -\frac{\varepsilon}{1+c_S}\begin{bmatrix}1 & 1 & \cdots & 1\end{bmatrix}^T \quad (14.5\text{-}18)$$

where ε is a small positive number, and the dimensionality is consistent the number of metrics (which for this problem will be 2). The advantage of this approach is that we need not keep track of the total number of constraints imposed which is convenient when studying the impact of various constraints on the design by simply adding and removing them.

If we find all constraints satisfied, then we will take the fitness to be the reciprocal of the size and loss, in particular,

$$f = \begin{bmatrix}\frac{1}{S_L} & \frac{1}{P_l}\end{bmatrix}^T \quad (14.5\text{-}19)$$

At this point, it is appropriate to detail the steps required to evaluate the fitness. In particular, the steps to evaluate the fitness are as follows:

1) As the first step, we initialize the constraints satisfied c_S and the constraints imposed c_I to zero.
2) Next, we perform the geometrical calculations using (14.2-28).
3) After the geometrical calculations are preformed, the MEC is setup using (14.3-59).
4) At this point, reluctances constraints $c_1 - c_3$ are evaluated using (14.5-9)–(14.5-11) with three executions of (14.5-16)–(14.5-17). At this point, if $c_S < c_I$, then the fitness is calculated using (14.5-18), and the fitness evaluation is complete. If this is not the case, we will proceed to Step 5.
5) In this step, we evaluate the magnetic analysis summarized by (14.4-30). On the basis of the convergence constraint c_4, we execute (14.5-16)–(14.5-17). If $c_S < c_I$, then the fitness is calculated using (14.5-18) and the fitness evaluation is complete. If this is not the case, we will proceed to Step 6.
6) Next, we evaluate inductance constraints $c_5 - c_7$ using (14.5-13)–(14.5-15). Again, we perform three executions of (14.5-16)–(14.5-17). At this point, if $c_S < c_I$, then the fitness is calculated using (14.5-18) and the fitness evaluation is complete. If this is not the case, we will proceed to Step 7.
7) At this point, all constraints have been satisfied and so we have a viable design. We compute the size S_L using (14.5-3)–(14.5-4). Next, we compute the resistive loss, where (10.7-2) is used to compute the resistivity, (13.6-45) for phase resistance, and finally (14.5-6) for resistive loss. To find proximity effect loss, we find γ using (14.3-53), and then the proximity effect loss using (14.5-7). The total loss is then found from (14.5-8). Finally, the fitness of the viable design is found from (14.5-19).

In the next section, we will use this process in a case study.

14.6 Case Study

In the previous section, we considered a design paradigm for a three-phase I-core-based inductor. We will now use that process in a design example. In particular, we will consider the design of a three-phase inductor for a 5 kW, 400 V dc to 230 V 60 Hz ac converter, with an assumed switching frequency of 19.98 kHz (20 kHz, but rounded to an integer number of switching cycles within a

Table 14.1 Case Study Operating Point Data

Symbol	Description	Value
v_{dc}	dc voltage	400 V
V_U	Utility voltage (line-to-neutral, rms)	$230/\sqrt{3}$ V
ω_e	Fundamental frequency	$2\pi 60$ rad/s
θ_{e0}	Time zero position of electrical angle	0
I_l	rms value of fund. component of inductor current	12.55 A
ϕ_{il}	Angle of fund. component of inductor current	0
L_q	q-axis inductance	1 mH
L_d	d-axis inductance	1 mH
r_l	Assumed phase resistance	0.1 Ω
f_{sw}	Switching frequency	19.98 kHz
t^*_{mx}	Time to simulate	1/60 s
h^*	Nominal time step	20 μs
d_{lm}	Maximum duty cycle	0.99
α	Convergence factor	0.5
e_a	Allowed error	10^{-5}

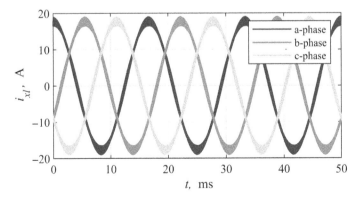

Figure 14.12 Case study assumed current waveforms.

Table 14.2 Three-Phase I-Core Inductor Specifications

Symbol	Description	Value
S	Operating point data (based in Table 14.1)	
t^*_{clc}	Desired coil-to-coil insulation thickness	1 mm
t^*_{ll}	Desired layer-to-layer insulation thickness	1 mm
k_d	Winding build factor	1.05
k_{bd}	Cond. bending radius constant (bend rad./strand rad.)	6
$\mathbf{D}_{CD}$	Structure of conductor data. Based on copper.	
$\mathbf{D}_{CR}$	Structure of core data. Based on a representative nanocrystalline material.	
T_w	Assumed winding temperature	100°C
L^*	Target inductance	1 mH
k_R	Allowed reluctance factor	20
k_S	Allowed saliency factor	1.05
k_M	Allowed mutual inductance factor	0.05
K	Number of test points	72

fundamental cycle). Operating point data is listed in Table 14.1, and based on this data one obtains the assumed current waveforms as shown in Figure 14.12.

Additional design specifications are set forth in Table 14.2. Of particular interest is the use of a "representative" nanocrystalline material, which can be used to form tape-wound core sections. This is a material with very thin laminations. It has a saturation flux density somewhat less than silicon steels, but can be used at much higher frequencies as compared to silicon steels (though not nearly so high as ferrites). The assumed parameters are listed with the silicon steels in the appendix, though it is not a silicon–steel material.

Table 14.3 Three-Phase I-Core Inductor Design Space

Par.	Description	Min.	Max.	Enc.	Gene
g	Air gap (m)	10^{-4}	10^{-2}	log	1
w_{lc}	Leg center width (m)	10^{-4}	10^{-1}	log	2
l_c	Core length (m)	10^{-4}	1	log	3
r_{bl}	Base depth to leg width ratio	0.5	1.5	log	4
N^*_{tl}	Desired turns per layer	0.51	100.4	log	5
N^*_l	Desired number of layers	0.51	5.49	log	6
N^*_{pr}	Number of parallel conductors	0.51	20.49	log	7
J^*_L	Desired current density (A/m²)	10^5	10^7	log	8
c_{cc}	Coil-to-coil clearance (m)	10^{-2}	$5 \cdot 10^{-2}$	log	9
c_{cr}	Coil-to-core end clearance (m)	10^{-3}	$5 \cdot 10^{-2}$	log	10
c_{le}	Leg-to-core end clearance (m)	10^{-3}	$5 \cdot 10^{-2}$	log	11

The search space is set forth in Table 14.3. Therein, many of the parameters are set via experimentation—making sure that the gene values do not become clustered on limits unless there is a physical limit. Others are based on what is deemed reasonable based on the experience. For example, the minimum coil-to-coil clearance was set to allow some provision for air flow between the coils.

Figure 14.13 illustrates the parameter distribution after a run. In this study, there were 1000 individuals and 2000 generations. The parameter numbers in Figure 14.13 correspond to the parameter order in (14.5-1) and gene number in Table 14.3. As can be seen, many of the parameters vary with size of the inductor, which increases from left to right. This is the most evident in the target current density J_L^* (gene 8). Observe that c_{cc} (gene 9) was tightly clustered to its minimum allowed value that was established to permit air flow. The number of parallel turns N_{pr}^* (gene 7) and coil-to-core clearance c_{cr} (gene 10) also hit their lower limits, but only for the smallest inductors. Interestingly, the leg-to-end clearance c_{le} (gene 11) does not approach its minimum value though this clearance increases mass. This is because it helps control saliency.

Figure 14.14 depicts the pareto-optimal front between loss and size. As we have seen many times in our studies, there is clear trade-off between the competing objectives of size and loss. Within Figure 14.14, Design 100 is highlighted. We will explore this design in some detail.

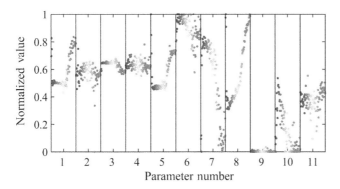

Figure 14.13 Three-Phase I-Core inductor parameter distribution.

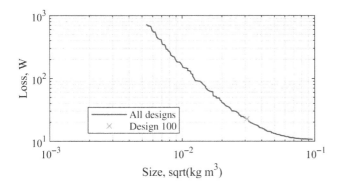

Figure 14.14 Three-Phase I-Core inductor pareto-optimal front.

Table 14.4 Design 100

Aggregate Inductor Dimensions------------------	Winding Data---
Overall Length: 6.36 cm	Wire Diameter: 1.516 mm
Overall Width: 11.26 cm	Wire Insulation Thickness: 33.02 um
Overall Depth: 8.252 cm	Bending Radius: 9.098 mm
Encapsulating Volume: 0.5911 L	Number of Layers: 3
Total Mass: 1.619 kg	Turns per Layer: 13
Size: 0.9784 (kg L)$^{1/2}$	Number of Turns: 39
	Number of Parallel Conductors: 2
Core Data---	Conductor Length (Each): 5.608 m
I-core Leg Width: 1.727 cm	Cl-Cr Ins.: 0.76 mm
I-core Leg Depth: 4.658 cm	Lyr-to-Lyr Ins.: 0.76 mm
I-core Leg Length: 3.281 cm	Winding Mass (1 of 3): 0.1647 kg
I-core Leg Mass (1 of 3): 0.1506 kg	
I-core Base depth: 1.666 cm	Electrical Model---
I-core Base width: 10.8 cm	Assumed Winding Temperature: 100°C
I-core Base length: 3.281 cm	Phase Res. at Assumed Wind. Temp.: 37.43 mΩ
I-core Base Mass (1 of 2): 0.3368 kg	Min. Incr. Inductance: 1.011 mH
	Max. Incr. Inductance: 1.051 mH
Spacing---	Max. Incr. QD Mutual Ind.: 12.09 µH
Air Gap: 1.308 mm	
Clearances:	Full Load Loss--
Leg-Core to Leg-Core: 2.411 cm	Total: 22.66 W
Leg-Core to Base Core End: 0.3989 cm	Center Leg Current Density: 3.479 A/mm^2
Coil-to-Coil: 1 cm	Resistive: 17.73 W
Coil-to-Core End: 0.1 cm	Proximity Effect: 2.178 W
	Core-Hysteresis: 0.7937 W
	Core-Eddy Current: 1.96 W

Features of Design 100 are set forth in Table 14.4. As can be seen, the design occupies roughly 0.59 L and has a mass of about 1.6 kg. Observe in the winding data that the coils are wound with two turns in hand—the use of parallel conductors reduces both bending radius and proximity effect loss. The ratio of maximum-to-minimum inductance is 1.04—near the limit. Clearly, the allowed saliency is a constraint to the design. Observe that the resistive loss dominates the total loss in this case.

A profile and top view of the inductor are set forth in Figure 14.15. Therein, the leg I-cores (lighter) and base I-cores (darker) are differentially shaded to make the top view clearer. Note that in the top view, the upper base core has been removed.

Looking at the design, the impact of the bending radius is clearly evident in the top view. Because of the significant bending radius, the end conductors are somewhat distant from the core, which will reduce proximity effect loss in this region at the expense of modestly greater resistive losses and a somewhat larger inductor.

At this point, it is appropriate to reflect on the design process. The three chief additions needed for the design approach are (i) the addition of a thermal model, (ii) the addition of a capacitance model,

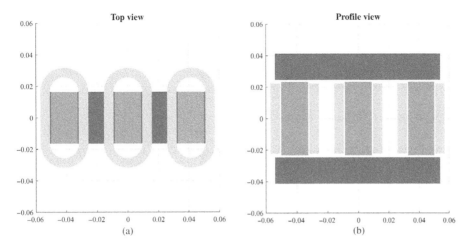

Figure 14.15 Design 100. Dimensions in meters.

and (iii) consideration of multiple operating points. These additions can be readily made using the material from Chapters 10, 12, and 13.

References

1 P. Krause, O. Waynczuk, S. Sudhoff, and S. Pekarek, *Analysis of Electric Machines and Drives Systems*, 3rd edition. Hoboken, NJ: IEEE Press and John Wiley & Sons, Inc., 2013.
2 O. Wasynczuk, S. Sudhoff, T. Tran, D. Clayton, and H. Hegner, A voltage control strategy for current-regulated PWM inverters, *IEEE Transactions on Power Electronics*, vol. **11**, no. 1, pp. 7–15, 1996.
3 A. Singh, M. Chinthavali, S. Sudhoff, K. Bennion, K. Prabakar, X. Feng, Z. Wang, and S. Campbell, Development and validation of a SiC based 50 kW grid-connected PV inverter, in *2018 IEEE Energy Conversion Congress and Exposition (ECCE)*, Portland, OR, 2018.
4 V. Dupalli and S. Sudhoff, Power density comparison of three-phase AC inductor architectures, in *2017 IEEE Electric Ship Technologies Symposium (ESTS)*, Arlington, VA, 2017.
5 S. Sudhoff, R. Swanson, A. Kasha, and V. Duppalli, Magnetic analysis of symmetrical three-phase Y-core inductors, *IEEE Transactions on Energy Conversion*, vol. **33**, no. 2, pp. 576–583, 2018.
6 S. D. Sudhoff, "*MATLAB codes for Power Magnetic Devices: A Multi-Objective Design Approach*", 2nd edition, [Online]. Available: http://booksupport.wiley.com

Problems

1 Using (14.1-12), (14.1-14), and (14.1-17) and the trigonometric identities in Appendix F show (14.1-18).

2 Starting with (14.1-26), and using (14.1-29) show (14.1-30) and (14.1-32).

3 Using (14.1-36) and (14.1-60) derive (14.1-61) and (14.1-62).

4 Using geometry, derive (14.3-19).

5 Show (14.3-20).

6 Using (14.3-20) derive (14.3-21) and (14.3-22).

7 Using (14.3-24) and (14.3-25) derive (14.3-26).

8 Using (14.3-24) and (14.3-25) derive (14.3-27), (14.3-28), and (14.3-29).

9 Using the path shown in Figure 14.9(b) derive (14.3-31), (14.3-32), and (14.3-33).

10 Using the path as shown in Figure 14.10, derive (14.3-37)–(14.3-41).

11 Starting with (14.3-57), show that we obtain (14.3-58).

12 Show that using (14.4-8) and (14.4-9) yields a Euclidean distance between any two of the points $\left(i^s_{qst1,k}, i^s_{dst1,k}\right)$, $\left(i^s_{qst2,k}, i^s_{dst2,k}\right)$, $\left(i^s_{qst,k-1}, i^s_{dst2,k-1}\right)$ equal to h.

13 Consider the data in Table 14.1. For the case study, and using code from [6], determine i^s_{qlrms}, i^s_{dlrms}, $\overline{\left(pi^s_{ql}\right)^2}$, $\overline{\left(pi^s_{dl}\right)^2}$, and $\overline{\left(pi^s_{ql}pi^s_{dl}\right)^2}$. Now, using the data from Table of 14.1, but replacing the inductor phase resistance with the value listed in Table 14.4, repeat the calculation of i^s_{qlrms}, i^s_{dlrms}, $\overline{\left(pi^s_{ql}\right)^2}$, $\overline{\left(pi^s_{dl}\right)^2}$, and $\overline{\left(pi^s_{ql}pi^s_{dl}\right)^2}$. Compute the percent change in each value.

15

Common-Mode Inductor Design

In the previous chapter, we designed a three-phase inductor as part of a dc-to-ac converter. While the three-phase inductor is the primary magnetic component of such a system, in many cases, there is another magnet component we have yet to consider—a common-mode inductor (CMI). As we will see, the CMI is a very interesting device used to reduce electromagnetic compatibility and electromagnetic interference problems. It is substantially different from anything we have looked at thus far.

15.1 Common-Mode Voltage and Current

When considering the performance of an electrical device, we often focus on the differential-mode voltage and current. For example, consider a device fed through a pair of wires such as the dc input terminals of the inverter studied in Chapter 14. The dc input voltage in that discussion was a differential-mode voltage. It was the voltage of the upper input terminal relative to the lower input terminal. In Chapter 14, we assumed that the dc input current went into the upper input terminal and returned to some source through the lower input terminal. The input current we thought of was the differential-mode current. The differential-mode current can be thought of as the intended current. Unfortunately, in reality, there will be another component of the current, which will cause the current going into the upper terminal to be different from that returning through the lower terminal. This difference is attributed to the presence of common-mode current, which is often associated with parasitic paths that do not appear on circuit diagrams but which are present in reality. CMIs are used to reduce these unintended currents.

Common-mode voltages and currents can lead to several system problems. First, these currents often find their way outside of a designated electrical system and into a grounding system or even a building or vehicle structure. This enables them to interfere with the proper operation of other systems. Because common-mode currents often follow unintended paths, the loop areas associated with the paths can be large, creating spatially dispersed magnetic fields throughout a building or vehicle, which can again interfere with other systems, particularly low-voltage electronics used for controls.

It is appropriate to begin our discussion of CMI design with the definitions of common-mode voltage and common-mode current. We will begin with common-mode voltage. Suppose we have a set of nodes X, and let x denote a member of that set, that is, $x \in X$. Let the voltage of

Power Magnetic Devices: A Multi-Objective Design Approach, Second Edition. Scott D. Sudhoff.
© 2022 The Institute of Electrical and Electronics Engineers, Inc. Published 2022 by John Wiley & Sons, Inc.
Companion website: www.wiley.com/go/sudhoff/Powermagneticdevices

node x relative to some point p be denoted v_x. Then, the common-mode voltage of that set of nodes is defined as

$$v_{cm,X} = \frac{1}{\|X\|} \sum_{x \in X} v_x \qquad (15.1\text{-}1)$$

where $\|X\|$ denotes the number of elements in X. The common-mode voltage is the average of the node voltages relative to node p.

Next, suppose we have a set of conductors denoted Y, and let $y \in Y$ denote a conductor within that set. The common-mode current associated with that set of conductors is defined as

$$i_{cm,Y} = \sum_{y \in Y} i_y \qquad (15.1\text{-}2)$$

The common-mode current is the sum of the currents in the conductors of the set.

In the two-conductor case, we often speak of the differential-mode voltage and current. If we have two nodes α and β, the differential-mode voltage associated with these nodes is

$$v_{dm} = v_\alpha - v_\beta \qquad (15.1\text{-}3)$$

If we have two conductors γ and δ, the differential-mode current is defined as

$$i_{dm} = \frac{1}{2}\left(i_\gamma - i_\delta\right) \qquad (15.1\text{-}4)$$

From (15.1-2), for the two-conductor case, $i_{cm} = i_\gamma + i_\delta$. With this observation and (15.1-4), we can show that $i_\gamma = i_{dm} + i_{cm}/2$ and $i_\delta = -i_{dm} + i_{cm}/2$, as shown in Figure 15.1. This reinforces the notion of the differential-mode current as the intended current and the common-mode current as an extraneous component of the current.

In the two-node/conductor case, normally, the differential-mode quantities are those components of the voltages and currents that are intended and the common- mode is unintended. In the three-phase case, one can view the q- and d-axis quantities as intended and the common-mode quantities (at least the common-mode current) as unintended.

Physically speaking, when running cables interconnecting power components, we typically bundle conductors so that the sum of the currents should be zero. Consider a three-phase system with currents i_a, i_b, and i_c. Ideally, the sum of the phase currents is zero, that is,

$$i_{cm} = i_a + i_b + i_c = 0 \qquad (15.1\text{-}5)$$

An advantage of such an approach is that if the sum of the currents is zero, the magnetic fields surrounding the cables fall off very rapidly with distance from the cable, and so the time rate of change of the fields will not induce currents and voltages in the surrounding conductive material,

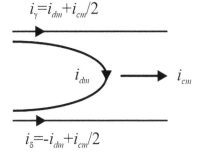

Figure 15.1 Differential- and common-mode currents.

be it circuits or structures. However, in many cases, even if no return path is evident, the sum of the currents is not zero. This is because the sum of the currents, that is, the common-mode current, follows an unintended parasitic path back to the source. This is particularly evident for the higher-frequency components of the currents.

Common-mode current is generally undesirable. It often leads to electromagnetic noise, which can degrade the performance of sensors and cause low-voltage control hardware to fail. Under some circumstances, it can even pose a safety hazard. CMIs are one tool used to mitigate common-mode current.

15.2 System Description

The focus of this chapter is the design of CMIs, which are used to reduce common-mode currents. In order to provide context to our discussion, let us consider the system shown in Figure 15.2.

Proceeding from left to right, the first system component shown is a generic DC device, which could be a battery, a photovoltaic (PV) array, or some other source or energy-storage mechanism. The next component is a CMI and then a dc link capacitor with capacitance C_{dc}. The power block consists of a two-level three-phase bridge such as that considered in Chapter 14, a more detailed view of which appears in Figure 15.3. Then, we have a three-phase inductor, again from Chapter 14. The output of the inductor is assumed to be connected to a three-phase wye-connected capacitor bank with a grounded neutral. This capacitor bank is tied in shunt with an ac system, which may be the utility grid, an ac microgrid, or simply an ac load.

Other components in Figure 15.2 include capacitances C_{dv} that are physically present to provide a low-impedance high-frequency path from the DC device rails to ground and large (in an Ohmic sense) resistors of value R_c whose purpose is to center the dc rail voltages symmetrically with respect to the ground bus.

Nodes "a" through "j" in Figure 15.2 are indicated, and the voltage of these nodes relative to node "g_i," that is, the inverter ground bus will be denoted v_x, where "x" is replaced by the symbol for the node. In other words, we are arbitrarily choosing our reference node to be the inverter ground. Formally, the reference node can only be a point since there will be a variation in voltage along any conductor; we will assume that the variation in voltage is negligible herein. Note that at very high frequencies (in the tens of MHz region and above), such an assumption is not warranted.

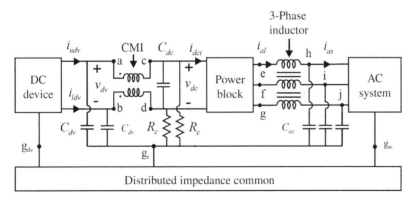

Figure 15.2 Dc-to-ac converter.

There are two other ground nodes in Figure 15.2. These include the ground node of the DC device, g_{dv}, and the AC system ground, g_{ac}. These grounds are all at different potentials relative to the reference node and are connected together in a somewhat nebulous fashion in a path that, for example, in a PV system, may include soil.

Other voltages and currents in Figure 15.2 include the differential DC device voltage, v_{dv}, the differential voltage, and the dc input to the power block (inverter), denoted v_{dc}, the a-phase current out of the power block i_{al}, and the a-phase AC system current, i_{as}. Other phase quantities are analogously defined.

Figure 15.3 depicts the topology of the power block (inverter). Therein, the devices are shown in insulated-gate bipolar transistors (IGBTs) but could be other devices such as metal–oxide–semiconductor field-effect transistors (MOSFETs). Nodes "r" and "s" in Figure 15.3 represent the bottom rail and heat sink, respectively. The capacitances C_1 and C_2 are parasitic; C_1 is the capacitance from each of the rails to the heat sink, and C_2 is the capacitance from each phase terminal to the heat sink. Herein, it is assumed that the heat sink is not connected to the inverter ground but is instead left floating. Connection of the heat sink to ground does not void the proposed approach if the parasitic capacitances are sufficiently small.

15.3 Common-Mode Equivalent Circuit

Our goal for this chapter is to design a CMI to reduce the common-mode current produced in the system of Figure 15.3. The first step to finding a solution is to understand the problem, and a good way to understand the problem of common-mode currents is through the use of a common-mode equivalent circuit. The idea of a common-mode equivalent circuit is that if we write the equations relating the common-mode voltages and currents, those equations correspond to an electric circuit diagram known as the common-mode equivalent circuit. In many cases, if we are fortunate, the common-mode equivalent circuit is independent of the differential-mode equivalent circuit. This

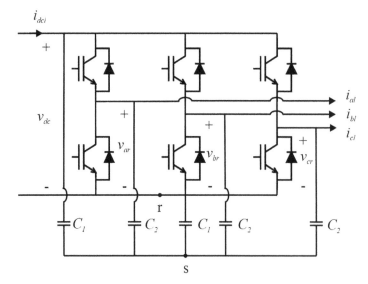

Figure 15.3 Power block.

15.3 Common-Mode Equivalent Circuit

is typically the case if our original circuit has appropriate symmetry. We will also find that many elements of the original circuit do not appear in the common-mode equivalent circuit because they do not conduct common-mode current. In any case, the common-mode equivalent circuit is a useful tool in understanding a system's common-mode behavior. In this section, we will develop the common-mode equivalent circuit corresponding to Figure 15.2 one component at a time.

DC Source and Capacitors

Let the currents coming out of the dc rails of the DC device be denoted i_{udv} and i_{ldv} as in Figure 15.2. From (15.1-1) and (15.1-2), the common-mode voltage and current associated with the DC device may be expressed as

$$v_{cmdv} = \frac{1}{2}(v_a + v_b) \tag{15.3-1}$$

$$i_{cmdv} = i_{udv} + i_{ldv} \tag{15.3-2}$$

Now, let us consider the capacitors with value C_{dv}. The current into each capacitor may be expressed as

$$i_{xcdv} = C_{dv} p v_x \tag{15.3-3}$$

where $x \in [a, b]$, and p is the Heaviside notation for the time derivative operator as previously used. From (15.3-3) and our common-mode definitions, we have that

$$i_{cmcdv} = 2 C_{dv} p v_{cmdv} \tag{15.3-4}$$

Together, (15.3-1), (15.3-2), and (15.3-4) lead to the left-most parts of the common-mode equivalent circuit shown in Figure 15.4.

Common-Mode Inductor

We will next turn our attention to the CMI. Observe that this inductor is a two-coil device and could be viewed as a transformer with an equal number of turns on the primary and secondary.

The voltage across the upper and lower coils may be expressed as

$$v_a - v_c = r_{cmi} i_u + p \lambda_u \tag{15.3-5}$$

$$v_b - v_d = r_{cmi} i_l + p \lambda_l \tag{15.3-6}$$

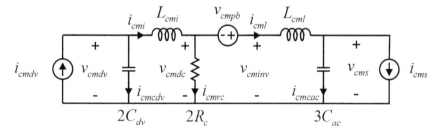

Figure 15.4 Common-mode equivalent circuit.

where i_u and i_l are the currents into the upper and lower windings from nodes "a" to "c" and "b" to "d," respectively, r_{cmi} is the winding resistance, and λ_u and λ_l are the winding flux linkages.

We will define the CMI voltage as

$$v_{cmi} = v_{cmdv} - v_{cmdc} = \frac{1}{2}r_{cmi}i_{cmi} + p\lambda_{cmi} \qquad (15.3\text{-}7)$$

where v_{cmdc} is the arithmetic mean of the respective node "c" and "d" voltages in accordance with (15.1-1). The CMI common-mode current is given by

$$i_{cmi} = i_u + i_l \qquad (15.3\text{-}8)$$

and the CMI common-mode flux linkage is defined as

$$\lambda_{cmi} = \frac{1}{2}(\lambda_u + \lambda_l) \qquad (15.3\text{-}9)$$

From (15.3-7) to (15.3-9), we see that the multi-winding CMI appears as a simple single-winding inductor in the common mode, as depicted in Figure 15.4.

For design purposes, we will use a nonlinear magnetic model to represent the relationship between flux linkage and current. However, for the purposes of insight, let us briefly consider a magnetically linear representation of the CMI. Using a linear model, the coil flux-linkage equations may be expressed as

$$\lambda_u = L_{cmil}i_u + L_{cmim}(i_u + i_l) \qquad (15.3\text{-}10)$$
$$\lambda_l = L_{cmil}i_l + L_{cmim}(i_u + i_l) \qquad (15.3\text{-}11)$$

where L_{cmil} and L_{cmim} are the leakage and magnetizing inductances, respectively. Note that the inductor is wound such that the flux produced by the two coil currents i_u and i_l adds.

Using the definitions of common-mode flux linkage and current, we can readily show that (15.3-10) and (15.3-11) become

$$\lambda_{cmi} = L_{cmi}i_{cmi} \qquad (15.3\text{-}12)$$

where

$$L_{cmi} = \frac{1}{2}L_{cmil} + L_{cmim} \qquad (15.3\text{-}13)$$

Power Block Input Network

The power block refers to the power semiconductor portion of the inverter, as shown in Figure 15.3. This is the same configuration studied in Chapter 14: however, in Figure 15.3, parasitic capacitances are shown. The power block input network includes the capacitor C_{dc} and the balancing resistors of value R_c, which are used to center the rail voltages about the ground bus.

With regard to the dc link capacitor, observe that the sum of the currents going from nodes "c" and "d" into the two terminals of the capacitor must be zero; thus

$$i_{ccdc} + i_{dcdc} = 0 \qquad (15.3\text{-}14)$$

As a result, there is no common-mode current associated with the capacitor, and it does not appear in the common-mode equivalent circuit.

The currents into the balancing resistors may be expressed as

$$i_{xrc} = \frac{1}{R_c} v_x \tag{15.3-15}$$

where $x \in [c, d]$. Thus, the common-mode current into the balancing resistor may be expressed as

$$i_{cmrc} = \frac{2}{R_c} v_{cmdc} \tag{15.3-16}$$

Power Block

The power block shown in Figure 15.3 is the central element of the inverter and is also the root source of our common-mode problems. Physically, the power block is often a single package with six transistors and six diodes or sometimes a collection of three phase-leg packages (each with two transistors and two diodes). Rarely discrete components are used. The parasitic capacitances shown in Figure 15.3 are effective values between different regions of the semiconductors and the heat sink to which the semiconductors are invariably attached. The impact of the capacitances varies considerably depending upon the electrical connection of the heat sink. One option is that the heat sink is "floating" and not intentionally connected to anything, although there will be parasitic capacitive connection to the case of the power converter, which is typically grounded. If the heat sink is left floating, the parasitic capacitances will not affect the common-mode equivalent circuit, though they will certainly impact the internal operation of the power semiconductors. This is the situation we will address. A second possibility is that the heat sink is grounded. This may have a safety advantage but can be detrimental in terms of common-mode noise issues.

Using Kirchhoff's voltage laws, we may write the voltage equations

$$v_{dc} = v_c - v_d \tag{15.3-17}$$

$$v_e = v_d + v_{ar} \tag{15.3-18}$$

$$v_f = v_d + v_{br} \tag{15.3-19}$$

$$v_g = v_d + v_{cr} \tag{15.3-20}$$

Manipulating (15.3-17)–(15.3-20), we can show that

$$v_{cminv} = v_{cmdc} + v_{cmpb} \tag{15.3-21}$$

where v_{cminv} is the arithmetic mean of node "e," "f," and "g" voltages in accordance with (15.1-1) and v_{cmpb} is the common-mode voltage produced by the power block, which is defined as

$$v_{cmpb} = \frac{1}{3}(v_{ar} + v_{br} + v_{cr}) - \frac{1}{2} v_{dc} \tag{15.3-22}$$

The term v_{cmpb} is very important to our work, as it is the dominant driving source for common-mode current. We will find that it has three components of interest in different frequency regimes. The first is at relatively low frequency, namely, the third harmonic of the fundamental component of the ac voltage. The second frequency regime of interest is at the switching frequency and harmonics thereof. This range will be of particular interest in our work. The third component is at the edge rate of the switching of the power electronic devices.

3-Phase Inductor

Let us next turn our attention to the three-phase filter inductor. The phase voltages across the three-phase inductor may be expressed as

$$v_{al} = v_e - v_h \tag{15.3-23}$$

$$v_{bl} = v_f - v_i \tag{15.3-24}$$

$$v_{cl} = v_g - v_j \tag{15.3-25}$$

Summing (15.3-23)–(15.3-25), the common-mode voltage across the three-phase inductor may be expressed as

$$v_{cml} = v_{cminv} - v_{cms} \tag{15.3-26}$$

where v_{cml} is the arithmetic mean of the ac inductor phase voltages and v_{cms} is the arithmetic mean of the node "h," "i," and "j" voltages.

Using the three-phase inductor phase voltage equations given by (14.1-24), we have

$$v_{cml} = \frac{1}{3} r_l i_{cml} + p\lambda_{cml} \tag{15.3-27}$$

where the common-mode flux linkage is given by

$$\lambda_{cml} = \frac{1}{3}(\lambda_{al} + \lambda_{bl} + \lambda_{cl}) \tag{15.3-28}$$

Using the linear flux-linkage equations given by (14.1-26), if

$$L_e + M_{ee} = L_c + M_{ec} \tag{15.3-29}$$

then, the common-mode flux linkage may be expressed as

$$\lambda_{cml} = L_{cml} i_{cml} \tag{15.3-30}$$

where

$$L_{cml} = \frac{1}{3}(L_e + M_{ec} + M_{ee}) \tag{15.3-31}$$

As an aside, in Chapter 14, expressions for L_q and L_d were given by (14.1-31) and (14.1-32), respectively. As a design requirement for the three-phase inductor design, we placed a constraint on saliency. From (14.1-31) and (14.1-32), it can be shown that the only way for $L_q = L_d$ and (15.3-29) to be simultaneously satisfied is if $L_e = L_c$ and $M_{ee} = M_{ec}$, that is, that the three-phase inductor is symmetrical. In [1], one topology for a truly symmetrical three-phase inductor is set forth. Alternately, in [2], methods of treating common-mode analysis of asymmetrical circuit topologies are considered.

The impact of asymmetry is that the differential-mode currents will create a voltage term in the common-mode equivalent circuit (and vice versa). From a pragmatic point of view, however, this effect can often be neglected because the common-mode voltage source v_{cmpb} dominates as a common-mode source. We will assume that this is the case herein.

3-Phase Capacitor and AC System

The currents into the ac capacitors may be expressed as

$$i_{xcac} = C_{ac} p v_x \tag{15.3-32}$$

where $x \in [h, i, j]$. From our definitions of common-mode voltage, current, and flux linkage, together with (15.3-32), we have

$$i_{cmcac} = 3C_{ac}pv_{cms} \tag{15.3-33}$$

Finally, referring to Figure 15.2, the common-mode current into the ac system may be expressed as

$$i_{cms} = i_{as} + i_{bs} + i_{cs} \tag{15.3-34}$$

Simplified Common-Mode Equivalent Circuit

At this point, all the common-mode representations of the system shown in Figure 15.2 and which appear in the common-mode equivalent circuit of Figure 15.4 have been developed. It is now appropriate to simplify the equivalent circuit.

To this end, we will make several assumptions. First, we will assume that in the absence of an external excitation, the common-mode currents i_{cmdv} and i_{cms}, out of the DC device and into the ac system, respectively, are zero. Further, we will assume that the capacitances C_{dv} and C_{ac} are such that the common-mode current arising from the power block is shunted away from the DC device and the AC system. This removes the two current sources from Figure 15.5. Next, we will assume that balancing resistance R_c is very large, which it must be to avoid excessive power loss. Thus, it can be neglected. With these assumptions, the common-mode equivalent circuit shown in Figure 15.4 reduces to the common-mode equivalent circuit of Figure 15.5(a).

To perform further simplification, we must consider our goals. As mentioned previously, there are three aspects to the common-mode problem. One is at low frequency and is a result of the fact that for some modulation strategies, such as space-vector modulation or extended sine-triangle modulation [3], v_{cmpb} contains a third-harmonic component (of the ac-side fundamental frequency). This component of the current is not readily blocked with a CMI because it is of very low frequency. Instead, we rely on the capacitive elements of Figure 15.5(a) to block this component. Observe that the two capacitors in series yield an effective capacitance of

$$C_{eff} = \frac{6C_{dv}C_{ac}}{2C_{dv} + 3C_{ac}} \tag{15.3-35}$$

and that the two inductors in series yield a net inductance of

$$L_{eff} = L_{cmi} + L_{cml} \tag{15.3-36}$$

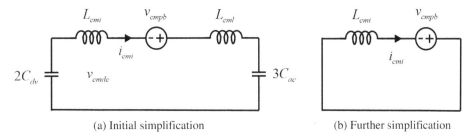

(a) Initial simplification (b) Further simplification

Figure 15.5 Simplified common-mode equivalent circuits.

If the effective capacitance is used to block the third-harmonic component of v_{cmpb}, then the low-frequency resonance at

$$f_{lfr} = \frac{1}{2\pi\sqrt{L_{eff}C_{eff}}} \qquad (15.3\text{-}37)$$

must be well above (ideally a factor of 10 or more) the third-harmonic frequency so that the system impedance looks capacitive to this frequency component. If this condition is met, then the system impedance at the frequency of the third harmonic will be capacitive. The capacitors C_{ac} and C_{dv} must be such that the effective capacitance C_{eff} is small enough (i.e., its impedance is high enough) to sufficiently limit the third-harmonic component of the common-mode current.

At the same time, this low-frequency resonance must be well below the switching frequency (ideally by a factor of 10 or more). This is because at the switching frequency, the impedance presented by the capacitors is typically very low; so, the inductors block the common-mode current. Thus, it is important that the system impedance transitions from being capacitive to inductive well before the switching frequency.

Clearly, if we are to block the low-frequency common-mode current with the capacitance, we must have that the low-frequency resonance given by (15.3-37) is both well above (ideally by a factor of 10 or more) the third harmonic of the fundamental and well below (ideally by a factor of 10 or more) the switching frequency.

Herein, we will take an alternate approach of using a modulation strategy which does not generate a third-harmonic component—for example standard (nonextended) sine-triangle modulation [3] as considered in the previous chapter. With that, the next problem is limiting the switching-frequency components of the common-mode current. Under the scenario of an appropriate modulation strategy (one not generating low-frequency components), we can choose capacitors such that at the switching frequency and above, the capacitors appear as short circuit, that is, we will choose the capacitor such that the switching frequency is at least an order of magnitude higher than the resonant frequency indicated by (15.3-37). In this case, we may treat the capacitances as short circuits in Figure 15.5(a). We will also assume that the common-mode inductance of the CMI is much greater than that of the common-mode inductance of the three-phase inductor, which is generally an excellent (and conservative) approximation. One reason that this is true is that many three-phase inductors are such that the magnetizing component of the inductance cancels in the common mode: thus, their common-mode inductance is primarily leakage. With these simplifications, the common-mode equivalent circuit of Figure 15.5(a) becomes that of Figure 15.5(b). It is this equivalent circuit upon which we will base our CMI design.

Before concluding this section, it should be noted that while the switching-frequency components of the common-mode current often dominate the rms value of the common-mode current, very high-frequency components associated with the edge rates of the switching of the power electronics devices also exist. While this can be very important, herein we will focus on the mitigation of the switching-frequency components of the common-mode current.

15.4 Common-Mode Inductor Specification

In this section, we consider what our design specifications for the inductor should be, which is of course necessary for any design process, optimization-based, or otherwise. The approach we will take here is a simple one—we will put a bound on the rms value of the common-mode current i_{cmi}. However, we must specify the conditions under which we impose this bound. To this end, we will also specify the common-mode flux-linkage waveform. This will be the focus of the rest of this section.

Common-Mode Flux Linkage

In order to establish the common-mode flux-linkage waveform and effectively design the CMI, we will make several approximations. The first of these is the switching frequency. In general, the dynamics of the inverter are a-periodic. If $mf_{sw} = nf_e$, where m and n are integers that do not share any common factor, then the system dynamics are periodic with a period $1/(nf_e)$. In order to avoid the need to analyze a-periodic behavior or behavior spanning more than one cycle of the fundamental, we will analyze the system based on an approximate switching frequency given by

$$f_{swa} = f_e \operatorname{round}\left(\frac{f_{sw}}{f_e}\right) \tag{15.4-1}$$

so that the system is periodic in the fundamental. This approximation works well in practice. For example, if the fundamental frequency is 60 Hz, and the switching frequency is 20 kHz, we would analyze the system (including time-domain simulations) at a frequency of 19.98 kHz, an error of 0.1%. If such an error makes a significant difference, it is doubtful that such a system would be operationally acceptable in any case. However, by analyzing the system as if the switching frequency were 19.98 kHz, we can perform an analysis based on a single switching cycle.

The next approximation we will make is to neglect the CMI resistance. This approximation is good one in that the CMI resistance is normally small (so as to present negligible loss to the system, as the device must carry the full differential-mode current) and also a conservative one, as the common-mode resistance should reduce the common-mode current. With this approximation, from Figure 15.5(b), we have that the common-mode flux linkage is governed by

$$p\lambda_{cmi} = v_{cmpb} \tag{15.4-2}$$

Therefore, the steady-state common-mode flux linkage is given by

$$\lambda_{cmi}(t) = \int_0^t v_{cmpb} d\tau - f_e \int_0^{1/f_e} v_{cmpb} d\tau \tag{15.4-3}$$

where τ is a dummy variable for time and f_e is the fundamental frequency. The second term is the initial condition that is calculated such that the common-mode flux linkage has no dc component. This follows from the assumption that there is no dc component to the common-mode current since the common-mode current normally involves a series capacitive element, which is often parasitic in nature.

To use (15.4-2), we calculate the line-to-bottom-rail voltages as described in Section 14.1 (neglecting semiconductor voltage drops) and then calculate v_{cmpb} using (15.3-22). We will explore this process in the following example.

Example 15.4A In this example, let us predict the common-mode flux-linkage waveform for a simple system. Let us assume that we have an inverter with an input voltage of $v_{dc} = 400$V and we wish to produce an ac waveform at 60 Hz with a fundamental component of 230 V, l–n, rms, using a switching frequency of 20 kHz. This corresponds to a peak value of 187.8 V, which, from our work in Section 14.1 or 13.5 of [3], we find we need $d = 0.9390$. From (15.4-1), we will analyze the system at a frequency of 19.98 kHz.

To compute the common-mode flux linkage, we first compute the phase-duty cycles from (14.1-6), where we take $\theta_c = 2\pi f_e t$ and $d_3 = 0$ since we will use standard (nonextended) sine-triangle modulation in order to avoid a third-harmonic ($3f_e$) component of the common-mode voltage. Next, we use (14.1-8) to compute the switching states at every instant of time. Then, the line-to-rail voltages are found using (14.1-4), neglecting the semiconductor switch drops. The power block common-mode voltage is then found from (15.3-22) and, finally, the common-mode flux linkage from (15.4-3).

Figure 15.6 shows the resulting common-mode voltage produced by the power block and the common-mode flux-linkage waveforms on multiple time scales. The first trace is the common-mode voltage over a narrow time slice, and the center trace illustrates the corresponding common-mode flux-linkage waveform. Here, the inflections at switching events are evident. The lower trace depicts the long-term trend in the common-mode flux-linkage waveform; observe the variation in the amplitude over one-sixth of a fundamental cycle.

Worst-Case Common-Mode Flux Linkage

Not surprisingly, the common-mode flux-linkage waveform is a function of duty cycle, and so it is interesting to identify how the common-mode flux-linkage waveform changes with duty cycle. We can intuit that the peak common-mode flux linkage will be a function of the dc voltage v_{dc}, the duty cycle d, and the switching frequency f_{swa} because these are the variables that determine the inverter voltage waveforms. Before proceeding to develop this relationship, let us consider a numerical example to guide our development.

Example 15.4B In this example, we will explore how the peak value of flux-linkage varies with duty cycle. We will consider a process very similar to Example 15.4 A but with the following changes. First, we will take $v_{dc} = 1$ V so that our results become normalized to v_{dc}. Next, instead of considering a single value of duty cycle d, we will vary it, and for each value of d, we will use the process described in Example 15.4 A to compute the peak value λ_{cmi}, analogous to the peak value seen in Figure 15.6, which in that case was 1.3 mVs. In this study, our peak value will be normalized to v_{dc}. The results of the study are shown in Figure 15.7. Clearly, for this sine-triangle modulation strategy, the peak common-mode flux linkage decreases linearly with duty cycle.

It is important to emphasize that this is not always the case and that small changes to modulation scheme can create a large difference in results. For example, consider displaced carrier sine-triangle modulation. This is very similar to what we have already considered, save for one difference. The carrier (triangle) wave for the b- and c-phases is displaced by that of the a-phase by 1/3 and 2/3 of a switching period, respectively. As it turns out, such a change will drastically reduce the peak common-mode flux linkage and also cause it to rise linearly with duty cycle, though it is always less than with a single carrier. The disadvantage of such an approach is that the q- and d-axis current ripples increase.

In Example 15.4B, we saw that the dc voltage-normalized peak common-mode flux linkage decreases linearly with duty cycle. We also expect that the peak common-mode flux linkage would decrease with the reciprocal of switching frequency. This is because the nature of the waveform does not change from switching cycle to switching cycle—but the duration of this cycle is the reciprocal of switching frequency. Considering these effects leads to

$$\lambda_{cmipk} = \frac{v_{dc}}{8 f_{swa}} \left(1 - \frac{1}{2} d\right) \qquad (15.4\text{-}4)$$

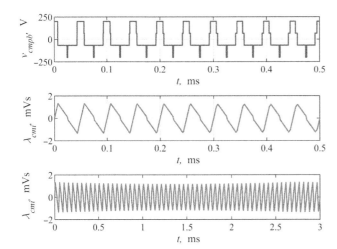

Figure 15.6 Common-mode flux-linkage waveform.

where the term in parentheses represents the best fit of the quantity $8\lambda_{cmipk}f_{swa}/v_{dc}$ versus d to a line and which just happens to be exactly represented with rational numbers, suggesting that a purely analytical derivation of (15.4-4) is possible. No doubt, this would be an excellent examination question.

From (14.1-11), we can see that for balanced conditions, the rms value of the fundamental component of the line-to-line voltage produced by the power block is given by

$$v_{pb,ll-rms} = \frac{\sqrt{6}}{4}v_{dc}d \tag{15.4-5}$$

Manipulating (15.4-4) and (15.4-5),

$$\lambda_{cmipk} = \frac{1}{8f_{swa}}\left(v_{dc} - \frac{\sqrt{6}}{3}v_{pb,ll-rms}\right) \tag{15.4-6}$$

From (15.4-6), we see that the worst-case operating point will be the one for which the difference between the dc voltage and $\sqrt{6}/3$ times the rms value of the fundamental component of the line-to-line power block voltage is the largest. This result is modulation-strategy-specific.

Proxy Common-Mode Flux Linkage

Now that we can identify a "worst-case" condition for our analysis, note that the common-mode flux-linkage waveform is typically rather involved, as seen in Example 15.4A and Figure 15.6. Rather than working with this waveform exactly, let us consider a simplified proxy flux-linkage waveform of the form

$$\lambda_p = \lambda_{p1}\cos\theta_{sw} + \lambda_{p3}\cos\left(3\theta_{sw} + \phi_{p3}\right) \tag{15.4-7}$$

where we take

$$\theta_{sw} = \omega_{sw}t \tag{15.4-8}$$

and where $\omega_{sw} = 2\pi f_{swa}$ and λ_{p1}, λ_{p3}, and ϕ_{p3} are constants used to characterize the worst-case common-mode flux linkage. The advantage of (15.4-7) over the original data is that it is periodic over a

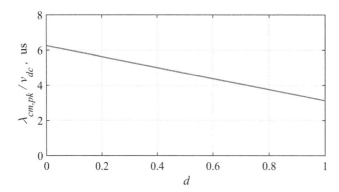

Figure 15.7 Peak common-mode flux linkage versus duty cycle.

single switching cycle, whereas, as can be seen from the final trace of Figure 15.6, the original data is not periodic when viewed over a large time window. Instead, the original data has an amplitude that is modulated at six times the fundamental frequency. The proxy waveform chosen includes a fundamental and third-harmonic component, which together can serve as a rough approximation to a quasi-triangular waveform, as seen in the second trace of Figure 15.6.

The importance of using this proxy waveform will become evident when we consider the magnetic analysis, where we will take an approach different from that before. In particular, in this chapter, we will use a time-domain simulation as part of the magnetic analysis, utilizing the extended Jiles–Atherton (EJA) studied in Chapter 6. The use of a proxy waveform will allow us to simulate a much shorter time span, thus increasing the speed of the fitness evaluation.

To determine λ_{p1}, λ_{p3}, and ϕ_{p3}, we will put three requirements on the proxy waveform. First, we will require that the peak value of (15.4-7) matches the peak value of the common-mode flux-linkage waveform, a value we will denote $\lambda_{cmi,pk}$. Secondly, we will require that the rms value of (15.4-7) matches the rms value of λ_{cmi}, denoted $\lambda_{cmi,rms}$. Third, we will require that the rms value of the time derivative of (15.4-7) matches that of λ_{cmi}, a value we will denote $p\lambda_{cmi,rms}$. The purpose of the first requirement is addressing magnetic saturation. The motivation for the second requirement is that, in the presence of magnetic linearity, the same rms flux linkage should result in the same rms current and the same resistive losses associated with that current. The third requirement is also motivated by losses as many loss mechanisms are proportional to the square of a time derivative.

From the latter two requirements, we can readily show that

$$\lambda_{p1}^* = \frac{1}{2}\sqrt{9\lambda_{cmi,rms}^2 - \frac{p\lambda_{cmi,rms}^2}{\omega_{sw}^2}} \tag{15.4-9}$$

$$\lambda_{p3}^* = \frac{1}{2}\sqrt{\frac{p\lambda_{cmi,rms}^2}{\omega_{sw}^2} - \lambda_{cmi,rms}^2} \tag{15.4-10}$$

where "*" denotes a candidate value. The reason we must specify a candidate value is if $\lambda_{p1} + \lambda_{p3} < \lambda_{cmi,pk}$, then our first requirement cannot be met while meeting the latter two requirements. For this case, in order to meet the first requirement and be conservative with the latter two, we calculate λ_{p1} and λ_{p3} as

$$\lambda_{p1} = \begin{cases} \lambda_{p1}^* \dfrac{\lambda_{cmi,pk}}{\lambda_{p1}^* + \lambda_{p3}^*} & \lambda_{p1}^* + \lambda_{p3}^* \leq \lambda_{cmi,pk} \\ \lambda_{p1}^* & \lambda_{p1}^* + \lambda_{p3}^* > \lambda_{cmi,pk} \end{cases} \tag{15.4-11}$$

$$\lambda_{p3} = \begin{cases} \lambda_{p3}^* \dfrac{\lambda_{cmi,pk}}{\lambda_{p1}^* + \lambda_{p3}^*} & \lambda_{p1}^* + \lambda_{p3}^* \leq \lambda_{cmi,pk} \\ \lambda_{p3}^* & \lambda_{p1}^* + \lambda_{p3}^* > \lambda_{cmi,pk} \end{cases} \qquad (15.4\text{-}12)$$

If $\lambda_{p1}^* + \lambda_{p3}^* \leq \lambda_{cmi,pk}$, then from (15.4-11) and (15.4-12), we will have that $\lambda_{p1} + \lambda_{p3} = \lambda_{cmi,pk}$, and from (15.4-7), we conclude that $\phi_{p3} = 0$. Otherwise, we must find the value of ϕ_{p3} such that the peak value of λ_p is equal to $\lambda_{cmi,pk}$. Thus, we have

$$\phi_{p3} = \begin{cases} 0 & \lambda_{p1}^* + \lambda_{p3}^* \leq \lambda_{cmi,pk} \\ \text{solution of } f_p(\phi_{p3}) = 0 & \lambda_{p1}^* + \lambda_{p3}^* > \lambda_{cmi,pk} \end{cases} \qquad (15.4\text{-}13)$$

where

$$f_p(\phi_{p3}) = \max_{\theta_{sw}} \left(\lambda_{p1} \cos \theta_{sw} + \lambda_{p3} \cos(3\theta_{sw} + \phi_{p3}) \right) - \lambda_{cmi,pk} \qquad (15.4\text{-}14)$$

For the case that $\lambda_{p1}^* + \lambda_{p3}^* > \lambda_{cmi,pk}$, $f_p(\phi_{p3}) = 0$ is solved numerically for ϕ_{p3}.

Example 15.4C In this example, let us find the proxy flux-linkage waveform corresponding to Example 15.4A. From the waveforms of Figure 15.6, we can readily find that $\lambda_{cmi,pk} = 1.328$ mVs, $\lambda_{cmi,rms} = 0.7739$ mVs, and $p\lambda_{cmi,rms} = 111.3$ V. From (15.4-9) and (15.4-10), we obtain $\lambda_{p1}^* = 1.073$ and $\lambda_{p3}^* = 0.2163$ mVs. Since $\lambda_{p1}^* + \lambda_{p3}^* < \lambda_{cmipk}$, we scale these values slightly using (15.4-11) and (15.4-12) to obtain $\lambda_{p1} = 1.105$ and $\lambda_{p3} = 0.2228$ mVs, and from (15.4-13), we obtain $\phi_{p3} = 0$ rad. The original common-mode flux-linkage waveform and its proxy are depicted in Figure 15.8. Note that while out of a sense of esthetics the zero crossings of the two waveforms have been aligned in Figure 15.8, there is no computational reason to do so.

To conclude this section, we have now succeeded in formulating one of our chief CMI specifications. That is, given λ_{p1}, λ_{p3}, and ϕ_{p3}, make sure that the rms common-mode current is less than an allowed rms value of $i_{cm,arms}$.

15.5 UR-Core Common-Mode Inductor

Now that we have our chief design specification, we can consider our CMI in more detail. To this end, let us begin with Figure 15.9, which shows a circuit diagram for the CMI. This was also shown as part of Figure 15.2, but here, we consider a view that is both more detailed and more generic. Relating the two figures, $v_{in} = v_{dv}$ and $v_{out} = v_{dc}$.

From this figure, we expect an inductor with two windings around a common core. With this in mind, we must decide on a choice of topology. Just to consider something different from that previously looked at, we will consider a UR-core CMI. In this arrangement, two UR-cores are used. A UR-core is similar to a U-core, except that the posts are round. The base of the UR core is of rectangular cross section. In this section, we will first define the topology and then summarize the geometrical calculations.

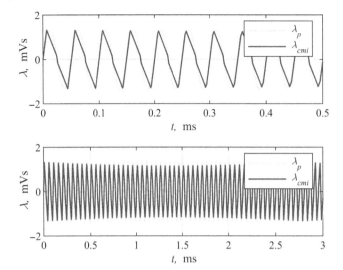

Figure 15.8 Original and proxy common-mode flux-linkage waveforms.

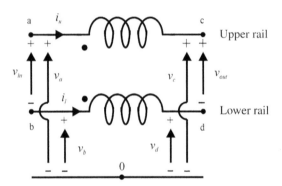

Figure 15.9 CMI circuit diagram.

Topology

The geometry of a UR-core-based CMI is shown in Figure 15.10. Therein, two UR cores are placed together to complete the magnetic path. Each core has a slot depth, d_s, a slot width w_s, a base core depth d_c, a base width w_c, and a post radius $r_c = w_c/2$.

There is a gap g between the cores. Unlike other inductors studied, it will often be that this gap is a result of surface roughness of placing two cores together rather than a specific desired air gap. The air gap is often on the order of 10 μm. The desire for a minimal air gap is a result of the fact that the common-mode component of the current density in a CMI is very low because typically the allowed common-mode current is a small fraction of the intended differential-mode current. The low current density of the common-mode component of the current, which is the component of the current that will drive the field in the core, leads to the desire for a large net permeance and the minimum possible air gap. While no air gap (and thus a single core) is attractive (and often used), the use of two cores makes the inductor easier to assemble in high-power applications.

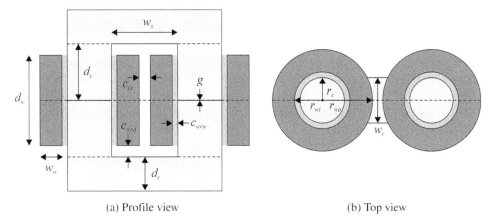

(a) Profile view   (b) Top view

Figure 15.10 UR-core common-mode inductor.

Figure 15.11 Winding cross section.

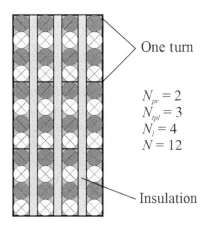

Other parameters in Figure 15.10 include the coil-to-coil clearance c_{cc}, winding-to-core clearance in the width direction c_{wcw}, and winding-to-core clearance in the depth direction c_{wcd}.

As can be seen, two coils are shown in Figure 15.10. One would think that each of these coils would correspond to a winding in Figure 15.9. This is a possibility. However, we will consider a different arrangement wherein we interleave the windings in each coil. This leads to the winding cross section shown in Figure 15.11. Therein, each turn of a coil is seen to consist of N_{pr} parallel strands and that the strands from the upper rail winding and those of the lower rail winding are interleaved. Other parameters of interest in Figure 15.11 are the number of turns per layer N_{tpl}, number of layers N_l, and number of turns per coil N. A numerical example of these quantities is included with the figure.

The reason for interleaving the winding is twofold. In addition to providing inductance in the common mode, the leakage inductance of the device will also act as an inductor in the differential mode. If the differential-mode inductance is too large, it can result in a stability problem [4]. A second advantage of interleaved windings is that they reduce the possibility of the undesired localized saturation of the core. That said, successful designs with noninterleaved windings are certainly possible and have advantages such as the opportunity for greater winding-to-winding voltage isolation.

The upper-rail windings of the one-coil structure may be connected in series or in parallel with the upper-rail windings of the second coil structure. Throughout this work, we will assume a series connection. With this geometry, the bending radius tends not to be active as a design constraint, so there is often no need to have a large number of conductors in parallel. Since the common-mode fields tend to be modest, large conductors are not problematic from a proximity effect loss point of view. Thus, given that we can tolerate larger conductors, a series connection tends to be favorable in terms of layer-to-layer capacitance, which often dominates the capacitance of multilayer designs.

Now that we have identified the inductor topology, we must characterize that geometry. One possible choice of independent variables is

$$\mathbf{G}_I = \begin{bmatrix} g & r_c^* & d_c & N_{tpl}^* & N_l^* & N_{pr}^* & a_{wcpr}^* & t_{ll}^* & c_{cc} & c_{wcd} & c_{wcw} & \mathbf{D} \end{bmatrix} \quad (15.5\text{-}1)$$

In (15.5-1), g, d_c, c_{cc}, c_{wcd}, and c_{wcw} are all defined in Figure 15.10. The variable r_c^* is the desired post radius; the actual value r_c is also defined in Figure 15.10. The variables N_{tpl}^*, N_l^*, N_{pr}^*, a_{wcpr}^*, and t_{ll}^* are all as in Section 13.6. The variable $\mathbf{D}$ is a structure of design data, which include a substructure $\mathbf{D}_{CD}$ of conductor parameters defined by (13.6-3), a substructure $\mathbf{D}_{CR}$ of core parameters, a build factor k_b, and a bending radius constant (allowed bending radius over the strand diameter) k_{bnd}.

The geometrical calculations begin with winding information. Given the desired layer-to-layer insulation thickness t_{ll}^*, we use (13.6-4) to find t_{ll}. From the desired number of turns per layer N_{tpl}^*, number of layers N_l^*, and number of parallel strands N_{pr}^*, we find the actual quantities N_{tpl}, N_l, N_{pr}, using (13.6-6), (13.6-6.7), and (13.6-8), respectively. Next, N is found from (13.6-9). Given the total (all parallel strands) of a conductor a_{wcpr}^*, we use (13.6-10), (13.6-11), and (13.6-12) to find the area of a single strand a_{wc}, radius of a single strand of conductor (without insulation) r_{wc}, and radius of a single strand with insulation r_w. Next, for the interleaved winding, the width and depth of a coil may be expressed as

$$w_w = 2N_l r_w k_b + (N_l - 1)t_{ll} \quad (15.5\text{-}2)$$

$$d_w = 4r_w(N_{tpl} + 1)N_{pr}k_b \quad (15.5\text{-}3)$$

where the $N_{tpl} + 1$ is used rather than N_{tpl} in (15.5-3) in order to accommodate completing the last turn.

We can now begin to calculate the dimensions of the core. The post radius is first calculated as

$$r_c = \max\left(r_c^*, 2k_{bnd}r_w - c_{wcw}\right) \quad (15.5\text{-}4)$$

Observe that the post radius is bounded by the bending radius of the conductor. From the post radius, we have

$$w_c = 2r_c \quad (15.5\text{-}5)$$

Next, we find the slot depth and width as

$$d_s = \max\left(0, \frac{1}{2}d_w - \frac{1}{2}g + c_{wcd}\right) \quad (15.5\text{-}6)$$

$$w_s = 2w_w + 2c_{wcw} + c_{cc} \quad (15.5\text{-}7)$$

Finally, the material volume and mass of one of the two cores may be expressed as

$$V_{cr} = \pi r_c^2(2d_s + d_c) + d_c w_c(w_s + 2r_c) \quad (15.5\text{-}8)$$

15.5 UR-Core Common-Mode Inductor

$$M_{cr} = V_{cr}\rho_{cr} \tag{15.5-9}$$

The length of a turn on the *i*th layer may be expressed as

$$l_{t,i} = \sqrt{4\pi^2(r_c + c_{wcw} + r_{wc}k_b(2i-1) + t_{ll}(i-1))^2 + (4r_{wc}k_bN_{pr})^2} \tag{15.5-10}$$

where the first term within the root corresponds to the circumference around the post and the second relates to the skew of the helical turn (which we have often previously neglected). In terms of the turn length, the total length of one strand of conductor in one coil may be expressed as

$$l_{wc} = N_{tpl}\sum_{i=1}^{N_l} l_{t,i} \tag{15.5-11}$$

whereupon the resistance of one winding in two-series-tied coils normalized to the conductivity (by multiplying by it) may be expressed as

$$r'_{cmi} = r_{cmi}\sigma = \frac{2l_{wc}}{a_{wc}N_{pr}} \tag{15.5-12}$$

This normalization is convenient when adding a thermal model.

At this point, the mass of both windings in one coil may be expressed as

$$M_{wc} = 2a_{wc}l_{wc}\rho_{cd}N_{pr} \tag{15.5-13}$$

and the total inductor mass can be expressed as

$$M_L = 2M_{wc} + 2M_{cr} \tag{15.5-14}$$

The overall inductor length, width, and depth may be expressed as

$$l_L = 2(r_c + c_{wcw} + w_w) \tag{15.5-15}$$
$$d_L = 2(d_c + d_s) + g \tag{15.5-16}$$
$$w_L = w_s + 4r_c + 2c_{wcw} + 2w_w \tag{15.5-17}$$

whereupon the cuboidal circumscribing volume may be expressed as

$$V_L = l_L d_L w_L \tag{15.5-18}$$

and the geometric size, defined as the geometric mean of volume and mass, may be computed as

$$S_L = \sqrt{V_L M_L} \tag{15.5-19}$$

Before concluding this section, it will be convenient for our magnetic analysis to compute the magnetic path length and cross sections for use in our magnetic analysis of the next section. The magnetic cross section, magnetic path length, and magnetic volume of a post (two posts per core and four posts per CMI) are readily expressed as

$$a_{mcp} = \pi r_c^2 \tag{15.5-20}$$
$$l_{mpp} = d_s + 0.5d_c \tag{15.5-21}$$
$$V_{mp} = a_{mcp}l_{mcp} \tag{15.5-22}$$

The magnetic cross section, magnetic path length, and magnetic volume of a base (with one base per core and two bases per CMI) are given by

$$a_{mcb} = d_c w_c \tag{15.5-23}$$

$$l_{mpb} = w_s + 2r_c \tag{15.5-24}$$

$$V_{mb} = a_{mcb}l_{mcb} \tag{15.5-25}$$

As we have done earlier in the work, it will be convenient to summarize the calculations of this section with a function call

$$\mathbf{G}_D = \mathrm{F}_G(\mathbf{G}_I, \mathbf{D}) \tag{15.5-26}$$

where $\mathbf{G}_I$ has already been defined, $\mathbf{D}$ is the structure with design information including the build factor k_b and bending ratio constant k_{bnd}, and

$$\mathbf{G}_D = \begin{bmatrix} w_w & d_w & r_c & w_c & d_s & w_s & V_{cr} & M_{cr} & \mathbf{l}_{t,i} & l_{wc} & r'_{cmi} & M_{wc} & M_L \\ l_L & d_L & w_L & V_L & S_L & a_{mcp} & l_{mpp} & V_{mp} & a_{mcb} & l_{mpb} & V_{mb} \end{bmatrix} \tag{15.5-27}$$

Finally, as we have done elsewhere, we can amalgamate the dependent and independent geometrical variables into a single structure

$$\mathbf{G} = [\mathbf{G}_I \ \mathbf{G}_D] \tag{15.5-28}$$

15.6 UR-Core Common-Mode Inductor Magnetic Analysis

We will now consider the magnetic analysis of the CMI. As discussed in Section 15.4, one of our chief design constraints will be to make sure that the rms value of the common-mode current $i_{cmi,rms}$ is less than a specified value for a proxy flux-linkage waveform parameterized by λ_{p1}, λ_{p3}, ϕ_{p3}, and ω_{sw}. This analysis will be quite different from any of our previous magnetic analysis because the nature of the CMI problem is much different from other inductors.

The CMI current has two components: the differential-mode current, which from (15.1-4) is given by $i_{dm,cmi} = (i_u - i_l)/2$, and the common-mode current, which we simply denote as $i_{cmi} = i_u + i_l$, as defined by (15.3-8). The differential-mode current, which is dominated by a dc component, is typically two orders of magnitude greater in magnitude than the common-mode current.

Observe that because of this large difference in levels, $i_{dm,cmi} \approx i_u \approx -i_l$. The current densities associated with the conductors carrying the upper and lower rail currents are "normal" when compared to other magnetic applications, for example, 5 A/mm^2 (acknowledging a considerable variance here). However, assuming a two-order-of-magnitude separation between the common-mode and differential-mode currents, the current density of the common-mode component of the current may only be 0.05 A/mm^2. Thus, the common-mode current density driving the inductor is very low, mainly because the windings have to carry the differential-mode current density that the core does not see. This extremely low current density tends to lead to low proximity and skin effect losses, which are proportional to current squared. However, the low current density also leads to an inductor with a relatively large amount of magnetic material relative to the amount of current driving that material. As a result, the time-domain waveform of the common-mode current and the rms value of that current are strongly influenced by magnetic hysteresis, much more so than other devices we have thus far considered. Thus, our magnetic analysis includes consideration of magnetic hysteresis, using our work in Section 6.8.

Let us begin our analysis by considering Figure 15.12, which depicts a magnetic equivalent circuit superimposed on a drawing of the core. Therein, F_p and F_b denote the MMF drop across one post, and one base section, and R_g is the reluctance of one air gap. Note that each coil contains N turns of

Figure 15.12 UR core magnetic analysis.

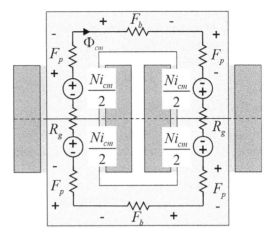

the upper rail winding and N turns of the lower rail winding for a total MMF of $N(i_u + i_l)$ or Ni_{cmi}, which has been split across the air gap out of a sense of esthetics. From Figure 15.12, we have

$$-2Ni_{cmi} + 2F_b + 4F_p + 2R_g \Phi_{cmi} = 0 \tag{15.6-1}$$

where the air-gap permeance (or reciprocal of reluctance) may be expressed as

$$P_g = \frac{1}{R_g} = \frac{\mu_0 \pi r_c^2}{g} + 2\mu_0 r_c \ln\left(1 + \frac{\pi \max(2d_s, w_s)}{2g}\right) \tag{15.6-2}$$

In (15.6-2), the first term represents a direct path and the second term represents a crude estimate of the fringing flux around the air gap based on "unrolling" the post into a linear segment of length $2\pi r_c$. Here, we can accept a crude estimate because the direct path dominates because of the typically very small air gap, which will be on the order of tens of microns. This is explored in the exercises (see Problem 5).

To proceed, we will assume we are given the parameters of the proxy flux-linkage waveform. We will create a time vector **t** with N_{pt} points whose elements span $[0 \ T_{sw}]$. Then, using (15.4-7) and (15.4-8), we will create a proxy flux-linkage vector, whose elements are the value of the proxy flux-linkage waveform λ_p at times corresponding to **t**. At those elements of time, we can compute the flux in the CMI

$$\Phi_{cmi} = \frac{1}{2N}\lambda_p \tag{15.6-3}$$

and then compute the flux density in the post and base

$$\mathbf{B}_p = \frac{1}{a_{mcp}}\Phi_{cmi} \tag{15.6-4}$$

$$\mathbf{B}_b = \frac{1}{a_{mcb}}\Phi_{cmi} \tag{15.6-5}$$

Next, analytically taking the time derivative of (15.4-7), we can compute a vector with values of the time derivative of proxy flux-linkage waveform $p\lambda_p$ and use analogs to (15.6-4)–(15.6-5) to find the time derivates of the flux linkage in the post and the base, denoted $p\mathbf{B}_p$ and $p\mathbf{B}_b$, respectively.

Once we have these derivatives, we will use a time-domain hysteresis model to compute the corresponding power loss density and field intensity in the post and the base. For example, if we use the EJA model encapsulated by (6.8-34),

$$[p_p, \mathbf{H}_p] = f_{EJA}(\mathbf{t}, \mathbf{B}_p, p\mathbf{B}_p, \mathbf{D}_{CR}) \tag{15.6-6}$$

$$[p_b, \mathbf{H}_b] = f_{EJA}(\mathbf{t}, \mathbf{B}_b, p\mathbf{B}_b, \mathbf{D}_{CR}) \tag{15.6-7}$$

where p_p and p_b are the power loss densities in the post and base, respectively, $\mathbf{H}_p$ and $\mathbf{H}_b$ are the time histories of the field intensities, and $\mathbf{D}_{CR}$ is a substructure of $\mathbf{D}$, which has core data, including the parameters of the hysteresis model.

The MMF drops across the post and base can then be found using

$$\mathbf{F}_p = \mathbf{H}_p l_{mpp} \tag{15.6-8}$$

$$\mathbf{F}_b = \mathbf{H}_b l_{mpb} \tag{15.6-9}$$

whereupon the time history of the common-mode current may be expressed as

$$\mathbf{i}_{cmi} = \frac{1}{N}\left(\mathbf{F}_b + 2\mathbf{F}_p + R_g \mathbf{\Phi}_{cmi}\right) \tag{15.6-10}$$

The rms common-mode current is then calculated as

$$i_{cmi,rms} = \sqrt{\frac{1}{T_{sw}} \int_0^{T_{sw}} i_{cmi}^2 \, dt} \tag{15.6-11}$$

which is evaluated numerically using points stored in $\mathbf{t}$ and $\mathbf{i}_{cmi}$.

Finally, we can compute the core loss as

$$P_c = 4p_p V_{mp} + 2p_b V_{mb} \tag{15.6-12}$$

It is convenient to encapsulate our magnetic analysis as

$$\mathbf{M} = F_M(\mathbf{G}, \mathbf{D}) \tag{15.6-13}$$

where $\mathbf{M}$ is a structure of magnetic analysis data given by

$$\mathbf{M} = \begin{bmatrix} i_{cmi,rms} & P_c & p_p & p_b \end{bmatrix} \tag{15.6-14}$$

Here, it is convenient to include p_p and p_b in addition to P_c in order to support thermal analysis.

15.7 Common-Mode Inductor Design Paradigm

In this section, we will develop a design paradigm for the CMI. Like our work in the previous chapter, we will not consider the inclusion of either a thermal model or capacitance. While these aspects of the design are critically important, this extension is left to the reader in the interest of brevity. In order to facilitate this extension, note that a complete capacitance model for the UR-core CMI is set forth in [5]. We will begin our discussion with the design space.

Design Space

A first step in the design process is to identify the degrees of freedom. In this work, we will focus primarily on the geometry and make material choices a priori in order to simplify the exposition. With this in mind, one choice of independent variables is

$$\boldsymbol{\theta} = \begin{bmatrix} g & r_c^* & d_c & N_{tpl}^* & N_l^* & N_{pr}^* & J^* & t_{ll}^* & c_{cc} & c_{cr} & c_{le} \end{bmatrix} \quad (15.7\text{-}1)$$

The reader will notice that the variables in $\boldsymbol{\theta}$ are all elements of $\mathbf{G}_I$ as defined by (15.5-1), except for J^*, the desired current density. We will relate J^* to the total desired conductor cross section a_{wcpr}^* by

$$a_{wcpr}^* = \frac{i_{dm,rms}}{J^*} \quad (15.7\text{-}2)$$

where $i_{dm,rms}$ is the rms value of the differential-mode current, which will be part of the design specifications stored in a structure $\mathbf{D}$. Since the differential-mode current is normally much larger than the common-mode current, it serves as an estimate for the rms winding current. The fields of $\mathbf{D}$ include

$$\mathbf{D} = \begin{bmatrix} f_{swa} & \lambda_p & \lambda_{p3} & \phi_{p3} & N_{tp} & i_{cm,arms} & k_{bd} & k_b & T_w & \mathbf{D}_{CD} & \mathbf{D}_{CR} \end{bmatrix} \quad (15.7\text{-}3)$$

Recall that f_{swa} (or $\omega_{swa} = 2\pi f_{swa}$), λ_p, λ_{p3}, and ϕ_{p3} are used to specify the proxy common-mode flux linkage. The variable N_{tp} is the number of points used to represent the proxy flux-linkage waveform in the time domain, and $i_{cm,arms}$ is the allowed rms current. The variables k_{bd} and k_b are the conductor bending ratio (bending radius/strand diameter) and build factor, and the assumed temperature of the winding is T_w. Finally, $\mathbf{D}_{CD}$ and $\mathbf{D}_{CR}$ are structures of conductor and core data, respectively. From a coding point of view, $\mathbf{D}$ may also contain other data such as wire gauge and insulating paper data.

Metrics

As in the previous chapter, in this chapter, we will take the geometric size defined by (15.5-19) to be our first metric to minimize. Our second metric to minimize will be loss. To this end, from the definition of the common-mode current (15.1-2) and differential-mode current (15.1-4), one can readily show that the current in the upper rail of the CMI (see Figure 15.9) is given by

$$i_u = i_{dm} + \frac{1}{2} i_{cmi} \quad (15.7\text{-}4)$$

where i_{dm} is the differential-mode current in the CMI. Assuming that the differential-mode current (predominantly dc) and the common-mode current (ac) are not correlated, the rms value of the current in the upper rail may be expressed as

$$i_{u,rms} = \sqrt{i_{dm,rms}^2 + \frac{1}{4} i_{cmi,rms}^2} \quad (15.7\text{-}5)$$

whereupon the total resistive loss (both windings of both coils) may be expressed as

$$P_r = 2 r_{cmi} i_{u,rms}^2 \quad (15.7\text{-}6)$$

Using the core loss from the magnetic analysis, the total loss may be expressed as

$$P_l = P_c + P_r \quad (15.7\text{-}7)$$

The reader may have issues with the fact that neither skin effect nor proximity effect losses have been included. While these could readily be added, the common-mode current density is very low, and so these components of loss tend to be very small compared to the dc-resistive loss, and so they are not included here for brevity.

Constraints

Now that we have set forth metrics, let us consider constraints. Because of the way we have constructed the geometry, for the problem as posed, we will only a single constraint, where the rms value of the common-mode current is less than the specified value, that is,

$$c_1 = \text{lte}(i_{cmi,rms}, i_{cmi,arms}) \tag{15.7-8}$$

If the approach here were extended to include the calculation of capacitance and to include a thermal equivalent circuit, it would be appropriate to place constraints on the resonant frequency (making sure it was well past the switching frequency) and the winding temperature.

Calculation of Fitness

In calculating the fitness, we will take the same approach as in the previous chapter and described by (14.5-16)–(14.5-19) without modification. At this point, it is appropriate to detail the steps required to evaluate the fitness. In particular, the steps to evaluate the fitness are as follows:

1) As a first step, we initialize the constraints satisfied c_S and the constraints imposed c_I to zero. Of course, with one constraint, this is probably excess formalism. However, it sets the stage for more comprehensive work.
2) Next, we calculate the desired wire cross section using (15.7-2).
3) The third step is the calculation of the geometrical calculations encapsulated by (15.5-26).
4) The magnetic analysis is then performed using (15.6-13).
5) If $c_S < c_I$, then the fitness is calculated using (14.5-18) and the fitness evaluation is complete. If this is not the case, we proceed to step 6.
6) Next, we compute the rms current in the winding using (15.7-5), the resistive loss using (15.7-6), and the total loss using (15.7-7).
7) Finally, we calculate fitness using the same fitness function as in Chapter 14, as given by (14.5-19).

In the next section, we will use this process in a case study.

15.8 Common-Mode Inductor Case Study

In the previous section, we considered a design paradigm for a UR-core common inductor. We will now use that process in a design example. In particular, we will consider the design of a UR-core CMI for a 5 kW, 400 V dc to 230 V l–l rms, 60 Hz ac inverter, with an assumed switching frequency of 19.98 kHz (20 kHz, but rounded to an integer number of switching cycles within a fundamental cycle). This case study is closely related to that of Chapter 14.

Design specifications are set forth in Table 15.1. Therein, the parameters of the proxy flux-linkage waveform come from Example 15.4C. The number of points used to represent the waveform in the

Table 15.1 UR-Core Common-Mode Inductor Specifications

Symbol	Description	Value
f_{swa}	Approximate switching frequency	19.98 kHz
λ_{p1}	Fund. component of proxy flux-linkage waveform	1.105 mVs
λ_{p3}	Third-harmonic component of proxy flux linkage	0.2228 mVs
ϕ_{p3}	Phase of third harmonic of proxy flux linkage	0 rad
N_{tp}	Number of temporal points used to represent proxy flux-linkage waveform	1000
$i_{cmi,arms}$	Allowed rms common-mode current	50 mA
$i_{dm,rms}$	RMS differential-mode current	13.5 A
k_{bd}	Cond. bend. radius constant (bend rad./strand diam.)	6
k_d	Winding build factor	1.05
T_w	Assumed winding temperature	100 °C
$\mathbf{D}_{CD}$	Structure of conductor data. Based on copper.	
$\mathbf{D}_{CR}$	Structure of core data. Based on MN67 ferrite.	

time domain, N_{pt}, is set to 1000 in order to well represent the waveform. As can be seen, the allowed rms common-mode current has been set to 50 mA. The assumed differential-mode current is 13.5 A; this represents 1 A of headroom over 12.5 A, which would correspond to 5 kW at 400 V. The assumed bending radius, build factor, and winding temperature are as those used in our case study of the previous chapter. The conductor data $\mathbf{D}_{CD}$ is based on copper; the insulation thickness t_{wi} is taken to be 32 μm. The core material is assumed to be MN67 ferrite, which was chosen rather arbitrarily; adding the core material to the search space is a rather natural extension of this example. Note that here we will need the parameters of the EJA model, as set forth in [5].

The search space is set forth in Table 15.2. Therein, many of the parameters are set via experimentation—making sure that the gene values do not become clustered on limits unless there is a physical limit. For example, the minimum air gap is set based on a reasonably worst case for air gap due to surface roughness. Other ranges are based on what is deemed reasonable based on experience. For example, the minimum coil-to-coil clearance was set to allow some provision for air flow between the coils. Before conducting a study, we may suspect that the thickness of the layer-to-layer insulation and clearances will go their minimum allowed value—in way of a spoiler, this will be the case. However, if we included capacitance calculations, there is a motivation to increase these quantities to decrease capacitance. If we included thermal analysis, there would be motivation to decrease the layer-to-layer insulation for better thermal transfer. As can be seen, the trade-offs can become quite convoluted.

Figure 15.13 illustrates the parameter distribution after a run. In this study, there were 1000 individuals and 1000 generations. The parameter numbers in Figure 13.12 correspond to the element number in (15.7-1) and gene number in Table 15.2. As can be seen, the air gap g (gene 1) went to the smallest allowed value. The desired post radius r_c^* (gene 2), the depth of core d_c (gene 3), the desired number of turns per layer N_{tpl}^* (gene 4), and the desired number of parallel conductors N_{pr}^* (gene 6) vary along the Pareto-optimal front. The desired number of layers N_l^* (gene 5) is loosely clustered—the loose clustering being due to the rounding operations. The target current density J_L^* (gene 7) is, not surprisingly, highly correlated with size. The thickness of the layer-to-layer insulation t_{ll}^*

15 Common-Mode Inductor Design

Table 15.2 UR-Core Three-Phase Inductor Design Space

Par.	Description	Min.	Max.	Enc.	Gene
g	Air gap (m)	10^{-5}	10^{-3}	log	1
r_c^*	Desired core (post) radius (m)	10^{-3}	10^{-1}	log	2
d_c	Depth of core (m)	10^{-3}	10^{-1}	log	3
N_{tpl}^*	Desired turns per layer	0.51	100.4	log	4
N_l^*	Desired number of layers	0.51	5.49	log	5
N_{pr}^*	Number of parallel conductors	0.51	20.49	log	6
J^*	Desired current density (A/m²)	10^5	10^7	log	7
t_{ll}^*	Desired thick. of the layer-to-layer ins. (m)	10^{-6}	10^{-2}	log	8
c_{cc}	Clearance from coil-to-coil (m)	10^{-2}	$5 \cdot 10^{-2}$	log	9
c_{wcw}	Clear. from coil-to-core in width (m)	10^{-4}	$5 \cdot 10^{-2}$	log	10
c_{wcd}	Clear. from coil-to-core in depth (m)	10^{-3}	$5 \cdot 10^{-2}$	log	11

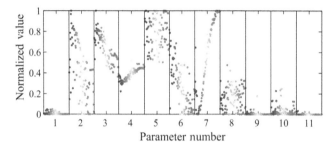

Figure 15.13 UR-core common-mode inductor parameter distribution.

(gene 8) is clustered toward lower values; the loose clustering is again because of the rounding operation. The various clearances all tend toward their minimum allowed value, as one would expect.

Figure 15.14 depicts the Pareto-optimal front between loss and size. As we have seen many times in our studies, there is clear trade-off between the competing objectives of size and loss. Within Figure 15.14, design 75 is highlighted. We will explore this design in some detail.

Details of design 75 are set forth in Table 15.3. As can be seen, the design occupies 0.18 l and has a mass of 0.52 kg. Observe that resistive loss dominates the total loss in this particular design. A profile and top view of the inductor are set forth in Figure 15.15.

The BH trajectory for this design, based on the proxy waveforms, is given in Figure 15.16. The reader will observe that the flux density is very low. This is associated with the undesired air gap, the very low common-mode current density (although it is "normal" in differential mode), and being a relatively low-loss design.

While the design is based on the proxy current waveform, it is interesting to investigate the response to the original flux-linkage waveform (and its derivative), which we considered in example 15.4A. This is shown in Figure 15.17. We see the same modulation in the common-mode current as we did in the common-mode flux linkage. With this excitation, the common-mode current has an

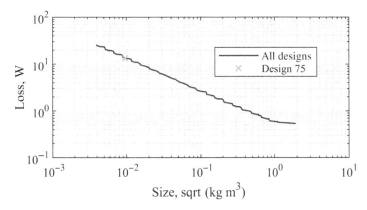

Figure 15.14 UR-core inductor Pareto-optimal front.

Table 15.3 Design 75

Aggregate Inductor	Winding Data
Dimensions	Wire diameter: 1.895 mm
Overall length: 3.070 cm	Wire insulation thickness: 33.02 μm
Overall width: 7.139 cm	Number of layers: 2
Overall depth: 8.149 cm	Turns of one winding per layer: 12
Encapsulating volume: 0.1786 l	Number of turns of one winding in one coil: 24
Total mass: 0.5230 kg	Number of parallel conductors per winding: 1
Size: 0.3056 sqrt (kg l)	Length of a strand of a winding in a coil: 2.006 m
	Layer-to-layer insulation thickness: 0 mm
Core Data	Winding mass per winding (both coils): 0.09369 kg
Core post radius: 1.037 cm	
Core base depth: 1.478 cm	Electrical Model
Slot depth: 2.596 cm	Assumed winding temperature: 100°C
Slot width: 1.996 cm	Winding resistance (r_{cmi}) at assumed temperature: 33.69 mΩ
UR-core mass (one of two): 0.1678 kg	
Spacing	Operating Data
Air gap: 10 μm	Common-mode current: 50 mA rms
Clearance: coil-to-coil: 1 cm	Winding current density: 5.139 A/mm² rms
Clearance: coil-to-core in width: 0.1 cm	Total loss: 13.45 W
Clearance: coil-to-core in depth: 0.01 cm	Resistive loss: 12.28 W
	Core hysteresis loss: 1.167 W
	Constraints
	Constraint report
	Constraint 1 (RMS current constraint) is satisfied
	Value/maximum allowed value = 99.9968%

rms value of 48.5 mA, somewhat smaller than the 48.9 mA calculated based on the proxy current waveform.

At this point, it is appropriate to reflect on the design process. The three chief additions needed for the design approach are (i) the addition of a thermal model, (ii) the addition of a capacitance model, and (iii) consideration of multiple operating points. These additions can be readily made using the material from Chapters 10, 12, and 13.

570 | 15 Common-Mode Inductor Design

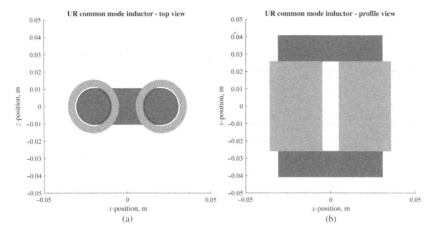

Figure 15.15 Design 75.

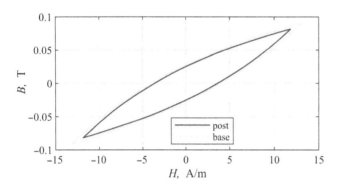

Figure 15.16 Design 75 BH trajectories based on proxy waveform.

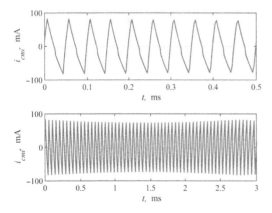

Figure 15.17 Design 75 BH trajectories using original flux-linkage waveform.

References

1 S. Sudhoff, R. Swanson, A. Kasha and V. Duppalli, "Magnetic analysis of symmetrical three-phase Y-core inductors," *IEEE Transactions on Energy Conversion*, vol. **33**, no. 2, pp. 576–583, 2018.

2 T. J. Donnelly, S. D. Pekarek, D. R. Fudge and N. Zarate, "Thévenin equivalent circuits for modeling common-mode behavior in power electronic systems," *IEEE Open Access Journal of Power and Energy*, vol. **7**, pp. 163–172, 20202.

3 P. Krause, O. Waynczuk, S. Sudhoff and S. Pekarek, *Analysis of Electric Machines and Drives Systems*, 3rd ed. Hoboken, NJ: IEEE Press and John Wiley & Sons, Inc., 2013.

4 S. D. Sudhoff, K. A. Corzine, S. F. Glover, H. J. Hegner and H. N. Robey, "DC link stabilized field oriented control of electric propulsion systems," *IEEE Transactions on Energy Conversion*, vol. **13**, no. 1, pp. 27–33, 1998.

5 S. D. Sudhoff, H. Singh, V. S. Duppalli and R. R. Swanson, "Capacitance of UR-core and C-core common mode inductors," *IEEE Power and Energy Technology Systems Journal*, vol. **6**, no. 2, pp. 113–121, 2019.

6 S. D. Sudhoff, "MATLAB Codes for Power Magnetic Devices: A Multi-Objective Design Approach, 2nd ed." [Online]. Available: http://booksupport.wiley.com.

Problems

1 Using (15.3-17)–(15.3-20) and the definitions of v_{cminv} and v_{cmdc}, show (15.3-21) and (15.3-22).

2 From (15.3-28) and (14.1-26), establish (15.3-29)–(15.3-31).

3 From (15.4-7), derive (15.4-9) and (15.4-10).

4 Derive (15.6-2).

5 Consider design 75 of Section 15.8. Assume that the core material is magnetically linear and has a relative permeability of $2.7 \cdot 10^4$. Further, suppose that the air gap is 10 μm. What fraction of the total reluctance of the magnetic path is due to the air-gap regions? What fraction of the air-gap permeance is associated with fringing?

6 Generate results analogous to Figure 15.7, (15.4-4), and (15.4-6) for sine-triangle modulation with displaced carrier waves.

7 Using the code from [6], repeat the case study of Section 15.8 if the air gap is completely removed. Compare the Pareto-optimal front of the case study with that obtained when the air gap is removed.

8 Using the code from [6], repeat the case study of Section 15.8 using a switching frequency of 60 kHz.

9 Starting with the code from [6], add a capacitance model based on [5]. Define the common-mode inductor's inductance as the peak value of the common-mode flux linkage over the peak

value of the common-mode current. Calling the resulting value of inductance L_{cmi} and the total capacitance C_{cmi}, add a constraint that the resonant frequency of L_{cmi} and C_{cmi} be at least 10 times higher than the switching frequency. Repeat the case study and compare the Pareto-optimal front so-obtained to that obtained from the case study.

16

Finite Element Analysis

In this book, much of the design work focused on the use of computationally efficient magnetic analysis such as magnetic equivalent circuits. However, sometimes it is difficult to apply such methods. Moreover, because such methods involve numerous approximations, it is important that any design is validated numerically before being constructed. Therefore, in this section, we will consider what is perhaps the most common method of numerical magnetic modeling—finite element analysis (FEA). There are many excellent references and books on this topic such as Salon [1], Silvester and Ferrari [2], and Aliprantis and Wasynczuk [3], to name a few. The goal here is to give the reader an introduction to, and flavor of, the approach. To this end, we will make several simplifications. First, we only consider the static (non-time-varying) conditions. Second, we only consider the two-dimensional (2D) case. Third, we restrict our attention to a formulation and implementation in Cartesian coordinates.

16.1 Maxwell's and Poisson's Equations

The goal of this section is to set the stage for numerical analysis of magnetic and electric fields. To this end, we start by considering Maxwell's equations that govern electromagnetics. For static conditions, these may be expressed as

$$\nabla \cdot \mathbf{D} = \rho_v \tag{16.1-1}$$

$$\nabla \cdot \mathbf{B} = 0 \tag{16.1-2}$$

$$\nabla \times \mathbf{E} = 0 \tag{16.1-3}$$

$$\nabla \times \mathbf{H} = \mathbf{J} \tag{16.1-4}$$

In (16.1-1)–(16.1-4), **B** and **H** are the magnetic flux density and field intensity, **D** and **E** are electric flux density and field intensity, **J** is current density, and ρ_v is the electric charge density. The operators • and × denote dot and cross products, respectively, and ∇ is the del operator which can be expressed in Cartesian coordinates as

$$\nabla = \left[\frac{\partial}{\partial x} \mathbf{a}_x \quad \frac{\partial}{\partial y} \mathbf{a}_y \quad \frac{\partial}{\partial z} \mathbf{a}_z \right]^T \tag{16.1-5}$$

where $\mathbf{a}_x$, $\mathbf{a}_y$, and $\mathbf{a}_z$ are unit vectors in the x-, y-, and z-directions, respectively.

We also consider the following constituatory relationships

$$\mathbf{B} = \mu \mathbf{H} \tag{16.1-6}$$

$$\mathbf{D} = \varepsilon \mathbf{E} \tag{16.1-7}$$

where μ is permeability, which may be a function of $\mathbf{B}$ or $\mathbf{H}$ as discussed in Chapter 2. It is also a function of position as materials change (from a core region to a free-space region). Note that as we saw in Sections 6.7 and 6.8, $\mathbf{B}$ is not actually a function of $\mathbf{H}$ in a mathematical sense because of the magnetic hysteresis, and so finite-element analysis has its own set of approximations.

As stated in the introduction, let us only consider the magnetostatic case in which the fields are temporally invariant. Based on (16.2-2), a vector magnetic potential $\mathbf{A}$ is defined such that

$$\mathbf{B} = \nabla \times \mathbf{A} \tag{16.1-8}$$

Substituting the right-hand side of (16.1-8) for $\mathbf{B}$ in (16.1-2) it can be seen that (16.1-2) is inherently satisfied.

Combining (16.1-4), (16.1-6), and (16.1-8), we obtain

$$\nabla \times (\gamma \nabla \times \mathbf{A}) = \mathbf{J} \tag{16.1-9}$$

where

$$\gamma = \frac{1}{\mu} \tag{16.1-10}$$

Expanding (16.1-9) in Cartesian coordinates and extracting the z-axis component we have

$$\frac{\partial}{\partial x}\left(\gamma\left(-\frac{\partial A_z}{\partial x} + \frac{\partial A_x}{\partial z}\right)\right) - \frac{\partial}{\partial y}\left(\gamma\left(\frac{\partial A_z}{\partial y} - \frac{\partial A_y}{\partial z}\right)\right) = J_z \tag{16.1-11}$$

where subscripts x, y, and z denote the x-axis, y-axis, and z-axis components of the respective vector quantity.

As mentioned in the introduction, we will consider the special case of a 2D analysis. In particular, we will assume that the geometry and fields are invariant in the z-axis, whereupon all partial derivatives with respect to z are zero, so that (16.1-11) reduces to

$$\frac{\partial}{\partial x}\left(\gamma \frac{\partial A_z}{\partial x}\right) + \frac{\partial}{\partial y}\left(\gamma \frac{\partial A_z}{\partial y}\right) = -J_z \tag{16.1-12}$$

If γ is spatially uniform (16.1-12) may be expressed in the form of Poisson's equation given by:

$$\frac{\partial^2 A}{\partial x^2} + \frac{\partial^2 A}{\partial y^2} = -\mu J \tag{16.1-13}$$

In (16.1-13) and henceforth in this discussion, we will omit the z subscript and it is understood that A and J are the z-axis component of the vector magnetic potential and current density, respectively. At this point, although A was introduced as a vector magnetic potential, since we are only looking at the z-axis component, it is a scalar field.

While the focus of this work is magnetics, in Chapter 12, we considered the impact of parasitic capacitance, and thus electrostatic analysis is also be of interest. To this end, we observe that for the electrostatic case the electric potential, or voltage, is related to the electric field by:

$$\mathbf{E} = -\nabla V \tag{16.1-14}$$

Substitution of (16.1-14) into (16.1-7) and the result into (16.1-3), it can be seen that (16.1-3) is satisfied. Substitution of (16.1-14) into (16.1-7) and the result into (16.1-1) we have

$$\frac{\partial^2 V}{\partial x^2} + \frac{\partial^2 V}{\partial y^2} + \frac{\partial^2 V}{\partial z^2} = -\frac{\rho_v}{\varepsilon} \tag{16.1-15}$$

Again, we arrive at Poisson's equation, though in this case in three dimensions.

As a third example, from (10.1-16), we see that for steady-state conditions, and using scaling, the heat equation can also be put into this form.

In solving Poisson's equation, we must also consider the boundary conditions. Two common types of boundary conditions are the Dirichlet boundary condition in which the scalar field value along a boundary is specified and the Neumann boundary condition in which the derivative of scalar field is specified.

In the case of magnetic systems, let us consider a Dirichlet condition, where $A = 0$ along some boundary. From the definition of vector magnetic potential, this implies that along the boundary, the component of **B** normal to the boundary is zero.

To see this, let us consider a local 2D coordinate system where $\mathbf{a}_n$ and $\mathbf{a}_t$ denote normal and tangential unit vectors at some point along a boundary. One may think of the normal direction as being a local x-direction and the tangential component as being a local y-direction. From (16.1-8) we have:

$$\mathbf{B} = \frac{\partial A}{\partial t}\mathbf{a}_n - \frac{\partial A}{\partial n}\mathbf{a}_t \tag{16.1-16}$$

In (16.1-16), t is position in the tangential direction, not time, as we are considering the static case. Likewise, n is the position in the normal direction. Both of these quantities are measured from some point of interest along the boundary.

From (16.1-16), if A is constant along a boundary, then its partial derivative in the direction parallel to the boundary, that is the tangential direction, is zero. Otherwise, it could not be constant along the boundary. Therefore, **B** cannot have a normal component and must be tangential to the boundary. In summary, for a Dirichlet condition, where $A = 0$ along a boundary, then along that boundary **B** only has a tangential component. We will refer to this case as a homogenous Dirichlet condition. For the electrostatic case, the Dirichlet condition involves assigning a known voltage to a boundary.

Now let us consider the Neumann boundary condition. For the magnetostatic case, we will consider the common version in which the normal derivative of A is set to zero along the boundary. We will also refer to this special case as a homogenous Neumann condition. From (16.1-16), we can see that for this boundary condition **B** will only have a normal component.

16.2 Finite Element Analysis Formulation

The finite element method can be derived via multiple ways. Herein, given that this book is heavily vested in optimization, we will derive the method as an optimization problem. We will focus on the static 2D electromagnetic case, but the reader understands the method is readily applied in three-dimensions over a wide-range of disciplines for both the static and dynamic (time-varying) analysis. There are numerous subtleties, and many careers have been devoted to the subject. The goal of this chapter is only to provide the reader with a brief introduction.

To develop the method, we pose our problem in several different ways. The first is the differential form where, from Section 16.1, we require

$$\frac{\partial}{\partial x}\left(\gamma \frac{\partial A}{\partial x}\right) + \frac{\partial}{\partial y}\left(\gamma \frac{\partial A}{\partial y}\right) = -J \text{ in } \Omega \tag{16.2-1}$$

with the boundary conditions as

$$A = 0 \text{ on } \Gamma_D \tag{16.2-2}$$

$$\frac{\partial A}{\partial n} = 0 \text{ on } \Gamma_N \tag{16.2-3}$$

In (16.2-1), Ω denotes the domain of the problem which is bounded by Γ. In (16.2-2) and (16.2-3) Γ_D and Γ_N denote those parts of the boundary Γ upon which the homogeneous Dirichlet and Neumann boundary conditions are employed. We will have $\Gamma = \Gamma_D \cup \Gamma_N$ and $\Gamma_D \cap \Gamma_N = \emptyset$, which is to say that every point on the boundary is subject to either a homogeneous Dirichlet or homogeneous Neumann condition, but not both simultaneously.

The boundary conditions are used in several ways. Suppose we are analyzing rectangular–cuboid core inductor as shown in Figure 16.1. Physically, for our analysis to be useful, the depth of this core into the page must be sufficient that what happens at the end of the core (in the directions into the page and out of the page) is negligible, since at the end regions, the fields are three dimensional in nature. If it were not for this requirement, the arrangement shown in Figure 16.1 could represent an I-core inductor; however, we will refrain from such a description since the aspect ratio required for a 2D field analysis is not consistent with a "normally" proportioned I-core.

Because of symmetry, we will only analyze the upper right-hand portion of the inductor. In this case, we expect that **B** will be directed vertically (tangentially) through the left/right centerline of the inductor, and so impose a Dirichlet condition along the dashed vertical line. Far from the inductor, we may take $A = 0$ and so also employ Dirichlet conditions along the dotted lines. On the top/bottom centerline of the inductor, we expect **B** is be normal to the solid horizontal line, and so here we impose a Neumann condition. Readers wishing to see an illustration of the field lines may peek ahead at Figure 16.6, though this will spoil the plot.

We will search for a solution of (16.2-1)–(16.2-3) such that A is within the Hilbert space of functions, and has (*i*) a square that can be integrated over the domain with a finite result, (*ii*) has a first derivative whose square can be integrated over the domain with a finite result, and (*iii*) satisfies the boundary conditions. We will call this space of functions F. Thus, $A \in F$.

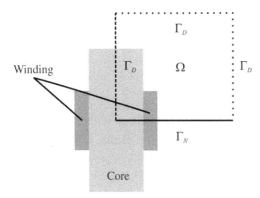

Figure 16.1 Selection of boundary conditions.

Our next step will be to derive the variational form of the problem. This will set the stage to turn the solution of the fields problem into an optimization problem. To this end, let us multiply (16.2-1) by a test function $v \in F$ and integrate over our domain. This yields:

$$\int_\Omega v \left[\frac{\partial}{\partial x}\left(\gamma \frac{\partial A}{\partial x}\right) + \frac{\partial}{\partial y}\left(\gamma \frac{\partial A}{\partial y}\right) \right] d\Omega = -\int_\Omega vJ\, d\Omega \quad (16.2\text{-}4)$$

Now, let us break our domain into two subdomains, Ω_1 and Ω_2, respectively, with boundary Γ_B between them. This is done so that γ is constant over each subdomain. Clearly, this could be extended to any number of subdomains, but for purposes of illustration we consider just two. We will require $\Omega = \Omega_1 \cup \Omega_2$ and $\Omega_1 \cap \Omega_2 = \emptyset$. In domain 1, we will have $\gamma = \gamma_1$ and in domain 2, we will have $\gamma = \gamma_2$, where γ_1 and γ_2 are constants. Domains 1 and 2 correspond to two regions of different but constant permeability. With this division, and simplifying using our requirement that γ is constant in each domain (16.2-4) becomes

$$\int_{\Omega_1} \gamma_1 v \nabla^2 A\, d\Omega + \int_{\Omega_2} \gamma_2 v \nabla^2 A\, d\Omega = -\int_\Omega vJ\, d\Omega \quad (16.2\text{-}5)$$

where $\nabla^2 A = \nabla \cdot \nabla A$ denotes the Laplacian of A.

Let us denote Γ_1 and Γ_2 as the boundary of each of the two subdomains. Further Γ_{1D} and Γ_{1N} will be those parts of Γ_1 on which Dirichlet and Neumann conditions are imposed, respectively. Similar nomenclature is used for subdomain 2. The boundary between the two domains will be denoted as Γ_B. These boundaries are illustrated in Figure 16.2.

From the chain rule, if we have two scalar fields g and f, then:

$$f\nabla^2 g = \nabla \cdot (f\nabla g) - \nabla f \cdot \nabla g \quad (16.2\text{-}6)$$

Applying (16.2-6) to (16.2-5) yields:

$$\begin{aligned}
&\int_{\Omega_1} \gamma_1 \nabla \cdot (v\nabla A) d\Omega - \int_{\Omega_1} \gamma_1 \nabla v \cdot \nabla A\, d\Omega + \\
&\int_{\Omega_2} \gamma_2 \nabla \cdot (v\nabla A) d\Omega - \int_{\Omega_2} \gamma_2 \nabla v \cdot \nabla A\, d\Omega = -\int_\Omega vJ\, d\Omega
\end{aligned} \quad (16.2\text{-}7)$$

Figure 16.2 Two domain problem.

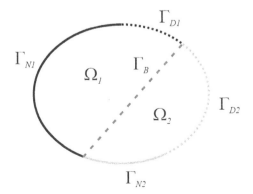

The divergence theorem relates volume and surface integrals of a field. In particular, the divergence theorem may be expressed as:

$$\int_V \nabla \cdot \mathbf{X}\, dV = \int_S \mathbf{X} \cdot d\mathbf{S} \tag{16.2-8}$$

In (16.2-8), V is a volume enclosed by surface $\mathbf{S}$, dv and $d\mathbf{S}$ are incremental volume and incremental surface, respectively, and $\mathbf{X}$ is a vector field.

Applying the divergence theorem to the first and third terms in (16.2-7) with $\mathbf{X} = v \nabla A$, noting that the normal component of $v \nabla A$ is $v\partial A/\partial n$, that the surface integral is an integral over the boundary, and that in the 2D case our surface integral becomes a line integral, we have:

$$\int_{\Gamma_{1D}+\Gamma_{1N}+\Gamma_B} \gamma_1 v \frac{\partial A}{\partial n} d\Gamma - \int_{\Omega_1} \gamma_1 \nabla v \cdot \nabla A\, d\Omega +$$
$$\int_{\Gamma_{2D}+\Gamma_{2N}+\Gamma_B} \gamma_2 v \frac{\partial A}{\partial n} d\Gamma - \int_{\Omega_2} \gamma_2 \nabla v \cdot \nabla A\, d\Omega = -\int_\Omega vJ\, d\Omega \tag{16.2-9}$$

Let us consider the boundary integral terms in (16.2-9). First, those portions of the boundary integrals along Γ_{1D} and Γ_{2D} must be zero because $v \in F$ and must satisfy the Dirichlet condition. Next, the portion of the integrals along Γ_{1N} and Γ_{2N} must be zero because of the Neumann boundary condition. Finally, using (16.1-10), (16.1-16), applying Ampere's law around a thin loop closely straddling Γ_B, and remembering that Γ_B is traversed in opposite directions when going around Ω_1 and Ω_2 (for example, so the integration is in the counterclockwise direction in both the subdomains), it can be shown that

$$\int_{\Gamma_B} \gamma_1 v \frac{\partial A}{\partial n} d\Gamma + \int_{\Gamma_B} \gamma_2 v \frac{\partial A}{\partial n} d\Gamma = 0 \tag{16.2-10}$$

Showing this is the topic of Problem 4 at the end of the chapter. Using (16.2-10), (16.2-9) reduces to:

$$\int_\Omega \gamma \nabla v \cdot \nabla A\, d\Omega = \int_\Omega vJ\, d\Omega \tag{16.2-11}$$

Next, let us denote the left-hand side of (16.2-11) by:

$$a(u,w) = \int_\Omega \gamma \nabla u \cdot \nabla w\, d\Omega \tag{16.2-12}$$

and the right-hand side by:

$$b(w) = \int_\Omega wJ\, d\Omega \tag{16.2-13}$$

From (16.2-11)–(16.2-13), we can pose the variational formulation to our problem, namely, find a function $A \in F$ such that

$$a(A,v) = b(v) \quad \forall v \in F \tag{16.2-14}$$

16.2 Finite Element Analysis Formulation

The variational formulation to our problem can be shown to be equivalent to the following minimization problem. In particular, find $A \in F$ such that

$$f(A) < f(v) \quad \forall v \in F/A \tag{16.2-15}$$

where

$$f(u) = \frac{1}{2}a(u,u) - b(u) \tag{16.2-16}$$

In order to see that (16.2-14) and (16.2-15) are equivalent, let us first show that (16.2-14) implies (16.2-15). To this end suppose A satisfies (16.2-14). Now let

$$v = A + w \tag{16.2-17}$$

where $w \in F$. From (16.2-16),

$$f(v) = \frac{1}{2}a(A+w, A+w) - b(A+w) \tag{16.2-18}$$

which, from (16.2-12) and (16.2-13) and manipulating, can be expressed as

$$f(v) = \left[\frac{1}{2}a(A,A) - b(A)\right] + [a(A,w) - b(w)] + \frac{1}{2}a(w,w) \tag{16.2-19}$$

The first term in brackets is $f(A)$. From (16.2-14), the second term in brackets must be zero. From (16.2-12), the final term is greater than or zero for $w \neq 0$. Thus,

$$f(v) > f(A) \tag{16.2-20}$$

for all $v \neq A$, and so we have established that if A satisfies (16.2-14), then it also satisfies (16.2-15).

Our next step is to show that the converse is true, which is to say that if A satisfies (16.2-15), then it will also satisfy (16.2-14). To this end, let us define a scalar function

$$g(\varepsilon) = f(A + \varepsilon v) \tag{16.2-21}$$

where A satisfies (16.2-15), v is any function in F, and ε is a real scalar. From (16.2-16), (16.2-12), and (16.2-13) we can show that

$$g(\varepsilon) = \frac{1}{2}a(A,A) + \varepsilon a(A,v) + \frac{1}{2}\varepsilon^2 a(v,v) - b(A) - \varepsilon b(v) \tag{16.2-22}$$

from which it follows that

$$\frac{dg(\varepsilon)}{d\varepsilon} = a(A,v) + \varepsilon a(v,v) - b(v) \tag{16.2-23}$$

From (16.2-15), we know that $g(\varepsilon)$ has a minimum at $\varepsilon = 0$. Therefore,

$$\frac{dg(0)}{d\varepsilon} = a(A,v) - b(v) = 0 \tag{16.2-24}$$

Therefore, from (16.2-24), we can see that (16.2-15) implies (16.2-14). Since we previously showed that (16.2-14) implies (16.2-15), the two formulations are equivalent.

The solution to (16.2-15) and (16.2-16) represents an exact solution to the original partial differential equation. In the next section, we will seek an approximate solution wherein we break up our domain Ω into a large number of triangular subdomains. This step is the heart of the FEA.

16.3 Finite Element Analysis Implementation

Rather than solving the exact minimization problem (16.2-15), let us solve a discretized equivalent. In particular let us, as an approximate solution, find $A_{pl} \in F_{pl}$ such that

$$f(A_{pl}) < f(\nu_{pl}) \quad \forall \nu_{pl} \in F_{pl} \tag{16.3-1}$$

where F_{pl} is the subset of F consisting of piecewise linear functions.

To this end, let us break our domain Ω into N_t triangular-shaped subdomains. The kth domain will be denoted Ω_k. The vertices of these triangles are nodes, and we will require that no node of one triangle can lie on the side of the another. The number of nodes denoted will be N_n. The value of A_{pl} (or ν_{pl}) on any one of these triangles will be the linear interpolate of its node values. We will also require that J and γ to be constant over every Ω_k. Thus, on over domain Ω_k their values will be denoted as J_k and γ_k, respectively.

With the division of the domain into subdomains, we may calculate $f(\nu_{pl})$ as

$$f(\nu_{pl}) = \sum_{k=1}^{N_T} f_k(\nu_{pl}) \tag{16.3-2}$$

where

$$f_k(\nu_{pl}) = f_{ak}(\nu_{pl}) - f_{bk}(\nu_{pl}) \tag{16.3-3}$$

In (16.3-3), from (16.2-12) and (16.2-13) we have

$$f_{ak}(\nu_{pl}) = \frac{1}{2}\int_{\Omega_k} \gamma_k \nabla \nu_{pl} \cdot \nabla \nu_{pl} d\Omega \tag{16.3-4}$$

and

$$f_{bk}(\nu_{pl}) = \int_{\Omega_k} \nu_{pl} J_k d\Omega \tag{16.3-5}$$

Before proceeding further, it is perhaps beneficial to pause in our development to consider how we will organize our problem from a data structure point of view. First consider Figure 16.3. Therein triangular domain Ω_k is shown. The domain has three vertices named L, M, and N. Vertex L will correspond to node l of a node list, vertex M to node m of that list, and vertex N to the node n of the same list. The coordinates of node l are (x_l, y_l), and the coordinates of the other nodes are

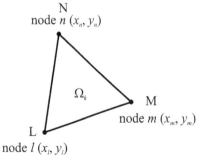

Figure 16.3 Triangular element.

Figure 16.4 Simple domain.

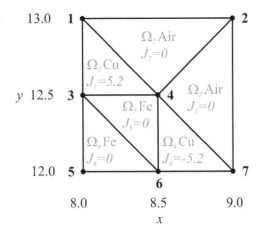

Table 16.1 Domain list for Figure 16.4

k	n	l	m	Material	J (A/mm²)
1	1	3	4	Cu	5.2
2	1	4	2	Air	0
3	4	7	2	Air	0
4	4	6	7	Cu	−5.2
5	3	6	4	Fe	0
6	3	5	6	Fe	0

denoted likewise. We will maintain a list or data structure of domains, and for each domain Ω_k, we will specify nodes l, m, n, the material that each domain is made of (essentially specifying γ_k), as well as the current density in that domain J_k. We will also maintain a second data structure which is a node list. For each node, the x- and y-coordinates of that node will be specified, and its status as a free node will be specified (it is free if not specified by a homogenous Dirichlet condition).

As an example, consider Figure 16.4, which shows a very simple domain. The domain or triangle list for this system is shown in Table 16.1. Therein, Cu and Fe denote copper and iron. The node list is set forth in Table 16.2. As can be seen, in this elementary example, nodes 1, 2, 3, and 5 are fixed to zero by a Dirichlet condition.

With this arrangement of our data, let us consider the evaluation of (16.3-4) and (16.3-5) over a domain Ω_k. On this domain, we will represent our piecewise linear function as

$$\nu_{\mathrm{pl}}(x,y) = \nu_l \alpha_{Lk}(x,y) + \nu_m \alpha_{Mk}(x,y) + \nu_n \alpha_{Nk}(x,y) \tag{16.3-6}$$

where ν_l, ν_m, and ν_n are the values of ν_{pl} at the nodes corresponding to the three vertices, and $\alpha_{Lk}(x, y)$ is a linear basis function which takes on a value of 1 at vertex L corresponding to node l, but is zero at the two vertices of Ω_k. The other two basis functions have analogous properties.

It is convenient to express the basis function $\alpha_{Ok}(x, y)$ in the form

$$\alpha_{Ok}(x,y) = \frac{1}{2\Delta_k}(p_{Ok} + q_{Ok}x + r_{Ok}y) \tag{16.3-7}$$

Table 16.2 Node list for Figure 16.4

n	x_n	y_n	Free
1	8.0	13.0	0
2	9.0	13.0	0
3	8.0	12.5	0
4	8.5	12.5	1
5	8.0	12.0	0
6	8.5	12.0	1
7	9.0	12.0	1

Table 16.3 Basis function coefficients

$p_{Lk} = x_m y_n - x_n y_m$	$q_{Lk} = y_m - y_n$	$r_{Lk} = x_n - x_m$
$p_{Mk} = x_n y_l - x_l y_n$	$q_{Mk} = y_n - y_l$	$r_{Mk} = x_l - x_n$
$p_{Nk} = x_l y_m - x_m y_l$	$q_{Nk} = y_l - y_m$	$r_{Nk} = x_m - x_l$

where $O \in [L, M, N]$ and

$$\Delta_k = \frac{1}{2}[(x_m - x_l)(y_n - y_l) - (y_m - y_l)(x_n - x_l)] \tag{16.3-8}$$

The absolute value of Δ_k is the area of the triangular domain. Showing this is the subject of Problem 5. The values of p_{Ok}, q_{Ok}, and r_{Ok} are specified in Table 16.3.

In order to derive these values, consider vertex L. We require $\alpha_{Lk}(x_l, y_l) = 1$, $\alpha_{Lk}(x_m, y_m) = 0$, and $\alpha_{Lk}(x_n, y_n) = 0$, from which we obtain three linear equations in three unknowns, p_{Lk}, q_{Lk}, and r_{Lk}. Solving the resulting system using Cramer's rule yields the values of p_{Lk}, q_{Lk}, and r_{Lk} given in Table 16.3. This is the topic of Problem 6. The values of p, q, and r for the other vertices may be likewise established, or alternatively established through appropriate relabeling of indices.

From (16.3-6) and (16.3-7) we have

$$\nabla \nu_{pl} = \frac{(\nu_l q_{Lk} + \nu_m q_{Mk} + \nu_n q_{Nk})\mathbf{a}_x + (\nu_l r_{Lk} + \nu_m r_{Mk} + \nu_n r_{Nk})\mathbf{a}_y}{2\Delta_k} \tag{16.3-9}$$

Substitution of (16.3-9) into (16.3-4) and evaluating yields

$$f_{ak} = \frac{1}{2}\boldsymbol{\nu}_{lmn}^T \mathbf{S}_k \boldsymbol{\nu}_{lmn} \tag{16.3-10}$$

In (16.3-10), $\mathbf{S}_k$ is the elementary stiffness matrix given by

$$\mathbf{S}_k = \frac{\gamma_k}{4\Delta_k}\begin{bmatrix} q_{Lk}q_{Lk} + r_{Lk}r_{Lk} & q_{Lk}q_{Mk} + r_{Lk}r_{Mk} & q_{Lk}q_{Nk} + r_{Lk}r_{Nk} \\ q_{Mk}q_{Lk} + r_{Mk}r_{Lk} & q_{Mk}q_{Mk} + r_{Mk}r_{Mk} & q_{Mk}q_{Nk} + r_{Mk}r_{Nk} \\ q_{Nk}q_{Lk} + r_{Nk}r_{Nk} & q_{Nk}q_{Mk} + r_{Nk}r_{Nk} & q_{Nk}q_{Nk} + r_{Nk}r_{Nk} \end{bmatrix} \tag{16.3-11}$$

and

$$\boldsymbol{\nu}_{lmn}^T = \begin{bmatrix} \nu_l & \nu_m & \nu_n \end{bmatrix}^T \tag{16.3-12}$$

Next, substitution of (16.3-6) into (16.3-5) yields

$$f_{bk} = w_k [1 \quad 1 \quad 1] \boldsymbol{\nu}_{lmn} \tag{16.3-13}$$

where

$$w_k = \frac{J_k |\Delta_k|}{3} \tag{16.3-14}$$

Combining (16.3-3), (16.3-10), and (16.3-13) we have

$$f_k = \frac{1}{2} \boldsymbol{\nu}_{lmn}^T \mathbf{S}_k \boldsymbol{\nu}_{lmn} - w_k [1 \quad 1 \quad 1] \boldsymbol{\nu}_{lmn} \tag{16.3-15}$$

If we substitute (16.3-15) into (16.3-2) we obtain the system

$$f = \frac{1}{2} \boldsymbol{\nu}^T \mathbf{S} \boldsymbol{\nu} - \mathbf{W}^T \boldsymbol{\nu} \tag{16.3-16}$$

where $\mathbf{v}$ is the vector of all node potentials.

The algorithm to determine $\mathbf{S}$ and $\mathbf{W}$ is straightforward. We begin by initializing $\mathbf{S} = \mathbf{0}$ and $\mathbf{W} = \mathbf{0}$ where $\mathbf{S}$ is a N_n by N_n matrix and $\mathbf{W}$ is N_n by 1 vector. Then, for $k = 1$ to $k = N_n$ we do the following. First, we look up the l, m, and n node numbers for domain k from the triangle list. Next, we increment the l,l element of $\mathbf{S}$ by the L,L element of $\mathbf{S}_k$, the l, m element of $\mathbf{S}$ by the L,M element of $\mathbf{S}_k$, and so forth, incrementing 9 elements in all. Then, the l, m, and n elements of $\mathbf{W}$ are incremented by w_k.

After the procedure of the previous paragraph, we may impose a Dirichlet condition. To apply a Dirichlet condition on node $\mathbf{v}_j$ the jth column of $\mathbf{S} = \mathbf{0}$ is set to zero. Then $\mathbf{S}_{jj}$ is set to one, and the jth row (element) of $\mathbf{W}$ is set to zero. This procedure is repeated for all nodes fixed by Dirichlet conditions, and will cause the fixed potentials to be equal to zero when the system (16.3-16) is solved. This can be seen by noting that the procedure just described decouples $\mathbf{v}_j$ from the rest of the node potentials in (16.3-16), and therefore minimizing (16.3-16) will require $\mathbf{v}_j = 0$. All boundaries upon which Dirichlet conditions are not imposed can be shown to automatically satisfy the Neumann boundary condition.

Once, we have formed $\mathbf{S}$ and $\mathbf{W}$, our goal is to minimize f in (16.3-16). We recognize that (16.3-16) is a quadratic form. Thus, its solution is given by solving

$$\mathbf{S} \boldsymbol{\nu} = \mathbf{W} \tag{16.3-17}$$

where upon the solution $\mathbf{v}$, which represents the z-axis component of the vector magnetic potential of the nodes, is known.

There are several points of interest in this process. First, for computational reasons, it should be recognized that $\mathbf{S}$ is symmetric. Secondly, it is important to observe that $\mathbf{S}$ will be sparse and so sparse matrix data representation and sparse matrix solvers should be used.

The description of this method has thus far assumed linearity. However, the nonlinear case is readily addressed. In this case, the method becomes iterative. The most straightforward approach is using a Gauss–Seidel iteration wherein the FEA is conducted with constant permeability. Then, based on the flux density in each element, the permeability of that element is adjusted, and the process is repeated until the solution converges in every element. Since, the Gauss–Seidel technique converges slowly in the neighborhood of the solution, other techniques (such as the Newton–Raphson method) are often used.

Once (16.3-17) has been solved, postprocessing commences to calculate quantities of interest. As an example, it is often desired to compute the flux lines. Flux lines are parallel to the flux density. To understand the computation of flux lines, from (16.1-8) with A only having a z-axis component,

$$\mathbf{B} = \frac{\partial A}{\partial y}\mathbf{a}_x - \frac{\partial A}{\partial x}\mathbf{a}_y \tag{16.3-18}$$

Now, consider a level set of A. A level set satisfies

$$A(x,y) = A_o \tag{16.3-19}$$

where A_o is a constant corresponding to level set o. Let $(x_o(s), y_o(s))$ represent parameterized curve in s which describes the level set. Taking the derivative of (16.3-19) with respect to s we have:

$$\frac{\partial A}{\partial x_o}\frac{\partial x_o}{\partial s} + \frac{\partial A}{\partial y_o}\frac{\partial y_o}{\partial s} = 0 \tag{16.3-20}$$

The tangent vector to the level set is given by:

$$\mathbf{T} = \frac{\partial x_o}{\partial s}\mathbf{a}_x + \frac{\partial y_o}{\partial s}\mathbf{a}_y \tag{16.3-21}$$

Next, let us take the cross product of **B** and **T**. From (16.3-18) and (16.3-21), we have

$$\mathbf{B} \times \mathbf{T} = \frac{\partial A}{\partial x_o}\frac{\partial x_o}{\partial s} + \frac{\partial A}{\partial y_o}\frac{\partial y_o}{\partial s} \tag{16.3-22}$$

which, from (16.3-20) is zero. We conclude that **B** and **T** are parallel and so flux lines are given by the level sets of A.

Computationally then, to draw a level set, we go through the triangle list and find those domains for which min $(v_l, v_m, v_n) \leq A_o \leq$ max (v_l, v_m, v_n). For each of those domains, a line segment representing the value of the level set is drawn. Expressions for the end points of these line segments can be derived from (16.3-6) and (16.3-7).

Example 16.3A In this example, we plot the field lines for a rectangular-cuboid core inductor such as is depicted in Figure 16.1. The rectangular cuboid is 10 cm in length (the vertical direction in the figure), and 4 cm width (the horizontal direction in the figure), and is constructed of MN67 ferrite (see Appendix B). Around the core is a winding bundle, each side of which occupies 4 cm in the vertical direction and 1 cm in the horizontal direction. The current density in the winding bundle is 5 A/mm^2. Because of the symmetry, only the first quadrant of the problem is considered. Part of the simple mesh is shown in Figure 16.5. The mesh has 1600 and 841 nodes. The mesh extends to 0.2 m in both the x- and y-directions, but the figure has been cropped for better visibility.

From the symmetry of the problem, we expect the field lines to be parallel to the y-axis at $x = 0$. Therefore, a Dirichlet condition is imposed on the line $x = 0$. A Dirichlet condition is also imposed on lines at $y = 0.2$ m and $x = 0.2$ m, where the field values should have decayed. Again by symmetry, we would expect the field lines to be normal to the x-axis. Thus, we impose a Neumann condition on the boundary $y = 0$. The resulting field lines, calculated using the method just described, are shown in Figure 16.6. Observe that they are tangent to the line at $x = 0$ and normal to the line at $y = 0$, which is consistent with the imposed boundary conditions.

Besides the computation of field lines and field values, another common postprocessing step is the computation of inductance. Let us consider the problem of using FEA to find the self-inductance of an inductor. There are several ways this can be done. Regardless of the approach,

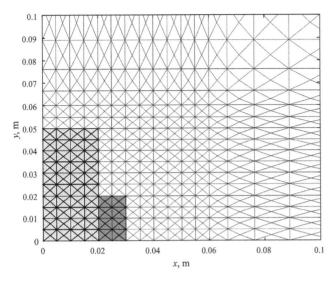

Figure 16.5 Rectangular–cuboid core inductor mesh.

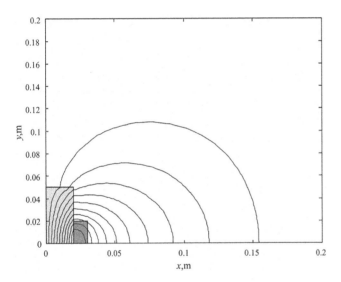

Figure 16.6 Rectangular–cuboid core inductor fields.

it is useful to note that for a given excitation current, the resulting flux density in each triangular subdomain is readily computed using (16.3-18) and is spatially constant within that domain.

Let us consider an energy-based approach. It is instructive to begin with the magnetically linear case. To use this approach, let us excite the system with a test current i_t which results in a flux density magnitude $B_{t,k}$ in domain k. From (2.7-16), we note that the volumetric energy density stored in the kth domain is given by

$$w_{f,k} = \int_0^{B_{t,k}} H_k dB_k \qquad (16.3\text{-}23)$$

Here, we have used the $w_{f,k}$ notation to remind ourselves that this is field energy (see Chapter 4). For the magnetically linear case $H_k = \gamma_k B_k$ and so the total field energy may be expressed as

$$W_f = \frac{1}{2}\sum_k |\Delta_k| \gamma_k B_{t,k}^2 \tag{16.3-24}$$

Observe that in the 2D case, this is actually field energy per unit length. From Chapter 4 and in particular (4.7-8) and (4.7-10), we have that for magnetically linear conditions

$$W_f = \frac{1}{2} L i_t^2 \tag{16.3-25}$$

Thus, comparing (16.3-24) and (16.3-25), the inductance (per unit length) may be computed as

$$L = \frac{\sum_k (|\Delta_k| \gamma_k B_{t,k}^2)}{i_t^2} \tag{16.3-26}$$

Now, let us consider the nonlinear case. To this end, using the definition of co-energy from Chapter 4 and following a derivation analogous to that of (2.7-16) it can be readily established that the co-energy density of the kth domain may be expressed as

$$w_{c,k} = \int_0^{H_{t,k}} B_k dH_k \tag{16.3-27}$$

where $H_{t,k}$ is the magnitude of the field intensity in domain k due to current i_t.

With this definition, consider the following experiment. We conduct a number of FEA analysis gradually increasing the current from 0 to i_t. As we do this, the magnitude of the flux density and field intensity in the kth domain gradually increases from 0 to $B_{t,\,k}$ and from 0 to $H_{t,\,k}$, respectively. The total field and co-energy may then be expressed as

$$W_f = \sum_k \left(|\Delta_k| \int_0^{B_{t,k}} H_k dB_k \right) \tag{16.3-28}$$

$$W_c = \sum_k \left(|\Delta_k| \int_0^{H_{t,k}} B_k dH_k \right) \tag{16.3-29}$$

The integrations in (16.3-28)–(16.3-29) are numerically evaluated over the field trajectories which occur as the current is increased from 0 to i_t. From (4.4-8), the flux linkage corresponding to the test current may be calculated as

$$\lambda_t = \frac{W_f + W_c}{i_t} \tag{16.3-30}$$

From this the inductance may be expressed as

$$L_t = \frac{\lambda_t}{i_t} \tag{16.3-31}$$

While we have formulated this process as if we were finding the flux linkage and inductance at a single point, by consideration of the process we can, as a biproduct, establish the flux linkage

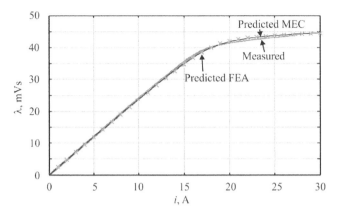

Figure 16.7 FEA predictions for $\lambda - i$ characteristic.

and inductance for every value of current considered in our trajectory of increasing the current from 0 to i_t.

Example 16.3B In this example, we will once again consider our UI-core inductor whose parameters are given in Figure 2.4. In particular, the $\lambda - i$ characteristics are determined using a 2D FEA analysis. The results are shown in Figure 16.7. As can be seen, the predictions of the 2D FEA analysis (shown as x's in the figure) are virtually identical to the predictions of the Magnetic Equivalent Circuit (MEC), and both models are reasonably close to the observed behavior. In this example, the device is relatively long, and so the results of a three-dimensional (3D) FEA were similar to those of the 2D FEA, except where the device begins to saturate. In this region, numerical issues arose with the (3D) model.

16.4 Closing Remarks

In this chapter, the FEA of magnetic fields was set forth. This chapter served to merely introduce the method; the reader is referred to the works of Salon [1], Silvester and Ferrari [2], and Aliprantis and Wasynczuk [3] for a thorough treatment of this topic. It should be observed that this method is supported by many commercial as well as open-source codes.

In this text, we have focused on using FEA as a means of design validation (particularly using 3D FEA), one should mention that it is certainly possible to use FEA as the computational engine for design, especially for problems wherein the magnetic analysis is particularly difficult. One such example, involving an internal rotor flux-modulation, PM machine is set forth in [4]. The penalty is computational speed, which has the net impact that the number of degrees of freedom must sometimes be reduced. Therefore, such an approach will invariably require the use of parallel computing, which is facilitated by the use of population-based optimization.

There are alternative methods of numerical magnetic analysis. For example, for 2-D analysis, the method of moments is particularly promising. A few examples of this method are set forth in [5] [6].

References

1 S. J. Salon, *Finite Element Analysis of Electrical Machines, first edition*. United States: Springer US, 1995.
2 P. P. Silvester and R. L. Ferrari, *Finite Elements for Electrical Engineers*, Cambridge: Cambridge University Press, 1983.
3 D. Aliprantis and O. Wasynczuk, *Electric Machines - Theory and Analysis Using the Finite Element Method*, Cambridge: Cambridge University Press, In Press.
4 J. M. Crider and S. D. Sudhoff, An Inner Rotor Flux-Modulated Permanent Magnet Synchronous Machine for Low-Speed High-Torque Applications, *IEEE Transactions on Energy Conversion*, vol. **30**, no. 3, pp. 1247–1254, 2015.
5 R. Howard and S. Pekarek, Two-Dimensional Galerkin Magnetostatic Method of Moments, *IEEE Transactions on Magnetics, vol. 53, no. December*, pp. 1–6, 2017.
6 D. Horvath, S. Pekarek and R. Howard, Analysis and Design of Electric Machines Using 2D Method of Moments, *IEEE International Electric Machines & Drives Conference (IEMDC)*, San Diego, CA, USA, 2019.

Problems

1 Using expansion in Cartesian coordinates, show that (16.1-8) satisfies (16.1-2).

2 From (16.1-9) show that we obtain (16.1-11).

3 From (16.1-8) show that we obtain (16.1-16).

4 Show (16.2-10).

5 Show that the area of a triangle domain k is given by (16.3-8).

6 Using (16.3-7) and (16.3-8) show that we obtain the first row of Table 16.3.

7 Consider the domain list in Table 16.1 and the corresponding node list in Table 16.2. Determine **S** and **W** after the first domain has been processed using the procedure described in the first full paragraph following (16.3-16).

Appendix A

Conductor Data and Wire Gauges

Data on conductor materials is listed in Table A.1. This is based on data compiled from references [1–5]. Each materials' conductivity may be approximated as

$$\sigma = \frac{\sigma_0}{1 + \alpha_T(T - T_0)} \tag{A-1}$$

where T is the temperature in °C, σ_0 is the resistive at $T = T_0$, and α_T is the temperature coefficient of resistivity. All the parameters σ_0, T_0, and α_T are listed in Table A.1. Therein, ρ is the mass density of the material and k is the thermal conductivity.

Wire gauges were originally established by the number of times the conductor was passed through a die. Each successive die lowered its diameter; and thus as the gauge goes up, the wire diameter goes down. There are several wire standards, including the American wire gauge (AWG) and the standard wire gauge (SWG). Conductor diameters for the AWG are listed in Table A.2, which is adapted from Herrington [6]. Therein, d_c denotes the diameter of the conductor.

Table A.1 Conductor Data

Material	σ_0, M℧/m	T_0, °C	α_T 1/°C	ρ, kg/m³	k, W/°km
Copper	59.6	20	$3.93 \cdot 10^{-3}$	8890	385
Aluminum	37.7	20	$3.94 \cdot 10^{-3}$	2705	205
Silver	63.0	20	$3.80 \cdot 10^{-3}$	10499	406
Gold	45.2	20	$3.40 \cdot 10^{-3}$	19301	314

Table A.2 American Wire Gauge Wire Diameters

Gauge	d_c (mm)
4/0	11.68
3/0	10.40
2/0	9.266
1/0	8.252
1	7.348

(*Continued*)

Power Magnetic Devices: A Multi-Objective Design Approach, Second Edition. Scott D. Sudhoff.
© 2022 The Institute of Electrical and Electronics Engineers, Inc. Published 2022 by John Wiley & Sons, Inc.
Companion website: www.wiley.com/go/sudhoff/Powermagneticdevices

Appendix A Conductor Data and Wire Gauges

Table A.2 (Continued)

Gauge	d_c (mm)
2	6.543
3	5.827
4	5.189
5	4.620
6	4.115
7	3.665
8	3.264
9	2.906
10	2.588
11	2.304
12	2.052
13	1.829
14	1.628
15	1.450
16	1.290
17	1.151
18	1.024
19	0.9119
20	0.8128
21	0.7239
22	0.6452
23	0.5740
24	0.5105
25	0.4547
26	0.4039
27	0.3607
28	0.3200
29	0.2870
30	0.2540
31	0.2261
32	0.2032
33	0.1803
34	0.1600
35	0.1422
36	0.1270
37	0.1143
38	0.1016
39	0.08890
40	0.07874

References

1 D. G. Fink and H. W. Beaty, *Standard Handbook for Electrical Engineers*, 13th edition. New York: McGraw-Hill, 1993.
2 D. E. Lide, *CRC Handbook of Chemistry and Physics, 84th edition*. Boca Raton, FL: CRC Press, 2003.
3 R. A. Serway, *Principles of Physics, 2nd edition*. Fort Worth: Saunders College Publishing, 1998.
4 H. B. Stauffer, *Engineer's Guide to the National Electric Code*. Boston: Jones and Bartlett Publishers, 2008.
5 H. D. Young, *University Physics, 7th edition*. Reading, MA: Addison-Wesley, 1992.
6 D. E. Herrington, *Handbook of Electronic Tables and Formulas*, Indianapolis, IN: Howard and Sams & Company, 1959.

Appendix B

Selected Ferrimagnetic Core Data

This Appendix lists some performance data for the selected ferrite materials. It should be understood that these materials were not selected for any reason other than availability within the author's laboratory. The data listed are intended for use in educational design examples. Table B.1 lists mass density that was measured by volumetric and mass measurements on toroidal samples. Table B.2 lists the data that characterize the $\mu_B()$ function as described in Section 2.3. This data was taken from Shane and Sudhoff [1] using the procedure and apparatus described therein. The corresponding B–H characteristics of the material are depicted in Figure B.1. MSE loss parameters for the materials are given in Table B.3.

Table B.1 Ferrite Mass Density Data

Material	MN8CX	MN60LL	MN67	MN80C	3C90
ρ (kg/m^3)	4612	4819	4795	4710	4743

Table B.2 Ferrite $\mu_{>B}()$ Data

Material	μ_r	α_k	β_k	γ_k
MN8CX	6326.0	0.41961	269.1083	0.42322
		0.18748	983.769	0.45413
		0.010849	8.69827	0.60172
		0.0081058	155.8723	0.38529
MN60LL	15042	2.1948	306.0455	0.453
		0.021436	998.3986	8.2682
		0.0078737	180.4701	0.40563
		0.001005	6.824599	0.38732
MN67	27085	1.0588	74.31812	0.55342
		0.021505	9.633883	0.7891
		0.021505	48.52124	0.4303

(Continued)

Power Magnetic Devices: A Multi-Objective Design Approach, Second Edition. Scott D. Sudhoff.
© 2022 The Institute of Electrical and Electronics Engineers, Inc. Published 2022 by John Wiley & Sons, Inc.
Companion website: www.wiley.com/go/sudhoff/Powermagneticdevices

Appendix B Selected Ferrimagnetic Core Data

Table B.2 (Continued)

Material	μ_r	α_k	β_k	γ_k
		0.001	106.7594	0.3785
MN80C	12402	0.76926	85.6801	0.49503
		0.048976	16.7675	0.58504
		0.01336	97.5145	0.41129
		0.0028669	2.98234	1.0098
3C90	22341	1.1542	431.1763	0.4742
		0.049742	2.29503	2.7955
		0.049644	15.04824	0.59862
		0.041155	74.28908	0.43996

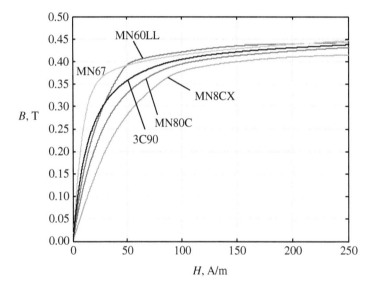

Figure B.1 B–H characteristics of the selected ferrites.

Table B.3 Ferrite MSE Parameters (B_b = 1T, f_b = 1 Hz)

Material	MN8CX	MN60LL	MN67	MN80C	3C90
$k_h(\text{W/m}^3)$	164	40.1	67.6	152	99.5
α	0.985	1.04	1.03	1.02	1.01
β	2.72	2.31	1.65	2.11	1.92

Reference

1 G. M. Shane and S. D. Sudhoff, Refinements in anhysteretic characterization and permeability modeling, *IEEE Transactions on Magnetics*, vol. **46**, no. 11, pp. 3834–3843, 2010.

Appendix C

Selected Magnetic Steel Data

This appendix lists some performance data for selected materials. It should be understood that these materials were not selected for any reason other than availability within the author's laboratory. The data listed is intended for use in educational design examples. Table C.1 list mass density which was measured by volumetric and mass measurements on toroidal samples. It also suggests a maximum allowed flux density. This is based on the point where the absolute relative permeability drops to 1000. Table C.2 lists the data which characterizes the μ_B () function described in Chapter 2. This data was taken from [1] using the procedure and apparatus described therein. The corresponding $B - H$ characteristics of the material are depicted in Fig. C.1. It should be understood that steel grades have performance which is a function of history and processing. Thus, not all M19 steel has the same magnetic characteristics. It should be noted that the material characterizations are homogenized in the sense that the predicted flux density is the average value within a stacked sample. Within the actual steel, the flux density will be higher because of the stacking factor. In other words, for the purposes of use, the stacking factor has already been included. Finally, note that some data for a representative nanocrystalline material has also been included. Loss data is presented in Table C.3.

Table C.1 Silicon Steel Data

Material	M19	M36	M43	M47	Hiperco50	Nanocrystalline
ρ (kg/m^3)	7402	7018	7291	7585	7845	5704
B_{mxa} (T)	1.39	1.34	1.39	1.49	2.07	0.974

Table C.2 Silicon Steel μ_B() Data

Material	μ_r	α_k	β_k	γ_k
M19	32686	0.098611	69.73973	1.399
		0.0014823	1.949541	2.1619
		0.001435	162.2767	1.2475
		0.001435	3.598553	2.0377
M36	26673	0.22599	271.8443	1.35065
		0.043195	97.31738	10

(*Continued*)

Power Magnetic Devices: A Multi-Objective Design Approach, Second Edition. Scott D. Sudhoff.
© 2022 The Institute of Electrical and Electronics Engineers, Inc. Published 2022 by John Wiley & Sons, Inc.
Companion website: www.wiley.com/go/sudhoff/Powermagneticdevices

Table C.2 (Continued)

Material	μ_r	α_k	β_k	γ_k
		0.031118	42.29465	1.32415
		0.0043748	0.8058081	5.38168
M43	24892	0.072885	33.9743	1.4138
		0.0039565	1.14108	4.2848
		0.0025477	3.37941	9.9998
		0.001	48.9292	1.4731
M47	9875	0.050919	18.03839	1.5613
		0.03933	20.61242	3.3712
		0.001	0.9336389	5.2715
		0.001	115.8623	1.3996
Hyperco50	43372	0.43708	17.13367	2.2836
		0.0003068	2.139356	1.3692
		0.00026279	163.4348	1.6772
		0.00024516	1.476588	3.494
Nanocrystalline	605910	0.75179	196.7667	0.98737
		0.18254	762.4804	0.97094
		0.011726	2.162413	3.79520
		0.001475	27.94608	0.90287

Table C.3 Silicon Steel MSE Parameters (B_b = 1T, f_b = 1 Hz)

Material	M19	M36	M43	M47	Hiperco50
Gauge	26	26	26	26	29
Thickness (mm)	0.66	0.66	0.66	0.66	0.33
ρ ($\mu\Omega$-cm)	52	44	42	30	41
k_h (W/m^3)	50.7	64.1	85.0	149	74.5
α	1.34	1.34	1.28	1.26	1.08
β	1.82	1.80	1.75	1.69	1.86
k_e (Am/V)	$2.65 \cdot 10^{-2}$	$4.05 \cdot 10^{-2}$	$4.14 \cdot 10^{-2}$	$26.1 \cdot 10^{-2}$	$3.02 \cdot 10^{-2}$

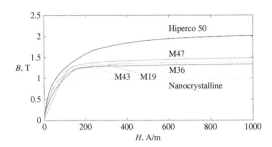

Figure C.1 *B–H* characteristics of selected steels.

Reference

1 G. M. Shane and S. D. Sudhoff, Refinements in anhysteretic characterization and permeability modeling, *IEEE Transactions on Magnetics*, vol. **46**, no. 11, pp. 3834–3843, 2010.

Appendix D

Selected Permanent Magnet Data

Table D.1 lists magnet data for a range of magnetic material. The data from this table were taken from Krizan and Sudhoff [1], which was in turn compiled from other sources. The data is for room temperature. It should be remembered that material parameters, especially permanent magnet parameters, are temperature dependent.

Table D.1 Selected Magnet Data

Material	B_r (T)	H_{ci} (kA/m)	χ_m	ρ (kg/m^3)
NdFeB N35	1.19	−867	0.09	7500
NdFeB N50	1.43	−836	0.36	7500
NdFeB Plastic	0.66	−577	0.24	5700
SmCo R20	0.90	−2400	0.02	8400
SmCo R32	1.15	−1350	0.10	8300
Ferrite AC-12	0.40	−318	0.10	4900
AlNiCo 8H	0.74	−151	1.50	7250

Reference

1 J. D. Krizan and S. D. Sudhoff, A design model for salient permanent-magnet machines with investigation of saliency and wide-speed-range performance, *IEEE Transactions on Energy Conversion*, vol. **28**, no. 1, pp. 95–105, 2013.

Appendix E

Phasor Analysis

The objective of this appendix is to provide a brief background in an ac steady-state circuit analysis using phasors. Phasor analysis is based upon two properties. The first of these is that the derivative of an exponential term is also exponential. In particular,

$$\frac{d}{du} e^{au} = a e^{au} \tag{E-1}$$

The second relationship upon which phasor analysis depends is Euler's identity:

$$e^{j\phi} = \cos\phi + j\sin\phi \tag{E-2}$$

where $j = \sqrt{-1}$.

Now suppose we wish to compute the steady-state solution to the differential equation

$$a_0 y + a_1 \frac{dy}{dt} + a_2 \frac{d^2 y}{dt^2} + \cdots = b_0 x + b_1 \frac{dx}{dt} + b_2 \frac{d^2 x}{dt^2} + \cdots \tag{E-3}$$

which may be expressed as

$$\sum_{n=0}^{N} a_n \frac{d^n y}{dt^n} = \sum_{m=0}^{M} b_m \frac{d^m x}{dt^m} \tag{E-4}$$

In (E-4), N is the order of the differential equation, and we will assume that the equation is proper so that $M \leq N$. The coefficients a_n and b_m are real. The input to the differential equation is x, which will be sinusoidal in time. We wish to find y for steady-state conditions. We will assume that the coefficients are such that a steady-state solution exists and is unique. Under these conditions, y will also be sinusoidal in time and have the same frequency as x.

Let us take

$$f = \sqrt{2} A_f \cos\left(\omega t + \phi_f\right) \tag{E-5}$$

as a general form for x and y, which is to say that (E-5) describes x with f replaced by x or with f replaced by y. In (E-5), ω is the radian frequency of the source excitation, A_f is the rms value of the waveform, and ϕ_f is the phase. With the form for these variables established, the problem of solving the differential equation becomes one of solving for A_y and ϕ_y given knowledge of A_x, ϕ_x, ω, and the coefficients a_n and b_m.

Power Magnetic Devices: A Multi-Objective Design Approach, Second Edition. Scott D. Sudhoff.
© 2022 The Institute of Electrical and Electronics Engineers, Inc. Published 2022 by John Wiley & Sons, Inc.
Companion website: www.wiley.com/go/sudhoff/Powermagneticdevices

In the process of solving (E-4), let us also consider the solution of a phase-shifted problem given by

$$\sum_{n=0}^{N} a_n \frac{d^n y_p}{dt^n} = \sum_{m=0}^{M} b_m \frac{d^m x_p}{dt^m} \tag{E-6}$$

This phase-shifted problem is the same as the original problem, except that the applied input x, and hence the output y is shifted by $\pi/2$. Thus, for our phase-shifted solution we have

$$f_p = \sqrt{2} A_f \sin(\omega t + \phi_f) \tag{E-7}$$

This phase-shifted problem is not of direct interest, but will be useful in solving the original problem.

Observe that x, x_p, y, and y_p are all real-valued time-varying quantities. We will now define composite variables of the form

$$\underline{f} = f + j f_p \tag{E-8}$$

As a result of the definition (E-8), the composite input and output variables, $\underline{x}$ and $\underline{y}$, are complex-valued time-varying quantities.

Observe that from (E-8) it follows that

$$f = \text{real}(\underline{f}) \tag{E-9}$$

and

$$f_p = \text{imag}(\underline{f}) \tag{E-10}$$

where $\text{real}(\underline{f})$ and $\text{imag}(\underline{f})$ are functions that return the real and imaginary part of a complex number. Using (E-9) and (E-10), the original and phase-shifted variables can readily be obtained in terms of the composite variables.

Substitution (E-5) and (E-7) into (E-8) yields

$$\underline{f} = \sqrt{2} A_f \left(\cos(\omega t + \phi_f) + j \sin(\omega t + \phi_f) \right) \tag{E-11}$$

By Euler's identity, (E-11) becomes

$$\underline{f} = \sqrt{2} A_f e^{j(\omega t + \phi_f)} \tag{E-12}$$

Manipulating (E-12), we obtain

$$\underline{f} = \sqrt{2} A_f e^{j\phi_f} e^{j\omega t} \tag{E-13}$$

We are now in a position to define what is meant by a phasor representation of a quantity. We will define a phasor as

$$\tilde{f} = A_f e^{j\phi_f} \tag{E-14}$$

which is often also denoted as

$$\tilde{f} = A_f \angle \phi_f \tag{E-15}$$

From (E-14) and (E-2) we can show that

$$A_f = |\tilde{f}| \tag{E-16}$$

and that

$$\phi_f = \text{angle}\left(\tilde{f}\right) \tag{E-17}$$

where $\text{angle}\left(\tilde{f}\right)$ is a function that returns the angle of its complex argument (with the real part treated as the independent variable and the imaginary part as the dependent variable). The phasor is a complex number, but it is not time-varying. From (E-13) and (E-14) a composite variable may be expressed as

$$\underline{f} = \sqrt{2}\tilde{f}\,e^{j\omega t} \tag{E-18}$$

At this point, we will return to the solution of our differential equation. Adding our original differential equation (E-4) to j times the phase-shifted equation (E-6), one obtains

$$\sum_{n=1}^{N} a_n \frac{d^n \underline{y}}{dt^n} = \sum_{m=1}^{M} b_m \frac{d^m \underline{x}}{dt^m} \tag{E-19}$$

In order to solve (E-19), note that from (E-1) we have

$$\frac{d^n}{dt^n} e^{j\omega t} = (j\omega)^n e^{j\omega t} \tag{E-20}$$

Multiplying (E-20) by $\sqrt{2}\tilde{f}$, we obtain

$$\frac{d^n}{dt^n} \underline{f} = (j\omega)^n \underline{f} \tag{E-21}$$

which may also be written as

$$\frac{d^n}{dt^n} \underline{f} = (j\omega)^n \sqrt{2}\tilde{f}\,e^{j\omega t} \tag{E-22}$$

Next, utilizing (E-22) in (E-19) yields

$$\sqrt{2}\tilde{y}e^{j\omega t} \sum_{n=1}^{N} a_n (j\omega)^n = \sqrt{2}\tilde{x}e^{j\omega t} \sum_{m=1}^{M} b_m (j\omega)^m \tag{E-23}$$

Canceling out the common $\sqrt{2}e^{j\omega t}$ term in (E-23)

$$\tilde{y} \sum_{n=1}^{N} a_n (j\omega)^n = \tilde{x} \sum_{m=1}^{M} b_m (j\omega)^m \tag{E-24}$$

Manipulating (E-24), we obtain

$$\tilde{y} = \tilde{x} \frac{\sum_{m=1}^{M} b_m (j\omega)^m}{\sum_{n=1}^{N} a_n (j\omega)^n} \tag{E-25}$$

The expression (E-25) allows us to calculate $\tilde{y}$ from $\tilde{x}$, ω, and the coefficients of the differential equation. This effectively yields the steady-state solution of the differential equation since A_y and ϕ_y are readily obtained from $\tilde{y}$ using (E-16) and (E-17). Alternately, from (E-9) and (E-18) we obtain

$$y = \text{real}\left(\sqrt{2}\tilde{y}e^{j\omega t}\right) \tag{E-26}$$

which reduces to

$$y = \sqrt{2}A_y \cos\left(\omega t + \phi_y\right) \tag{E-27}$$

It should be emphasized that the introduction of phase-shifted and composite variables was in order to show why the phasor approach to solving differential equations works. These variables are not used nor even thought about in practice.

Let us now consider how our results apply to circuit analysis. The voltage across and current through a resistor, denoted v_r and i_r, respectively, are related by

$$v_r = R i_r \tag{E-28}$$

where the resistor resistance is denoted R. This is a zero-order differential equation of the form (E-4), so (E-24) applies. Thus,

$$\tilde{v}_r = R \tilde{i}_r \tag{E-29}$$

It is convenient to define the impedance as the ratio of the phasor representation of the voltage over that of the current. Thus, the impedance of a resistor may be expressed

$$Z_r = \frac{\tilde{v}_r}{\tilde{i}_r} = R \tag{E-30}$$

Next, let us consider a magnetically linear constant geometry inductor. Denoting the voltage across the inductor v_L and the current through the inductor i_L, we may write

$$v_L = L \frac{di_L}{dt} \tag{E-31}$$

where L is the inductance. This is a first-order differential equation of the form (E-4), so (E-25) applies. We have

$$\tilde{v}_L = j\omega L \tilde{i}_L \tag{E-32}$$

From (E-32), the impedance is given by

$$Z_L = \frac{\tilde{v}_L}{\tilde{i}_L} = j\omega L \tag{E-33}$$

The current flowing into a capacitor, i_c, may be expressed in terms of the capacitor voltage v_c as

$$i_c = C\frac{dv_c}{dt} \tag{E-34}$$

where C is the capacitance. This is again a first-order differential equation of the form (E-4). Thus,

$$\tilde{i}_c = j\omega C \tilde{v}_c \tag{E-35}$$

The impedance looking into the capacitor may therefore be written as:

$$Z_c = \frac{\tilde{v}_c}{\tilde{i}_c} = \frac{1}{j\omega C} \tag{E-36}$$

At this point, we have established how common circuit elements are represented by phasors and impedances. It should be noted that our results are readily extendible to any linear equation. Suppose:

$$\sum_{o=1}^{O} c_o z_o = 0 \tag{E-37}$$

In (E-37), O is the number of terms, and c_o and z_o are a coefficient and arbitrary variable associated with the oth term of a linear equation. Using the same techniques, we used to solve the differential equation (E-4), we can show

$$\sum_{o=1}^{O} c_o \tilde{z}_o = 0 \tag{E-38}$$

The implications of (E-37) and (E-38) are that all circuit analysis techniques one might use for dc resistive circuits also work for steady-state ac problems since all linear equations apply to the corresponding phasor relationships. This includes nodal analysis, mesh analysis, and voltage and current division, since all of these techniques originate in linear equations.

As a specific example, consider a series connection of a resistor, inductor, and capacitor connected to a source v_S. We may write

$$v_R + v_L + v_C = v_S \tag{E-39}$$

From (E-38), we have

$$\tilde{v}_R + \tilde{v}_L + \tilde{v}_C = \tilde{v}_S \tag{E-40}$$

Because of the series connections of elements

$$i_R = i_L = i_C = i \tag{E-41}$$

it follows from (E-38) that

$$\tilde{i}_R = \tilde{i}_L = \tilde{i}_C = \tilde{i} \tag{E-42}$$

Relating the element voltages to the currents using (E-29), (E-32), and (E-35) and manipulating, we can readily solve for the phasor representation of the current, that is,

$$\tilde{i} = \frac{\tilde{v}_S}{R + j\omega L + \frac{1}{j\omega C}} \tag{E-43}$$

Thus, given the amplitude and phase of the source, (E-43) could be used to find the amplitude and phase of the resulting current. Let us consider the particular case of the example in which

$$v = \sqrt{2}\,200\cos(500t + 1) \tag{E-44}$$

where the 1 is 1 radian, $R = 5\,\Omega$, $L = 1$ mH, and $C = 1$ mF. Comparing (E-44) with (E-5) and (E-15), we have

$$\tilde{v} = 200e^{j1} = 200\angle 1 = 108 + j168 \tag{E-45}$$

and

$$\omega = 500 \tag{E-46}$$

From (E-43), we obtain

$$\tilde{i} = 38.3\angle 1.29 \tag{E-47}$$

where the angle is in radians. Thus,

$$i = \sqrt{2}\,38.3\cos(500t + 1.29) \tag{E-48}$$

The reader is referred to any introductory text on circuit analysis for a more detailed review.

Appendix F

Trigonometric Identities

$$e^{jx} = \cos x + j \sin x$$
$$a \cos x + b \sin x = \sqrt{a^2 + b^2} \cos(x + \phi) \quad \phi = \text{angle}(a - jb)$$
$$\cos^2 x + \sin^2 x = 1$$
$$\sin 2x = 2 \sin x \cos x$$
$$\cos 2x = \cos^2 x - \sin^2 x = 2 \cos^2 x - 1 = 1 - 2 \sin^2 x$$
$$2 \cos x \cos y = \cos(x + y) + \cos(x - y)$$
$$2 \sin x \sin y = \cos(x - y) - \cos(x + y)$$
$$2 \sin x \cos y = \sin(x + y) + \sin(x - y)$$
$$\cos(x \pm y) = \cos x \cos y \mp \sin x \sin y$$
$$\sin(x \pm y) = \sin x \cos y \pm \cos x \sin y$$
$$\cos x + \cos\left(x - \frac{2\pi}{3}\right) + \cos\left(x + \frac{2\pi}{3}\right) = 0$$
$$\sin x + \sin\left(x - \frac{2\pi}{3}\right) + \sin\left(x + \frac{2\pi}{3}\right) = 0$$
$$\cos^2 x + \cos^2\left(x - \frac{2\pi}{3}\right) + \cos^2\left(x + \frac{2\pi}{3}\right) = \frac{3}{2}$$
$$\sin^2 x + \sin^2\left(x - \frac{2\pi}{3}\right) + \sin^2\left(x + \frac{2\pi}{3}\right) = \frac{3}{2}$$
$$\sin x \cos x + \sin\left(x - \frac{2\pi}{3}\right)\cos\left(x - \frac{2\pi}{3}\right) + \sin\left(x + \frac{2\pi}{3}\right)\cos\left(x + \frac{2\pi}{3}\right) = 0$$
$$\sin x \cos y + \sin\left(x - \frac{2\pi}{3}\right)\cos\left(y - \frac{2\pi}{3}\right) + \sin\left(x + \frac{2\pi}{3}\right)\cos\left(y + \frac{2\pi}{3}\right) = \frac{3}{2}\sin(x - y)$$
$$\sin x \sin y + \sin\left(x - \frac{2\pi}{3}\right)\sin\left(y - \frac{2\pi}{3}\right) + \sin\left(x + \frac{2\pi}{3}\right)\sin\left(y + \frac{2\pi}{3}\right) = \frac{3}{2}\cos(x - y)$$
$$\cos x \cos y + \cos\left(x - \frac{2\pi}{3}\right)\cos\left(y - \frac{2\pi}{3}\right) + \cos\left(x + \frac{2\pi}{3}\right)\cos\left(y + \frac{2\pi}{3}\right) = \frac{3}{2}\cos(x - y)$$

Power Magnetic Devices: A Multi-Objective Design Approach, Second Edition. Scott D. Sudhoff.
© 2022 The Institute of Electrical and Electronics Engineers, Inc. Published 2022 by John Wiley & Sons, Inc.
Companion website: www.wiley.com/go/sudhoff/Powermagneticdevices

Index

a

AC conductor losses. *See* Conductor, losses
AC electrical frequency   279
Air-gap field intensity   276
   radial component of   276
Alnico (aluminum, nickel, cobalt) magnet   96
Ambient temperature   375, 386, 388, 391
American Wire Gauge   109
   wire diameters   589
Ampere's law   43, 65, 71, 74, 276, 401, 403, 405, 420, 423
Anhysteretic behavior, measuring   188–196
Annealing   187
Anti-ferromagnetic materials   49
Atomic magnetic moment, arrangements   49
Average power loss density   175
Axial heat flow   363–364
   spatial averages, temperature   363

b

Backiron segment   285
Balanced set
   transformation   291–292
      three phase quantities, identical amplitudes   291
Bending radius   525, 533, 538, 540
Bessel functions   406, 407
B–H characteristics
   anhysteretic   165, 188, 194, 196
   ferromagnetic and ferrimagnetic materials   50
   MN80C   173
   M47 silicon steel
      anhysteretic   51
   selected ferrites   173, 594
   selected steels   596
B–H trajectories   173, 176, 205
   energy transferred to device during   177
   minor loop behavior   176
Biological genetics   7–10
   deoxyribonucleic acid   7
Boundary condition
   Dirichlet   575
   Neumann   575, 576, 578, 583
Branch fluxes   135
Bridge, three-phase   503, 504
Build factor   533
Buried-magnet rotor   303

c

Canceled conductors   269–270, 318
   number defined   270
Canonical genetic algorithm   12
Capacitance
   of curved plates   437–438
   differential   438, 439
   dynamic coil-to-core   488
   dynamic layer-to-layer   488
   dynamic turn-to-turn   445–446, 488
   of insulated conductor   438–440
   of isolated conductive cylinders   440–442
   layer-to-core   433
   layer-to-layer   433, 438, 449–453
   measuring   455–459
   in multi-winding systems   452–455
   parallel plate   435–437
   of parallel plate capacitor   436

Capacitance (cont'd)
  parasitic   433–458
  static turn-to-turn   443, 448, 450
  turn-to-turn   433, 438, 442–446,
      448–451
Capacitors
  ceramic   474
  electrolytic   474
  film   474
Carrier wave   505, 512
Carter's coefficient   282–284
  concept of   282
  stator tooth width   282
  tooth geometry   284
Carter's method   282
Case studies   113, 336–344
  design of a single-phase core-type
      transformer   251–259
    core flux density waveforms   258
    design space   253
    electrical parameters   257
    no-load flux density waveforms   259
    operating point performance   258
    parameter distribution   254
    primary flux linkage vs. current   258
    rated base primary/secondary currents   251
    rated load impedance   251
    transformer design Pareto-optimal
        front   254
    transformer fixed parameters   253
    transformer specifications   252
  design of UI-core inductor
    B-H characteristics   116
    conductor count vs. mass   117
    conductor materials   113
    current density   116
      vs mass   116
    design space, values   114
    Ferrite MN60LL   116
    final population gene distribution   115
    inductor specifications   114
    material properties   113, 589–590
    Pareto-optimal front   115, 116
  EI-core electromagnet
    genetic algorithm-based optimization   157

    conductor count vs volume   159
    core widths vs. volume   160
    current density vs. volume   159
    design space   157
    effects of wire size descritization on   162
    electromagnet specifications   156
    gene distribution plot   158
    geometrical parameters vs. volume   160
    Pareto-optimal front   157, 162
  machine design
    component mass vs. electromagnetic
        mass   341
    cross section   343, 344
    current-related parameters vs.
        electromagnetic mass   341
    design specifications   338
    flux density vs. rotor position   344
    Hiperco50   341
    machine parameters vs. electromagnetic
        mass   342
    material selection vs. electromagnetic
        mass   340
    parameter distribution   339
    parameter ranges   338
    Pareto-optimal front   339, 343
    power loss components vs. electromagnetic
        mass   340
  thermal model of electromagnet   380–382
    air-gap thermal resistance   384
    electro-thermal analysis   387–389
      thermal analysis of electromagnet design,
          390–396
    thermal equivalent circuit
        architecture   384–387
    thermal representation of a rounded
        corner   382
    winding to core resistance   383–384
Chain rule   577
Chromosome   8
  crossover   14
  pairing   8
  replication   8
  segregation   14
  sex   8
Circuit, thermal-equivalent   472, 484, 486, 487

Co-energy   128–132, 134
  associated with electromechanical system   130
  force from   132–133
  mathematical expression   129
Combined loss modeling   181–183
Common-mode current   543–545
Common-mode voltage   543–545
Conductivity   34, 49, 69, 106, 167, 181, 290
  thermal   354, 362, 369–372, 390, 395, 412, 589
Conductors   34, 43–44, 77, 105, 106, 116, 142, 153, 167, 266
  canceled   270, 318
  cylindrical   405
    impedance characteristic   408
    skin effect   405–409
  density   267
    a-phase stator   267
    position, sinusoidal function   267
    positive, negative   267
  distributions, condition on   267
  end   269–270
  losses   399
    in rotating machinery   422–426
    in UI-core inductor   426–431
  mass   110
  number   266
  parallel   232, 253, 266, 268, 270, 289, 318, 320, 343, 372, 387
  proximity effect (See Proximity effects)
  radius   106, 109, 153, 236, 372, 383, 393, 417, 428
  rectangular, exposed to magnetic field   409
  round   414
    exposed to a magnetic field   410
  size, actual   109
  strip   400
    skin effect in   399–405
  symbol   270
Conservative fields, conditions for   133
Constituatory relationships   574
Constraints   2, 26, 27, 31–33, 35, 111, 247–249, 392
  aspect ratio   119
  current   307
  design   111, 334, 391
  functions   35
  thermal   111, 154
  torque   335
  voltage   307, 312
Convective heat transfer   374–375
Converter
  angle   505
  buck   461–501
  dc-to-ac   504, 529, 543, 545
Core loss   244–245
  high-frequency   529, 531–532
  low-frequency   529, 531
Core, nanocrystalline   446, 456
Core, tape-wound   186–187
Core-type transformer design   245
  calculation of fitness   249–251
  design space   245–246
  metrics and constraints   246–249
Core-type transformer MEC   238–244
Coulomb force   48
Cross section, magnetic   481, 482, 518
Curie temperature   96
Current density   111, 476, 477, 492, 495, 497
  design   37
  distribution in strip conductor   404
Current flow   44, 47, 171, 307, 412
Current ripple
  dc   514, 515
  inductor   467, 468, 471, 473, 488, 491
  input inductor   467, 491
Current source operation   310–312. See also Permanent magnet ac (PMAC) machine
  current source fed
    PMAC machine characteristics   312
  d-axis current   310, 311
  inverse transformation, use of
    abc variable current   310
  inverter current regulation   309
  line-to-line voltage   311
    peak value   311
  nonsalient machine   310
  power out of machine   310

Current source operation (*cont'd*)
  q-axis current   310, 312
  rms value, applied voltage   310
  salient machines   311
  torque
    desired   310
    maximum   312
Cylindrical conductors   405
  impedance characteristic   408
  skin effect   405

d

DC bus voltage   514, 515
DC coil resistance   105–107
DC inductor design   108
  calculation of fitness function   113
  geometrical parameters   108
  parameter vector to characterize design   108
  problem formulation   108–113
  UI-core inductor   108
Decoding algorithm   11
Demagnetization   95, 97, 320, 327, 330, 333, 334, 343
Deoxyribonucleic acid (DNA)   7
  helix structure   8
  hydrogen bonds   8
  mutation   18
  nonfunctional   8
  purines   8
  pyrimadines   8
Design approach   1–3
Design equations   1, 2
Design metrics   2, 32, 108, 110, 118, 152, 153, 333–334
Design problems   1, 7, 108, 112, 117, 152, 245
Design process
  manual   2
  optimization-based   2
Design space   564, 565, 568
Device parameters   1
Device performance   1
Device specifications   1
Diamagnetic materials   48

Differential-mode current   543, 544, 550, 553, 558, 562, 565, 567
Differential-mode voltage   543, 544
Diode
  PN   461, 470
  power   461
  Schottky   461, 471, 473
Discrete event handling   514
Distributed windings   263–265. *See also* Winding
  continuous description   267
    conductor density   267
  converting   268
  discrete description   265–269
  resistance of   289
  slots, large number   267
  turns, number   271
  B–H trajectory, determining   205
  hysteron behavior   202
  incremental magnetization   204
  Jiles–Atherton model   205
  magnetization states   203
  material magnetization   201
  normalized hysteron density function   202
  normalized magnetization   202
  Preisach model   201
  reversible magnetization   201
  saturated magnetization   202
Divergence theorem   578
Doudle-sheet tester   184-185
Duty cycle   461, 465, 473, 489, 490, 492, 499
  phase   505

e

Eddy current losses   165–172
  calculation   166
  defined   165
  instantaneous power dissipation   166
  normalized power loss density   167
  prediction power loss in aluminum alloy   167–172
    flux density waveform   169, 170
    normalized power loss *vs.* frequency   169
    top view of test configuration   168
Edge rate   549, 552

EI-core electromagnet   141–143
  design   151
    actual conductor size   152
    calculation of fitness function   156
    calculation of volume   153
    choice for parameter vector   152
    conductor radius   152
    constraint   153, 154
    packing factor   153
    power loss   154
    problem formulation   152–155
    ratios defined   152
    use of MEC   155
    winding width and depth, computation   153
  electric analysis   142
    steady-state dc current   143
  force analysis   147–151
    core made of 3C90 ferrite   149–151
    flux density   149
  magnetic analysis   141–147
    calculation of fringing permeances   144–145
    EI-core magnetic equivalent circuit   144–146
    energy associated with path   146
    horizontal/vertical slot leakage permeances   144–145
    permeance of center gap   145
    reduced magnetic equivalent circuit   148
    slot leakage path   146
    specifying reluctances of core pieces   143
Electrical position   265
Electric circuit   43, 46, 48, 56
  phasor analysis   221
  resistive   80
  translation of magnetic circuits to   59–64
Electric field   434–438, 440, 444, 449
Electric machinery   290
  equations, sets   290
    flux linkage equations   290
    torque equation   290
    voltage equations   290
Electromagnet   1. *See also various case studies*
  constraints   247
  design   141
  thermal equivalent circuit   385
Electromagnet architectures   141
  bobbin core   142
  E-core arrangement   141–143
  solenoid   141, 142
  U-core arrangements   141, 142
Electromagnetic mass   110
Electromagnetic system   43, 50, 59
Electromagnetic torque   135, 139, 264, 290, 297, 298, 306, 326
Electromechanical devices
  energy storage   123–125
    mathematical derivations   124–125
Electromechanical system   43
Electrostatics   433–442
Elementary stiffness matrix   582
Elitism operator   37
Empirical modeling of core loss   177
  Steinmetz equation   177
    generalized   179–180
    modified   178–179
End conductors   269–270
End leakage inductance   285
End leakage permeance
  calculation of   287
  end winding   288
  exterior path   288
  flux flow   288
  machine end   288
End winding segment, dimensions   288
Energy storage factor   206, 208
Epstein Frame   183–184, 188–190, 194, 197
Equivalent circuit
  common-mode   546–552
  differential-mode   546
Equivalent electrical angles   264
Extended Jiles-Atherton model   205–211

f

Faraday's law   61–64
Fast-average   506, 510
FEA. *See* Finite element analysis (FEA)
Ferrimagnetic materials (ferrites)   49
  B–H characteristic   50

Ferrimagnetic materials (ferrites) (*cont'd*)
  hysteresis loss   165
Ferrite mass density data   593
Ferrite MSE parameters   594
Ferromagnetic field analysis   327–332
  ferromagnetic portions, machine
    flux density waveforms, determining   327
    permanent magnet field intensity   332
      minimum field intensity, computing of   332
      minimum stator MMF   332
    rotor flux   330–332
    stator backiron flux   329
    stator core loss   329–330
      flux density waveform   330
      resulting core loss   330
      time domain waveforms   330
      time vector, calculation of   329
    stator tooth flux   328–329
Field distribution, periodicity   282
Field, electric   434–438, 440, 444, 449
Field energy   130, 135, 137
  calculation   125–127
    multi-input system   126
    single input electrical system   126
  force from   127–128
Field intensity, anhysteretic   189, 206
Filter inductor, three-phase   550
Finite element analysis (FEA)   573–588
  2-D FEA analysis   587
  3-D FEA analysis   587
  predictions   587
  vs. MEC analysis   587
Fitness functions   12, 35
  compute   249, 337
  for design problems   31
    formulation   31–33
  mathematical   12
  scaled   21
  single-and multi-objective optimization   35
Fitness *vs.* generation   36
Flux
  air-gap flux   304
  calculation   47
  fringing (*See* Fringing flux in magnetic circuits)
  incremental   281

  magnetic   46
  magnetizing ratio   112
  rotor (*See* Rotor flux) source expression   58
  sum of fluxes into (or out of) any node   48
  through open surface, calculation   46–47
Flux density   37, 38, 43, 46, 47, 50, 55, 278–279
  air gap   280
  circumferential   72
  core flux density waveforms   258
  current normalized   415
  electric   434, 436, 437
  fringing   419
  imposed   412
  maximum tooth   328
  radial variation   282
  ratio   53
  residual   50
  rotor backiron   330
  saturation   116
  sinusoidal   171
  solution for radial   321, 325
  tangential   331, 335
  time-varying   172
  waveform   169, 170
Flux linkage   43, 59, 60, 63, 112, 123, 135, 280–282, 295
  anhysteretic   191, 193
  calculation of   280–281
  common-mode   548, 550, 552–557
  contribution of   281
  magnetizing   61, 219, 231, 298, 326
  vertical leakage   426
  vs. current   60, 62–63
    dc offset   63
    hysteresis loop   64
    measurement   62
    predicted characteristics   63
  winding   61
Flux lines, computing   584
Flux linkage equations   123, 127, 130, 133, 135, 290, 293, 297, 306, 326–327
  machines   306
  rotor reference frame   295
  transformation   293–294

permanent magnet synchronous machine   293
    rotor-position-dependent inductance matrix   293
Flux-linkage waveform, proxy   555, 557, 562, 563, 565, 566
Flux permeance
    exterior adjacent conductor leakage   75–78
    exterior isolated conductor leakage   78–79
    magnetic equivalent circuit with leakage   80
    slot leakage   73–75
Frequency response   433, 446, 449, 452, 455–458
Fringing flux in magnetic circuits   64–68
    calculation   66
    field intensities   65
    fringing face permeance   68
    Gauss's law   65
    incorporated into MEC   65
    MMF drop   66
    relative permeability   65
    reluctance   68
    total air gap permeance of the UI-core inductor   68

g

Gametes
    diversity   9
    formation   8
GAs. *See* Genetic algorithms (GAs)
Gauss's law   46, 47, 65
    for electrostatics   434, 435
Generic electrical machine   263
    position measurements, definition   264
Generic position   265
Genetic algorithms (GAs)   1, 7
    canonical   10
        computing fitness of member of population   13
        crossover point   13
        effectiveness   16
        genetic operators, use of   12
        illustration   12
        initialization   12
        mating pool, determined by selection process   12
        next generation   12
        n-way tournament selection   13
        pseudo-code   13, 14
        roulette wheel selection   13
        selection process   13
        stopping criterion, checking   12
        vs. traditional optimization algorithms   10
    computing fitness   12
    decoding algorithm   11
    denoting population   10
    fitness function   11
    genetic code
        decoding function   11
        for ith individual organized as   10, 11
        role of crossover   9
Genetic code   9–11
Geometric size   534
GOSET code   37
Gray code scheme   15

h

Heat energy   349–351
    mean temperature, material   349
    spatial average   349
    thermal capacitance   350
    thermal energy density   349
        expression   349
        in material sample   349
Heat equation   349–351
    derivation   351, 352
    elemental cuboid   350
        electrical power dissipated   351
    thermal equivalent circuits, development of   352
    time derivative, thermal energy   351
Heat flow   349–351
    Fourier's law   350
    heat flux   350
        expression   350
    heat transfer rate   350
    net heat transfer   350
        expression   351
Heat sink   472–474
Heaviside notation   547
Horizontal averaging   188

Hysteresis 50
  average power loss density in material due to 175
  characteristic 175
  effects 50
  estimate losses 50
  loop 189
  loss in ferrites 165
  loss in magnetic materials 172
    toroid 173–174
  Steinmetz relationship 50
Hysteresis model, time-domain 564

i

IEC_60404, 183
Impedance capacitor 604
  characteristics of cylindrical conductor 408
  characteristics of a strip conductor 405
  dc impedance 404
  internal, of a conductor 408
  load 222, 226, 248, 251
  magnetizing branch 226
Inductance 59–64, 280–282
  absolute 60
  associated with strip conductor 403
  axis 343
  calculation of 278, 280
  constraint 35, 111
  incremental 61, 112, 119, 529–531, 533, 535
  leakage (See Leakage inductance)
    machine 307, 310
  magnetizing 223, 225, 228, 231, 282
  self-and mutual 61, 282, 284
Induction motor 263
  machine 264, 270
  stator core 263
  stator housing 263
Inductor
  architectures 103–105
  aspect ratio 111
  common-mode 547–548, 552–570
  dimensions vs mass 118
  effective permeance 71
  height, width, and length 111
  stick-like 111
  three-phase 503–541
  total mass 110
Inductor design, common mode 547–570
Inhomogeneous regions 367–373
  conductor current 372
  conductor region 368
  dimensions
    air/potting material region 368
    conductor region 368
    insulator region 368
  effective anisotropic material 371
  effective material 371
    conductivities 372
    parameters 369
  effective thermal conductivity 371
  effective value
    density 371
    specific heat capacity 371
  heat transfer rates 372
  homogenized material, determine basic strategy 368
  homogenized region, representation 368
    determine 372
    heterogeneous region 368, 371
    homogenized material 368, 371
    thermal resistance calculation, arrangement for 368, 371
  homogenized thermal conductivities 369
  spatial temperature dependence 373
  thermal equivalent circuit, cuboidal element 372
  thermal resistance
    computing 369
    effective value 370
    expression 370
    net 370
    series/parallel connection, treatment of 370
    x-direction 369, 370
    y-direction 369, 370
    z-direction 370, 371
  winding bundle
    air/potting material 367
    aspect ratio 369
    conductor material 368

cross section   368
   insulation material   368
   thermal conductivity   368
Insulated-gate bipolar transistor (IGBT)   461
Insulating layer, coil-to-core   484
Insulation
   coil-to-core   478–480, 484
   layer-to-layer   477–480, 484, 486, 495
Inverter   543, 545, 546, 548, 549, 553, 554, 566
Iron oxide magnetite   49

j

Jiles-Atherton model   201, 205–211

k

Kirchhoff's current law   48, 376
Kirchhoff's flux law   46, 48, 55, 82, 85, 138
Kirchhoff's voltage law   43

l

Laminations   171, 172, 183, 185–187
Laminations, die stamped   187
Laser cutting, impact of   187
Leakage flux in magnetic circuits   68–70
Leakage flux linkage   280
Leakage inductance   80, 242, 280, 284–289
   air gap   286
   calculation of   280
   end conductors
      cross section of   285
      leakage flux linkage   284
   machine, length   286
   machine, longitudinal cross section of end conductors   285
   magnet depth   286
   parasitic   280
   slot conductors   285
      cross section of   285
      slot leakage inductance contributions of   285
Leakage permeance   68, 70, 155
   associated with the vertical component of field   145
   calculation   76
      end leakage permeance   287
      slot leakage permeance   317
      total end leakage permeance   289
   expression for horizontal slot   79, 145, 146
   horizontal   73, 74
   primary coil   232
   relating mean-squared field and   416–417
   slot leakage permeance due to paths   286, 287
   vertical slot   145
Leakage reluctance   80
Linearity, magnetic   535
Loss (es)
   characterization, with spatially varying fields   199–201
   core   479, 482–484, 491, 498
   modeling, time-domain   201–205
   models, characterizing   197–201
   proximity-effect   498, 513, 520, 522, 526, 532, 534
   switching   469–471, 473, 489
   turn-on   470
Lumped parameters   326–327. *See also* Transformer performance
   electrical parameters   327
   end-flux paths, machine   327
   flux linkage equations, associated parameters   326
      abc variable   326
      flux density expression   326
      qd0 leakage flux   326
      leakage flux linkage   326
      expression   326
   machine model   326
   magnetizing flux linkages   326
   Park's transformation   326
   stator resistance   326, 327

m

Machine   267
   developed diagram   267
   large air gaps   282
   multi-pole   282
   slot conductors
      front, back end   269
   symmetry of   277

Machine geometry 312–317
  dependent variables as vector 317
  independent variables 314
  inert region 313
  leakage inductance calculations 316
  material types 313
  parameters, calculation 313
  rotor backiron 315
  slot leakage permeance, calculation 316–317
  slot/tooth region 313
    dimensions 314
    rectangular slot approximation 316
    slot area 315
    slot depth 316
    slot width 316
    tooth base area 315
    tooth fraction 314
    tooth tip area 315
    tooth tip depth 316
    tooth tip fraction 314
    tooth tip width 315
  stator backiron 312
  stator teeth 315
  surface mounted permanent magnet synchronous machine
    cross section 312
    variables 313
  variables calculation 315
  vector-valued function 317
Machine model, QD variables 305–306.
    See also Permanent magnet ac (PMAC) machine
  electromagnetic torque 306
  flux linkage equations 306
  3-phase electric machinery 305
    delta-connection 305
    ideal machine 305
    wye-connection 305
  steady-state conditions 306
  voltage equations, rotor reference frame 305
Magnetically linear systems 134–135
  energy storage 70–73
  energy storage in 70–73
Magnetic circuits 43, 46, 48
  Ampere's law for 43–46

  calculating force using 135–138
    branch flux 137
    branch MMF drop 137
    excitation of single branch 136
    flux linkages 136
    Kirchhoff's flux law, use of 43–46, 138
      allowable combination and outcome 138
      mechanical degree of freedom 137
  to electric circuits, translation of 59
  fringing flux in 64–68
  leakage flux in 68–70
  magnetomotive force in 46–48
  nonlinear (See Nonlinear magnetic circuits)
Magnetic equivalent circuit (MEC) 56–59
  application to UI-Core inductor 103–104
  computing reluctances 59
  construction 57
  nonlinear analysis 90–94
  simple nonlinear 88
  UI-core inductor 56
    air gap g 57
    location of nodes 57
Magnetic flux 46
Magnetic hysteresis 165, 172, 206
Magnetic materials 48–55
  characterization of loop 50, 183–197
  conductive materials 48
  diamagnetic effect 48
  hysteresis 50
    estimate losses 50
    Steinmetz relationship 50
  magnetic moments 50
Magnetic moment 48, 50, 172
Magnetic permeability 50, 51
Magnetization 51, 52, 95, 178, 202, 303, 332
  irreversible 207, 208
  reversible 207, 208
Magnetizing flux linkage 280–282
  radial flux flow 280
Magnetizing inductance 280
  distributed windings calculation of 280
Magnet materials
  permanent 95–98, 320
    magnet configuration 98
    magnet MEC element 97

permeability   286
plastic encapsulated NdFeB   342
selected permanent magnet properties   96
Magnetomotive force (MMF)   43
  air gap   276–278
    Ampere's law   276
    backiron, sum   278
    calculate, fields   278
  creation of   280
  drop   45
    around a closed loop   45
    defined   45
  integration, path   276
  radial component, teeth   276
  radial direction   276
  reluctances   58
  rotating   278–280
  sinusoidal, space   279
  sources, windings   44
  use of Kirchhoff's law   45
Magnetomotive force sources   43
Manual design approach   2
Material boundaries   373–376
  contact resistance   374
    contact heat transfer coefficient   374
    expression   374
    heat transfer rates, across boundaries   374
    surface roughness, related to   374
  convective heat transfer   374–375
    coefficient   374
    heat transfer rate, solid to liquid   374
    natural convection   375
    thermal resistance, expression   374
  radiation   375–376
    Celsius   375
    heat transfer rate   375
    Kelvin   375
    materials, emissivity   376
    net heat transfer rate   375
    Stefan–Boltzmann constant   375
    temperature, linearized   375
Material properties, anhysteretic   183
MATLAB-based genetic optimization   36
  toolbox   36
Maxwell's equations   573–575

Maxwell's equations, integral form   435
  mean path length, magnetic   481
Mean-squared field, for select geometries   417
  exterior adjacent conductors   418
  exterior isolated and non-gapped closed-slot conductors   418
  gapped closed-slot conductors   419–422
  open-slot conductors   418–419
MEC. *See* Magnetic equivalent circuit (MEC)
Meiosis   8
Mendel, Gregory   8
Metallic metals   96
Metamodel   476, 477, 495
Metal-oxide-semiconductor field-effect transistor (MOSFET)   461, 471, 495
MMF. *See* Magnetomotive force (MMF)
MN67, 567, 584
MN80C   173, 192, 198
M47 silicon steel
  anhysteretic B–H characteristic   51
Multi-objective optimization   1, 25–27
  results   38
  using genetic algorithms   27–31
Mutations   10, 14
  beneficial   9
  chromosome   14
  DNA   18
  impact in meiosis   9
Mutual inductance factor   538
Mutual inductance, specification of   535
Mutual inductor, compensating   190
Mutual leakage inductance   280, 284

n

Nelder–Mead simplex method   5, 37
Neutral point   504
Newton's method   5, 6
  apply to find minimizer of function   6
  defining operators   5
  defining function minimizers   5
  Hessian of derivative   5
  limitations   7
  population-based optimization algorithm   7
Nickel ferrite   49
Nickel–zinc ferrite   49

Nonlinear magnetic circuits   80–81
  numerical solution   80–81
    mesh analysis   85–88
    nodal analysis   82–85
    standard branch   81
Nonlinear magnetic equivalent circuit   53
Nonlinear magnetic systems   88
  magnetic material characteristics, reluctances   88
Newton–Raphson method   89
  nodal analysis vs. mesh analysis   88–90
  permeability calculation   89
  reluctance terms, expression   89
  residual
    for nodal analysis   90
    using mesh analysis   91
Numerical analysis   1
  in manual design process   2

o

Objective functions   2
  mathematical properties   3–5
    defining parameter vector   3
    definition of convex function   5
    definition of convex set   5
    discrete element   3
    elements, real numbers   3
    geometrical parameter   3
    global minimizer   3
    local extrema, existence of   4
    local minima, existence of   4
    scalar variables   3
    search space   3
Ohmic resistance   166
Ohm's law   48, 55–56
  implicit equation for flux   59
  MMF drop   55
Operating point, worst-case   555
Optimization algorithms   3
Optimization-based approach, advantages   2
Optimization-based design   3

p

Packing factor of coil   110
Path length, magnetic   518

Paramagnetic materials   49
Parameter identification   191
Pareto-optimal front   25–27, 339
  calculation with ε-constraint method   27
Pareto-optimal set   26–27
Park's transformation   290–291, 326
  geometric interpretation of   291
    direct axis   291
    rotating coordinate system, axis of   291
  three-phase quantity   290
  zero sequence axis
    phase variables, value   291
Permanent magnet ac (PMAC) machine
  material parameters   320
    conductor type   320
    permanent magnet material   320
    permanent magnet type   320
    rotor steel type   320
    stator steel type   320
    temperature-dependent properties   320
    vectors of   320, 321
  operating characteristics   305–312
    current source operation   310–312
    machine model, QD variables   305–306
    three-phase bridge inverter   306–308
    voltage source operation   308–309
Permanent magnet materials   95, 120
  applying Ampere's law   98
  B–H and M–H characteristic   95–96
  ceramic/ferrite magnetic materials   96
  Curie temperature   96
  demagnetization   97
  field intensity in magnet material, 98
  magnet MEC element   97
  metallic metals   96
  permanent magnet configuration   98
  rare-earth combinations   96
  selected permanent magnet properties   96
  sintering   96
Permanent magnet synchronous machines   303–304
  air-gap flux flow   304
  interior magnets   304
    buried-magnet rotor   303

spoked magnet rotor   303
2-pole surface-mounted radial flux   303–304
   magnetization, direction of   303
   V-shape magnet arrangement   303
Permeability functions   52
Permittivity   434–436, 438, 446, 448, 450, 451
   of free space   434
   relative   434, 438, 446, 448, 451
Phase currents   515, 528
Phasor analysis   601–606
   composite variable, expression   602
   current flowing, into a capacitor   605
   Euler's identity   601
   impedance   604
      looking into the capacitor   605
   properties   601
   solution of a phase-shifted problem   602
   steady-state solution   601
Phasors   510
Pinning sites   165, 172, 207, 208
Poisson's equation   573–575
Positive flux   264
   magnetic material   264
P-pole machines   265
Potting material   517, 533
Power block   545, 546, 548–549, 551, 554, 555
Power loss density, measuring   181, 197
Proximity effect loss, gamma-normalize   534
Proximity effects   399
   in a group of conductors   413
      dynamic resistance   415–416
      loss in terms of flux density   413–414
      multi-winding systems   415–416
      round conductors   414
   independence of skin and   411–413
   in single conductor   409–411
Pseudo-code
   calculation of the fihecul function   250, 337
   check of constraints satisfied against imposed   251, 337
   Kung's method   28
   mating   13

q

Quasi-Newton methods   5

r

Radial field analysis   321–325
   air-gap MMF drop   323
      Carter's coefficient   323
      field intensity and flux density   322
      quasi-reluctance   323
   analytical field solution   321
      key assumption of   321
   large effective air gaps   321
   magnetic analysis, machine   321
MMF drop   321, 322
   permanent magnet MMF   322–323
      flux density and field intensity, relationship   323
      MMF drop, obtaining expression for   324
      MMF source   325
      radial magnetization, illustration   324
      spatial dependence   324
      square wave function   325
   radial field variation   323
      flux density, arbitrary radius   323
      Gauss's law   323
   radial flux density   322, 325
      solution for   325
   stator MMF   322–323
      conductor density distribution   322
      expression   322
      a-phase winding function   322
      qd variables   322
      winding functions, determining   322
Radial field variation, impact of   190–194
Radial heat flow   364–367
   circuit derivation   365, 366
   extremum point   367
   heat transfer rates   367
   peak temperature, calculation of   366, 367
   in radial direction   363
   spatial averages   363
      exterior node temperatures   366
   thermal equivalent circuit   366
   thermal resistances   368
Radiation   375–376
Real-coded genetic algorithms   15–25
   crossover   16–18
      multi-point simple blend crossovers   18

Real-coded genetic algorithms (*cont'd*)
   single-point simple-blend crossover   18
   vector crossover   17
  death   23
  deterministic search   23
  diversity control   22
  elitism   22
  encoding   15–16
  enhanced   23–25
  evolution   20
  local search   23
  migration   22–23
  mutation   18–21
   absolute vector mutation   18
   fitness and gene values   20
   modified gene value   18
   mutated value calculation   18
   relative vector mutation   18
  scaling   21–22
   linear scaling methods   21
   scaled fitness calculation   21
Reference frame theory   290–294, 506, 507
  Park's transformation   290–291
  transformation
   of balanced set   291–292
   of flux linkage equations   293–294
   of power   294
   of voltage equations   292–293
Reluctance   55
  factor   538
  leakage   80
Resistance   289–290
  ac   483
  axial distance
   machine laminations, end   289
  calculated, phase   290
  cross-sectional conductor area   289
  dynamic   522, 523, 526–528
  effective series   461, 475
  parasitic   461
  thermal   472, 473, 485–487, 495
Ring mapped index   270
Ripple
  input filter   466–467
  output voltage   468

  peak-to-peak   466
Root mean square   464
Rotating machinery   265
Rotating rotor   263
Rotor   263–264
  field intensity   278
  structures   265
Rotor flux   330–332
  backiron flux   330
  backiron region, machine   331
  Gausses' law   330
  maximum radial flux density   330
  peak radial flux density, rotor   330
  peak tangential flux density   330–331
  peak values   331
   permanent magnet   331
   stator MMF   331
  radial component, flux density   331
  rotor backiron   330, 331
   flux density   330, 331
  rotor fields   330
   calculation   330
  stator flux density waveforms, tooth   330
   calculation   330
  stator MMF   331
Runge–Kutta method   463

S

Saliency   514–515, 535, 539, 540
  factor   538
Scalar variables   3
Scheduling   514
Schema theory   14
Selected magnet data   599
Semiconductors, power   549
Sense winding   167, 183–186, 188, 190, 192, 193, 200
Silicon steel   456
Silicon steel data   595
Silicon steel MSE parameters   596
Simulation, time domain   511–514
Single-objective optimization
  study   36
  using Newton's method   5–7
Single-phase transformers   215

transformer performance   223
  calculation of lumped parameters   223–224
  inrush current   230–231
  lumped parameter   223–224
  magnetizing characteristics   225
  no-load, full-load, and overload analysis   230
  operating point analysis   225–226
  circuit solution   226
  convergence evaluation   229
  final calculations   229–230
  initialization   226–227
  lumped circuit parameter update   227–229
  nonlinear magnetizing current   227
  regulation   224–225
Single-sheet tester   184–185, 188, 194
Sinusoidal turns distribution   279
SI units   43
Skin effect   498
  in cylindrical conductors   405–409
  in strip conductors   399–405
Slot depth   109
Slot effects   282–284
Slot leakage
  permeance   73–75, 286
  inductance   285
Spatial temperature dependence   373
Spoked magnet rotor   304
Stacking factor   172, 186, 187
Stationary circuit transformation   507
Stationary reference frame   508, 510–512
Stator
  distributed winding   264
  field intensity 278
  machine   265
  reference axis   263–264
  slots   263
  transformation   507
  uniformly rotating poles   264
  windings   266
    mutual leakage inductance   280
    4-pole 36-slot machine   271
    slots
      locations   266–267
      structure   266
Stator backiron flux   329

backiron flux calculation   329
flux density   329
  calculation of   329
Stator currents, control philosophy   320–321
  current
    phase angle   321
    rms value   321
  machine design considerations   320
Stator tooth flux   328–329
  air-gap radial flux   328
  ith tooth flux, expression of   328
  maximum tooth flux density   328
  rotor   328
    position values   328
    simple rotation   328
  permanent magnet ac machine   328
    operation   328
  tooth flux values, matrix   328
  tooth flux vector
    synthesis   328
Stator winding   317–319
  conductor density   317
    fundamental amplitude   317
    given by   317
  conductors   317
    canceled   318
    cross-sectional area   318
    diameter   319
    largest area   319
    total number, given by   318
  constructing a machine   317
  end conductor distribution   318
    winding   318
  end winding bundle   319
    dimension approximation   319
  end winding offset   319
  magnetic analysis   318
  packing factor   318
  slot conductor distribution   318
  stator geometry   319
    independent variables, winding   319
  winding depth, computing of   315
  wire cross-sectional area   318
Superparamagnetic materials   49
Surface-mounted PM machine   286

Switching frequency   461, 473, 488, 490, 492–494, 497–499
Switching period   461
Switch, power electronic   461
Synchronous machines   280, 291
  permanent magnet   291
  wound-rotor   291
Systems, distributed parameter   456

t

Taylor series   6, 91, 171
Temperature, junction   469, 472, 473
T-equivalent circuit model   217–220
Thermal conductivity   517, 533
Thermal equivalent circuit, cuboidal region   358–361
  boundary conditions   358
    temperature   358
  circuit illustration   360
  development   358
    assumptions   358
      heat equation, cuboidal region   358
      heat transfer rates, calculation   359
    dynamic portion, obtaining   359
    extremum points   360
    input conditions   358
    spatial mean temperature   358
    temperature   360
      peak value   361
    temperature distribution
      steady-state   359
Thermal equivalent circuit, cylindrical region   361–367
  axial heat flow (See Axial heat flow)
  development   363–364
  geometry, cylindrical region   362
  heat flow equation   362
    cylindrical region   362
  radial heat flow (See Radial heat flow)
  spatial average   364
  thermal conductivity, assumptions   362
Thermal equivalent circuit networks   376–380
  graphical element representation   379
    concise circuit symbols   380
    cuboidal element   379
    cylindrical element   379
    graphical shorthand notation   379
    one-dimensional element   379
    standard circuit symbols   379
  interconnections of   376
  laws   376–377
    application   377
  nodal network analysis   378–379
    heat transfer rates   378
    nodal temperature, vector   378
    thermal nodal analysis formulation algorithm   379
  standard branch   377–378
    degree of temperature dependence   377
    dependent source, purpose of   377
    heat transfer rate   378
    illustration   377
    power dissipation   377
    resistivity   377
    source term   377
    thermal conductance   377
Thermal equivalent circuit, one-dimensional heat flow   352–358
  circuit, illustration of   354
  development   352
    assumptions   358
    boundary conditions   358
    Euler integration algorithm   355
    extremum point   355
    heat flux   353
    heat transfer rate   353
      vs. time   355, 356
    mean temperature   352
      vs. time   355
    one-dimensional heat flow
      example of   352
      special case   358
    partial differential equation (PDE)   355
    peak temperature   356
    spatial average   354
      temperature, second derivative   354
    steady-state portion   354
    steady-state temperature profile   355

temperature distribution   352
  expression   352
  parabolic   356
  steady-state   352
temperature profile   355
  vs. time   356, 357
thermal analysis   355
thermal resistance   353, 357
total power dissipation
  in a region   353
  vs. partial differential equation (PDE) solution   355
Third harmonic   549, 551–553, 556
Three-phase bridge inverter   306–308. *See also* Permanent magnet ac (PMAC) machine
  constraints   307
    current   307
    voltage   307
  high-frequency voltage ripple   307
  illustration   307
  inverter power loss   307
    mechanisms   307
  line-to-line voltage   307
  neutral voltages   307
  a-phase current   307
    equation   307
    instantaneous power loss, given by   308
  semiconductor power loss   308
  switching frequencies   307
    range   307
  three phase legs   307
    three phases of machines, connected to   307
  voltage drop   307
Three-phase inductor   543, 545, 550, 552, 568
Toroidal tester   186, 188, 190, 197
Time-domain simulation   463
Torque   135, 294–299
  electromagnetic   135, 290
  energy based approach, calculation   294–297
    coupling field   294
    electrical input energy, expression   295
    stator phase flux linkages, time rate change   295
  field approach to calculation   297–299

Transformation of power   294
  instantaneous power three-phase machine   294
Transformation, QD   506–508
Transformer architectures   215–217
  core-type single-phase   215
  shell-type single-phase   215
  UI-and EI-core arrangements   216
Transformer, core-type   452
Transformer performance   223
  calculation of lumped parameters   223–224
  inrush current   230–231
  lumped parameter   223–224
  magnetizing characteristics   225–226
  no-load, full-load, and overload analysis   230
  operating point analysis   226
    circuit solution   226
    convergence evaluation   229
    final calculations   299–230
    initialization   226–227
    lumped circuit parameter update   227–229
    nonlinear magnetizing current   227
  regulation   224–225
Transistor, power   461
Trigonometric identities   607

U

UI-core design   37
UI-core inductor   1, 33, 34
  incorporation of leakage flux permeances into   79–80
  leakage flux path   69
  nonlinear MEC   94
  simplified magnetic equivalent circuit   59

V

Vector magnetic potential   574, 575, 583
Voltage, blocking   470, 471, 473, 490
Voltage equations   290
  applied voltage resistive drop   290
  change, time rate flux linkages   290
  transformation   292–293
    stator voltage equations   292
    three-phase electric machinery   292

Voltage ripple
  input capacitor   465, 467, 491
  output capacitor   465, 491
Voltage source operation   308–309. *See also* Permanent magnet ac (PMAC) machine
  applied voltage, fundamental component given by   308
  desired voltage, obtaining   308
  equations   308
  electrical rotor position, sensing of   308
  important operating characteristics   309
  inverters, act as   308
  three-phase balanced set   308
  voltage source fed PMAC machine characteristics   309
Volume, magnetic   481, 518

W

Winding
  arrangements   270–271
    coil   270–271
    concentric   270–271
    consequent pole   270–271
    double-layer winding   270
    lap   270–271
    wave   271
  bundle   57, 77, 78, 241, 287, 319, 367, 381, 383, 384, 419
  depth   57, 109

  distributed, description   263–265
    induction motor   263
    P-pole machines   265
  flux linkage   280
  flyback   449
  interleaved   559
  leakage inductance   280
  multiple sets, phases   265
  non-interleaved   559
  orthocyclic   445
  orthogonal   443, 449
  resistance, calculation   269
  rotating machinery   265
  skew   481
  standard   449
  symmetry conditions, on conductor distributions   267
  three-phase stator   279
  volume located
    end-turns   289
    slots   289
Winding functions   271–275
  calculation   273
  conductor distribution and   275
  continuous representation   275
  discrete representation   273, 275
  manipulation of yields   273
  symmetry conditions   274
  usages   271
Wire gauge standards   109

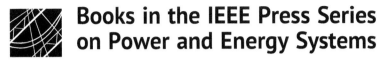
Books in the IEEE Press Series on Power and Energy Systems

Series Editor: ME El-Hawary, Dalhousie University, Halifax, Nova Scotia, Canada.

The mission of the IEEE Press Series on Power and Energy Systems is to publish leading-edge books that cover a broad spectrum of current and forward-looking technologies in the fast-moving area of power and energy systems including smart grid, renewable energy systems, electric vehicles and related areas. Our target audience includes power and energy systems professionals from academia, industry and government who are interested in enhancing their knowledge and perspectives in their areas of interest.

1. *Electric Power Systems: Design and Analysis, Revised Printing*
   Mohamed E. El-Hawary

2. *Power System Stability*
   Edward W. Kimbark

3. *Analysis of Faulted Power Systems*
   Paul M. Anderson

4. *Inspection of Large Synchronous Machines: Checklists, Failure Identification, and Troubleshooting*
   Isidor Kerszenbaum

5. *Electric Power Applications of Fuzzy Systems*
   Mohamed E. El-Hawary

6. *Power System Protection*
   Paul M. Anderson

7. *Subsynchronous Resonance in Power Systems*
   Paul M. Anderson, B.L. Agrawal, J.E. Van Ness

8. *Understanding Power Quality Problems: Voltage Sags and Interruptions*
   Math H. Bollen

9. *Analysis of Electric Machinery*
   Paul C. Krause, Oleg Wasynczuk, and S.D. Sudhoff

10. *Power System Control and Stability, Revised Printing*
    Paul M. Anderson, A.A. Fouad

11. *Principles of Electric Machines with Power Electronic Applications*, Second Edition
    Mohamed E. El-Hawary

12. *Pulse Width Modulation for Power Converters: Principles and Practice*
    D. Grahame Holmes and Thomas Lipo

13. *Analysis of Electric Machinery and Drive Systems*, Second Edition
    Paul C. Krause, Oleg Wasynczuk, and S.D. Sudhoff

Power Magnetic Devices: A Multi-Objective Design Approach, Second Edition. Scott D. Sudhoff.
© 2022 The Institute of Electrical and Electronics Engineers, Inc. Published 2022 by John Wiley & Sons, Inc.
Companion website: www.wiley.com/go/sudhoff/Powermagneticdevices

14. *Risk Assessment for Power Systems: Models, Methods, and Applications*
    Wenyuan Li

15. *Optimization Principles: Practical Applications to the Operations of Markets of the Electric Power Industry*
    Narayan S. Rau

16. *Electric Economics: Regulation and Deregulation*
    Geoffrey Rothwell and Tomas Gomez

17. *Electric Power Systems: Analysis and Control*
    Fabio Saccomanno

18. *Electrical Insulation for Rotating Machines: Design, Evaluation, Aging, Testing, and Repair*
    Greg C. Stone, Edward A. Boulter, Ian Culbert, and Hussein Dhirani

19. *Signal Processing of Power Quality Disturbances*
    Math H.J. Bollen and Irene Y. H. Gu

20. *Instantaneous Power Theory and Applications to Power Conditioning*
    Hirofumi Akagi, Edson H. Watanabe and Mauricio Aredes

21. *Maintaining Mission Critical Systems in a 24/7 Environment*
    Peter M. Curtis

22. *Elements of Tidal-Electric Engineering*
    Robert H. Clark

23. *Handbook of Large Turbo-Generator Operation and Maintenance,* Second Edition
    Geoff Klempner and Isidor Kerszenbaum

24. *Introduction to Electrical Power Systems*
    Mohamed E. El-Hawary

25. *Modeling and Control of Fuel Cells: Distributed Generation Applications*
    M. Hashem Nehrir and Caisheng Wang

26. *Power Distribution System Reliability: Practical Methods and Applications*
    Ali A. Chowdhury and Don O. Koval

27. *Introduction to FACTS Controllers: Theory, Modeling, and Applications*
    Kalyan K. Sen, Mey Ling Sen

28. *Economic Market Design and Planning for Electric Power Systems*
    James Momoh and Lamine Mili

29. *Operation and Control of Electric Energy Processing Systems*
    James Momoh and Lamine Mili

30. *Restructured Electric Power Systems: Analysis of Electricity Markets with Equilibrium Models*
    Xiao-Ping Zhang

31. *An Introduction to Wavelet Modulated Inverters*
    S.A. Saleh and M.A. Rahman

32. *Control of Electric Machine Drive Systems*
    Seung-Ki Sul

33. *Probabilistic Transmission System Planning*
    Wenyuan Li

34. *Electricity Power Generation: The Changing Dimensions*
    Digambar M. Tagare

35. *Electric Distribution Systems*
    Abdelhay A. Sallam and Om P. Malik

36. *Practical Lighting Design with LEDs*
    Ron Lenk, Carol Lenk

37. *High Voltage and Electrical Insulation Engineering*
    Ravindra Arora and Wolfgang Mosch

38. *Maintaining Mission Critical Systems in a 24/7 Environment, Second Edition*
    Peter Curtis

39. *Power Conversion and Control of Wind Energy Systems*
    Bin Wu, Yongqiang Lang, Navid Zargari, Samir Kouro

40. *Integration of Distributed Generation in the Power System*
    Math H. Bollen, Fainan Hassan

41. *Doubly Fed Induction Machine: Modeling and Control for Wind Energy Generation Applications*
    Gonzalo Abad, Jesús López, Miguel Rodrigues, Luis Marroyo, and Grzegorz Iwanski

42. *High Voltage Protection for Telecommunications*
    Steven W. Blume

43. *Smart Grid: Fundamentals of Design and Analysis*
    James Momoh

44. *Electromechanical Motion Devices, Second Edition*
    Paul Krause, Oleg Wasynczuk, Steven Pekarek

45. *Electrical Energy Conversion and Transport: An Interactive Computer-Based Approach, Second Edition*
    George G. Karady and Keith E. Holbert

46. *ARC Flash Hazard and Analysis and Mitigation*
    J.C. Das

47. *Handbook of Electrical Power System Dynamics: Modeling, Stability, and Control*
    Mircea Eremia, Mohammad Shahidehpour

48. *Analysis of Electric Machinery and Drive Systems, Third Edition*
    Paul C. Krause, Oleg Wasynczuk, S.D. Sudhoff, and Steven D. Pekarek

49. *Extruded Cables for High-Voltage Direct-Current Transmission: Advances in Research and Development*
    Giovanni Mazzanti, Massimo Marzinotto

50. *Power Magnetic Devices: A Multi-Objective Design Approach*
    S.D. Sudhoff

51. *Risk Assessment of Power Systems: Models, Methods, and Applications, Second Edition*
    Wenyuan Li

52. *Practical Power System Operation*
    Ebrahim Vaahedi

53. *The Selection Process of Biomass Materials for the Production of Bio-Fuels and Co-Firing*
    Najib Altawell

54. *Electrical Insulation for Rotating Machines: Design, Evaluation, Aging, Testing, and Repair, Second Edition*
    Greg C. Stone, Ian Culbert, Edward A. Boulter, and Hussein Dhirani

55. *Principles of Electrical Safety*
    Peter E. Sutherland

56. *Advanced Power Electronics Converters: PWM Converters Processing AC Voltages*
    Euzeli Cipriano dos Santos Jr., Edison Roberto Cabral da Silva

57. *Optimization of Power System Operation, Second Edition*
    Jizhong Zhu

58. *Power System Harmonics and Passive Filter Designs*
    J.C. Das

59. *Digital Control of High-Frequency Switched-Mode Power Converters*
    Luca Corradini, Dragan Maksimoviæ, Paolo Mattavelli, and Regan Zane

60. *Industrial Power Distribution, Second Edition*
    Ralph E. Fehr, III

61. *HVDC Grids: For Offshore and Supergrid of the Future*
    Dirk Van Hertem, Oriol Gomis-Bellmunt, and Jun Liang

62. *Advanced Solutions in Power Systems: HVDC, FACTS, and Artificial Intelligence*
    Mircea Eremia, Chen-Ching Liu, and Abdel-Aty Edris

63. *Operation and Maintenance of Large Turbo-Generators*
    Geoff Klempner, Isidor Kerszenbaum

64. *Electrical Energy Conversion and Transport: An Interactive Computer-Based Approach*
    George G. Karady, Keith E. Holbert

65. *Modeling and High-Performance Control of Electric Machines*
    John Chiasson

66. *Rating of Electric Power Cables in Unfavorable Thermal Environment*
    George J. Anders

67. *Electric Power System Basics for the Nonelectrical Professional*
    Steven W. Blume

68. *Modern Heuristic Optimization Techniques: Theory and Applications to Power Systems*
    Kwang Y. Lee, Mohamed A. El-Sharkawi

69. *Real-Time Stability Assessment in Modern Power System Control Centers*
    Savu C. Savulescu

70. *Optimization of Power System Operation*
    Jizhong Zhu

71. *Insulators for Icing and Polluted Environments*
    Masoud Farzaneh, William A. Chisholm

72. *PID and Predictive Control of Electric Devices and Power Converters Using MATLAB®/Simulink®*
    Liuping Wang, Shan Chai, Dae Yoo, Lu Gan, Ki Ng

73. *Power Grid Operation in a Market Environment: Economic Efficiency and Risk Mitigation*
    Hong Chen

74. *Electric Power System Basics for Nonelectrical Professional, Second Edition*
    Steven W. Blume

75. *Energy Production Systems Engineering*
    Thomas Howard Blair

76. *Model Predictive Control of Wind Energy Conversion Systems*
    Venkata Yaramasu, Bin Wu

77. *Understanding Symmetrical Components for Power System Modeling*
    J.C. Das

78. *High-Power Converters and AC Drives, Second Edition*
    Bin Wu, Mehdi Narimani

79. *Current Signature Analysis for Condition Monitoring of Cage Induction Motors: Industrial Application and Case Histories*
    William T. Thomson, Ian Culbert

80. *Introduction to Electric Power and Drive Systems*
    Paul Krause, Oleg Wasynczuk, Timothy O'Connell, and Maher Hasan

81. *Instantaneous Power Theory and Applications to Power Conditioning, Second Edition*
    Hirofumi, Edson Hirokazu Watanabe, Mauricio Aredes

82. *Practical Lighting Design with LEDs, Second Edition*
    Ron Lenk, Carol Lenk

83. *Introduction to AC Machine Design*
    Thomas A. Lipo

84. *Advances in Electric Power and Energy Systems: Load and Price Forecasting*
    Mohamed E. El-Hawary

85. *Electricity Markets: Theories and Applications*
    Jeremy Lin, Jernando H. Magnago

86. *Multiphysics Simulation by Design for Electrical Machines, Power Electronics and Drives* Marius Rosu, Ping Zhou, Dingsheng Lin, Dan M. Ionel, Mircea Popescu, Frede Blaabjerg, Vandana Rallabandi, David Staton

87. *Modular Multilevel Converters: Analysis, Control, and Applications*
    Sixing Du, Apparao Dekka, Bin Wu, and Navid Zargari

88. *Electrical Railway Transportation Systems*
    Morris Brenna, Federica Foiadelli, and Dario Zaninelli

89. *Energy Processing and Smart Grid*
    James A. Momoh

90. *Handbook of Large Turbo-Generator Operation and Maintenance, 3rd Edition*
    Geoff Klempner, Isidor Kerszenbaum

91. *Advanced Control of Doubly Fed Induction Generator for Wind Power Systems*
    Dehong Xu, Dr. Frede Blaabjerg, Wenjie Chen, Nan Zhu

92. *Electric Distribution Systems, 2nd Edition*
    Abdelhay A. Sallam, Om P. Malik

93. *Power Electronics in Renewable Energy Systems and Smart Grid: Technology and Applications*
    Bimal K. Bose

94. *Distributed Fiber Optic Sensing and Dynamic Rating of Power Cables*
    Sudhakar Cherukupalli, and George J Anders

95. *Power System and Control and Stability, Third Edition*
    Vijay Vittal, James D. McCalley, Paul M. Anderson, and A.A. Fouad

96. *Electromechanical Motion Devices: Rotating Magnetic Field-Based Analysis and Online Animations, Third Edition*
    Paul Krause, Oleg Wasynczuk, Steven D. Pekarek, and Timothy O'Connell

97. *Applications of Modern Heuristic Optimization Methods in Power and Energy Systems*
    Kwang Y. Lee and Zita A. Vale

98. *Handbook of Large Hydro Generators: Operation and Maintenance*
    Glenn Mottershead, Stefano Bomben, Isidor Kerszenbaum, and Geoff Klempner

99. *Advances in Electric Power and Energy: Static State Estimation*
    Mohamed E. El-hawary

100. *Arc Flash Hazard Analysis and Mitigation, Second Edition*
     J.C. Das

101. *Maintaining Mission Critical Systems in a 24/7 Environment, Third Edition*
     Peter M. Curtis

102. *Probabilistic Power System Expansion Planning with Renewable Energy Resources and Energy Storage Systems*
     Jaeseok Choi and Kwang Y. Lee

103. *Power Magnetic Devices: A Multi-Objective Design Approach*, Second Edition.
     Scott D. Sudhoff.

CPSIA information can be obtained
at www.ICGtesting.com
Printed in the USA
LVHW060230141121
703217LV00003B/51